VOLUME 1

MODERN ELECTROCHEMISTRY

SECOND EDITION

Ionics

VOLUME 1

MODERN ELECTROCHEMISTRY

SECOND EDITION

Ionics

John O'M. Bockris
Distinguished Professor of Chemistry
Texas A&M University
College Station, Texas

and

Amulya K. N. Reddy
President
International Energy Initiative
Bangalore, India

Plenum Press • New York and London

Library of Congress Cataloging-in-Publication Data

Bockris, J. O'M. (John O'M.), 1923–
Modern electrochemistry: ionics / John O'M. Bockris and Amulya K. N. Reddy. --2nd ed.
p. cm.
Includes bibliographical references and index.
ISBN 0-306-45554-4 (v. 1). -- ISBN 0-306-45555-2 (v. 1: pbk.)
1. Electrochemistry. I. Reddy, Amulya K. N. II. Title.
QD553.B63 1998
541.3'7--DC21
97-24151
CIP

ISBN 0-306-45554-4 (Hardbound)
ISBN 0-306-45555-2 (Paperback)

A Division of Plenum Publishing Corporation
233 Spring Street, New York, N. Y. 10013

http://www.plenum.com

10 9 8 7 6 5 4 3 2 1

Printed in the United States of America

To P. Debye and E. Hückel

PREFACE TO THE FIRST EDITION

This book had its nucleus in some lectures given by one of us (J.O'M.B.) in a course on electrochemistry to students of energy conversion at the University of Pennsylvania. It was there that he met a number of people trained in chemistry, physics, biology, metallurgy, and materials science, all of whom wanted to know something about electrochemistry. The concept of writing a book about electrochemistry which could be understood by people with very varied backgrounds was thereby engendered. The lectures were recorded and written up by Dr. Klaus Muller as a 293-page manuscript. At a later stage, A.K.N.R. joined the effort; it was decided to make a fresh start and to write a much more comprehensive text.

Of methods for direct energy conversion, the electrochemical one is the most advanced and seems the most likely to become of considerable practical importance. Thus, conversion to electrochemically powered transportation systems appears to be an important step by means of which the difficulties of air pollution and the effects of an increasing concentration in the atmosphere of carbon dioxide may be met. Corrosion is recognized as having an electrochemical basis. The synthesis of nylon now contains an important electrochemical stage. Some central biological mechanisms have been shown to take place by means of electrochemical reactions. A number of American organizations have recently recommended greatly increased activity in training and research in electrochemistry at universities in the United States. Three new international journals of fundamental electrochemical research were established between 1955 and 1965.

In contrast to this, physical chemists in U.S. universities seem—perhaps partly because of the absence of a modern textbook in English—out of touch with the revolution in fundamental interfacial electrochemistry which has occurred since 1950. The fragments of electrochemistry which are taught in many U.S. universities belong not to the space age of electrochemically powered vehicles, but to the age of thermodynamics and the horseless carriage; they often consist of Nernst's theory of galvanic cells (1891) together with the theory of Debye and Hückel (1923).

Electrochemistry at present needs several kinds of books. For example, it needs a textbook in which the whole field is discussed at a strong theoretical level. The most pressing need, however, is for a book which outlines the field at a level which can be understood by people entering it from different disciplines who have no previous background in the field but who wish to use modern electrochemical concepts and ideas as a basis for their own work. It is this need which the authors have tried to meet.

The book's aims determine its priorities. In order, these are:

1. Lucidity. The authors have found students who understand advanced courses in quantum mechanics but find difficulty in comprehending a field at whose center lies the quantum mechanics of electron transitions across interfaces. The difficulty is associated, perhaps, with the interdisciplinary character of the material: a background knowledge of physical chemistry is not enough. Material has therefore sometimes been presented in several ways and occasionally the same explanations are repeated in different parts of the book. The language has been made informal and highly explanatory. It retains, sometimes, the lecture style. In this respect, the authors have been influenced by *The Feynman Lectures on Physics*.

2. Honesty. The authors have suffered much themselves from books in which proofs and presentations are not complete. An attempt has been made to include most of the necessary material. Appendices have been often used for the presentation of mathematical derivations which would obtrude too much in the text.

3. Modernity. There developed during the 1950s a great change in emphasis in electrochemistry away from a subject which dealt largely with solutions to one in which the treatment at a molecular level of charge transfer across interfaces dominates. This is the "new electrochemistry," the essentials of which, at an elementary level, the authors have tried to present.

4. Sharp variation is standard. The objective of the authors has been to begin each chapter at a very simple level and to increase the level to one which allows a connecting up to the standard of the specialized monograph. The standard at which subjects are presented has been intentionally variable, depending particularly on the degree to which knowledge of the material appears to be widespread.

5. One theory per phenomenon. The authors intend a *teaching book*, which acts as an introduction to graduate studies. They have tried to present, with due admission of the existing imperfections, a simple version of that model which seemed to them at the time of writing to reproduce the facts most consistently. They have for the most part refrained from presenting the detailed pros and cons of competing models in areas in which the theory is still quite mobile.

In respect to references and further reading: no detailed references to the literature have been presented, in view of the elementary character of the book's contents, and the corresponding fact that it is an introductory book, largely for beginners. In the "further reading" lists, the policy is to cite papers which are classics in the development of the subject, together with papers of particular interest concerning recent developments, and in particular, reviews of the last few years.

It is hoped that this book will not only be useful to those who wish to work with modern electrochemical ideas in chemistry, physics, biology, materials science, etc., but also to those who wish to begin research on electron transfer at interfaces and associated topics.

The book was written mainly at the Electrochemistry Laboratory in the University of Pennsylvania, and partly at the Indian Institute of Science in Bangalore. Students in the Electrochemistry Laboratory at the University of Pennsylvania were kind enough to give guidance frequently on how they reacted to the clarity of sections written in various experimental styles and approaches. For the last four years, the evolving versions of sections of the book have been used as a partial basis for undergraduate, and some graduate, lectures in electrochemistry in the Chemistry Department of the University.

The authors' acknowledgment and thanks must go first to Mr. Ernst Cohn of the National Aeronautics and Space Administration. Without his frequent stimulation, including very frank expressions of criticism, the book might well never have emerged from the Electrochemistry Laboratory.

Thereafter, thanks must go to Professor B. E. Conway, University of Ottawa, who gave several weeks of his time to making a detailed review of the material. Plentiful help in editing chapters and effecting revisions designed by the authors was given by the following: Chapters IV and V, Dr. H. Wroblowa (Pennsylvania); Chapter VI, Dr. C. Solomons (Pennsylvania) and Dr. T. Emi (Hokkaido); Chapter VII, Dr. E. Gileadi (Tel-Aviv); Chapters VIII and IX, Prof. A. Despic (Belgrade), Dr. H. Wroblowa, and Mr. J. Diggle (Pennsylvania); Chapter X, Mr. J. Diggle; Chapter XI, Dr. D. Cipris (Pennsylvania). Dr. H. Wroblowa has to be particularly thanked for essential contributions to the composition of the Appendix on the measurement of Volta potential differences.

Constructive reactions to the text were given by Messers. G. Razumney, B. Rubin, and G. Stoner of the Electrochemistry Laboratory. Advice was often sought and accepted from Dr. B. Chandrasekaran (Pennsylvania), Dr. S. Srinivasan (New York), and Mr. R. Rangarajan (Bangalore).

Comments on late drafts of chapters were made by a number of the authors' colleagues, particularly Dr. W. McCoy (Office of Saline Water), Chapter II; Prof. R. M. Fuoss (Yale), Chapter III; Prof. R. Stokes (Armidale), Chapter IV; Dr. R. Parsons (Bristol), Chapter VII; Prof. A. N. Frumkin (Moscow), Chapter VIII; Dr. H. Wroblowa, Chapter X; Prof. R. Staehle (Ohio State), Chapter XI. One of the authors (A.K.N.R.) wishes to acknowledge his gratitude to the authorities of the Council of Scientific and Industrial Research, India, and the Indian Institute of Science, Bangalore, India, for various facilities, not the least of which were extended leaves of absence. He wishes also to thank his wife and children for sacrificing many precious hours which rightfully belonged to them.

PREFACE

The textbook *Modern Electrochemistry* by Bockris and Reddy originated in the needs of students at the Energy Conversion Institute of the University of Pennsylvania in the late 1960s. People trained in various disciplines from mathematics to biology wanted to understand the new high-energy-density storage batteries and the doubling of the efficiency of energy conversion offered by fuel cells over heat engines. The task was to take a group that seemed to be above average in initiative and present electrochemistry well enough to meet their needs.

The book turned out to be a great success. Its most marked characteristic was—is—lucidity. The method used was to start off at low level and then move up in a series of very small steps. Repetition is part of the technique and does not offend, for the lesson given each time is the same but is taught differently.

The use of the book spread rapidly beyond the confines of energy conversion groups. It led to the recognition of *physical* electrochemistry—the electrochemical discipline seen from its roots in physics and physical chemistry, and not as a path to superior chemical analysis. The book outlined electrochemical science for the first time in a molecular way, paying due heed to thermodynamics as bedrock but keeping it as background. The success of the effort has been measured not only by the total sales but by the fact that another reprinting had to be made in 1995, 25 years after the first one. The average sales rate of the first edition is even now a dozen copies a month!

Given this background, the challenge of writing a revised edition has been a memorable one. The changes in the state of electrochemical science in the quarter century of the book's life have been broad and deep. Techniques such as scanning tunneling microscopy enable us to *see atoms* on electrodes. Computers have allowed a widespread development of molecular dynamics (MD) calculations and changed the balance between informed guesses and the timely adjustment of parameters in force laws to enable MD calculations to lead to experimental values. The long-postponed introduction of commercial electric cars in the United States has been realized and is the beginning of a great step toward a healthier environment. The use of the new room-temperature molten salts has made it possible to exploit the advantage of working with pure liquid electrolytes—no solvent—without the rigors of working at 1000 °C.

All the great challenges of electrochemistry at 2000 A.D. do not have to be addressed in this second edition for this is an *undergraduate* text, stressing the teaching of *fundamentals* with an occasional preview of the advancing frontier.

The basic attributes of the book are unchanged: lucidity comes first. Since the text is not a graduate text, there is no confusing balancing of the merits of one model against those of another; the most probable model at the time of writing is described. Throughout it is recognized that theoretical concepts rise and fall; a theory that lasts a generation is doing well.

These philosophies have been the source of some of the choices made when balancing what should be retained and what rewritten. The result is quite heterogeneous. Chapters 1 and 2 are completely new. The contributions from neutron diffraction measurements in solutions and those from other spectroscopic methods have torn away many of the veils covering knowledge of the first 1–2 layers of solvent around an ion. Chapter 3 also contains much new material. Debye and Huckel's famous calculation is two generations old and it is surely time to move toward new ideas. Chapter 4, on the other hand, presents much material on transport that is phenomenological—material so basic that it must be presented but shows little variation with time.

The last chapter, which is on ionic liquids, describes the continuing evolution that is the result of the development of low-temperature molten salts and the contributions of computer modeling. The description of models of molten silicates contains much of the original material in the first edition, for the models described there are those still used today.

A new feature is the liberal supply of problems for student solution—about 50 per chapter. This idea has been purloined from the excellent physical chemistry textbook by Peter Atkins (W. H. Freeman). There are exercises, practice in the use of the chapter's equations; problems (the chapter's material related to actual situations); and finally, a few much more difficult tasks which are called "microresearch problems," each one of which may take some hours to solve.

The authors have not hesitated to call on colleagues for help in understanding new material and in deciding what is vital and what can be left for the literature. The authors would particularly like to thank John Enderby (University of Bristol) for his review of Chapter 2; Tony Haymet (University of Sydney) for advice on the weight to be given to various developments that followed Debye and Hückel's ground-breaking work and for tutoring us on computational advances in respect to electrolytic ion pairs. Michael Lyons (University of Dublin) is to be thanked for allowing the present authors use of an advanced chapter on transport phenomena in electrolytes written by him. Austin Angell (Arizona State University of Tempe) in particular and Douglas Inman

(Imperial College) have both contributed by means of criticisms (not always heeded) in respect to the way to present the material on structure in pure electrolytes.

Many other electrochemists have helped by replying to written inquiries.

Dr. Maria Gamboa is to be thanked for extensive editorial work, Ms. Diane Dowdell for her help with information retrieval, and Mrs. Janie Leighman for her excellence in typing the many drafts.

Finally, the authors wish to thank Ms. Amelia McNamara and Mr. Ken Howell of Plenum Publishing for their advice, encouragement, and patience.

ACKNOWLEDGMENTS

In writing a book of this type, the authors have accessed the advice of many colleagues, often by telephone discussions and sometimes in written exchanges.

A few individuals, however, deserve mention for having done much more than normal collegial cooperation implies.

Thus, the chapter on solvation was greatly helped by consultation and correspondence with Professor J. E. Enderby (University of Bristol).

In respect to Chapter 3, advice was sought from and given by Professor Harold Friedman (University of New York at Stony Brook) and Professor J. C. Rasaiah (University of Maine).

Professor Antony Haymet (University of Sydney, Australia) was particularly helpful in the giving of his latest work, sometimes unpublished, and the giving of advice, both in writing and in telephone discussions.

Chapter 4 is rewritten to a lesser degree than the other chapters but the new material has been discussed with Professor B. E. Conway (University of Ottawa).

Chapter 5 was greatly improved by discussions and several letter exchanges with Professor Austin Angel (Arizona State University at Tempe) and to some extent with Professor Douglas Inman (Imperial College of Science and Technology, London University).

CONTENTS

CHAPTER 2

ION–SOLVENT INTERACTIONS

CHAPTER 3

ION–ION INTERACTIONS

CHAPTER 4

ION TRANSPORT IN SOLUTIONS

CHAPTER 5

IONIC LIQUIDS

NOMENCLATURE

Symbol	*Name*	*SI unit*	*Other units frequently used*
GENERAL			
A	area of an electrode–solution interface	m^2	
a_i	activity of species i		
c_i	concentration of species i	mol m^{-3}	M, N
c^0	bulk concentration	mol m^{-3}	M, N
E	total energy	J	
F	force	N	
f_c	frictional coefficient	J	
f_i	partition function of species i		
$g_{i,j}$	radial pair distribution function		
k, ν	frequency	s^{-1}	
k	wave vector ($= 2\pi/\lambda$)	m^{-1}	
m	mass	kg	
n	number of moles		
N	number of molecules		
p	momentum		
p_r	radial momentum		
P	pressure	Pa	atm

For each symbol the appropriate SI unit is given. Multiples and submultiples of these units are equally acceptable and are often more convenient. For example, although the SI unit of concentration is **mol m**$^{-3}$, concentrations are frequently expressed in **mol dm**$^{-3}$ (or M), **mol cm**$^{-3}$, or **mmol dm**$^{-3}$.

Symbol	*Name*	*SI unit*	*Other units frequently used*
P_i^*	vapor pressure of species i	Pa	atm
r, d, l	distance	m	Å
T	thermodynamic temperature	K	°C
t	time	s	
U	potential energy	J	
V	volume	m^3	
v	velocity	$m\ s^{-1}$	
W	work	J	
w_i	weight fraction of species i		
x_i	molar fraction of species i		
$\overline{\mu}$	reduced mass		
ρ	density	$kg\ m^{-3}$	
τ	relaxation time	s	
λ	wavelength	m	
$\tilde{v}$	wavenumber	m^{-1}	
θ	angle	°	

ION– AND MOLECULE-RELATED QUANTITIES

a	distance of closest approach	m	Å
p	quadrupole moment of water	debye cm^2	
q	Bjerrum parameter	m	
z_i	charge number of an ion i		
α'	symmetry factor		
α_i	polarizability of species i	m^3 $molecule^{-1}$	
$\alpha_{d,i}$	distortion polarizability of species i	m^3 $molecule^{-1}$	
β	compressibility	Pa^{-1}	
μ	dipole moment	debye	esu cm
κ^{-1}	Debye–Hückel reciprocal length	m	

THERMODYNAMICS OF A SIMPLE PHASE

E_{pzc}, *p.z.c.*	potential of zero charge	V
E_a	energy of activation	$J\ mol^{-1}$
ΔG	molar Gibbs free-energy change	$J\ mol^{-1}$
ΔH	heat or enthalpy change	$J\ mol^{-1}$
K	equilibrium constant of the reaction	

Symbol	*Name*	*SI unit*	*Other units frequently used*
K_A, K_D	association constant and dissociation constant		
ΔS	entropy change	$J\ K^{-1} mol^{-1}$	eu
X	electric field	$V\ m^{-1}$	
μ_i	chemical potential of species *i*	$J\ mol^{-1}$	
$\bar{\mu}_i$	electrochemical potential of species *i*	$J\ mol^{-1}$	
ρ	charge density C	m^3	
ε	dielectric constant		
$\hat{\varepsilon}(\omega)$	complex permittivity		
ψ	electrostatic potential between two points	V	

ACTIVITIES IN ELECTROLYTIC SOLUTIONS AND RELATED QUANTITIES

$\alpha_{\pm}$	mean activity
γ_i, f_i	activity coefficient of species *i*
$f_{\pm}$, $\gamma_{\pm}$	stoichiometric mean molar activity coefficient
f_{IP}	activity coefficient for the ion pairs
I	ionic strength

MASS TRANSPORT

D_i	diffusion coefficient of species *i*	$m^2\ s^{-1}$
J_B	flux density of species B	$mol\ m^{-2} s^{-1}$
Re	Reynolds number	
v_d	drift velocity	$m\ s^{-1}$
η	viscosity	poise

CHARGE TRANSPORT PROPERTIES OF ELECTROLYTES

q_i	charge of species *i*	C
R	resistance of the solution	Ω
$1/R$	conductance	S, Ω^{-1}
σ	specific conductivity	$S\ m^{-1}$ or $\Omega^{-1}\ m^{-1}$
Λ_m	molar conductivity of an electrolyte	$S\ m^2\ mol^{-1}$ or $\Omega^{-1}\ m^2\ mol^{-1}$
Λ	equivalent conductivity	$S\ m^2\ mol^{-1}\ eq^{-1}$ or $\Omega^{-1}\ m^2\ mol^{-1}\ eq^{-1}$
t_i	transport number of ionic species *i* in an electrolytic solution	number

Symbol	*Name*	*SI unit*
$\bar{\mu}_{abs}$	absolute mobility	$m\ s^{-1}\ N^{-1}$
μ_i, $(\mu_{conv})_i$	conventional (electrochemical) mobility of species i	$m^2\ V^{-1}\ s^{-1}$

KINETIC PARAMETERS

I	electric current	A
j	current density	$A\ m^{-2}$
$\vec{k}$, $\overleftarrow{k}$	rate constants	
ν_i	stoichiometric number of species i	

STATISTICS AND OTHER MATHEMATICAL SYMBOLS

$C(t)$	autocorrelation function
$G_{A,B}(r)$	pair correlation function
g_o	probability distribution coefficient
I	moment of inertia
P_r	probability
$\langle x \rangle$	average value of variable x
$\langle x^2 \rangle$	mean square value of variable x
x_{rms}	root-mean-square value of variable x
$\bar{y}$	Laplace transform of y
θ	fraction number, e.g. fraction of ions associated
erf(y)	error function
$x_{\pm}$	mean value of variable x

USEFUL CONSTANTS

Symbol	*Name*	*Value*
c_0	velocity of light	2.998×10^{10} cm s^{-1}
e_0	electron charge	1.602×10^{-19} C
F	Faraday's constant = $e_0 N_A$	9.649×10^5 C mol^{-1}
h	Planck's constant	6.626×10^{-34} J s
k	Boltzmann's constant	1.380×10^{-23} J K^{-1}
m_e	mass of electron	9.110×10^{-31} kg
m_p	mass of proton	1.673×10^{-27} kg
N_A	Avogadro's number	6.022×10^{23} mol^{-1}
ε_0	permittivity of free space	8.854×10^{-12} $C^2\ N^{-1} m^{-2}$
0 K	absolute zero of temperature	–273.15 °C
π	pi	3.14159 . . .

Useful Unit Conversion Factors

Potential	*Length*	*Volume*	*Mass*	*Force*	*Pressure*	*Energy*
1 V	1 m	1 m^3	1 g	1 N	1 Pa	1 J
1 J C^{-1}	100 cm	1000 dm^3	10^{-3} kg	10^5 dynes	1 N m^{-2}	10^7 ergs
	1000 mm	1000 liters			10^{-5} bar	0.239 cal
	10^6 μm				9.872×10^{-6} atm	6.242×10^{18} eV
	10^9 nm				7.502×10^{-3} mmHg	
	10^{10} Å				7.502×10^{-3} torr	
	10^{12} pm					

CHAPTER 1

ELECTROCHEMISTRY

1.1. A STATE OF EXCITEMENT

Electrochemistry was born from a union between biochemistry and electricity and is the essential discipline among the chemical sciences needed to prepare society for near-future times. The birth of electrochemistry happened over 200 years ago (1791) in Bologna, Italy, where Luigi Galvani was dissecting a frog: "One of those who was assisting me touched lightly and by chance the point of his scalpel to the internal crural nerves of the frog (an electric machine was nearby), then suddenly all the muscles of its limbs were seen to be contracted . . ."

Galvani's discovery was followed nine years later by that of his compatriot, Volta, who communicated to the Royal Society in London an amazing thing: If one used a pasteboard membrane to separate silver plates from zinc plates, and wetted the ensemble with salt water, an electric current flowed. Volta called his device "the artificial electric organ."

These past events in Italy resonate in a modern decision of the California state legislature. In 2002, California will begin limiting the number of "emitting" vehicles that may be sold in the state, with sales of these being completely eliminated by 2017. Volta's discovery is the basis for the development by U.S. automakers of an emission-free vehicle—one that is electrochemically powered.

In 1923, Debye and Hückel wrote a paper describing for the first time a credible theory of the properties of ionically conducting solutions. In 1994, Mamantov and Popov edited a book in which the first chapter is called "Solution Chemistry: A Cutting Edge in Modern Technology." The book describes some frontiers of the electrochemistry of today: chloraluminate-organic systems that make room-temperature molten salts the basis of high-energy electricity storers; the use of vibrational spectroscopy to study ion–ion interactions; and the application of the molecular dynamic technique to the ionic solutions unfolded by Debye and Hückel just one lifetime before.

Take another gigantic leap along the timeline of electrochemical discovery and application. Consider Michael Faraday,[1] that London superstar who in 1834 discovered the relation between the amount of electricity consumed and the amount of metal produced in solid form from some invisible particles in solution. In 1995, more than a century later, Despic and Popov wrote an article that described electroforming of (almost) anything from its ion in solution: powders or dendrites, whiskers or pyramids, in laminar shapes of any chosen composition (including that of semiconductors) or indeed in nanometer sizes. This is what has become of Faraday's electrodeposition at the cutting edge, as well as in practical applications such as electrodissolution to shape metal parts in the making of Rolls Royce cars.

Of all these jumps in electrochemistry, each separated by around a century, there is one that best of all shows how electrochemistry is both deep-rooted and at the frontier of the twenty-first century. It was the pedant Julius Tafel who found, in 1905, that electric currents passing across metal–solution interfaces could be made to increase exponentially by changing the electric potential of the electrode across the surface of which they passed. In this way, he complemented the finding in ordinary kinetics made by Arrhenius 16 years earlier. Arrhenius's equation tells us that an increase of temperature increases the rate of chemical reaction exponentially:

$$\text{Velocity}_{\text{thermal}} \propto A e^{-E_a/RT} \tag{1.1}$$

where A and E_a are constants, with E_a representing the activation energy.

Corresponding to this, Tafel found a similar equation:

$$\text{Velocity}_{\text{echem}} = B e^{-\alpha VF/RT} \tag{1.2}$$

where B, α, and F are constants. The term αVF (a constant α times a potential V times a quantity of electricity F) represents energy; one can see that the two equations are related.

To what did Tafel's discovery lead in our lifetime? To the first moon landing in 1969! In the 64 years from Tafel's discovery of how electrochemical reaction rates vary with potential, it had become possible to take his equation and use it in the development of an electrochemical fuel cell that produced electricity from chemicals directly without moving parts. This was firstly done by Groves in 1939. Those 64 years following Tafel's discovery had been well used, particularly at Cambridge University in England, by Tom Bacon (a descendant of the seventeenth-century Bacon, Baron Verulam, whom some see as the founder of Science as we know it). Tom Bacon established in practice what theorists had for long reasoned: that the electrochemical fuel cell produced electrical energy from chemical fuels at twice the efficiency of a

---

[1]Despite his achievements, Faraday was subject to the whip of religious discipline. When he accepted an invitation to visit Buckingham Palace to receive an award from Queen Victoria, and missed a religious service, his church told him that he had preferred Mammon to God and ought to leave the church!

heat engine driving a generator. Thus, when the NASA pioneers turned to the design of the first space vehicles—low weight being at a premium—they chose the electrochemical fuel cell (which provides the same amount of energy as conventional cells at half the weight) as the source of auxiliary power in space. These cells are used in all U.S. space vehicles and will be likely to power the first mass produced electric cars.[2]

These few examples grew out of the chest of treasures opened up by Galvani and by Volta. Diabetics will soon be able to check their glucose levels by glancing at a wrist meter that measures sugar content electrochemically. Tritium, an essential component of nuclear weaponry, may be made electrochemically at a fraction of the cost of its production in a nuclear reactor.

Holding off dielectric breakdown in water by means of electrochemically formed coatings can allow condenser plates to store gargantuan energies for powering the lasers of the Star Wars weaponry. Electrochemistry can be used to consume domestic wastes with no noxious effluents reaching the air. The North Sea oil platforms are protected by corrosion inhibitors that slow down the electrochemical reactions that deteriorate the metal in the rigs.

In this book, an attempt will be made to present the basis of all this new technology, but in a way in which the first consideration is a lucid explanation. Before we look closely at the individual parts of the territory, it is good to have a look at the whole country from above.

1.2. TWO KINDS OF ELECTROCHEMISTRY

According to the philosophers, all science is one, but that is not how it seems in the case of electrochemistry. The two main types do not at first seem to be strongly connected (see Fig. 1.1). These are the physical chemistry of ionically conducting solutions (ionics) and the physical chemistry of electrically charged interfaces (electrodics).

This text discusses four aspects of ionic electrochemistry: ion–solvent interactions, ion–ion interactions, ion transport in solution, and ionic liquids.

The physical chemistry of ionic solutions deals with ions and solvents and how ions interact dynamically with water as they move about in solutions. The study of ion–ion interactions tells us how ions associate, sometimes even forming polymers in solution. These interactions are important for the new spectroscopic techniques, neutron diffraction and infrared spectroscopy; and for molecular dynamics (MD).

The study of transport covers diffusion and conductance of ions in solution, where much of the basis is phenomenological.

[2]Batteries carry the active material, the reaction of which produces electricity. Fuel cells store the fuel used to produce electricity in a fuel tank. Batteries limit the range of electric cars to ~150 miles. However, electric cars powered by fuel cells are limited only by the size of the fuel tank.

IONICS	**ELECTRODICS**
Concerns ions in solution and in the liquids arising from melting solids composed of ions.	Concerns the region between an electronic and an ionic conductor and the transfer of electric charges across it.

Fig. 1.1. A way to divide the two quite different aspects of the field of electrochemistry. In this book, the point of view is presented that the electrodic area should be the realm associated with electrochemistry. Ionics is a necessary adjunct field (just as is the theory of electrons in metals and semiconductors, which is adequately dealt with in books on the solid state).

The last part of ionic electrochemistry, ionics, is about "pure electrolytes." A few decades back this electrochemistry would have been all about high-temperature liquids (liquid common salt at 850 °C was the role model). However, this has changed, and the temperatures for eliminating the solvent have decreased considerably. Some molten salts are now room temperature liquids. At the other end of the temperature scale are the molten silicates, where large polyanions predominate. These are important not only in the steel industry, where molten silicate mixtures form blast furnace slags, but also in the corresponding frozen liquids, the glasses.

The other half of electrochemistry, electrodics, in vol. 2 has surpassed ionics in its rate of growth and is coming into use in enterprises such as the auto industry, to obtain electrochemical power sources for transportation. Such a change in the way we power our cars is seen by many as the only way to avoid the planetary warming caused by the CO_2 emitted by internal combustion engines.

Our discussion of electrodics starts with a description of the *interfacial region* between the metal and solution phases. This is the stage on which the play is to be performed. It involves the kinetics of electrons moving to and fro across areas with immensely strong electric field strengths (gigavolts per meter) that are unavailable in the laboratory. This is the heart of electrochemistry—the mechanism of electrically controlled surface reactions.

Electrons are quantal particles and much basic electrochemistry in the past few decades has been quantal in approach, so a simplified description of the current state of this field is given.

After this, the text moves to the main applications of electrochemistry. There is the conversion of light to electricity and the photosplitting of water to yield pure hydrogen as a storage medium for electricity produced from solar light. Some organic reactions are better carried out under electrochemical conditions, because one can vary the energy of the available electrons so easily (i.e., by changing V in Tafel's law). The stability of materials (corrosion protection) is indeed a vast area, but the basic mechanisms are all electrochemical and deserve a whole chapter. There are two other important and growing areas of electrochemistry: bioelectrochemistry and the most active and expanding of all, the electrochemistry of cleaner environments.

What is the connection between the two main areas in electrochemistry—the science of solutions (ionics) and that of charge transfer across solid–solution interfaces (electrodics)? There is indeed a close connection. The interfacial region at electrodes (and all wet surfaces, including the surface of plants undergoing photosynthesis) is surrounded by ions in solution (or in the moisture films on surfaces). Thus it is important that we know all about them. The electrode is the stage; the solution is the theater and the audience. It is also the place that supplies the players—ions and solvent—while electrons are clearly supplied from resources in the wings.

1.3. SOME CHARACTERISTICS OF ELECTRODICS

Electron transfer between two phases is the fundamental act of electrochemistry (see Fig. 1.2) and governs much in nature. Until well past the middle of this century, there was no knowledge of the breadth of interfacial charge transfer. It used to be thought of as something to do with metals. All that is changed. Now we know it involves semiconductors and insulators, also, insofar as these bodies are in contact with ion-containing liquids. For example, proteins undergo electron charge transfer when they are in contact with glucose in solution.

The fundamental act in electrochemistry (the simple act of Fig. 1.2) is prevalent in nature, and that is why electrodics is such an important part of science. It is a vast

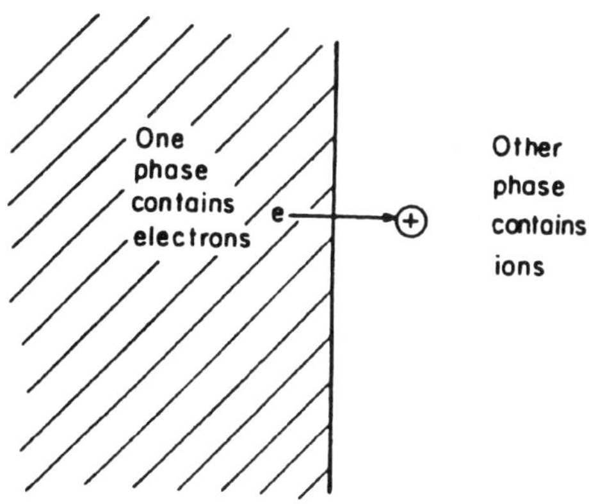

Fig. 1.2. The fundamental act in electrochemistry. Often, the electron-containing phase is a metal and the ion-containing phase is an aqueous solution. However, germanium in contact with a molten salt would also involve electron- and ion-containing systems.

field underlying all those phenomena in chemistry, biology, and engineering that involve real (and therefore moist or wet) surfaces.

1.4. PROPERTIES OF MATERIALS AND SURFACES

It is now well recognized that the properties of most materials are controlled by the properties of their surfaces. Furthermore, most surfaces are "wet." The quotation marks are used because "wet" does not here mean "placed in a solution." It can also mean "covered by a film of moisture." How pervasive are—in this sense—wet surfaces! Indeed, most outdoor and some indoor surfaces are in contact with invisible films of moisture, and the presence of these films is a required element in the corrosion and decay of metals. Thoroughly dry surfaces, removed from contact with air, do not corrode.

Wet surfaces have interfacial regions. One is in the solid, the other is in the solution. Across these regions are super-intense electric fields that accelerate or decelerate the passage of electrons from solid to solution and vice versa.

Corrosion—the gradual decay of materials—occurs in many ways, all involving electrochemical surface reactions. The essence of it is the electrochemical dissolution of atoms in the surface into the ion-containing film that is in contact with the corroding metal. However, such dissolution has to be accompanied by a counter-reaction and this is often the electrochemical decomposition of water to form hydrogen on the metal surface. If that occurs, the H in the form of minute protons, H^+, may enter the metal, diffuse about, and cause a weakening of metal–metal bonds and hence stress-corrosion cracking.

Local mechanical stress also plays an important part in determining the behavior of materials: combined with electrochemical corrosion, it may lead to bridges collapsing and ships splitting in half.[3]

Is friction electrochemical also? At least on moist surfaces, the distance between surface promontories—the protrusions of the metal–metal contacts—is controlled by the repulsion of like charges from ions adsorbed from electrolyte-containing moisture films onto surfaces. Indeed, if a pendulum swings on a fulcrum containing a metal–metal contact, its rate of decay (which is increased by the friction of the contact) maximizes when the interfacial excess electrical charge is a minimum; the friction therefore is a maximum (because the metal contacts, unrepelled by charges, are in closer contact).

1.4.1. Interfaces in Contact with Solutions Are Always Charged

Even when a solid–solution interface is at equilibrium (i.e., nothing net happening), electron transfer occurs at the same speed in each direction, for there are excess

[3]This can happen when the front and back of a ship are momentarily suspended on the peaks of waves in rough weather.

electric charges on both sides of the interface. Consider the interior of a metal. It consists of a lattice of metallic ions populated by electrons in the plasma state that are mobile and moving randomly at about 10^3 km s^{-1}.

Now, in a thought experiment, an extremely thin, sharp knife cuts the metal in half at great speed. Moreover, this imaginary act occurs under a solution of ions. Consider only one of the two surfaces formed by the knife. The electrons near the new surface are suddenly confronted with a boundary, which they overshoot or undershoot. Within 10^{-9} s, ions in the solution nearest to the metal arrange themselves to present a whole load of possible receiver or donor states for electrons. Depending on how the balance of tendencies goes, the electrons will either depart from the metal and head for the receiver states in solution leaving the surface of the metal positively charged, or take on board a load of electrons from the ions that have turned up from the solution, making the metal surface negatively charged. Whichever way it happens, the surface of the metal now has an excess positive or negative electric charge. The interior remains electroneutral (see Fig. 1.3).

Now, this argument can be generalized. It indicates that an uncharged metal or electron conductor in an ionic solution always manifests an excess surface electric charge, and the gigavolt per meter field, which results from having this sheet of excess electric charge on the metal facing a layer of opposite charge on the solution layer in contact with it, has extremely far-reaching consequences for the properties of the interface and eventually of the material beneath it.

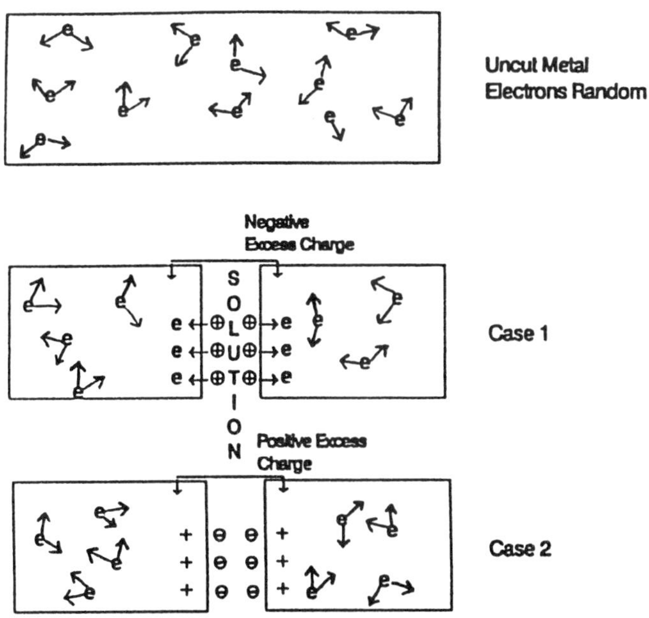

Fig. 1.3. Surfaces in solution carry a net excess charge.

1.4.2. The Continuous Flow of Electrons across an Interface: Electrochemical Reactions

It has been argued in the preceding section that all surfaces carry an excess electric charge, i.e., that surfaces in contact with ionic solutions are electrified. However, the argument was made by considering an isolated piece of material unconnected to a source or sink of electrons.

Suppose now that the metal, an electronic conductor, is connected to a power supply,[4] i.e., to a source of electrons so large in capacity that, say, 10^{19} to 10^{20} electrons[5] drawn from the source leave it unaffected in any significant way. To make the discussion specific, assume that the electronic conductor is a platinum plate and the ionically conducting phase is an aqueous solution of HI.

Then, by connecting the electrical power source to the platinum plate, it becomes possible for electrons to flow from the source to the surface of the plate. Before this was done, the electrified platinum–solution interface was in equilibrium. Under these equilibrium conditions, the platinum plate had a net surface electric charge, and the ionically conducting solution had an equal excess electric charge, though opposite in sign. Furthermore, the passage of electrons across the interface, which is associated with electron-transfer reactions, is occurring at an equal rate in the two directions. What happens when a disequilibrating shower of extra charges from the power source arrives at the surface of the platinum? The *details* of what happens, the mechanism, is a long story, told partly in the following chapters. However, the essence of it is that the new electrons overflow, as it were, the metal plate and *cross the metal–solution interface* to strike and neutralize ions in the layer of solution in contact with the metal, e.g., hydrogen ions produced in solution from the ionization of HI in the solution phase. This process can proceed continuously because the power source supplying the electrons can be thought of as infinite in capacity and the ionic conductor also has an abundance of ions in it; these tend to migrate up to the metal surface to capture there some of the overflowing electrons.

What is being described here is an *electrochemical reaction*; i.e., it is a *chemical transformation involving the transfer of electrons across an interface*, and it can be written in familiar style as

$$2\,H^+ + 2\,e \rightarrow H_2$$

The hydrogen ions are "discharged" (neutralized) on the electrode and there is an evolution of hydrogen outside the solution, as a gas.

---

[4]Actually, a power supply has two terminals and one must also consider how the metal–electrolyte interface is connected to the other terminal; however, this consideration is postponed until the next section.

[5]An Avogadro number (~10^{24}) of electrons deposited from ions in solution produces 1 gram-equivalent (g-eq) of metal passed across an interface between metal and solution; hence 10^{19} to 10^{20} electrons produce 10^{-5} to 10^{-4} g-eq of material.

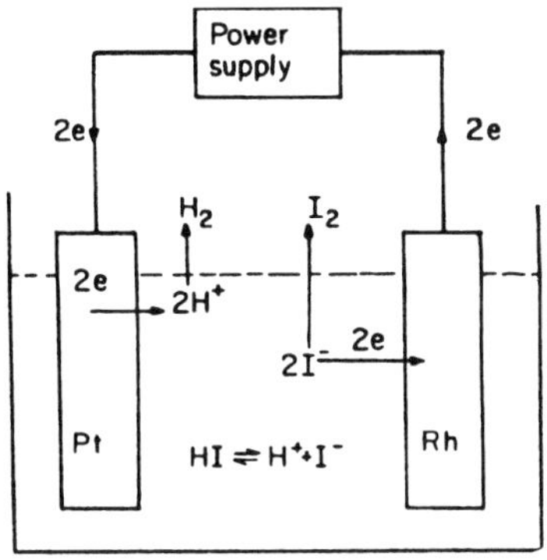

Fig. 1.4. The electrochemical reactor.

The simplicity of such a formulation should not obscure the fact that what has been described is a remarkable and distinctive part of chemistry. An electric current, a controllable electron stream, has been made to react in a controlled way with a chemical substance and produce another *new chemical substance.*[6] That is what a good deal of electrochemistry is about—*it is about the electrical path for producing chemical transformations.* Much of electrochemistry is also connected with the other side of this coin, namely, the production of electric currents *and therefore electric power* directly from changes in chemical substances. This is the method of producing electrical energy without moving parts (see fuel cells, Chapter 13 in vol II).

1.4.3. Electrochemical and Chemical Reactions

There is another aspect of the electrochemical reaction that has just been described. It concerns the effect on the iodide ions of hydrogen iodide, which must also have been present in the HI solution in water. Where do they go while the hydrogen ions are being turned into hydrogen molecules?

The I^- ions have not yet appeared because only half of the picture has been shown. In a real situation, one immerses another electronic conductor in the same solution (Fig. 1.4). Electrical sources have two terminals. The assumption of a power source pumping electrons into a platinum plate in contact with an ionic solution is essentially a thought experiment. In the real situation, one immerses another electronic conductor in the same solution and connects this second electronic conductor to the other terminal of the power source. Then, whereas electrons from the power source pour into the platinum plate, they would flow *away from* the second electronic conductor (made, e.g., of rhodium) and back to the power source. It is clear that, if we want a system that can operate for some time with hydrogen ions receiving electrons from the

[6]There is not much limit on the kind of chemical substance; for example, it does not have to be an ion. $C_2H_4 + 4\,H_2O \rightarrow 2\,CO_2 + 12\,H^+ + 12\,e$ is as much an electrochemical reaction as is $2\,H^+ + 2\,e \rightarrow H_2$.

platinum plate, then iodide ions have to give up electrons to the rhodium plate at the same rate as the platinum gives up electrons. Thus, the whole system can function smoothly without the loss of electroneutrality that would occur were the hydrogen ions to receive electrons from the platinum without a balancing event at the other plate. Such a process would be required to remove the negatively charged ions, which would become excess ones once the positively charged hydrogen ions had been removed from the solution.

An assembly, or system, consisting of one electronic conductor (usually a metal) that acts as an electron source for particles in an ionic conductor (the solution) and another electronic conductor acting as an electron sink receiving electrons from the ionic conductor is known as an *electrochemical cell*, or *electrochemical system*, or sometimes an *electrochemical reactor*.

We have seen that electron-transfer reactions can occur at one charged plate. What happens if one takes into account the second plate? There, the electron transfer is from the solution to the plate or electronic conductor. Thus, if we consider the two electronic conductor–ionic conductor interfaces (namely, the whole cell), there is no *net* electron transfer. The electron outflow from one electronic conductor equals the inflow to the other; that is, a *purely chemical reaction* (one not involving net electron transfer) *can be carried out in an electrochemical cell*. Such net reactions in an electrochemical cell turn out to be formally identical to the familiar thermally induced reactions of ordinary chemistry in which molecules collide with each other and form new species with new bonds. There are, however, fundamental differences between the ordinary chemical way of effecting a reaction and the less familiar electrical or electrochemical way, in which the reactants collide not with each other but with separated "charge-transfer catalysts," as the two plates which serve as electron-exchange areas might well be called. One of the differences, of course, pertains to the facility with which the rate of a reaction in an electrochemical cell can be controlled; all one has to do is electronically to control the power source. This ease of control arises because the electrochemical reaction rate is the rate at which the power source pushes out and receives back electrons after their journey around the circuit that includes (Figs. 1.4 and 1.5) the electrochemical cell.

Thus, one could write the electrochemical events as

$$2\,\mathrm{HI} \xrightarrow[\text{solution}]{\text{in}} 2\,\mathrm{H}^+ + 2\,\mathrm{I}^-$$

$$2\,\mathrm{H}^+ + 2\,\mathrm{e} \xrightarrow[\text{plate}]{\text{at the Pt}} \mathrm{H}_2\uparrow \tag{1.3}$$

$$2\,\mathrm{I}^- \xrightarrow[\text{plate}]{\text{at the Rh}} \mathrm{I}_2 + 2\,\mathrm{e}$$

$$\text{net reaction: } 2\,\mathrm{HI} \rightarrow \mathrm{H}_2 + \mathrm{I}_2$$

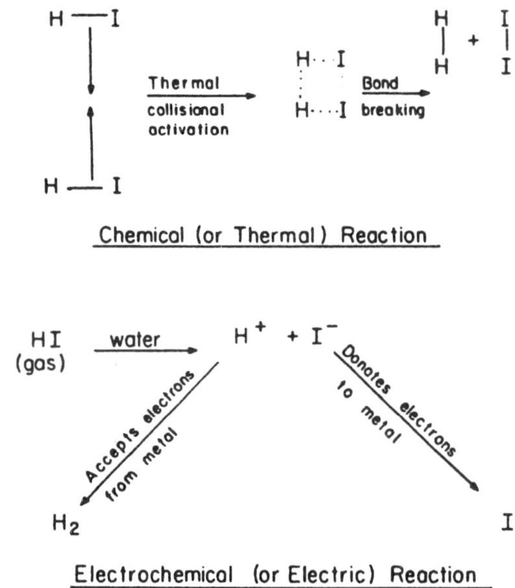

Fig. 1.5. The chemical and electrochemical methods of carrying out reactions. In the electrochemical method, the particles do not collide with each other but with separated sources and sinks of electrons.

Thus, from an overall point of view (not thinking of the molecular-level mechanism), this net cell reaction is identical to that which would occur if one heated hydrogen iodide and produced hydrogen and iodine by a purely chemical, or thermal, reaction.

There is another way in which the electrical method of carrying out chemical reactions is distinct from the other methods for achieving chemical changes (Fig. 1.5). The ordinary reactions of chemistry, such as the homogeneous combination of H_2 and I_2 or the heterogeneous combinations of H_2 and O_2, occur because thermally energized molecules occasionally collide and, during the small time they stay together, change some bonds to form a new arrangement. Correspondingly, photochemical reactions occur when photons strike molecules and give them extra energy so that they break up and form new compounds. In a similar way, the high-energy particles emanating from radioactive substances can energize molecules, which then react. The electrical method of causing chemical transformations is different from the other three methods of provoking chemical reactivity in that the overall electrochemical cell reaction is composed of two separate electron-transfer reactions that occur at *spatially separated*

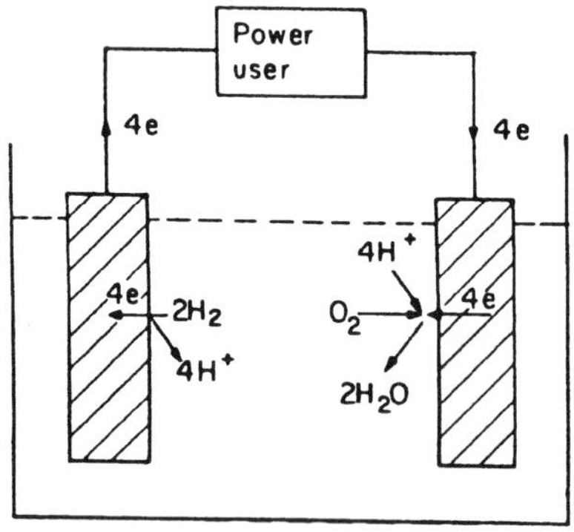

Fig. 1.6. Grove was the first to obtain electric power directly from a chemical reaction.

electrode–electrolyte interfaces and are each susceptible to electrical control as far as rate (electron flow) is possible.

If electrical energy provokes and controls chemical reactions, chemical reactions working in the other direction can presumably give rise to a flow of electricity. Thus, two reactant substances may be allowed to undergo spontaneous electron transfer at the separated (electrode) sites, which is characteristic of the electrochemical way of bringing about chemical reactions, and then the electrons transferred in the two reactions at the two interfaces will surge spontaneously through an electrical load, for example, the circuit of an electric motor (Fig. 1.6). In this reverse process, also, there is a unique aspect when one compares it with the production of available energy from thermally induced chemical reactions. It can be shown (see Chapter 13 in vol 2) that the fraction of the total energy of the chemical reaction that can be converted to mechanical energy is intrinsically much greater in the electrical than in the chemical way of producing energy. This is a useful property when one considers the economics of running a transportation system by means of a fuel cell–electric motor combination rather than by means of the energy produced in the combustion of gasoline.

1.5. THE RELATION OF ELECTROCHEMISTRY TO OTHER SCIENCES

1.5.1. Some Diagrammatic Presentations

Let us look at Fig. 1.7 to see something of the parentage of conventional electrochemistry, in both the ionic and electrodic aspects. We could also look at these relations in a different way and make the central thought a charge transfer at interfaces, while stressing the interdisciplinary character of the fields involved in studying it. Such

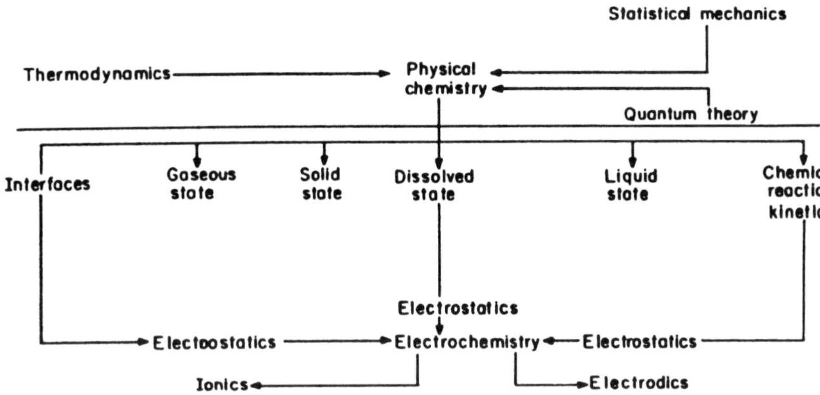

Fig. 1.7. Physical chemistry and electrochemistry.

a view is implied in Fig. 1.8, where space limits the number of disciplines mentioned that are associated with the study of electrified interfaces.

Apart from the large number of areas of knowledge associated with modern electrochemistry, there are many areas to which it contributes or in which it plays an essential role. Thus, much surface chemistry under real conditions involves moisture; hence the electrified interfaces for which electrochemical concepts are relevant are as wide in application as practical surface chemistry itself. This, together with the fact that the subject embraces interactions between electric currents and materials (i.e., between two large areas of physics and chemistry), implies a widespread character for the phenomena subject to electrochemical considerations (Fig. 1.8).

1.5.2. Some Examples of the Involvement of Electrochemistry in Other Sciences

1.5.2.1. *Chemistry.* There are many parts of mainline chemistry that originated in electrochemistry. The third law of thermodynamics grew out of observations on the temperature variations of the potential of electrochemical reactions occurring in cells. The concepts of pH and dissociation constant were formerly studied as part of the electrochemistry of solutions. Ionic reaction kinetics in solution is expressed in terms of the electrochemical theory developed to explain the "activity" of ions in solution. Electrolysis, metal deposition, syntheses at electrodes, plus half of the modern methods of analysis in solution depend on electrochemical phenomena. Many biomolecules in living systems exist in the colloidal state, and the stability of colloids is dependent on the electrochemistry at their contact with the surrounding solution.

1.5.2.2. *Metallurgy.* The extraction of metals from their compounds dissolved in molten salts, the separation of metals from mixtures in solution, and the

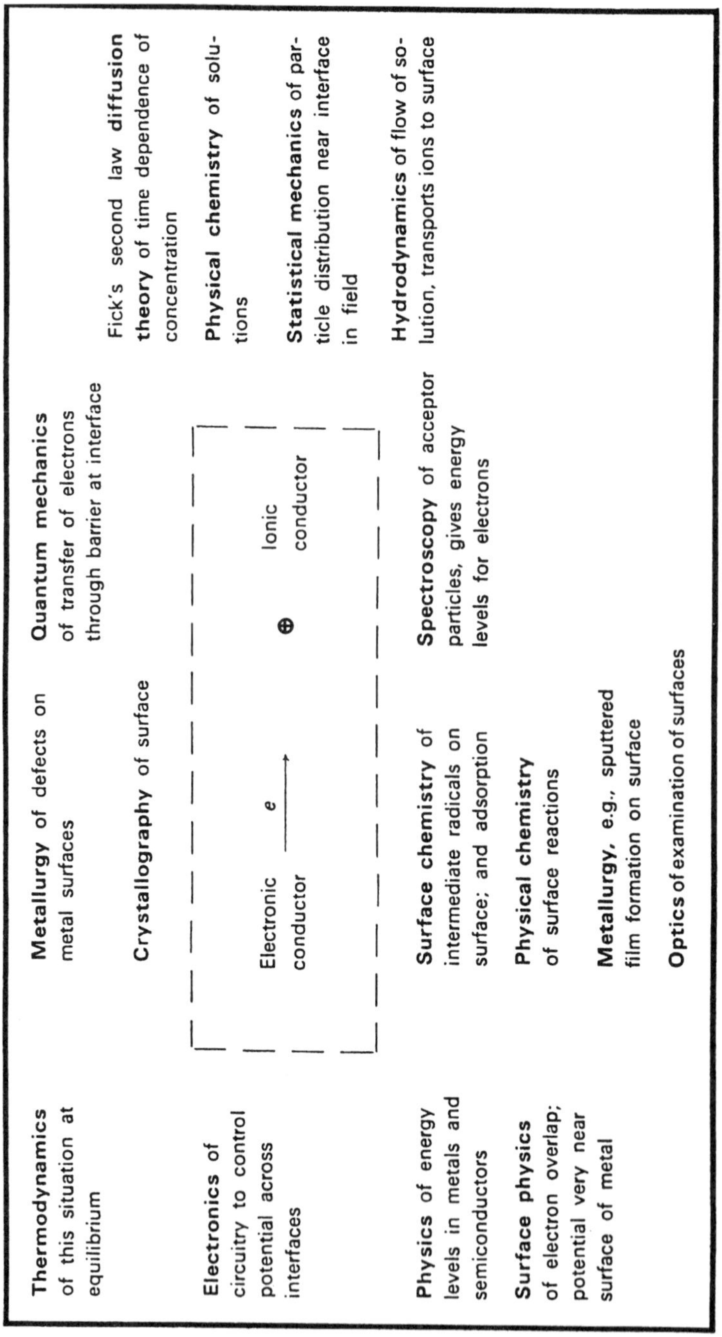

Fig. 1.8. Some disciplines involved in the study of charge transfer at interfaces.

protection of metals from corrosive decay are among the many applications of electrochemistry in metallurgy.

1.5.2.3. *Engineering.* Electrochemical engineering is the basis of a large portion of the nonferrous metals industries, in particular, the production of aluminum by deposition from a molten salt containing aluminum oxide. As noted earlier, electrochemical energy converters, fuel cells, provide the on-board power for space vehicles, and there are prospects of evolving from the thermal to the electrochemical method of utilizing the energy of chemical reactions. One of the most important applications is the prospect of clean, pollution-free electrical on-board power for automobiles. Environmental issues, in general, as well as the real threat of global warming from CO_2 buildup, favor clean electrochemical processes.

1.5.2.4. *Biology.* Food is converted to energy by biochemical mechanisms that have an efficiency much greater than that of some corresponding forms of energy conversion involving the heat-engine principle. Such high efficiency in energy conversion involves electrochemical reactions in the mitochondrion, a part of the biological cell. Correspondingly, the transmission of impulses through nerves, as well as the stability of blood and the functioning of many of the macromolecules involved in biological processes, depends on aspects of electrochemistry that concern electrochemical charge transport and the repulsion between bodies bearing the same electrical charge. The formation of blood clots and the resulting heart attack are influenced by the electrical charge on the arterial wall and that on the colloidal particles in blood.

1.5.2.5. *Geology.* An example of electrochemistry in geology concerns certain types of soil movements. The movement of earth under stress depends on its viscosity as a slurry; that is, a viscous mixture of suspended solids in water with a consistency of very thick cream. Such mixtures of material exhibit thixotropy,[7] which depends on the interactions of the double layers between colloidal particles. These in turn depend on the concentration of ions, which affects the field across the double layer and causes the colloidal structures upon which the soil's consistency depends to repel each other and remain stable. Thus, in certain conditions the addition of ionic solutions to soils may cause a radical increase in their tendency to flow.

1.5.3. Electrochemistry as an Interdisciplinary Field, Distinct from Chemistry

All fields in chemistry (e.g., that of the liquid state or of reaction kinetics) are connected to each other and, indeed, fields treated under chemistry tend, as time goes on, to move toward the more sophisticated level attained in physics. Chemists undertake approximate treatments of relatively complicated problems that are not yet

---

[7]Thus, certain soils, when appropriately agitated, suddenly become much less viscous and start to flow easily, a dangerous thing if there is a house on top.

simplified enough for physicists to approach them in a more exact way. However, the connections between chemistry and physics (e.g., the study of liquids and gaseous reaction kinetics) largely occur through the parent areas of chemistry (see Fig. 1.7), for example, statistical and quantum mechanics. A direct connection to areas just outside chemistry does not immediately follow, e.g., liquids and reaction kinetics.

In *electro*chemistry, however, there is an immediate connection to the physics of current flow and electric fields. Furthermore, it is difficult to pursue interfacial electrochemistry without knowing some principles of theoretical structural metallurgy and electronics, as well as hydrodynamic theory. Conversely (see Section 1.5.2), the range of fields in which the important steps are controlled by the electrical properties of interfaces and the flow of charge across them is great and exceeds that of other areas in which physical chemistry is relevant.[8] In fact, so great is the range of topics in which electrochemical considerations are relevant that a worker who is concerned with the creation of passive films on metals and their resistance to environmental attack is scarcely in intellectual contact with a person who is interested in finding a model for why blood clots or someone seeking to solve the quantum mechanical equations for the transfer of electrons across interfaces.

This widespread involvement with other areas of science suggests that in the future electrochemistry will be treated increasingly as an interdisciplinary area as, for example, materials science is, rather than as a branch of physical chemistry.

At the same time, there is a general tendency at present to break down the older formal disciplines of physical, inorganic, and organic chemistry and to make new groupings. That of materials science—the solid-state aspects of metallurgy, physics, and chemistry—is one. Energy conversion—the energy-producing aspects of nuclear fission, electrochemical fuel cells, photovoltaics, thermionic emission, magnetohydrodynamics, and so on—is another. Electrochemistry would be concerned with the part played by electrically charged interfaces and interfacial charge transfers in chemistry, metallurgy, biology, engineering, etc.

1.6. THE FRONTIER IN IONICS: NONAQUEOUS SOLUTIONS

Studies of ionic solutions have been overwhelmingly aqueous in the hundred years or so in which they have been pursued. This has been a blessing, for water has a dielectric constant, ε, of ~80, about ten times larger than the range for most nonaqueous solvents. Hence, because the force between ions is proportional to $1/\varepsilon$, the tendency of ions in aqueous solutions to attract each other and form groups is relatively small, and structure in aqueous solutions is therefore on the *simple* side. This enabled a start to be made on the theory of ion–ion attraction in solutions.

---

[8]As apart from areas of basic science (e.g., quantum mechanics) that primarily originate in physics and underlie all chemistry, including, of course, electrochemistry.

If one goes to a nonaqueous solvent system (an organic one such as acetonitrile, CH_3CN; or a pure electrolyte such as the KNO_3-$NaNO_3$ eutectic), the dielectric constant is more in the range of 2–20, and there is a greater tendency than that in aqueous solutions for ions of opposite sign to get together and stay together. Further, the bonding that develops is not purely Coulombic but may involve solvent and H-bond links, some of them unexpected. Figure 1.9 shows dimethyl sulfoxide (DMSO) with the molecular distances in picometers.

Electrochemical measurements (mainly conductances) have been made in both the organic and the pure electrolyte kind of nonaqueous solution for at least two generations. Why, then, is there talk of nonaqueous electrochemistry as one of the *frontiers* of the field?

One reason is that much better methods of detecting impurities (parts per billion) now exist and hence of purifying solvents (and keeping them pure—they all tend to pick up water). However, there is a greater reason and that is the emergence of several new methods for determining structure. These are

1. X-ray diffraction measurements *in solution* (a development of the earliest method of structural determination)
2. X-ray absorption measurements of fine structure in solution (EXAFS)
3. Neutron diffraction
4. Infrared (IR) and Raman spectroscopy

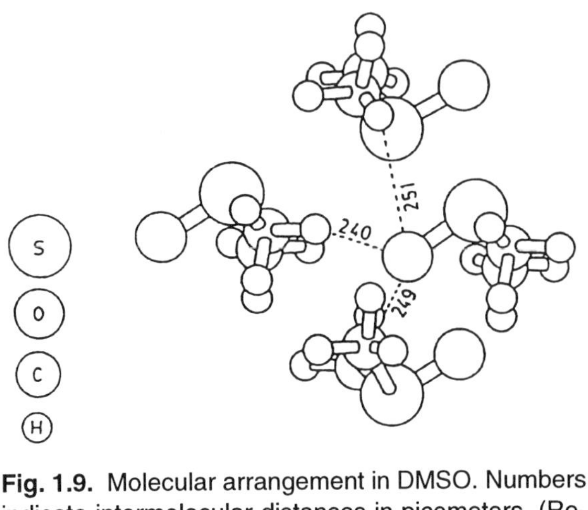

Fig. 1.9. Molecular arrangement in DMSO. Numbers indicate intermolecular distances in picometers. (Reprinted from G. Mamantov and A. I. Popov, *Chemistry of Nonaqueous Solutions: Current Progress*, p. 188. VCH Publishers, New York, 1992.)

Later in this book we will discuss these new tools, how they work, and what they bring to electrochemistry. They have provided the ionics electrochemist with a new kind of microscope.

Although this is still introductory material, three rather general points may be made.

1. Until the 1980s, the major methods of investigation in ionics were nonspectroscopic. For example, conductance results were used to infer the existence of complex ions. Alternatively and typically, the change of the dielectric constant of a solution as a function of the concentration of ions (measurements at various frequencies) was interpreted in terms of structural hypotheses about ion–solvent interactions.

The new optical and spectroscopic methods are more discerning, more definite in what they reveal. For example, in solutions of $AgNO_3$ in water and in dimethylformamide, one used to speak of ion-pair formation, $Ag^+NO_3^-$. By now it is known that there are several kinds of ion pairs. For example, the ions may be in direct contact or they may be separated by a solvent molecule. The concentration of the free ions (if they give vibration spectra) can be followed. In general, an enormous increase in detail (corresponding to an increase in knowledge of the variety of particles present in the nonaqueous systems) has become available.

2. Amazing bonds have been revealed by the new methods. For example, Perelygin found that thiocyanates of the alkali metals form groups in acetonitrile. Valence theory is sometimes hard put to interpret the unusual forms found:

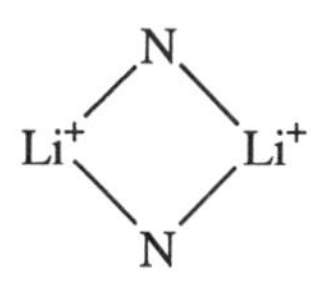

3. Few measurements of the so-called "driving force,"[9] the Gibbs free energy, $\Delta G°$, have become available as yet; however, for a number of reactions in organic nonaqueous solutions it is *entropy* driven, that is, $\Delta G°$ is driven to a negative value over the positive (endothermic) $\Delta H°$ by a positive $\Delta S°$ and its influence as $T\Delta S°$ in the basic thermodynamic equation: $\Delta G° = \Delta H° - T\Delta S°$.

Necessity does seem to be the mother of invention. This nonaqueous electrochemistry has great practical value, for example, in new high-energy-density batteries and fuel cells—just the things needed for electricity storage and production (respectively) on board nonpolluting electric cars.

[9]Clearly, a standard free *energy* difference cannot be a driving *force*. However, the larger the negative value of $\Delta G°$, the more will be the tendency of a reaction, with reactants and products in their standard states, to proceed.

1.7. A NEW WORLD OF RICH VARIETY: ROOM-TEMPERATURE MOLTEN SALTS

The image in most chemists' minds of a "molten salt" is probably liquid NaCl at 900 °C. There are many such "pure electrolytes" that have the great advantage of a large electrochemical "window"; that is, one can carry out electrode processes in them over a much greater range of cell potentials than is possible in aqueous solutions, where the range is limited by the evolution of H_2 if the potential becomes too negative and of O_2 if it becomes too positive.

This situation has been radically altered during the past 20 years or so by work that has been led (in separate and individual ways) by three U.S. electrochemists, namely, Osteryoung, Hussey, and Wilkes, respectively. Thus, the first truly room-temperature pure electrolyte (i.e., a system consisting of ions without a solvent) was due to Osteryoung et al. in 1975. This is 1-(1-butylpyridinium chloride) (BupyCl):

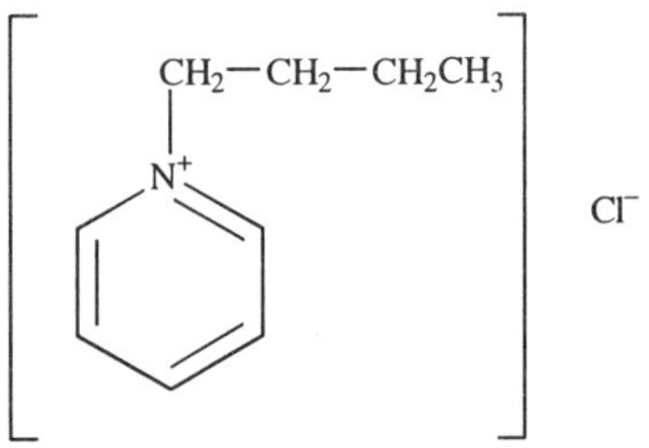

Wilkes, in particular, has developed the use of 1-methyl-3-ethylimidazolonium chloride (MeEtInCl):

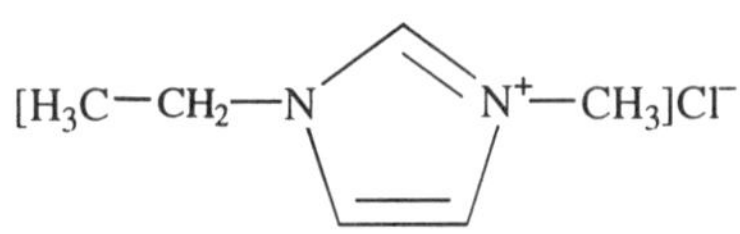

Compounds of these two electrolytes are leading members in the extensive development of this new chemistry. For example, it has been possible to investigate tetrachlorobenzoquinone in $AlCl_3$-BupyCl at an electrode consisting of glassy carbon; changing the potential of this electrode changes the oxidation state of the quinone. The resulting absorption spectrum is shown in Fig. 1.10. Reactions involving Cu, Ag, Au, Co, Rh, Ir, Mo, W, etc. have all been investigated in room-temperature molten salts.

A fertile field exists here for batteries and fuel cells: a rechargeable couple involving the considerable electrical energy that can be stored in Al and O_2 can be developed. In the first aqueous Al cell, developed by Solomon Zaromb in 1960, the product of the anodic dissolution of Al was the insoluble $Al(OH)_3$, and no electrical recharge was possible.

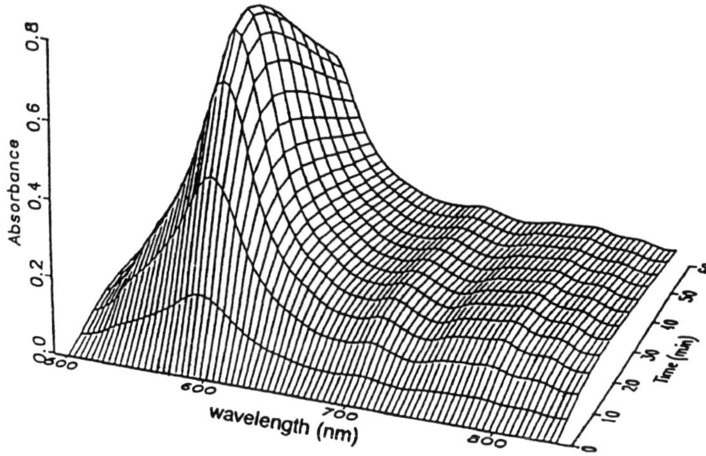

Fig. 1.10. Absorption spectra of a 4.59-m*M* solution of perylene in 44.4–55.6 mol% $AlCl_3$-MeEtimCl melt taken at 5-minute intervals during 60-min electrolysis at a platinum screen optically transparent electrode (OTE). The applied potential was −1.85 V, and the OTE path length was 0.10 cm. (Reprinted from J. E. Coffield, G. Mamantov, S. P. Zingg, G. P. Smith, and A. C. Buchanan, *J. Electrochem. Soc.* **139**: 355, 1992.)

1.8. ELECTROCHEMICAL DETERMINATION OF RADICAL INTERMEDIATES BY MEANS OF INFRARED SPECTROSCOPY

Electrochemical reactions on electrodes involve consecutive reactions with several steps. Knowledge of the reactants and products in each step may provide a valuable piece of evidence by means of which the pathway—and sometimes even the rate-determining step—can be identified.

Two difficulties exist before this can be done. First, the concentrations of surface species are, at most, 10^{-9} mol cm^{-2}, about 10^{15} atoms cm^{-2}, depending on the size of the radical or intermediate molecule to be observed. However, light for any spectroscopic method has to go through a layer of solution before it strikes the electrode. Now a square centimeter of a 0.1 *M* solution 1 mm thick contains about 10^{19} ions. The signal from the radicals on the electrode has to compete with much stronger signals from this layer.

To overcome these hurdles, one has to have a supersensitive measurement and then some way of separating the *surface* signal from competing signals of the same frequency produced by molecules or ions in the adhering solution.

How this could be done was first shown by Neugebauer and co-workers in 1981 (but it was developed particularly by Alan Bewick and Stan Pons in the 1980s). In spectroscopy, in general, it is possible to enhance and elicit a given line by repeating its spectra many hundreds of times. The signal is enhanced when the spectrum is repeated if it always occurs at the same frequency. Any false blips in the intensity–wavelength relation are determined by random fluctuations—they won't be enhanced by repetition because they don't always occur at the same wavelength. To separate the surface signal from the solution spectrum Greenler's theorem was used. According to this theorem, if the incident angle of the beam from the vertical is very high (e.g., 88–89°), there is a radical difference between the information carried by the parallel and vertical components of the polarized light beam reflected through the solution from the electrode. The parallel component carries both the surface and the solution information; the vertical carries only the solution information. Hence, if the polarities of beams reflected from the surface of an electrode in solution are alternated from parallel to vertical and then vertical to parallel several hundred thousand times per second, and the strength of the signals of the vertical components is subtracted from that of the parallel ones at various wavelengths in the IR region, there should remain (according to Greenler's theorem) the lines characteristic of the surface species only.

Of course, this is a rough outline of a sophisticated and complex technique. The solution layer in contact with the electrode should be very thin to reduce competition

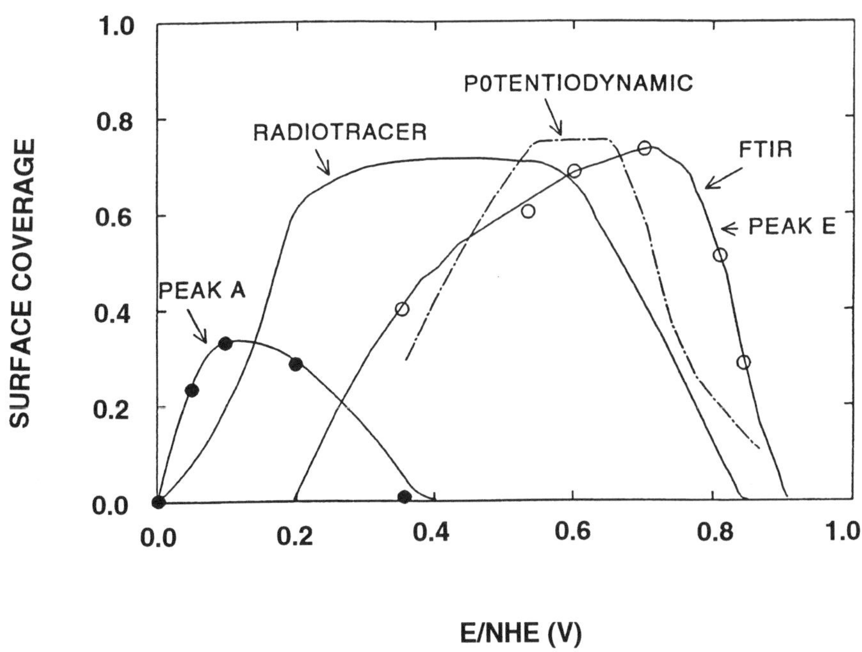

Fig. 1.11. Relative areas of two peaks (A and E) as a function of electrode potential. (Reprinted from K. Chandrasekaran, J. C. Wass, and J. O'M. Bockris, *J. Electrochem. Soc.* **137**: 519, 1990.)

from the ions and molecules in solution (separating parallel and vertical beam information by subtracting the vertical from that of the parallel signal is only an approximation for metal–solution systems). This technique is called Fourier transform IR absorption spectroscopy (FTIRAS). The Fourier transform is used in the mathematics of calculating the results. While the process is complex, the results are relatively simple and helpful. One of these is shown in Fig. 1.11.

This methodology reveals the tell-tale signal of bonds present in intermediate radicals on the electrode surface if the electrochemical reaction is operating in a steady state, i.e., the reaction is in its final pathway with the final rate-determining step. The exciting possibility of advanced versions of the FTIRAS technique is that really rapid (millisecond) changes in the spectra of the surface radicals can be recorded. Then, changes in the nature and concentration of radicals that may occur during the switching-on phase of the electrical current could be measured. This can provide much information on the buildup and structure of the pathway of the reaction occurring on the surface. When this happens, fast-reacting surface spectroscopic measurements will become a principal aid to studies of mechanisms in electrochemistry.

1.9. RELAY STATIONS PLACED INSIDE PROTEINS CAN CARRY AN ELECTRIC CURRENT

The body contains several thousand different enzymes that are catalysts for specific bioreactions. Without them, biochemical pathways in the body would not function.

Diabetics have a need for a glucose meter that would show the glucose concentration in the blood at any moment without having to take a sample. If the glucose builds up too much, the meter would send out a signal telling the wearer of the need for insulin.

Glucose gets oxidized with the cooperation of an enzyme called glucose oxidase, which has a molecular diameter of 86 Å. Suppose we could immobilize glucose oxidase on an organic semiconductor such as polypyrrole, and the electrons produced when glucose in the blood is oxidized could be brought out through the glucose oxidase to work a meter on the wrist: our aim would be achieved, and diabetics could monitor their condition at any time by a glance at the wrist.[10]

One could carry this idea further. Since enzymes are so specific in reacting to only certain molecules, one could imagine a future "general diagnoser," a plaque with a series of enzymes adsorbed on microelectrodes that are exposed to the blood, with each able to pick out the molecules that indicate the presence of specific diseases. Oxidation reactions would provide electrons and an electrical signal would indicate the disease through the circuitry, which would identify the enzyme from which the current originated. For such an idea to work, one must have electronic conductivity of

[10]A wrist glucose meter? Not yet, but several are in development.

the enzyme because eventually the electrons from the reaction on its surface (in contact with the blood) must get through the 10 nm of enzyme to the metallic circuitry of the wrist meter.

Now, most enzymes are centered on a specific redox atom (e.g., Fe) and in order to be oxidized or reduced, the electron, the effector of the act, must travel through the enzyme to the so-called heme group, the vital Fe-containing group, as in hemoglobin, for example.

Enzymes are complex organic substances and are not expected to be good electron conductors at all. If an electron is going to get to and from the heme group to an outside contact, the best hope is quantum mechanical tunneling. However, there is a limit to the jump length in tunneling; it is about 2 nm. Supposing the heme group is in the middle of the enzyme glucose oxidase; then, as it is ~9 nm across, the electron would have to jump ~4 nm, which is not possible.

Adam Heller in 1986 devised and achieved the solution to this, which is illustrated in Fig. 1.12. With his associate, Y. Degani, Heller introduced extra redox centers into

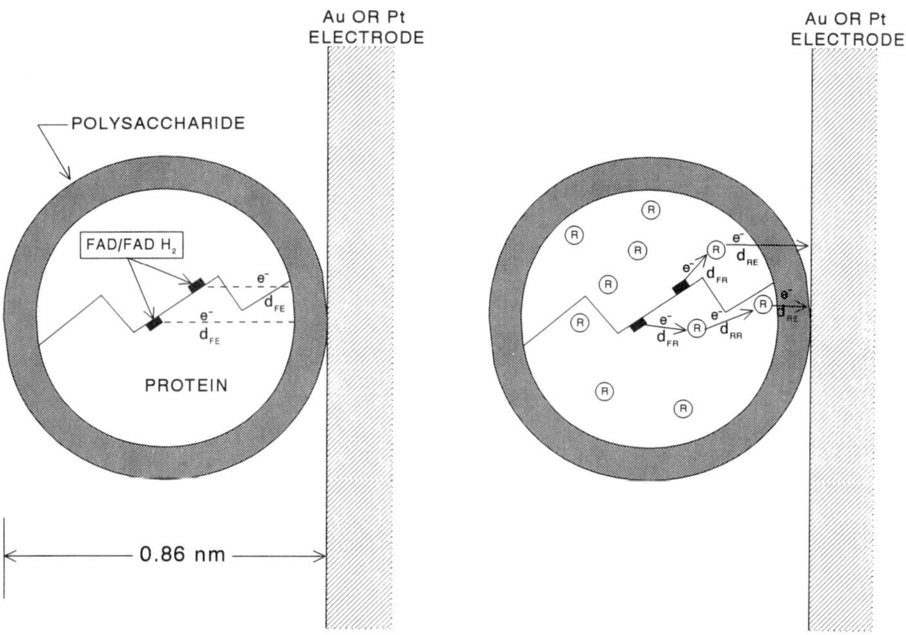

Fig. 1.12. Schematic drawing of the glucose oxidase molecule, showing the electron-transfer distances involved in the various steps of moving an electron from its two flavin adenine dinucleotide/reduced flavin adenine dinucleotide ($FAD/FADH_2$) centers to a metal electrode. Left: The enzyme before modification. Right: The modified enzyme, after chemical attachment of an array of electron transfer relays ("R"). (Reprinted from Y. Degani and A. Heller, *J. Phys. Chem.* **91**: 1286, 1987.)

glucose oxidase. The result was duly electrifying—the enzyme became radically more conducting than before.

1.10. SPECULATIVE ELECTROCHEMICAL APPROACH TO UNDERSTANDING METABOLISM

We eat and various biochemical processes produce glucose from some of the food molecules. We breathe and the oxygen changes the glucose to gluconic acid, and finally by way of the ubiquitous enzymes, to CO_2. On the way, we get mechanical energy to operate our muscles, including the heart pump, which drives the fuel-carrying blood around the body circuit. This is metabolism, and one thing about it that is not well understood is why the energy conversion (chemical energy of the oxidation of food to mechanical energy of the body) is so efficient (50%), compared with the efficiency of a normal heat engine (~25%).

The reaction that gives chemical energy to heat engines in cars (hydrocarbon oxidation) has to obey the Carnot cycle efficiency limitation $(1 - T_{\text{low}}/T_{\text{high}})$. With T_{low} around body temperature (37 °C), the metabolic T_{high} would have to be 337 °C to explain this metabolic efficiency in terms of a heat engine. Thus, the body energy conversion mechanism cannot use this means to get the energy by which it works.

However, there are electrochemical energy converters (fuel cells), such as the one shown in Fig. 1.6. An electrochemical energy converter is not restrained by the Carnot efficiency of 25% and can have efficiencies up to $\Delta H/\Delta G$ for the heat content and free energy changes in the oxidation reactions involved in digesting food. This ratio is often as much as 90% (cf. Chapter 13).

Hence, to explain the high metabolism of > 50%, we are forced into proposing an electrochemical path for metabolism. How might it work? Such a path was proposed by Felix Gutmann in 1985 and the idea is shown in a crude way in Fig. 1.13.

Mitochondria are tiny systems found in every biological cell, and they are known to be the seat of the body's energy conversion. Suppose one could identify certain organic groups on the mitochondria as electron acceptors and other groups as electron donors—microelectrodes, in fact. Glucose diffuses into the cell and becomes oxidized at the electron acceptors. At the electron donor groups, O_2 is reduced using the electrons provided by the glucose. The mitochondrion has made millions of micro fuel cells out of its two kinds of electrodic groups and now has electrical energy from these cells to give—and at an efficiency typical of fuel cells of ~50%.

There is much more to the story—how energy in living systems is stored, for example, and finally how it is transported to all the body parts which use it—this is discussed in Chapter 14. This electrochemical (and vectorial) approach to metabolism, which was proposed in 1986, is not yet widely accepted by tradition-bound biologists, but it has one tremendous thing going for it—it solves the problem of why the efficiency of the conversion of the chemicals in food to mechanical energy is so much higher than it can be in alternative energy conversion pathways.

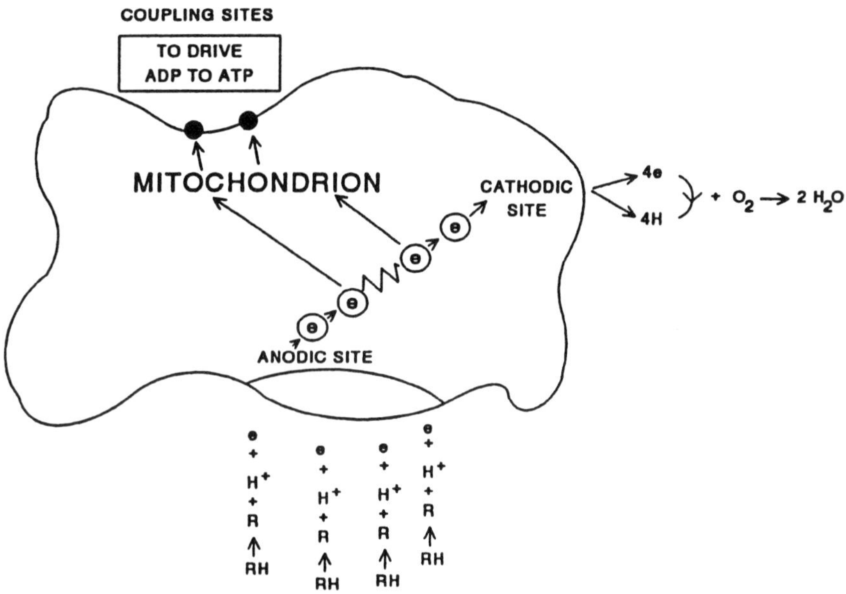

Fig. 1.13. Schematic representation of the principle of the biological fuel cell concept. R and RH represent the oxidized and reduced form of a bio-molecule. ADP is adenosine diphosphate; ATP is adenosine triphospate. (Reprinted from J. O'M. Bockris and S. U. M. Khan, *Surface Electrochemistry*, p. 699. Plenum Press, New York, 1993.)

1.11. THE ELECTROCHEMISTRY OF CLEANER ENVIRONMENTS

Many of the problems of the environment are caused by the fact that the main method by which we obtain the energy to run our civilization is still by means of the combustion of fossil fuels. This has two unacceptable effects. The first—and little realized—is that inhaling gasoline vapors has proved to be carcinogenic to laboratory animals, and there is an indication that it is one of the causes of the increased spread of cancer among humans. The second is that the CO_2 injected into the atmosphere by combustion is causing global warming.

Electrochemical Technology is involved in a major way in the only alternative to oil and gasoline as an energy source which also avoids the hazards of nuclear power. Thus, the collection of solar light and its conversion by means of photovoltaics to electricity and then the electrochemical splitting of water to yield clean hydrogen, would provide an inexhaustible replacement for fossil fuels (see Chapter 15). A pilot plant to do this now exists at Neunburg vorm Wald, in southern Germany.

However, electrochemistry will have to play a broader role than this if sustainable and clean energy is to become a reality. It provides a general approach that is superior to conventional chemical approaches for two reasons.

1. Concentration and temperature determine the rate at which chemical processes take place. Electrochemical processes are equally affected by these variables, but also are controlled in selectivity of reaction by the electrical potential of the electrode (see Section XXX). Hence, there is an extra variable that controls chemical processes that occur electrochemically rather than chemically. Moreover, this variable is applied easily by turning the knob on an electrical power source.

2. When electrochemical processes are used to clean up carbonaceous material, the only gas produced is CO_2—there are no noxious products of partial combustion, such as NO and CO, to be injected into the atmosphere. When hydrogen is used in a fuel cell to produce electric power, it is made by splitting water and it produces water right back again as a by-product of the power generation.

This book contains several examples of electrochemical clean-up processes (see Chapter 15), but one is briefly described here. It is in the cleanup of wastewater, defined as water having impurities in the range of 5–500 ppm. There is a problem in using an electrochemical approach because of the low electrical conductance of the water. However, in one of the several electrochemical companies developed around Texas A&M University, Duncan Hitchens has solved the problem as shown in Fig. 1.14.

A proton exchange membrane on the right draws off protons cathodically; the platinized iridium in the middle presents a large area for anodic oxidation. The arrangement makes the current pathway so small that the low conductivity of the

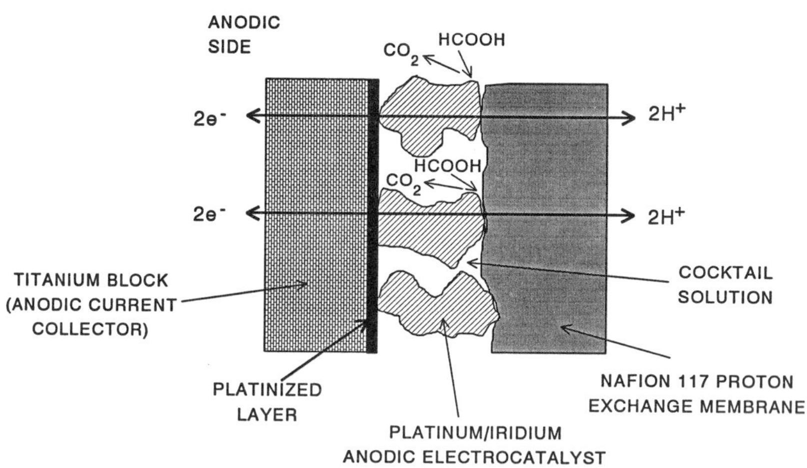

Fig. 1.14. Schematic representation of wastewater treatment process. (Reprinted from O. J. Murphy, G. D. Hitchens, L. Kaba, and C. E. Verostko, *Water Res.* **26**: 443, 1992.)

polluted water does not matter. By flowing wastewater through an electrochemical reactor of this type, impurities are reduced about a thousand times.

1.12. SCIENCE, TECHNOLOGY, ELECTROCHEMISTRY, AND TIME

1.12.1. Significance of Interfacial Charge-Transfer Reactions

It is informative in this chapter to make some attempt to place electrochemistry among the sciences and obtain the relative measure of its significance. It is difficult to do this for a field, the modern phase of which dates back only 50 years. Let us try, but let us realize that what we are doing here is speculating, although we will give some of the reasoning that supports our conclusions.

First, we can ask a relative question: Is interfacial electrochemistry simply a special aspect of reaction kinetics, somewhat analogous to photochemistry? In photochemistry, one might say, one studies the effect of energy packets (photons) striking molecules; in electrochemistry, one studies the effect of striking molecules dissolved in solution with electrons emitted from electrically charged conductors.

Or does interfacial electrochemistry have a greater significance? The evidence for tending to this latter view is as follows:

1. Interfacial electrochemistry has a high degree of prevalence in the practical world compared with other branches of knowledge outside physics. A way to appreciate this is to realize how often one is concerned with electrochemical phenomena outside the laboratory. For example, one starts a car and listens to its radio on battery power; television pictures are transmitted from space vehicles to the earth by fuel-cell power; a sports car may be made of electrochemically extracted aluminum; the water that is in one's coffee may be obtained by electrochemical deionization from impure or brackish water; some persons ride in electrically powered cars; one wears clothes of nylon produced from adiponitrile, which is electrochemically synthesized; one adds an inhibitor to the radiator fluid of a car to reduce electrochemical corrosion. Finally, one thinks by using electrochemical mechanisms in the brain, and one's blood remains liquid as long as the electrochemical potential at the interface between the corpuscles and their solution remains sufficiently high and of the same sign.

2. This ubiquitous role for electrified interfaces throughout many aspects of science suggests that electrochemistry should not be regarded as only a *branch* of chemistry. Rather, while most chemists have concentrated upon *thermally activated* reactions and their mechanisms, with electrochemical reactions as some special academic subcase, there is a parallel type of chemistry based not on the collisions of molecules and the energy transfers that underlie these collisions but on interfacial electron transfers. It is this latter chemistry that seems to underlie much of what goes on in the world around us, for example, in photosynthesis, in metabolism, and in the

decay of metallic structures. The examples of alternative thermal and electrical approaches to common chemical events given in Table 1.1 may illustrate this.

1.12.2. The Relation between Three Major Advances in Science, and the Place of Electrochemistry in the Developing World

If one stands sufficiently removed from one's specialization in the sciences and looks as far back as the nineteenth century, then three great scientific contributions stand out as measured by their impact on science and technology. They are

1. The electromagnetic theory of light due to Maxwell (nineteenth century)
2. The theory of relativistic mechanics due to Einstein (nineteenth century)
3. The theory of quantum mechanics originating with the work of Planck, Einstein, and Bohr and developed by Schrödinger, Heisenberg, Born, and Dirac (twentieth century)

It is important to recall why these contributions to physics and chemistry are regarded as so outstanding. Maxwell's theory provides the basis for the transfer of energy and communication across distances and the delivery of mechanical power on command as a consequence of the controlled application of electric currents causing magnetic fields. The significance of relativistic mechanics is that it helps us understand the universe around us—great bodies, far away, traveling very fast. Quantum mechanics is the basis of solid-state devices and transistor technology. It led to a revolution in thinking; for example, we now know that macroscopic and microscopic systems behave in fundamentally different ways. It allows us to understand how small particles can penetrate barriers otherwise insurmountable (a realization that helps us understand the functioning of fuel cells). A fourth contribution, the discovery of *over unity* processes—machines that seem to produce more energy than was put in—hold promise for great advances. However, this work is still in a very early stage and we cannot yet gauge its technological significance.[11]

How is the eventual magnitude of a contribution in science weighed? Is it not the degree to which the applications that arise from it eventually change everyday life? Is not the essence of our present civilization the attempt to control our surroundings?

It is in this light that one may judge the significance of the theory of electrified interfaces and thus electrochemistry. It is of interest to note how interfacial charge-transfer theories are based on a combination of the electric currents of Maxwell's theory and the quantum-mechanical tunneling of electrons through energy barriers.

---

[11]Over-unity machines are claimed by their inventors to be emerging, as seen in the late 1990s. Their mechanisms are far from clear as yet. They seem to involve nuclear reactions under very low temperature situations, or alternatively, they are machines that are claimed to convert the zero point energy of their surroundings to electricity. If (as seems likely) they become commercialized in the early decades of the new century, the cost of electricity will fall and a great augmentation of the electrical side of chemistry will occur.

TABLE 1.1

Examples of Alternative Thermal and Electrochemical Paths in Chemical Events

Phenomenon or Process	Thermal	Electrochemical
The determination of free-energy changes and equilibrium constants in chemical reactions	Determine equilibrium constant and use $\Delta G^\circ = -RT \ln K$	Determine thermodynamic cell potential and use $\Delta G^\circ = -nFE$
Synthesis, e.g., water from hydrogen and oxygen	Occurs heterogeneously presumably by non-charge-transfer collisional processes $H_2 + \frac{1}{2}O_2 \rightarrow H_2O$	Occurs in electrochemical cell by reactions $H_2 \rightarrow 2\,H^+ + 2\,e$ $\frac{1}{2}O_2 + 2H^+ + 2\,e \rightarrow H_2O$ $H_2 + \frac{1}{2}O_2 \rightarrow H_2O$
Biochemical digestion	Series of enzyme-catalyzed chemical reactions	Some enzymatic reactions may act through electrochemical mechanisms analogous to the local cell theory of corrosion
Many so-called chemical reactions, e.g., chemical synthesis of Ti	$TiCl_4 + 2\,Mg \rightarrow Ti + 2\,MgCl_2$ (apparently a thermal collisional reaction)	$2\,Mg \rightarrow 2\,Mg^{2+} + 4e$ $Ti^{4+} + 4\,e \rightarrow Ti$ $4\,Cl^- \rightarrow 4\,Cl^-$
Production of electrical energy	$H_2 + \frac{1}{2}O_2$ explodes, produces heat, expands gas, causes piston to move, and drives generator	H_2 and O_2 ionize on electrodes, as above in this column, and produce current
Storage of electrical energy	Electricity pumps water up to height and allows it to fall on demand to drive generator	Allow to cause some electrochemical change, e.g., $Cd^{2+} + 2\,e \rightarrow Cd$, $Ni^{2+} \rightarrow Ni^{4+} + 2\,e$, which will be reversed on demand
Synthesis of inorganic and organic material, e.g., Al, adiponitrile	$2\,Al_2O_3 + 3\,C \rightarrow 3\,CO_2 + 4\,Al$; Tetrahydrofuran $\rightarrow$ 1,4-dichlorobutane $\rightarrow$ adiponitrile	$Al^{3+} + 3\,e \rightarrow Al$ $H_2C{=}HC-CN \xrightarrow[\text{coupling}]{\text{cathodic}} NC-(CH_2)_4-CN$
Spreading of cracks through metal	Amount of stress at the apex of a crack per unit area is so high that the crack is propagated into the metal bulk	Bottom of crack dissolves anodically, obtaining current from local cell formed with surface (which is an electron donor, probably to O_2 from air)

A number of illustrations have been given to support the statement that electrochemical mechanisms are relevant to many fields of science. The nineteenth century contributed to physics the theory of electromagnetism. The twentieth century contributed to physics the relativistic theory and the quantum theory. In the twenty-first century, it seems reasonable to assume that the major preoccupations will be in the direction of working out how we can make a sustainable world that continues to have an abundant supply of energy, and that does not suffocate in its own refuse or become too hot to live on because of the continued use of oil and coal as fuels.

Two very general types of probable advances can be expected. One is in the direction noted; for example the development of practical devices for the production of cheap electricity, or the development of prosthetic devices operated by the electric circuits of the body.

The other type will be those developments necessary as a result of the interference with nature over the past 50 years; for example, electrically powered vehicles to avoid increasing the CO_2 content of the atmosphere and processes to reduce pollution in our water supply.

Electrochemistry is the core science upon which many of these electrically oriented advances of the next few decades will depend for their practical execution (see Fig. 1.15). It underlies electrically powered synthesis, extractions (including fresh from brackish water), machining, stabilization of materials, storage of energy in the form of electricity, efficient conversion to electricity of the remaining fossil fuels, and

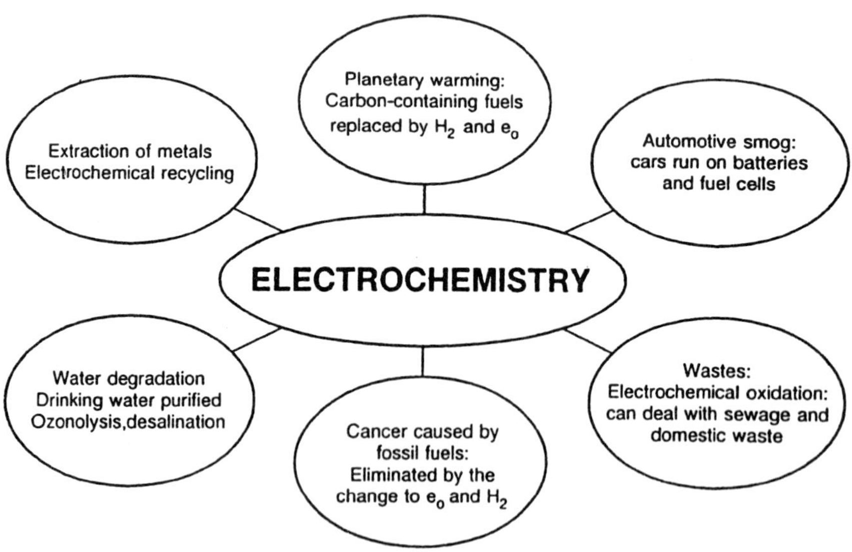

Fig. 1.15. Electrochemistry as a core science for the development of a sustainable society.

the basis for understanding much in molecular biology. The fixing of CO_2 and its photoconversion to methanol (a fuel) is also a prospect.

It is worthwhile thinking also that urban areas are likely to develop as a function of the availability of electricity from solar or nuclear sources. There will also be an increasing need to invest resources in preventing the exhaustion of many vital metals.[12]

The forms in which energy will be transmitted will be electricity with hydrogen as a storage medium. Towns in the 21st century will tend to be self-contained. Little material mass will leave or enter them. The processes on which towns will run will be all electrical, and those involving matter therefore electrochemical. Transportation will use energy stored in hydrogen and in condensers. Manufacturing and machining processes and recovery of materials used or discarded will all be electrochemical. Polluted liquids will be cleaned in packed bed electrolysers (Fig. 1.16). Wastes will be processed electrochemically in molten salts. Medical electronics—the electronic-electrodic combination in medical research—will be highly developed toward various combinations of humans and machines.

Thus, it seems reasonable to expect the achievement of several electrochemically based innovations by 2050: the provision of cheap heat electrically from storage units charged during off-peak times; electrochemically powered vehicles, including ships; an economical and massive solar conversion system; hydrogen storage and transmission to avoid systems that add further CO_2 to the atmosphere; extensive use of electrochemical machining and electrochemically based tools; an internal fuel-cell-powered heart; and electrometallurgical extractions of materials on a large scale from the moon (their transfer to earth will be easy because of the moon's low gravity). An immediately developable area lies in the electrochemical aspects of molecular biology (the replacement of electrically functioning body parts) and in the development of circuitry that will join the brain and its electrochemical mechanisms to artificial limbs with their electrochemical functions and perhaps even to circuits not connected to the body. Cyborgs, the person–machine combination, will become a part of life.

Let us therefore read this book with some sense of where we are on the scale of time, in the development of that great revolution begun in the eighteenth century. For it was then that we discovered how to make heat give mechanical power. However, this great discovery, and all that it has produced, has brought with it an unacceptablc penalty—the pollution and planetary warming caused by the use of heat to produce mechanical power. We are just at the point on the time scale where we must wean ourselves away from oil (that mother's milk) and other CO_2-producing fuels that ran the first century of technology and find how to support the population of our overburdened planet by the use of fusion energy (from the sun, itself, or perhaps from the benign energy of low-temperature nuclear reactors). However, as we move away from pollution, CO_2, and planetary warming, it is certain that a greatly enhanced

[12]Only iron and aluminum are present in amounts to last hundreds of years. Unless they are recycled, many metals will be exhausted in the twenty-first century.

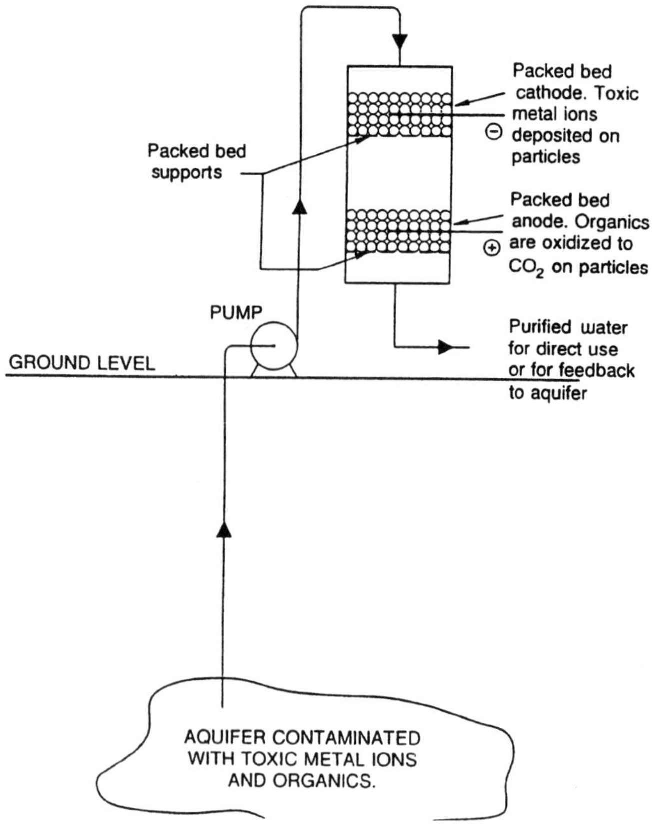

Fig. 1.16. Electrochemical treatment of a contaminated aquifer.

dependence on electricity as a clean source of energy will have to occur. This book stands as a guide on how to achieve this great transition to a sustainable world.

Further Reading

Books

1. D. B. Hibbert, *Dictionary of Electrochemistry*, Wiley, New York (1984).
2. A. J. Fry and W. E. Britton, eds., *Topics in Organic Electrochemistry*, Plenum Press, New York (1986).
3. J. Goodisman, *Electrochemistry: Theoretical Foundations, Quantum and Statistical Mechanics, Thermodynamics, Solid State*, Wiley, New York (1987).
4. O. Murphy, S. Srinivasen, and B. E. Conway, *Electrochemistry in Transition: 20–21st Century*, American Chemical Society, Washington D.C. (1988).

5. K. Scott, *Electrochemical Reaction Engineering*, Academic Press, New York (1993).
6. V. S. Bagotsky, *Fundamentals of Electrochemistry*, Plenum Press, New York (1993).
7. N. Masuko, O. Tetsuya, and Y. Fukunaka, *New Trends and Approaches in Electrochemical Technology*, Kodansha, Tokyo (1993).
8. J. Lipkowski and P. N. Ross, *Structure of Electrified Interfaces*, VCH Publishers, New York (1993).
9. J. O'M. Bockris and S. U. M. Khan, *Surface Electrochemistry*, Plenum Press, New York (1993).
10. J. Wang, *Analytical Electrochemistry*, VCH Publishers, New York (1994).
11. *Electrochemical Engineering in the Environment*, Institution of Chemical Engineers, Rugby, UK (1994).
12. C. A. C. Sequeira, *Environmentally Oriented Electrochemistry*, Elsevier, Amsterdam (1994).
13. R. C. Alkire and D. M. Kolb, Eds., *Advances in Electrochemical Science and Engineering*, V.C.H. Weinheim, New York (1995).
14. F. Goodridge and K. Scott, *Electrochemical Process Engineering: A Guide to the Design of Electrolytic Plant,* Plenum Press, New York (1995).
15. J. A. G. Drake, *Electrochemistry and Clean Energy*, Royal Society of Chemistry, Cambridge (1995).
16. I. Rubenstein, *Physical Electrochemistry: Principles, Methods, and Applications*, M. Dekker, New York (1995).
17. R. J. D. Miller, G. L. McLendon, A. J. Nozik, W. Schmickler, and F. Willig, *Surface Electron Transfer Processes*, VCH Publishers, New York (1995).
18. W. Schmickler, *Interfacial Electrochemistry*, Oxford University Press, Oxford (1996).

Monograph Series

1. J. O'M. Bockris, B. E. Conway, and E. Yeager, eds., *Comprehensive Treatise of Electrochemistry, Vol. 10*, Plenum Press, New York (1986).
2. J. O'M. Bockris, B. E. Conway, and R. White, eds., *Modern Aspects of Electrochemistry, Vol. 29*, Plenum Press, New York (1995).
3. A. J. Bard, ed., *Journal of Analytical Chemistry: The Series of Advances, Vol. 17*, M. Dekker, New York (1991).
4. H. Gerischer and C. W. Tobias, eds., *Advances in Electrochemistry and Electrochemical Engineering, Vol. 13*, Wiley, New York (1984).
5. H. Gerischer and C. W. Tobias, eds., *Advances in Electrochemical Science and Engineering, Vol. 3*, VCH Publishers, New York (1994).
6. D. Pletcher, ed., *Electrochemistry, Vol. 10*, The Royal Society of Chemistry, London (1985).
7. A. J. Bard, ed., *Encyclopedia of Electrochemistry of the Elements, Vol. 15*, M. Dekker, New York (1984).
8. *The Electrochemical Society Series*, Wiley, New York.
9. J. Lipkowski and N. P. Ross, eds., *Frontiers of Electrochemistry*, VCH Publishers, New York (1996).

10. International Union of Pure and Applied Chemistry (IUPAC), *Monographs in Electroanalytical Chemistry and Electrochemistry*, M. Dekker, New York (1995).
11. J. Braunstein, G. Mamantov, and G. P. Smith, eds., *Advances in Molten Salt Chemistry*, Vol. 6, Plenum Press, New York (1987).

Journals

1. Electrochemical themes are often treated in other journals of physical chemistry (e.g., *Journal of Physical Chemistry*) and occasionally in journals of chemical physics.
2. Journal of the Electrochemical Society, P. A. Kohl, ed., The Electrochemical Society, Inc., Pennington, NJ.
3. *Interface*, L. P. Hunt, ed., The Electrochemical Society, Inc., Pennington, NJ.
4. *Electrochimica Acta*, R. D. Armstrong, ed., Elsevier Science Ltd. for the International Society of Electrochemistry (ISE), Oxford.
5. *Journal of Applied Electrochemistry*, A. A. Wragg, ed., Chapman and Hall, London.
6. *Langmuir*, W. A. Steel, ed., The American Chemical Society, Washington, DC.
7. *Journal of Colloid and Interface Science*, Darsh T. Wasan, ed., Academic Press, Orlando, FL.
8. *Russian Journal of Electrochemistry (Elektrokhimiya)*, Yakov M. Kolotyrkin, ed., Interperiodica Publishing, Moscow.
9. *Journal of Electroanalytical Chemistry*, R. Parsons, ed., Elsevier Science, S.A., Lausanne, Switzerland.
10. *Corrosion Science*, J. C. Scully, ed., Elsevier Science Ltd., Oxford.
11. *Corrosion*, J. B. Lumsden, ed., National Association of Corrosion Engineers International, Houston, TX.
12. *Bioelectrochemistry and Bioenergetics*, H. Berg, ed., Elsevier Science, S.A., Lausanne, Switzerland.

CHAPTER 2

ION–SOLVENT INTERACTIONS

2.1. INTRODUCTION

Aristotle noted that one could separate water from a solution by means of evaporation. Some two millennia later, Fourcroy in 1800 focused attention on the *interaction* of a solute with its solvent.

These early observations serve to introduce a subject—the formation of mobile ions in solution—that is as basic to electrochemistry as is the process often considered its fundamental act: the transfer of an electron across the double layer to or from an ion in solution. Thus, in an electrochemical system (Fig. 2.1), the electrons that leave an electronically conducting phase and cross the region of a solvent in contact with it (the interphase) must have an ion as the bearer of empty electronic states in which the exiting electron can be received (electrochemical reduction). Conversely, the filled electronic states of these ions are the origin of the electrons that enter the metal in the

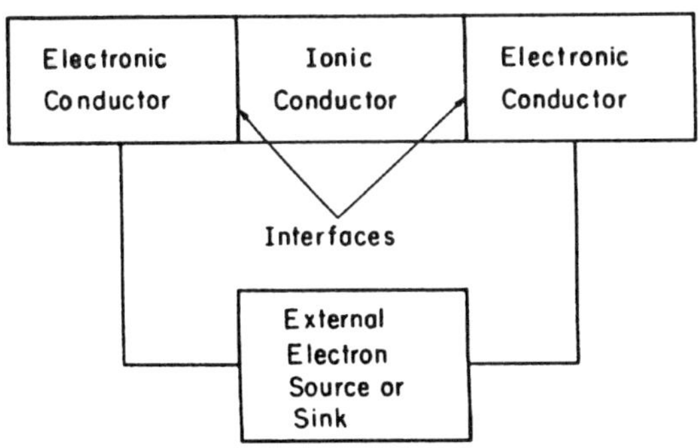

Fig. 2.1. The essential parts of an electrochemical system.

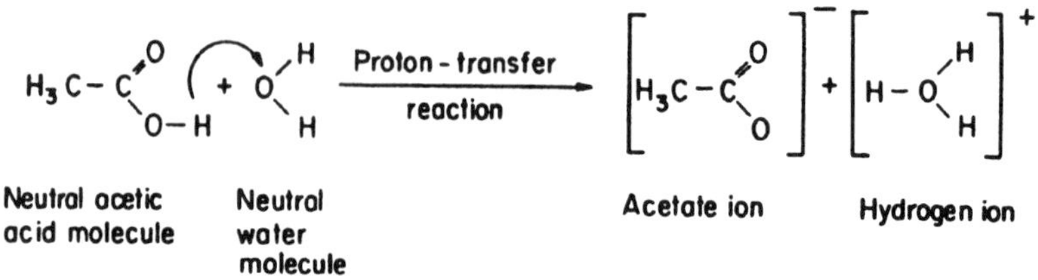

Fig. 2.2. The chemical method of producing ionic solutions.

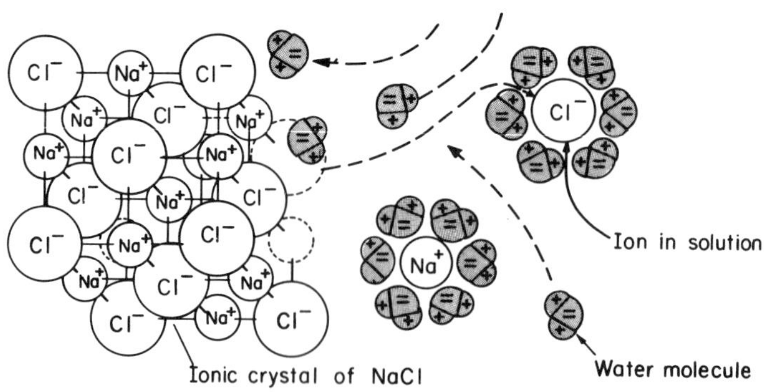

Fig. 2.3. Dissolution of an ionic crystal by the action of a solvent.

reverse electron-transfer reactions involved in electrochemical oxidation reactions. These basic electrochemical processes are shown in Fig. 2.1.

There are two distinct ways that mobile ions form in solution to create the ionically conducting phases that make up the solution side of an electrode–solution system (one half of the electrochemical system shown in Fig. 2.1).

The first one is illustrated in Fig. 2.2. It applies to ion formation in a solvent where the solute is a neutral molecule; in Fig. 2.2 it is acetic acid, CH_3COOH. The figure shows the reaction between solute and solvent that forms the ions in the solvent, and therefore the solution. A characteristic of solutions formed in this way is that usually[1] the ionic concentration is a rather small fraction (e.g., about 0.1%) of the solute molecules are ionized to give ions (Table 2.1).

The second (perhaps more frequently applicable) method of forming mobile ions in solution is quite different: it involves the dissolution of a solid lattice of ions such as the lattice of the often-cited sodium chloride. Some attempt to show what happens in this type of ion formation is reproduced in Fig. 2.3. It is as though the solvent, colliding with the walls of the crystal, gives the ions in the crystal lattice a better deal

[1]However, HCl is a molecule and when this molecule ionizes by reacting with a water molecule, (it does so rather fully), nearly 100% of the molecule can become ionized.

TABLE 2.1

Ionic Concentrations in Pure Water, Pure Acetic Acid, and Acetic Acid Solution

	Ionic Concentration (g-ions dm^{-3} at 298 K)
Pure water	10^{-7}
Pure acetic acid	$10^{-6.5}$
0.1 *N* acetic acid solution	10^{-3}

energetically than they have within the lattice. It entices them out of the lattice and into the solution.[2]

Of course, this implies that there is a considerable energy of interaction between the lattice ions and the solvent molecules. It is this ion–solvent interaction, the immediate cause of the formation of conducting ionic solutions from salts, that is the subject of this chapter.

2.2. BREADTH OF SOLVATION AS A FIELD

The wide range of areas affected by solvation can be seen when one considers the basic role hydration plays in; for example, geochemistry, and indeed in the whole hydrosphere. The pH of natural waters, with all the associated biological effects, is affected by the dissolution of $CaCO_3$ from river beds; and the degree of this dissolution, like any other, is determined by the solvation of the ions concerned. Alternatively, consider the modern environmental problem of acid rain. The basic cause is the formation of atmospheric SO_2 as a result of burning fossil fuels. The pH reached in naturally occurring water is a result of the dissolution of SO_2 in rain and the subsequent creation of the sulfuric acid; H_2SO_4, because the stability of the H^+ and SO_4^{2-} ions that arise is determined by their hydration. The acidity of natural waters then depends upon the original concentration of SO_2 in the air as well as the action of various associated ionic reactions which tend to counter the pH change the SO_2 causes, but which, because they involve ions, themselves depend for their energetics on the ions' solvation energies.

It has already been implied that ion–solvent interactions have widespread significance in electrochemistry, and some of the ramifications of this were discussed in

[2] In this process (and up to a certain concentration) all the ions in the lattice salt become mobile solution ions, although when the ionic concentration gets high enough, the negative and positive ions (*anions* and *cations*, respectively) start to *associate* into nonconducting ion pairs. The specific conductance of salt solutions therefore passes through a maximum, if plotted against concentration.

[3] In this chapter, *solvation* and *hydration* will both be used to describe the interaction of an ion with its surroundings. Clearly, solvation is the general term but most cases of it are in fact hydration.

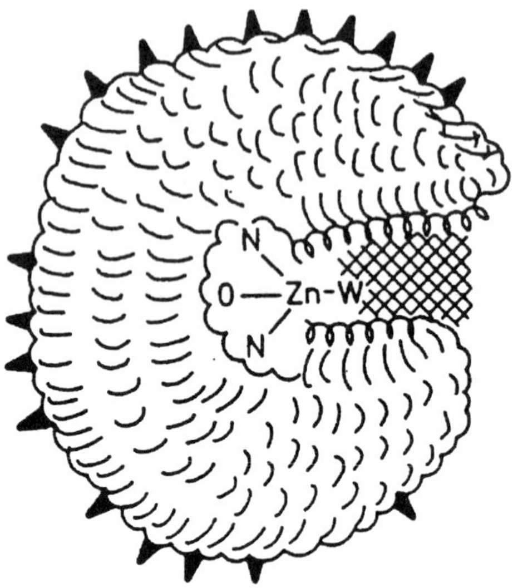

Fig. 2.4. Schematic view of carboxypeptidase A. The coordination sphere around Zn at the bottom of the groove is shown (W = water). The curly lining of the pocket symbolizes hydrophobic residues and the hatched area indicates "organized" water molecules. The small pointed figures at the outer surface refer to polar side groups solvated by the external solvent. (Reprinted from R. R. Dogonadze, A. A. Kornyshev, and J. Ulstrup "Theoretical Approach in Solvation," in *The Chemical Physics of Solvation,* Part A, R. R. Dogonadze, E. Kalman, A. A. Kornyshev, and J. Ulstrup, eds., Elsevier, New York, 1985.)

Chapter 1. However, ionic hydration[3] also plays a leading role in biology. Figure 2.4 shows how the structure of an enzyme depends upon hydration. The diagram indicates how there are effects from "organized water" (i.e., water molecules associated with the Zn nucleus of the carboxypeptidase A) as well as some hydrophobic effects.

This mention of hydrophobicity serves to introduce a lesser mentioned field, that in which solute–solute interactions combine to force *out* solvent molecules, in direct contrast to the more normal ionic hydration effects in which the ions draw solvent molecules into themselves. Such hydrophobic effects occur, for example, when there are large solute groups present (e.g., neutral hydrocarbons in water) and sometimes with large charged groups such as NR_4^+ ions, the size of which can cause attractive–dispersive interactions between the ions and organic molecules to compete with ion–solvent attraction.

TABLE 2.2
Some Approaches to the Investigation of Solvation

From the Properties of Solutions	From Spectroscopic and Other Approaches[a]
Mobility of ions	NMR
Individual partial molar volumes	Raman
Activities of the solvent	EXAFS
Compressibilities of solutions	Neutron diffraction
Vibration potentials	Model calculations using statistical mechanics
Dielectric constants	Monte Carlo
Individual ionic entropies	Molecular dynamics

[a]NMR, nuclear magnetic resonance; EXAFS, extended X-ray absorption fine structure. Neutron diffraction deals with momentum transfer: spectroscopic methods with energy transfer.

2.3. A LOOK AT SOME APPROACHES TO SOLVATION DEVELOPED MAINLY AFTER 1980

There are three main approaches to the study of solvation that have been developed largely since the first edition of this book was published in 1970 (Table 2.2).

2.3.1. Statistical Mechanical Approaches

Statistical mechanical approaches apply mainly to deductions about structure and are the basis of interpretations of the entropy of ions in solution and the solution's heat capacity. The entropy of a system can be calculated if the partition functions of the ions and the water molecules surrounding them are known.[4] The partition functions (translatory, rotational, and vibrational) can be obtained from textbook material by assuming a structure of the ion–solvent complex. By comparing calculations based on various assumptions about structure with the values obtained from experiments, certain structures can be shown to be more likely (those giving rise to the calculations that match the experiment), others less probable, and some so far from the experimental values that they may be regarded as impossible.

2.3.2. What Are Monte Carlo and Molecular Dynamics Calculations?

The easy availability of computers of increasing power has greatly encouraged the two approaches described in this section. In the Monte Carlo (MC) approach,

[4]Although these statistical mechanical approaches have been used increasingly in recent times, a paper by Eley and Evans written as early as 1938 is the origin of partition functional treatments of solutions.

random movements of ions in the solution (and the waters near them) are tested by calculating the energy changes they would bring about were they to occur. The ones that happen with the lowest energy (with a negative free-energy change) are those taken to be occurring in reality.

The molecular dynamics (MD) calculations are different from the Monte Carlo ones. Instead of using assumptions about random movements of the ions and solvent molecules and calculating which of the movements is good (lowers energy) or bad (raises it), the molecular dynamics approach works out the potential energy of the molecular entities as they interact with each other. Then, by differentiating these energies with respect to distance, one can derive the force exerted on a given particle at each of the small time intervals (mostly on a femtosecond scale). As a result of such computations, the dynamics of each particle, and hence the distribution function and eventually the properties of the system, can be calculated. The critical quantities to know in this approach are the parameters in the equation for the intermolecular energy of interaction. To compensate for the fact that only interactions between nearest neighbors are taken into account (no exact calculations can be made of multibody problems), these parameters are not calculated from independent data, but an assumption is made that the two-body-problem type of interaction is acceptable and the parameters are computed by using a case for which the answer is known. The parameters thus obtained are used to calculate cases in which the answer is unknown.

The ability of these computational approaches to predict reality is good. A limitation is the cost of the software, which may amount to many thousands of dollars. However, some properties of solutions can be calculated more cheaply than they can be determined experimentally (Section 2.5). Increasing computer power and a lowering of the cost of the hardware indicates a clear trend toward the ability to calculate chemical events.

2.3.3. Spectroscopic Approaches

In the latter half of this century physicochemical approaches have increasingly become spectroscopic ones. Infrared (IR), Raman, and nuclear magnetic resonance (NMR) spectroscopic approaches can be used to register spectra characteristic of the ion–solvent complex. The interpretation of what molecular structures give these spectra then suggests structural features in the complex. On the other hand, the spectroscopic technique does not work well in dilute solutions where the strength of the signal emitted by the ion–solvent interaction is too small for significant determination. This limits the spectroscopic approach to the study of solvation because it is only in dilute solutions that the cations and anions are sufficiently far apart to exert their properties independently. Thus, spectroscopic methods are still only a partial help when applied to solvation. Spectroscopic and solution property approaches are summarized in Table 2.2.

Two other points must be made here:

1. Water is a very special solvent in respect to its structure (Section 2.4) and the fact that nearly all of our knowledge of ions in solutions involves water arises from its universal availability and the fact that most solutions met in practice are aqueous. However, studying the hydration of ions rather than their solvation limits knowledge, and a welcome modern trend is to study nonaqueous solutions as well.

2. Modern theoretical work on solution properties often involves the use of *mean spherical approximation*, or MSA. This refers to models of events in solution in which relatively simple properties are assumed for the real entities present so that the mathematics can be solved analytically and the answer obtained in terms of an analytical solution rather than from a computer program. Thus, it is assumed that the ions concerned are spherical and incompressible. Reality is more complex than that implied by the SE approximations, but they nevertheless provide a rapid way to obtain experimentally consistent answers.

2.4. STRUCTURE OF THE MOST COMMON SOLVENT, WATER

One can start by examining the structure of water in its gaseous form. Water vapor consists of separate water molecules. Each of these is a bent molecule, the H–O–H angle being about 105° (Fig. 2.5). In the gaseous oxygen atom, there are six electrons in the second shell (two 2s electrons and four 2p electrons). When the oxygen atoms enter into bond formation with the hydrogen atoms of adjacent molecules (the liquid phase), there is a blurring of the distinction between the s and p electrons. The six electrons from oxygen and the two from hydrogen interact. It has been found that four pairs of electrons tend to distribute themselves so that they are most likely to be found in four approximately equivalent directions in space. Since the motion of electrons is described by quantum mechanics, according to which one cannot specify precise orbits for the electrons, one talks of the regions where the electrons are likely to be found as orbitals, or blurred orbits. The electron orbitals in which the electron pairs are likely to be found are arranged approximately along the directions joining the oxygen atom to the corners of a tetrahedron (Fig. 2.6). The eight electrons around the oxygen are neither s nor p electrons; they are sp^3 hybrids. Of the four electron orbitals, two are used for the O–H bond, and the remaining two are as free as a lone pair of electrons.

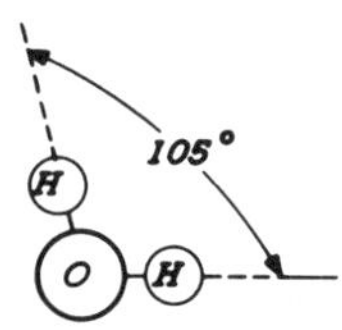

Fig. 2.5. A water molecule is nonlinear.

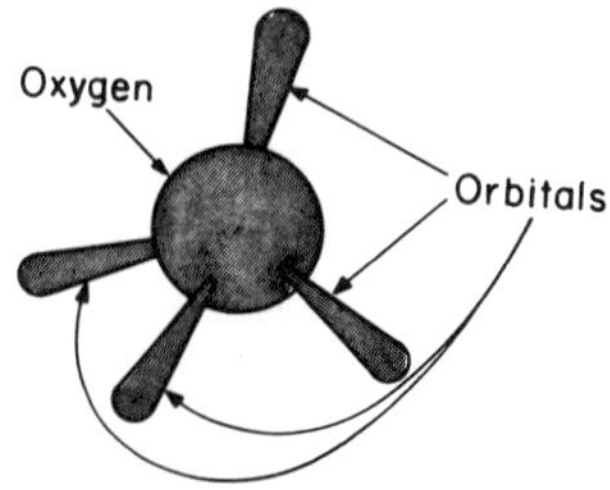

Fig. 2.6. The hybrid orbitals of an oxygen atom.

Because of the repulsion of the electron pairs, the H–O–H angle is not exactly equal to the tetrahedral angle (109° 28′), but is a few degrees less than that.

Water molecules vibrate and rotate in the gas phase. In the liquid phase, the rotation is hindered (*libration*) because of the structure (intermolecular H bonds) and the vibrations are modified (Section 2.11.2).

The free orbitals in which are found the electron lone pairs confer an interesting property on the water molecule. The center of gravity (Fig. 2.7) of the negative charge in the water molecule does not coincide with the center of gravity of the positive charge. In other words, there is a separation within the electrically neutral water molecule: it is thus called an electric *dipole*. The *moment* of a dipole is the product of the electrical charge at either end times the distance between the centers of the electrical charge, $q \cdot d$. The dipole moment of water is 1.87 D in the gas phase (but becomes larger when the water molecule is associated with other water molecules) (Section 2.4.2).

In fact, although water can be treated effectively as a dipole (two equal and opposite charges at either end of a straight line), a more accurate representation of the electrical aspects of water is to regard the oxygen atom as having two charges and each hydrogen atom as having one. This model will be studied in the theory of hydration heats (Section 2.15).

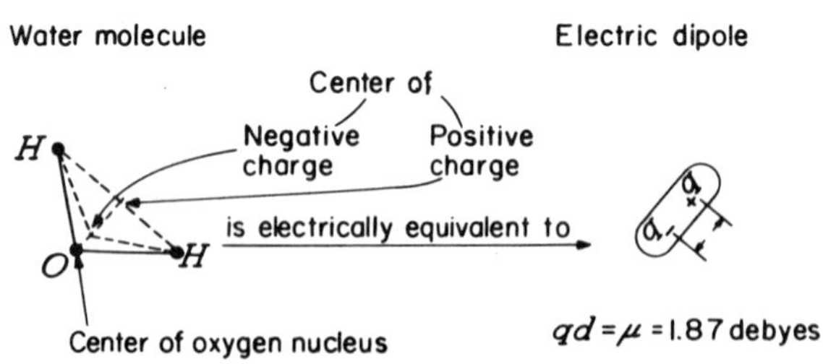

Fig. 2.7. A water molecule can be considered electrically equivalent to a dipole.

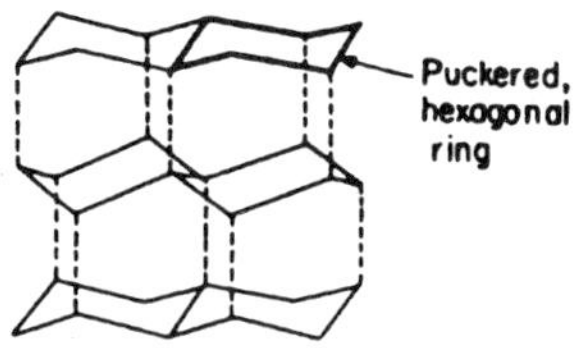

Fig. 2.8. The oxygen atoms in ice, which are located at the intersections of the lines in the diagram, lie in a network of open puckered hexagonal rings.

The availability of the free orbitals (with lone electron pairs) on the oxygen atom contributes not only to the dipolar character of the water molecule but also to another interesting consequence. The two lone pairs can be used for electrostatic bonding to two other hydrogen atoms from a neighboring water molecule. This is what happens in a crystal of ice. The oxygen atoms lie in layers, with each layer consisting of a network of open, puckered hexagonal rings (Fig. 2.8). Each oxygen atom is tetrahedrally surrounded by four other oxygen atoms. In between any two oxygen atoms is a hydrogen atom (Fig. 2.9), which provides a hydrogen bond. At any instant the hydrogen atoms are not situated exactly halfway between two oxygens. Each oxygen has two hydrogen atoms near it (the two hydrogen atoms of the water molecule) at an estimated distance of about 175 pm. Such a network of water molecules contains interstitial regions (between the tetrahedra) that are larger than the dimensions of a water molecule (Fig. 2.10). Hence, a free, nonassociated water molecule can enter the interstitial regions with little disruption of the network structure.

This important property of water, its tendency to form the so-called *H bonds* with certain other atoms, is the origin of its special characteristic, the netting up of many water molecules to form large groups (Fig. 2.10). *H bonds* may involve other types of

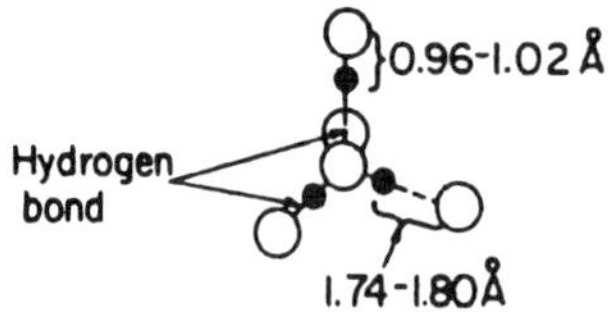

Fig. 2.9. The hydrogen bond between two oxygen atoms (the oxygen and hydrogen atoms are indicated by ○ and •, respectively).

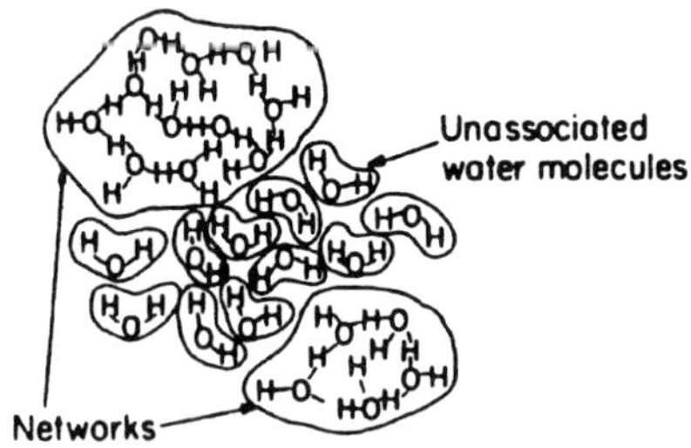

Fig. 2.10. Schematic diagram to show that in liquid water there are networks of associated water molecules and also a certain fraction of free, unassociated water molecules.

atoms, such as Cl, F, O, and S when they are in solution. There is nothing mysterious about the energy behind H bonding; it derives from the positive charge on the proton and the negative one on F^-, for example.

However, H bonding (the value of the bond strength is small, only 10–40 kJ mol^{-1}) does affect the properties of water and is responsible for water's anomalously high boiling point. If one extrapolates the boiling points of the hydrides of the elements in group VI of the Periodic Table to the expected value for the hydride of O, it turns out to be ~ 215 K. The fact that it is actually 158 K higher than that is undoubtedly because individual water molecules are not free to evaporate as the temperature is increased. Many of them bond to each other through the H bonds. The thermal stability of water has had an important effect on the structure of the earth, for if there were no H bonds, the seas would never have formed (they would have remained in the vapor phase) and it is doubtful if life would have begun.

Structural research on water, which originated in a classic paper by Bernal and Fowler, has shown that under most conditions liquid water is best described as a rather broken-down, slightly expanded (Table 2.3) form of the ice lattice (Fig. 2.8). Thus, X-ray and other techniques indicate that in water there is a considerable degree of short-range order that is characteristic of the tetrahedral bonding in ice. Thus, liquid water partly retains the tetrahedral bonding and resulting network structure characteristic of crystalline ice. In addition to the water molecules that are part of the network, some structurally free, nonassociated water molecules can be present in interstitial regions of the network (Fig. 2.10). When a network water molecule breaks its hydrogen bonds with the network, it can move as an interstitial water molecule that can rotate freely. The classification of the water molecules into network water and free (or interstitial) water is not a static one. It is dynamic. As argued in a classic paper by Frank and Wen, clusters of water molecules cooperate to form networks and at the same time the networks can break down. A water molecule may be free in an interstitial

TABLE 2.3
The Structure of Ice and Liquid Water

	Ice	Liquid Water
Mean O–O distance	276 pm	292 pm
Number of oxygen nearest neighbors	4	4.4–4.6

position at one instant and in the next instant it may become held as a unit of the network.[5]

In the 1970s and 1980s, calculational approaches (in addition to the X-ray studies) were added to the tools for the attack on the structure of water. In the molecular dynamics approach, classical mechanics is used to calculate the successive movements of molecules in the structure. Such an approach is dependent on the correctness of the equation that represents the energies of interaction between the particles. The basic equation for these interactions is the "Lennard-Jones 6–12" potential.

$$U_r = -\frac{A}{r^6} + \frac{B}{r^{12}} \tag{2.2}$$

The first term represents the attraction between two molecules and the second the repulsion that also occurs between them.

One of the results of these calculations is that the number of water molecules in an ordered structure near a given water molecule drops away rapidly from the original molecule considered. The similarity of the liquid to solid ice does not remain too far from a given water molecule, i.e., the long-range order present in the solid is soon lost in the liquid.

---

[5]*Radial distribution functions* are met along the path between the results of X-ray and neutron diffraction examinations of water and the deriving of structural information, which is more difficult to do with a liquid than with a solid. Radial distribution functions are, e.g.,

$$\int_0^a 4\pi r^2 g(r) \frac{\sin Sr}{Sr} dr \tag{2.1}$$

and can be seen as proportional to the intensity of the reflected X-ray beam as a function of the incident angle, θ. Thus, in Eq. (2.1) $S = (4\pi/\lambda) \sin \theta/2$. The significance of $4\pi r^2 g(r)\, dr$ allows one to calculate the number of oxygen atoms between r and $r + dr$. In this way one can derive the number of nearest neighbors in the liquid from any central O. The value is 4 for ice and, rather curiously, increases as the temperature is increased (it is 4.4 at 286 K). This may be due to the disturbing effect of pure H_2O molecules, which increase in number with temperature. Their presence would *add* to the intensity due to the regular pattern and account for values greater than 4.2.

Knowing the distribution functions, $g(r)$, as a function of r from experiment or calculation from a central particle, it is possible to calculate how the water molecules spread out from a central particle. With knowledge of this function, it is possible to calculate various properties of liquid water (e.g., its compressibility), and these can then be compared with experimental values as a test of the calculation.

2.4.1. How Does the Presence of an Ion Affect the Structure of Neighboring Water?

The aim here is to take a microscopic view of an ion inside a solvent. The central consideration is that ions orient dipoles. The spherically symmetrical electric field of the ion may tear water dipoles out of the water lattice and make them point (like compass needles oriented toward a magnetic pole) with the appropriate charged end toward the central ion. Hence, viewing the ion as a point charge and the solvent molecules as electric dipoles, one obtains a picture of ion–dipole forces as the principal source of ion–solvent interactions.

Owing to the operation of these ion–dipole forces, a number of water molecules in the immediate vicinity of the ion (the number will be discussed later) may be trapped and oriented in the ionic field. Such water molecules cease to associate with the water molecules that remain part of the network characteristic of water (Section 2.4.3). They are immobilized except insofar as the ion moves, in which case the sheath of immobilized water molecules moves with the ion. The ion and its water sheath then become a single kinetic entity (there is more discussion of this in Section 2.4.3). Thus, the picture (Fig. 2.11) of a hydrated ion is one of an ion enveloped by a solvent sheath of oriented, immobilized water molecules.

How about the situation far away from the ion? At a sufficient distance from the ion, its influence must become negligible because the ionic fields have become

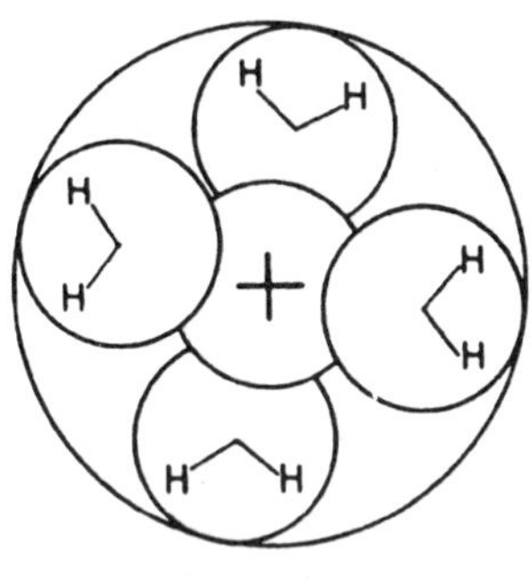

Fig. 2.11. An ion enveloped by a sheath of oriented solvent molecules.

attenuated to virtually zero. The normal structure of water has been re-attained: it is that of bulk water.

In the region between the solvent sheath (where the ionic influence determines the water orientation) and the bulk water (where the ionic influence has ceased to dominate the orientation of water molecules), the ion still has some orienting influence on the water network: it tries to align the water dipoles parallel to the spherically symmetrical ionic field, and the water network tries to convert the water in the in-between region into a tetrahedral arrangement (Fig. 2.12). Caught between the two types of influences, the in-between water adopts some kind of compromise structure that is neither completely oriented nor yet fixed back into the undisturbed water structure shown in Fig. 2.10. The compromising water molecules are not close enough to the ion to become oriented perfectly around it, but neither are they sufficiently far away from it to form part of the structure of bulk water; hence, depending on their distance from the ion, they orient out of the water network to varying degrees. In this intermediate region, the water structure is said to be partly broken down.

One can summarize this description of the structure of water near an ion by referring to three regions (Fig. 2.12). In the *primary*, or structure-enhanced, region next to the ion, the water molecules are oriented out of the water structure and immobilized by the ionic field; they move as and where the ion moves. Then, there is a *secondary*, or structure-broken (SB), region, in which the normal bulk structure of

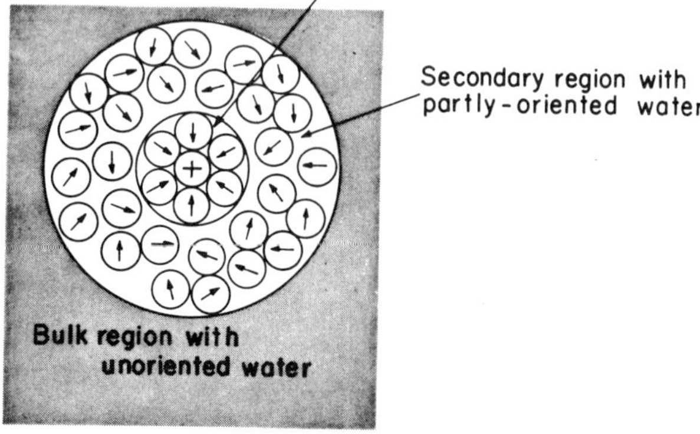

Fig. 2.12. The neighborhood of an ion may be considered to consist of three regions with differing solvent structures: (1) the primary or structure-forming region, (2) the secondary or structure-breaking region, and (3) the bulk region.

water is disturbed to varying degrees. The in-between water molecules, however, do not partake of the translational motion of the ion. Finally, at a sufficient distance from the ion, the water structure is unaffected by the ion and again displays the tetrahedrally bonded networks characteristic of bulk water.

The three regions just described differ in their degree of sharpness. The primary region (discussed in greater detail later), in which there are (at least for the smaller cations) water molecules that share the translational motion of the ion, is a sharply defined region.

In contrast, the secondary region, which stretches from the termination of the primary region to the resumption of the normal bulk structure, cannot be sharply defined; the bulk properties and structure are asymptotically approached.

These structural changes in the primary and secondary regions are generally referred to as solvation (or as hydration when, as is usual, water is the solvent). Since they result from interactions between the ion and the surrounding solvent, one often uses the term "solvation" and "ion–solvent interactions" synonymously; the former is the structural result of the latter.

2.4.2. Size and Dipole Moment of Water Molecules in Solution

In discussions of the *radius of a molecule*, the only well-defined, exact part of the answer lies in the internuclear distances (e.g., between H and O for water), but this distance comprises only part of the radius, albeit the main part. All other measures of the radius of a molecule are affected by some aspect of the occupancy of space that is connected with the packing of the molecule, which varies with circumstances.

In ice (where X-ray measurements of internuclear distances, d, are more exact than those in the liquid), the O–O internuclear distance is accurately known: 280 pm. The distance between the nucleus of oxygen and that of hydrogen (see a space-fitting picture of water, Fig. 2.9) is 138 pm.

The so-called *van der Waals radius*[6] of H can be taken to be 52.8 pm (this is the Bohr radius for the ground state of H). The van der Waals radius of O is ~115 pm and from these values 168 pm is obtained for the radius of water.

Now, several other radii between ca. 140 and 193 pm can be obtained by equating $4/3\ \pi r^3$ to certain volumes (e.g., the close-packed volume of water, or a volume based on the molecular dimensions). Each of these radii is "applicable" as the circumstances dictate.

Thus, the "radius of water" varies, according to the method used to estimate it, between 168 (the model shown in Fig. 2.9) and 193 pm, which arises from measurements on the density of water (and the resulting molar volume of 18 cm^3). The larger

---

[6]The phrase "van der Waals radius" arises as a distinction from "internuclear distance radii." Thus, from the van der Waals equation for the P-V relation in gases (an improvement on the simple gas law $PV = nRT$), a quantity b can be found which refers to the space taken up out of the whole gas volume V by the molecules themselves.

value is too high because it includes the free space in liquid water. The lower value is too low, because it neglects the volume needed to cover some water movements in the liquid. The mean, 181 ± 13 pm, is a regrettably imprecise figure for such an important quantity, but, as can be understood, it all depends upon what one takes into account.

The dipole moment of water in the gas phase is well known as 1.87 D but is 2.42 D in water at 298 K. The reason for the difference is that in liquid water there is electrostatic pull on a given water molecule from the surrounding ones, and this lengthens the distance in the dipoles (Fig. 2.7).

2.4.3. The Ion–Dipole Model for Ion–Solvent Interactions

The preceding description of the solvent surrounding an ion was used as the basis of a structural treatment of ion–solvent interactions initiated by Bernal and Fowler (1933). Their paper is a seminal one for much else in the structural picture of hydration (Section 2.4).

Bernal and Fowler thought of the changes of free energy during solvation as being largely electrostatic in nature. In their picture, the passage of an ion began in the gas phase, and here they took (for most of their calculations) the ions to have a zero potential energy of interaction. Transferring an ion mentally to the interior of the solvent, they thought this to be associated with three energy changes:

1. The ion–dipole interaction caused by the ion's attracting water dipoles in the first layer around the ion (ion–dipole interaction). (*cf.* Appendix 2.2)
2. Then there would be the energy needed to break up the water structure, which occurs when ions enter it.
3. There would be a further allowance for the ion–water interactions "further out" from the first layer to the rest of the solvent surrounding the ion. Such an interaction might be thought to be relatively small, not only because of the increasing distance but because of the increasing dielectric constant; small (near to 6) near the ion but rapidly attaining the value of about 80 (at 25 °C), as near as 1000 pm from the ion. Both these factors would diminish the individual ion–dipole interaction energy.

Bernal and Fowler's calculation remains famous because it grappled for the first time with the structure of water and with ion–solvent interactions on a molecular basis. Better theories have been developed, but most have their roots in the Bernal and Fowler work of 1933.

The modern developments of solvation theory will not be discussed at this point because nearly all the tools for investigating the ion–solvent interaction have become available since 1933. One has to see some of the information they have provided before ion–solvent interactions can be worked out in a more quantitative way.

Further Reading

Seminal

1. M. Born, "Free Energy of Solvation," *Z. Phys.* **1**: 45 (1920).
2. J. D. Bernal and R. H. Fowler, "The Structure of Water," *J. Chem. Phys.* **1**: 515 (1933).
3. R. W. Gurney, *Ionic Processes in Solution*, McGraw-Hill, New York (1953).
4. H. S. Frank and W. Y. Wen, "Water Structure Near to Ions," *Faraday Discuss., Chem. Soc.* **24**: 133 (1957).

Monograph

1. B. E. Conway, *Ionic Hydration in Chemistry and Biology*, Elsevier, New York (1981).

Papers

1. C. Sanchez-Castro and L. Blum, *J. Phys. Chem.* **93**: 7478 (1989).
2. E. Guardia and J. A. Padro, *J. Phys. Chem.* **94**: 6049 (1990).
3. B. Guillot, P. Martean, and J. Ubriot, *J. Chem. Phys.* **93**: 6148 (1990).
4. S. Golden and T. R. Tuttle, *J. Phys. Chem.* **93**: 4109 (1990).
5. D. W. Mundell, *J. Chem. Ed.* **67**: 426 (1990).
6. F. A. Bergstrom and J. Lindren, *Inorg. Chem.* **31**: 1525 (1992).
7. Y. Liu and T. Ichiye, *J. Phys. Chem.* **100**: 2723 (1996).
8. A. K. Soper and A. Luzar, *J. Phys. Chem.* **100**: 1357 (1996).
9. B. Madan and K. Sharp, *J. Phys. Chem.* **100**: 7713 (1996).
10. G. Hummer, L. R. Pratt, and A. E. Garcia, *J. Phys. Chem.* **100**: 1206 (1996).

2.5. TOOLS FOR INVESTIGATING SOLVATION

2.5.1. Introduction

The more recently used methods for investigating the structure of the region around the ion are listed (though not explained) in Table 2.2. It is convenient to group the methods shown there as follows.

Several methods involve a study of the properties of solutions in equilibrium and are hence reasonably described as *thermodynamic*. These methods usually involve thermal measurements, as with the heat and entropy of solvation. Partial molar volume, compressibility, ionic activity, and dielectric measurements can make contributions to solvation studies and are in this group.

Transport methods constitute the next division. These are methods that involve measurements of diffusion and the velocity of ionic movement under electric field gradients. These approaches provide information on solvation because the dynamics of an ion in solution depend on the number of ions clinging to it in its movements, so that knowledge of the facts of transport of ions in solution can be used in tests of what entity is actually moving.

A third group involves the *spectroscopic approaches*. These are discussed in Section 2.11.

Finally, computational approaches (including the Monte Carlo and molecular dynamic approaches) are of increasing importance because of the ease with which computers perform calculations that earlier would have taken impractically long times.

2.5.2. Thermodynamic Approaches: Heats of Solvation

The definition of the heat of solvation of a salt is the change in heat content per mole for the imaginary[7] transition of the ions of the salt (sufficiently far apart so that they have negligible energies of interaction between them) from the gas phase into the dissolved state in solution. Again, a simplification is made: the values are usually stated for dilute solutions, those in which the interaction energies between the ions are negligible. Thus, the ion–solvent interaction is isolated.

The actual calorimetric measurement that is made in determining the heat of hydration of the ions of a salt is not the heat of hydration itself, but the heat of its dissolution of the salt in water or another solvent (Fajans and Johnson, 1942). Let this be ΔH_{soln}. Then one can use the first law of thermodynamics to obtain the property which it is desired to find, the heat of hydration. What kind of thought process could lead to this quantity? It is imagined that the solid lattice of the ions concerned is broken up and the ions vaporized to the gaseous state (heat of sublimation). Then one thinks of the ions as being transferred from their positions far apart in the gas phase to the dissolved state in dilute solution (heat of hydration). Finally, the cycle is mentally completed by imagining the dissolved ion reconstituting the salt-lattice ($-\Delta H_{\text{soln}}$). It is clear that by this roundabout or cyclical route[8] the sum of all the changes in the cycle should be equal to zero, for the initial state (the crystalline salt) has been re-formed.

This process is sketched in Fig. 2.13.

Thus,

$$L_{\text{sub}} + \Delta H_{s,\text{salt}} - \Delta H_{\text{soln}} = 0 \tag{2.3}$$

[7]It *is* an imaginary transition because we don't know any actual way of taking two individual ions from the gas phase and introducing them into a solution without passing through the potential difference, $\Delta\chi$, which occurs across the *surface* of the solution. There is always a potential difference across an interphase, and were ions *actually* to be transferred from the gas phase to the interior of a solution, the energy, $z_i e_0 \Delta\chi$, would add to the work one calculates by the indirect method outlined here. It's possible to add this energy of crossing the interphase to the heats of solution of Table 2.4 and the resultant values are consequently called "real heats of solvation," because they represent the actual value that would be obtained if one found an experimental way to go from the vacuum through the interface into the solution.

[8]When, as here, an imaginary cycle is used to embrace as one of its steps a quantity not directly determinable, the process used is called a "Born–Haber cycle." The sum of all the heats in a cycle must be zero.

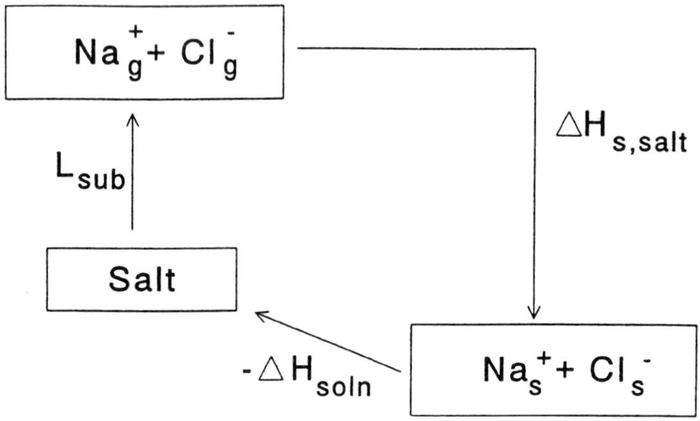

Fig. 2.13. An imaginary cycle–route process.

TABLE 2.4

Heats of Hydration of the Alkali Halides and Quantities Used in Their Derivation[a]

Salt	U_o (calc.)[b]	U^b (T – 291 K)[b]	Initial Heat of Solution, $\Delta H_{soln,o}$	$\Delta H_{s,+} + \Delta H_{s,-}$
Li-F	–245.5	–247.7	1.1	–245.9
–Cl	–201.8	–202.7	–8.6	–211.3
–Br	–191.0	–191.4	–11.1	–202.5
–I	–177.7	–177.6	–14.8	–192.4
Na-F	–217.2	–218.2	0.6	–217.6
–Cl	–185.0	–185.5	1.3	–184.2
–Br	–177.4	–177.8	0.2	–177.6
–T	–165.5	–165.6	–1.4	–167.0
K-F	–194.0	–194.7	–4.1	–198.8
–Cl	–169.2	–169.6	4.4	–165.2
–Br	–162.6	–162.8	5.1	–157.7
–I	–153.8	–153.9	5.1	–148.8
Rb-F	–184.5	–184.9	–5.8	–190.7
–Cl	–163.3	–163.4	4.5	–158.9
–Br	–157.4	–157.4	6.4	–151.0
–I	–149.3	–149.2	6.5	–142.7

[a]Units are kcal mol^{-1}. 1 calorie = 4.184 J.
[b]Lattice energy.

Now, only ΔH_{soln} (the heat of solution) is experimentally measured, and hence the evaluation of the heat of solvation of the salt, $\Delta H_{s,\text{salt}}$, by means of Eq. (2.3) involves trusting the reliability of information on the heat of sublimation of the salt, L_{sub}. In some cases, L_{sub} is known reliably (±1%) from calculations. Alternatively, it can be determined experimentally. From L_{sub}, then, and the measured heat of dissolution, ΔH_{soln}, the heat of hydration of the salt concerned in the solvent can be deduced.

Fajans (1962) was the first scientist to put these thoughts into practice. One finds that the determined heats of solvation are relatively small and endothermic (+) for some salts but exothermic (–) for others. However, lattice energies are known to be in the region of several hundreds of kilojoules mol^{-1}, so that, in rough terms [Eq. (2.3)], heats of solvation should not be more than a dozen kilojoules mol^{-1} (numerically) different from lattice energies. In Table 2.4 a compilation is given of the quantities mentioned earlier in the case of the alkali halides.[9] Now, the method described here gives the sum of the heat of hydration of the ions of a salt. The question of how to divide this sum up into individual contributions from each of the ions of a salt requires more than the thermodynamic approach that has been used so far. The way this is done is described in later sections (e.g., in Section 2.6.2 or 2.15.9).

2.5.3. Obtaining Experimental Values of Free Energies and Entropies of the Solvation of Salts

In the preceding section it was shown how to obtain, a little indirectly, the heat of solvation of a salt. However, it is the *free energies* of the participants in a chemical reaction that determine the state of equilibrium so that one cannot leave the situation with only the ΔH determined. Free energy and entropy changes have to be dealt with also.

How can the free energy of a solution be obtained? Consider a saturated solution of a 1:1 salt of the type MA. Because the solid salt lattice is in equilibrium with its ions in solution, the chemical potential of the salt, μ_{MA}, can be expressed in terms of its individual chemical potentials and activities (a_{M^+} and a_{A^-}) in solution,

$$\mu_{\text{MA,crystal}} = \mu^{\circ}_{\text{MA,crystal}} = \mu_{\text{M}^+\text{A}^-,\text{soln}}$$

$$= \mu^{\circ}_{\text{M}^+,\text{soln}} + RT \ln(a_{\text{M}^+})_{\text{soln,sat}} + \mu^{\circ}_{\text{A}^-,\text{soln}} + RT \ln(a_{\text{A}^-})_{\text{soln,sat}} \tag{2.4}$$

In thermodynamic reasoning, there has to be a *standard state*. The standard state for the solid crystal is the substance in its pure state at 298 K. It follows that the standard chemical potential of solution is:

$$\Delta\mu^{\circ}_{\text{soln}} = \mu^{\circ}_{\text{M}^+,\text{soln}} + \mu^{\circ}_{\text{A}^-,\text{soln}} - \mu^{\circ}_{\text{MA,crystal}}$$

$$= -RT \ln\,[a_{\text{M}^+,\text{soln,sat}}\, a_{\text{A}^-\text{soln,sat}}] = -2\,RT \ln\, a_{\pm(\text{soln,sat})} \tag{2.5}$$

---

[9]The alkali halides are chosen as good examples because the lattice energy is particularly well known and reliable.

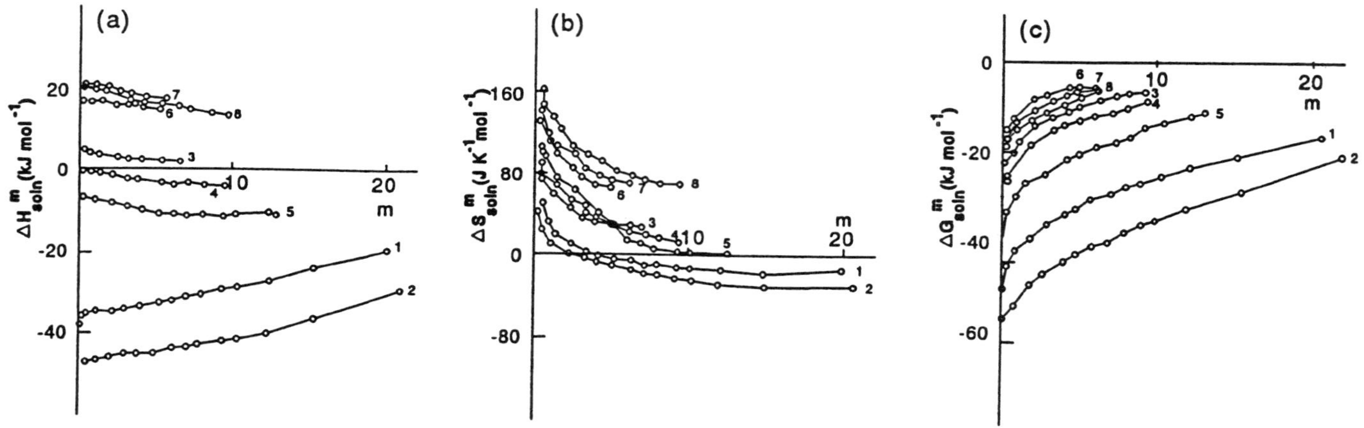

Fig. 2.14. Dependence of $\Delta H_{m,\text{soln}}$, $\Delta S_{m,\text{soln}}$, and $\Delta G_{m,\text{soln}}$ on electrolyte concentration in aqueous solutions for some alkali metal halides at 298.15 K: 1, LiCl; 2, LiBr; 3, NaCl; 4, NaBr; 5, NaI; 6, KCl; 7, KBr; 8, KI. (Reprinted from G. A. Krestov, *Thermodynamics of Solvation*, Ellis Harwood, London, 1991.)

where a is activity of the ion in solution and $a_{\pm(\text{soln,sat})}$ is the mean ionic activity of the solution defined as $a_{\pm} = (a_+a_-)^{1/2}$.

Thus, if one knows the mean activity (see Section 2.9.1) of the electrolyte for the condition at which the solution is saturated and in equilibrium with the salt, one has the standard free energy change of solution. To obtain the standard free energy of solvation (hydration) from this, one has also to know the free energy of the salt lattice at 298 K. This is easily obtainable (see any physicochemical text) if there are data on the specific heat of the given salt as a function of temperature, so that the entropy of the salt in its standard state can be determined in the usual way of integrating the $\Delta C_{\text{p}} - T$ relation (where C_{p} is specific heat) to obtain entropies. Knowing then the standard free energy of solution and that of the salt lattice, one applies reasoning similar to that used earlier for the heats [Eq. (2.3)]. One can thus obtain free energies of hydration of salts. Knowing the $\Delta G°$ and $\Delta H°$ for the hydration process, one may calculate the standard entropy of solvation of the salt from the well-known thermodynamic equation $\Delta G_s^{\circ} = \Delta H_s^{\circ} - T\Delta S_s^{\circ}$. Values of ΔS_s° will be discussed further in Section 2.15.12, which covers the process of splitting up ΔS_s° into its component parts for the individual ions concerned.

Why should one bother with these thermodynamic quantities when the overall aim of this chapter is to determine the structure of liquids near ions? The answer is the same as it would be to the generalized question: What is the utility of thermodynamic quantities? They are the quantities at the base of most physicochemical investigations. They are fully *real*, no speculations or "estimates" are made on the way (at least as far as the quantities for salts are concerned). Their numerical modeling is the challenge that the theoretical approaches must face. However, such theoretical approaches must assume some kind of structure in the solution and only a correct assumption is going to lead to a theoretical result that agrees with experimental results. Thus, such agreement indirectly indicates the structure of the molecules.

Finally, this section ends with a reminder that heats, entropies, and free energies of hydration depend on concentration (Fig. 2.14) and that there are significant changes in values at very low concentrations. It is the latter values that are the desired quantities because at high concentrations the heats and free energies are influenced not only by ion–solvent interactions (which is the objective of the venture) but also by interionic forces, which are much in evidence (Chapter 3) at finite concentrations.

2.6. PARTIAL MOLAR VOLUMES OF IONS IN SOLUTION

2.6.1. Definition

The molar volume of a pure substance can be obtained from density measurements, i.e., ρ = (molecular weight)/(molar volume). The volume contributed to a solution by the addition of 1 mole of an ion is, however, more difficult to determine. In fact, it has to be measured indirectly. This is because, upon entry into a solvent, the ion changes the volume of the solution not only by its own volume, but by the change due, respectively, to a breakup of the solvent structure near the ion and the compression of the solvent under the influence of the ion's electric field (called *electrostriction*; see Section 2.22).

The effective ionic volume of an ion in solution, the partial molar volume, can be determined via a quantity that *is* directly obtainable. This is the apparent molar volume of a salt, $V_{m,2}$, defined by

$$V_{m,2} = \frac{V - n_1 V_{m,1}}{n_2} \tag{2.6}$$

where V is the volume of a solution containing n_1 moles of the solvent and n_2 moles of the solute and $V_{m,1}$ is the molar volume of the solvent. It is easily obtained by measuring the density of the solution.

Now, if the volume of the solvent were not affected by the presence of the ion, $V - n_1 V_{m,1}$ would indeed be the volume occupied by n_2 moles of ions. However, the complicating fact is that the solvent volume is no longer $V_{m,1}$ per mole of solvent; the molar volume of the solvent is affected by the presence of the ion, and so $V_{m,2}$ is called the *apparent* molar volume of the ion of the salt. Obviously, as when $n_2 \to 0$, the apparent molar volume of the solvent in the solution must become the real one, because the disturbing effect of the ion on the solvent's volume will diminish to zero. Hence, at *finite* concentration, it seems reasonable to write the following equation:

$$\tilde{V} = \tilde{V}^0 + n_2 \left(\frac{\partial V_{m,2}}{\partial n_2} \right)_{T,P,n_1} \tag{2.7}$$

This equation tells one that the density of the solution that gives $V_{m,2}$ for a series of concentrations gives the partial molar volume $\tilde{V}$ at any value of n_2. Knowing $V_{m,2}$ from ρ_{soln} and $\rho°$, Eq. (2.7) can be used to obtain $\tilde{V}$ as a function of n_2. Extrapolation of $\tilde{V}$ to $n_2 = 0$ gives the partial molar volume of the electrolyte *at infinite dilution*, $\tilde{V}^0$ (i.e., free of interionic effects).

Once partial molar volumes are broken down into the individual partial ionic volumes (see Section 2.6.2), the information given by partial molar volume measurements includes the net change in volume of the solvent that the ion causes upon entry and hence it provides information relevant to the general question of the structure near the ion, that is, its solvation.

2.6.2. How Does One Obtain Individual Ionic Volume from the Partial Molar Volume of Electrolytes?

From an interpretive and structural point of view, it is not much use to know the partial molar volumes of electrolytes unless one can separate them into values for each ion. One way of doing this might be to find electrolytes having ions with the same

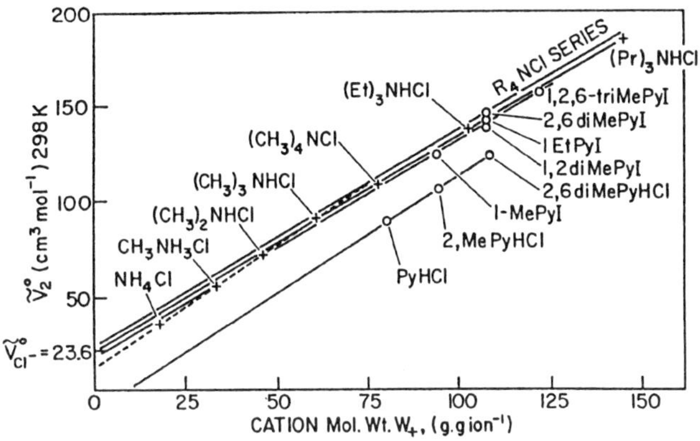

Fig. 2.15. Plot of infinite-dilution partial molar volumes of homologous R_4N^+ chlorides in water against the cation molecular weight, allowing extrapolation to obtain $\tilde{V}^{0}_{Cl^-}$. (Reprinted from B. E. Conway, *Ionic Hydration in Chemistry and Biophysics*, Elsevier, New York, 1981.)

crystallographic radii (the standard example is KF) and allot to each ion one half of the partial molar volume of the electrolyte. However, this method does not give results in agreement with those of other methods, which agree among themselves. Why is this? It is because electrostriction and the breakdown of the solvent structure in the neighborhood of the ion are not purely Coulombic (depending on simple distance laws), but are also specific (depending to some degree on chemical bonding, like hydrogen bond formation between ion and solvent).

Correspondingly, objections can be made to making a plot of the values of $\tilde{V}$ for the electrolyte against $1/\tilde{V}_{\text{cation}}$, and extrapolating to $1/\tilde{V}_{\text{cation}} = 0$. At first sight, one thinks this should give the value of the partial molar volume for an anion which is the partner of each of the various cations of increasing size in the data that would make up the plot.[10] However, questions of the specificity of some interactions, the absence of allowance for dead space, etc., make this approach too flawed to be acceptable.

2.6.3. Conway's Successful Extrapolation

Conway has suggested a method that seems to give results in agreement with those of a second entirely different method, the *ionic vibration* method (see later discussion). Conway found that plotting the partial molar volume of a series of electrolytes involving large cations (e.g., a tetraalkylammonium series) and a constant smaller

[10]Of course, if one obtains reliably the value of $\tilde{V}^{0}_{i}$ for one ion, then knowing the partial molar volumes for a series of electrolytes containing that one known ion enables the $\tilde{V}^{0}_{i}$ for the counterions of a series of electrolytes to be known.

anion, Cl^-, against the *molecular weight* of the cation (instead of the reciprocals of the cations' volumes) is useful. Extrapolation of this plot to zero cation molecular weight should give the partial gram-ionic volume for the partner anion. Conway's plot is given in Fig. 2.15. The method works because the tetraalkylammonium ions are large and hence cause little electrostriction (i.e., compression of the surrounding solvent), which is the reason for the apparent lack of agreement of the other extrapolations. The reason for the apparent absence of other specific effects, such as the structure breaking which the big tetraalkylammonium cations would be expected to produce, is less obvious. The basis for the success claimed by Conway's method is the agreement (particularly for $\tilde{V}^0_{H^+}$) of the values it gives with those of an entirely different method, the ionic vibration approach (Section 2.7). The values for $\tilde{V}^0_{Cl^-}$ and $\tilde{V}^0_{H^+}$ from the present method are found to be 23.6 and –5.7 cm^3 mol^{-1}, respectively. Why is one of these values negative? It can only mean that addition of H^+ to the solution causes more contraction among the surrounding solvent molecules than the volume added by the cation (which in this case is small).

2.7. COMPRESSIBILITY AND VIBRATION POTENTIAL APPROACH TO SOLVATION NUMBERS OF ELECTROLYTES

2.7.1. Relation of Compressibility to Solvation

In 1938, Passynski made the following argument, which relates the compressibility of a solution to the sum of the *primary* solvation numbers of each ion of an electrolyte. Primary here means ions that are so compressed by the ions' field that they themselves have zero compressibility.

Passynski measured the compressibility of solvent (β_0) and solution (β), respectively, by means of sound velocity measurements. The compressible volume of the solution is V and the incompressible part, v ($v/V = \alpha$). The compressibility is defined in terms of the derivative of the volume with respect to the pressure, P, at constant temperature, T. Then,

$$\beta_0 = -\frac{1}{V-v}\left(\frac{\partial(V-v)}{\partial P}\right)_T = -\frac{1}{V-v}\left(\frac{\partial V}{\partial P}\right)_T \tag{2.8}$$

$$\beta = -\frac{1}{V}\left(\frac{\partial V}{\partial P}\right)_T \tag{2.9}$$

Therefore

$$\frac{\beta}{\beta_0} = \frac{V-v}{V} = \frac{V-\alpha V}{V} \tag{2.10}$$

and

$$\alpha = 1 - \frac{\beta}{\beta_0} = \frac{v}{V} \tag{2.11}$$

Let the g-moles of salt be n_2; these are dissolved in n_1 g-moles of solvent. Then, there are $\alpha n_1/n_2$ g-moles of incompressible solvent per g-mole of solute. This was called by Passynski (not unreasonably) the *primary solvation number* of the salt, although it involves the assumption that water held so tightly as to be incompressible will qualify for primary status by traveling with the ion.

To obtain *individual* ionic values, one has to make an assumption. One takes a large ion (e.g., larger than I^-) and assumes its primary solvation number to be zero,[11] so that if the total solvation number for a series of salts involving this big anion is known, the individual hydration numbers of the cations can be obtained. Of course, once the hydration number for the various cations is determined by this artifice, each cation can be paired with an anion (this time including smaller anions, which may have significant hydration numbers). The *total* solvation numbers are determined and then, since the cation's solvation number is known, that for the anion can be obtained.

In Passynski's theory, the basic assumption is that the compressibility of water sufficiently bound to an ion to travel with it is zero. Onori thought this assumption questionable and decided to test it. He used more concentrated solutions (1–4 mol dm^{-3}) than had been used by earlier workers because he wanted to find the concentration at which there was the beginning of an overlap of the primary solvation spheres (alternatively called Gurney co-spheres) of the ion and its attached primary sheath of solvent molecules.

Figure 2.16 shows the plot of the mean molar volume of the solution $\overline{V}_m$ multiplied by the compressibility of the solution β as a function of the molar fraction of the NaCl solute x_2. At $x_2 \approx 0.07$ (~4 mol dm^{-3}), the values for the three temperatures become identical. Onori arbitrarily decided to take this to mean that $\beta\overline{V}_m$ has no further temperature dependence, thus indicating that *all the water in the solution* is now in the hydrated sphere of the ions and these, Onori thought, would have a β with no temperature dependence (for they would be held tight by the ion and be little dependent on the solvent temperature).

These assumptions allow the compressibility of the hydration sheath itself to be calculated (Passynski had assumed it to be negligible). To the great consternation of some workers, Onori found it to be significant—more than one-tenth that of the solvent value.

[11]Thus, whether molecules move off with an ion is determined by the struggle between the thermal energy of the solution, which tends to take the water molecule away from the ion into the solvent bulk, and the attractive ion–dipole force. The larger the ion, the less likely it is that the water molecule will remain with the ion during its darting hither and thither in solution. A sufficiently large ion doesn't have an adherent (i.e., primary) solvation shell, i.e., $n_s = 0$.

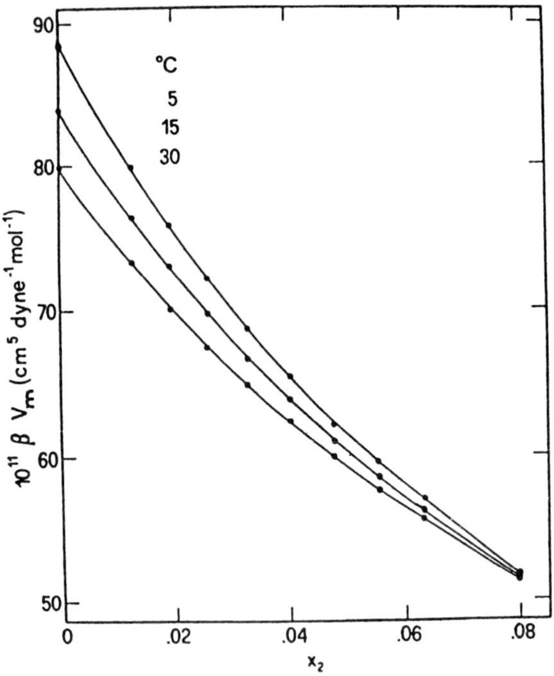

Fig. 2.16. Adiabatic compressibility of aqueous sodium chloride solutions as a function of salt mole fraction, x_2, at various temperatures. The solid line is the calculated compressibility. The expected error is on the order of the dot size. (Reprinted from G. Onori, *J. Chem. Phys.* **89:** 510, 1988.)

Taking into account a finite compressibility of the hydration waters led Onori to suggest solvation numbers that differed from those of Passynski with his assumption of zero compressibility of the inner region of the solvation shell. For example, for a 1.5 *M* solution, Onori has the value 19 for the sum of the solvation numbers of Na^+ and Cl^-, whereas the Passynski at 0.05 *M* solution is 6! However, later on (Section 2.22), when electrostriction is discussed in detail, Onori's estimate will be shown to be unlikely.

2.7.2. Measuring Compressibility: How It Is Done

The most convenient way to measure the compressibility of a liquid or solution is from the velocity of sound in it. A well-known equation derived in physics texts states that

$$\beta = \frac{1}{c_0^2 \rho} \tag{2.12}$$

where c_0 is the velocity of an ultrasonic wave in the medium and ρ is the density.

Note that β determined by this equation is an *adiabatic* β, not an isothermal one, because the local compression that occurs when the ultrasound passes through the solution is too rapid to allow an escape of the heat produced.[12]

A word about a particularly clever sound velocity measurement technique is justified. It is due in its initial form to Richards et al. One creates an ultrasonic vibration by bringing a piezoelectric crystal with oscillations in the megahertz range into contact with a fixed transducer. The latter has one face in contact with the liquid and sends out a beam of sound through it. Another transducer (the receiver) is not fixed and its position is varied with respect to that of the first transducer over distances that are small multiples of the wavelengths of the sound waves (~5×10^{-3} cm). A stepping motor is used to bring about exact movements, and hence positions, of the receiver transducer. As the movable transducer passes through nodes of the sound waves, the piezoelectric crystal on the receiver transducer reacts and its signal is expressed through an electronic circuit to project Lissajous figures (somewhat like figures of eight) on the screen of a cathode ray oscilloscope. When these figures attain a certain configuration, they indicate the presence of a node (the point in a vibration where the amplitude is negligible) and by counting the number of nodes observed for a given distance of travel of the receiver transducer, the distance between two successive nodes—the wavelength of sound in the liquid (λ)—is obtained. The frequency of piezoelectric crystal used (e.g., barium titanate), ν (e.g., 5 MHz), is known. Because

$$\nu\lambda = c_0 \tag{2.13}$$

where c_0 is the velocity of sound needed in Eq. (2.12) for the compressibility, the latter can be found from Eq. (2.13).

2.8. TOTAL SOLVATION NUMBERS OF IONS IN ELECTROLYTES

Total solvation numbers arise directly from the discussion following Passynski's Eq. (2.11), which requires measurements only of the compressibility (β) of the solution (that of the solvent usually being known). Bockris and Saluja used this method in 1972 to obtain the total solvation numbers of both ions in a number of salts (Table 2.5).

In the next section it will be shown how these total solvation numbers for salts can be turned into individual solvation numbers for the ions in the salt by the use of information on what are called "ionic vibration potentials," an electrical potential

---

[12]A study of the temperature dependence of β shows that it is positive for tetraalkylammonium salts.

TABLE 2.5

Adiabatic Compressibility and Total Solvation Number of Electrolytes as a Function of Concentration at 298 K

c (mol/liter)	Adiabatic Compressibility $\times 10^6$ (bar^{-1})	Total Solvation Number of Salt	c (mol/liter)	Adiabatic Compressibility $\times 10^6$ (bar^{-1})	Total Solvation Number of Salt
	NaF			$BaBr_2$	
1.0	38.72	7.5 ± 0.0	3.29	20.39	7.5 ± 0.0
0.5	41.57	7.9 ± 0.0	1.21	33.60	10.7 ± 0.0
0.1	44.04	8.5 ± 0.3	0.50	39.59	12.4 ± 0.0
0.05	44.35	9.1 ± 0.1	0.14	43.09	14.6 ± 0.2
			0.05	44.1	15.2 ± 0.5
	NaCl				
5.1	26.75	3.9 ± 0.0		BaI_2	
2.0	36.64	4.7 ± 0.0	2.50	26.00	9.1 ± 0.0
0.1	40.04	5.7 ± 0.0	0.50	39.86	11.7 ± 0.0
0.1	44.20	6.5 ± 0.3	0.25	41.99	13.2 ± 0.1
0.05	44.45	6.9 ± 0.5	0.05	44.05	15.4 ± 0.4
			0.025	44.40	16.2 ± 1.0
	NaBr				
5.2	26.93	3.7 ± 0.0		$MgCl_2$	
2.0	35.99	5.1 ± 0.0	4.22	21.61	6.1 ± 0.0
1.1	39.75	5.7 ± 0.0	2.11	29.74	8.4 ± 0.0
0.1	44.23	6.1 ± 0.3	1.05	35.94	10.1 ± 0.0
0.05	44.47	6.3 ± 0.5	0.09	43.88	12.3 ± 0.3
			0.05	44.21	12.8 ± 0.6
	NaI				
5.0	28.78	3.2 ± 0.0		$CaCl_2$	
2.1	36.83	4.4 ± 0.0	4.05	22.90	5.9 ± 0.0
1.0	40.84	4.6 ± 0.0	2.02	30.94	8.0 ± 0.0
0.1	44.28	5.5 ± 0.3	1.00	36.91	9.4 ± 0.0
0.05	44.48	6.1 ± 0.5	0.10	43.82	11.1 ± 0.3
			0.05	44.22	12.1 ± 0.5
	LiCl			$SrCl_2$	
5.0	29.93	3.3 ± 0.0			
2.0	37.15	4.5 ± 0.0	3.11	27.38	6.2 ± 0.0
1.0	40.60	4.9 ± 0.0	1.50	33.91	8.5 ± 0.0
0.1	44.30	5.3 ± 0.3	0.80	38.42	9.5 ± 0.0
0.05	44.49	6.0 ± 0.4	0.10	44.04	8.8 ± 0.3
			0.03	44.45	13.2 ± 0.8

TABLE 2.5
Continued

c (mol/liter)	Adiabatic Compressibility $\times 10^6$ (bar^{-1})	Total Solvation Number of Salt	c (mol/liter)	Adiabatic Compressibility $\times 10^6$ (bar^{-1})	Total Solvation Number of Salt
	KCl				
4.0	29.88	4.0 ± 0.0		$LaCl_3$	
1.0	39.84	5.7 ± 0.0	1.68	28.11	11.6 ± 0.0
0.17	43.90	6.2 ± 0.1	0.34	40.51	15.4 ± 0.0
0.05	44.46	6.6 ± 0.5	0.03	44.23	18.3 ± 0.7
			0.02	44.47	18.9 ± 1.4
	RbCl				
3.5	31.84	4.0 ± 0.0		$CeCl_3$	
1.7	38.13	4.9 ± 0.0	1.68	29.74	9.7 ± 0.0
0.9	41.17	4.9 ± 0.0	0.34	40.73	14.6 ± 0.0
0.1	44.30	5.0 ± 0.2	0.03	44.14	21.5 ± 0.7
0.05	44.50	5.5 ± 0.5	0.02	44.42	22.4 ± 1.3
	CsCl			$ThCl_4$	
5.0	28.15	3.2 ± 0.0	1.25	29.39	14.5 ± 0.0
1.0	42.03	3.2 ± 0.0	0.25	40.90	18.8 ± 0.0
0.1	44.43	3.6 ± 0.3	0.11	42.84	21.0 ± 0.3
0.05	44.52	5.1 ± 0.5	0.05	43.72	22.9 ± 0.5
			0.03	44.19	26.6 ± 1.1
	$BaCl_2$				
1.65	30.90	9.9 ± 0.0			
0.74	37.65	11.6 ± 0.0			
0.50	39.44	12.8 ± 0.0			
0.1	43.65	13.5 ± 0.0			
0.05	44.14	14.6 ± 0.5			

Source: Reprinted from J. O'M. Bockris and P. P. S. Saluja, *J. Phys. Chem.* **76**: 2140, 1972.

difference that can be detected between two electrodes sending sound waves to each other.

2.8.1. Ionic Vibration Potentials: Their Use in Obtaining the Difference of the Solvation Numbers of Two Ions in a Salt

Sound waves pass through matter by exerting a force on the particles in the path of the beam. Transmission of the sound occurs by each particle giving a push to the next particle, and so on.

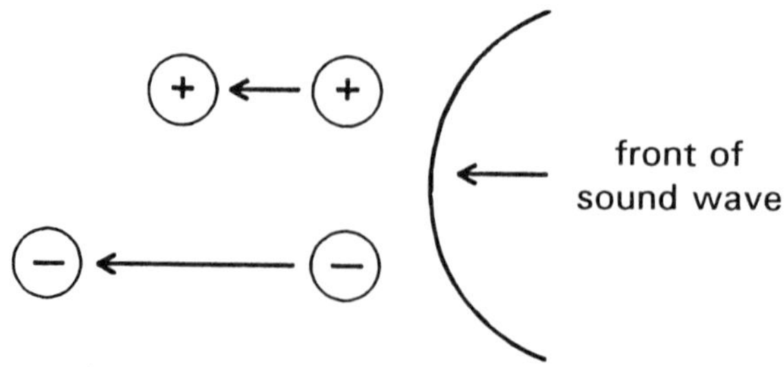

Fig. 2.17. Debye vibronic potential. Schematic on origin. In one phase of the sound waves, both ions are accelerated toward the left. However, small cations tend to be heavier than small anions because the cations carry a larger load of primary hydration water.

When a beam of sound is emitted from a transducer onto a solution that contains cations and anions, each having a different mass, these masses undergo a different degree of displacement per cycle, for while each receives the same pulse from the sound, each has a different inertia. Figure 2.17 shows that the displacement of each ion in one phase of the cycle is canceled in the next. However, there is a net difference in position of the cation and anion which remains and this gives rise to a degree of nonelectroneutrality that can be measured in the form of an ionic vibration potential, usually found to be in the range of 1–5 μV cm^{-1} s.

In 1933 Peter Debye formulated a sophisticated theory about all this.[13] He assumed, as is also intuitively obvious, that the "supersonic emf," that is, the ionic vibration potential produced by the ultrasonic beam, would be proportional to the *difference* of the masses of the moving ions. Debye's expression can be reduced to

$$\Delta E = 1.55 \times 10^{-7} a_0 \left(\frac{t_+}{z_+} m_{m,+} - \frac{t_-}{z_-} m_{m,-} \right) \tag{2.14}$$

where a_0 is the velocity amplitude[14] of the ultrasonic wave; t_+ and t_- are the respective transport numbers of cation and anion; and z_+ and z_- are the corresponding charges on cation and anion. The *apparent molar mass*, $m_{m,i}$, of the moving ion is defined as the mass of the solvated ion minus the mass of the solvent displaced, $V_1\rho_1$, where ρ_1 is the density of the solvent and V_1 its volume.

---

[13]It is of interest to note that the lengthy and complex calculation Debye made was published in the same (first) edition of the *Journal of Chemical Physics* as an article by Bernal and Fowler, who first suggested several seminal concepts about the structure of water that are now commonly accepted in solution theory.
[14]The velocity amplitude is measured in cm s^{-1}. It is the ratio of the pressure of the ultrasonic wave to the characteristic acoustic impedance of the media.

Now the mass of the solvated ion $(M_i)_s$ is the sum of the mass of the bare ion, M_i, and that of the primary solvation water, $n_i M_1$, where n_i is the solvation number of the ion and M_1 is the molar weight of the solvent.

Thus:

$$m_{m,i} = M_i + n_i M_1 - V_1 \rho_1 \tag{2.15}$$

If one takes this definition of $m_{m,i}$ and uses it in the above expression for ΔE, one obtains the rather lengthy expression:

$$\left(\frac{t_+}{z_+}\right) n_+ - \left(\frac{t_-}{z_-}\right) n_- = \frac{\Delta E / a_0}{M_1 \; 1.55 \times 10^{-7}} - \frac{1}{M_1}\left(\frac{t_+}{z_+} M_+ - \frac{t_-}{z_-} M_-\right)$$

$$+ \frac{\rho_1}{M_1}\left(\frac{t_+}{z_+}(V_1) - \frac{t_-}{z_-}(V_s)\right) \tag{2.16}$$

Now, for the aqueous case, Eq. (2.16) reduces to

TABLE 2.6

Relevant Parameters to Obtain Individual Solvation Numbers

Electrolyte	t_+ (Infinite Dilution)	t_- (Infinite Dilution)	$(V_s)_+$, cm^3 mol^{-1}	$(V_s)_-$, cm^3 mol^{-1}	A	ψ_0/a_0, μV cm^{-1} s: 0.05 N	0.1 N	1.0 N	5.0 N
NaF	0.476	0.524	115.8	110.1	−0.20	1.0	1.0	0.9	Insol
NaCl	0.396	0.604	115.8	92.4	0.14	1.1	1.0	0.7	(0.7)
NaBr	0.392	0.608	115.8	90.7	1.66	−2.8	−2.8	−2.6	(−2.6)
NaI	0.395	0.605	115.8	91.5	3.23	−6.4	−6.4	−5.7	(−5.7)
LiCl	0.336	0.664	140.7	92.4	0.40	0.1	0.1	0.05	(0.0)
KCl	0.490	0.510	91.5	92.4	−0.20	1.7	1.7	1.7	(1.7)
RbCl	0.511	0.489	89.9	92.4	−1.42	5.2	5.2	(5.2)	(5.2)
CsCl	0.500	0.500	89.9	92.4	−2.78	8.1	8.1	(8.1)	(8.1)
$BaCl_2$	0.454	0.546	166.4	92.4	−1.35	4.2	4.2	3.6	(3.6)
$BaBr_2$	0.448	0.552	166.4	90.7	−0.03	−1.8	1.6	−0.8	(0.8)
BaI_2	0.452	0.548	166.4	91.5	1.45	−1.7	1.8	−2.0	(2.03)
$MgCl_2$	0.410	0.590	197.9	92.4	0.12	1.1	1.0	1.0	(1.0)
$CaCl_2$	0.438	0.562	176.5	92.4	−0.12	1.1	1.1	1.0	0.8
$SrCl_2$	0.438	0.562	176.5	92.4	−0.70	2.6	2.5	2.1	1.8
$LaCl_3$	0.476	0.524	233.1	92.4	−0.82	3.9	3.7	3.2	(3.2)
$CeCl_3$	0.460	0.540	233.1	92.4	−0.91	3.8	3.8	3.0	2.5
$ThCl_4$	0.485	0.515	330.9	92.4	−0.95	3.4	3.2	2.9	(2.9)

Source: Reprinted from J. O'M. Bockris and P. P. S. Saluja, *J. Phys. Chem.* **76**: 2140, 1972.

TABLE 2.7

Absolute Values of Ionic Solvation Number (SN) vs. Concentration at 298 K

Electrolyte	Concentration	SN of Cation	SN of Anion	Electrolyte	Concentration	SN of Cation	SN of Anion
NaF	1.0	4.0	3.5	$BaBr_2$	3.29	5.7	0.9
	0.5	4.3	3.6		1.21	7.9	1.4
	0.1	4.6	3.9		0.50	9.2	1.6
NaCl	5.1	2.8	1.1		0.14	11.1	1.8
	2.0	3.2	1.5		0.05	11.6	1.8
	1.0	3.8	1.9	BaI_2	2.50	7.4	0.9
	0.1	4.4	2.1		0.50	9.2	1.3
	0.05	4.7	2.2		0.25	10.3	1.5
NaBr	5.2	3.0	0.7		0.05	11.9	1.8
	2.0	3.8	1.3		0.025	12.6	1.8
	1.1	4.2	1.5	$MgCl_2$	6.22	5.1	0.5
	0.1	4.4	1.7		2.11	6.8	0.8
	0.05	4.5	1.8		1.05	8.1	1.0
NaI	5.0	3.1	0.1		0.09	9.7	1.3
	2.1	3.8	0.6		0.05	10.1	1.4
	1.0	4.0	0.6	$CaCl_2$	4.05	4.5	0.7
	0.1	4.3	1.2		2.02	5.0	1.0
	0.05	4.6	1.5		1.00	7.1	1.2

LiCl	5.0	2.6	0.7		0.10	8.3	1.4
	2.0	3.4	1.1		0.05	9.1	1.5
	1.0	3.7	1.2	$SrCl_2$	3.11	4.4	0.9
	0.1	3.9	1.4		1.50	6.0	1.3
	0.05	4.4	1.6		0.80	6.9	1.3
KCl	4.0	2.4	1.6		0.10	6.6	1.1
	1.0	3.3	2.4		0.03	9.8	1.7
	0.17	3.6	2.6	$LaCl_3$	1.68	9.4	0.7
	0.05	3.8	2.8		0.34	12.3	1.0
RbCl	3.5	2.4	1.6		0.03	14.8	1.2
	1.7	2.8	2.1		0.02	15.3	1.2
	0.9	2.8	2.1	$CeCl_3$	1.68	7.5	0.7
	0.1	2.9	2.1		0.34	11.6	1.0
	0.05	3.1	2.4		0.03	17.4	1.4
CsCl	5.0	1.7	1.5		0.02	18.1	1.4
	1.0	1.7	1.5	$ThCl_4$	1.25	11.9	0.6
	0.1	1.9	1.7		0.25	15.4	0.8
	0.05	2.7	2.4		0.11	17.4	0.9
$BaCl_2$	1.65	6.9	1.5		0.05	18.4	1.1
	0.74	8.2	1.7		0.03	22.0	1.1
	0.50	9.0	1.9				
	0.10	9.7	1.9				
	0.05	10.5	2.0				

Source: Reprinted from J. O'M. Bockris and P. P. S. Saluja, *J. Phys. Chem.* **76**: 2140, 1972.

$$\left(\frac{t_+}{z_+}\right)n_+ - \left(\frac{t_-}{z_-}\right)n_- = \frac{\Delta E/a_0}{2.79} + A \tag{2.17}$$

where A depends on the electrolyte.

Bockris and Saluja applied these equations derived by Debye to a number of electrolytes and, using data that included information provided by Conway and by Zana and Yeager, calculated the difference of the solvation numbers of the ions of salts. The relevant parameters are given in Table 2.6.

Now, since the ultrasound method gives the difference of the hydration numbers, while the compressibility method gave the sum, individual values can be calculated (Table 2.7).

2.9. SOLVATION NUMBERS AT HIGH CONCENTRATIONS

2.9.1. Hydration Numbers from Activity Coefficients

A nonspectroscopic method that has been used to obtain hydration numbers at high concentrations will be described here only in qualitative outline. Understanding it quantitatively requires a knowledge of ion–ion interactions, which will be developed in Chapter 3. Here, therefore, are just a few words of introduction.

The basic point is that the mass action laws of chemistry ([A][B]/[AB] = constant) do not work for ions in solution. The reason they do not work puzzled chemists for 40 years before an acceptable theory was found. The answer is based on the effects of electrostatic interaction forces between the ions. The mass action laws (in terms of concentrations) work when there are no charges on the particles and hence no long-range attraction between them. When the particles are charged, Coulomb's law applies and attractive and repulsive forces (dependent on $1/r^2$ where r is the distance between the ions) come in. Now the particles are no longer independent but "pull" on each other and this impairs the mass action law, the silent assumption of which is that ions are free to act alone.

There are several ways of taking the interionic attraction into account. One can work definitionally and deal in a quantity called "*activity*," substituting it for concentration, whereupon (by definition) the mass action law works. Clearly, this approach does not help us understand why charges on the particles make the concentration form of the mass action law break down.

From a very dilute solution (10^{-4} mol dm^{-3}) to about 10^{-2} mol dm^{-3}, the ratio of the activity to the concentration (a_i/c_i or the *activity coefficient*, γ_i) keeps on getting smaller (the deviations from the "independent" state increase with increasing concentration). Then, somewhere between 10^{-2} and 10^{-1} M solutions of electrolytes such as NaCl, the activity coefficient (the arbiter of the deviations) starts to hesitate as to which direction to change with increasing concentration; above 1 mol dm^{-3}, it turns around and *increases* with increasing concentration. This can be seen schematically in Fig. 2.18.

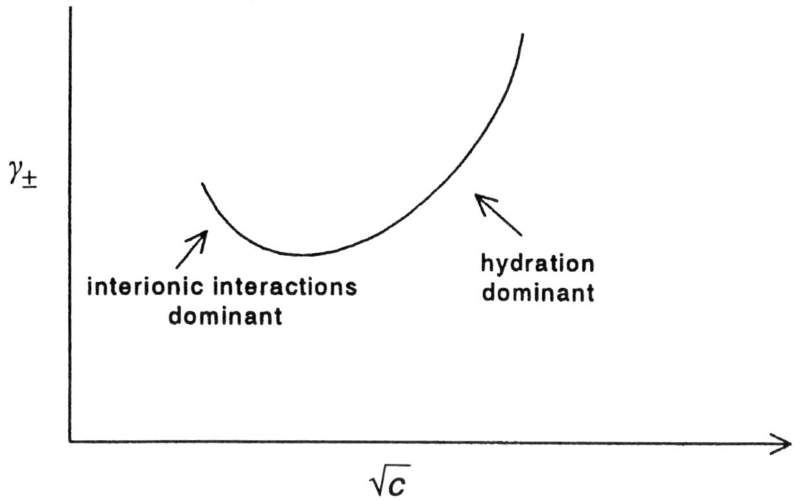

Fig. 2.18. Schematic of the observed trend of $\gamma_\pm$ versus $\sqrt{c}$ curve for salts showing a minimum.

Why does it do this? There may be more than one reason. A reason that was suggested long ago by Bjerrum and developed intensively by Stokes and Robinson in the 1950s is concerned with solvation and has more than historical interest. This is how they argued.

In 1 liter (1 dm^3) of pure water, there are 55.5 mol of water [(1000 g)/(18 g/mol)], all of it available to solvate ions. As ions are added, the water can be divided into two types, the so-called *free water* (unattached to any ions) and the *water associated* with (i.e., hydrating) ions.

The idea is that the water adhering to ions is out of commission as far as functions of the free water go. Only a fraction of the free water is available to solvate the added ions. That is, the effective concentration is increased compared with that which one would calculate if one assumed all the water was active. Thus, in the "concentration" calculation (so and so many moles of ions per liter of water), it is implicitly assumed that all the water molecules are "active." Suppose half the water molecules are temporarily associated with ions; then the *effective* concentration (i.e., the activity) is doubled.

Of course, in this simplified presentation, one assumes that by the time the concentration is so high that the activity coefficient–concentration relation turns upward (Fig. 2.18), the interionic interaction effects, although still there, have been overwhelmed by the effect of ions in reducing free waters. In reality, both this effect and the interionic effects that dominated at lower concentrations (below the minimum) should be taken into account.

By now, the reader will begin to see the point, and how all this is related to determining hydration number. In Chapter 3, a quantitative expression between this activity coefficient and the concentration will be developed. In it, the unknown is n_h, the *hydration number* of the electrolyte. The sum of the hydration number of the cation and anion will be found and so a measurement of the activity coefficient at a particular concentration (from, e.g., 0.1 mol dm^{-3} up to 10 mol dm^{-3}) will yield the hydration number at that concentration.

A second difficulty is more subtle. The activity coefficient is determined not only by water that is adhering to ions, but also by increasing interionic effects, and our ability to allow for these at very high concentrations such as 5 mol dm^{-3} is not good. Spectroscopy tells us also that ionic association is occurring in these high ranges, but there is not much information on this association for ions that do not have IR spectra (e.g., Na^+ and Cl^-). These matters will be discussed again quantitatively in Chapter 3.

2.10. TRANSPORT

2.10.1. The Mobility Method

The mobility method is a rough-and-ready method for obtaining information on the number of solvent molecules that accompany an ion in motion. Its basic theory is really quite simple. One equates the electrostatic force pulling the ion forward, $z_i e_0 X$ ($z_i e_0$ = charge of the ion; X = electric field gradient), to the viscous resistance to the ion's flow. This view neglects all interionic interactions (Chapter 4) but would apply at sufficiently high dilution. This viscous resistance is given by Stokes' law, $6\pi r \eta v$, where r is the radius of the entity moving through a liquid of viscosity η and at a velocity v. The bulk viscosity is used. However, in reality an ion breaks up the solvent near it as it darts from here to there in the solution, so that a viscosity less than that of the undisturbed bulk water should be used with Stokes' law. Again, to determine the radius of the ion plus the adherent solvent and to determine how many water molecules fit in, one has to know the volume of water attached to the ion. This is not the ordinary bulk volume but a compressed value arising from the effect of the ion's field on normal water.

Finally, the validity of using Stokes' law to find the force of viscous resistance against movement in a liquid has to be questioned. In the original derivation of this formula, the model used was that of a *solid sphere* passing in a stately way in a straight line through a viscous fluid like molasses. The extrapolation to atomic-sized particles that move randomly in a solvent which itself has innumerable complex, dynamic movements might be thought to stretch the equation so far from its original model that it would become inapplicable. Nevertheless, tests (Chapter 5) show that Stokes' law *does* apply, although, depending on the shape of the particle, the 6 (which is valid for spheres) might have to be modified for other shapes (e.g., 4 for cylinders).

Now, from the equivalent conductivity of an electrolyte Λ (Chapter 4, Section 4.3.7) at concentrations low enough so that the ions are virtually free from the influence of interionic forces, or ion-pair formation, individual ions (none associated into pairs) can be related to the mobility of the ions, u_+ and u_- by the equation (Section 4.4.5).

$$\Lambda = F(u_+ + u_-) \tag{2.18}$$

The transport number (Section 4.5.2) of the cation, t_+, is given by

$$t_+ = \frac{u_+}{u_+ + u_-} = \frac{u_+ F}{\Lambda} \tag{2.19}$$

and

$$u_i = \frac{t_i \Lambda}{F} \tag{2.20}$$

In electrostatics, a charge q under the influence of a field X experiences a force Xq. Here,

$$z_i e_0 X = 6\pi r_S \eta \, u_i \tag{2.21}$$

where r_S is r_{Stokes} (the radius of the ion and its primary solvation shell) and this allows one to use independent knowledge of the mobility, u_i, to obtain the radius of the moving solvated ion (Chapter 3). One also knows the crystallographic radius of the bare ion inside the solvation sheath, r_c. Hence:

$$V_{\text{hydrated water}} = \frac{4}{3}\pi(r_S^3 - r_c^3) \tag{2.22}$$

$$n_s = \frac{r_S^3 - r_c^3}{r_{H_2O}^3} \tag{2.23}$$

TABLE 2.8

Nearest Integer Hydration Number of Electrolytes from the Mobility Method and the Most Probable Value from Independent Experiments

Salt	Hydration Number (nearest integer)	Hydration Number from Other Experimental Methods
LiCl	7	7 ± 1
LiBr	8	7 ± 1
NaCl	4	6 ± 2
KCl	2	5 ± 2
KI	3	4 ± 2

TABLE 2.9

Primary Hydration Numbers from Ionic Mobility Measurements

Ion	Li^+	Na^+	Mg^{2+}	Ca^{2+}	Ba^{2+}	Zn^{2+}	Cd^{2+}	Fe^{2+}	Cu^{2+}	Pb^{2+}
V_s	31.8	15.0	95.2	67.8	48.5	92.3	91.2	92.3	95.0	34.3
n_{min}	3.5	2	10.5	7.5	5	10	10	10	10.5	4
n_{max}	7	4	13	10.5	9	12.5	12.5	12.5	12.5	7.5

Source: Reprinted from B. E. Conway, *Ionic Hydration in Chemistry and Biophysics*, Elsevier, New York, 1981.

is a rough expression for the volume occupied by water molecules that move with one ion. The expression is very approximate (Tables 2.8 and 2.9), because of the uncertainties explained earlier.

On the other hand, the transport or mobility approach to determining the primary hydration number does give a value for what is wanted, the number of water molecules that have lost their own degrees of translational freedom and stay with the ion in its motion through the solution. This approach has the advantage of immediately providing the individual values of the solvation number of a given ion, and not the sum of the values of those of the electrolyte.

Why bother about these hydration numbers? What is the overall purpose of chemical investigation? It is to obtain knowledge of invisible structures, to see how things work. Hydration numbers help to build up knowledge of the environment near ions and aid our interpretation of how ions move.

2.11. SPECTROSCOPIC APPROACHES TO OBTAINING INFORMATION ON STRUCTURES NEAR AN ION

2.11.1. General

There is nothing new about spectroscopic approaches to solvation, the first of which was made more than half a century ago. However, improvements in instrumentation during the 1980s and 1990s, and above all the ready availability of software programs for deconvolving spectra from overlapping, mixed peaks into those of individual entities, have helped spectra give information on structures near an ion. This is not to imply that they supersede alternative techniques, for they do carry with them an Achilles heel in that they are limited in sensitivity. Thus, by and large, only the more concentrated solutions (>0.1 mol dm^{-3}) are open to fruitful examination. This is not good, for in such concentrations, interionic attraction, including substantial ion pairing and more, complicates the spectral response and makes it difficult to compare information obtained spectroscopically with that obtained using the partial molar volumes and vibration potential methods (Section 2.7).

One point should be noted here: the importance of using a 10% D_2O mixture with H_2O in IR spectroscopic measurements because of the properties of HOD, which contributes a much more clearly resolved spectrum with respect to O-D. Thus, greater clarity (hence information) results from a spectrum in the presence of HOD. However the *chemical* properties (e.g., dipole moment) of HOD are very similar to those of H_2O.

Raman spectra have a special advantage in analyzing species in solution. This is because the integrated intensity of the spectral peaks for this type of spectroscopy is proportional to the concentration of the species that gives rise to them.[15] From observations of the intensity of the Raman peaks, equilibrium constants K can be calculated and hence $\Delta G°$s from the thermodynamic equation $K = e^{-\Delta G°/RT}$ can be derived. Furthermore, if one carries out the Raman experiment at various temperatures, one can determine both the heat and the entropy of solution. Since $\Delta G° = \Delta H° - T\Delta S°$, a plot of ln K against $1/T$ gives the enthalpy of solvation from the slope and the entropy from the intercept. This provides much information on the various relations of ions to water molecules in the first one or two layers near the ion. In particular, the use of a polarized light beam in the Raman experiments provides information on the *shape* of complexes present in a solution.

2.11.2. IR Spectra

In obtaining information on solvation that can be deduced from IR spectra, the first thing that must be understood is that the raw data, the peaks and their frequencies, seldom speak directly but need to be decoded. Spectra in the IR region are mainly messages fed back from the solvent, and it is from the interpretation of evidence for *changes* in the solvent's libration and rotation when ions are introduced (rather than any new peaks) that information on solvation may sometimes be drawn. One has to take the spectrum of the solvent, then that of the solution, and subtract them to obtain the effect of the solute (Fig. 2.19). Vibration spectra have frequencies in the region of 10^{14} s^{-1} but it is usual to refer to the inverse of the wavelength, that is, the *wavenumber*, $\bar{\nu} = 1/\lambda$. Since $c_0 = \nu\lambda$, then $1/\lambda = \nu/c_0$. It turns out then that the wavenumbers of most covalent bonds are numerically in the thousands.[16]

Intramolecular effects can be detected in the near infrared or high-frequency region ($\bar{\nu} \sim 1000$ cm^{-1}). *Intermolecular* effects are seen in the far infrared or low-frequency region ($\bar{\nu}$ down to 100 cm^{-1}). Early measurements showed that ions can cause new peaks to arise that are at distinctly higher wavenumbers than those in pure water. The explanation proposed is that some of the hydrogen bonds present in pure

---

[15]This tends to be the case for all spectra. For other spectra it involves sensitivity factors or nonlinearity at higher concentrations; that is, it is approximate.

[16]A typical value for ν is 9×10^{13} s^{-1} and $c_0 = 3 \times 10^{10}$ cm s^{-1}. Hence, the wave number, $\bar{\nu}$, is $\bar{\nu} = 9 \times 10^{13}/3 \times 10^{10} = 3 \times 10^3$ cm^{-1}.

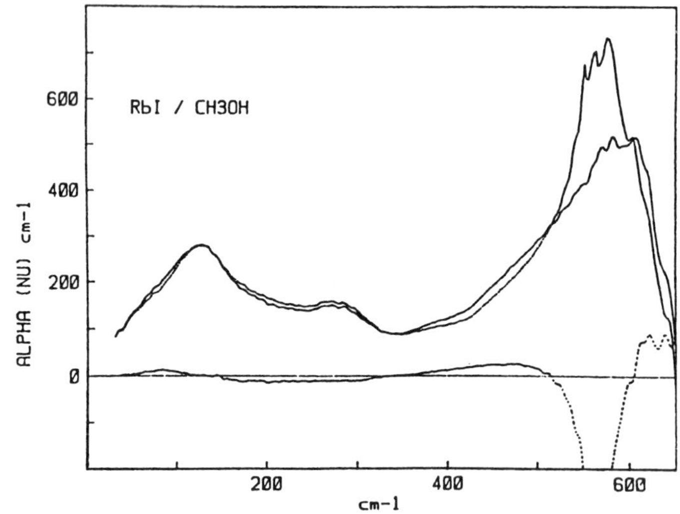

Fig 2.19. The absorption coefficients of pure methanol (Top line) and a solution of 0.4 *M* RbI in CH_3OH (Lower line). The bottom curve represents the difference between the solution absorption band and the solvent absorption band. (Reprinted from B. Guillot, P. Marteau, and J. Obriot, *J. Chem. Phys.* **93**: 6148, 1990.)

water have been broken due to the presence of ions, thus producing some quasi-free water molecules, which are the source of the new peaks.

There is plenty of spectroscopic evidence for a *structure-breaking* effect of ions in aqueous solution (although there is also evidence for formation of new structures). Because an increase in temperature also causes structure breaking, there has arisen the concept of "structural temperature" to describe ionic effects that produce the same degree of damage that would be produced by increasing the temperature. This structure breaking occurs in the "secondary" water (i.e., that outside the primary hydration sphere) because the first sheath of water around the ion is structure forming. Anions, which are usually bigger than cations, have consequently a less tightly held primary co-sphere and their large size makes them more responsible for structure breaking in solution than the corresponding cations.

The far IR (lower energy) spectra (100 cm^{-1}) show intermolecular effects in which the spectra reflect the effect of ions on the movements of the whole water molecule and (in contradistinction to vibrational movements within individual molecules)[17] are dynamically dominated by the mass of oxygen and not that of hydrogen.

[17]For a water molecule, the reduced mass is $1/\overline{\mu} = 1/M_O + 1/2M_H$ and from Eq. (2.24) the vibrational spectrum is dominated by M_H.

From an interpretation of peaks in the far IR spectral region, one can obtain knowledge of hindered translations among water molecules in ionic solutions. Somewhat surprisingly, the force constants associated with such movements are *lowered* by the presence of ions because ions free some water molecules from the surrounding solvent structures. Thus, force constants are given by $\partial^2 U/\partial r^2$, where U is the potential energy of particle–particle interaction. The librative frequencies (Section 2.4) of water also show up in this region and decrease in the order KF > KCl > KBr > KI. Thus, ions of smaller radii (higher field and force) give higher librative frequencies, as expected because of the equation

$$\nu = (2\pi)^{-1}\sqrt{k/\tilde{\mu}} \tag{2.24}$$

where k is the force constant (a function of the energy of ion–water interaction) and $\tilde{\mu}$ is the reduced mass of the vibrating entities.

James and Armitage have analyzed the far IR spectra of some ionic solutions and attempted to distinguish waters in the primary hydration shell (those waters that *stay* with the ion as it moves) from waters ("secondary hydration") which, although affected by the ion, are not attracted by it enough to move with it.

Studies that provide more illumination arise from IR measurements in the work of Bergstrom and Lindgren (Table 2.10). They have made IR studies of solutions containing Mn^{2+}, Fe^{2+}, and Co^{2+}, which are transition-metal ions, and also certain lanthanides, La^{3+}, Nd^{3+}, Dy^{3+}, and Yb^{3+}, at a concentration of 0.2–0.3 mol dm^{-3}. They find that the O–D stretching vibrations in HOD (2427 cm^{-1} in the absence of ions) are affected by the presence of these substances (Fig. 2.20). Both the transition-metal ions and the lanthanide elements perturb the HOD molecule in a similar way. However, even the trivalent lanthanide ions (for which stronger effects are expected because of the higher ionic charge) only perturb the nearest-neighbor water molecules (the first

TABLE 2.10

Bergstrom and Lindgren's Determination of Primary Hydration Number from IR Measurements (Transition Metal-Ions and Lanthanides)

Ion	Hydration No.
Mn^{2+}	6.5
Fe^{2+}	7.3
Co^{2+}	6.5
La^{3+}	7.8
Nd^{3+}	8.0
Dy^{3+}	8.7
Yb^{3+}	8.8

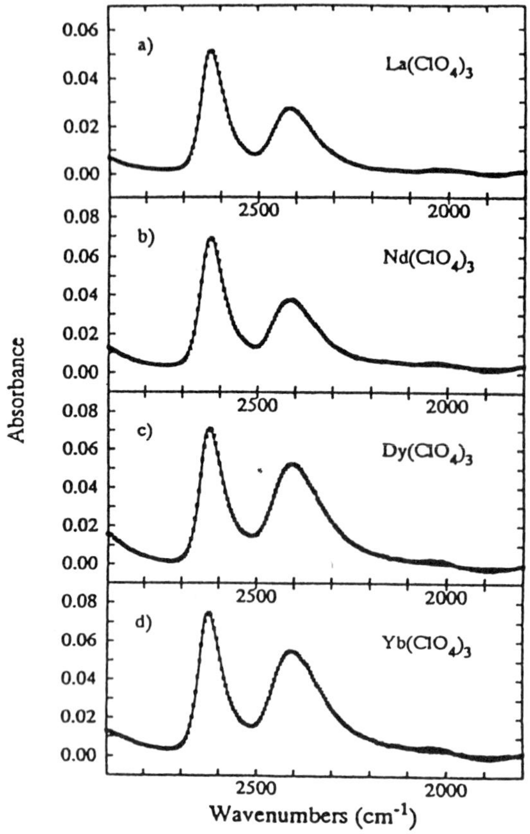

Fig. 2.20. O–D stretching spectra of aqueous solutions of (a) 0.236 *M* $La(ClO_4)_{3+}$ and (b) 0.342 *M* $Nd(ClO_4)_{3+}$ (c) 0.366 *M* $Dy(ClO_4)_{3+}$ and (d) 0.375 *M* $Yb(ClO_4)_{3+}$. The crosses are the observed points, and the solid line is the function fitted to the spectra. See Table 1 for band parameters. T = 20.0 °C, path length = 0.0444 mm, and c(HDO) = 6.00 mol%. (Reprinted from P. A. Bergstrom, *J. Phys. Chem.* **95**: 7650, 1991).

shell around the ion) and there is no effect detectable in the IR spectra on the second and other layers. The hydration numbers thus obtained are given in Table 2.10.

The hydration numbers in the table, deduced from IR spectroscopy, are much lower than values given by nonspectroscopic methods. The latter give hydration numbers for two- and three-valent ions 1.3 times greater than the spectroscopic values. It seems particularly surprising that in the spectroscopic results (Table 2.10) the three-valent ions give hydration numbers little different from those of charge two.

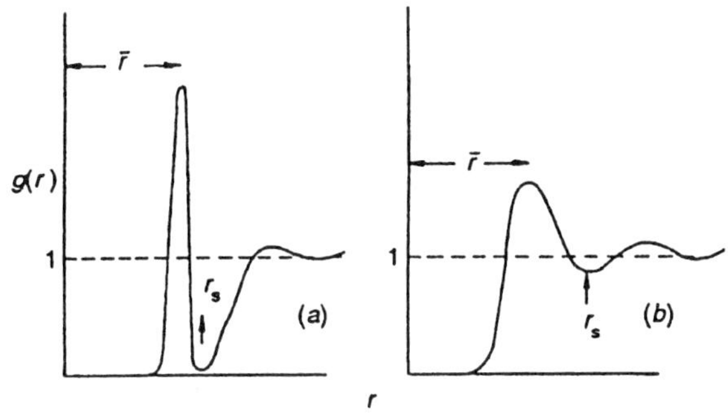

Fig. 2.21. Generic radial distribution functions for (a) a well-coordinated liquid with a long-lived coordination shell and (b) a weakly coordinated liquid. (Reprinted from J. E. Enderby, *Chem. Soc. Rev.* **24**: 159, 1995.)

Evidently the first layer is about the same when filled for both the two- and the three-valent ions, and perhaps the IR spectrum does not register effects given by a second layer of water. One might indeed well expect the first layer of water around an ion to be full, and therefore expect the same for the two- and three-valent cations of similar radii.

The presence of electrolytes in solution is often ill characterized spectroscopically by vague shoulders or bumps that make the interpretation hazardous and lacking in quantitative information on the ion–solvent structure. However, far IR does allow one to understand the spectra and obtain knowledge of the ion–solvent structure. For example, it is important to distinguish contact ion pairs (CIP) in the spectrum. These must be clearly identified and their effects allowed for before the spectrum can be used to obtain knowledge of ion–solvent interactions (Fig. 2.21).

2.11.3. The Neutron Diffraction Approach to Solvation

The seminal event in the foundation of solid-state science was the realization by von Laue that the ordered structure of atoms in a crystalline solid might act as a diffraction grating for X-rays. The corresponding formula by Bragg,

$$n\lambda = 2d \sin \theta \tag{2.25}$$

This represents the path-length difference with θ as the glancing angle (see Section 5.2.3) allowed d, the distance between atoms, to be determined for the first time. The path to structure determination was open.

X-ray analysis works when there are indeed ordered rows of atoms in a crystal. It also works in an examination of the structure of molten salts (Chapter 5), when there is short-range order; the molten salts are somewhat like a disordered solution containing much open space, which some call holes. However, X-ray analysis does not work well for ionic solutions because the ordered elements (solvated ions) turn up only occasionally and are interrupted by relatively large distances of disordered solvent molecules.

Although, as will be seen in this section, the developments of neutron diffraction during the 1980s (Enderby and Neilson) have led to substantial advances in determining the coordination numbers of water molecules around ions, there are still some hurdles to clear before one can use this powerful method:

1. It is necessary to have a *neutron source* and that, in turn, means that only laboratories that have access to a nuclear reactor can do such work.[18]
2. It is desirable to work with D_2O rather than H_2O because of the sharper diffraction patterns obtained are the former. This can become a cost burden.
3. (1) and (2) can be overcome but the last hurdle is too high and must be accepted: the method is limited (as with most of the spectroscopic methods) to solutions of 1 mol dm^{-3} or more,[19] whereas most of the data in the literature concern dilute solutions (e.g., 10^{-3} mol dm^{-3}) where the univalent salts are about 12 nanometers apart (one can, then, picture an isolated ion in its solvation shell).

Having given the obstacles to attainment, let it be said that there are two ways in which neutron diffraction can be used to obtain information on the structure around an ion. In the first (Soper et al., 1977), the objective is the distribution function in the equation:

$$dn_\alpha = 4\pi \rho_\beta g_{\alpha\beta}(r) r^2 dr \tag{2.26}$$

where $\rho_\beta = N_\beta/V$.

This equation refers to a reference ion α and gives the average number of β particles that exist in a spherical shell of radius r and thickness dr (of course, on a time average). The symbol $g_{\alpha\beta}(r)$ is the so-called *distribution function*. For multicomponent systems, one can generalize $g_{\alpha\beta}(r)$ to include all the possible combinations of atoms [e.g., NaCl would have a total of 10 $g_{\alpha\beta}(r)$ values].

---

[18]In U.S. universities, this means that the professor concerned must write a proposal to ask for time on a reactor at, for example, Brookhaven or Oak Ridge or Argonne. Such proposals wait in line until they are evaluated and then, if accepted, the professor's team must move to the reactor for some days of intense activity—probably shift work to make 24-hr per day use of the time allotted.

[19]It is easy to show that the average distance between ions in a solution of 1:1 salt is $(1000/2N_A c)^{1/3}$, where c is the concentration in mol dm^{-3}. For a 1 M solution, this comes to 1.2 nm. However, for an ion of 0.1-nm radius and layer of two waters, the radius is about $0.1 + 4 \times 0.16$ nm and therefore the internuclear distance between two ions in contact is about 1.4 nm. For the 10^{-3} mol dm^{-3} case, the distance apart is 12.2 nm; i.e., the ions are isolated. Thus, it is very desirable to try to get spectroscopic measurements at these dilute solutions. Only then can the results of spectroscopic work be compared directly with deductions made about solvation (a concentration-dependent property) from measurements of solution properties.

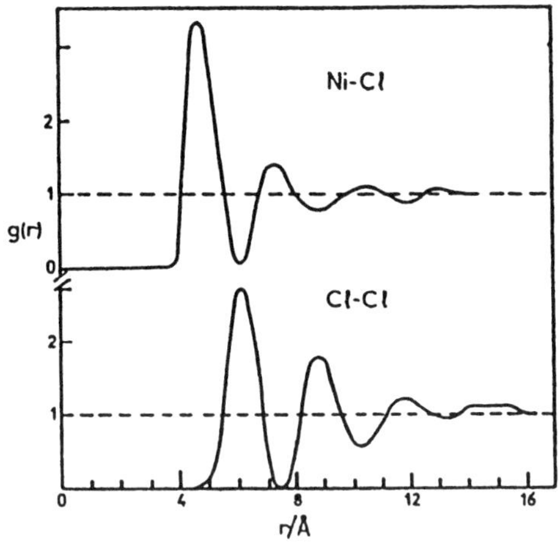

Fig. 2.22. The radial distribution functions g_{NiNi} and $g_{ClCl}(r)$ for a 4.35 *M* solution of $NiCl_2$ in D_2O (neutron results). (Reprinted from J. E. Enderby, "Techniques for the Characterization of Electrodes and Electrochemical Processes," in R. Varma and J. R. Selman, eds., *The Electrochemical Society Series,* Wiley, New York, 1991, p. 110).

It follows that

$$4\pi\rho_\beta \int_0^{r_s} g_{\alpha\beta}(r)\, r^2\, dr$$

is the coordination number, where r_s is the radius of the first shell of water molecules around the ion.[20]

Two extreme types of results are obtained from this approach (which is limited to solutions of C > 0.1 m), and they are shown in Fig. 2.21. The left-hand diagram shows the kind of result in which the molecules are strongly coordinated in a first shell with a lifetime that could vary from, say, 10^{-5} s to hundreds of hours (in exceptional cases). The second diagram is typical of a more weakly coordinated ion in which the

[20]The actual experimental determination (analogous to the Bragg determination of d, the distance apart of atoms in crystals) is of $g_{\alpha\beta}$. This is what is done with neutron diffraction. It can also be calculated using molecular dynamics in which the basic assumption is that there is a certain force field between the particles. As explained elsewhere (Section 2.17.4), this is given in terms of the corresponding energy of interaction of pairs of particles. Although the attraction energy is always given by $z_i z_j / \varepsilon r$, there are various versions of the repulsion potential.

lifetime of water molecules is short, perhaps only 10^{-11} s. For multicomponent systems it is possible to have several internuclear distances and, correspondingly, a number of distribution functions. By applying Eq. (2.26) to the data for $g_{\alpha\beta}(r)$, such as that shown in the diagram, one can find out how many water molecules are in the first shell, as in the case of concentrated $NiCl_2$ solutions (Fig. 2.22).

There are always water molecules located around a stationary ion and the structure of these waters will be dominated by the field of the ion (rather than the pull back into the structure of the water). This dominance is stronger the smaller the ion because the ion–solvent interaction is inversely proportional to $1/r^2$. As the schematics of a typical distribution function suggest, there may be a second layer in addition to the first shell of solvent associated with the ion. In this layer the structure is not yet that of bulk water, though such second solvation layers are usually more prominent with divalent ions (they are even more so with 3+ and 4+ ions) and are not seen for univalent ions, in the company of which hydration waters stay for very short times.

Now, the question is how to get information on the more subtle quantity, the hydration *numbers*. Some confusion arises here, for in some research papers the coordination number (the average number of ions in the first layer around the ion) is also called the hydration number! However, in the physicochemical literature, this latter term is restricted to those water molecules that spend at least one jump time with the ion, so that when its dynamic properties are treated, the effective ionic radius seems to be that of the ion plus one or more waters. A startling difference between co-ordination number and solvation number occurs when the ionic radius exceeds about 0.2 nm (Fig. 2.23a).

It is important, then, to find out the time that waters stay with the ion. Thus, one can make an order-of-magnitude calculation for the jump time, by a method shown in Section 4.2.17. It comes to approximately 10^{-10} s.

One could conclude that if a water molecule stays with its ion for more than about 10^{-10} s, it has accompanied the ion in a jump. That is, during the time the water is associated with the ion, it is likely to have made one move with the ion in its sporadic random movements (and therefore counts as a hydration number rather than a static or equilibrium coordination number).[21] Figure shows the ratio of the solvation number to the coordination number. The ratio $\tau_{\text{ion-wait}}/\tau_{\text{water orient}}$ is an important quantity

[21]In the case of ions for which the ion–water binding is very strong (the transition-metal ions particularly), the hydration number may be greater than the coordination number, because more than one shell of waters moves with the ion and the hydration number will encompass *all* the water molecules that move with it, while the coordination number refers to the ions in just the first shell.

However, with larger ions, which have weaker peripheral fields, there is less likelihood that a water molecule will stay for the time necessary to accomplish an ion movement (e.g., $> 10^{-10}$ s). Thus, for larger ions like Cl^- and large cations such as $N(C_2H_5)_4^+$, the coordination number will be 6 or more, but the hydration number may tend to be 0. The hydration number is a dynamic concept; the coordination number is one of equilibrium: it does not depend on the lifetime of the water molecules in the shell but measures their time-averaged value.

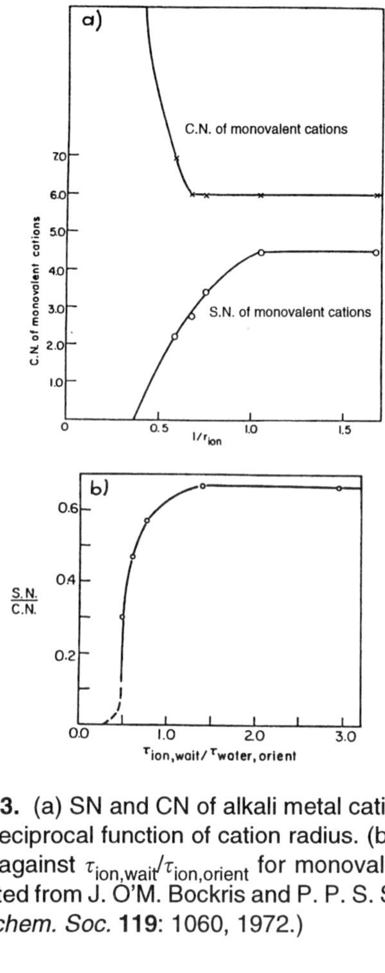

Fig. 2.23. (a) SN and CN of alkali metal cations plotted as reciprocal function of cation radius. (b) SN/CN plotted against $\tau_{\text{ion,wait}}/\tau_{\text{ion,orient}}$ for monovalent ions. (Reprinted from J. O'M. Bockris and P. P. S. Saluja, *J. Electrochem. Soc.* **119**: 1060, 1972.)

because it emphasizes the dynamic character of the solvation number. Thus, $\tau_{\text{ion-wait}}$ represents the time taken for an ion to remain at a given site in its movement into the solution. Correspondingly, $\tau_{\text{water orient}}$ represents the time for a water molecule (at first fixed within the water structure) to break out of this and orient towards the ion to a position of maximum interactions upon the ions arrival at a given site. In the case (Fig. 2.23b) where this ratio is big enough, the ratio of the solution number to the co-ordination number will be above 0.5.

There is a second way to use neutrons to investigate the structure of ions, particularly with respect to the time of movement of the water molecules. A remarkable advance was accomplished by Hewich, Neilson, and Enderby in 1982. They used *inelastic neutron scattering*. Upon analysis of their results, they found that they could obtain D_H, the diffusion coefficient for displacement of water. The special point they

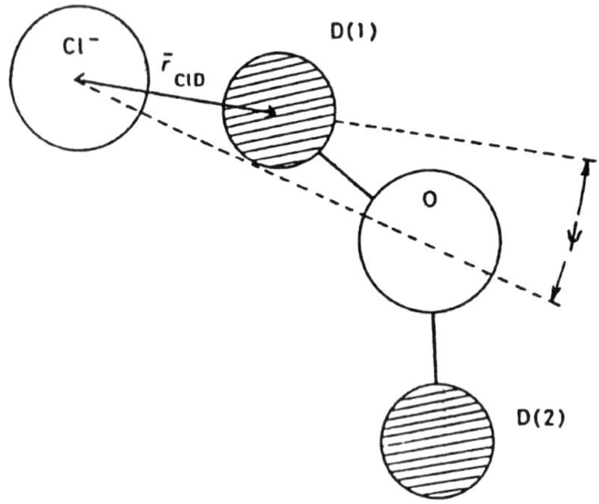

Fig. 2.24. Geometric arrangement of a D_2O molecule close to a chloride ion. (Reprinted from D. H. Powell, A. C. Barnes, J. E. Enderby, G. W. Neilson, and P. H. Salmon, *Faraday Disc. Chem. Soc.* **85**: 137, 1988.)

discovered was that certain functions that should have varied in a Lorentzian way with the frequency of the neutrons used were a poor fit to the expected variation and were better deconvoluted into two *D*s, one for water in the inner sphere and one for water in the outer sphere. Thus, this was direct evidence that during the movement of some ions (Ni is the one cited) there is an inner layer of about 6 but an outer layer of about 15 that also moves with the ion. Here, then, the coordination number would be 6 (number of molecules in the inner sphere) and the hydration number would be 6 + 15 = 21.

Finally, in this very general account of neutron interference and scattering applied to ions in solution, it is interesting to note that the tilt angle of the water molecule to the ion can be obtained (Fig. 2.24). Again, Enderby and Neilson are the progenitors of this kind of information and an example (together with one for the wag angle ψ and its variation with concentration) is given here.†

Ferrous and ferric ions have been examined in respect to their solvation shells, particularly by NMR methods. For Fe^{2+}, the value obtained for the number of water

†John Enderby is Professor of Physics in the H. H. Wills Physics Laboratory of the University of Bristol. Together with his colleague, G. W. Neilson, he has made, arguably more contributions to specific, quantitative, knowledge of the region close to an ion in solution than that of any other worker since 1950. Thus, the developments of the neutron diffraction methods at Bristol have gone far to making it possible (for concentrated solutions at least) to distinguish between waters remaining with an ion during its movements and those which are simply affected by an ion as it passes by.

molecules in a first shell was 5.85 and the lifetime of these waters was relatively long, about 2×10^{-7} s. Thus, because this time is much greater than that needed for a diffusive movement (10^{-10} s), the waters in the first shell are certainly a part of the hydration number. By analogy with Ni^{2+}, it seems likely that the total hydration number is greater than this coordination number because of a second shell containing water with a lifetime greater than 10^{-11} s, which would then also qualify (because its lifetime greatly exceeds a jump time) as contributing to a hydration number. Similar remarks apply to Fe^{3+} but the lifetime of the inner shell is greater (about 6×10^{-3} s) than that for ferrous ions because of the stronger binding to water with ferric rather than ferrous ions.

The Cr^{3+} ion has a special place in the history of solvation because it was the first ion for which the lifetime of the water in its hydration shell was measured. This work was done by Hunt and Taube in 1957. The exchange of water between the hydration shell and the surroundings was slow, so the change in the concentration of the isotope could be measured. The lifetime found is 1.6×10^7 s (about 6 months). There is evidence of outer-sphere water, so the value for the total hydration water that travels with the ion is much higher than the value of about 6 found for the first shell in transition-metal ions.

The hydration number of the Li^+ ion has been measured by a number of nonspectroscopic methods and a value of 5 ± 1 represents the range of results. The difference method of neutron diffraction gives 6 with reasonable consistency. The lifetime is 3×10^{-9} s, so the value is clearly a hydration number (i.e., it is much greater than 10^{-11} s, so water travels with the ion).

An interesting result is obtained for Cl^-. Neutron diffraction work shows the value to be 6, which is in strong disagreement with nonspectroscopic methods, which give the much lower value of 1 or 2. However, light is at once thrown on what seems a breach difficult to mend when the lifetime of the 6 waters around the Cl^- is found to be only 10×10^{-11} s; this is on the borderline as viable for a dynamic hydration number. It seems likely, then, that the Cl^- moves on average without its hydration water, because its lifetime is barely enough for the time needed ($> 10^{-11}$ s) for wares to travel one jump distance. Thus, the 1 or 2 hydration numbers of other measurements is understandable. It is possible to regard the 6 value as coordination water, but few of the waters manage to stay with the ion when it moves from site to site.

2.11.4. To What Extent Do Raman Spectra Contribute to Knowledge of the Solvation Shell?

Raman spectra have a more involved origin than do IR absorption spectra. They concern the *scattering* of light. This is a subject that was studied in the nineteenth century when Rayleigh showed that the elastic scattering of light (no absorption) was proportional in intensity to λ^{-4}, where λ is the wavelength of the scattered light. In conditions that pertain to the application of this formula, the photon is assumed to "bounce" off the molecules it strikes with the same energy latterly as it had initially.

Smekal predicted a radically different type of scattering in 1923. He suggested that if a beam of light were incident on a solution, the scattered light would contain (apart from the elastically scattered light of unchanged frequency) two more frequencies, one greater and one lesser than the frequency of the incident beam. These would result from a collision of a photon with a molecule in which the photon would indeed (in contrast to Rayleigh's idea) exchange energy with the molecule it struck and bring about a transition in the molecule's vibrational band. Smekal thought that two kinds of energy exchange would occur. The photon might excite the molecule to higher states and the resulting frequency of the scattered light would be lower than the incident frequency (because some energy had been removed from the light). Alternatively, the molecule might give energy to the photon and the emerging photon would have a higher frequency than that of the initial beam. These nonelastic scattering effects predicted by Smekal in 1923–1924 were first experimentally observed by the Indian scientists Raman and Krishnan in 1928.

Thus, in the Raman spectra, it is a frequency shift, a $\Delta\nu$, that is observed. The value of such shifts in the frequency of nonelastically scattered light is not dependent on the exciting frequency,[22] but on the structure of the molecule with which the photon interacts. Thus, the Raman shifts reflect *vibrational* transitions, as do the IR spectra.

One other aspect of Raman spectra must be explained before one can understand how information on structures around ions in solution can be extracted from positive and negative Raman shifts. In Rayleigh scattering, the oscillating dipoles may radiate in all directions with the frequency of the exciting beam. However, in Raman spectra, the radiation depends on the polarizability of the bond. In some molecules, polarizability does not have the same value in all directions and is called nonisotropic. Let it emit light in the x-y plane only. Then the intensity ratio (light in the xy plane)/(light in the incident beam) is called the *depolarization factor*. For Raman scattering, this depolarization factor gives information on structure, for example, on the degree of symmetry of the entity in the solution, the Raman spectra of which are being observed (Tanaka et al., 1965).

2.11.5. Raman Spectra and Solution Structure

One type of Raman study of solutions concentrates on water–water bonding as it is affected by the presence of ions. Hydrogen bonds give Raman intensities, and the variation of these with ionic concentration can be interpreted in terms of the degree and type of structure of water molecules around ions. The ClO_4^- ion has often been used in Raman studies to illustrate structure-breaking effects because it is a relatively large ion.

In studies of the spectra of intramolecular water and how they are affected by ions, new Raman peaks can be interpreted in terms of the model of solvation suggested by

---

[22]This must, however, have a frequency greater than that of any of the transitions envisaged, but less than that which would cause an electronic transition.

Bockris in 1949 and by Frank and Evans in 1957. According to these workers, there are two regions. One is in the first (and for polyvalent cations, the second) layer near the ions, where water molecules are tightly bound and give rise to new frequencies. Such waters accompany the ion in its movements in the solution.

There is also a broken-structure region outside the first one to two layers of water molecules around the ion. Here the solvating waters are no longer coordinated, as in the bulk, by other waters, because of the ion's effect, but they are outside the *primary* hydration shell, which moves with the ion. Such intermediate waters, though partly broken out of the bulk water structure, do not accompany an ion in its diffusional motion.

Studies consistent with these ideas were first performed by Walfren in 1971 in H_2O–D_2O mixtures. Concentrations of 1 to 4 mol dm^{-3} were employed to get measurable effects. Thus, at 4 mol dm^{-3}, some 40–50% of the water present is at any moment in the primary hydration sheath!

Intermolecular effects in ionic solutions can also be studied in the Raman region between 200 and 1000 cm^{-1}. Librational modes of water show up here. The intensity of such peaks changes linearly with the ionic concentration. The ν_2 bending mode in the Raman spectra of alkali halides in water was studied by Weston. They may also be interpreted in terms of models of primary hydration (water staying with the ion in motion) and a secondary disturbed region.

Raman spectra can also be used to determine the degree of dissociation of some molecules, namely, those that react with the solvent to give ions (e.g., HCl). If the Raman frequency shifts for the dissociated molecules are known, then they can be used to calculate equilibrium constants at different temperatures. Then, once the temperature coefficient of the equilibrium constant K is known, one can determine ΔH° and ΔS° of the dissociation reaction (Section 2.13).

The study of the NO_3^- ion falls into the category of Raman studies that concentrate on interpreting spectra caused by the solute. It illustrates the use of Raman spectra to give structural conclusions via the study of symmetry. The free NO_3^- ion should have what is termed "D_{3h} symmetry" and give rise to three Raman bands corresponding to two degenerate asymmetric stretching modes (ν_3 and ν_4) and one symmetric stretching mode.

Irish and Davis studied the effect of solvation on the spectra of this NO_3^- ion and the splitting of the ν_3 band. They found that H bonding removes the degeneracy of this mode. The symmetry change would be from D_{3h} to C_{2h} and this is interpreted to mean that hydration effects have brought about a nonequivalence of the O atoms in NO_3^-, a most unexpected effect.

2.11.6. Information on Solvation from Spectra Arising from Resonance in the Nucleus

It has been often suggested that nuclear resonance might be used to gain information in solvation studies. Thus, in hydroxylic solvents, electron shielding around the proton should be affected by ions and thus, in terms of changes in nuclear resonance frequencies, solvation-bound water and free water should be distinguishable. It turns out that this is a nonevent for water because the rapid exchange between the two types, bound and unbound, gives rise to only a broad peak.

In order to obtain information from nuclear resonance, the proceedings must be a bit complicated. One adds a paramagnetic ion to a solution in which the solvation of a diamagnetic entity is to be measured. Then, two types of water around, for example, an Al^{3+} ion, bound and unbound waters, can be distinguished by observing the resulting nuclear magnetic resonance spectra of ^{17}O. The nuclear spin in the ^{17}O interacts with the electron spin vector of the paramagnetic ion added as a helpful auxiliary ion, and this changes the field on the ^{17}O nucleus. This shift in the NMR spectra of ^{17}O between water attached to the ion and bulk water has to be sufficiently large, and this in turn may allow a separation to be made between water bound to the diamagnetic ion and free water. In this rather complex and devious way, it is possible to obtain estimates of the number of waters in the first layer next to an ion.

However, disappointingly, again the values obtained from this NMR spectroscopic approach (e.g., 6 for Al^{3+}, Ga^{3+}, and Be^{2+}) are less than the values obtained for these ions (e.g., 14 for Be^{2+}) from the relatively self-consistent values of mobility, entropy, and compressibility. Is this simply because the nonspectroscopic measurements are usually done at high dilutions (e.g., 10^{-3} mol dm^{-3}) to diminish interionic effects, and the spectroscopic ones have to be done at 0.5 mol dm^{-3} or greater concentrations, because the spectroscopic shifts are relatively insensitive, and hence need the high concentration to score a detectable effect?

Swift and Sayne used concepts similar to those of Bockris and Saluja: if a molecule stays associated with an ion for more than the time needed for a diffusional jump, it "counts" as a primary hydration number. This approach yields approximately 4 solvation molecules for Mg^{2+} and Ca^{2+}, and 5 for Ba^{2+} and Sr^{2+}, whereas nonspectroscopic methods for these systems yield values that are two to three times larger. Does NMR measure only water arranged in a first, octahedral layer in the first shell near the ion and is it insensitive to the rest of the water structure near an ion?

Further Reading

Seminal

1. P. Debye, "The Vibrational Potential in Solution," *J. Chem. Phys.* **1**: 13 (1933).
2. D. D. Eley and M. G. Evans, "Statistical Mechanics of Ions in Solution," *Trans. Faraday Soc.* **34**: 1093 (1938).
3. M. Passynski, "Compressibility and Solvation," *Acta Phys. Chim. USSR* **8**: 385 (1938).
4. K. Fajans and O. Johnson, "Heats of Hydration," *J. Am. Chem. Soc.* **64**: 668 (1942).
5. R. H. Stokes and R. A. Robinson, "Hydration Numbers from Activity Measurements," *J. Am. Chem. Soc.* **70**: 1870 (1948).
6. J. B. Hasted, D. M. Ritson, and C. H. Collie, "Dielectric Constants of Solutions," *J. Chem. Phys.* **16**: 1 (1948).
7. J. O'M. Bockris, "Primary and Secondary Solvation," *Quart. Rev. Chem. Soc. Lond.* **3**: 173 (1949).

8. A. M. Azzam, "Theoretical Calculation of Hydration Numbers," *Z. Phys. Chem. (N.F.)* **33**: 320 (1962).
9. R. Zana and E. Yeager, "Solvation Numbers from Partial Molar Volumes," *J. Phys. Chem.* **71**: 521 (1967).
10. G. E. Walfren, "Spectroscopic Evidence on Hydration," *J. Chem. Phys.* **55**: 768 (1971).
11. J. O'M. Bockris and P. P. S. Saluja, "Hydration Numbers for Partial Molar Volumes and Vibrational Potentials from Solutions," *J. Phys. Chem.* **76**: 2140 (1972).
12. A. K. Soper, G. W. Neilson, J. E. Enderby, and R. A. Howe, "Neutron Diffraction and Coordination Numbers in Solution," *J. Phys. Soc.* **10**: 1793 (1977).
13. N. A. Hewich, G. W. Neilson, and J. E. Enderby, "Deconvolution within Neutron Diffraction Data and the Structure of the Hydration Sheath," *Nature* **297**: 138 (1982).
14. H. Friedman, "Time of residence of water molecules near ions," *Chem. Scripta* **25**: 42 (1985).

Review

1. G. A. Krestov, *Thermodynamics and Structure in Solvation*, Ellis Harwood, New York (1990).

Papers

1. S. Koda, J. Goto, T. Chikusa, and H. Nomura, *J. Phys. Chem.* **93**: 4959 (1989).
2. G. Omori and A. Santucci, *J. Chem. Phys.* **93**: 2939 (1990).
3. M. Jukiewicz and M. Figlerowicz, *Ultrasonics* **28**: 391 (1990).
4. B. Wilson, R. Georgiadis, and J. A. Bartmess, *J. Am. Chem. Soc.* **113**: 1762 (1991).
5. I. Howell, G. W. Neilson, and P. Chieuk, *J. Mol. Struct.* **250**: 281 (1991).
6. P. A. Bergstrom and J. Lindgren, *Inorg. Chem.* **31**: 1529 (1992).
7. P. A. Bergstrom, J. Lindgren, M. Sandrum, and Y. Zhou, *Inorg. Chem.* **31**: 150 (1992).
8. B. Guillot, P. Martineau, and J. O. Grist, *J. Chem. Phys.* **93**: 6148 (1993).
9. M. Maroncelli, V. P. Kumer, and A. Papazyan, *J. Phys. Chem.* **97**: 13 (1993).
10. J. M. Alia, H. G. M. Edwards, and J. Moore, *Spectrochim. Acta* **16**: 2039 (1995).
11. J. Barthell, *J. Mol. Liquids* **65**: 177 (1995).
12. Y. Tominaga, Y. Wang, A. Fujiwara, and K. Mizoguchi, *J. Molec. Liquids* **65**: 187 (1995).
13. M. J. Shaw and W. E. Geiger, *Organometallics* **15**: 13 (1996).
14. A. E. Johnson and A. B. Myers, *J. Phys. Chem.* **100**: 7778 (1996).
15. A. S. L. Lee and Y. S. Li, *Spectrochim. Acta* **52**: 173 (1996).
16. G. W. Neilson and J. E. Enderby, *J. Phys. Chem.* **100**: 1317 (1996).

2.12. DIELECTRIC EFFECTS

2.12.1. Dielectric Constant of Solutions

The electric field created between two plates of a parallel-plate condenser in a vacuum is greater than the field that exists between the same plates with the same

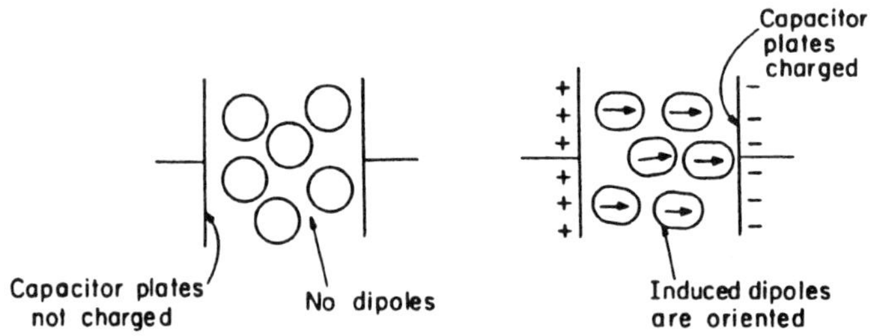

Fig. 2.25. The charge on the capacitor plates induces dipoles in the molecules of a dielectric.

charge on them but separated by a material medium, a so-called dielectric. The model by which dielectrics affect fields is easy to understand at a qualitative level. The molecules of the medium either contain a permanent dipole moment in their structure or have one induced by the field between the plates. Of course, when the field on the condenser plates is switched on, the dipoles orient against it (Fig. 2.25) and cause a counter electric field. The result is that the net electric field between the plates is less than it is when there is no medium between them (Fig. 2.26).

The counter field and the resulting net field can be calculated in mathematical form, but historically a more empirical way has been used; the field in the presence of a dielectric is simply expressed by dividing the field in its absence by an empirical "dielectric constant." The greater the counter field set up by the medium between the plates (Fig. 2.26), the greater the dielectric constant and the less the net electric field.

With this simple background model, then, it is easy to see that there will be a *decrease* of the dielectric constant of solutions (compared with that of the original

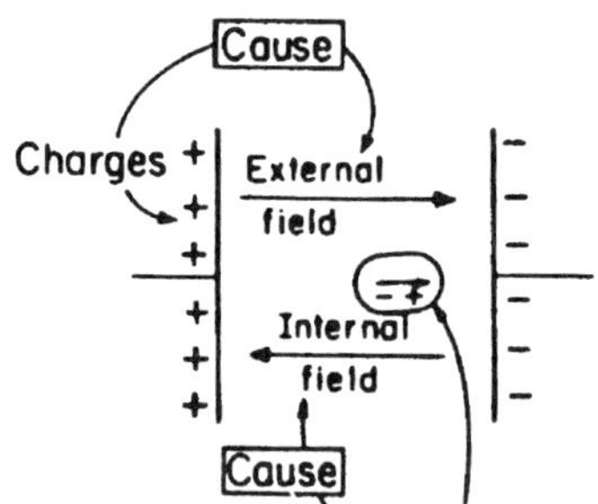

Fig. 2.26. The orientation of dipoles in the dielectric sets up an internal field that is directed counter to the external field produced by the charges on the plates.

TABLE 2.11

Hydration Numbers from Dielectric Increment Measurements

Ion	n	Ion	n
H^+	10	Rb^+	4
Li^+	6	Mg^{2+}	14
Na^+	4	Ba^{2+}	14
K^+	4	La^{3+}	22

Source: From J. B. Hasted, *Aqueous Dielectrics*, Chapman and Hall, London, 1973.

liquid) if ions are added to them. Thus, the ions undergo solvation and, to some critical distance from the ions' center, hold the solvent molecules tightly against the tendency of the field to orient them to oppose the applied field.

Those water molecules that are prevented from orienting ("irrotationally bound") to oppose the field will be withdrawn from those contributing to the counter field and hence the dielectric constant of the ionic solution will be reduced from what the solvent would have without the ions.

There are a number of publications in the field of the structure of ionic solutions that are particularly seminal and although published half a century ago have great influence on present concepts. One of them is the paper by Bernal and Fowler in which the associated structure of water was first established (in 1933) from the interpretation of the original X-ray data on liquids. However, another paper of great importance is that by Hasted, Ritson, and Collie in 1948, for it was here that the dielectric properties of solutions were first recorded on a large scale. In subsequent publications the relation of the dielectric constant of the solution and solvation was first investigated.

Some of the facts that Hasted et al. established are shown in Table 2.11. They found that the lowering of the dielectric constant of 1 *M* solutions is in the range of 10–20%. This can be nicely explained by taking the water in contact with the ion as dielectrically saturated (unable to orient on the demand of the external field), but still having a dielectric constant of only 6,[23] compared with the value of 80 for bulk water unaffected by ions. The table shows the number of water molecules per ion pair that one has to assume are saturated (i.e., irrotationally bound in the vicinity of each ion) to make the above model come out right (i.e., reproduce the measured dielectric constants of solutions). This model leads to a very simple equation for the dielectric constant of a solution:

[23]This value (6) is the dielectric constant of water under conditions of dielectric saturation. The ion's field not only stops the water orienting under the influence of the ac field exerted on the solution, it also breaks up the associated water structure (which made the dielectric constant of water so huge compared with that of other liquids). The 6 represents the counter field offered by the distortion of the positions of the nuclei in H and O and of the electron shells of these atoms.

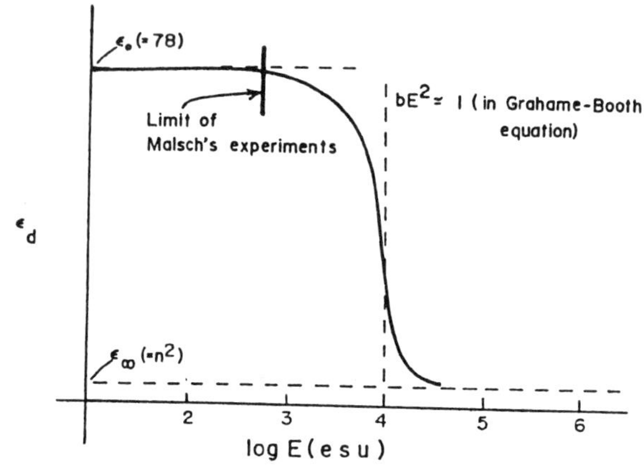

Fig 2.27. Differential dielectric constant ε_d as a function of field that is near an ion. (Reprinted from B. E. Conway, *Ionic Hydration in Chemistry and Biophysics*, Elsevier, New York, 1981.)

$$\varepsilon_{\text{soln}} = 80\left(\frac{55 - c_i n_s}{55}\right) + 6\,\frac{c_i n_s}{55} \tag{2.27}$$

where n_s is the total number of water molecules held by the ion, and c_i is the ionic concentration in mol dm^{-3}. Here, the first term represents the contribution from the bulk water molecules and the second term that from the bounded waters. Thus, measurements of the dielectric constants of ionic solutions provide a way to determine primary hydration numbers, the number of water molecules that stay with an ion while it diffuses in a solution.

Of course the assumption that in ionic solutions there are just two dielectric constants, one at 6 and the other at 80, is a simplification. There must be an intermediate region in the first two or three layers out near the ion in which the dielectric constant varies quite rapidly as one passes from the 6 of the first layer to the 80 a few layers further out.

This broken-down region near the ion was the subject of mathematical discussion by Webb as early as 1926, by Conway et al., and by Booth, whose paper also can be considered seminal. Grahame made an attempt to simplify Booth's equation for the dielectric constant as a function of field strength, and a diagram due to him is shown in Fig. 2.27.

Although the dielectric constant shown here is in terms of the *field* near the ion, not the distance from it, it is fairly[24] simple to find the distance that corresponds to those fields in the diagram and thus know what the dielectric constant is as a function of distance.

---

[24]Only fairly simple because the field itself depends on ε_r, the quantity one is trying to find. There is thus a catch to obtaining the distance corresponding to a certain field. An early solution to the problem was given by Conway, Bockris, and Ammar in 1951.

The main purpose of this section is to give the basis of how measurements of the dielectric constants of ionic solutions can give information on solvation, particularly primary hydration numbers. However, dielectric measurements *as a function of frequency* also give information on the dynamic behavior of water by allowing us to determine the relaxation time of water in ionic solutions and expressing the changes in terms of the number of water molecules bound to the ion.

Dielectric measurements of ionic solutions are also important for another topic that will be dealt with in Section 2.22, namely, electrostriction, the study of the compressive effect of the very strong electric fields produced by ions on the surrounding medium. When one looks into the effect of ions on the frequency at which water undergoes relaxation (i.e., when water no longer reacts to an applied field), it is found that the cation has a greater effect than the anion. The reason is shown in Fig. 2.28. For the cation, the two protons of a solvated water stick out from the ion and are bound to other waters, which restricts their libration and hence their contribution to the dielectric constant. Anions orient the protons of the hydration waters to themselves and away from binding by waters outside the first shell, thus having less binding effect on their movement than do the cations.

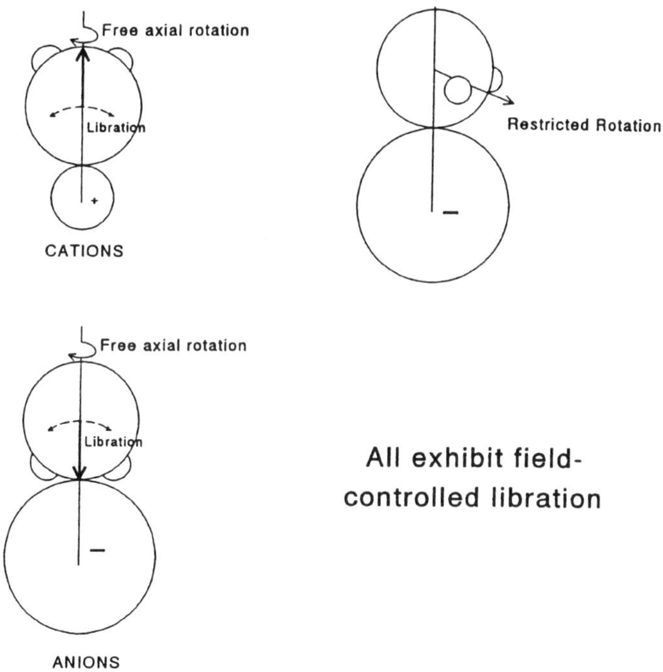

Fig. 2.28. Orientation and rotation possibilities for H_2O dipoles at anions and cations. (Reprinted from B. E. Conway, *Ionic Hydration in Chemistry and Biophysics*, Elsevier, New York, 1981.)

2.12.2. How Does One Measure the Dielectric Constant of Ionic Solutions?

There is much to think about here. If one wishes to measure the dielectric constant of a *liquid*, not a conducting ionic solution, one simply uses an alternating current (ac) bridge containing a capacitor in one of the arms. Then the capacitance is measured in the presence of the liquid, the dielectric constant of which is to be measured, and then without it, i.e., in the presence of air. Since the dielectric constant is near to unity in the latter case, this gives rise to knowledge of the dielectric constant of the liquid because the capacitance of the cell in the bridge arm increases as the dielectric constant increases.

However, when the liquid is a conducting solution, this approach breaks down. The conductance of the solution contributes to the impedance of the cell, which (depending on frequency) may no longer be overwhelmingly capacitative. Hence, the dielectric constant of a conducting liquid cannot be simply measured, because of the conductive components of the impedance.

Two approaches can be used to avoid this difficulty. In the one, which is best used for frequencies well below the relaxation frequency of water, one measures the *force* between two plates that have the conducting liquid between them. This force is independent of the conductance of the liquid or the direction of the field. If d is the distance between the plates, it is easy to show that the force is

$$\text{force} = \frac{E^2 A \varepsilon}{8\pi d^2} \tag{2.28}$$

where E is the potential difference between the plates, A is their area, and ε is the dielectric constant (Fig. 2.26). To obtain measurements that avoid significantly changing the temperature of the solution, d should be minimal. The field between the two plates is small (10^3 V cm^{-1}). Vibrations must also be minimized (e.g., by placing the apparatus on a stone slab supported on an inflated inner tube). An optical lever magnifies the very small movement of the plates.[25] Because everything is known in Eq. (2.28) except ε, the latter can be measured.

If it is desired to measure the dielectric constant at high frequencies (e.g., 10^7–10^{10} Hz), a different technique is necessary, partly because one may be working in the range of the relaxation time of water. The measurement of this, and the degree to which it is affected by the presence of ions, provides yet another way of finding how many molecules of water are bound to cation and to anion. The technique involves the use of a *wave guide* consisting of a coaxial liquid-filled cell, containing a probe that is moved about until the interference signal between it and an alternator is found to be

---

[25]A movement of 0.1 mm in the plate can be magnified more than 100 times by projecting a beam of light reflected from a mirror attached to the plate over a distance of 25 m.

zero. From this (for details of how to do this, see Conway, 1982), it is possible to find the ε of the solution as a function of frequency (and hence the relaxation time of water as a function of the presence of ions). This quantity clearly depends on how many water molecules have been withdrawn from the free solvent, where they can relax, and how many are attached to ions, where they cannot. It therefore leads to primary hydration numbers

Studies of the dielectric constant of solutions and the relaxation times of water in the presence of ions have been refined since the 1980's and indeed difficulties do turn up if one looks at data from measurements over large frequency ranges. The variation of the dielectric constant with frequency has been studied particularly by Winsor and Cole, who used the Fourier transform of time domain reflectometry to obtain dielectric constants of aqueous solutions and the relaxation times in them. Their frequency ranges from over 50 MHz to 9 GHz.

The problem of making significant dielectric constant measurements in these ranges is to separate the relaxation effects of the ionic atmosphere around the ions (Chapter 3) from effects connected with ion–solvent interactions. At low concentrations (< 0.01 mol dm^{-3}), the former effects are less important, but at such concentrations the decrements in the dielectric constant are too small for accurate measurement. Theoretical work makes it clear that a series of measurements over a large range of frequencies (e.g., 1 Hz to 1 GHz) are needed to separate dielectric effects from those due to relaxation of the ionic atmosphere.

Nevertheless, in spite of these warnings, values of dielectric decrements have a sufficiently clear basis to allow their use in discussing the elusive solvation numbers.

2.12.3. Conclusion

Measurements of dielectric constants of solutions allow the deduction of not only how many waters are taken up and held irrotationally by ions, but also how the ions affect the frequency of the movements of molecules near them. This will help a person interested in electrostatic effects calculate the local pressure near an ion (Section 2.22.1).

Further Reading

Seminal

1. J. J. Webb, "Electric Field near an Ion in Solution," *J. Am. Chem. Soc.* **48**: 2589 (1926).
2. J. B. Hasted, D. M. Ritson, and C. H. Collie, "The Dielectric Constants of Ionic Solutions," *J. Chem. Phys.* **16**: 1 (1948).
3. F. Booth, "Dielectric Constant As a Function of the Applied Field," *J. Chem. Phys.* **19**: 1451 (1951).
4. D. C. Grahame, "Electric Field and Dielectric Constant near an Ion," *J. Phys. Chem.* **11**: 1054 (1951).

5. J. O'M. Bockris and J. Bowler-Reed, "A Technique for Measuring Dielectric Constants in Conducting Solution," *J. Appl. Phys.* **2**: 74 (1951).

Papers

1. D. Bertolini, M. Cassetari, E. Tombari, and S. Verenesi, *Rev. Sci. Instrum.* **61**: 450 (1990).
2. R. S. Drago, D. C. Feris, and N. Wong, *J. Am. Chem. Soc.* **112**: 8953 (1990).
3. S. Safe, R. K. Mohr, C. J. Montrose, and T. A. Litovitz, *Biopolymers* **31**: 1171 (1991).
4. M. Bruehl and J. T. Hynes, *J. Phys. Chem.* **96**: 4068 (1992).
5. J. Z. Bao, M. L. Swicord, and C. C. Davis, *J. Chem. Phys.* **104**: 4441 (1996).
6. J. L. Buck, *IEEE Transactions* **45**: 84 (1996).

2.13. IONIC HYDRATION IN THE GAS PHASE

Now that some methods for investigating the structure of the ion–solvent complex in solution have been described, it is time to learn systematically what is known about it. One can start by considering systems that avoid the complexity of liquid water. By varying the partial pressure of water vapor while keeping it low (0.1–10 kPa), it is possible to find the equilibrium constant between water vapor and the entities represented by a number of ion–solvent aggregates, $M(H_2O)_n^+$, in the gas phase (Kebarle and Godbole, 1968).

Thus, if the equilibrium constant K for one of these equilibria is known, ΔG° can be derived from the well-known thermodynamic relation $K = \exp(-\Delta G^\circ/RT)$. If K (and hence ΔG°) is known as a function of T, ΔH° can be obtained from the slope of an $\ln K - 1/T$ plot and ΔS° from the intercept.

The seminal work in this field was carried out by Kebarle and it is surprising to note the gap of 30 years between the foundation paper by Bernal and Fowler on solvation in solution and the first examination of the simpler process of hydration in the gas phase. A series of plots showing the concentrations of various hydrate complexes for $Na(H_2O)_n^+$ as a function of the total pressure of water vapor is given in Fig. 2.29.[26]

Now, an interesting thing can be done with the ΔH values obtained as indicated earlier. One takes the best estimate available for the primary hydration number in solution (see, e.g., Tables 2.7 and 2.11). One then calculates the corresponding heat of hydration in the gas phase for this number and compares it with the corresponding individual heat of hydration of the ion in solution. The difference should give the residual amount of interaction heat outside the first layer (because in the gas phase there is no "outside the first layer").

The hydration energy for the outer shell turns out to be 15% of the whole for cations and about 30% for anions. Thus, in hydration of the alkali and halide ions,

---

[26]In Fig. 2.29 a non-SI unit, the torr, is used. The unit is named for Torricelli, who first discussed the partial vacuum above mercury contained in a tube and found it to be ~1/760 of an atmosphere.

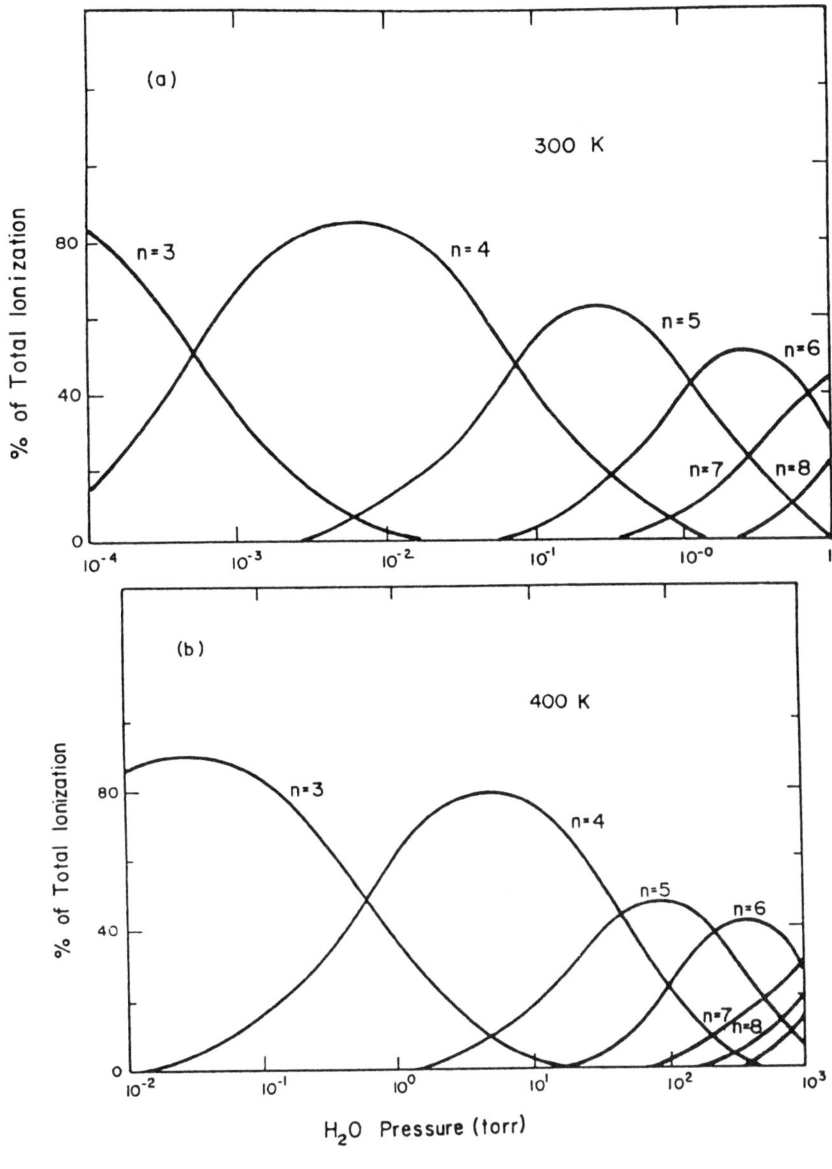

Fig. 2.29. Relative concentrations of Na^+ ion hydrates, $Na(H_2O)_n^+$, in the gas phase at (a) 300 K and (b) 400 K. (Reprinted from B. E. Conway, *Ionic Hydration in Chemistry and Biophysics*, Elsevier, New York, 1981.)

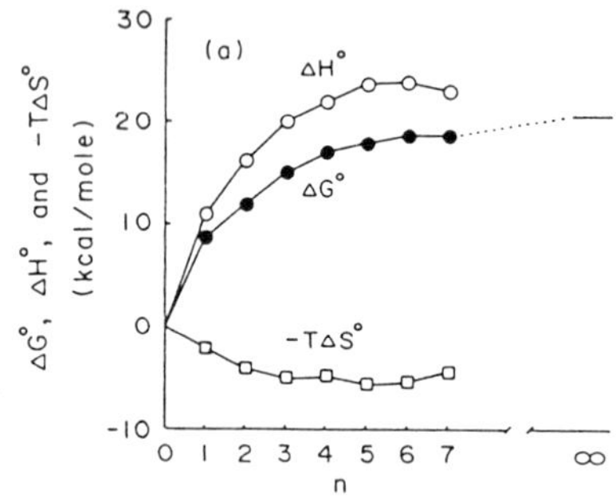

$F^-(CH_3CN)_n + Cl^- = Cl^-(CH_3CN)_n + F^-$

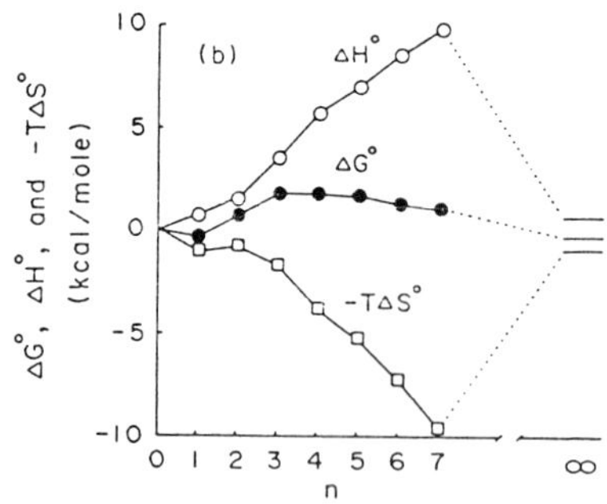

$Cl^-(CH_3CN)_n + Br^- = Br^-(CH_3CN)_n + Cl^-$

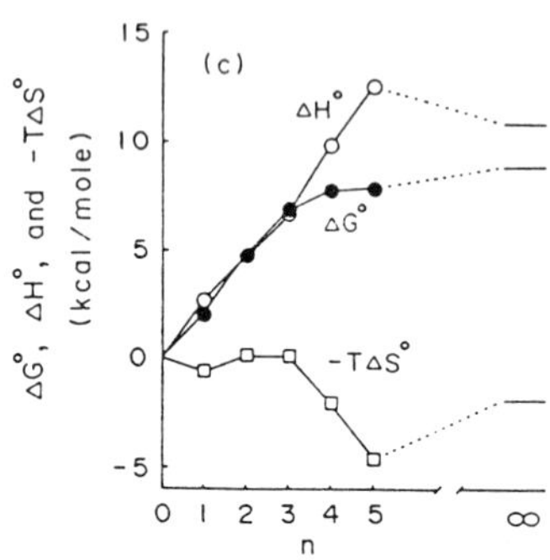

$Cl^-(CH_3CN)_n + I^- = I^-(CH_3CN)_n + Cl^-$

Fig. 2.30. The *n* dependence of differential integrated thermochemical values, $\Delta G°$, $\Delta H°$, and $-T\Delta S°$, for three different solvation reactions. The values with $n = \infty$ correspond to the relative thermochemical values for the bulk hydration of the ions (1 cal = 4.184 J). (Reprinted from K. Hiraoka, S. Mizuse, and S. Yamabe, *J. Phys. Chem.* **92**: 3943, 1988.)

between 70 and 85% of the heat of hydration of ions comes from the first shell. One reason for the interest in this result lies in the electrode process field where the traditional theory of the energetics of electron transfer has in the past stressed the outer and not the inner shell solvation as having a major influence on electron transfer, although the present material makes the waters in the first layer those which exert the major control on the ion–solvent interaction energy.

Kebarle used a pulsed electron beam to produce ions for injection into a mass spectrograph that contained the water molecules at a determined, but variable partial pressure. Hiraoka et al. found that $-\Delta G°$ decreases more rapidly with a change in the numbers of water molecules attached to Cl^- than for Br^- and crosses over at $n_h = 4$. Evidence for some degree of covalent bonding occurs for $F^-(CH_3CN)$. The completion of the first solvation shell does not occur until $n_h = 8$ in this case, a surprisingly high number. The trends found are diagrammed in Fig. 2.30.

Hiraoka et al. have also discussed how the results of their measurements on solvation in the gas phase are related to the more usually discussed liquid phase solvation. The first water molecules go onto the ion and are structure-forming.

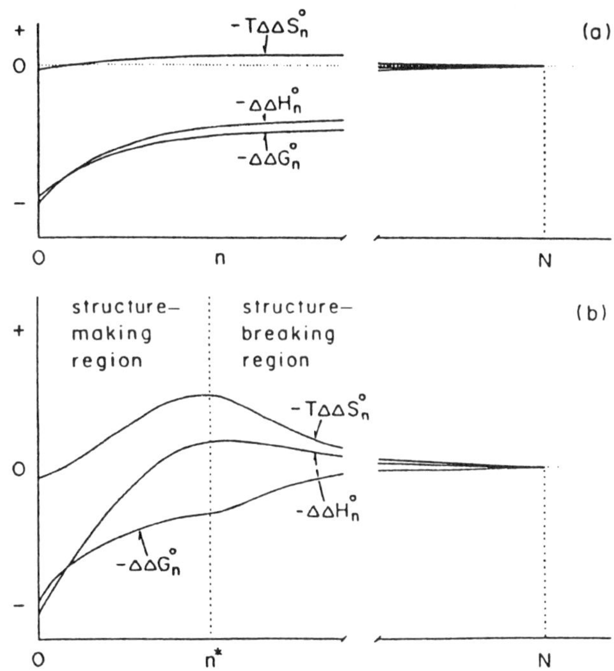

Fig. 2.31. Schematic representation of n dependence of $-\Delta\Delta G^\circ_n$, $-\Delta\Delta H^\circ_n$, and $-T\Delta\Delta S^\circ_n$ for reactions (a) $Cl^-(H_2O)_{n-1} + H_2O = Cl^-(H_2O)_n$ and (b) $Cl^-(CH_3CN)_{n-1} + CH_3CN = Cl^-(CH_3CN)_n$. (Reprinted from K. Hiraoka, S. Mizuse, and S. Yamabe, *J. Phys. Chem.* **92**: 3943, 1988.)

However, as the shell builds up and breaks more and more H bonds in the surrounding solvent, the effect of ions on liquid water begins to become more structure-breaking (Fig. 2.31).

So far, only singly charged ions have been mentioned and there is a good reason for this. A flow of univalent ions can be generated by electron pulses in an atmosphere of water vapor. If the electron energy is sufficiently high to bring about the second ionization of the metal atoms to form M^{2+}, ionization of water also occurs and hence experiments are rendered useless. The results of measurements of equilibria become too complex to interpret. However, in 1984, Yamashita and Fenn introduced a technique that *sprays* ions already in solution into a mass spectroscope. It is possible to spray into the instrument any ions that exist in solution. Such electrospray techniques have opened up an exciting new area of possibilities in gas phase solvation studies, but the database in the 1990s is as yet too sparse to support broader conclusions.

2.14. INDIVIDUAL IONIC PROPERTIES

2.14.1. Introduction

It is usually relatively easy to find the solvation-related property of an *electrolyte* (as, e.g., the heat of hydration, Section 2.5.2) or the partial molar volume (Section 2.6.2) of a salt in solution. However, experiments that reflect the properties of *individual ions* are difficult to devise, the only simple, direct one being the transport number of an ion (Section 2.10) and the associated individual ionic mobility (Section 2.10.1).

It is important to separate the two contributions to ionic solvation in a salt such as NaCl. Thus, the degree of hydration depends first upon the ion–solvent distance. The crystallographic radii of cations are less than those of the parent atom, but those of anions are larger than those of the parent. Cations, therefore, tend to be more hydrated than anions because the attraction rendered by the ion is inversely proportional to its radius. However, when, as with K^+ and F^-, the crystallographic radii are essentially identical, there is even then a nonequal heat of hydration for each ion. This is because the dipole moment of the water molecule is not symmetrically distributed with respect to its geography; its center, which determines the degree of interaction with the ion, is nearer the positive than the negative ion, so that the former is again favored in respect to solvation compared with the latter.

Obtaining the individual properties of ions with solvation numbers from measurements of ionic vibration potentials and partial molar volumes is not necessary in the study of gas phase solvation (Section 2.13), where the individual heats of certain hydrated entities can be obtained from mass spectroscopy measurements. One injects a spray of the solution under study into a mass spectrometer and investigates the time of flight, thus leading to a determination of the total mass of individual ions and adherent water molecules.

2.14.2. A General Approach to Individual Ionic Properties: Extrapolation to Make the Effects of One Ion Negligible

Let it be assumed that the value of the interaction energy of an ion with a solvent is an inverse function of the ion–first water shell distance, r. Then, if one has a series of salts (R_1A, R_2A, . . .) where R is, say, a tetraalkylammonium ion, and the anion is constant, the electrolyte property (e.g., the heat of hydration) can be plotted for the series of RAs, against $1/r_i^n$ (where r represents the cation radius), and the extrapolated value for $1/r_i^n = 0$ is then the individual heat of hydration for the common anion, A^-.

If an accepted value of the property for this one anion can be derived, then, of course, it can be coupled with data for various electrolytes containing this anion. If the data pertain to dilute solutions, avoiding the interfering effects of ion–ion interactions, it is possible to derive the individual value for the heat of hydration of the cations.

This method sounds simple at first. However, there are certain difficulties. One has to decide on a value of n in the plot of $1/r_i^n$ and this may not always be unity or simple. Various terms that affect the calculation of the heat of hydration of ions depend on r^{-2}, r^{-3}, and r^{-4}. Against which one should one plot?

Because the appropriate n is uncertain, it may be a better tactic to make a different extrapolation and plot the property of the electrolyte against the molecular weight of the cation and then extrapolate to zero, as with partial molar volumes, which was illustrated in Fig. 2.15.

Conway and his associates have been foremost in studying individual ionic properties and have published a weighted analysis of many suggested methods for obtaining individual ionic properties (see the reading lists). Among the methods chosen by Conway et al. as excellent, two have been discussed in this chapter so far, namely, extrapolation against cation molecular weight and combining partial molar volume with data on ionic vibration potential to determine individual solvation numbers. Another method with good reliability involves measurements of the heat produced in reversible electrolytic cells, which can be used to deduce individual, ionic entropies, as will be explained in Section 2.15.8. First, though, it is desirable to describe one particular method used for the individual heat of hydration of the proton, clearly a most fundamental quantity.

2.15. INDIVIDUAL HEAT OF HYDRATION OF THE PROTON

2.15.1. Introduction

A particular method of obtaining this fundamental quantity was given by Halliwell and Nyburg in 1963 and although there have been several reexaminations of the process,[27] changes of only about 1% in a value first calculated in 1963 have been made.

---

[27]These include information on the dynamics of proton hydration.

It seems reasonable therefore to examine the process that gave such a hardy number so reliably.

2.15.2. Relative Heats of Solvation of Ions on the Hydrogen Scale

Consider the unambiguous experimental value ΔH_{HX-H_2O}, the heat of hydration of HX. It is made up of the heats of hydration of H^+ ions and X^- ions.[28]

$$\Delta H_{HX} = \Delta H_{H^+} + \Delta H_{X^-} \tag{2.29}$$

The heat of solvation of X^- ions relative to that of H^+ ions [i.e., ΔH_{X^-}(rel)] can be defined by considering ΔH_{H^+} as an arbitrary zero in Eq. (2.29), that is,

$$\Delta H_{X^-}(\text{rel}) = \Delta H_{X^-}(\text{abs}) + \Delta H_{H^+}(\text{abs}) = \Delta H_{HX} \tag{2.30}$$

Thus, the notation (rel) and (abs) has been inserted to distinguish between the *relative* ΔH_{X^-} value of X^- ions on an arbitrary scale of $\Delta H_{H^+}(\text{abs}) = 0$ and the *absolute* or true ΔH_{X^-} values. From Eq. (2.30),

$$\Delta H_{X^-}(\text{rel}) = \Delta H_{HX} \tag{2.31}$$

and since ΔH_{HX} (the heat of hydration of an electrolyte) can be experimentally obtained to a precision determined only by measuring techniques (i.e., it includes no structural assumptions), one can see that the relative heat of solvation, ΔH_{X^-}(rel), is a clear-cut experimental quantity.

Relative heats of solvation can also be defined for positive ions. One writes

$$\Delta H_{MX} = \Delta H_{M^+}(\text{abs}) + \Delta H_{X^-}(\text{abs}) \tag{2.32}$$

and substitutes for ΔH_{X^-}(abs) from Eq. (2.30). Thus, one has

$$\Delta H_{X^-}(\text{abs}) = \Delta H_{X^-}(\text{rel}) - \Delta H_{H^+}(\text{abs}) = \Delta H_{HX} - \Delta H_{H^+}(\text{abs}) \tag{2.33}$$

which, when inserted into Eq. (2.32), gives

$$\Delta H_{MX} = \Delta H_{M^+}(\text{abs}) + \Delta H_{HX} - \Delta H_{H^+}(\text{abs}) \tag{2.34}$$

or

---

[28]One should, strictly speaking, write

$$\Delta H_{HX-H_2O} = \Delta H_{H^+-H_2O} + \Delta H_{X^--H_2O}$$

but the $-H_2O$ will be dropped in the subsequent text to make the notation less cumbersome.

TABLE 2.12

Relative Heats of Hydration of Individual Ions, ΔH_{H^+} (abs) = 0

Ion	Relative Heat of Hydration
Li+	+136.34
Na^+	+163.68
K^+	+183.74
Rb^+	+188.80
Cs^+	+194.60
F^-	−381.50
Cl^-	−347.50
Br^-	−341.00
I^-	−331.20

$$\Delta H_{M^+}(\text{abs}) - \Delta H_{H^+}(\text{abs}) = \Delta H_{MX} - \Delta H_{HX} \tag{2.35}$$

Taking ΔH_{H^+}(abs) as an arbitrary zero in this equation permits the definition of the relative heat ΔH_{M^+}(rel) for the heat of solvation of positive ions

$$\Delta H_{M^+}(\text{rel}) = \Delta H_{M^+}(\text{abs}) - \Delta H_{H^+}(\text{abs}) = \Delta H_{MX} - \Delta H_{HX} \tag{2.36}$$

Since ΔH_{MX} and ΔH_{HX} are unambiguous experimental quantities, so are the relative heats of solvation, ΔH_{M^+}(rel), of positive ions.

On this basis, a table of relative heats of solvation of individual ions can be drawn up (Table 2.12). These relative heats can be used, as will be promptly shown, to examine the degree of truth in the assumption that ions of equal radii and opposite charge have equal heats of solvation.

2.15.3. Do Oppositely Charged Ions of Equal Radii Have Equal Heats of Solvation?

Consider two ions M_i^+ and X_i^- of *equal radius r_i but opposite charge*. If their *absolute* heats of solvation are equal, one expects that

$$\Delta H_{M_i^+}(\text{abs}) - \Delta H_{X_i^-}(\text{abs}) = 0 \tag{2.37}$$

But, from the definition of the *relative* heats of solvation of positive ions [Eq. (2.36)] and of negative ions [Eq. (2.33)], one has by subtraction

$$\Delta H_{M_i^+}(\text{abs}) - \Delta H_{X_i^-}(\text{abs}) = [\Delta H_{M_i^+}(\text{rel}) - \Delta H_{X_i^-}(\text{rel})] + 2\Delta H_{H^+}(\text{abs}) \tag{2.38}$$

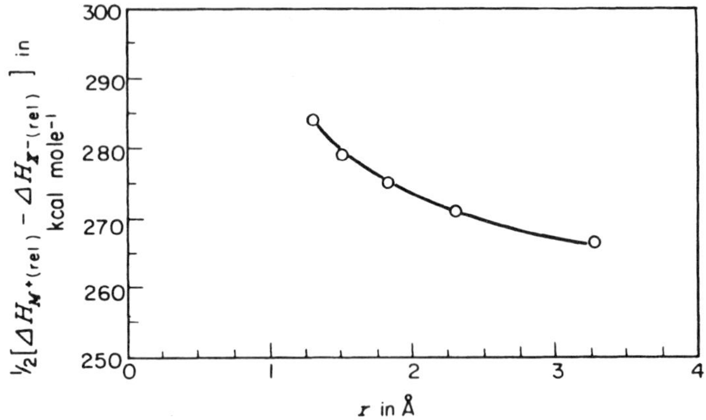

Fig. 2.32. Plot of the difference between the relative heats of hydration of oppositely charged ions with equal radii vs. ionic radius. Values are in kilocalories (1 cal = 4.184 J).

If, therefore, the left-hand side is zero, then one should find, since ΔH_{H^+}(abs) is a constant, that

$$\Delta H_{M_i^+}(\text{rel}) - \Delta H_{X_i^-}(\text{rel}) = \text{a constant} \tag{2.39}$$

This prediction can easily be checked. One makes a plot of the experimentally known relative heats of solvation of positive and negative ions as a function of ionic radius. By erecting a perpendicular at a radius r_i, one can get the difference $[\Delta H_{M_i^+}(\text{rel}) - \Delta H_{X_i^-}(\text{rel})]$ between the relative heats of solvation of positive and negative ions of radius r_i. By repeating this procedure at various radii, one can make a plot of the differences $[\Delta H_{M^+}(\text{rel}) - \Delta H_{X^-}(\text{rel})]$ as a function of radius. If oppositely charged ions of the same radius have the same absolute heats of hydration, then $[\Delta H_{M^+}(\text{rel}) - \Delta H_{X^-}(\text{rel})]$ should have a constant value independent of radius. It does not (Fig. 2.32).

2.15.4. The Water Molecule as an Electrical Quadrupole

The structural approach to ion–solvent interactions has been developed so far by considering that the electrical equivalent of a water molecule is an idealized dipole, i.e., two charges of equal magnitude but opposite sign separated by a certain distance. Is this an adequate representation of the charge distribution in a water molecule?

Consider an ion in contact with the water molecule; this is the situation in the primary hydration sheath. The ion is close enough to "see" one positively charged region near each hydrogen nucleus and two negatively charged regions corresponding

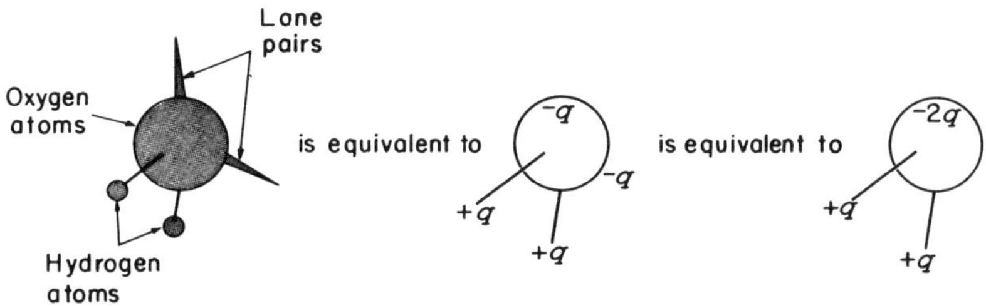

Fig. 2.33. The electrical equivalence between a water molecule and a quadrupole.

to the lone pairs near the oxygen atom. In fact, from this intimate viewpoint, the charge distribution in the water molecule can be represented (Fig. 2.33) by a model with four charges of equal magnitude q: a charge of $+q$ near each hydrogen atom, and two charges each of value $-q$ near the oxygen atom. Thus, rather than consider that the water molecule can be represented by a dipole (an assembly of two charges), a better approximation, suggested by Buckingham (1957), is to view it as a *quadrupole*, i.e., an assembly of four charges. What may this increase in realism of the model do to the remaining discrepancies in the theory of ion–solvent interactions?

2.15.5. The Ion–Quadrupole Model of Ion–Solvent Interactions

The structural calculation of the heat of ion–solvent interactions involves the following cycle of hypothetical steps: (1) A cluster of $n + 1$ water molecules is removed from the solvent to form a cavity; (2) the cluster is dissociated into $n + 1$ independent water molecules; (3) n out of $n + 1$ water molecules are associated with an ion in the gas phase through the agency of ion–dipole forces; (4) the primary solvated ion thus formed in the gas phase is plunged into the cavity; (5) the introduction of the primary solvated ion into the cavity leads to some structure breaking in the solvent outside the cavity; and (6) finally, the water molecule left behind in the gas phase is condensed into the solvent. The heat changes involved in these six steps are W_{CF}, W_D, $W_{I\text{-}D}$, W_{BC}, W_{SB}, and W_C, respectively, where, for $n = 4$.

$$W = W_{CF} + W_D + W_{SB} + W_C = +20 \text{ for positive ions}$$

$$= +30 \text{ for negative ions} \tag{2.40}$$

$$W_{I\text{-}D} = -\frac{4N_A z_i e_0 \mu_W}{(r_i + r_W)^2} \tag{2.41}$$

$$W_{BC} = -\frac{N_A (z_i e_0)^2}{2(r_i + 2r_W)} \left(1 - \frac{1}{\varepsilon_W} - \frac{T}{\varepsilon_W{}^2} \frac{\partial \varepsilon_W}{\partial T} \right) \tag{2.42}$$

and the total heat of ion–water interactions is

$$\Delta H_{I\text{–}H_2O} = W + W_{I\text{–}D} + W_{BC} \tag{2.43}$$

If one scrutinizes the various steps of the cycle, it will be realized that for only one step, namely, step 3, does the heat content change [Eq. (2.41)] depending upon whether one views the water molecule as an electrical dipole or quadrupole. Hence, the expressions for the heat changes for all steps *except* step 3 can be carried over as such into the theoretical heat of ion–water interactions, $\Delta H_{I\text{–}H_2O}$, derived earlier. In step 3, one has to replace the heat of ion–dipole interactions, $W_{I\text{–}D}$ [Eq. (2.41)], with the heat of ion–quadrupole interactions (Fig. 2.34).

What is the expression for the energy of interaction between an ion of charge $z_i e_0$ and a quadrupole? The derivation of a general expression requires sophisticated mathematical techniques, but when the water molecule assumes a symmetrical orientation (Fig. 2.35) to the ion, the ion–quadrupole interaction energy can easily be shown to be (Appendix 2.3)

$$E_{I\text{–}Q} = -\frac{z_i e_0 \mu_W}{r^2} \pm \frac{z_i e_0 p_W}{2r^3} \tag{2.44}$$

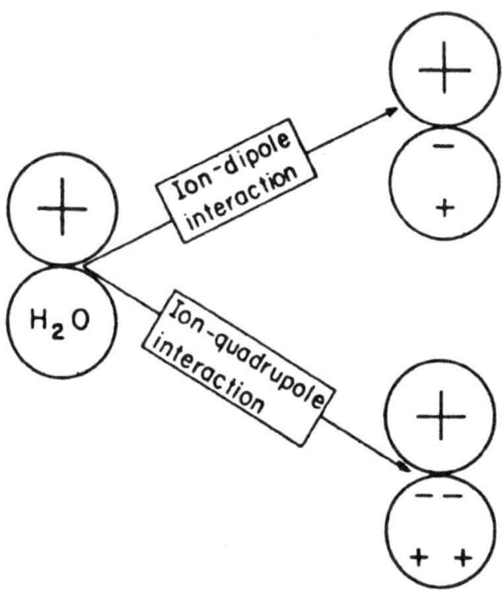

Fig. 2.34. Improvement in the calculation of the ion–water molecule interactions by altering the model of the water molecule from a dipole to a quadrupole.

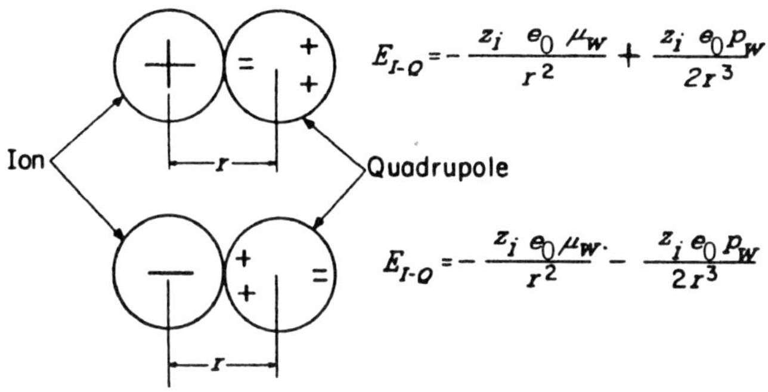

Fig. 2.35. The symmetrical orientation of a quadrupole to an ion.

where the + in the ± is for positive ions, and the – is for negative ions, and p_W is the quadrupole moment (3.9×10^{-26} esu) of the water molecule. It is at once clear that a difference will arise for the energy of interaction of positive and negative ions with a water molecule, a result hardly foreseeable from the rudimentary dipole viewpoint and hence probably accountable for the result shown in Fig. 2.32.

The first term in this expression [Eq. (2.44)] is the dipole term, and the second term is the quadrupole term. It is obvious that with increasing distance r between ion and water molecule, the quadrupole term becomes less significant. Or, in other words, the greater the value of r, the more reasonable it is to represent the water molecule as a dipole. However, as the ion comes closer to the water molecule, the quadrupole term becomes significant, i.e., the error involved in retaining the approximate dipole model becomes more significant.

When the ion is in contact with the water molecule, as is the case in the primary solvation sheath, expression (2.44) for the ion–quadrupole interaction energy becomes

$$E_{I-Q} = -\frac{z_i e_0 \mu_W}{(r_i + r_W)^2} \pm \frac{z_i e_0 p_W}{2(r_i + r_W)^3} \tag{2.45}$$

The quantity E_{I-Q} represents the energy of interaction between one water molecule and one ion. If, however, four water molecules surround one ion and one considers a mole of ions, the heat change W_{I-Q} involved in the formation of a primary solvated ion through the agency of ion–quadrupole forces is given by

$$W_{I-Q} = 4N_A E_{I-Q} = -\frac{4N_A z_i e_0 \mu_W}{(r_i + r_W)^2} \pm \frac{4N_A z_i e_0 p_W}{2(r_i + r_W)^3} \tag{2.46}$$

where, as before, the + in the ± refers to positive ions and the – to negative ions.

Substituting this expression for W_{I-Q} in place of W_{I-D} in expression (2.43) for the heat of ion–water interactions, one has

$$\Delta H_{I\text{-}H_2O} = 20 - \frac{4N_A z_i e_0 \mu_W}{(r_i + r_W)^2} + \frac{4N_A z_i e_0 p_W}{2(r_i + r_W)^3} - \frac{N_A (z_i e_0)^2}{2(r_i + 2r_W)} \times \left(1 - \frac{1}{\varepsilon_W} - \frac{T}{\varepsilon_W{}^2}\frac{\partial \varepsilon_W}{\partial T}\right) \tag{2.47}$$

for positive ions and

$$\Delta H_{I\text{-}H_2O} = 30 - \frac{4N_A z_i e_0 \mu_W}{(r_i + r_W)^2} - \frac{4N_A z_i e_0 p_W}{2(r_i + r_W)^3} - \frac{N_A (z_i e_0)^2}{2(r_i + 2r_W)} \times \left(1 - \frac{1}{\varepsilon_W} - \frac{T}{\varepsilon_W{}^2}\frac{\partial \varepsilon_W}{\partial T}\right) \tag{2.48}$$

for negative ions (values in kilocals (gram ion)$^{-1}$.

2.15.6. Ion–Induced Dipole Interactions in the Primary Solvation Sheath

At this level of sophistication, one wonders whether there are other subtle interactions that one ought to consider.

For instance, when the water molecule is in contact with the ion, the field of the latter tends to distort the charge distribution in the water molecule. Thus, if the ion is positive, the negative charge in the water molecule tends to come closer to the ion, and the positive charge to move away. This implies that the ion tends to induce an extra dipole moment in the water molecule over and above its permanent dipole moment. For small fields, one can assume that the induced dipole moment μ_{ind} is proportional to the inducing field X

$$\mu_{\text{ind}} = \alpha X \tag{2.49}$$

where α, the proportionality constant, is known as the *deformation polarizability* and is a measure of the "distortability" of the water molecule along its permanent dipole axis.

Thus, one must consider the contribution to the heat of formation of the primary solvated ion (i.e., step 3 of the cycle used in the theoretical calculation presented earlier), which arises from interactions between the ion and the dipoles induced in the water molecules of the primary solvent sheath. The interaction energy between a dipole and an infinitesimal charge dq is $-\mu\, dq/r^2$, or, since dq/r^2 is the field dX due to this charge, the interaction energy can be expressed as $-\mu\, dX$. Thus, the interaction energy between the dipole and an ion of charge $z_i e_0$, exerting a field $z_i e_0/r^2$, can be found by, performing the integration $-\int_0^{z_i e_0/r^2} \mu \, dX$. In the case of permanent dipoles, μ does not depend on the field X and one gets the result (Appendix 2.2):

$$-\int_0^{z_i e_0/r^2} \mu \, dX = -\frac{z_i e_0}{r^2}\mu \tag{2.50}$$

For induced dipoles, however, $\mu_{\text{ind}} = \alpha X$, and, hence,

$$-\int_0^{z_i e_0/r^2} \mu \, dX = -\int_0^{z_i e_0/r^2} \alpha X \, dX = -\left[\frac{\alpha X^2}{2}\right]_0^{z_i e_0/r^2} = \alpha \frac{(z_i e_0)^2}{2r^4} \tag{2.51}$$

Considering a mole of ions and four water molecules in contact with an ion, the heat of ion–induced dipole interactions is

$$-\frac{4N_A \alpha (z_i e_0)^2}{2(r_i + r_W)^4}$$

Introducing this induced dipole effect into the expression for the heat of ion–solvent interactions [Eqs. (2.47) and (2.48)], one has

$$\Delta H_{I\text{-}H_2O} = 20 - \frac{4N_A z_i e_0 \mu_W}{(r_i + r_W)^2} + \frac{4N_A z_i e_0 p_W}{2(r_i + r_W)^3} - \frac{N_A (z_i e_0)^2}{2(r_i + 2r_W)}\left(1 - \frac{1}{\varepsilon_W} - \frac{T}{\varepsilon_W^2}\frac{\partial \varepsilon_W}{\partial T}\right) - \frac{4N_A \alpha (z_i e_0)^2}{2(r_i + r_W)^4} \tag{2.52}$$

for positive ions, and

$$\Delta H_{I\text{-}H_2O} = 30 - \frac{4N_A z_i e_0 \mu_W}{(r_i + r_W)^2} - \frac{4N_A z_i e_0 p_W}{2(r_i + r_W)^3} - \frac{N_A (z_i e_0)^2}{2(r_i + 2r_W)}\left(1 - \frac{1}{\varepsilon_W} - \frac{T}{\varepsilon_W^2}\frac{\partial \varepsilon_W}{\partial T}\right) - \frac{4N_A \alpha (z_i e_0)^2}{2(r_i + r_W)^4} \tag{2.53}$$

for negative ions.

2.15.7. How Good Is the Ion–Quadrupole Theory of Solvation?

A simple test for the validity of these theoretical expressions (2.52) and (2.53) can be constructed. Consider two ions M_i^+ and X_i^- of equal radius but opposite charge. The difference $\Delta H_{M_i^+}(\text{abs}) - \Delta H_{X_i^-}(\text{abs})$ in their absolute heats of hydration is obtained by subtracting Eq. (2.53) from Eq. (2.52). Since the signs of the dipole term,

$$\frac{4N_A z_i e_0 \mu_W}{(r_i + r_W)^2}$$

the Born charging term,

$$\frac{N_A (z_i e_0)^2}{2(r_i + 2r_W)}\left(1 - \frac{1}{\varepsilon_W} - \frac{T}{\varepsilon_W{}^2}\frac{\partial \varepsilon_W}{\partial T}\right)$$

and the induced dipole term,

$$\alpha \frac{4N_A (z_i e_0)^2}{2(r_i + r_W)^4}$$

are invariant with the sign of the charge of the ion, they cancel out in the subtraction (as long as the orientation of a dipole near a cation is simply the mirror image of that near an anion). The quadrupole term, however, does not cancel out because it is positive for positive ions and negative for negative ions. Hence, one obtains[29]

$$\Delta H_{M_i^+}(\text{abs}) - \Delta H_{X_i^-}(\text{abs}) = -41 + \frac{4N_A z_i e_0 p_W}{(r_i + r_W)^3} \tag{2.54}$$

It is seen from this equation that the quadrupolar character of the water molecule would make oppositely charged ions of equal radii have radius-dependent differences in their heats of hydration (Fig. 2.32). Further, Eq. (2.38) has given

$$\Delta H_{M_i^+}(\text{abs}) - \Delta H_{X_i^-}(\text{abs}) = \Delta H_{M_i^+}(\text{rel}) - \Delta H_{X_i^-}(\text{rel}) + 2\Delta H_{H^+}(\text{abs}) \tag{2.38}$$

By combining Eqs. (2.38) and (2.54), the result is

$$\Delta H_{M_i^+}(\text{rel}) - \Delta H_{X_i^-}(\text{rel}) = -2\Delta H_{H^+}(\text{abs}) - 41 + \frac{4N_A z_i e_0 p_W}{(r_i + r_W)^3} \tag{2.55}$$

Thus, the ion–quadrupole model of ion–solvent interactions predicts that if the experimentally available differences $\Delta H_{M_i^+}(\text{rel}) - \Delta H_{X_i^-}(\text{rel})$ in the relative heats of solvation of oppositely charged ions of equal radii r_i are plotted against $(r_i + r_W)^{-3}$, one should get a straight line with a slope $+4N_A z_i e_0 p_W$. From Fig. 2.36, it can be seen that

[29]Expression (2.54) is based on the assumption of the radius independence of $W_+ - W_- = 84 - 125 = 41$ kJ (gram ion)$^{-1}$ and the constancy of n with radius over the interval concerned.

the experimental points do give a straight line unless the ionic radius falls below about 13 nm. Further, the theoretical slope (4.51 kJ mol^{-1} nm^3) is in fair agreement with the experimental slope (3.80 kJ mol^{-1} nm^3).

It can therefore be concluded that by considering a quadrupole model for the water molecule, one can not only explain why oppositely charged ions of equal radius have differing heats of hydration (Fig. 2.32), but can also quantitatively predict the way these differences in the heats of hydration will vary with the radius of the ions concerned.

What are we seeking in this section? The objective is a method to unscramble the individual heats of hydration from values known for the salt, i.e., for at least two individual ions.

An elegant method of obtaining such experimental values is now at hand. Starting from the experimentally proved linearity of $\Delta H_{M_i^+}(\text{rel}) - \Delta H_{X_i^-}(\text{rel})$ vs. $(r_i + r_W)^{-3}$ (Fig. 2.36), one can take Eq. (2.55) and, following Halliwell and Nyburg, extrapo-

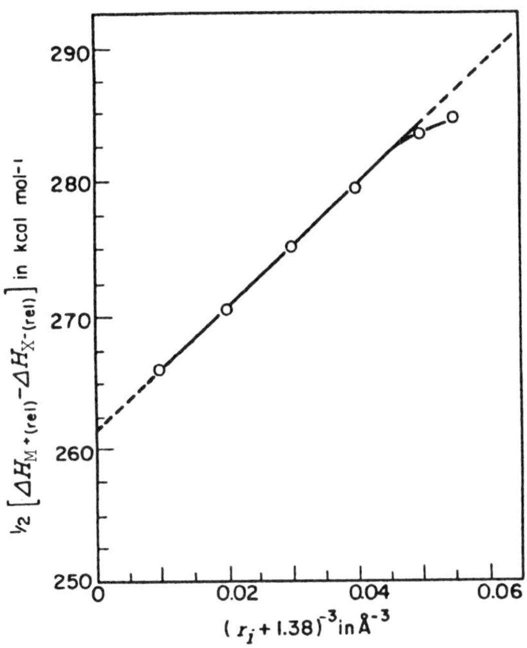

Fig. 2.36. The plot of half the difference in the relative heats of hydration of positive and negative ions of the same radii versus $(r_i + 1.38)^{-3}$. The solid line is through experimental points and the dotted line is the extrapolation of the straight line going through the experimental points (1 cal = 4.184 J; 1 Å = 100 pm).

TABLE 2.13
Quasi-Experimental Absolute Heats of Hydration of Various Individual Ions

Ion	Absolute Heat of Hydration (kcal. (gram ion)$^{-1}$)
Li^+	−129.7
Na^+	−102.3
K^+	−82.3
Rb^+	−77.2
CS^+	−71.4
F^-	−115.5
Cl^-	−81.5
Br^-	−75.0
I^-	−65.2

late the $\Delta H_{M_i^+}(\text{rel}) - \Delta H_{X_i^-}(\text{rel})$ vs. $(r_i + r_W)^{-3}$ plot to infinite radius, i.e., to $(r_i + r_W)^{-3} \rightarrow 0$. The intercept, which is 2184 kJ mol^{-1}, is then equal to $-2\Delta H_{H^+}(\text{abs}) - 41$, or $\Delta H_{H^+}(\text{abs}) = -1113$ kJ mol^{-1}.

Once one has thus obtained the experimental absolute heat of hydration of the proton, one can use it in combination with experimental values of H^+-containing electrolytes (e.g., HCl) to obtain absolute values of the heat of hydration of other ions, (Table 2.13). Such an approach has a better basis than an older method in which the salt considered had equal radii and the absolute heat of solvation of each was equal to half that of the whole. However, this assumption of the equality of the heats of hydration of ions with the same radius is just not true (Fig. 2.32).

Conway and Soloman corrected Halliwell and Nyburg's (1963) value for the numerical value of the dipole moment of water and got −1117 kJ mol^{-1}. Then, Lister, Nyburg, and Doyntz reattacked the calculation using a different series of ions of the same size. They got −1096 kJ mol^{-1}.

The best value in the late 1990s is −1110 kJ mol^{-1} (*cf.* the original value of −1113). As stated earlier, knowledge of the absolute heat of hydration of H^+ allows individual heats of hydration of anions to be obtained from the corresponding heats for an electrolyte, HX. Knowing the individual heats for some typical X^- ions allows one to couple these values with the heats of hydration of MXs and obtain individual values for M^+ ions. A few values are given in Table 2.13.

2.15.8. How Can Temperature Coefficients of Reversible Cells Be Used to Obtain Ionic Entropies?

In Section 2.13, it was shown how it is possible to obtain the entropy of solvation for an electrolyte, but that left open the separation into the individual entropies of solvation for each ion.

A well-known thermodynamic expression for the change in entropy $\Delta S°$ of a process is

$$\Delta S = -\frac{\partial \Delta G}{\partial T} \tag{2.56}$$

with ΔG as the free-energy change.

Furthermore, when an electrochemical cell works in a thermodynamically reversible way (see Vol. 2, Chapter 7),

$$\Delta G° = -nFE \tag{2.57}$$

where n is the number of electrons for one step of the overall reaction.

It follows from Eq. (2.56) that

$$\Delta S = nF\frac{\partial E}{\partial T} \tag{2.58}$$

Now, this ΔS of a *cell* reaction must be composed of the entropies of at least two different ions in solution (because two electrode reactions are involved, one at each electrode), so that Eq. (2.58) cannot lead directly to an individual ionic entropy.

However, in 1941, Lee and Tai considered the potential and temperature coefficients of the following cells:

1. $Hg(ecm) \mid Hg_2Cl_2(sat), KCl(a_{Cl^-} = 1) \mid Hg$ $\quad E_1$
2. $H_2 \mid Hg_2Cl_2(sat), HCl(a_{H^+} = 1) \mid Hg$ $\quad E_2$
3. $Hg(ecm) \mid HCl(a_{H^+} = 1) \mid H_2$ $\quad E_3$

The suffix (ecm) for cells 1 and 3 represents the term "electrocapillary maximum" and can be regarded (Vol. 2, Chapter 6) as a potential at which the electrode has zero charge.

Lee and Tai assumed that the potential at which the excess charge on the electrode is zero also indicates an interfacial potential difference of zero. This would not be consistent with the viewpoint of workers in the late 1990s (Vol. 2, Chapter 6), but let it be assumed to be so for now and follow Lee and Tai's reasoning.

Contemplating then cell (3),[30] if the Hg electrode does not contribute to the temperature coefficient of the potentials measured, then E_3 of the cell (being equivalent to $\Delta S°$) must yield here the entropy difference ($\frac{1}{2}S_{H_2} - S_{H^+}$) of the ion undergoing a reversible equilibrium reaction at Pt (right-hand electrode, cell 3) with H_2 in the gas phase and H^+ in solution at unit activity.

---

[30]In fact, Lee and Tai made measurements on cells (1) and (2) and obtained data on cell (3) by observing that $E_3 = E_1 - E_2$.

TABLE 2.14

Absolute Standard Partial Gram-Ionic Entropies of H^+ and Cl^- Ions[a]

Author(s)	$\tilde{S}^{o}_{Cl^-}$	Discrepancy between Values with Various Electrolytes	Value of $\tilde{S}^{o}_{H^+}$
Eastman (NaCl, KCl, HCl)	18.5 (288 K)	0.2	-5.0 ± 1.87
Crockford and Hall (NaCl, KCl, NH_4Cl, HCl)	19.8 (285.5 K)	1.8	-6.3 ± 3.8
Lange and Hesse (HCl)	18.2 (298 K)		−4.7
Lee and Tai	Value obtained using electrode at ecm		−5.4

Source: Reprinted from B. E. Conway, *Ionic Hydration in Chemistry and Biophysics*, Elsevier, New York, 1981.
[a]Units are cal K^{-1} mol^{-1}. 1 cal = 4.184 J.

However, the entropy of H_2 in the gas phase is well known and hence $S^{o}_{H^+}$ can be obtained. Lee and Tai, in fact, obtained -22.6 J K^{-1} mol^{-1} for $S^{o}_{H^+}$, the absolute standard entropy of H^+ in solution.

What of Lee and Tai's assumption that a charge-free surface involves no potential contribution to the cell? In fact, work done much later suggests that the missing temperature coefficient is only 0.01, so that the error Lee and Tai introduced by their outmoded assumption is indeed negligible.

Other work on the temperature coefficient of cells gave rise to a more complex analysis, but produced essentially the same result as that of Lee and Tai. Thus, Table 2.14 can be taken to indicate a result of -20.9 J K^{-1} mol^{-1} for this important quantity, $S^{o}_{H^+}$.

Having obtained the individual value of the gram-ionic entropy of the hydrogen ion in solution, the individual *entropy of hydration* can be obtained by a straightforward calculation of the value of $(S_{H^+})_{gas}$ from statistical mechanical reasoning.

To use this value of S_{H^+} to obtain the individual ionic entropies of other ions in solution, it is necessary to know values for the entropy of hydration of a number of electrolytes containing H^+. Thereafter, the value of the entropy of the counterion can be obtained. It can then be used in conjunction with entropies of hydration of electrolytes containing the counterion to determine the absolute entropies of partner ions in the electrolyte containing the constant anion. Of course, in all cases, the value of the entropy of the ion in the gaseous state must be subtracted from that of the ion in solution to give the entropy of hydration [i.e., $\Delta S_{hyd} = (S_i)_{soln} - (S_i)_{gas}$].

These considerations of individual entropies take it for granted that values of $(S)_{hyd}$ for a group of electrolytes are known (Table 2.15). This is acceptable pedagogically because in Section 2.5.3 one learned how to obtain ΔH_s and ΔG_s. So, the equation

TABLE 2.15

Entropies of Hydration of Individual Ions at 298 K and Quantities Used in Their Derivation[a,b]

Ion	Conventional $\tilde{S}_i^\circ$	$\tilde{S}_i$ (assuming $\tilde{S}^\circ_{H^+} = -5.0$)	$S^\circ_{i,g}$	$\Delta S^\circ_{s,i}$ [c]
H^+	0	−5.0	26.0	−31.0
Li^+	4.7	−0.3	31.8	−32.1
Na^+	14.0	9.0	35.3	−26.3
K^+	24.2	19.2	36.9	−17.7
Rb^+	28.7	23.7	39.3	−15.6
NH_4^+	26.4	21.4	—	—
Ag^+	17.5	12.5	39.9	−27.4
Tl^+	30.5	25.5	41.9	−16.4
Mg^{2+}	−31.6	−41.6	35.5	−77.1
Ca^{2+}	−11.4	−21.4	37.0	−58.4
Sr^{2+}	−7.3	−17.3	39.3	−56.6
Ba^{2+}	2.3	−7.7	40.7	−48.4
Zn^{2+}	−25.7	−35.7	38.5	−74.2
Cd^{2+}	−16.4	−26.4	40.1	−66.5
Fe^{2+}	−25.9	−25.9	38.0	−73.9
Cu^{2+}	−26.5	−36.5	38.5	−75.0
Sn^{2+}	−4.9	−14.9	40.3	−55.2
Pb^{2+}	3.9	−6.1	41.9	−48.0
Al^{4+}	−76.0	−91.0	35.9	−126.8
F^-	−2.3	3.0	34.8	−31.8
Cl^-	13.5	18.5	36.7	−18.2
Br^-	19.7	24.7	39.1	−14.4
I^-	25.3	30.3	40.4	−10.1
OH^-	−2.5	2.5	—	
NO_3^-	35.0	40.0	—	—
ClO_4^-	43.6	48.6	—	—
SO_4^{2-}	4.4	5.6	—	—
CO_3^{2-}	−13.0	−3.0	—	—
PO_4^{3-}	−45.0	−30.0	—	—

Source: Reprinted from B. E. Conway, *Ionic Hydration in Chemistry and Biophysics*, Elsevier, New York, 1981.

[a] Based on Latimer's scale of conventional ionic entropies. Units are cal K^{-1} mol^{-1}.
[b] 1 cal = 4.184 J.
[c] Ionic entropy of hydration as $\Delta S^\circ_{s,i} = \tilde{S}^\circ_i - S^\circ_{i,g}$ when $S^\circ_{i,g}$ is the gas-phase ionic entropy.

$\Delta S_s = (\Delta H_s - \Delta G_s)/T$ can be used to obtain ΔS_s, the entropy of hydration of an electrolyte.

However, there is a direct method to obtain ΔS_s, via the thermodynamics of reversibly behaving cells *without* any special assumptions. Thus, consider a cell that is assumed to be run in a thermodynamically reversible way:

$$H_2 \mid HCl \mid Hg_2Cl_2 \mid Hg$$

The overall reaction is

$$H_2 + Cl_2 \rightarrow 2\,H^+ + 2\,Cl^-$$

Now:

$$\Delta S = -nF\frac{\partial E}{\partial T} \tag{2.59}$$

However, the entropies of H_2 and Cl_2 in the gas phase can easily be calculated so that the sum of the entropies of H^+ and Cl^- in solution can be obtained. The value H^+ in solution is known (2.15.8). It is fairly easy to devise these kinds of cells, which have been used in obtaining much data.

2.15.9. Individual Ionic Properties: A Summary

In Section 2.15, methods for obtaining the properties of individual ions (their hydration numbers, heats, and entropies) have been considered. Starting with a general method—extrapolation to eliminate the effect of a partner on the value of the (easily obtainable) corresponding electrolyte property—two special cases were dealt with: how one obtains the individual values of the heat of hydration of the proton and then its entropy.

Values of the thermal properties of individual ions do not have the same status as thermodynamic properties (for these are assumption-free). For thermodynamic properties, only the accuracy of the experimental determination can be questioned. For example, in electrochemical cells, are they being operated in equilibrium as required? Thus, determining the values of the properties of individual ions always involves some assumption—that it is appropriate to extrapolate according to a certain $1/r^n$ law, for example—and therefore the values will always be open to improvement. Some of the newer values increasingly refer to nonaqueous solutions.

2.15.10. Model Calculations of Hydration Heats

In the 1970s Bockris and Saluja developed models incorporating and extending ideas proposed by Eley and Evans, Frank and Wen, and Bockris and Reddy. Three basic models of ionic hydration that differ from each other in the structure in the first coordination shell were examined. The features of these models are given in Table 2.16. The notations chosen for the models were 1A, 1B, 1C; 2A, 2B, 2C; and 3A, 3B, 3C, where 1, 2, and 3 refer to three basic hydration models, and A, B, and C refer to the subdivision of the model for the structure-broken (SB) region. These models are all defined in Table 2.16. A model due to Bockris and Reddy (model 3 in Table 2.16 and Fig. 2.37) recognizes the distinction between *coordination number* (CN) and *solvation number* (SN).

TABLE 2.16
Models of Ionic Solvation

<table>
<tr><td colspan="3">Model 1
First coordination shell
a. CN = 4 for all monovalent ions
b. No distinction between SN and CN
Second layer
(SB region)</td></tr>
<tr><td>Model A</td><td>Model B</td><td>Model C</td></tr>
<tr><td>Molecules in the first shell are H bonded to those in the SB region; rest of the solvent as bulk</td><td>SB region consists of dimers formed from monomers drawn from liquid water</td><td>SB region consists of monomers, some of which librate with respect to molecules in the first shell</td></tr>
<tr><td colspan="3">Model 2
First coordination shell
a. CN ranges from 6 to 8 for monovalent ions
b. No distinction between SN and CN
Second layer
(SB region)</td></tr>
<tr><td>Model A</td><td>Model B</td><td>Model C</td></tr>
<tr><td colspan="3">Model 3
First coordination shell
a. CN ranges from 6 to 8 for monovalent ions
b. Distinction between SN and CN
Second layer
(SB region)</td></tr>
<tr><td>Model A</td><td>Model B</td><td>Model C</td></tr>
</table>

In the concept of nonsolvated coordinated water, it is assumed that an ion exists in two states in the solution. It is either stationary or engaged in diffusional movement. As an ion arrives at a new site after a movement, its situation in respect to coordination by water is of two types. Clearly, the water molecules must always have some degree of orientation in the direction of their dipoles in the first layer around the ion. It is practical to divide these molecules into two types. There will be those oriented so that these dipoles interact maximally with the ion, and another type that is oriented so that the dipoles are oriented at 90° to the ion and do not interact with it at all. The coordination number refers to the total number of water molecules in the first layer around the ion, independently of how they are oriented.

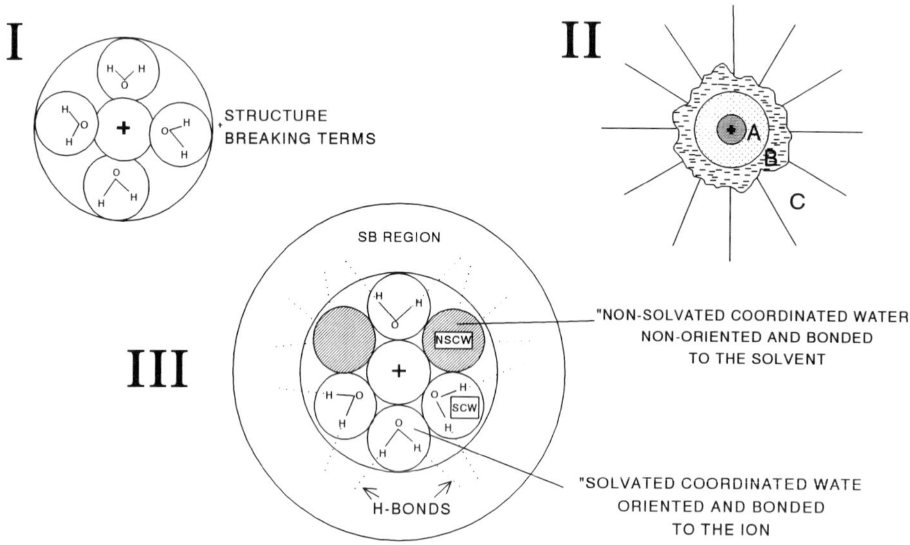

I BERNAL AND FOWLER (1933) ELEY AND EVANS (1938):
(i) All Monovalent Ions are 4-Coordinated
(ii) C.N. = S.N.
(iii) Structure Breaking Terms

II FRANK AND WEN (1957):
(i) Two Layer Model
(ii) All Molecules in Region "A" are immobilized, i.e., S.N. = C.N.
(iii) A Structure Broken Region "B"

III BOCKRIS AND REDDY (1970):
(i) S.N. Distinguished from C.N.
(ii) Structure Broken Region (About a Monolayer)

Fig. 2.37. Models for the region near an ion. (Reprinted from J. O.'M. Bockris and P. P. S. Saluja, *J. Phys. Chem.* **76**: 2298, 1972.)

The *dynamic* solvation number (as distinguished from the *static* coordination number) is the number of these water molecules *that remain with the ion for at least one diffusive movement*. When an ion arrives at a new site, it may remain there long enough to influence a number of the surrounding water molecules to come out of the water structure and become part of the primary solvation shell that moves with the ion. Conversely, it may remain at a given site for a time so brief that it does not have an effect on the waters that are relatively stable and fixed tightly in the water structure that surrounds it. In the solvent structure these latter may still be thought of as part of the coordinating waters of the ion, but their dipoles have not had sufficient time to rotate into the attractive position ($\cos \theta = 1$), such as that of the solvationally coordinated waters represented by the letters SCW.

2.15.11. Heat Changes Accompanying Hydration

The following steps in a cycle (Fig. 2.38 and Tables 2.17 and 2.18) are carried out in calculating the heat of hydration (ΔH_h): (1) the evaporation of n water molecules from liquid water to the gas phase ($\Delta H_1 = n\Delta H_{\text{evap}}$), where n is the number of molecules in the first shell in a given model and ΔH_{evap} is the heat of evaporation (42.2 kJ mol^{-1}); (2) the gain on interaction of an ion with n water molecules in the gas phase ($\Delta H_2 = n\Delta H_{\text{ion-}w}$), where $\Delta H_{\text{ion-}w}$ is the ion–water interaction and is negative in sign; (3) a lateral interaction term: $\Delta H_3 = \Sigma\Delta H_{\text{lat}}$; (4) the phenomenon known as *Born charging* in which the ion with its coordination shell is returned to the solvent ($\Delta H_4 = \Delta H_{BC}$) (see Appendix 2.1); and (5) the interaction among the molecules in the first shell with molecules in the outer layer and, consequently, structural alterations in the structure-broken region making a net contribution ($\Delta H_5 = \Sigma\Delta H_{SB}$).

The heat of ionic hydration ΔH_h can now be calculated as

$$\Delta H_h = \sum_{z=1}^{5} \Delta H_x \tag{2.60}$$

When all the molecules in the coordination shell are treated as identical, ΔH_h will be given by the cycle in Fig. 2.38.

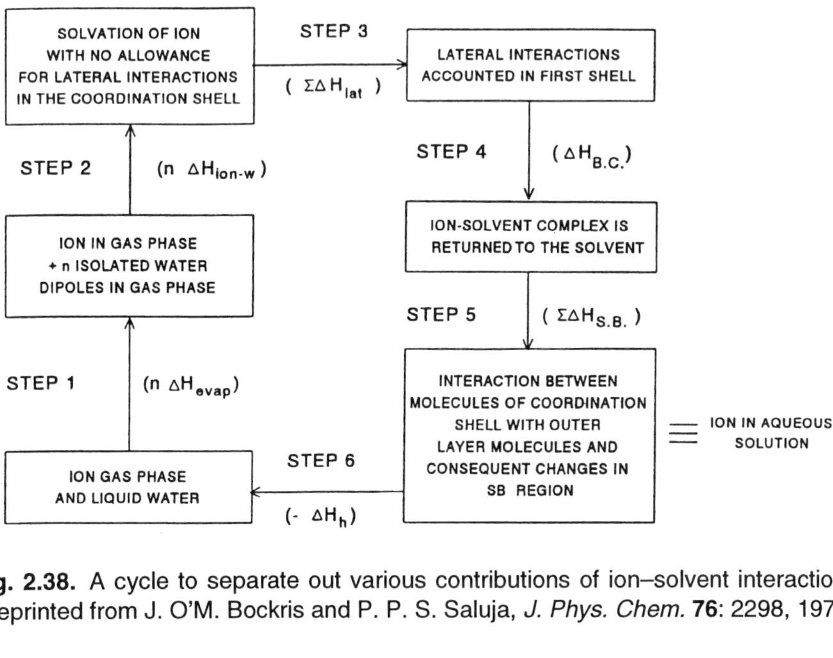

Fig. 2.38. A cycle to separate out various contributions of ion–solvent interactions. (Reprinted from J. O'M. Bockris and P. P. S. Saluja, *J. Phys. Chem.* **76**: 2298, 1972.)

TABLE 2.17

Relevant Quantities for Predicting Heat of Hydration, ΔH_h

	Li^+	Na^+	K^+	Rb^+	Cs^+	F^-	Cl^-	Br^-	I^-
τ_{ion}, Å	0.60	0.95	1.33	1.48	1.69	1.36	1.81	1.95	2.16
$\Delta H_{ion\text{-}w} = H_{ion\text{-}SCW}$, kcal mol^{-1}	−29.92	−20.50	−14.75	−13.17	−11.38	−27.23	−18.67	−16.82	−14.51
$\Delta H_{ion\text{-}NSCW}$, kcal mol^{-1}	1.67	2.03	1.41	1.06	1.64	−10.94	−6.84	−5.99	−4.88
n_{CW} (X-ray)	(6)	(6)	(6)	(6)	(7)	(6)	(7)	(7)	(8)
n_{SCW}	(4)	(4)	(3)	(3)	(2)	(4)	(2)	(2)	(1.5)
n_{NSCW}	(2)	(2)	(3)	(3)	(5)	(2)	(5)	(5)	(6.5)
$(\Sigma E_{lat})_{CN\text{-}4}$, kcal	−2.3	−0.8	1.5	1.5	1.5	7.7	4.8	4.2	3.4
$(\Sigma E_{lat})_{CN\text{-}nCW}$, kcal	−20.7	−2.6	2.6	3.2	5.0	21.7	20.0	17.5	22.0
$(\Sigma E_{lat})_{SN \neq CN}$, kcal	−2.3	−0.8	0.0	0.0	0.0	13.75	0.0	0.0	0.0
ΔH_{BC}, kcal	−47.32	−42.91	−38.96	−37.59	−35.84	−38.68	−34.90	−33.87	−32.84
n_{SB}	23.6	28.7	34.8	37.4	41.1	35.3	43.4	46.0	50.2
$\Sigma\Delta H_{SB}$ (model 1A), kcal	−20.2	−20.2	−20.2	−20.2	−20.2	−20.2	−20.2	−20.2	−20.2
$\Sigma\Delta H_{SB}$ (model 2A), kcal	−30.3	−30.3	−30.3	−30.3	−35.4	−30.3	−35.4	−35.4	−40.4
$\Sigma\Delta H_{SB}$ (model 3A), kcal	−35.4	−35.4	−37.9	−37.5	−48.0	−32.8	−48.0	−48.0	−56.8
$\Sigma\Delta H_{SB}$ (model 1–3B), kcal	−64.5	−73.0	−83.0	−87.0	−92.75	−83.75	−96.25	−100.25	−106.7
$\Sigma\Delta H_{SB}$ (model 1C), kcal	−29.7	−31.7	−34.3	−35.3	−36.8	−34.4	−37.8	−38.8	−40.5
$\Sigma\Delta H_{SB}$ (model 2C), kcal	−40.0	−41.7	−44.3	−45.3	−51.8	−44.4	−52.8	−53.8	−60.5
$\Sigma\Delta H_{SB}$ (model 3C), kcal	−44.7	−46.8	−51.8	−52.3	−64.3	−46.9	−65.3	−65.8	−76.7

TABLE 2.18
Ionic Solvation Numbers Obtained in Dilute Solution

Ion	Vibration Potentials + Partial Molar Volumes	Passynski Compressibility	Ulich Entropy	Ulich Mobility	Zana and Yeager	Averages Nearest Integer
Li^+	4.5	5–6	5	3.5	5–6	5 ± 0.4
Na^+	4.5	6–7	4	2.0	7.0	5 ± 0.6
K^+	3.8	6–7	3		6.0	5 ± 1.6
Rb^+	3.0	—	3		—	3 ± 0.0
Cs^+	2.5	—	—		—	—
F^-	4.0	2.0	5		2.0	3 ± 1.2
Cl^-	2.2	0–1	3		1	2 ± 0.8
Br^-	1.8	0	2		0–1	2 ± 0.4
I^-	1.5	0	1		0	1 ± 0.1
Mg^{2+}	10.0	16	13	10.5	15.0	13 ± 2
Ca^{2+}	9.0	—	10	7.5	—	9 ± 0.8
Ba^{2+}	11.0	16	8	5.0	13.0	11 ± 4

$$(\Delta H_h)_{SN=CN} = n\Delta H_{\text{evap}} + n\Delta H_{\text{ion-}w}$$

$$+ (\sum \Delta H_{\text{lat}})_{SN=CN} + \Delta H_{BC} + (\sum \Delta H_{SB})_{SN=CN} \tag{2.61}$$

When, as suggested by Bockris and Reddy, the number of water molecules that stay with the ion (SN) is distinguished from the number that simply surround the first shell (CN), $\Delta H_{\text{ion-}w}$ will differ for molecules referred to as solvational coordinated water and for those referred to as nonsolvational coordinated water (NSCW). Then ΔH_h will be given by

$$(\Delta H_h)_{SN \neq CN} = n\Delta H_{\text{evap}} + n_{SCW}\,\Delta H_{\text{ion-}SCW}$$

$$+ n_{NSCW}\,\Delta H_{\text{ion-}NSCW} + (\sum \Delta H_{\text{lat}})_{SN \neq CN} + \Delta H_{BC} + (\sum \Delta H_{SB})_{SN \neq CN} \tag{2.62}$$

Let's now evaluate the different energy terms involved in calculating the heat of hydration of the ions [Eqs. (2.61) and (2.62)] and compare the various models described in Table 2.16.

2.15.11.1. $\Delta H_{\text{ion-water}}$. The energy of an ion interacting with a water molecule oriented along the ionic field is

$$\Delta H_{\text{ion-}w} = \Delta H_{\text{ion-dipole}} + \Delta H_{\text{ion-quadrupole}} + \Delta H_{\text{ion-induced dipole}} \tag{2.63}$$

or

$$\Delta H_{\text{ion-}w} = -\frac{|z|e\mu}{\varepsilon r^2} \pm zep_w \frac{1}{2\varepsilon r^3} - \frac{1}{2}\frac{\alpha_w (zr)^2}{r^4} \tag{2.64}$$

where μ_w, p_w, and α_w are the dipole moment, quadrupole moment, and polarizability of the water molecule. The ion–water distance is symbolized by r $(= r_i + r_w)$, where r_i is crystallographic ionic radius and r_w is the radius of the water molecule. The ± sign in the ion–quadrupole term refers to cation and anion, respectively.

Using values of 1.84×10^{-18} esu cm, 3.9×10^{-26} esu cm^2, and 1.444×10^{-24} cm^3 molecule^{-1} for μ_w, p_w, and α_w, respectively, the term $\Delta H_{\text{ion-}SCW}$ can be rewritten as

$$\Delta H_{\text{ion-}SCW}\ (\text{kJ mol}^{-1}) = 60.24\left[\frac{-8.8357}{r^2} \pm \frac{9.3639}{r^3} - \frac{16.6487}{r^4}\right] \tag{2.65}$$

2.15.11.2. $\Delta H_{\text{ion-}NSCW}$. When SN ≠ CN, there will be two interaction energies: (1) the interaction energy of a solvationally coordinated water with the ion, $\Delta H_{\text{ion-}SCW}$; and (2) the interaction energy of nonsolvationally coordinated water (NSCW) with the ion, $\Delta H_{\text{ion-}NSCW}$. The $\Delta H_{\text{ion-}SCW}$ energy is given by Eq. (2.65). An NSCW does not give any average preferred orientation.[31] Before it has time to orient toward the ion, the latter has left the site. Thus, NSCW is still a part of the water structure, but on the average has an H-bonding position blocked by the ion. Thus, its net ion–dipole interaction energy is zero. The interaction energy of an NSCW with the ion arises from the ion–induced dipole, the ion quadrupole, and the dispersion interactions.

Thus,

$$\Delta H_{\text{ion-}NSCW} = \Delta H_{\text{ion-induced dipole}} + \Delta H_{\text{ion-quadrupole}} + \Delta H_{\text{dip}} \tag{2.66}$$

and

$$\Delta H_{\text{ion-}NSCW} = -\frac{1}{2}\frac{\alpha_w (ze)^2}{r^4} \pm \frac{zep_w}{2r^3} - \frac{3}{4}\frac{h\nu\alpha_{\text{ion}}\alpha_w}{r^6} \tag{2.67}$$

[31]Of course, in reality, there will not only be these two positions but also all possible intermediate positions, in which the ion–solvent interaction is given for one solvent molecule by $z_i e_0 \mu \cos\theta / \varepsilon_r r^2$. The value of θ is zero for a dipole oriented directly to the ion ($\cos 0° = 1$); and 90° for a distant water molecule ($\cos 90° = 0$). However, there will be all different values of θ in between, and the model presented here does not deal with them. As a simplification; it takes the two extreme situations (all oriented and not at all oriented) and pretends any water molecule belongs to one group or the other.

where α_{ion} is the polarizability of the ion. Values of $\Delta H_{\text{ion-}NSCW}$ calculated from Eq. (2.67) are listed in Table 2.17.

2.15.11.3. ΔH_{lat}. The lateral interactions in the coordination shell for a CN of 4 and 6 can be calculated. A $\Sigma\Delta H_{\text{lat}}$ vs. CN plot obtained for CNs of 4 and 6 can be extrapolated. Thus, $\Sigma\Delta H_{\text{lat}} \rightarrow 0$ when CN $\rightarrow$ 1, i.e., there is no other molecule to interact with; and for CN = 6, the extrapolation is made to follow the shape of the curve of $\Sigma\Delta H_{\text{lat}}$ vs. CN, obtained from the equation

$$\sum \Delta H_{\text{lat}} = \frac{D_n \mu_w^2}{r^3} \tag{2.68}$$

where D_n is a geometrical factor depending on the CN. The $(\Sigma\Delta H_{\text{lat}})$ arises only from the SCW. Thus, for large ions, which have a low SN, $\Sigma\Delta H_{\text{lat}}$ obtained from the $\Sigma\Delta H_{\text{lat}}$ vs. CN plot is negligible. Therefore, for large ions ($r_i > 250$ pm), $\Sigma\Delta H_{\text{lat}}$ has been neglected.

2.15.11.4. ΔH_{BC} This term can be determined from

$$\Delta H_{BC} = -\frac{N(ze)^2}{2(r_i + 2r_w)}\left[1 - \frac{1}{\varepsilon} - \frac{T}{\varepsilon^2}\left(\frac{\partial \varepsilon}{\partial T}\right)\right] \tag{2.69}$$

Substituting in Eq. (2.69) the appropriate values for ε and $(\partial\varepsilon/\partial T)$, the following equation is obtained:

$$\Delta H_{BC}\,(\text{in kcal mol}^{-1}) = -\frac{6.73z^2}{(r_i + 2r_w)} \tag{2.70}$$

where r_i and r_w are in picometer (pm) units.

2.15.11.5. $\Sigma\Delta H_{SB}$ *(Model A).* This model (Table 2.16) considers H bonding between the molecules in the first layer and those in the region of structure breaking (SB) (Fig. 2.37). The solvated coordinated water (SCW), which is oriented toward the ion, has two H-bonding positions blocked by the ion and offers the remaining two sites for H bonding to the solvent molecules in the SB region (Fig. 2.37). The NSCW is still attached predominantly to the solvent structure but on the average has an H-bonding position blocked by the ion and thus can offer three H-bonding sites to molecules in the SB region. Thus, the net gain in energy from H bonding between molecules in the B region and those in the first layer will be $(2/2)E_{\text{H-bond}}$ for the SCW and $(3/2)E_{\text{H-bond}}$ for NSCW, i.e.,

$$\sum \Delta H_{SB}\,(\text{modelA}) = n_{CW}[(2/2)E_{\text{H-bond}}] \tag{2.71}$$

for the case where the solvation number and the coordination number is the same (SN = CN), and

$$\sum \Delta H_{SB}\,(\text{modelA}) = n_{SCW}[(2/2)E_{\text{H-bond}}) + n_{SCW}[(3/2)E_{\text{H-bond}}] \tag{2.72}$$

when the solvation number differs from the coordination number (SN ≠ CN). The $E_{\text{H-bond}}$ is the energy of one H bond, equal to –21 kJ mol^{-1}. The numerical values are listed in Table 2.17.

2.15.11.6. $\Sigma\Delta H_{SB}$ *(Model B).* This model (Table 2.16) considers the structure-breaking region as consisting of dimers, which are formed from monomers from liquid water. If the lowering in the potential energy due to formation of a dimer from two monomers is $\Delta E_{m\to d}$, then

$$\sum \Delta H_{SB}\,(\text{modelB}) = \frac{n_{SB}}{2}\,(\Delta E_{m\to d}) \tag{2.73}$$

where n_{SB} is the number of molecules in the SB region per ion, and $\Delta E_{m\to d}$ is taken as the energy of one H bond, i.e., –21 kJ mol^{-1}. Values of $\Sigma\Delta H_{SB}$ (model B) are listed in Table 2.17.

2.15.11.7. $\Sigma\Delta H_{SB}$ *(Model C).* In this model (Table 2.16) the SB region is considered to consist basically of monomers. Some of these monomers are H bonded to the molecules in the first shell. The SCW and NSCW participate in two and three H bonds with molecules in the SB region. The formation of an SB region of monomers can be treated according to the cycle shown in Fig. 2.39. Therefore, one can write

$$\Delta H_{SB} + \Delta H_{\text{evap},SBW} + \Delta H_{\text{cond},\,w} = 0 \tag{2.74}$$

or

$$\Delta H_{SB} = -\Delta H_{\text{evap},SBW} - \Delta H_{\text{cond}\,w} \tag{2.75}$$

or

$$\Delta H_{SB} = \Delta H_{\text{evap},w} - \Delta H_{\text{evap},SBW} \tag{2.76}$$

$\Delta H_{\text{evap},SBW}$ clearly cannot be taken as the heat of evaporation for liquid water. It may be possible to get some rough value for it. In some theories of liquids, a property of interest is that of the free volume, that is, the volume occupied by the liquid minus the volume of the actual molecules considered to occupy "cells." One such theory gives

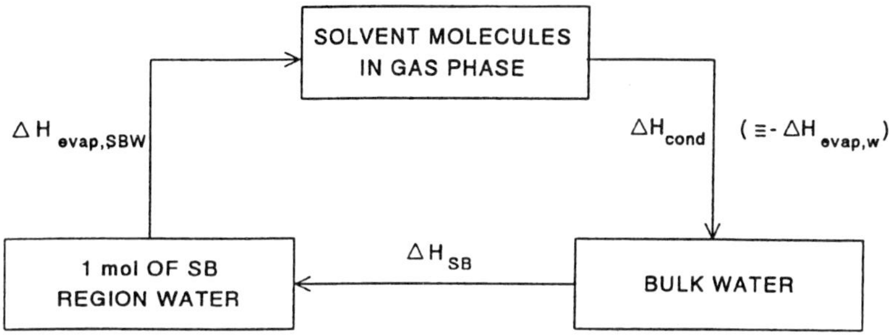

Fig. 2.39. The formation of an SB region of monomers can be treated according to the cycle shown. (Reprinted from J. O'M. Bockris and P. P. S. Saluja, *J. Phys. Chem.* **76**: 2298, 1972.)

$$\ln V_{f,w} = 1 + \ln \frac{RT}{P_w^*} - \frac{\Delta H_{\text{evap},w}}{RT} \tag{2.77}$$

and

$$\ln V_{f,SBW} = 1 + \ln \frac{RT}{P_{SBW}^*} - \frac{\Delta H_{\text{evap},SBW}}{RT} \tag{2.78}$$

Here the P_w^* terms refer to the vapor pressure of bulk water, while P_{SBW}^* is that of the structure-breaking region.

From Eqs. (2.77) and (2.78),

$$\ln \frac{V_{f,SBW}}{V_{f,w}} = -\left(\frac{\Delta H_{\text{evap},SBW} - \Delta H_{\text{evap},w}}{RT}\right) \tag{2.79}$$

on the first-approximation assumption that the vapor pressures of the two forms of water are not significantly different. Using a value of $V_{f,SBW}$ of 0.20 cm^3 mol^{-1} for water in the SB region and 0.40 cm^3 mol^{-1} for $V_{f,w}$

$$\ln \frac{0.20}{0.40} = -\left(\frac{\Delta H_{\text{evap},SBW} - \Delta H_{\text{evap},w}}{RT}\right) \tag{2.80}$$

or

$$\Delta H_{\text{evap},w} - \Delta H_{\text{evap},SBW} = -RT \ln 2 \tag{2.81}$$

Substitution of values from Eq. (2.81) into Eq. (2.76) gives

$$\Delta H_{SB} = -RT \ln 2 = -1.72 \text{ kJ mol}^{-1} \tag{2.82}$$

Thus, for Li^+, where n_{SB} (the number of molecules in the structure-breaking region per ion; see later discussion) is 24, the net contribution of ΔH_{SB} will be −41 kJ. Similar calculations can be carried out for other ions. The net contribution from $\Sigma\Delta H_{SB}$ (model C) comes to

$$\sum \Delta H_{SB}(\text{model C}) = n_{SB}[(2/2)E_{\text{H-bond}}] \tag{2.83}$$

for SN = CN, and

$$\sum \Delta H_{SB}\,(\text{model C}) = n_{SB}\Delta H_{SB} + n_{SCW}[(2/2)E_{\text{H-bond}}] + n_{SCW}\,[(3/2)E_{\text{H-bond}}] \tag{2.84}$$

for SN ≠ CN, where ΔH_{SB} in the first term is obtained from Eq. (2.82). The values of $\Sigma\Delta H_{SB}$ (model C) are listed in Table 2.17.

2.15.11.8. n_{SB}. The total number of molecules in the SB region can be calculated by consideration of the close packing of water molecules in the area of a sphere consisting of the ion plus the first layer. Thus, the number of molecules of cross-sectional area πr_w^2 will be given as

$$n_{SB} = \frac{4\pi(r_i + 2r_w)^2}{\pi r_w^2} \tag{2.85}$$

Values of n_{SB} are listed in Table 2.17.

2.15.11.9. *Numerical Evaluation of* ΔH_h. The heats of hydration of monovalent ions have been calculated for the various models by using Eqs. (2.61) and (2.62) and the parameters listed in Table 2.17. The results are shown in Figs. 2.40 and 2.41. From Figs. 2.40 and 2.41 it can be seen that the experimental data for cations fit model 3C of Table 2.16 best, while for the anions, the best fit is with model 3A.

Both models of best fit assume that there is a distinction between the coordination shell and the solvation shell. The difference between cations and anions is that the anion calculations are more consistent with H bonding from the first coordination shell around the ions and the water in the structure-broken region; with the cations, the better model fit is to stress the librating properties of water in the structure-broken region. This difference may arise from the smaller peripheral field strengths of the anions (larger radius) so that there is more time for orientation and H bonding with the (larger amount of) structure-broken waters after an ion arrives in a given region.

All these conclusions must be tempered by continuous reminders that solvation is a dynamic matter and that water molecules are constantly being attracted by the

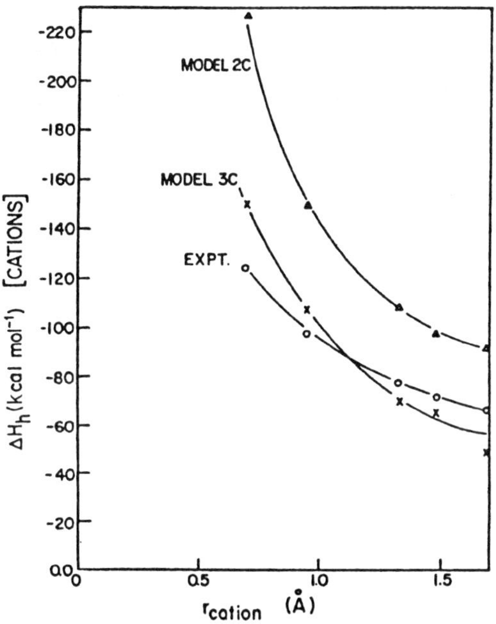

Fig. 2.40. Heat of hydration against radius of cations for various competing models (1 Å = 100 pm; 1 kcal = 4.184 kJ). (Reprinted from J. O'M. Bockris and P. P. S. Saluja, *J. Phys. Chem.* **76**: 2298, 1972.)

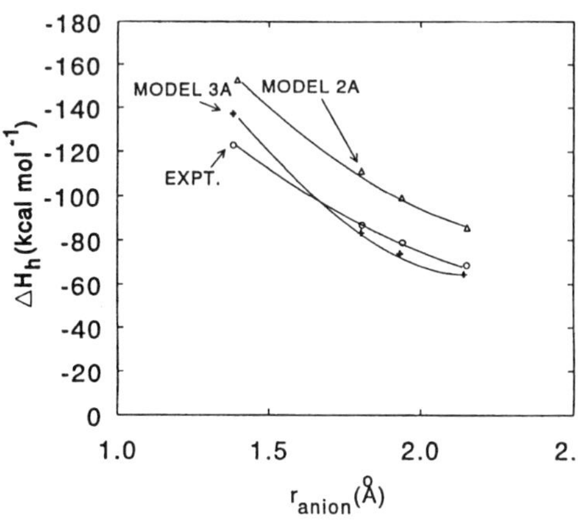

Fig. 2.41. Heat of hydration against radius of anions, for various models (1 Å = 100 pm; 1 kcal = 4.184 kJ). (Reprinted from J. O'M. Bockris and P. P. S. Saluja, *J. Phys. Chem.* **76**: 2298, 1972.)

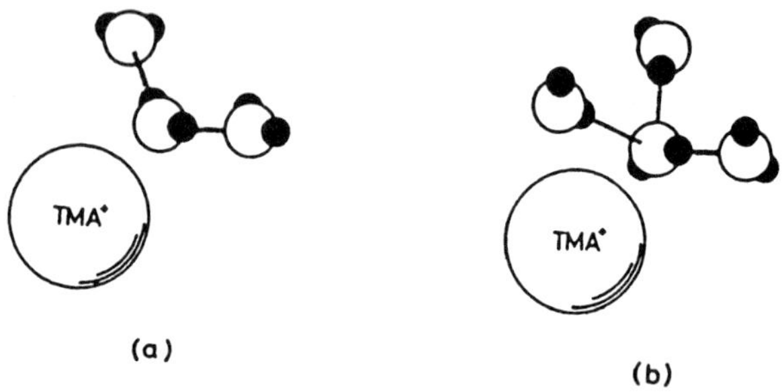

Fig. 2.42. Orientational change of a water molecule in the first hydration shell of a tetramethylammonium ion (TMA^+). (a) A water molecule in the first hydration shell is attached to the TMA^+ ion by charge–electric dipole interaction. (b) Orientational change of a water molecule in the first hydration shell is complemented by formation of another hydrogen bond to an outer water molecule. (Reprinted from Y. Nagano, H. Mizuno, M. Sakiyama, T. Fujiwara, and Y. Kondo, *J. Phys. Chem.* **95**: 2536, 1991.)

ion–dipole forces; but they are also shaken about by the thermal forces, by lateral repulsion between waters, and by the breaking and formation of hydrogen bonds. Orientational changes of waters in the first shell of a tetramethylammonium cation are shown in Fig. 2.42.

2.15.12. Entropy of Hydration: Some Possible Models

The entropy of solvation values reflect solvational structure near an ion. The following discussion of models that are more in agreement with the experimental values of solvational entropies follows the seminal treatment due to Bockris and Saluja in 1982. Models for the region near an ion are shown in Fig. 2.37. The entropy of hydration from the model with no SB region was 170 to 250 J K^{-1} mol^{-1} lower (more negative) than the experimental values, and was therefore not pursued further.

2.15.13. Entropy Changes Accompanying Hydration

The ΔS_h is the entropy change that accompanies the transition of an ion in the gas phase to an ion in solution (with the arbitrary neglect of any change in crossing the gas–liquid interface).[32] Thus,

[32]This does not introduce an error, due to the way in which solvation entropies are calculated. Thus, one calculates the processes of ordering in solution when an ion becomes hydrated and then subtracts the ion's entropy in the gas phase. This is precisely what the experimental value reflects.

$$\Delta S_h = S_{i,\text{soln}} - S_{i,g} \tag{2.86}$$

The second term in the right-hand side of Eq. (2.86) is the entropy of the ion in the gas phase and the first term represents the total entropy change caused upon the entry of the ion into the solution. $S_{i,\text{soln}}$ will be given as

$$[S_{i,\text{soln}}]_{SN=CN} = S_{i,tr} + n_{CW}(S_{i\text{-}CW} - S_{w\text{-}w}) + \Delta S_{BC} + \Delta S_{SB} \tag{2.87}$$

$$[S_{i,\text{soln}}]_{SN\neq CN} = S_{i,tr} + n_{SCW}(S_{i\text{-}SCW} - S_{w\text{-}w}) + n_{NSCW}(S_{i\text{-}NSCW} - S_{w\text{-}w}) + \Delta S_{BC} + \sum \Delta S_{SB} \tag{2.88}$$

where $S_{i,tr}$ is the translational entropy of the ion, $S_{i\text{-}CW}$ is the entropy of coordinated water in the first shell around the ion, $S_{w\text{-}w}$ is the entropy of water in liquid water, ΔS_{BC} is Born-charging entropy, $S_{i\text{-}SCW}$ and $S_{i\text{-}NSCW}$ are the entropy of solvationally coordinated water and nonsolvationally coordinated water, respectively, near the ion, and $\Sigma\Delta S_{SB}$ is the net entropy contribution of the structure-broken region around the ion.

The substitution into Eq. (2.86) of the values of $S_{i,\text{soln}}$ from Eqs. (2.87) and (2.88) gives the entropy of hydration. Relevant terms will be evaluated next.

2.15.13.1. $S_{i,g}$. One uses the Sackur–Tetrode equation for the translational entropy of a monatomic ion in the gas phase and it gives

$$S_{i,g} = \frac{3}{2} R \ln m_{\text{ion}} + 26.0 \tag{2.89}$$

where m_{ion} is the mass of the ion (Table 2.19).

2.15.13.2. ΔS_{BC}. The Born-charging entropy is given by

$$\Delta S_{BC} = \frac{N_A (z_i e_0)^2}{2(r_i + 2r_w)} \frac{1}{2}\left(\frac{\partial \varepsilon}{\partial T}\right) \tag{2.90}$$

which at 298 K becomes

$$\Delta S_{BC} = -\frac{4.09.5 z_i^2}{(r_i + 2r_w)} \quad (\text{J K}^{-1}\ \text{mol}^{-1}) \tag{2.91}$$

where r_i and r_w are in nanometers (Table 2.19).

2.15.13.3. $S_{i,tr}$. The partition function corresponding to "translation" of a particle in a liquid can be written as

TABLE 2.19

Relevant Parameters in Calculations of the Entropy of Ionic Hydration

	Li^+	Na^+	K^+	Rb^+	Cs^+	F^-	Cl^-	Br^-	I^-
$S_{i,g}$	31.8	35.3	36.9	39.2	40.6	34.8	36.7	39.1	40.6
ΔS_{BC}	−2.9	−2.6	−2.4	−2.3	−2.2	−2.4	−2.1	−2.1	−2.0
$\Delta S_{BC} - S_{i,g}$	−34.7	−37.9	−39.3	−41.5	−42.8	−37.2	−38.8	−41.2	−42.6
$S_{\text{rot}}(SCW)$	5.62	6.12	6.55	6.70	6.90	5.74	6.24	6.38	6.57
$S_{\text{vib}}(SCW)$	0.64	1.65	2.39	2.77	3.08	2.16	2.96	3.33	3.64
$S\ (SCW)$	6.26	7.77	8.94	9.47	9.98	7.90	9.20	9.71	10.21
$S_{\text{rot}}(NSCW)$	9.43	9.18	9.66	10.03	10.50	6.94	7.57	7.74	8.02
$S_{\text{vib}}(NSCW)$	0.89	2.30	3.36	3.85	4.33	3.14	4.27	4.74	5.16
$S\ (NSCW)$	10.32	11.48	13.02	13.88	14.83	10.08	11.84	12.48	13.18
$\Sigma\Delta S_{SB}$ (model 1A)	10.24	10.24	10.24	10.24	10.24	10.24	10.24	10.24	10.24
$\Sigma\Delta S_{SB}$ (model 2A)	15.36	15.36	15.36	15.36	17.92	15.36	17.92	17.92	20.48
$\Sigma\Delta S_{SB}$ (model 3A)	17.92	17.92	19.20	19.20	24.32	16.64	24.32	24.32	28.80
$\Sigma\Delta S_{SB}$ (model 2C)	32.8	40.4	49.6	53.5	58.6	50.3	62.1	65.9	72.2
$\Sigma\Delta S_{SB}$ (model 3C)	32.3	39.9	48.9	52.8	57.5	50.1	61.0	64.8	71.0

$$f_{tr,\text{ion}} = \frac{(2\pi m_{\text{ion}} kT)^{3/2}}{h^3} V_{f,\text{ion}} \tag{2.92}$$

where $V_{f,\text{ion}}$ is the *free volume available to the ion in solution* described in Section 2.15.11. Therefore, the entropy of translation is

$$S_{i,tr} = R \ln f_{tr,\text{ion}} + RT \frac{\partial \ln f_{tr,\text{ion}}}{\partial T} \tag{2.93}$$

Also,

$$V_{f,\text{soln}} = x_{\text{salt}} V_{f,\text{salt}} + (1 - x_{\text{salt}}) V_{f,\text{water}} \tag{2.94}$$

where $V_{f,\text{soln}}$, $V_{f,\text{salt}}$, and $V_{f,\text{water}}$ are the free volumes of solution, salt, and water, respectively, and x_{salt} is the mole fraction of the salt. Equation (2.94) can be rearranged as

$$V_{f,\text{soln}} = V_{f,\text{water}} + x_{\text{salt}}(V_{f,\text{salt}} - V_{f,\text{water}}) \tag{2.95}$$

and a plot of $V_{f,\text{soln}}$ vs. x_{salt} can then be used to calculate $V_{f,\text{salt}}$.

The $V_{f,\text{soln}}$ can be calculated from the velocity of sound by the expression

$$V_{f,\text{soln}} = \left(\frac{v_{o,g}}{v_{o,\text{soln}}} \right)^3 V_{m,\text{soln}} \tag{2.96}$$

where $v_{o,g} = (\gamma RT/M)^{1/2}$ is the velocity in the gas phase given by kinetic theory, $v_{o,\text{soln}}$ is the velocity of sound in solution, $V_{m,\text{soln}}$ is the molar volume of the solution, $\gamma = C_p/C_v$, and M is the molecular weight. Equation (2.96) at 298.16 K becomes upon substitution of appropriate values:

$$V_{f,\text{soln}} = \left(\frac{18.63 \times 10^4}{v_{o,\text{soln}}} \right)^3 V_{m,\text{soln}} \tag{2.97}$$

The molar volume $V_{m,\text{soln}}$ of a binary solution is given by

$$V_{m,\text{soln}} = \frac{1}{\rho \sum_i (w_i/M_i)} \tag{2.98}$$

where w_i and M_i are the weight fraction and molar weight of the ith component, respectively, and ρ is the density of the solution. For example, for NaCl, Eq. (2.98) reduces to

$$V_{m,\text{soln}} = \frac{18}{(1 - 0.6923w_2)} \tag{2.99}$$

where w_2 is the weight fraction of NaCl in solution. M_{soln} needed to calculate $V_{o,\text{soln}}$ is given by

$$M_{\text{soln}} = x_{\text{salt}} M_{\text{salt}} + x_{\text{solv}} M_{\text{solv}} \tag{2.100}$$

where x_{salt} and x_{solv} are the mole fractions of salt and solvent in the solution. Using measurements of the velocity of sound, $V_{m,\text{soln}}$ from Eq. (2.99), and M_{soln} from Eq. (2.100), $V_{f,\text{soln}}$ can be calculated from Eq (2.97) for different mole fractions, x_{salt}, of the salt.

The $V_{f,\text{salt}}$ obtained from the slope of $V_{f,\text{soln}}$ vs. x_{salt} [*cf.* Eq. (2.95)] is near zero and it is inferred that this value indicates a value of zero for the translational entropy of the ion. A similar result was obtained for the solvated complex. Zero translational entropy for a solvated ion is a reasonable conclusion. Thus, for most of its time, the ion is still in a cell in the solution and only occasionally does it jump into a vacancy, or if it shuffles about, its movement is so constrained compared with that of a gas that it may approach zero.

2.15.13.4. $S_{i\text{-SCW}}$. The entropy of solvationally coordinated water is made up of librational ($S_{i,\text{lib}}$) and vibrational ($S_{i,\text{vib}}$) contributions. $S_{i,\text{lib}}$ can be calculated as follows.

$S_{i\text{-lib}}$. The partition function of a particle, f_{rot}, under an electric field is

$$f_{\text{lib}} = \frac{8\pi^2(8\pi^3 I_1 I_2 I_3 k^3 T^3)^{1/3}}{\sigma'_w h^3} \frac{\sinh E_r/kT}{E_r/kT} \tag{2.101}$$

where I_1, I_2, and I_3 are the moments of inertia of the water molecules about three mutually perpendicular axes and E_r is the ion–water interaction energy. The symbol σ'_w is the symmetry factor and is equal to 2 for water. Therefore (see physicochemical texts)

$$S_{i,\text{lib}} = R\left[\frac{\ln 8\pi^2(8\pi^3 I_1 I_2 I_3 k^3 T^3)^{1/3}}{\sigma'_w h^3} + \ln \sinh(E_r/kT) - E_r/kT + \coth(E_r/kT) + 5/2\right] \tag{2.102}$$

In the case of water molecules oriented near the ion, $E_r >> kT$ and Eq. (2.102) becomes

$$S_{i,\text{lib}} = R\left[\ln\frac{8\pi^2(8\pi^3 I_1 I_2 I_3 k^3 T^3)^{1/3}}{\sigma'_w h^3} - \ln E_r/kT - \ln 2 + 5/2\right] \tag{2.103}$$

Inserting numerical values for I_1, I_2, and I_3, S_{lib} at 298.16 K becomes

$$S_{i,\text{lib}} = R[5.57 - \ln E_r/kT] \tag{2.104}$$

Substituting $\Delta H_{\text{ion-}w}$ from Table 2.17 for E_r, $S_{i,\text{lib}}$ for the SCW molecules near ions can be calculated. The quantity E_r represents the sum of the ion–dipole, ion–quadrupole, and ion–induced dipole interactions.

In the computation of the rotational entropies of SCW and NSCW near an ion, the rotation is restricted to libration about the axis perpendicular to the dipole. The third rotation, i.e., about the dipole axis, does not change the orientation of the dipole and may be better calculated as if it were a free rotation. The partition function for this is

$$f_{\text{rot,free}} = \frac{2\pi}{\sigma'}\left(\frac{2\pi I_w kT}{h^2}\right)^{1/2} \tag{2.105}$$

where I_w is the moment of inertia of water molecules about the dipole axis and is 1.9187×10^{-40} g cm^2. Using this partition function in the general equation for entropy in terms of partition function, one calculates $S_{\text{rot,free}} = 14.35$ J K^{-1} mol^{-1}.

Thus,

$$S_{SCW,\text{rot}} = \frac{2}{3} S_{SCW,\text{lib}} + 3.43 \tag{2.106}$$

and

$$S_{NSCW,\text{rot}} = \frac{2}{3} S_{NSCW,\text{lib}} + 3.43 \tag{2.107}$$

$S_{i,\text{vib}}$. The term $S_{i,\text{vib}}$ can be calculated from

$$f_{i,\text{vib}} = \left(2 \sinh \frac{h\nu}{2kT}\right)^{-1} \tag{2.108}$$

and therefore

$$S_{i,\text{vib}} = R \ln\left(2 \sinh \frac{h\nu}{2kT}\right)^{-1} + R\left(\frac{h\nu}{2kT}\right)\coth\left(\frac{h\nu}{2kT}\right) \tag{2.109}$$

where

$$\nu = \frac{1}{2\pi}\sqrt{k/\bar{\mu}} \tag{2.110}$$

and k is the force constant and $\bar{\mu}$ is the reduced mass.

Now

$$k = \left(\frac{\partial^2 U_r}{\partial x^2}\right)_{r=r_{eq}} \tag{2.111}$$

and

$$U_r = -\frac{ze\mu}{r^2} + \frac{A}{r^9} \tag{2.112}$$

Now, at equilibrium, for the ion–water separation

$$\left(\frac{\partial U}{\partial r}\right)_{r=r_{eq}} = 0 \tag{2.113}$$

It follows that

$$A = \frac{2}{9} ze\mu r_{eq}^7 \tag{2.114}$$

From Eqs. (2.112) and (2.114) and expressing the displacements from the equilibrium separation, r_{eq}, as $(r_{eq} + x)$ and $(r_{eq} - x)$, one gets for k:

$$k = \frac{14ze\mu}{r_{eq}^4} \tag{2.115}$$

The numerical form turns out to be

$$k = \frac{123.7 \times 10^4 z}{r_{eq}^4} \quad \text{(in dyn cm}^{-1} \text{ if } r_{eq} \text{ is in angstroms)} \tag{2.116}$$

Thus, ν can be evaluated by inserting k from Eq. (2.116) into Eq. (2.110). The values of ν, when substituted in Eq. (2.109), give $S_{i,\text{vib}}$ (Table 2.19).

2.15.13.5. S_{NSCW}. The term S_{NSCW} is made up of two contributions, $S_{NSCW,\text{lib}}$ and $S_{NSCW,\text{vib}}$. The librational entropy of the NSCW is obtained from Eqs. (2.104) and (2.107) by inserting the value of $\Delta H_{\text{ion-}NSCW}$ for E_r from row 3 of Table 2.17 (see Table 2.19).

$S_{NSCW,\text{vib}}$ can be calculated from Eq. (2.109) if ν_{NSCW} is known. Waters that are not coordinated solvationally with the ion (NSCW) have as attractive force only an ion–induced dipole component (ΔH_h). Thus, the force constant k_{NSCW} can be worked out by using Eqs. (2.111)–(2.114):

$$k_{NSCW} = \frac{10(ze)^2}{r_{\text{eq}}^6} \tag{2.117}$$

or in numerical form:

$$k_{NSCW} = \frac{332.9 \times 10^4 z^2}{r_{\text{eq}}^6} \tag{2.118}$$

Substitution of k_{NSCW} in Eq. (2.110) gives ν_{NSCW}, which, when used in Eq. (2.109), gives the vibrational contribution to the entropy of an NSCW near ions (Table 2.19).

2.15.13.6. S_{SB} *(Model A)*. In this model, the hydrogen-bonded molecule of the SB region is seen as executing one free rotation about the H-bond axis. The two rotations about the axes perpendicular to the H-bond axis will be librational. $S_{\text{rot-H-bond axis}}$ can be obtained from Eq. (2.109) using a frequency represented by a wave-number of 60 cm^{-1} (or $1.80 \times 10^{12}\ \text{s}^{-1}$) assigned for the rotational band of a one-hydrogen-bonded species. Thus,

$$S_{\text{rot-H-bond-axis}} = 18.74\ \text{J K}^{-1}\ \text{mol}^{-1} \tag{2.119}$$

The entropy corresponding to two librations can be obtained by inserting the librational frequencies, $\nu_{1,\text{lib}}$ and $\nu_{2,\text{lib}}$, in Eq. (2.109). The assignments of these νs, based on the interpretation of the infrared spectrum for two libration frequencies, are 4.35×10^{12} and $5.25 \times 10^{12}\ \text{s}^{-1}$. Substituting these values in Eq. (2.109) gives

$$(S_{1,\text{lib}})_{SB} = 11.46\ \text{J K}^{-1}\ \text{mol}^{-1} \tag{2.120}$$

$$(S_{2,\text{lib}})_{SB} = 10.08\ \text{J K}^{-1}\ \text{mol}^{-1} \tag{2.121}$$

The three vibrations, corresponding to the three translational degrees of freedom, for a molecule involved in a single H bond are assigned frequencies of $1.80 \times 10^{12}\ \text{s}^{-1}$. The entropy for these vibrations follows from Eq. (2.109) as

$$S_{SB,\text{vib}} = 3(10.67) = 32.00\ \text{J K}^{-1}\ \text{mol}^{-1} \tag{2.122}$$

From Eqs. (2.119) to (2.122)

$$S_{SB,\text{H-bonded}} = 18.74 + 11.46 + 10.08 + 32.00 = 72.29\ \text{J K}^{-1}\ \text{mol}^{-1} \tag{2.123}$$

Thus

$$S_{SB}\ (\text{model A}) = n_{SB,\text{H-bonded}}(S_{SB,\text{H-bonded}} - S_{w,w}) \tag{2.124}$$

(Table 2.19).

2.15.13.7. S_{SB} *(Model C).* In this model, the molecules in the SB region, which are not H bonded to the first shell, are taken as *unbonded*, i.e., freely rotating monomers. These molecules are assigned three translational and three rotational degrees of freedom. The intramolecular bond stretching and bending modes are of high frequencies and do not contribute to the entropy of an unbonded molecule.

One has to know the free volume available to the unbonded molecule, $V_{f,SB}$, and its temperature dependence, so that the translational entropy may be calculated. The term $V_{f,SB}$ may be found from sound velocity measurements [Eq. (2.96)]. If one takes a number of unassociated liquids and plots their free volumes V_f obtained from the velocity of sound against their molar weights, an extrapolation through a molar weight of 18 will give $V_{f,SB}$ for a freely rotating monomer of water.[33] A value of 0.2 $cm^3\ mol^{-1}$ for $V_{f,SB}$ is obtained. The temperature dependence of $V_{f,SB}$ is obtained from the integration of the heat capacity

$$(C_v)_{SB} = RT^2\left(\frac{\partial^2 \ln V_{f,SB}}{\partial T^2}\right) \tag{2.125}$$

which gives

$$T\frac{\partial \ln V_{f,SB}}{\partial T} = \frac{(C_v)_{SB}}{R} - \frac{F'}{2} - 1.2 \tag{2.126}$$

where $(C_v)_{SB}$ is the heat capacity of the freely rotating monomers in the SB region, F' is the number of degrees of freedom, and the term -1.2 is the integration constant. $(C_v)_{SB}$ has a value of $(6/2)R$ arising from three translational and three rotational degrees of freedom. Thus, from Eq. (2.126)

$$T\frac{\partial \ln V_{f,SB}}{\partial T} = -1.2 \tag{2.127}$$

Using $V_{f,SB}$ of 0.20 $cm^3\ mol^{-1}$ and $T(\partial \ln V_{f,SB}/\partial T)$ of -1.2 in the general equation relating entropy to partition function,

$$S = Nk\left(\ln f - T\frac{\partial \ln f}{\partial T}\right)$$

[33]This is to be distinguished from the entropy of *water in water* for which, of course, free-volume values are available directly from measurements of sound velocity.

one finds

$$S_{SB,tr} = 29.29 \text{ J K}^{-1} \text{ mol}^{-1} \tag{2.128}$$

Furthermore, $S_{SB,\text{rot}}$ for three degrees of free rotations is calculated with the same general approach to be 43.85 J K^{-1} mol^{-1}. Adding these components together, one obtains a calculated entropy for the structure-broken part of model C, that is,

$$S_{SB,\text{freely rot}} = 29.29 + 43.85 = 73.14 \text{ J K}^{-1} \text{ mol}^{-1} \tag{2.129}$$

Thus, the total entropy for model C is

$$\begin{aligned}\sum \Delta S_{SB} \text{ (model C)} = n_{\text{H-bonded}} &(S_{SB,\text{H-bonded}} - S_{w,w}) \\ &+ (n_{SB} - n_{\text{H-bonded}})(S_{SB,\text{freely rot}} - S_{w,w})\end{aligned} \tag{2.130}$$

The values of ΔS_h of models A and C as well as their comparisons with experimental values are shown in Table 2.20 and in Figs. 2.43 and 2.44. It is necessary to reject model 1C because X-ray determinations of the CN indicate not 4, but numbers that vary from 6 to 8 as a function of the ion. The experimental results on solvation numbers have similar inference; a sharp distinction between the solvation numbers of

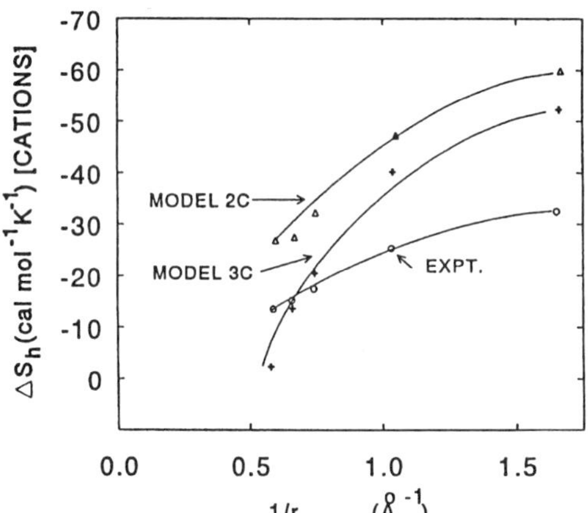

Fig. 2.43. Entropy of hydration against reciprocal of ionic radius for monovalent cations (1 Å = 100 pm; 1 cal = 4.184 J). (Reprinted from J. O'M. Bockris and P. P. S. Saluja, *J. Phys. Chem.* **76**: 2298, 1972.)

TABLE 2.20

Comparison of Calculated and Experimental Values for Entropy of Hydration (calories K^{-1})

Ion	Li^+	Na^+	K^+	Rb^+	Cs^+	F^-	Cl^-	Br^-	I^-
ΔS_h									
Model 1A	−63.42	−60.58	−57.30	−57.38	−56.64	−59.36	−55.76	−56.12	−55.52
Model 2A	−77.78	−71.92	−66.30	−65.32	−67.02	−70.44	−68.48	−67.31	−68.44
Model 3A	−67.10	−62.00	−50.20	−48.30	−36.30	−64.80	−48.90	−47.10	−40.80
Model 1C	−40.0	−29.6	−17.1	−13.3	−7.0	−13.4	−2.6	0.8	7.7
Model 2C	−60.4	−46.9	−32.1	−27.2	−26.4	−35.5	−24.3	−19.4	−17.9
Model 3C	−52.7	−40.0	−20.5	−14.6	−3.2	−31.4	−12.2	−6.6	0.7
ΔS_h (exptl.)[a]	−33.7	−26.2	−17.7	−14.0	−14.0	−31.8	−18.2	−14.5	−9.0

[a]The experimental values refer to $S^\circ_{H^+} = 5.3 \pm 0.3$ e.u.

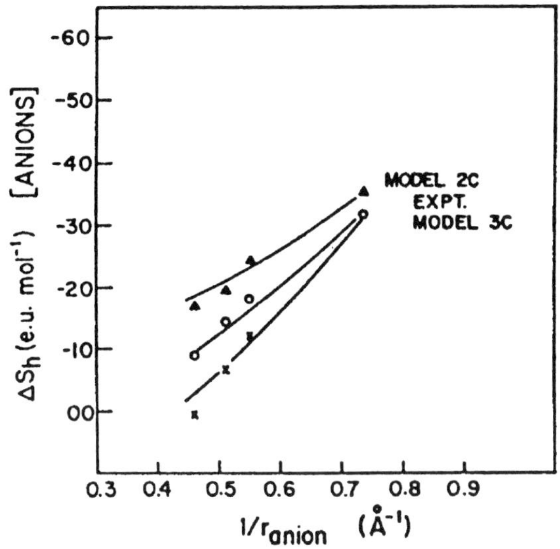

Fig. 2.44. Entropy of hydration against reciprocal of ionic radius for halide ions (1 Å = 100 pm; 1 e.u. = 4.184 J K^{-1} mol^{-1}). (Reprinted from J. O'M. Bockris and P. P. S. Saluja, *J. Phys. Chem.* **76**: 2298, 1972.)

larger and smaller ions is indicated (the SN of Na^+ is 4, and that of I^- is 1.5). Models 2A and 2C turn out, after the calculations have been made, to be the least consistent with the experiments (Figs. 2.43 and 2.44).

Thus, model 3C is the most experimentally consistent model. It is consistent with a model in which there is a difference between coordinated water and solvational water. Some of the waters in the structure-broken region are librating monomers. The entropy choice of 3C is the same as the choice in the heat calculation for cations. However, there is much deviation for the cations (Fig. 2.43) and only the anion model is more consistent with experiments (Figs. 2.43 and 2.44). The model in which two different kinds of coordinating waters in the first shell have been assumed (i.e., a solvational and a nonsolvational coordination number) gives numerically better consistency with experiments than models lacking this feature.

Conclusions for the monovalent ions can be drawn from this fairly detailed analysis. (1) A division of a region around the ion into two parts (Bockris, 1949; Frank and Wen, 1957) is supported. (2) In the first layer around the ion, one can distinguish two kinds of water molecules, referred to as "solvated" and "nonsolvated." (3) The second layer is also one water molecule thick and consists basically of monomers, some of which librate.

2.15.14. Is There a Connection between the Entropy of Solvation and the Heats of Hydration?

It is common in many properties of chemical systems to find that as the ΔH of the property varies in a series of elements, there is a corresponding $T\Delta S$ change that counteracts the ΔH change. The appropriate relations are shown for ionic hydration in Fig. 2.45. Because of the general relation $\Delta G = \Delta H - T\Delta S$, this compensation effect quiets down the variation of ΔG_{solv} among the ions.

An interpretation of this well-known compensation effect has been given by Conway. In this interpretation, ion–dipole attraction is the larger contributor to the ΔH_s values. As it increases (becomes more negative), the vibrational and rotational frequencies will increase and hence the entropy contribution will become more negative. Because $\Delta G = \Delta H - T\Delta S$, the increased negativity of $T\Delta S$ (i.e., a *positive* contribution to ΔG) will compensate for the increasingly negative ΔH_{solv}.

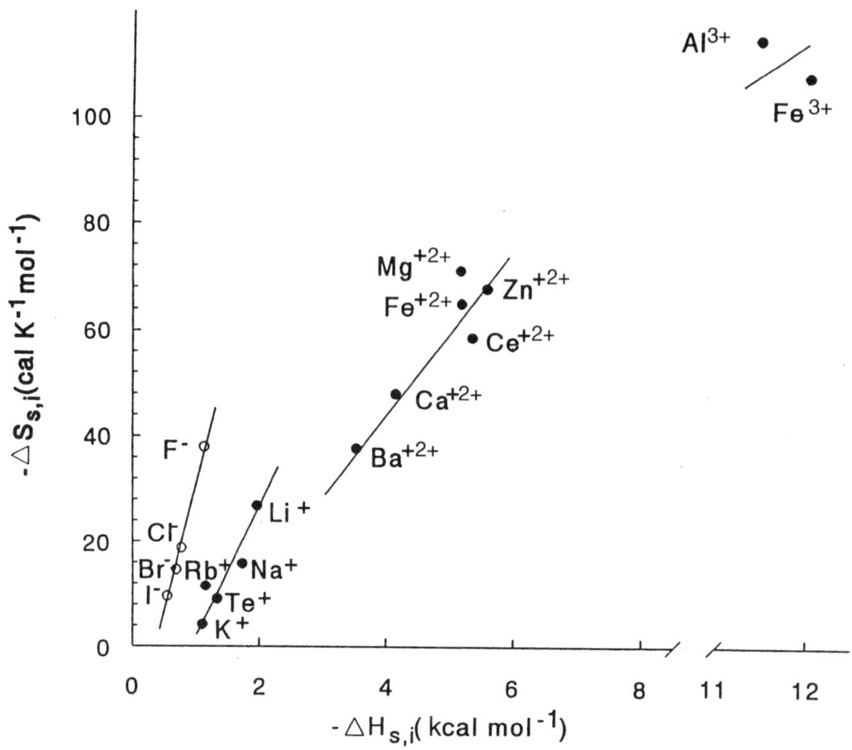

Fig. 2.45. Compensation plot of enthalpies vs. standard entropies of hydration for a number of ions. (Reprinted from D. D. Eley and H. G. Evans, *Trans. Faraday Soc.* **34**: 1093, 1938.)

2.15.15. Krestov's Separation of Ion and Solvent Effects in Ion Hydration

Krestov in 1993 broke down the solvent contribution into an "ordering" one—that is, a negative entropy due to the enhanced order brought about by the ordering of the solvent around the ion—and a disordering entropy (i.e., ΔS positive) caused by the solvent breakdown. Temperature affects the ordering part little, but the disordering contribution diminishes with an increase in temperature because the water is already broken up before entry of the ion.

2.16. MORE ON SOLVATION NUMBERS

2.16.1. Introduction

One of the challenges of solvation studies consists in separating effects among the ions of a salt (e.g., those due to the anion and those due to the cation) and this difficulty, that of determining the *individual* solvation heats (see Section 2.15), invades most methods devoted to the determination of individual ionic properties (Fig. 2.46). When it comes to the solvation number of an ion, an unambiguous determination is even more difficult because not all workers in the field understand the importance of distinguishing the coordination number (the nearest-neighbor first-layer number) from

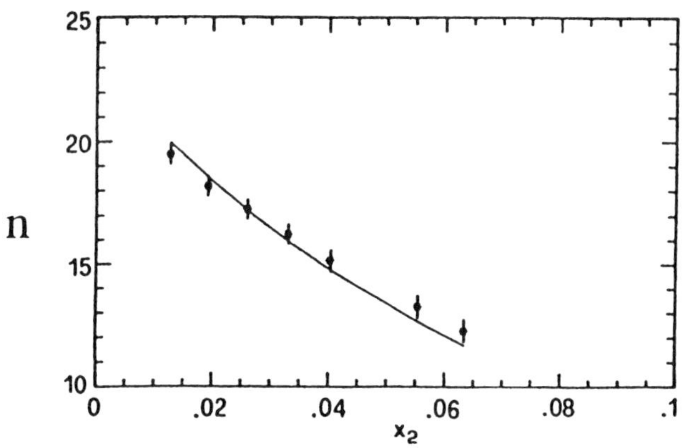

Fig. 2.46. "Hydration number" vs. sodium chloride molar fractions. (•) Experimental data; (–) fitting. (Reprinted from G. Onori and A. Santucci, *J. Chem. Phys.* **93**: 2939, 1990.) The large values arise as a consequence of dropping the assumption of an incompressible hydration sheath.

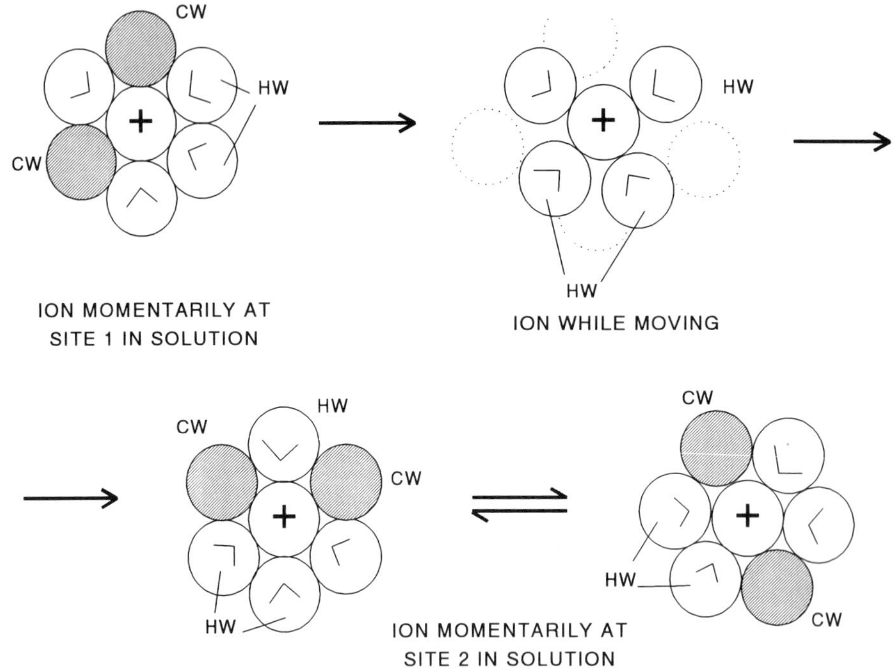

Fig. 2.47. Hydration number *n* in relation to coordination number CN in the motion of an ion in water and in relation to exchange of water molecules. CW, coordinated H_2O; HW, primary hydrating H_2O molecules. (Reprinted from J. O'M. Bockris and P. P. S. Saluja, *J. Phys. Chem.* **76**: 2298, 1972.)

the dynamic solvation number (the number of water molecules that remain with an ion for at least one movement (Fig. 2.47).

Thus, a major misunderstanding is committed by those who confuse solvation numbers with the number of solvent molecules in contact with an ion, the coordination number. It has already been implied and indeed spelled out that the term *solvation number* implies a dynamic concept. Solvation numbers reflect the dynamic situation of the ion as it moves around in the solution. Thus, two hydration numbers may be described, but only one of these is open to numerical determination. This is the so-called *primary hydration number*, that is, the number of water molecules that have lost their own freedom of translational motion and move along with the ion in its random movements in the solution. A *secondary hydration number* refers to the water molecules in the area around the ion that *are affected by* the ion's presence. Clearly, this second quantity depends entirely on the degree of the effect on the solvent molecules outside the first and second layers and hence on the sensitivity of the method being used. It includes waters in the structurally broken-down region out from the first layer of waters attached to the anion.

However, the term *primary solvation number*, although it has apparently a clear definition (see above), is open to further discussion. Thus it may be that a water molecule loses its degree of translational freedom and travels with a given ion as it moves from site to site. The question is, how long does it have to remain the ion's consort to count in the definition? There is no difficulty in accepting "travels with the ion in its movements" when the lifetime of the complex is clearly greater than that needed for registration in some experimental method. Thus, a relatively long lifetime, more than 10^{-6} s, say, would find a strong positive vote on the question of whether such an ion counts primarily as a hydrated molecule. However, it seems better to accept the limit of one jump as the necessary qualification. Thus, if the lifetime of water molecules in contact with an ion is only enough for one jump, it still means that the ion always has water molecules with it on jumping, and the number of these is reasonably taken as the primary hydration (solvation) number.

Another matter concerns the time of "reaction" between a set of water molecules after an ion has just pushed its way into the middle of them. Thus, if the lifetime of molecules in the primary solvation shell is sufficiently short, there must be some jumps in which the ion is bare or at least only minimally clothed. How is the hydration number affected by the time needed for the solvent molecules buried in the solvent layer to break out of that attachment and rotate so that their dipoles are oriented toward the ion to maximize the energy of interaction ($\cos\theta = 1$)?

2.16.2. Dynamic Properties of Water and Their Effect on Hydration Numbers

A critical question one might ask in obtaining a closer concept of the primary hydration number pertains to the average time during which a water molecule stays with an ion during its movements (*cf.* Fig. 2.23). One can begin by repeating yet again the distinction between coordination number and primary hydration water. The coordination number is the number of particles in contact with the ion (independently of whether any of the molecules concerned move with the ion and independently of how they orient in respect to it). It is a matter of geometry, space filling, etc. Now, when an ion moves from one site to the next, not all of the water molecules coordinating it at the beginning of the movement cling to it.

One can see at once that the free volume for coordinating waters is

$$\frac{4}{3}\pi[(r_i + 2r_w)^3 - r_i^3]$$

If water molecules fill this space without regard to directed valences and without accounting for repulsion between head-to-head dipoles, the coordination number could be obtained by dividing the above number by the volume of one water molecule.

For $r_i = 70$, 100, 150, and 200 pm, the calculated coordination numbers obtained on such a basis are shown in Tables 2.21 and 2.22. Real values should be less, because of lateral repulsion of molecules.

TABLE 2.21

Geometrically Calculated Coordination Numbers (Space Filling) for Univalent Ions[a]

Radius of Ion (Å)	Coordination Number (nearest integer)
0.7	9
1.0	11
1.5	15
2.0	20

[a]The values of the coordination number will be less because the molecules repel each other and the space is not filled.

The solvation numbers would be expected to be zero for sufficiently large ions. For these ions, the ion's field is too weak to hold the dipoles and rises toward (but never equals) the coordination number as the ion's size falls; therefore its Coulombic attraction upon the water molecule increases (Fig. 2.23).

2.16.3. A Reconsideration of the Methods for Determining the Primary Hydration Numbers Presented in Section 2.15

Most[34] of the methods employed to yield primary hydration numbers have already been presented (Section 2.15). Thus, one of the simplest methods, which has been applied to organic molecules as well as ions in solution, is that originated by Passynski. As with most methods, it purports to give the total solvation number for the salt, but this should not be counted as a difficulty. More to the point is Onori's criticism that the primary hydration shell is not incompressible. Onori has measured this while dropping Passynski's assumption, but the solvation numbers he gets seem unreasonably large and may be invalidated by his counter assumption, i.e., that solvation is temperature independent.

The mobility method (Section 2.10.1) has the advantage of yielding individual solvation numbers directly (as long as the transport numbers are known). However, this positive point is offset by the fact that the viscosity term used should be the *local* viscosity near the ion, which will be less than the viscosity of the solvent, which is

---

[34]Thus the entropy of solvation (Section 2.15.12) can be used to obtain hydration numbers. Knowing the value of ΔS (Section 2.5.3), it is necessary only to know the entropy change of one water molecule as it transfers from a position in a somewhat broken-up water lattice, where it has librative (and some limited translatory) entropy but ends up, after having been trapped by the ion, with only vibrational entropy. The value assigned to this change is generally 25 J K^{-1} mol^{-1} of water, so that $n_s = \Delta S_s/25$.

As a rough-and-ready estimate, this will do. However, the method is open to further development, particularly as to the broken-down character of water in the neighborhood of the ion and how this affects the value of 25 J K^{-1} mol^{-1} per water molecule.

TABLE 2.22

Comparison of Theoretically and Experimentally Obtained Solvation Numbers

Method	Be^{2+}	Mg^{2+}	Ca^{2+}	Sr^{2+}	Ba^{2+}	Zn^{2+}	Cd^{2+}	Fe^{2+}	Co^{2+}	Ni^{2+}	Sn^{2+}	Pb^{2+}	Cu^{2+}
Theoretical	15.6	13.5	12.3	11.4	11	13.3	12.4	13.1	13.3	13.5	12.9	11.3	13.4
average of: Mobility, Entropy, Compressibility		14	12.2	10.8	9	12.3	12	12.2					12.2

Source: Reprinted from B. E. Conway, *Ionic Hydration in Chemistry and Biophysics*, Elsevier, New York, 1981.

usually used in the calculation. It is interesting to note that the use of raw solvent viscosities does, however, give values that agree (±30%) with the mean of the other five or so nonspectroscopic approaches.

The use of hydration numbers calculated from the effect of ions on the dielectric constant of ionic solutions was seen (Section 2.12.1) at first to be relatively free of difficulties. However, the theory has become more sophisticated since the original conception, and it has been realized that in all but quite dilute solutions, interionic forces affect a straightforward interpretation of the relaxation time, making it important to have sets of data over a wide range of frequencies—1 mHz to 1 GHz. The methods of Yeager and Zana and Bockris and Saluja (Section 2.8.1), which use compressibility measurements to obtain the sum, and the ionic vibration potentials to get the difference, represent an approach to determining hydration number that has the least number of reservations (although there are questions about the degree of residual compressibility of the inner sheath).

Finally, there is the troubling matter that the spectroscopic methods of measurement generally gave results as much as 50% lower (Table 2.10) than the values obtained by the relatively concordant nonspectroscopic methods (compressibility,

TABLE 2.23

Hydration Numbers for Some Alkali Metal and Halide Ions Obtained from MD Calculations and X-Ray and Neutron-Diffraction Experiments[a]

Ion	Solute Salt	Hydration Number	
		MD (conc.)	Diffraction
Li^+	LiCl	5.7	4 ± 1
	LiBr		4 and 6
	LiI	7.1	
Na^+	NaCl	6.6	4
K^+	KF	5.3	2–4
	KCl	6	2 and 4
	KI	1.7–3.2	0.5–5
F^-	CsF	6.3	
	NH_4F		4.5
Cl^-	HCl		4
	LiCl	7.4	Various values in the range 6–9
	NaCl	6.7	6
		7.7 (0.55 M)	6.3
I^-	LiI	7.3	8.8

Source: Reprinted from B. E. Conway, *Ionic Hydration in Chemistry and Biophysics*, Elsevier, New York, 1981.

[a]It seems likely that although these values are called hydration numbers, they are, in fact, coordination numbers.

TABLE 2.24

Summary of Primary Hydration Numbers for Alkali Metal and Halide Ions

	Li^+	Na^+	K^+	Rb^+	F^-	Cl^-	Br^-	I
n[a]	5 ± 1	5 ± 1	4 ± 2	3 ± 1	4 ± 1	1 ± 1	1 ± 1	1 ± 1
No. of methods	5	5	4	4	3	3	3	2
n(calc.)	6	5	3	2	5	3	2	0

Source: Reprinted from B. E. Conway, *Ionic Hydration in Chemistry and Biophysics*, Elsevier, New York, 1981.
[a]Based on interpretation of various experimental ionic properties discussed in text.

activity, entropy, mobility, partial ionic volume, vibration potentials, and dielectric constant). There seem to be two interpretations for this marked discrepancy. The first is that the spectroscopic methods are relatively insensitive and for this reason are only applicable to concentrated solutions. However, here the number of water molecules available per ion markedly decreases (5×10^4 water molecules in a 10^{-3} *M* solution and 10 water molecules in a 5 *M* solution) so that there would be a mass tendency toward lower hydration numbers (see Fig. 2.46). Apart from this, in solutions as concentrated as those used, e.g., in neutron diffraction (> 1.5 mol dm^{-3} for 2:1 salts such as $NiCl_2$), the situation becomes complicated for two reasons: (1) the formation of various kinds of ion pairs and triplets and (2) the fact that so much of the water available is part of the hydration sheaths that are the object of investigation. Thus, ionic concentration, which refers conceptually to the number of *free* waters in which the hydrated ion can move, has a different meaning from that when the number of water molecules that are tied up is negligible.

Conversely, spectroscopic methods (particularly NMR and neutron diffraction) can be used to sense the residence time of the water molecules within the solvent sheaths around the ion. Thus, they could offer the most important data still required—a clean quantitative determination of the number of molecules that move with the ion. Unfortunately they only work in concentration regions far higher than those of the other methods. A summary of results from these methods is given in Tables 2.23 and 2.24.

2.16.4. Why Do Hydration Heats of Transition-Metal Ions Vary Irregularly with Atomic Number?

The theoretical discussion of the heats of ion–solvent interactions has been restricted so far to stressing the alkali metal and alkaline earth cations and halide anions. For these ions, a purely electrostatic theory (Section 2.15.10) provides fair coincidence with experiments. However, with the two- and three-valent transition-metal ions, where directed orbital interactions with water may have more influence,

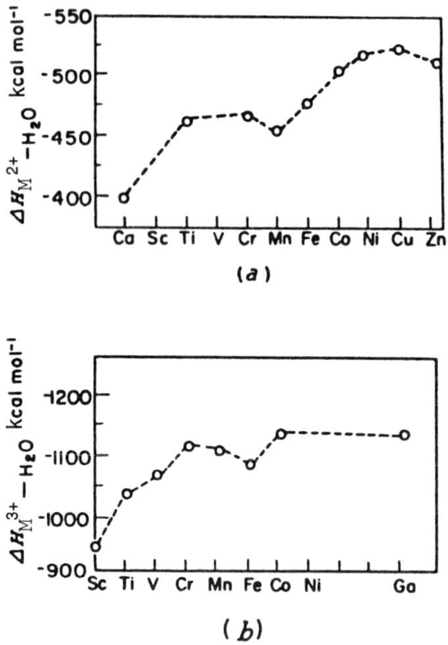

Fig. 2.48. Plot of the heats of hydration of transition-metal ions (and their immediate neighbors) vs. atomic number. (a) Divalent ions; (b) trivalent ions (1 cal = 4.184 J).

interesting and unexpected changes of the hydration heats occur with change of atomic number.

One way of seeing these changes is to plot the experimental heats of hydration of the transition-metal ions against their atomic number. It is seen that in the case of both divalent and trivalent ions, the heats of hydration lie on double-humped curves (Fig. 2.48).

Now, if the transition-metal ions had spherical charge distributions, then one would expect that with increasing atomic number there would be a decreasing ionic radius[35] and thus a smooth and monotonic increase of the heat of hydration as the atomic number increases. The double-humped curve implies therefore the operation

---

[35]The radius of an ion is determined mainly by the principal quantum number and the effective nuclear charge. As the atomic number increases in the transition-metal series, the principal quantum number remains the same, but the effective charge of the valence electrons increases; hence, the ionic radius should decrease smoothly with an increase in atomic number.

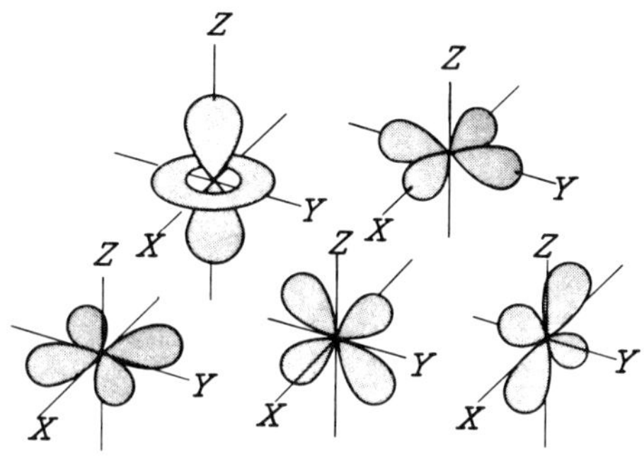

Fig. 2.49. The five 3d orbitals.

of factors that make transition-metal ions deviate from the behavior of charged spheres. What are these factors?

In the case of transition-metal ions, the 3d orbitals are not spherically symmetrical; in fact, they are as shown in Fig. 2.49. In a gaseous ion (i.e., a free unhydrated ion), all the 3d orbitals are equally likely to be occupied because they all correspond to the same energy. Now, consider what happens when the ion becomes hydrated by six[36] water molecules situating themselves at the corners of an octahedron enveloping the ion. The lone electron pairs of the oxygen atoms (of the water molecules) exert a repulsive force on the valence electrons of the ion (Fig. 2.50).

This repulsive force acts to the same extent on all the p orbitals, as may be seen from Fig. 2.51. The d orbitals, however, can be classified into two types: (1) those that are directed along the x, y, and z axes, which are known as the d_ε orbitals, and (2) those that are directed between the axes, which are known as the d_γ orbitals. It is clear (Fig. 2.52) that the repulsive field of the lone electron pairs of the oxygen atoms acts more strongly on the d_ε orbitals than on the d_γ orbitals. Thus, under the electrical influence of the water molecules of the primary solvation sheath, all the 3d orbitals do not correspond to the same energy. They are differentiated into two groups, the d_ε orbitals corresponding to a higher energy and the d_γ orbitals corresponding to a lower energy. This splitting of the 3d orbitals into two groups (with differing energy levels) affects the heat of hydration and makes it deviate from the values expected on the basis of the theory developed earlier in this chapter, which neglected interactions of the water molecules with the electron orbitals in the ion.

Thus, consider a free vanadium ion V^{2+} and a hydrated vanadium ion. In the case of the free ion, all the five 3d orbitals (the two d_ε and the three d_γ orbitals) are equally likely to be occupied by the three 3d electrons of vanadium. This is because in the free ion, all five 3d orbitals correspond to the same energy. In the hydrated V^{2+} ion,

---

[36]The figure of six, rather than four, is used because of the experimental evidence that transition-metal ions undergo six coordination in the first shell. Correspondingly, the hydration numbers are 10–15.

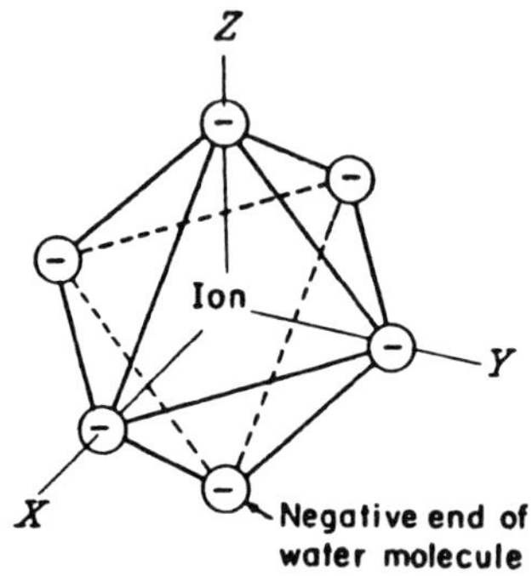

Fig. 2.50. Schematic diagram to show that the valence electrons of the positive ion are subject to the repulsion of the negative ends of the octahedrally coordinating water molecules. The negative charge arises from the presence of lone electron pairs on the oxygen atoms of the water molecules.

however, the d_γ orbitals, corresponding to a lower energy, are more likely to be occupied than the d_ε orbitals. This implies that the mean energy of the ion is less when the ion is subject to the electrical field of the solvent sheath than when it is free. Thus, the change in the mean occupancy of the various 3d orbitals, arising from the electrical

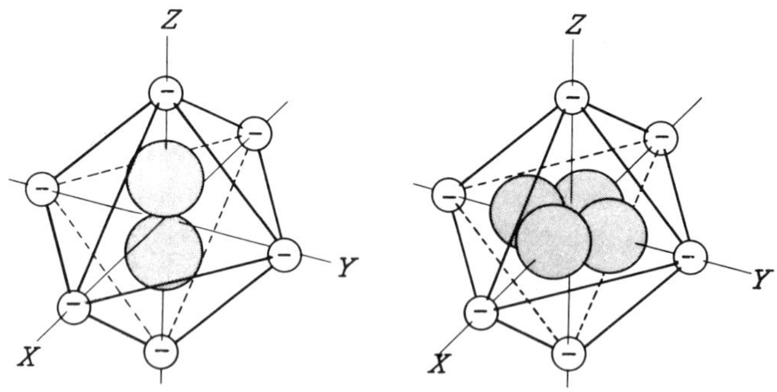

Fig. 2.51. Schematic diagram that shows that the three p orbitals (directed along the axes) are equally affected by the repulsive field of octahedrally coordinating water molecules.

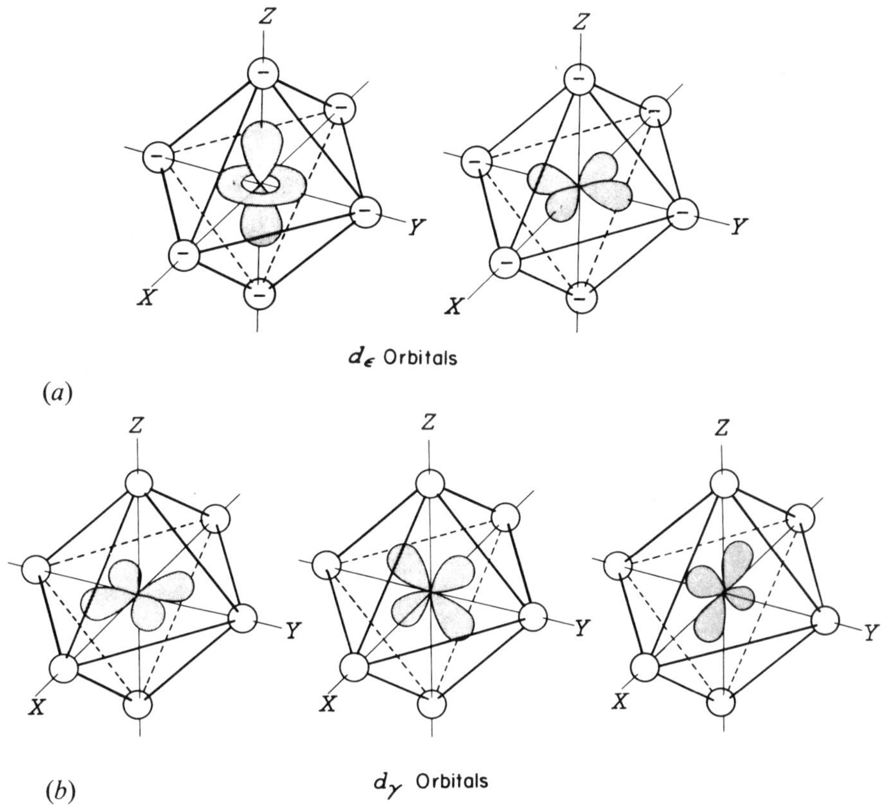

Fig. 2.52. Because the d_ε orbitals (a) are directed along the axes and toward the negative ends of the water molecules, they correspond to a higher energy than the d_γ orbitals (b), which are directed between the axes.

field of the water molecules coordinating the ion, has conferred an extra field stabilization (lowering of energy) on the ion–water system and because of this, the heat of hydration is made more negative.

In the case of the hydrated divalent manganese ion, however, its five 3d electrons are distributed[37] among the five 3d orbitals, and the decrease in energy of three electrons in the d_γ orbitals is compensated for by the increase in energy of the two electrons in the d_ε orbitals. Thus, the mean energy of the ion in the hydrated state is

---

[37]The five electrons tend to occupy five different orbitals for the following reason: In the absence of the energy required for electrons with opposite spins to pair up, electrons with parallel spins tend to occupy different oribtals because, according to the Pauli principle, two electrons with parallel spins cannot occupy the same orbital.

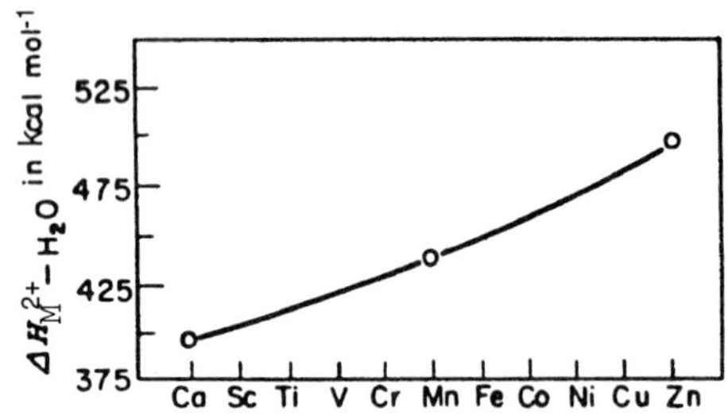

Fig. 2.53. A plot of the heat of hydration of Ca^{2+}, Mn^{2+}, and Zn^{2+} vs. atomic number.

the same as that calculated from a model that neglects interactions affecting the filling of orbitals. Similarly, for Ca^{2+} with no 3d electrons and Zn^{2+} with a completely filled 3d shell, the heat of hydration does not become more negative than would be expected from the electrostatic theory of ion–solvent interactions developed in Section 2.4.3. It can be concluded, therefore, that the experimental heats of hydration of these three ions should vary in a monotonic manner with atomic number as indeed they do (Fig. 2.53).

All the other transition-metal ions, however, should have contributions to their heats of hydration from the field stabilization energy produced by the effect of the field of the water molecules on the electrons in the 3d orbitals. It is these contributions that produce the double-humped curve of Fig. 2.54. If, however, for each ion, the energy[38] corresponding to the water-field stabilization is subtracted from the experimental heat of hydration, then the resulting value should lie on the same smooth curve yielded by plotting the heats of hydration of Ca^{2+}, Mn^{2+}, and Zn^{2+} versus atomic number. This reasoning is found to be true (Fig. 2.54).

The argument presented here has been for divalent ions, but it is equally valid (Fig. 2.55) for trivalent ions. Here, it is Sc^{3+}, Fe^{3+}, and Ga^{3+} which are similar to manganese in that they do not acquire any extra stabilization energy from the field of the water molecules acting on the distribution of electrons in their d levels.

Thus, it is the contribution of the *water-field stabilization energy* to the heat of hydration that is the special feature distinguishing transition-metal ions from the alkali-metal, alkaline-earth-metal, and halide ions in their interactions with the solvent.

This seems quite satisfying, but interesting (and apparently anomalous) results have been observed by Marinelli and Squire and others concerning the energy of interaction of successive molecules as the hydration shell is built up in the gas phase. Thus, it would be expected that the first hydrating water would have the greatest heat of binding, because there are no other molecules present in the hydration shell with

[38]This energy can be obtained spectroscopically.

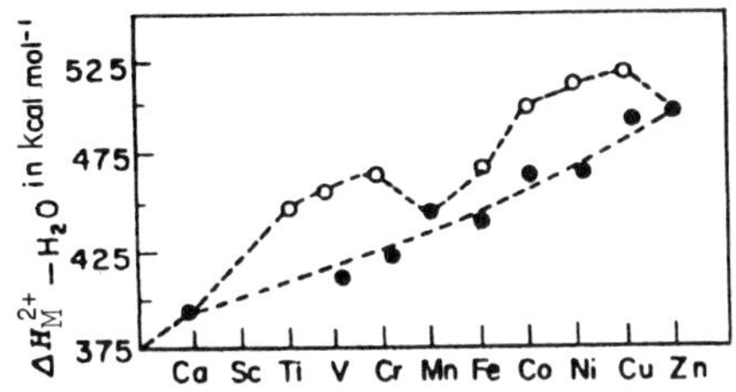

Fig. 2.54. A plot of the heat of hydration of divalent transition-metal ions vs. atomic number (○, experimental values; ●, values after subtracting water-field stabilization energy) (1 cal = 4.184 J).

which to experience lateral repulsion. On the other hand, it is found that with some transition-metal ions (e.g., V, Cr, Fe, and Co), the second water bonds with greater strength than the first!

Rosi and Bauschlicher have made detailed molecular-orbital calculations of the interaction of successive water molecules with transition-metal ions to interpret this anomaly. Their quantum-chemical calculations are able to reproduce the anomalous heats. Depending upon the ion, it is found (in agreement with experiments) that the binding strength of the second hydration water is greater than that of the first.

The anomalous results (the binding energy of the second water being greater than that of the first) can be explained even though the binding energy of hydration water in transition-metal ions is still largely electrostatic. The essential cause is changes in the occupancy of the metal-ion orbitals as a result of differences in repulsion between neighboring waters.

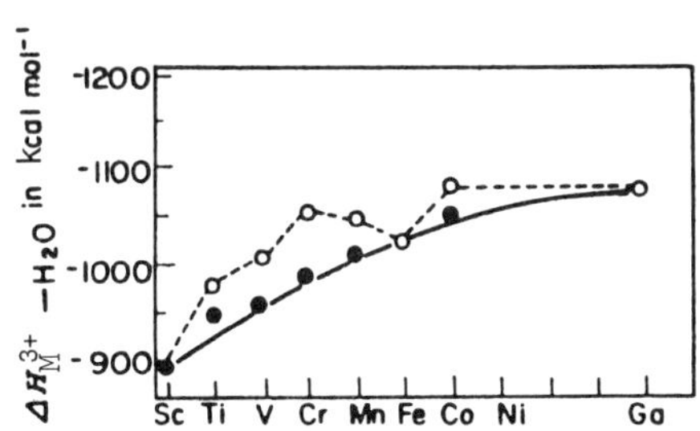

Fig 2.55. A plot of the heat of hydration for trivalent ions.

The key to understanding this surprising result is the interplay between the 4p and the 4s orbitals. The latter can mix in 4p character with its shoelace shape. Such an orbital (effectively an electron cloud) reduces the repulsion between water molecules 1 and 2. Such a reduction allows the bonding energy of the second water to be greater than that of the first. The degree of this mixing of 4p character into 4s (and the resulting effect on bonding) depends on the ion. If the effect is small, repulsion between water molecules 1 and 2 remains great enough for the second water to have a *lesser* bonding energy than the first, as would ordinarily be expected.

Further Reading

Seminal

1. D. D. Eley and M. G. Evans, "Statistical Mechanics in Ionic Solutions," *Trans. Faraday Soc.* **34**: 1093 (1938).
2. S. Lee and M. Tai, "Individual Ionic Entropy of the Proton," *J. Chinese Chem. Soc.* **9**: 60 (1941).
3. F. H. Halliwell and N. C. Nyburg, "Hydration Heat of the Proton," *Trans. Faraday Soc.* **59**: 1126 (1963).
4. J. O'M. Bockris and P. P. S. Saluja, "Model-Based Calculations of Hydration Energies," *J. Phys. Chem.* **76**: 2295 (1972).
5. B. E. Conway, "Individual Ionic Properties," *J. Solution Chem.* **7**: 721 (1978).
6. P. Kebarle and E. W. Godbole, "Hydration in the Gas Phase," *J. Chem. Phys.* **39**: 1131 (1983).

Reviews

1. R. R. Dogonadze, A. A. Kornyshev, and J. Ulstrup, "Theoretical Approaches to Solvation," in *The Chemical Physics of Solvation*, R. R. Dogonadze, E. Kalman, A. A. Kornyshev, and J. Ulstrup, eds., Part A, Elsevier, New York (1985).
2. G. A. Krestov, "Individual Ionic Properties," in *Thermodynamic Structure of Solvation*, Ellis Harwood, New York (1990).

Papers

1. R. C. Keese and A. W. Casheman, *J. Am. Chem. Soc.* **14**: 9015 (1989).
2. J. Marinelli and R. R. Squires, *J. Am. Chem. Soc.* **111**: 4101 (1989).
3. M. Rosi and C. W. Bauschlicher, *J. Chem. Phys.* **90**: 7264 (1989).
4. R. S. Drago, D. C. Feris, and N. Wang, *J. Am. Chem. Soc.* **112**: 8953 (1990).
5. A. D. Paynton and S. E. Feller, *J. Electrochem. Soc.* **137**: 183 (1990).
6. A. G. Sharpe, *J. Chem. Ed.* **67**: 309 (1990).
7. S. Golden and T. R. Tuttle, *J. Phys. Chem.* **95**: 4109 (1991).
8. H. Mizuno and M. Sakiyami, *J. Phys. Chem.* **95**: 2536 (1991).
9. T. F. Magnera, D. E. David, and J. Mich, *Chem. Phys. Lett.* **182**: 363 (1991).
10. P. M. Quereschi, S. Kamoonpuri, and M. Igbal, *J. Chem. Ed.* **68**: 109 (1991).

11. H. Ohtaki and T. Rednai, *Chem. Rev.* **93**: 1157 (1993).
12. Y. Marcus, *Faraday Trans.* **89**: 713 (1993).
13. B. E. Conway and D. P. Wilkinson, "Evolution of Single Ion Entropies," *Electrochim. Acta* **38**: 997 (1993).
14. T. Barthel, *J. Mol. Liquids* **65**: 177 (1995).
15. J. M. Alia, H. G. M. Edwards, and J. Moore, *Spectrochim. Acta* **16**: 2039 (1995).
16. P. Dangelo, A. Dinola, M. Mangani, and N. V. Pavel, *J. Chem. Phys.* **104**: 1779 (1996).

2.17. COMPUTER-SIMULATION APPROACHES TO IONIC SOLVATION

2.17.1. General

For about 95% of the history of modern science, since Bacon's work in the seventeenth century, the general idea of how to explain natural phenomena consisted of a clear course: collection of the facts, systemization of them into empirical laws, the invention of a number of alternative and competing intuitive models by which the facts could be qualitatively understood, and, finally, mathematical expression of the more qualitatively successful models to obtain sometimes more and sometimes less numerical agreement with the experimental values. The models that matched best were judged to describe a particular phenomenon better than the other models.

Until the 1960s, one of the difficulties in this approach was the lengthy nature of the calculations involved. Using only mechanical calculators, adequate numerical expression of a model's prediction would often have taken an impractical time.

Electronic calculating machines (the hardware) that can read instructions on how to carry out the calculations (the software) have made it much easier to select models that give the best prediction. However, not only has this technology transformed the possibilities of calculating the consequences of intuitive assumptions (reducing the time needed for calculation from weeks and months to minutes and hours),[39] but it has made possible another approach toward putting experimental results into a theoretical framework. Instead of making up intuitive suppositions as to what might be happening in the system concerned, and seeing how near to reality calculations with competing models can come, the alternative mode is to calculate the forces between the particles concerned and then to use classical mechanics to calculate the properties of the particles without prior assumptions as to what is happening. In physical chemistry, phenomena often result from a series of collisions among particles, and *that* is what can be calculated.

Computational chemistry can be applied to all parts of chemistry, for example, to the design of corrosion inhibitors that are not toxic to marine life. In this section, a

[39]This assumes that a program for the calculation concerned has already been written. If not, it may take an experienced specialist 6 months to write, and cost $50,000 to buy.

brief account will be given of how far this approach has gone in improving our understanding of ionic solvation.

2.17.2. An Early Molecular Dynamics Attempt at Calculating Solvation Number

Palinkas et al. were the first (1972) to calculate the expression:

$$n_s = 4\pi\rho_w \int_{r_i}^{\infty} g_{i-o}(r)\, r^2\, dr \tag{2.131}$$

where ρ_w is the mean density of water molecules, and g_{i-o} is the radial pair distribution function for the pair ion–oxygen. A plot of n_s against r leads to a series of maxima, the first much greater than the second, and the number of waters "under" this peak is the number in the first shell nearest to the ion.

These sophisticated calculations are impressive but they err in representing their results as solvation numbers. They are, rather, coordination numbers and grow larger with an increase in the size of the ion (in contrast to the behavior of the hydration numbers, which decrease as the ion size increases).

2.17.3. Computational Approaches to Ionic Solvation

In considering various computational approaches to solvation, it must first be understood that the ion–water association alone offers a great range of behavior as far as the residence time of water in a hydration shell is concerned. Certain ions form *hydrates* with lifetimes of months. However, for the ions that are nearly always the goal of computation (ions of groups IA and IIA in the Periodic Table and halide ions), the lifetime may be fractions of a nanosecond.

As indicated earlier in this chapter (see Section 2.3), there are three approaches to calculating solvation-related phenomena in solution: Quantum mechanical, Monte Carlo, and molecular dynamics.

The quantum mechanical approach, which at first seems the most fundamental, has major difficulties. It is basically a 0° K approach, neglecting aspects of ordering and entropy. It is suited to dealing with the formation of molecular bonds and reactivity by the formation in terms of electron density maps.[40] However, ionic solutions are *systems* in which order and entropy, its converse, are paramount considerations.

The most fruitful of the three approaches, and the one likely to grow most in the future, is the molecular dynamics approach (Section 2.3.2). Here, a limited system of ions and molecules is considered and the Newtonian mechanics of the movement of

[40]There is a more fundamental difficulty: the great time such calculations take. If they have to deal with more than ten electrons, *ab initio* calculations in quantum mechanics may not be practical.

all the particles in the system is worked out, an impractical task without computers. The foundation of such an approach is knowledge of the intermolecular energy of interaction between a pair of particles. The validity, and particularly the integrity, of the calculations is dependent upon the extent to which the parameters in the equations representing attraction and repulsion can be obtained *independently* of the facts that the computation is to calculate. Thus, if the energy U_r for the interaction of a particle with its surroundings is known, then $\partial U_r/\partial r$ is the force on the particle and hence the acceleration and final velocity can be calculated [e.g., every femtosecond (10^{-15}s)]. With the appropriate use of the equations of statistical mechanics, the properties of a system (particularly the dynamic ones such as diffusion coefficients and the residence times of water molecules) can then be calculated.

In spite of these confident statements, the computation of the properties of ionic solutions is truly difficult. This is partly because of the general limitations of molecular dynamics. Because it is based on classical mechanics, MD cannot deal with situations in which $h\nu/kT > 1$, i.e., quantal situations (e.g., molecular vibrations). Again, MD depends on potential and kinetic energy (as does quantum mechanics), but it does not account for entropy, which is an important characteristic of equilibrium conditions in systems.

Another problem is that long-range Coulombic forces, which are the principal actors in solvation, have to be subjected in practice to a cutoff procedure (thus, they tend to continue to be significant outside the volume of the few hundred particles in the system considered), and the effect of the cutoff on the accuracy of the final calculation is sometimes unclear. For these reasons, much of the computational work on solvation has been carried out with gas-phase clusters, where the essence of the solvational situation is retained but the complexities of liquids are avoided.

2.17.4. Basic Equations Used in Molecular Dynamics Calculations

The basis of MD calculations in solvation is pairwise interaction equations between the ion and the water molecule. The form of these equations depends greatly upon the water molecule model chosen; there are several possibilities.

For example, suppose one can choose a rigid three-point-charge model of water with an internal geometry of 109.47° and 100 pm for the HOH angle and OH distance, respectively. The interaction energy involves a "Lennard-Jones" 6–12 potential for electrostatic interactions between water–water and ion–water pairs, U_{pair}; a nonadditive *polarization energy*, U_{pol}; and a term that includes exchange repulsion for ion–water and water–water pairs, $U_{\text{3-body}}$:

$$U_{\text{total}} = U_{\text{pair}} + U_{\text{pol}} + U_{\text{3-body}} \qquad (2.132)$$

The pair additive potential is

$$U_{\text{pair}} = \sum_i \sum_j \frac{A_{ij}}{\mathbf{r}_{\mathbf{ij}}^{12}} - \frac{C_{ij}}{\mathbf{r}_{\mathbf{ij}}^{6}} + \frac{q_i q_j}{\mathbf{r}_{\mathbf{i,j}}} \tag{2.133}$$

the polarization energy is

$$U_{\text{pol}} = \frac{1}{2} \sum \mu_{\text{ind}} X_i^0 \tag{2.134}$$

and a three-body repulsion term is

$$U_{\text{3-body}} = A \exp(-\beta r_{12}) \exp(-\beta r_{13}) \exp(-\gamma r_{23}) \tag{2.135}$$

where μ_{ind} is the induced dipole moment and X_i^0 is the electrostatic field from the charges. The term $\mathbf{r}_{\mathbf{ij}}$ is the vector from atom j to atom i, and q_j is the charge on atom j. The distances r_{12} and r_{13} are ion–oxygen distances for the trimer, and r_{23} is the oxygen–oxygen distance for the two water molecules in the trimer. Finally, A, C, and γ are empirical constants.

An iterative approach is often taken to solve the equations. Iteration may be continued until the difference in the induced dipole for successive calculations is 0 to 0.1 D. Typically one uses a system of, say, 215 waters for one ion in a cubic cell with an 1860-pm side. The time step is 1 fs. Coulombic interactions may be cut off at as little as 800 pm. Each set of calculations involves computer software (the cost of which may be very high) and various mathematical procedures to solve the equations of motion.

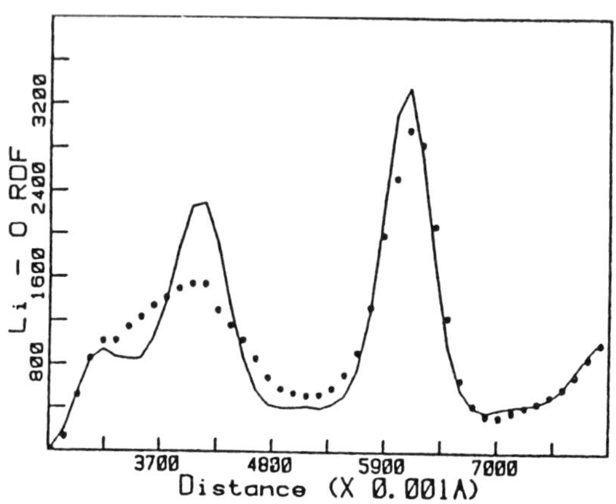

Fig. 2.56. Li-O radial distribution functions (RDFs). The curve represents results from the full-scale MD method. (Reprinted from S. B. Zhu, J. Lee, and G. W. Robinson, *J. Phys. Chem.* **94**: 2114, 1990.)

An example of an Li-O radial distribution function (calculated by MD) is shown in Fig. 2.56. From such calculations one can obtain relaxation features of the rotational and translational motion of the neighboring waters and find the effect of Li^+ or F^-, for example.

What is practical in the later 1990s depends upon whether software able to carry out the computation needed has already been written and is available for about $10,000. It is usually economically impractical for university groups to have a program written for each calculation, so that one finds oneself sometimes carrying out a calculation for which there is the software rather than obtaining the software to carry out the needed calculation.

2.18. COMPUTATION OF ION–WATER CLUSTERS IN THE GAS PHASE

Ion–water clusters have been examined by Dang et al., who calculated the interaction between $Na^+(OH_2)_4$, $Na^+(OH_2)_6$, and $Cl^-(H_2O)_4$. Their orientations and structures at various times in the stimulation are shown in Fig. 2.57.

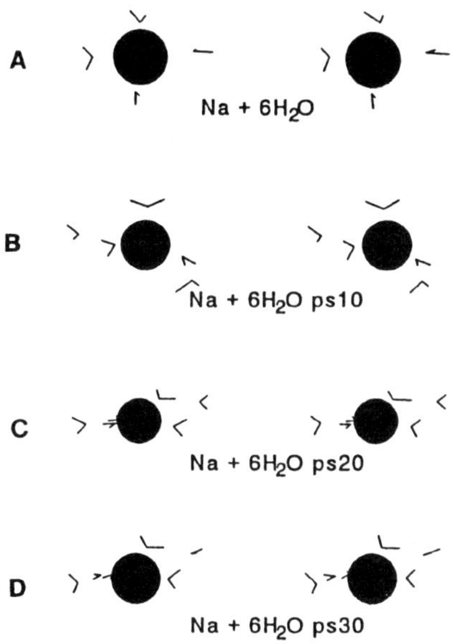

Fig. 2.57. Snapshots of water around the ions of the $Na^+(H_2O)_6$ complex during a simulation of dynamics. (a) 1 ps; (b) 10 ps; (c) 20 ps; (d) 30 ps. (Reprinted from L. X. Dang, J. F. Rice, J. Caldwell, and P. A. Kollman, *J. Am. Chem. Soc.* **113:** 2481, 1991.)

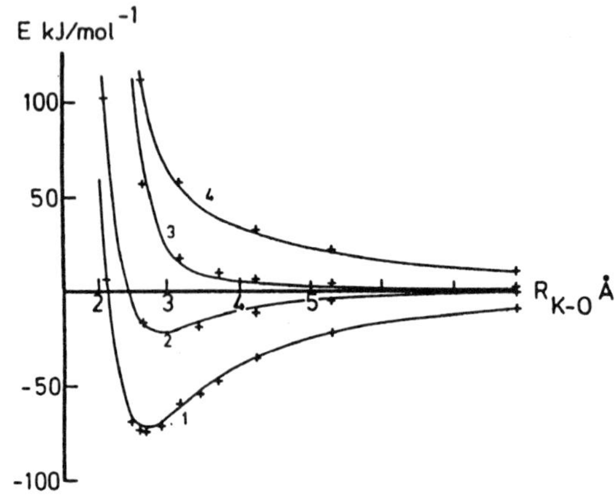

Fig. 2.58. Dependence of $K^+(OH_2)$ interaction energies on the K-O distance for different orientations of the H_2O molecule (1 Å = 100 pm). 1–4 refer to the orientations. (Reprinted from G. G. Malenkov, "Models for the Structure of Hydrated Shells of Simple Ions Based on Crystal Structure Data and Computer Simulation," in *The Chemical Physics of Solvation*, Part A, R. R. Dogonadze, E. Kalman, A. A. Kornyshev, and J. Ulstrup, eds., Elsevier, New York, 1985.)

H–O bonding is important for the Cl^- complex because the Hs are drawn toward the Cl^- and leave the O for H bonding with other waters. The converse is true for $Na^+(OH_2)$ interactions. During simulations, it is found that water molecules transfer from the sheath in contact with the ion to a second sheath in the cluster.

The $Na^+(OH_2)$ turns out to be best fitted with a 4 + 2 structure rather than an octahedral one. It seems likely that the coordination geometries for cluster water in the gas phase and water around the ions in solution differ significantly, but the gas-phase calculations provide an introductory step to the solution ones.

The dependence of the $K^+(OH_2)$ interaction on distance is shown in Fig. 2.58. The ion-O radial distribution functions for $Na^+(H_2O)$ and $K^+(H_2O)$ clusters are shown in Fig. 2.59. A histogram that illustrates the distribution of O–Na–O angles in an $Na(H_2O)_6$ cluster (simulated at 298 K) is shown in Fig. 2.60. Finally, Fig. 2.61 shows the number of water molecules in a sphere of radius r within the cluster.

These diagrams indicate the limit of the hydration shell in the gas-phase ion as the first minimum in the radial distribution function. It is well pronounced for K^+, which has 8 molecules as the calculated coordination number on the cluster; curiously, the sharpness of the definition for Na^+ is less at $N = 6$ (and sometimes 7). The influence

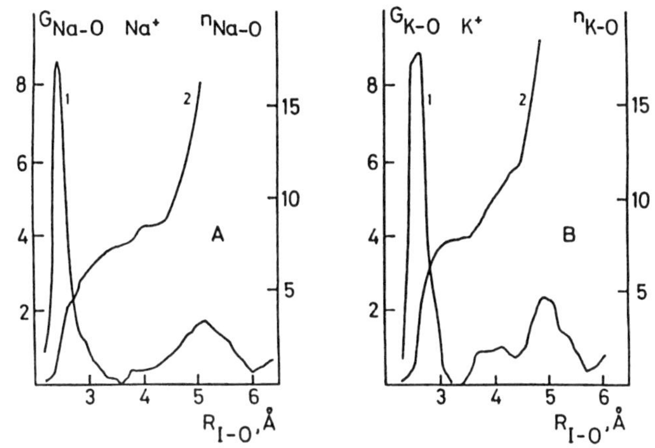

Fig. 2.59. Ion-O radial distribution functions in the ion–$(H_2O)_{199}$ cluster. (a): Na^+, (b) K^+. 1: $G_{T\text{–}O}$ (ordinate to the left). 2: Number of H_2O molecules in the sphere of radius R (ordinate to the right). (Reprinted from G. G. Malenkov, "Models for the Structure of Hydrated Shells of Simple Ions Based on Crystal Structure Data and Computer Simulation," in *The Chemical Physics of Solvation*, Part A, R. R. Dogonadze, E. Kalman, A. A. Kornyshev, and J. Ulstrup, eds., Elsevier, New York, 1985.)

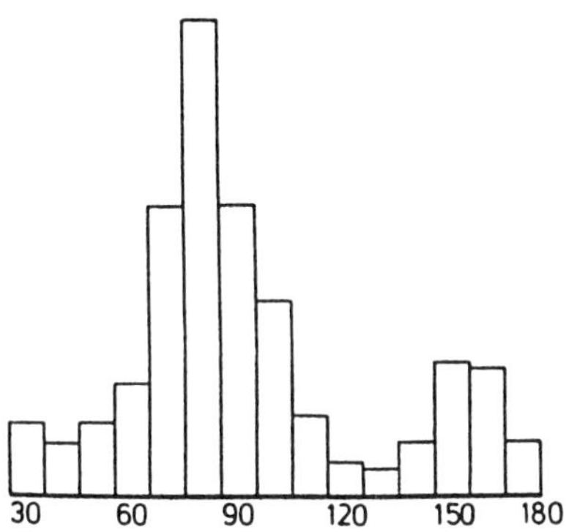

Fig. 2.60. Histogram showing the distribution of O–Na–O angles in a $Na(H_2O)_6$ cluster simulated at 298 K. (Reprinted from G. G. Malenkov, "Models for the Structure of Hydrated Shells of Simple Ions Based on Crystal Structure Data and Computer Simulation," in *The Chemical Physics of Solvation*, Part A, R. R. Dogonadze, E. Kalman, A. A. Kornyshev, and J. Ulstrup, eds., Elsevier, New York, 1985.)

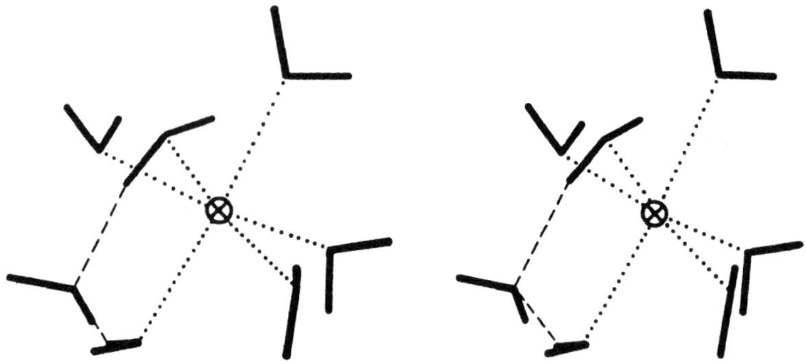

Fig. 2.61. Stereoscopic picture of a "frozen" $Na(H_2O)_7$ cluster. CN = 6. Na–O close contacts (242–245 pm) are shown by dotted lines, hydrogen bonds by dashed lines. (Reprinted from G. G. Malenkov, "Models for the Structure of Hydrated Shells of Simple Ions Based on Crystal Structure Data and Computer Simulation," in *The Chemical Physics of Solvation*, Part A, R. R. Dogonadze, E. Kalman, A. A. Kornyshev, and J. Ulstrup, eds., Elsevier, New York, 1985.)

of the ion on the cluster's structure in these computations becomes negligible at around 700 pm (Fig. 2.62).

If the minimum of the potential corresponds to an Na^+-O separation of 238 pm and a K^+-O separation of 278 pm, then the most probable distances in simulated clusters containing 6 water molecules are about 244 and 284 pm, respectively, at 300 K and about 2 pm less at 5 to 10 K. In Table 2.25, potential parameters that provide such results are given, and the dependence of the ion–water interaction energy on the ion–O distance is shown in Fig. 2.63.

Experimental mass spectrometric data on the hydration of ions in the gas phase that can be compared with calculations of small clusters are available. Full accordance of the computed results with these data is not expected, partly because the aim was to simulate the condensed phase, and the interaction potentials used may not adequately reproduce the properties of small systems in the gas phase at low pressures. However, mass spectrometric data provide reliable experimental information on the hydration of separate ions in the gas phase, and comparison of the results of simulation with these data is an important test of the reliability of the method.

In cluster calculations, an element essential in solution calculations is missing. Thus, intrinsically, gas-phase cluster calculations cannot allow for ionic movement. Such calculations can give rise to average coordination numbers and radial distribution functions, but cannot account for the effect of ions jumping from place to place. Since one important aspect of solvation phenomena is the solvation number (which is intrinsically dependent on ions moving), this is a serious weakness.

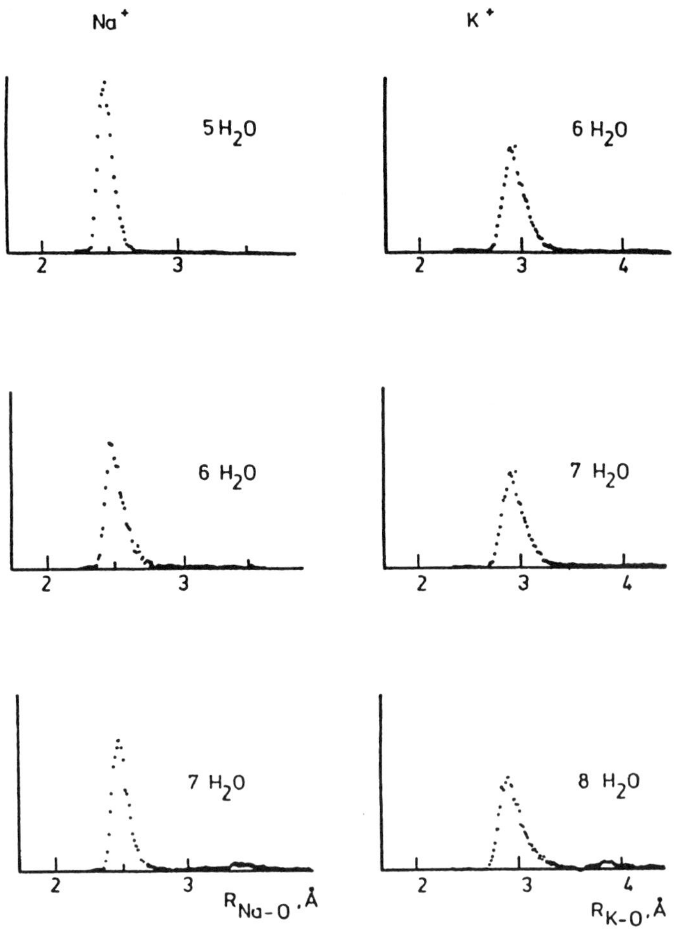

Fig. 2.62. Ion-O radial distribution functions for simulated ion $(H_2O)_n$ clusters (300 K). Ordinate, G_{T-O} in arbitrary units. (Reprinted from G. G. Malenkov, "Models for the Structure of Hydrated Shells of Simple Ions Based on Crystal Structure Data and Computer Simulation," in *The Chemical Physics of Solvation*, Part A, R. R. Dogonadze, E. Kalman, A. A. Kornyshev, and J. Ulstrup, eds., Elsevier, New York, 1985.)

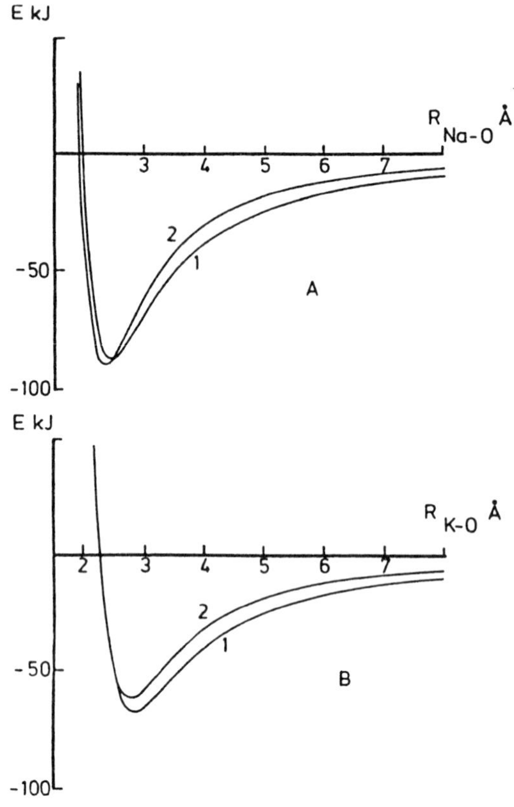

Fig. 2.63. Dependence of ion–H_2O interaction on ion-O distance. 1: Symmetric H_2O molecule orientation (ion-O vector coincides with the water dipole moment direction): 2: lone pair-to-ion orientation. (a) Na^+, (b) K^+ (1 Å = 100 pm). (Reprinted from G. G. Malenkov, "Models for the Structure of Hydrated Shells of Simple Ions Based on Crystal Structure Data and Computer Simulation," in *The Chemical Physics of Solvation*, Part A, R. R. Dogonadze, E. Kalman, A. A. Kornyshev, and J. Ulstrup, eds., Elsevier, New York, 1985.)

TABLE 2.25

Parameters of the Potential Function Used in Equations[a]

$$U_{I\text{-}W} = \sum_{i=1}^{4} \frac{q_i}{r_{O-I}} - \frac{A_{I-O}}{r_{I-O}^{6}} + B_{I-O}\exp(-C_{I-O}r_{I-O}) + \sum_{i=1}^{2} -\frac{A_{I-H}}{r_{I-H}^{6}} + B_{I-H}\exp(-C_{I-H}\times r_{I-H})$$

Interaction				
I	j	A_{i-j}	B_{i-j}	C_{i-j}
O	O	4169	34,330,500	4.33
O	H	159	4940	3.90
Na^+	O[b]	6285	4,400,000	4.18
Na^+	O[c]	6285	9,218,000	4.18
K^+	O[b]	8925	377,100	3.50
K^+	O[c]	12,980	569,000	3.50

[a] Distances in angstroms, energy in kJ/mol.
[b] Adjusted to quantum mechanical calculations.
[c] Improved by taking into account crystallographic data.

2.19. SOLVENT DYNAMIC SIMULATIONS FOR AQUEOUS SOLUTIONS

Of the models of water chosen as practical for MD simulation work, the central force model is the best because the effect of ions on intramolecular frequencies can be studied. Heinzinger and Palinkas first used such a model in 1982 to calculate the ion–water pair potentials as a function of the ion–water distance and orientations as shown in Fig. 2.64. Such computations were carried out using 200 water molecules, 8 cations, and 8 anions in a 2.2 M solution. The side of the cube in the computation was 2000 pm.

The time-average positions for Mg^{2+}, F^-, Cs^+, and I^- can be seen in Fig. 2.64. For Mg^{2+}, the arrangement is octahedral but for F^- there is only a small preference for octahedral coordination. On the other hand, Cs^+ and I^- are firmly octahedral. As one goes outward past 400 pm, the preferential orientation is gone except for Li^+ and this seems to form a second shell. It must be again stressed that the numbers are all time-averaged (coordination) numbers and have only a tenuous relation to the time-dependent hydration numbers.

It is possible to calculate diffusion coefficients by computing the mean square displacement distance and dividing by $6t$. [The basic relation here is the Einstein–Smoluchowski equation (Section 4.2.6)]. The values are surprisingly good and are shown in Table 2.26.

Both transition times, reorientation of water near the ion and translation, can be calculated. The value for the reorientation time of I^- is 5 ± 2 ps; this is a low value because of the weak field in the water arising from the large size of I^-. The hindered

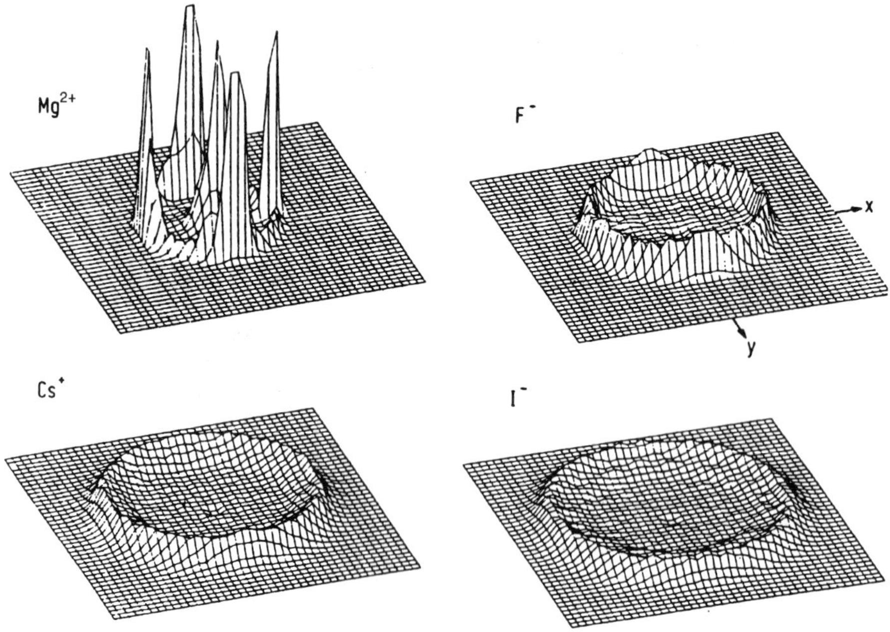

Fig. 2.64. Three-dimensional drawings of the projections of the oxygen atom positions of the six and eight nearest-neighbor water molecules around Mg^{2+} and F^-, and Cs^+ and I^-, respectively, onto the *xy* plane of a coordinate system. The drawings are calculated from MD simulations of 1.1 molal $Mg^{38}Cl_2$ as well as 2.2 molal $Cs^{18}F$ and $Li^{35}I$ solutions. (Reprinted from K. Heinzinger and G. Palinkas, "Computer Simulation of Ion Solvent Systems," in *The Chemical Physics of Solvation*, Part A, R. R. Dogonadze, E. Kalmar, A. A. Kornyshev, and J. Ulstrup, eds., Elsevier, New York, 1985.)

translation can also be calculated and the frequencies of movement are found to be in the range of the librational modes for water. I^- moves faster than Li^+ because it moves almost bare of water. Li^+ drags with it its hydration number waters.

The work of Guardia and Padro in applying MD to solvation is of particular interest because of the attention given to simulations oriented to the actual calculations

TABLE 2.26

Self-Diffusion Coefficients of Ions and Solvent Water (D_w) in a 2.2 molal LiI Solution Obtained from MD Simulation and Experiments at 305 K[a]

	D_w	D_{Li^+}	D_{I^-}
MD	2.48 ± 0.06	0.7 ± 0.3	1.40 ± 0.15
Experimental	2.35	1.0	1.47

Source: From J. A. Padro and E. Guardia, *J. Phys. Chem.* **94**: 2113, 1990.
[a]Units are 10^5 cm^2 s^{-1}.

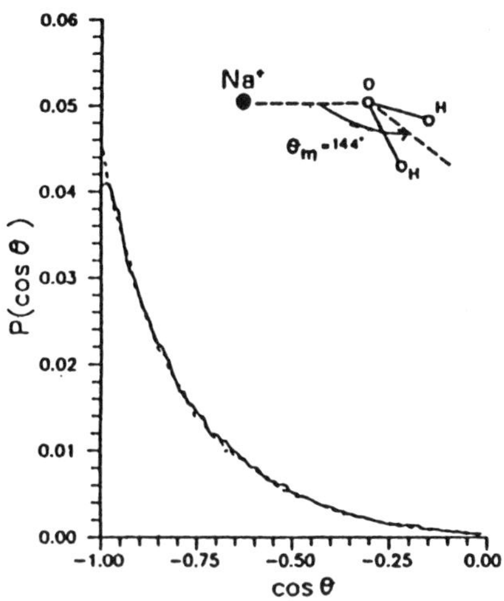

Fig. 2.65. Probability distribution of cos θ for water molecules in the first hydration shell of Na^+ ion. ____, Flexible water, . . ., rigid water. (Reprinted from E. Guardia and J. A. Padro, *J. Phys. Chem.* **94**: 6049, 1990.)

of hydration numbers (and not coordination numbers!) Figure 2.65 shows the flexible water model used.

Among the variations for the water molecule are a rigid and a flexible model, and Guardia and Padro found that significant changes arose with the introduction of flexibility in respect to orientation times. The most important aspect of this work is the computation of the residence times (*cf.* Section 2.11.3).

TABLE 2.27

Residence Times and Hydration Numbers

	Residence Times (picoseconds)		Computed Hydration Numbers	
	Na^+	F^-	Na^+	F^-
Flexible model	38.5	21.4	5.4	5.7
Rigid model	23.8	15.8	5.2	5.4
Computer simulation[a]	9.9	20.3	3.8	4.6
Experimental		5 ± 1	4 ± 1	

Source: From E. Guardia and J. A. Padro, *J. Phys. Chem.* **94**: 2113, 1990.
[a]Data correspond to computer simulations at different temperatures, i.e., 282 K for Na^+ and 278 K for F^-.

The basic idea of calculating a hydration number is to study the number of water molecules in the shell as a function of time. At $t = 0$, the number is the coordination number, and at $t = \tau_{i-w}$, all the waters have been replaced. Then the mean number during this time is taken as the solvation number:

$$n_h = (CN)_{t=0} e^{-\tau_{w-w}/\tau_{i-w}} \tag{2.136}$$

where τ_{w-w} is the residence time for water around water. Although this approach does not take into account the distribution of residence times for the various waters "caught" in different positions when the ions arrive, it does give reasonable values (Table 2.27).

Further Reading

Seminal

1. J. O'M. Bockris and P. P. S. Saluja, "The Time Dependence of Solvation Numbers," *J. Electrochem. Soc.* **119**: 1060 (1972).
2. G. Palinkas, W. O. Riede, and K. Heinzinger, "Calculation of Distribution Functions," *Z. Naturforsch.* **32**: 1137 (1972).
3. F. H. Stillinger and A. Rahman, "A Statistical Mechanical Approach to Ionic Solution," *J. Chem. Phys.* **60**: 1545 (1974).

Papers

1. A. Chandra and B. Bagchi, *J. Phys. Chem.* **93:** 6996 (1989).
2. E. Guardia and J. A. Padro, *J. Phys. Chem.* **94:** 2113 (1990).
3. P. Cieplak and P. Kollman, *J. Chem. Phys.* **92:** 6761 (1990).
4. P. A. Kollman, *J. Am. Chem. Soc.* **113:** 2681 (1991).
5. T. Yabe, S. Sankareman, and J. K. Kochi, *J. Phys. Chem.* **95:** 4147 (1991).
6. H. Yu, B. M. Pettitt, and M. Karplus, *J. Am. Chem. Soc.* **113:** 2425 (1991).
7. R. W. Rick and B. J. Berne, *J. Am. Chem. Soc.* **116**: 3949 (1994).
8. P. J. Rossky, K. P. Johnson, and P. B. Babuena, *J. Phys. Chem.* **100:** 2706 (1996).
9. C. C. Pye, W. Rudolph, and R. A. Pourier, *J. Phys. Chem.* **100:** 601 (1996).
10. T. Z. M. Denti, T. C. Beutler, W. F. Vangunsteeren, and F. Diedrich, *J. Phys. Chem.* **100:** 4256 (1996).

2.20. INTERACTIONS OF IONS WITH NONELECTROLYTES IN SOLUTION

2.20.1. The Problem

The picture that has emerged in this book so far is of ions interacting with a solvent and producing the interesting effects that go under the name "solvation." The solvent structure is not the same after an ion has entered it near the ion. Some of the water molecules are wrenched out of the quasi-lattice and appropriated by the ion as part of its primary solvation sheath. Further off, in the secondary solvation sheaths, the ions produce the telltale effects of structure breaking.

What happens if, in addition to ions and water molecules, molecules of nonelectrolytes are also present in the system? Or what will occur if ions are added to a solution already saturated with nonelectrolyte molecules?

One thing is certain: the fact that the ion takes the water out of circulation for a time means that there will be on average less free water to dissolve the nonelectrolyte. The nonelectrolyte molecules will find themselves suddenly having less water to associate with, and some of them will shun the loneliness imposed by the water's preference for the newly added ions and find it energetically favorable to go back to their parent lattice, i.e., precipitate out. This is the origin of a term from organic chemistry—*salting out.* It means causing a nonelectrolyte to precipitate out of a solution by adding an electrolyte to it to draw off available solvent molecules.

Occasionally, however, the ions are deviants and associate preferentially with the nonelectrolyte solute, shunning the water (hydrophobic effects). In the rare instances where these deviants appear, there is a rapid departure of the nonelectrolyte from the parent lattice and the solubility of the former is *enhanced* rather than decreased. The phenomenon is called *salting in.*

Two aspects of the theory of salting out are considered below. First, the effects of the primary solvation sheath have to be taken into account: how the requisition of water by the ions causes the nonelectrolyte's solubility to decrease. Second, the effects of secondary solvation (interactions outside the solvation sheath) are calculated. The two effects are additive.

2.20.2. Change in Solubility of a Nonelectrolyte Due to Primary Solvation

It is easy to calculate this change of solubility so long as there are data available for the solvation numbers of the electrolyte concerned. Let it be assumed that the normal case holds true and that ions are solvated entirely by water molecules (i.e., the organic present is pushed out). Then recalling that 1 liter of water contains 55.55 moles, the number of water molecules left free to dissolve nonelectrolytes after the addition of ions is $55.55 - c_i n_s$, where n_s is the solvation number of the electrolyte concerned.

Assuming that the solubility of the nonelectrolyte is simply proportional to the number of water molecules outside the hydration sheath, then

$$\frac{S}{S_0} = \frac{55.55 - c_i n_s}{55.55} \tag{2.137}$$

where S_0 is the solubility of the nonelectrolyte before the addition of electrolyte and S that after it.

Assume $n_s = 6$, $c_i = 1\ M$; then from Eq. (2.137), $(S/S_0) = 0.89$. For comparison, with $KClO_4$ as electrolyte, and 2,4-dinitrophenol as nonelectrolyte, the corresponding experimental value is about 0.7. Some further effects must be taken into account.

2.20.3. Change in Solubility Due to Secondary Solvation

It has been stressed that solvation is a far-reaching phenomenon, although only the coordination number and the primary solvation number can be determined. However, there are effects of ions on the properties of solutions that lie outside the radius of the primary hydration sheath. These effects must now be accounted for, insofar as they relate to the solubility of a nonelectrolyte. Let the problem be tackled as though no primary solvation had withdrawn water from the solution. One can write

$$n_{NE,r} = n_{NE,b}\, e^{-AG/RT} \tag{2.138}$$

where $n_{NE,r}$ is the number per unit volume of nonelectrolyte molecules at a distance r from the ion, $n_{NE,b}$ is the same number in the bulk, and the free-energy change ΔG is the work W_r done to remove a mole of water molecules and insert a mole of nonelectrolyte molecules at a distance r from the ion outside the primary solvation sheath.

A first-approximation calculation would be like this. The field X_r of the ion at a distance r falls off with distance according to Coulomb's law[41]

$$X_r = \frac{z_i e_0}{\varepsilon r^2} \tag{2.139}$$

If, now, a dipole (aligned parallel to the ionic field) is moved from infinity, where the field $X_r = 0$, through a distance dr to a point where the corresponding field is dX, the elementary work done is $-\mu\ dX$. Thus, the work to bring a mole of nonelectrolyte molecules from infinity to a distance r is

$$-N_A \int_0^{X_r} \mu_{NE}\, dX$$

and the work done to remove a water molecule to infinity is $N_A \int_0^{X_r} \mu_w\, dX$. The net work of replacing a water molecule by a nonelectrolyte molecule would therefore be given by

[41]Notwithstanding the considerations of Section 2.12.1, the use of the *bulk* dielectric constant of water for dilute solutions of nonelectrolyte is not very inaccurate in the region *outside* the primary solvation sheath. The point is that in this region (i.e., at distances > 500 to 1000 pm from the ion's center), there is negligible structure breaking and therefore a negligible decrease in dielectric constant from the bulk value.

$$W_r = N_A\left(\int_0^{X_r} \mu_w\, dX - \int_0^{X_r} \mu_{NE}\, dX\right) \tag{2.140}$$

Now, the reader will probably be able to see there is a flaw here. Where? The error is easily recognized if one recalls the Debye argument for the average moment of a gas dipole. What is the guarantee that a water dipole far from the ion is aligned parallel to the ionic field? What about the thermal motions that tend to knock dipoles out of alignment? So what matters is the *average* dipole moment of the molecules in the direction of the ionic field. Thus, one has to follow the same line of reasoning as in the treatment of the dielectric constant of a polar liquid and think in terms of the average moment $<\mu>$ of the individual molecule, which will depend in Debye's treatment on the interplay of electrical and thermal forces, and in Kirkwood's treatment also on possible short-range interactions and associations of dipoles. One has therefore

$$W_r = N_A\left(\int_0^{X_r} <\mu_w>\, dX - \int_0^{X_r} <\mu_{NE}>\, dX\right) \tag{2.141}$$

Further, with α now the orientation polarizability,

$$<\mu> = \alpha X \tag{2.142}$$

so that

$$W_r = \frac{N_A(\alpha_w - \alpha_{NE})X_r^2}{2} \tag{2.143}$$

This expression for the work of replacing a water molecule by a nonelectrolyte molecule at a distance r from an ion can now be introduced into Eq. (2.138) to give (in number of nonelectrolyte molecules per unit volume)

$$n_{NE,r} = n_{NE,b} \exp\left[\frac{(\alpha_{NE} - \alpha_w)X_r^2}{2kT}\right] \tag{2.144}$$

The exponent of Eq. (2.144) is easily shown to be less than unity at 298 K for most ions. Thus, for distances outside the primary hydration shell of nearly all ions, the field X will be sufficiently small (because of the large dielectric constant of bulk water), so one can expand the exponential and retain only the first two terms, i.e.,

$$n_{NE,r} = n_{NE,b}\left[1 + \frac{(\alpha_{NE} - \alpha_w)X_r^2}{2kT}\right] \tag{2.145}$$

Thus, the *excess number* per unit volume of nonelectrolyte molecules at a distance r from the ion is (in number of molecules per unit volume)

$$n_{NE,r} - n_{NE,b} = \frac{n_{NE,b}(\alpha_{NE} - \alpha_w)X_r^2}{2kT} \tag{2.146}$$

However, this is only the excess number of nonelectrolyte molecules per unit volume at a distance r from the ion. What is required is the total excess number per unit volume throughout the region outside the primary solvation sheath, i.e., in region 2. One proceeds as follows: the excess number, *not* per unit volume, but in a spherical shell of volume $4\pi r^2\, dr$ around the ion, is

$$(n_{NE,r} - n_{NE,b})4\pi r^2\, dr$$

and therefore the total excess number of nonelectrolyte molecules per ion in region 2 is (with r_h equal to the radius of the primary hydration sheath)

$$n_{NE,2} - n_{NE,b} = \int_{r_h}^{\infty} (n_{NE,r} - n_{NE,b})4\pi r^2 dr = \int_{r_h}^{\infty} \frac{n_{NE,b}(\alpha_{NE} - \alpha_w)X_r^2}{2kT}\, 4\pi r^2 dr \tag{2.147}$$

If one sets $X_r = z_i e_0/(\varepsilon r^2)$, then the excess number of nonelectrolyte molecules caused to be in solution per ion is

$$\frac{4\pi(z_i e_0)^2 n_{NE,b}(\alpha_{NE} - \alpha_w)}{2\varepsilon^2 kT} \int_{r_h}^{\infty} \frac{1}{r^2}\, dr = \frac{4\pi(z_i e_0)^2 n_{NE,b}}{2\varepsilon^2 kTr_h}(\alpha_{NE} - \alpha_w) \tag{2.148}$$

The excess number of nonelectrolyte molecules in a real solution containing c_i moles dm^{-3} of binary electrolyte is given by Eq. (2.148), multiplied by the number of moles per cubic centimeter in the solution, namely, $N_A c_i/1000$. The expression is[42]

$$n_{NE} - n_{NE,b} = \frac{N_A c_i}{1000}\left[\frac{4\pi(z_i e_0)^2 n_{NE,b}(\alpha_{NE} - \alpha_w)}{\varepsilon^2 kTr_h}\right] \tag{2.149}$$

where n_{NE} is the number of nonelectrolyte molecules per cubic centimeter in the solution after addition of the electrolyte.

Hence,

$$\frac{n_{NE} - n_{NE,b}}{n_{NE,b}} = \frac{N_A c_i}{1000}\left[\frac{4\pi(z_i e_0)^2(\alpha_{NE} - \alpha_w)}{\varepsilon^2 kTr_h}\right] \tag{2.150}$$

[42]A factor of 2 has been removed from the denominator of this equation, compared with Eq. (2.148), because there are two ions in the binary electrolyte, each of which is assumed to give the same effect on the solubility.

The terms n_{NE} and $n_{NE,b}$ are numbers of nonelectrolyte molecules per cubic centimeter; $N_A n_{NE} = S$ and $N_A n_{NE,b} = S_0$; these terms were defined in Section 2.20.2. It follows that

$$S = S_0 - \frac{S_0 N_A c_i}{1000}\left[\frac{4\pi(z_i e_0)^2(\alpha_w - \alpha_{NE})}{\varepsilon^2 kTr_h}\right] \tag{2.151}$$

This equation clearly shows the effect of the secondary solvation. It turns out that the orientation polarizabilities α_{NE} and α_w depend on the square of the permanent dipole moments of the molecules. Water has a higher dipole moment than most nonelectrolytes. When $\alpha_w > \alpha_{NE}$, S is less than S_0 and there is salting out. HCN is an example of a substance the dipole moment of which is greater than that of water. (It masquerades as a nonelectrolyte because it is little dissociated in aqueous solution.) Appropriately, HCN is often salted in.

2.20.4. Net Effect on Solubility of Influences from Primary and Secondary Solvation

Equation (2.137) can be written as

$$S_{\text{prim}} = S_0 - K_1 c_i \tag{2.152}$$

Equation (2.151) can be written as

$$S_{\text{sec}} = S_0 - K_2 c_i \tag{2.153}$$

The treatment of the effect of secondary solvation has assumed that the primary solvational effects do not exist. In fact, the secondary solvational effects work on the diminished concentration of nonelectrolyte which arises because of the primary solvation. Hence,

$$S = S_{\text{prim}} - K_2 c_i \tag{2.154}$$

$$= S_0 - K_1 c_i - K_2 c_i \tag{2.155}$$

$$= S_0 - \frac{S_0 c_i n_s}{55.55} - \frac{N_A c_i S_0}{1000}\left[4\pi(z_i e_0)^2 \frac{(\alpha_w - \alpha_{NE})}{\varepsilon^2 kTr_h}\right] \tag{2.156}$$

S as written in Eq. (2.154) has taken into account the primary and secondary solvation and can be identified with the solubility of the nonelectrolyte after addition of ions to the solution. Hence,

$$\frac{S_0 - S}{S_0} = \frac{n_s c_i}{55.55} + \frac{N_A c_i}{1000}\left[\frac{4\pi(z_i e_0)^2(\alpha_w - \alpha_{NE})}{\varepsilon^2 kTr_h}\right] \tag{2.157}$$

or:

$$\Delta S/S_0 = k_1 c_i \tag{2.158}$$

the k_1 is called Setchenow's constant (Fig. 2.66).

There is fair agreement between Eq. (2.158) and experiment. If the nonelectrolyte has a dipole moment less than that of water, it salts out. In the rare cases in which there is a dipole moment in the nonelectrolyte greater than that of water, the nonelectrolyte salts in.

Salting *out* has practical implications. It is part of the electrochemistry of everyday industrial life. One reclaims solvents such as ether from aqueous solutions by salting them out with NaCl. Salting out enters into the production of soaps and the manufacture of dyes. Detergents, emulsion polymerization (rubber), and the concentration of antibiotics and vitamins from aqueous solutions all depend in some part of their

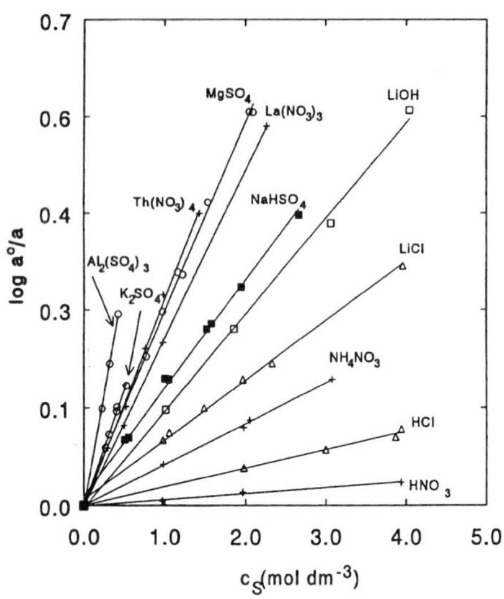

Fig. 2.66. Setchenow plots for oxygen at 310.2 K and 1 atm (101.325 kPa) oxygen partial pressure in aqueous solutions of some representative electrolytes. (Reprinted from W. Lang and R. Zander, *Ind. Eng. Chem. Fundam.* **25**: 775, 1986.)

manufacture upon salting in. However, the salting in with which they are associated is not the rare deviant phenomenon arising when the dipole moment of the nonelectrolyte is greater than that of water. It possesses the characteristic of always being associated with organic electrolytes in which the ions are *large*. Here, even when water has a dipole moment *greater* than that of the nonelectrolyte, which would be expected to give salting out, salting in occurs.

2.20.5. Cause of Anomalous Salting In

The picture given above seems satisfactory as a first approximation and for dilute solutions, but it has this one disturbing feature; namely, there are situations where theory predicts salting out but experiment shows salting in. In such cases, the theory seems to favor ions being surrounded by water whereas in fact they move away from water (hydrophobic effects). Now, it will be recalled that only ion–dipole forces have been reckoned with in treating the interactions among the particles populating the primary solvation shell. Thus, the ion–water and ion–nonelectrolyte forces have been considered to be of the ion–dipole type of directional forces which orient polar particles along the ionic field. Perhaps this restriction is too severe, for there are also nondirectional forces, namely, dispersion forces.

These dispersion forces can be seen classically as follows: The *time-average* picture of an atom may show spherical symmetry because the charge due to the electrons orbiting around the nucleus is smoothed out in time. However, an instantaneous picture of, say, a hydrogen atom would show a proton "here" and an electron "there"—two charges separated by a distance. Thus, every atom has an *instantaneous* dipole moment; of course, the time average of all these oscillating dipole moments is zero.

Now, an instantaneous dipole in one atom will induce an instantaneous dipole in a contiguous atom, and an instantaneous dipole–dipole force will arise. When these forces are averaged over all instantaneous electron configurations of the atoms and thus over time, it is found that the time-averaged result of the interaction is finite, attractive, and nondirectional. Forces between particles that arise in this way are called *dispersion forces*.

Dispersion forces give rise to an interaction energy in which the potential energy of interaction varies as r^{-6}, where r is the distance between the centers of the two substances interacting. Thus, the equation for the dispersive energy of interaction may be written as λ/r^6, where λ is a constant independent of r. The rapid decrease of such forces with increase of distance from the origin makes it unnecessary to consider dispersion interactions outside the primary solvation shell; by then, they have already decreased to an extent that they no longer warrant consideration. Inside the primary hydration sheath, the dispersion interaction can be treated in the same way as the ion–dipole interaction. That is, in the replacement of a water molecule by a nonelectrolyte molecule, one must take into account not only the difference in ion–dipole interactions but also the difference between the dispersion interactions. Thus, one must reconsider the situation in which the ion interacts with water molecules.

The picture given in the last three sections was that the water would usually be the entity to be predominantly attracted to the ion, so that the amount of water available for the dissolution of the nonelectrolyte would be reduced and its solubility consequently would fall. The only exceptions that were recognized for this were those unusual cases in which the nonelectrolyte had a dipole moment greater than that of water. As for the ion, only its radius and charge played a part in the matter; it did not influence the situation in any more structural way, e.g., in terms of its polarizability.

On what molecular features do dispersion forces depend? What relative attractiveness has a given ion on the one hand for a water molecule, and on the other for a nonelectrolyte? Equations for the dispersive force interactions have been worked out for interactions in the gas phase. A simple version of such equations would be

$$U = \frac{h\nu\alpha_{d\text{-ion}}\alpha_{d\text{-molec}}}{2R^6} \tag{2.159}$$

where the ν is the frequency of vibration of the electron in its lowest energy state, the αs are the *distortion polarizabilities* of the entities indicated, and R is their distance apart. It must be noted at once that the polarizability indicated here, the distortion polarizability, differs from that which enters into the equations for the dipole effects, which has simply been termed α. This latter α, which influences the theory of effects of secondary solvation upon salting out and salting in, is due to the orientation of dipoles against the applied field. The distortion polarizability α_d is due to the stretching of the molecule under the influence of a field. The former polarizability α is connected to the dipole moment of the molecule according to a formula such as Eq. (2.142). The α_d is more complexly connected to the size of the molecule. There is a parallelism with the radius and in spherical symmetry cases it is found that α_d approximately follows r^3, where r is the radius of the molecule.

Now, if the size of the nonelectrolyte (hence $\alpha_{d,\text{molec}}$) is greater than that of the water molecule (and for organic nonelectrolytes this is often so), it is clear from the above equation that the dispersive interaction of a given ion is going to be greater with the nonelectrolyte than with the water. This is a reversal of the behavior regarded as usual when only the permanent dipoles of the water and the nonelectrolyte are taken into account (for the dipole moment of water is higher than that of most nonelectrolytes). In view of the above situation it may be asked, why is not salting in (which happens when the nonelectrolyte outcompetes the water molecule in its attraction to the ion) the normal case? In this section, it is the dispersive interactions which have been the center of attention: they have been suddenly considered in isolation. One has to ask, however, whether the dispersive ion–water (and dispersive ion–nonelectrolyte) interactions will dominate over the ion–dipole interactions. If the ion–dipole interactions dominate the dispersive interactions, the considerations of those earlier sections are applicable and salting out is the norm, with salting in a rare exception. When the dispersive interactions predominate, it is the other way around; salting in (and hydrophobic effects) becomes the norm.

What factors of structure tend to make the dispersive forces dominate the situation? Clearly, they will be more likely to have a main influence upon the situation if the nonelectrolyte is large (because then the distortion polarizability is large), but there are many situations where quite large nonelectrolytes are still salted *out*. The dispersive interaction contains the *product* of the polarizability of both the ion and the nonelectrolyte (or the ion and water, depending upon which interaction one is considering), so that it is when *both* the ion *and* the nonelectrolyte are large (hence, both the αs are large) that the dispersive situation is likely to dominate the issue, rather than the ion–dipole interaction.

In accordance with this it is found that if one maintains the nonelectrolyte constant[43] and varies the ion in size, though keeping it of the same type, salting in begins to dominate when the ion size exceeds a certain value. A good example is the case of the ammonium ion and a series of tetraalkylammonium ions with increasing size, i.e., NR_4^+, where R is CH_3, C_2H_5, etc. Here, the salting in begins with the methylammonium ion (its α_d is evidently large enough), the ammonium ion alone giving salting out. The degree of salting in increases with an increase in the size of the tetraalkylammonium cation. Thus, the observations made concerning the salting in of detergents, emulsions, and antibiotics by organic ions are, in principle, verified. An attempt has been made to make these considerations quantitative.

This discussion of the effect of ions upon the solubility of nonelectrolytes is sufficiently complicated to merit a little summary. The field is divided into two parts. The first part concerns systems in which the dispersive interactions are negligible compared with the dipole interactions. Such systems tend to contain relatively small ions acting upon dissolved molecules. Here salting out is the expected phenomenon—the solubility of the nonelectrolytes is decreased—and the reverse phenomenon of salting in occurs only in the rare case in which the nonelectrolyte dipole moment is greater than the dipole moment of the solvent. In the other group of solubility effects caused by ions, the ions concerned tend to be large and because distortion polarizability increases with size, this makes the dispersive activity between these large ions and the nonelectrolyte become attractive and dominate this situation so that the organic molecule is pulled to the ions and the water is pushed out. Then salting in (solubility of the nonelectrolyte increases) becomes the more expected situation.

2.20.6. Hydrophobic Effect in Solvation

Solvation entropies (Section 2.15.13) are negative quantities. Since $\Delta S_{\text{solv}} = S_{i,\text{soln}} - S_{i,\text{gas}}$, this means that the disorder in solution (the entropy) is less than that of the ion

---

[43]For example, it may be benzoic acid, which is considered a nonelectrolyte because it dissociates to a very small degree.

in the gas phase. The interpretation of this is in terms of the structured presence of water molecules around the ion (low entropy). However, there must be another component in the events that make up the measured entropy, for the ion breaks the water structure; i.e., it increases entropy. This is called the "hydrophobic aspect of solvation." There is a large literature on this phenomenon and it can be seen by its effects on several properties of solutions, not only on ΔH_s and ΔS_s, but also on the partial molar volume, specific heat effect, etc.

Among the early discussions of hydrophobic effects were those of Frank et al. They studied the highly negative entropies of hydration of the rare-gas atoms. These might have been expected to give much less negative values because of the absence of tightly ordered hydration shells. To interpret the order indicated by the highly negative entropies, they suggested that when the normal structure of water was broken down by the dissolution of the rare-gas atoms, a new type of water structure—iceberglike groups—was formed. Such groups arise from the breakup of normal water and thus result from hydrophobicity.

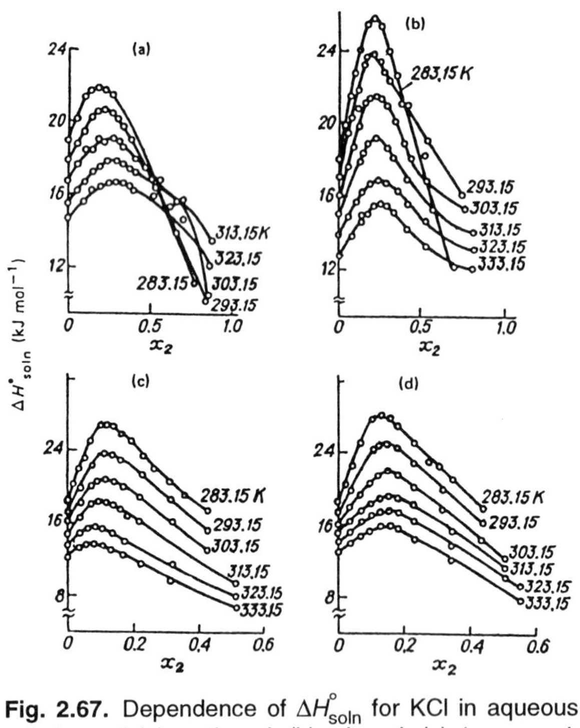

Fig. 2.67. Dependence of $\Delta H^{\circ}_{\text{soln}}$ for KCl in aqueous solutions of (a) methanol, (b) ethanol, (c) 1-propanol, and (d) 2-propanol on composition at different temperatures. (Reprinted from G. A. Krestov, *Thermodynamics of Solvation*, Ellis Harwood, London, 1991.)

Frank and Evans, in studying the numbers for the hydrational entropies of ordinary monatomic ions, found them *insufficiently* negative, indicating that, due to structure breaking, the entropy of the ion itself should be larger than that expected if only the ordering effect of the ion is considered.

An interesting variation of the heat of solution (e.g., for KCl) can be observed in water–alcohol mixtures. The position of the maxima of the curves shifts with increasing temperature along the ordinate in conjunction with the decrease of the endothermicity of the KCl dissolution. This is related to an increasing disruption of the water structure—a hydrophobic effect (Fig. 2.67).

The solubility of noble gases in various solutions (often aqueous–nonaqueous mixtures) gives indications of both *hydrophobic* and *hydrophilic* effects (Fig. 2.68). When substances exhibiting both effects are present, there is a maximum in the solubility of argon. Thus (Fig. 2.68, curve 1) in the system water–acetone, no hydrophilic effects are caused by the added solvent component, and the solubility increases. On the other hand, for systems in which urea is added, there are no hydrophobic effects and the solubility or the gas therefore decreases. In curve 2 of Fig. 2.68, hydrophilic and hydrophobic effects compete (due to the properties of acetamide in water) and there is a maximum on the curve.

Another source of hydrophobic effects arises from solute–solute attraction. The usual effects of interactions between ions of like sign are, of course, repulsive. However, if the ions are sufficiently large, attractive interactions will arise due to

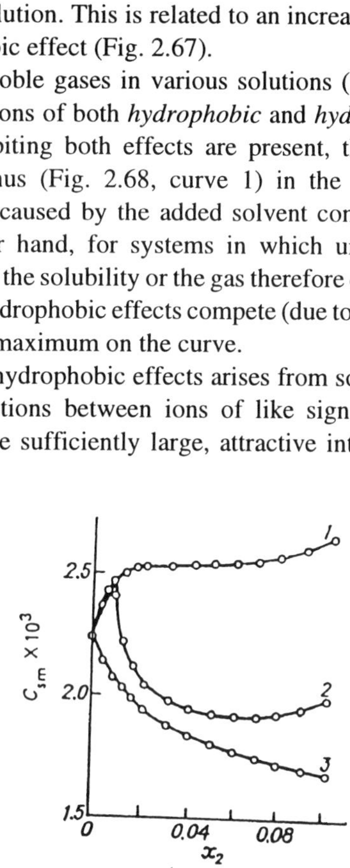

Fig. 2.68. Dependence of solubility of argon in aqueous solutions of (1) acetone, (2) acetamide, and (3) urea (NH_2CON_2H) on composition at 273.15 K. (Reprinted from G. A. Krestov, *Thermodynamics of Solvation*, Ellis Harwood, London, 1991.)

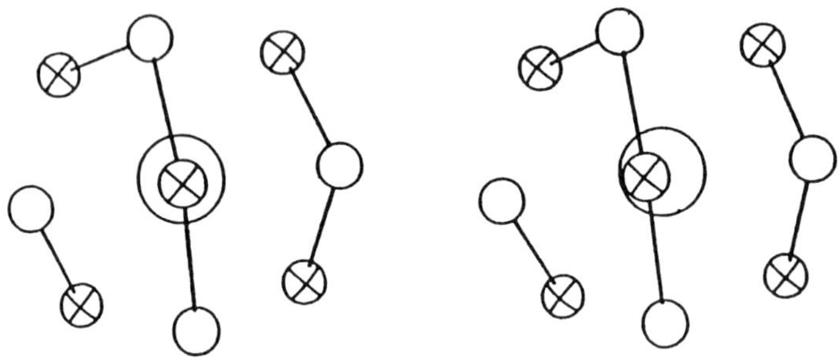

Fig. 2.69. Stereographic view of part of a possible solvation shell. Only a portion of the hydrogen bonds in the solvation shell is shown. Large sphere: nonpolar solute, $\sigma = 320$ pm. Cross-hatched spheres: nearest-neighbor water molecules. Plain spheres: a group of next-nearest-neighbor water molecules located in the solvation shell and surrounding one of the nearest-neighbor molecules. O–O distances range from 290 pm for hydrogen bonds to 320–350 pm otherwise. (Reprinted from E. Grunwald, *J. Am. Chem. Soc.* **108**: 5926, 1986.)

significant dispersion force interaction between the solute particles, and this tends to push water away from the ions; i.e., make it act *hydrophobically*.

These thermodynamic approaches to hydrophobic effects are complemented by spectroscopic studies. Tanabe (1993) has studied the Raman spectra manifested during the rotational diffusion of cyclohexane in water. The values of the diffusion coefficients are approximately half those expected from data for other solvents of the same viscosity, and the interpretations made are in terms of hindered rotation arising from the icebergs presumably formed (*cf.* Frank and Evans) around the cyclohexane.

Correspondingly, NMR studies of the rate at which CH_4 "tumbles" in mixtures of nonaqueous solvents with water show that it is moving approximately ten times faster than the solvent rotation; i.e., it moves independently of the solvent and thus acts hydrophobically.

Simulations of solutions have been used to study hydrophobic effects. Thus, Rossky and Zicki (1994) found that hydration shells of methane and neon remain intact in mixed solvents; this is understandable in terms of clathrate formation—an example of an unusual degree of disordering from the normal structure of water.

Hydrocarbons in water give rise to hydrophobic solvation shells in which the water structure is thoroughly disturbed though still forming a solvation shell around a spherical solute. An example of a calculated situation of this type is shown in Fig. 2.69.

Further Reading

Seminal

1. P. Setchenow, "Salting Out Coefficients," *Z. Phys. Chem.* **4**: 117 (1889).

2. P. Debye and W. McAuley, "Theory of Salting Out," *Z. Phys. Chem.* **26**: 22 (1925).
3. J. O'M. Bockris, J. Bowler-Reed, and J. A. Kitchener, "The Salting In Effect," *Trans. Faraday Soc.* **47**: 184 (1951).
4. R. McDevilt and F. Long, "Salting Out," *Chem. Rev.* **51**: 119 (1952).

Reviews

1. B. E. Conway, *Ionic Hydration in Chemistry and Biology*, pp. 444–465, Elsevier, New York (1981).
2. G. A. Krestov, *Thermodynamics of Solvation*, Ellis Harwood, New York (1991).

Papers

1. W. Lang and R. Zander, *Ind. Eng. Chem. Fundam.* **25**: 775 (1986).
2. J. Butz, P. H. Karpinski, J. Mydiarz, and J. Nyvil, *Ind. Eng. Chem. Prod. Res. Dev.* **25**: 657 (1986).
3. D. C. Leggett, T. F. Jenkins, and P. H. Miyeres, *J. Anal. Chem.* **62**: 1355 (1990).
4. G. Mina-Makarius and K. L. Pinder, *Can. J. Chem. Eng.* **69**: 308 (1991).

2.21. DIELECTRIC BREAKDOWN OF WATER

2.21.1. Phenomenology

Experiment shows that the magnitude of the electric field between two plates depends on the medium contained between them. The quantity *dielectric constant* is used as a measure of the effect of the medium in reducing the field that exists if nothing is there. Water has a particularly large dielectric constant (ca. 78 at 25 °C). This means that if the field between two plates separated by a vacuum amounts to X_0, the field would be reduced to $X_0/78$ if water were used to fill the space between the plates.

Dielectric constants depend little upon the strength of the applied field until extremely high fields are reached—fields greater than 10^6 V cm^{-1}. However, at some critical field strength, a complex phenomenon occurs (Figs. 2.70 and 2.71). It is called *dielectric breakdown*. It can be described in a general way by saying that a dielectric liquid subjected to a sufficiently high electric field suddenly ceases to behave in the customary field-reducing manner. At the same time, a number of characteristic phenomena (e.g., light emission) occur.

What are the phenomena characteristic of dielectric breakdown?[44] The first six in importance are:

---

[44]The phenomenon occurs for solids, liquids, and gases. Because this chapter is concerned with ionic solutions, the material here is limited to liquids and dilute solutions.

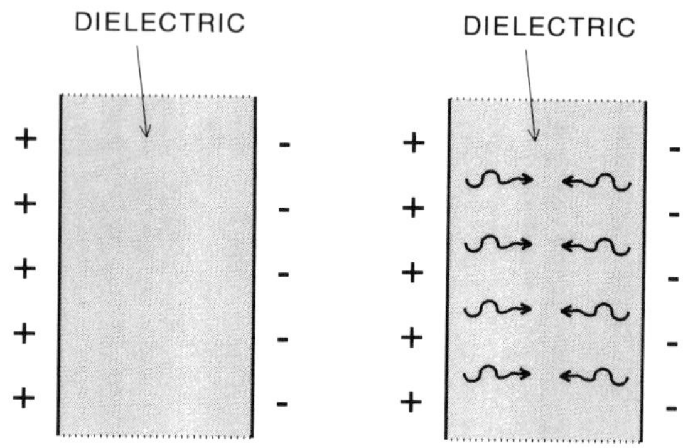

Fig. 2.70. (a) The field between the plates is normal and the water dielectric is chemically stable. Its property as an insulator depends on the concentration of ions but it is highly resistive. (b) When the field applied is sufficiently high, dielectric breakdown occurs; the liquid no longer supports the field and the charges flow away, forming "streamers."

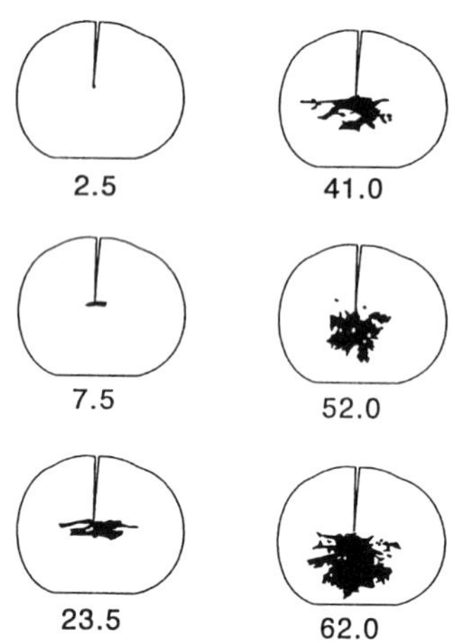

Fig. 2.71. Negative streamer growth in a white naphthenic mineral oil (Marcol 70). Gap: 1.27 cm; voltage step: 185 kV; time in microseconds. (Reprinted from R. E. Tobeozon, *The Liquid State and Its Electrical Properties*, Plenum Press, New York, 1988.)

1. The time between application of an electric field sufficiently high to cause the breakdown phenomena and their occurrence is about 1 μs.
2. Bright, white light is emitted.
3. The most singular and characteristic phenomenon is that called *streamers*. The electrodes concerned emit a piercing series of filaments (Fig. 2.71) that spread out across the solution from the electrode. The filaments turn out to be low-density tubules through the liquid and fade away rapidly after breakdown. These low-density regions in the dielectric remind one of the branchlike growth of trees (Fig. 2.71). The heads of the streamers travel at high velocity, reaching in some systems 1% of the velocity of light.
4. Although older books give tables of the *breakdown voltage* for a series of different liquids, it is not helpful to describe dielectric breakdown phenomena in terms of the voltage applied between two metal plates. The phenomenon depends on the strength of the *electric field* (volts per unit distance), not upon volts. It is also not enough to state a field strength as the volts between two electrodes immersed in the dielectric divided by the distance between the plates. This is because the phenomenon is known to be critically associated with the *interfacial regions* of the cathode and anode concerned. In such regions, however, there are huge discontinuities in field strength. A total of 30,000 V applied over 10 cm may cause dielectric breakdown, but this may not be directly related to the apparent 3000 V cm^{-1} field across the whole liquid between the plates. It may depend upon the 10^6 V cm^{-1} field very near the surface or even the 10^8 V cm^{-1} over a few nanometers, which can be calculated to occur at the tips of the spikelike micropromontories that exist on many real surfaces.
5. An increase in pressure applied to the liquid suppresses the formation of streamers, which are the most telling sign of breakdown.
6. The critical applied volts (and the associated electric field at the interfaces concerned) that cause breakdown depend upon the conductance of the solution (Fig. 2.72).

2.21.2. Mechanistic Thoughts

Dielectric breakdown is a phenomenon of practical importance; it is by no means only an academic puzzle. One can appreciate this through the analogy between mechanical strength and dielectric strength. Any mechanical structure collapses if too great a stress is placed upon it. Correspondingly, any substance can withstand an electric field only to a certain degree of field strength, after which it ceases to remain an insulator and opens the flood gates for electrons to come across; as a result, the resistance to the applied potential undergoes a catastrophic decline. For example, a condenser may be used with a water dielectric to store great amounts of electrical energy to activate electromagnetic weapons. It then becomes important to know, and

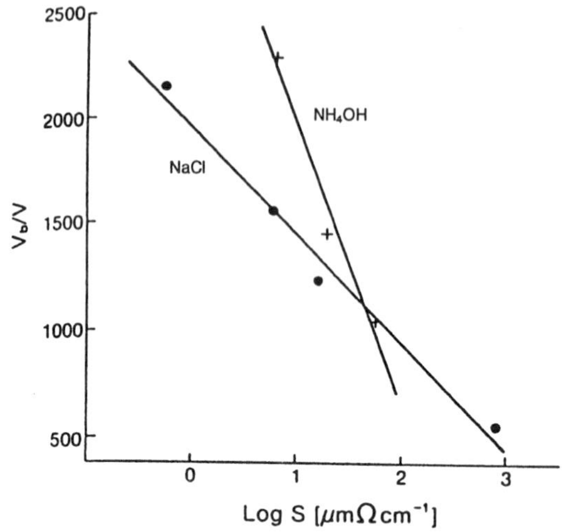

Fig 2.72. The dependence of the breakdown potential (V_b) on the log of conductivity (log S) of NaCl (•) and of NH_4OH (+) solutions. (Reprinted from M. Szklarczyk, R. Kainthla, and J. O'M. Bockris, *J. Electrochem. Soc.* **136**: 2512,

be able to control, the critical field strength, the exceeding of which would break down the dielectric and waste away the electrical energy stored within it.[45]

There is not yet a consensus as to the mechanism of dielectric breakdown in liquids. Research during the last decades of the twentieth century on the breakdown of liquid dielectrics was funded mainly to investigate dielectric breakdown in oils because transformers use oils as dielectrics and sometimes suffer catastrophic dielectric breakdown accompanied by sparking. Insulating oil then catches on fire and equipment costing millions of dollars is damaged beyond repair. Thus, not much laboratory work aimed at determining the mechanism of dielectric breakdown in water *at a molecular level* exists. For example, work in which all is maintained constant except for a systematic change in the nature of the electrode surfaces seems to be conspicuous by its absence.

The most obvious theory—one still held even today by engineers—is that a sufficiently high electric field existing in *the liquid* will eventually place an electrical force on the solvent dipoles, which will cause the force applied to break the chemical

[45]In the Star Wars program of the early 1990s, giant space-borne lasers were to zap enemy missiles as they left their silos. Their ground-based power was to he held in condensers with water or very dilute solutions as the dielectric. Transfer of electrical energy to the orbiting satellite was to be made by beaming it in the megacycle frequency range.

bond and hence destroy its dielectric properties. Thus, μX is the force on a dipole, where X is the electric field and μ is the dipole moment. If $\mu X > \Delta G^{\circ}_{\text{dissocn}}$ for the liquid ($\Delta G^{\circ}_{\text{dissocn}}$ is the standard free energy of a dissociation reaction), it should break down. The results of calculations along these lines for water show that more than 10^7 V cm^{-1} would be needed and this kind of electric field strength could only be attained at an interface. However, even then, breakdown by electrical *tearing apart* does not merit too much attention any more because it has been known since the 1950s that fields at interfaces between electrodes and solutions are on the order of 10^7 V cm^{-1} (Chapter 6), yet water there retains its chemical stability.

On the other hand, most chemists and physicists who have discussed this phenomenon in the 1980s and 1990s observed that the streamers *come from the electrode* and that light is emitted from the electrode. The interfacial region undoubtedly plays the determinative role in the dielectric breakdown of liquids.

One view has concentrated upon seeing water and dilute solutions thereof as if they were intrinsic semiconductors, i.e., semiconductors in which no impurities have been added to provide foreign atoms that could ionize and provide electrons to increase conductance. Such bodies are known to have three vital regions. In one, the electron

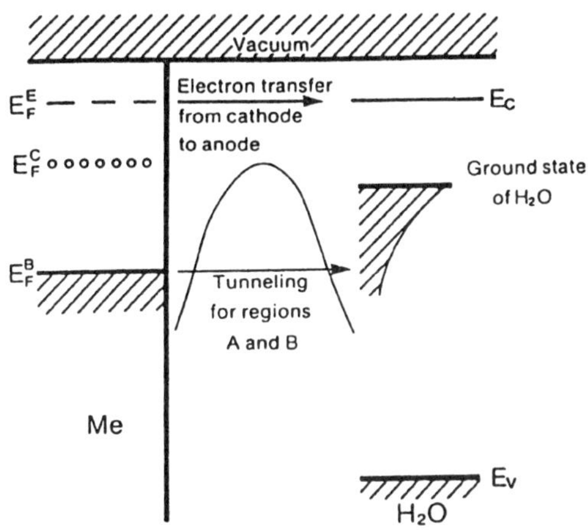

Fig. 2.73. A schematic representation of the Fermi level of the electrons in the electrode and different levels in H_2O. E_F^B, E_F^C, and E_F^E correspond to the position of the Fermi level, E_v and E_c represent the valence and conduction bands of water. The barrier for the electrons at the metal–solution interface is also shown. (Reprinted from M. Szklarczyk, R. Kainthla, and J. O'M. Bockris, *J. Electrochem. Soc.* **136**: 2512, 1989.)

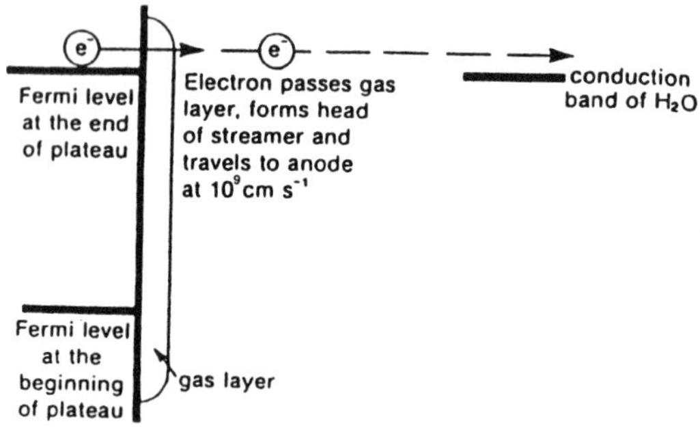

Fig. 2.74. Schematic representation of the transfer of electrons from metal to the conduction band of water before and after the breakdown. The current-potential relation leading to breakdown involves a plateau in which current hardly increases as potential approaches breakdown. (Reprinted from M. Szklarczyk, R. Kainthla, and J. O'M. Bockris, *J. Electrochem. Soc.*, **136**: 2512, 1989.)

energy levels are full and transport by electrons is therefore nonexistent. This *valence band* region is separated from the *conductance band* by what is called an *energy gap*; i.e., between the energy of the band of electron levels that conducts and the band that does not, there is a region in which no electrons exist. However, energetically above the top of the gap in the conduction band, electrons can conduct freely and with a mobility millions of times above that of ions in solution.

A theory put forward by Szklarczyk et al. suggests that the high applied potential (and high resulting field strength in solution) are secondary reflections of the effect they cause on the fundamental Fermi energy level of electrons in this metal. Here are most of the available electrons in a metal. Should it be possible to lift this electron-emitting level in the electrode to sufficiently high levels, the level of the conduction band in water (seen as a semiconductor) would be reached (Fig. 2.73). *This* would be the critical event for breakdown, for if electrons in the metal attained such an energy, they would enter the conduction band of water and behave as electrons in the conduction band of a semiconductor, traveling too rapidly to react chemically with

[46]Distance = (speed)(time). Suppose one allows twice the diameter of water (~ 600 pm) for the distance between the center of a water molecule and that in which an electron could affect a water chemically, and 0.1 of the speed of light for the velocity of passage of a streamer. Then $6 \times 10^{-8} = 3 \times 10^{9}\ \tau$, where τ is the time the water has to recognize an electron; 2×10^{-17} s leaves water unaffected as far as the vibrational and rotational levels are concerned. Even the electrons in water move only about 1/100 of a diameter in such a short time.

water,[46] but forming the head of the streamers and traveling across to the counter electrode. The great outburst of electrons arising from reaching the conduction band in water and flowing through the dielectric (opening of said flood gates, etc.) would precipitate dielectric breakdown (Fig. 2.74).

This section briefly describes an intriguing and practical phenomenon found in water and ionic solutions. A detailed comparison of this new model with experiment would take a disproportionate amount of space. One matter only is mentioned. Does the model stated explain the apparent avalanchelike effect shown in Fig. 2.71? Perhaps. For there are always particles in practical solution, solid particles and some metallic. The phenomena of breakdown are probably determined by many factors. A stream of electrons from the cathodes could cause collisional phenomena in the solution and thus secondary emissions from the particles struck by the electrons, which would then cause many more electron–particle collisions and eventually an avalanche of electrons.

2.22. ELECTROSTRICTION

Ions exert electrical forces on solvent molecules in their vicinity. Because pressure is defined as force per unit area, this means that ions exert a pressure on the solvent and/or other nearby ions. As shown below, this pressure is very high (it may exceed 10^9 Pa) compared with pressures normally encountered in the laboratory.[47]

Phenomena connected with this large pressure are referred to under the title of *electrostriction*. Molecules and ions are squeezed and decrease in size. Electrostriction is the reason that the partial molar volume of ions may become negative, for the volume-decreasing effect of adding them to a system can be greater than the volume increase caused by the addition of the ions themselves.

Effects of this kind are shown in Fig. 2.75. However, electrostriction has its limits. As seen in Fig. 2.75, the value of the compressibility itself is reduced as the electric field (and hence the local pressure) increases. Some details of this are worked out in the next section.

2.22.1. Electrostrictive Pressure near an Ion in Solution

The molar volume of water in the natural state is 18 cm^3 mol^{-1} but if water molecules are close-packed in the liquid state, this volume would become only 12 cm^3 mol^{-1}, a reduction of 33%. Thus, the volume available for the effects of electrical constriction by ions on water molecules is as much as 6 cm^3 mol^{-1}.

It is easy to calculate a typical pressure exerted by an ion on a water molecule in the first hydration shell. Thus, the energy of interaction of an ion of radius r_i on a water molecule of radius r_w is

---

[47]When a gas is compressed, the particles come into contact at pressures of a few thousand atmospheres. It is impractical to deal with gases at substantially higher pressures.

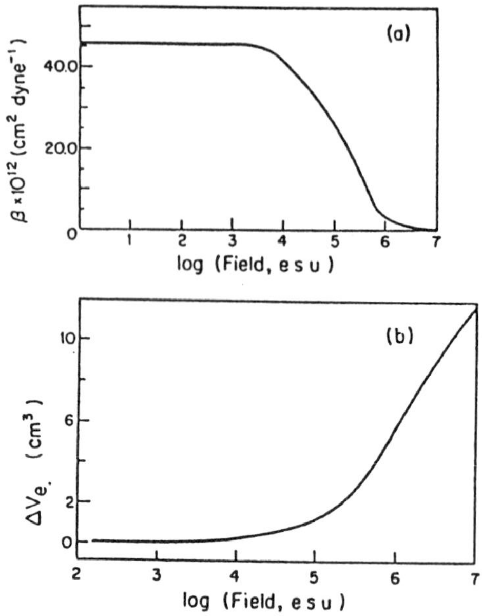

Fig. 2.75. (a) Local compressibility β of water as a function of log(field) near an ion. (b) Electrostriction, ΔV, in water as a function of log(field) near an ion. (Reprinted from B. E. Conway, *Ionic Hydration in Chemistry and Biophysics*, Elsevier, New York, 1981.)

$$U_r = \frac{e_0 \mu \cos\theta_r}{\varepsilon_r r^2} \tag{2.160}$$

Now,

$$F_r\, dr = -\, dU_r \tag{2.161}$$

where F_r is the force of the ion on the water molecule at a distance r from the ion.

Therefore,

$$F_r = \frac{2e_0 \mu \cos\theta_r}{\varepsilon_r r^3} \tag{2.162}$$

The force calculated here is upon a water molecule in the hydration shell. If one takes the cross section of the water as r_w^2, the pressure (pressure = force/unit area) at $r = (r_i + r_w)$ with $r_i = 200$ pm, $r_w = 138$ pm, $\mu_w = 1.8$ D, $\varepsilon_r = 6$, and $\theta_r = 0$ is $P = 10^{10}$ Pa.

2.22.2. Maximum Electrostrictive Decrease in the Volume of Water in the First Hydration Shell

The definition of compressibility is

$$\beta = -\frac{1}{V}\left(\frac{\partial V}{\partial P}\right)_T \tag{2.163}$$

Then,[48]

$$\beta \, dP = -\frac{dV}{V} \tag{2.164}$$

and after integration,

$$V_P = V^0 \exp(-\beta P) \tag{2.165}$$

Although β is known to decrease with increase of pressure, it is of interest to make a first approximation by taking β_{water} as constant ($\beta_{P^0} = 4.56 \times 10^{-11}$ Pa^{-1}) up to pressures on the order of 10^{10} Pa. Then, for $V^0 = 18$ cm^3 mol^{-1}, Eq. (2.165) becomes

$$V_P = 18 \exp(-4.56 \times 10^{-11} \times 10^{10}) = 11.5 \text{ cm}^3 \text{ mol}^{-1} \tag{2.166}$$

However, water cannot be compressed below the volume of the close-packed structure, 12 cm^3 mol^{-1}. Hence, when the compressibility is taken as pressure independent, the calculated effect of pressure becomes overestimated. What, then, is the relation of β to P?

2.22.3. Dependence of Compressibility on Pressure

It is known empirically that for water

$$\left(\frac{1}{\beta}\right)_P - \left(\frac{1}{\beta}\right)_{P0} = 7.5P \text{ (atm)} \tag{2.167}$$

Hence, at $P = 10^{10}$ Pa, $\beta = 1 \times 10^{-6}$ atm^{-1} = 1×10^{-11} Pa^{-1}. It can be seen that $\beta_P/\beta_{P^0} = 0.22$, a 22% decrease in β compared with that at the standard condition of 10^5 Pa.

---

[48]A pressure of 10^5 Pa = 1 bar ≃ 1 atm is the standard pressure for reporting data, and it is denoted by a superscript null after the variable, e.g., V^0.

TABLE 2.28
Calculated Electrostriction, ΔV_e (cm^3 mol^{-1}), of Water as a Function of Field (Intermediate Fields)

ΔV_e (cm^3 mol^{-1})	E (esu)[a]	ΔV_e (cm^3 mol^{-1})	E (esu)[a]
0	0	0.995	6.01×10^4
5×10^{-4}	4×10^2	1.36	1.09×10^5
0.010	2×10^3	1.82	1.82×10^5
0.030	3×10^3	2.36	2.70×10^5
0.085	5.51×10^3	3.17	4.06×10^5
0.201	9.51×10^3	5.08	8.16×10^5
0.536	2.31×10^4	5.71	1.00×10^6

Source: Reprinted from B. E. Conway, *Ionic Hydration in Chemistry and Biophysics*, Elsevier, New York, 1981.
[a] 1 esu = 3.33×10^{-10} coulombs.

This decrease in compressibility with increase in pressure explains the overlarge decrease in volume (44%) due to electrostriction calculated earlier, assuming β to be independent of pressure. It also supports Passynski's approximation $(\beta)_{\text{hydn shell}} = 0$.

In Fig. 2.75 the decrease of β is shown as a function of log X, where X is the electric field in the hydration shell. For an ion of radius 100 pm, this is[49]

$$X = k' \frac{e_0}{\varepsilon r^2} = k' \frac{1.6 \times 10^{-19}}{6(10^{-10})^2} = 2.40 \times 10^{10}\ \text{N C}^{-1} = 2.40 \times 10^{10}\ \text{Vm}^{-1} \tag{2.168}$$

If one compares this with the Conway relation of local pressure to field, one sees that this field is equivalent in electrostrictional pressure to about 4×10^9 Pa, at which (see above) the compressibility is about 22% of the value at the standard pressure.

Thus in Fig. 2.75(a), the decrease of the molecular volume of water in the hydration sheath of an ion, the radius of which is 100 pm, is about 2 cm^3 mol^{-1}, or about 11% (the first approximation was 44%).

For instance,

$$\text{at } P = 10^{10}\ \text{Pa}, \beta = 10^{-11}\ \text{Pa}^{-1}, \text{ and from Eq. (2.165)},$$

[49]The Système International of electrical units includes most of the common electrical units, such as volt (V), ampere (A), ohm (Ω), watt (W), and coulomb (C). In this system, Coulomb's law reads

$$F = k' \frac{|q_1, q_2|}{r^2}$$

where $k' = 1/4\pi\varepsilon_0 \sim 9.0 \times 10^9$ N m^2 C^{-2}. Useful conversions to remember are 1 J = 1 N m and 1 J/C = 1 V. In the cgs system (in which the basic units are the centimeter, gram, and second), the constant k' is defined to be unity, without unit. In this system the unit of electric charge is the *statcoulomb* or esu (electrostatic unit), which is still used in many scientific publications. The conversion factor is, to four significant figures, 1 C = 2.998×10^9 esu.

$$\Delta V = 18[\exp(-10^{-11} \times 10^{10}) - 1] = -1.71 \text{ cm}^3 \qquad (2.169)$$

Hence, the compression in the first layer of water molecules around an Na^+ ion would be about (1.71/18) or 10% of the average volume of water in the bulk.

2.22.4. Volume Change and Where It Occurs in Electrostriction

Calculation shows that most of the electrostrictional volume change due to the electrical pressure exerted by an ion in the surrounding solvent arises in the first shell of solvent around it. The calculated volume changes for typical fields are shown in Table 2.28.

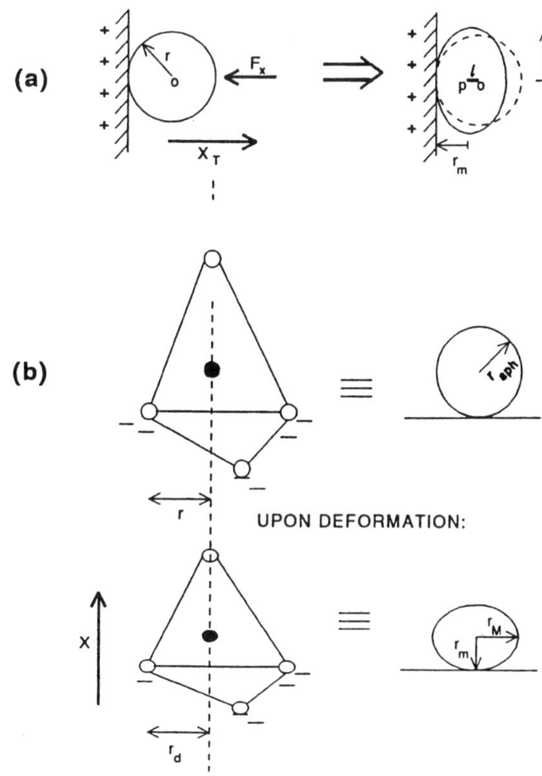

Fig. 2.76. (a) Deformation of the adsorbed ions due to the electric field at the metal–solution interphase. (b) Schematic representation of the bisulfate ion and its equivalent spherical structure before and after its deformation at the interphase. (Reprinted from J. O'M. Bockris, Maria Gamboa-Aldeco, and M. Szklarczyk, *J. Electroanalyt. Chem.* **339:** 335, 1992.)

2.22.5. Electrostriction in Other Systems

This section concentrates on the electrostrictional volume contraction of water molecules in the hydration shell. However, electrostriction is a general phenomenon and whenever there are electric fields on the order of $10^9 - 10^{10}$ V m^{-1}, the compression of ions and molecules is likely to be significant (although less so than occurs by using the compressibility figures usually given in books because these usually neglect the decrease of β with increase of pressure). Bockris and Gamboa have calculated the degree of compression of SO_4^{2-} and Cl^- ions near interfaces. The spherical shape is lost and the ions become lenslike in shape (Fig. 2.76).

Electrostriction in solids is important as the origin of piezoelectric effects. Von Sterkenberg has measured the electrostrictive coefficients in alkaline (earth) fluorides and found electrostriction there to be anisotropic.

Further Reading

Seminal

1. T. J. Webb, "Field and Pressure Near an Ion," *J. Am. Chem. Soc.* **48**: 2589 (1926).
2. H. S. Frank, "Theory of Electrostriction," *J. Chem. Phys.* **23**: 2033 (1955).
3. J. Padova, "Pressure Effects in Ionic Solutions," *J. Chem. Phys.* **39**: 1552 (1963).
4. J. E. Desnoyers, R. E. Verrall, and B. E. Conway, "Electrostriction in Electrolytes," *J. Chem. Phys.* **43**: 243 (1965).

Papers

1. S. W. P. von Sterkenberg, *J. Phys. Appl. Chem.* **17**: 69 (1982).
2. J. L. Ord, *J. Electrochem. Soc.* **138**: 2934 (1991).
3. J. O'M. Bockris, M. Gamboa-Aldeco, and M. Szklarczyk, *J. Electroanal. Chem.* **339**: 355 (1992).
4. S. W. P. von Sterkenberg and Th. Kwaitaal, *J. Appl. Phys.* **25**: 843 (1992).
5. M. Szklarzcyck, in *Modern Aspects of Electrochemistry*, Vol. 25, Ed. J. O'M. Bockris, B. E. Conway, and R. H. White, Plenum, New York (1993).
6. G. Kloos, *J. Appl. Phys.* **28**: 1680 (1995).
7. W. L. Marshall, *J. Solution Chem.* **22**: 539 (1993).
8. T. M. Letcher, J. J. Paul, and R. L. Kay, *J. Solution Chem.* **20**: 1001 (1991).

2.23. HYDRATION OF POLYIONS

2.23.1. Introduction

Most of the ion-exchange resins consist of polyions. A typical one is also the most well known: Nafion, the structure of which is shown in Fig. 2.77; the figure also shows the structure of some proteins (these are often polyelectrolytes).

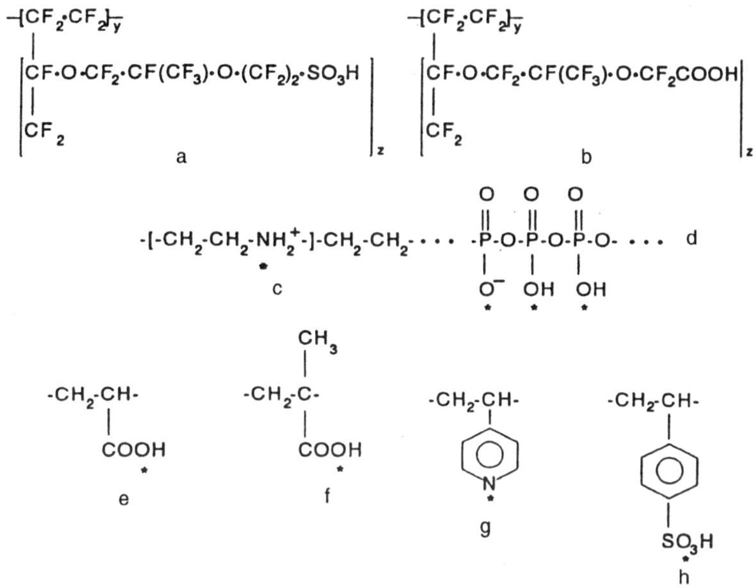

Fig. 2.77. Ion-exchange resins. a = nafion; b = the Asahi structure similar to nafion's; c = polyimines; d = polyphosphates; e = polyacrylic acid; f = polymethacrylic acid; g = polyvinylpyridine; h = polysterene sulphonic acid.

In all these structures the interaction of the solvent with the polyelectrolyte is critical for its stability (*cf.* Fig. 2.4). The ion concerned is usually nonspherical and a more typical configuration is cylindrical, as with the linear polyphosphates.

Thus, the ionizing group in many of these materials is in the side chain, as with polyacrylic acid [Fig. 2.77(e)]. Correspondingly, in the linear polyions there are structures of the kind shown in Fig. 2.77(c) and (d).

2.23.2. Volume of Individual Polyions

To obtain the partial molar volume of a series of polyvinyl acid salts containing a cation of tetraalkylammonium, one plots the volume V of the total electrolyte against $1/V_{\text{cation}}$ and obtains the partial molar volume by extrapolating the volume of the anion. Such volumes are given in Table 2.29.

2.23.3. Hydration of Cross-Linked Polymers (e.g., Polystyrene Sulfonate)

It is possible to study step-by-step hydration of biomolecules as was first done by Gluckauf. The hydration is localized and connected to ionic binding. Thus, Gregor et al. found that the order of adsorption onto polystyrene was $H^+ > Li^+ > Na^+ > K^+$.

TABLE 2.29

Individual Partial Molar Volumes of Macroions of Weak Polyelectrolytes at 298 K[a]

Weak polyelectrolyte[b] (α)	$\overline{V}_p^0$	V_I	V_{st}	V_e
KPMA (0.95)	43.5*	81.6	−9.1	−29.0
(0.75)	46.6*			−25.9
(0.6)	51.3*			−21.2
(0.4)	60.0*			−12.5
PEI (HCl) (0.8)	29.1	54.7	−21.5	−4.1
TP (HCl) (0.8)	24.9	54.0	−20.3	−8.8
TT (HCl) (0.8)	24.9	53.3	−19.6	−8.8
DT (HCl) (0.8)	24.5	52.0	−18.4	−9.1
ED (HCl) (0.8)	20.3	50.6	−19.2	−11.1

Source: Reprinted from B. E. Conway, *Ionic Hydration in Chemistry and Biophysics*, Elsevier, New York, 1981.

[a]Units are cm^3 $monomoles^{-1}$. $\overline{V}_p^0$, partial molar volume of the polyion; V_I, ionic volume; V_{st}, volume due to change in the volume of the surrounding water; and V_e, volume due to the electrostriction of the surrounding water.

[b]KPMA, potassium polymethacrylate; PEI, polyethyleneimines; TP, tetraethylene pentamine; TT, triethylene tetramine; DT, diethylene triamine; ED, ethylene diamine.

2.23.4. Effect of Macroions on the Solvent

Macroions, such as proteins, have an effect on the self-diffusion of a solvent. Thus, the presence of the macroion obstructs the solvent's movement. On the other hand, it binds the solvent and substantial fractions of it may be trapped within the molecule.

The fluidity of the solvent is changed because of the structure-breaking effects caused by the presence of macroions. Self-diffusion data on water in the presence of proteins allows one to measure their hydration. This is usually measured in water per gram of anhydrous protein.

2.24. HYDRATION IN BIOPHYSICS

Hydration of biopolymers is a mechanism for stabilizing these materials (Fig. 2.78). When proteins are completely dry, they tend to decompose. One way of evaluating hydration in polyions is to measure the dielectric constant of a solution containing a dissolved protein as a function of concentration at radio frequency. The dielectric constant falls with increase in concentration and the water per polyion can be calculated by assuming that water bound to the protein no longer makes any contribution to the dielectric constant. Thus, Buchanan calculated the irrotationally bound water from such experiments. Some of this water is hidden in cavities within the structure of the protein molecule.

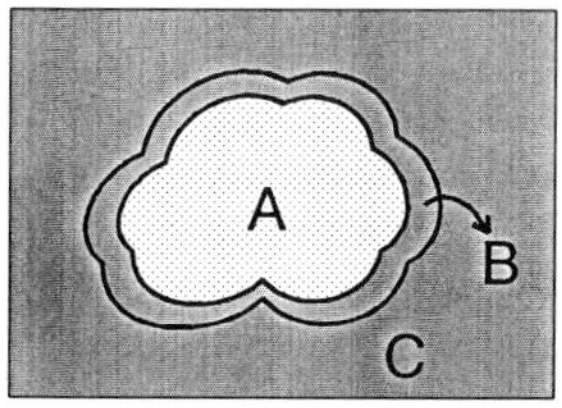

Fig. 2.78. A schematic drawing of protein and water. The regions of the protein and the excluded volume of it are represented as A and A + B, respectively. The regions of solute (low dielectric media) and solvent (high dielectric media) are represented as A and B + C, respectively. (Reprinted from M. Irisa, T. Takahashi, F. Hirata, and T. Yanagida, *J. Mol. Liquids* **65**: 381, 1995.)

As with normal electrolytes (Section 2.12.1), the dielectric constant depends linearly on concentration:

$$\varepsilon = \varepsilon_0(1 - \delta c) \tag{2.170}$$

where:

$$\delta = \frac{\beta}{100}\,[\tilde{V}(\varepsilon_{H_2O} - \varepsilon^{\infty}_{prot}) + N_{w,irr}(\varepsilon_{H_2O} - \varepsilon^{\infty}_{w})] \tag{2.171}$$

Here c is the concentration in gram percent, $\tilde{V}$ is the partial specific volume of the protein, $\varepsilon^{\infty}_{prot}$ is the high-frequency value of the dielectric constant of the protein (= 2) or of water (= 6) and ε^{∞}_{w} is that. The unknown here is $N_{w,irr}$, the number of irrotationally bound waters and this can then be calculated. It is usual to assume that the water is adjacent to charged groups.

2.24.1. A Model for Hydration and Diffusion of Polyions

For polyelectrolytes containing discrete charged groups, the usual configuration is cylindrical. Thus, the field midway along the chain is described by Conway (1982) as

$$X = \sum_{0}^{n} \frac{ze_0 r}{[r^2 + (2n-1)^2\lambda^2]^{3/2}\varepsilon} \tag{2.172}$$

TABLE 2.30

Free-Energy Changes for the Introduction of a Water Molecule into Pure Water and into Two Cavities in Sulfate-Binding Protein (SBP)[a]

Energy component	Theoretical ΔA for Pure Water	Experimental ΔG for Pure Water	Theoretical ΔA for Hydrated Cavity in SBP	Theoretical ΔA for "Dry" Cavity in SBP
Electrostatic	-8.4 ± 0.3	-8.7	-11.9 ± 0.9	-1.9 ± 0.9
Lennard-Jones	2.0 ± 0.1	2.4	-4.5 ± 0.0	-4.3 ± 0.2
Total	-6.4 ± 0.4	-6.3	-16.4 ± 0.9	-6.2 ± 1.1

[a]Units are kcal/mol.

As expected, ε decreases near the ions concerned. However, the structure of water is disturbed by the field from such charged bodies only 500–600 pm out into the solvent.

2.24.2. Molecular Dynamics Approach to Protein Hydration

Internal water plays a role in the structure of proteins. It is difficult to detect and measure these waters by means of X-rays and therefore statistical thermodynamic calculations may be helpful.

An example of this kind of calculation, due to Wade et al., is the computation of the hydration of two internal cavities in a sulfate-binding protein. The results are given in Table 2.30. The main difference between having a "dry" cavity and having a wet one is the hydration bond energy.

2.24.3. Protein Dynamics as a Function of Hydration

Proteins in the dry state are "frozen." They only open up and start moving if some water is added, as in nature. It turns out that protein movements in, e.g., lysozyme are activated only when there is 0.15 g of water per gram of protein, a good example of the effect of hydration on living processes. However, it is difficult to examine protein dynamics in solution because to make a satisfactory interpretation of the observations, one would have first to do the corresponding spectroscopy in the dry state; this is difficult because of the "frozen" state referred to and a tendency to decompose.

To avoid this difficulty, one technique is to use *reverse micelles*. These materials can host a protein in a small water pool. Reverse micelles are spherical aggregates formed by dissolving amphiphiles in organic solvents. The polar head of the amphiphilic molecule is in the interior of the aggregate and the hydrophobic tail is in the organic phase. The micellar suspension is transparent, and controlled amounts of water can be added.

TABLE 2.31
Results of Protein Hydration

Protein[a]	Concentration Range (mg/ml)	N_{total} (liters/protein)
Cc(A)	15–65	290 ± 30
Mb(A)	14–56	350 ± 40
Ov(dw)	12–49	870 ± 100
Ov(A)	14–60	860 ± 95
Ov(M)	15–58	840 ± 95
BSA(dw)	17–68	1300 ± 140
BSA(A)	11–43	1260 ± 150
BSA(M)	14–61	1410 ± 160
Hb(A)	17–66	1260 ± 140

Source: Reprinted from M. Suzuki, J. Shigematsu, and T. Kodama, *J. Phys. Chem.* **100**: 7282, 1996.
[a]Cc, cytochrome; Mb, myoglobin; Ov, ovalbumin; BSA, bovine serum albumin; Hb, hemoglobin; dw, distilled water; A, buffer A which contained 10 mM KCl and 10 mM sodium borate (pH = 8.0). M, buffer M, which contains 20 mM KCl, 5 mM $MgCl_2$ and 10 mM sodium borate (pH = 7.5).

One can then use the time course of the fluorescence decay. The change in frequency of the emitted light compared with that of the incident light contains information from which the rotational and internal dynamics can be studied.

The rotational frequency in lysozomes decreases with increase of water. This type of study throws some light on protein–micellar interactions and how they are affected by hydration.

X-ray and neutron diffraction measurements on polyion hydration give the number of water molecules involved per repeat group in the structures. About one water molecule per repeat group is the result for polymethyl methacrylate. The results of hydration for a variety of proteins are given in Table 2.31.

2.24.4. Dielectric Behavior of DNA

The rotation of large molecules in solution together with their hydration is the basis of many of the properties of solutions containing polyions. One has to ask questions, however, about the decrement of the dielectric constant in the case of linear polyelectrolytes, which include DNA. Substances such as this exhibit dielectric decrements as expected but it is difficult to account for their magnitude in terms of hydration. Thus, there might be a rotation but this cannot be about the long axis because in such a case δ (Section 2.24) should increase when the molecules are oriented perpendicularly to the electric field and this is not found to be the case.

Could this be behavior in terms of the Maxwell–Wagner dispersion,[50] which would arise through conductivity in the double layer near the polyanion. In support of this, the dielectric constant falls as the frequency increases (Fig. 2.79).

[50]This is the variation of the resistance or capacitance with frequency.

Now, ice has a dielectric dispersion over the range 10^2–10^4 Hz, and water over the range 10 Hz–10^2 Hz. Thus, the high rate of fall of dielectric constant with increased concentration for linear polyelectrolytes may come from the icelike structure of water in their vicinity.

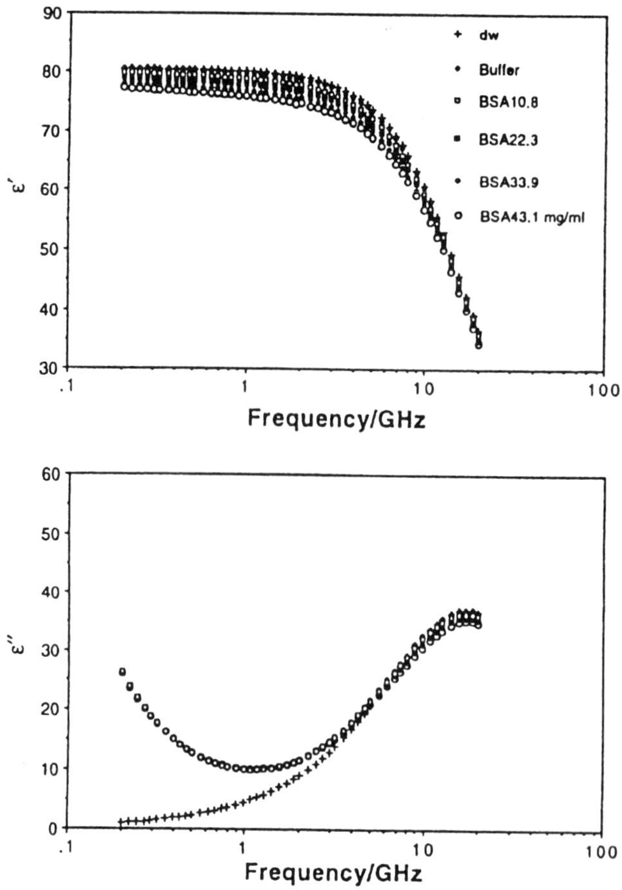

Fig. 2.79. Dielectric spectra of solutions of the protein bovine serum albumin (BSA) at 20.0 ± 0.01 °C. dw, buffer, and the numbers after BSA indicate distilled water, buffer A, and BSA concentrations in mg/ml in buffer A, respectively. (Reprinted from M. Suzuki, J. Shigematsu, and T. Kodama, *J. Phys. Chem.* **100**: 7279, 1996.)

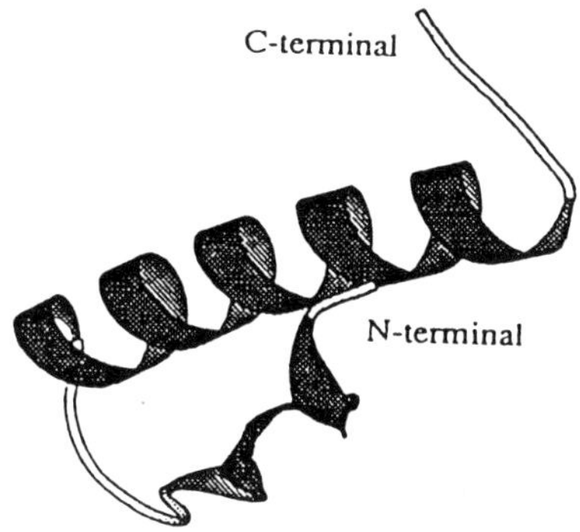

Fig. 2.80. A secondary structure of avian pancreatic polypeptide. (Reprinted from M. Irisa, T. Takahashi, F. Hirata, and T. Yanagida, *J. Mol. Liquids* **65**: 381, 1995.)

2.24.5. Solvation Effects and the α-Helix–Coil Transition

In many biological materials in solution, there is an equilibrium between a coil-like film and a helix[51] (Fig. 2.80). The position of the equilibrium between coil and helix-like forms in biomolecules in solution is determined by hydration. For example, in the solvent Cl–CH_2–CH_2OH, light scattering studies show that poly-γ-benzyl glutamate is found to consist of random coils, but in other solvents it forms a helix.

The form and structure of proteins demands hydration. Doty, Imahor, and Klemperer found that a typical structure was 15% α-helix in 0.14 M NaCl, but at pH 4 it changed to 50% α-helix. The change is a function of hydrogen bonding. The main difference between having a "dry" cavity and a wet one is the hydrogen bond energy.

2.25. WATER IN BIOLOGICAL SYSTEMS

2.25.1. Does Water in Biological Systems Have a Different Structure from Water *in Vitro*?

It has been suggested that water in biological systems has a different structure from that of free water. Thus, Szent-Györgyi (the discoverer of vitamin C) was the first to suggest that an icelike structure surrounded proteins and other biomolecules. Cope examined cell hydration and asked whether water is affected by *conformational*

[51]A helix is an elongated form of a coil.

[52]*Conformational* refers to spatial arrangements of molecules that can be obtained by rotating groups around bonds.

changes[52] in the protein structure. Much of the water present in a biological cell would be associated with the cell surface and hence have an oriented or rigid structure. Thus, the interior of biological cells contains an interconnecting network of cellular organelles: there is little space inside for free water.

Water in biological systems does indeed have a structure different from that of the bulk, as it does at all interfaces, and the appropriate way to look at the alleged difference in its structure is to realize that there is a large surface-to-volume ratio in cell-like structures so that the fraction of water associated with surfaces is large, and; by definition, under electric fields that are characteristic of the surface of double layers at surface–solution interfaces. Hence, the icelike structure that Szent-Györgyi saw as characteristic of biological systems would be a description of any near-surface water and becomes the norm for biological cell water because most water is near a surface and hence has abnormal structure (Ling and Drost-Hansen, 1975).

2.25.2. Spectroscopic Studies of Hydration of Biological Systems

Zundel reports on the hydration of thin polyelectrolyte ion-exchange membranes subject to progressive increases in water sorption. Spectroscopic observations of these systems reflect the hydration of polyion charge centers in the membranes but in the presence of associated counterions, H_3O^+, which in turn are progressively hydrated. Zundel worked with polysulfonates and found the spectra of unhydrated polymer salts: the cation is attached unsymmetrically to the SO_3^- groups. This mode of attachment leads to a loss of degeneracy[53] in the antisymmetric vibrations.

The fact that this is so allows one to follow the hydration of cation–polyanion association. This also affects the assignment of changes of band frequency in the water molecule as it dissociates with the counterion groups.

In the region of 2000–4000 cm^{-1}, it is possible to obtain spectroscopic results that reflect the bound water at polyions. These occur at 3000–3700 cm^{-1} in H_2O and 2200–2750 cm^{-1} for D_2O. As the radius of the bound counteranion and polystyrene sulfonate decreases, the OH stretching vibration of the hydration water molecule increases (e.g., from Ba^{2+} to Be^{2+} and from La^{3+} to Sc^{3+}).

These spectroscopic results are consistent with corresponding thermodynamic results obtained by Gluckauf and Kitt. They found that the greatest values of ΔH and ΔS for sorption occurred during the first water molecule aggregation per ion pair and decreased for subsequent ones.

2.25.3. Molecular Dynamic Simulations of Biowater

So far there have been few in-depth MD simulations of water at the interfaces of cells. As indicated earlier, the results of general studies and spectroscopic data suggest that much of the water in biological cells is affected by the interface. The fact that the

[53]Degeneracy = more than one state having the same energy.

surface-to-volume ratio is so large in biological cells means that most of the water there is affected by the surface forces. However, this is *not* supported by the MD calculations of Ahlstrom et al., who made an MD simulation of the Ca^{2+}-binding protein, parvalbumin. These workers found, contrary to the expectation arising from knowledge of the high surface-to-volume ratio, that relatively few ions were immobilized (Fig. 2.81) on the surface of the protein. However, they did conclude that the dipole of the waters in contact with the protein was indeed oriented perpendicularly to the protein surface. These findings contradict NMR data that indicate surface waters as having a longer relaxation time than in the bulk. The potential functions of the MD simulation may have been ill chosen. Electron density distributions of water in cells show order out to 1500 pm.

2.26. SOME DIRECTIONS OF FUTURE RESEARCH IN ION–SOLVENT INTERACTIONS

Traditionally, the interaction between particles in chemistry has been based upon empirical laws, principally on Coulomb's law. This law is also the basis of the attractive part of the potential energy used in the Schrödinger equation, but the resultant

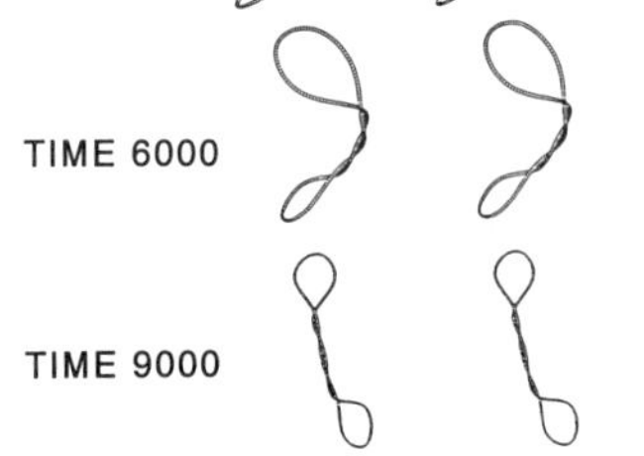

Fig. 2.81. Snapshots from a stochastic dynamics simulation of DNA supercoiling. (Reprinted from P. Ahlstrom, O. Teleman, and B. Jocsom, *J. Am. Chem. Soc.* **110**: 4198, 1988.)

energies of interaction obtained by solving it are then quantized. Such an approach was begun for ion–water interactions by Clementi in the 1970s.

A quantum mechanical approach to ion–water interactions has the up side that it is the kind of development one might think of as inevitable. On the other hand, there is a fundamental difficulty that attends all quantum mechanical approaches to reactions in chemistry. It is that they concern potential energy and do not account for the entropic aspects of the situation. The importance of the latter (*cf.* the basic thermodynamic equation $\Delta G = \Delta H - T\Delta S$) depends on temperature, so that at $T = 0$, the change in entropy in a reaction, ΔS, has no effect. However, in calculations of solvation at ordinary temperatures, the increase in order brought about by the effect of the ion on the water molecules is an essential feature of the situation. Thus, a quantum mechanical approach to solvation can provide information on the energy of individual ion–water interactions (clusters in the gas phase have also been calculated), but one has to ask whether it is relevant to solution chemistry.

Another problem in the quantal approach is that ions in solution are not stationary as pictured in the quantum mechanical calculations. Depending on the time scale considered, they can be seen as darting about or shuffling around. At any rate, they move and therefore the reorientation time of the water when an ion approaches is of vital concern and affects what is a solvation number (waters moving with the ion) and what is a coordination number (Fig. 2.23). However, the Clementi calculations concerned stationary models and cannot have much to do with dynamic solvation numbers.

Finally, Conway points out that the values of the energies obtained by solving the Schrödinger equation and by electrostatics are about the same. But this should be! The quantum mechanical calculations do not infer that the electrostatic ones (Section 2.15.11) are wrong. Indeed, the Schrödinger equations solved by Clementi involved the same energy that was used in the electrostatic method. It is a matter of whether the sophistication of quantization gives increasing insight into the behavior of a real system in solution. The answer in the 1990s is: not yet.

However, science evolves! More will be seen of the quantum mechanical approach to solvation and in particular in nonaqueous solutions when there is more chance of interactions involving overlap of the orbitals of transition-metal ions and those of organic solvent molecules. Covalent bond formation enters little into the aqueous calculations because the bonding orbitals in water are taken up in the bonds to hydrogen. With organic solvents, the quantum mechanical approach to bonding may be essential.

The trend pointing to the future is the molecular dynamics technique, and treatments of this kind were discussed in Section 2.17. The leaders here are Heinzinger and Palinkas (in the 1990's), and the major thing to note is that the technique provides needed information on the *number* of waters in the first shell. The MD method does this by calculating the distribution function (Section 2.17.2) and may also provide information on a second layer, lifetimes of the solvent molecules in the hydration sheath, and so on.

2.27. OVERVIEW OF IONIC SOLVATION AND ITS FUNCTIONS

2.27.1. Hydration of Simple Cations and Anions

The first thing that was dealt with in this chapter was the fundamental question of how things dissolve. Materials have stability in the solid lattice and therefore in the solution there must be energies of interaction that compensate for the lattice energies and thus make it comfortable for the ion to move into the solution. This energy of interaction of solvent and ion is called the *solvation energy*. Solvation and the corresponding water-related hydration are fields having a great breadth of application and indeed one can even see a relevance to environmental problems such as acid rain because the pH of natural waters depends on the stabilization of certain ions in solution, and this depends primarily upon their solvation energies.

It is useful to divide the methods of investigating hydration into four groups. First of all, there are the many methods in which one looks at the ionic solution in the equilibrium state; for example, one can study its compressibility or one can study heats and entropies of hydration and fit the values to various model systems.

Another approach involves measuring the transport of the ion concerned under an applied electric field, its mobility. This approach has a unique aspect: it provides the individual properties of ions directly so long as they are in dilute solution.

Spectra of all kinds (infrared, Raman, nuclear magnetic resonance, neutron diffraction) have been applied to electrolytic solutions for a long time, and the contrast between the indications from such measurements (which involve the grave disadvantage of having to be carried out at high concentrations) and those of other (concordant) methods of investigation (usually carried out in dilute solutions to diminish the effects of ion–ion interactions) in measuring solvation numbers is troubling. It seems likely that the spectroscopic results report the much smaller values obtained in solvation studies at higher ionic concentrations (i.e., lower water concentration). The thermodynamic methods (e.g., partial molar volumes and the Debye vibrational potential) give much higher results than the spectroscopic methods because they measure the situation at high concentrations of water (i.e., low ionic concentrations) or dilute solutions. However, increasingly neutron diffraction and MD methods are being able to provide information on the *time* spent by water molecules in the first and second layer near an ion. By comparing these times with those known for the time an ion spends between its movements in solution, it is becoming increasingly possible to use spectroscopic methods (particularly the recent neutron diffraction methods) to make a distinction between coordination numbers and numbers of waters which travel with the ion, the hydration number.

Finally, there is the theoretical method of approaching ionic solvation including the molecular dynamics simulations. These have become increasingly used because they are cheap and quick. However, MD methods use two-body interaction equations and the parameters used here need experimental data to act as a guide for the determination of parameters that fit.

Most of the data that one gets from the experimental point of view of hydration concerns the net effect of all the ions in the electrolytes, and it is necessary to pull out the quantities corresponding to each ion separately. This is tricky but possible. For example, one can couple a certain anion with a series of cations of increasing size and extrapolate a plot in such a manner that the effect of the cation becomes negligible and that of the anion is isolated. Then, the individual value for that anion can be used to obtain individual values for any cations that can form salts with this anion.

Apart from neutron diffraction, what other method distinguishes between the static or equilibrium coordination number and the dynamic solvation number, the number of solvent molecules that travel with an ion when it moves? One method is to obtain the sum of the solvation numbers for both cation and anion by using a compressibility approach, assuming that the compressibility of the primary solvation shell is small or negligible, then using the vibration potential approach of Debye to obtain the difference in mass of the two solvated ions. From these two measurements it is possible to get the individual ionic solvation numbers with some degree of reliability.

Another approach that can help in getting hydration numbers is the study of dielectric constants—both the static dielectric constants and the dielectric constant as it depends on frequency. Such measurements give a large amount of information about the surrounds of the ion but a good deal more has to be done before the theoretical interpretation can bear the weight of clear structural conclusions. Density, mobility, and entropy measurements may also be informative.

The material so far has all come from discussions of methods of examining hydration and solvation in solutions. It is now good to turn to a simpler field, the study of which began much later—hydration of the gas phase. It was not until high-pressure mass spectroscopy became available that such studies could be easily made. If we had had these studies 50 years ago, it would have been much easier to interpret the values obtained in solution. Thus, the numbers we get from the unambiguous gas-phase results are close to results for the first shell of water around ions in solution, and knowledge of this, and the energies that go with it, helps us greatly in building models in solution.

The models assume that the energies are entirely electrostatic and go about calculation in terms of ion–dipole forces for the first layer, together with correction for the quadrupole properties of water, and for the extra dipole that the ion induces in the water. There are two more steps. One is to take into account the interaction of the ion with distant water molecules, which we do by means of the Born equation, and then finally we take into account the structure-breaking effects of the ions on the surrounding solvent. The agreement between theory and experiment is good and this applies (though somewhat less well) to the corresponding calculation of the entropies of hydration.

A special case, but one of seminal importance, is the heat of hydration of the proton because so much depends upon it. A rather clever method was set out many years ago by Halliwell and Nyburg and although their approach has been reexamined many times, the numerical values have not changed by more than one percent. The method involves an application of the known *differences* of quadrupole interactions of ions which have the same size but different signs.

2.27.2. Transition-Metal Ions

Our overview of hydration so far is based upon ions which are simple, mostly those from groups IA and IIA. We haven't yet said anything about the ions with more complex electronic structures, for example, the transition-metal ions, two-valent and three-valent entities. There are many who would expect such ions to have valence-force interactions (orbital bond formation) with water molecules, but in fact they don't. It is possible to interpret their hydration heats in terms of electrostatic interactions with water, but one has to be more sophisticated and no longer regard the ions as simple spheres but take into account the shape and direction of their molecular orbitals and how these affect the electrostatics of the interactions with water molecules.

It turns out that the splitting of electron levels in the ions caused by the water molecules changes the electrostatic interaction of the ions with the waters. This in turn makes the course of a plot of individual hydration of transition-metal ions against the radii of the ions no longer a continuous one, which one would expect from first-approximation electrostatic studies, but takes into account some irregularities that can be quite reasonably accounted for. Now, there is also lurking among the literature on ion hydration in the gas phase the surprising (but clearly explainable) fact that the *second* water molecule sometimes has a larger heat of hydration than the first.

2.27.3. Molecular Dynamics Simulations

The last part of our treatment of the central aspects of the chapter concerned molecular dynamics. We showed the power of molecular dynamics simulations in ionic solutions and what excellent agreement can be obtained between, say, the distribution function of water molecules around an ion calculated from molecular dynamics simulation and that measured by neutron diffraction.

However, there has been in the past some semantic confusion in these calculations. Most of them are about distribution functions. Some of the workers concerned assumed that they were calculating hydration numbers when in fact they were calculating time-averaged coordination numbers. Nevertheless, a few groups have indeed been able to calculate the rate at which water molecules react when an ion is brought near them and the lifetime of the water molecules near the ion as it moves. From this, they have calculated hydration numbers that are in fair agreement with those determined experimentally.

2.27.4. Functions of Hydration

The rest of this chapter has been concerned with phenomena and effects that are connected with hydration. The first one (salting out) concerns the solubility of nonelectrolytes as affected by the addition of ions to the solution. Here two effects are to be dealt with. The first is salting out—the decrease of solubility that the ions cause. This is easily understood because of course the ions remove quite a lot of the waters from availability to the incoming solute by taking them off into temporary inactivity in the hydration shell so that the organic molecule has less water (per liter of solution) to dissolve in and its solubility is thus decreased. Salting in is a little bit more difficult to understand, especially anomalous salting in, which occurs when the equations for salting out indicate that there should be a decrease in the solubility of a nonelectrolyte upon the addition of ions but in fact there is an increase. It turns out that this is caused by dispersion force interactions by which the ions (large ions such as those of the NR_4^+ series are involved here) attract the organic molecules to themselves and push the water out, thus giving more water for the organic molecules to dissolve in and an increased solubility. Such phenomena provide some basis for an interpretation of hydrophobic effects in hydration.

Electrostriction is the study of the effects of squeezing of ions and molecules by the electrical forces that are exerted upon them by the ions we have been dealing with (Section 2.22). It is only recently that modelers have begun to take into account the shapes formed by these compressed bodies. In fact, they do become lenslike in shape (not spheres) and when this is taken into account, agreement between theory and experiment is improved.

Hydrophobic effects are on a list of special phenomena. They are closely tied to salting in because one of the reasons for hydrophobic effects (water pushing-out effects, one could say) is that the ions of the solute tend to attract each other or other nonelectrolytes present and push the water between them out. Structure breaking in a solution, some part of which rejects water in the rearrangements formed, also gives hydrophobic effects.

Polyelectrolytes occur in ion-exchange membranes and thus their study has great material value. They have a central importance in biology and the study of their electrochemistry, as ions, their natural interactions in solution, etc., are important although we are only able to give a short description of them in a chapter of restricted length.

Finally, the question of the structure of biological water is one of far-reaching importance. Some workers in the last few decades have suggested that water in biological systems is special but our answer is that this special structure is so readily explicable that no mystery exists. Biological cells are sized on the micron scale and contain much solid material. The surface-to-volume ratio inside such cells is very large. Most of the waters in cells are in fact surface waters. In this sense, biological water *is* special but only because it has lost the netted-up properties of bulk water and adopted the individual two-dimensional structure of water at all surfaces.

APPENDIX 2.1. THE BORN EQUATION

In the preceding text, particularly in Section 2.15.11, use is made of Born's equation, a famous classical equation first deduced in 1920. This equation is generally used to express that part of the free energy of the solvation of ions which arises from interactions outside the first, oriented, layers of dipoles near the ion. Thus, sufficiently near the ion, the structure of the water is fairly definite and can be used to write equations that express simple models close enough to reality to be credible (see Appendices 2.2 and 2.3). A molecular picture is more difficult to sustain outside these first one or two layers. It is argued that there it is better to work in terms of continuum electrostatics and to suppress questions concerned with structure in the solution, etc.

The basic model upon which Born's equation rests involves a mental image of a metallic sphere. It is argued that when such a sphere (at first grounded and charge free) is given an electric charge q, this charging process must be equivalent to some amount of energy.

The reasoning is that when a series of small amounts of charge are brought upon the ion, some work has to be done to put them there because after the first charge arrives, the rest of the charge bits (all positive, say) have to push against the repelling interaction between the positive charges themselves and the positive charge already building up on the metallic sphere.

Now, from electrostatics, the work done, W, when there is a change of charge Δq of a body of potential, ψ, is given by

$$dW = \psi \, \Delta q \tag{A2.1.1}$$

In the case of the conducting sphere upon which charge is building, the potential ψ depends upon the charge and so to avoid conceptual trouble [what ψ to use in Eq. (A2.1.1) as q changes], we take an infinitesimally small change of charge dq and argue that for such very very small changes of charge, ψ will be very very nearly constant.

To find the work W done in a real finite buildup of charge, one has to overcome a problem—that ψ itself depends on the degree of charge—and hence express ψ in terms of q.

It is easy to show that for a conducting sphere, the value of ψ is given by

$$\psi = \frac{q}{r} \tag{A2.1.2}$$

With this (and the assumptions) as background, one may write for the work to build up a charge q on the sphere:

$$W = \int_0^q \frac{q}{r}\, dq = \frac{q^2}{2r} \tag{A2.1.3}$$

Now, solvation energy—and Born's equation is usually proposed as giving at least some part of that—is the difference of free energy of an ion *in vacuo* and that of an ion in solution. If this work of charging which has been calculated above is then taken as the basis for a change of energy upon the transfer from vacuum to solution, one has ($q_{\text{ion}} = z_i e_0$):

$$\text{Energy of charging in vacuum: } (z_i e_0)^2 / r_i \tag{A2.1.4}$$

$$\text{Energy of charging in solution: } (z_i e_0)^2 / \varepsilon r_i \tag{A2.1.5}$$

where ε is the dielectric constant of the solution.

Hence, the work of charging in solution, $(z_i e_0)^2/2r_i\varepsilon$, minus the work of charging the solution, $(z_i e_0)^2/2r_i$, is argued to be a contribution to the solution energy. Therefore, the corresponding energy change or the so-called *Born term* is

$$\Delta G_{\text{Born}} = \frac{(z_i e_0)^2}{2r_i\varepsilon} - \frac{(z_i e_0)^2}{2r_i} = -\frac{(z_i e_0)^2}{2r_i}\left(1 - \frac{1}{\varepsilon}\right) \tag{A2.1.6}$$

This equation is used in most theories of solvation (Section 2.15.10) as though it represented, not the difference in the energy of charging up a conducting sphere *in vacuo* and then in solution, but the energy of interaction of an ion with a solvent.

There are a number of fundamental difficulties with the Born equation, which is still presented in this text because of the prominent part it plays in most theories (it accounts for around one-third of the hydration energy calculated for simple ions).

1. Behind the idea of the work of charging is an assumption that the charging occurs slowly, so that all parts of the system concerned are arranged in their equilibrium configuration. This is arguable. Any real change of charge on an ion in solution occurs in a time of about 10^{-15} s, so fast, indeed, that atomic motions in molecules (e.g., vibrations) are taken to be stationary in comparison (Franck–Condon principle).

2. The thinking behind the deduction of Born's equation is pre-quantal. In reality, atoms do not charge up by the aggregation of a series of infinitesimally small amounts of charge. They become charged by means of the sudden transfer of one electron. The energy of charging an atom to a positive ion is called the ionization energy and is a known quantity. The energy of charging an atom to be a negative ion is called the electron affinity and is also known. Both these energies differ significantly from the Born energy of charging.

3. Although the Born charging energy differs from either the ionization energy or the electron affinity, its values for simple ions are not unreasonable. However, when one comes to apply Born's concepts to protons and electrons, irrational energies result. For example, the Born energy of a proton in the gas phase is ~1000 times the normal range of chemical energies. The corresponding self-energy for an electron alone is greater than $m_e c^2$!

So, Born's equation remains a controversial part of the theory of solvation although there have been many recent attempts striving to justify it. The difficulty resides in the avoidance of molecular-level arguments and in applying continuum electrostatics, which clearly involves fundamental limitations when it comes to atomic and subatomic charged bodies. However, there is a greater doubt. What is required in calculations of solvation energy is the ion–solvent interaction energy. Does the Born equation measure the difference of two self-energies, which is not a quantity to be used in solvation calculations at all?

APPENDIX 2.2. INTERACTION BETWEEN AN ION AND A DIPOLE

The problem is to calculate the interaction energy between a dipole and an ion placed at a distance r from the dipole center, the dipole being oriented at an angle θ to the line joining the centers of the ion and dipole (Fig. A2.2.1). (By convention, the direction of the dipole is taken to be the direction from the negative end to the positive end of the dipole.)

The ion–dipole interaction energy U_{I-D} is equal to the charge $z_i e_0$ of the ion times the potential ψ_r due to the dipole at the site P of the ion

$$U_{I-D} = z_i e_0 \psi_r \tag{A2.2.1}$$

Thus, the problem reduces to the calculation of the potential ψ_r due to the dipole. According to the law of superposition of potentials, the potential due to an assembly of charges is the sum of the potentials due to each charge. Thus, the potential due to a dipole is the sum of the potentials $+q/r_1$ and $-q/r_2$ due to the charges $+q$ and $-q$, which constitute the dipole and are located at distances r_1 and r_2 from the point P. Thus,

$$\begin{aligned}\psi_r &= \frac{q}{r_1} - \frac{q}{r_2} \\ &= q\left(\frac{1}{r_1} - \frac{1}{r_2}\right)\end{aligned} \tag{A2.2.2}$$

From Fig. A2.2.2, it is obvious that

$$r_1^2 = Y^2 + (z+d)^2 \tag{A2.2.3}$$

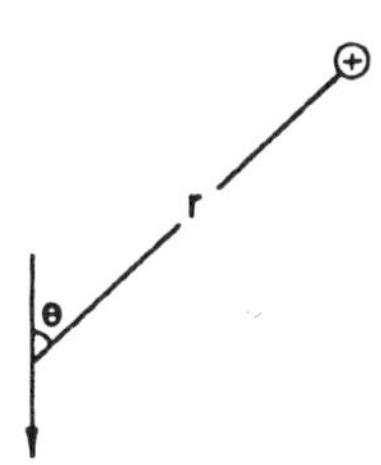

Fig. A2.2.1

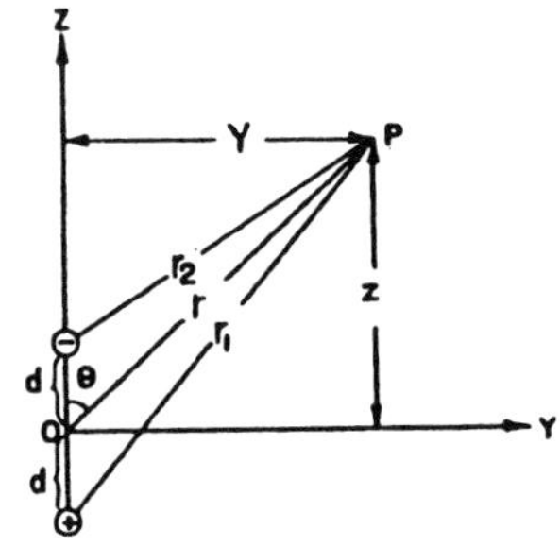

Fig. A2.2.2

and therefore,

$$\frac{1}{r_1} = [Y^2 + (z + d)^2]^{-1/2}$$

$$= [(Y^2 + z^2) + d^2 + 2zd]^{-1/2}$$

$$= (r^2 + d^2 + 2zd)^{-1/2}$$

$$= \frac{1}{r}\left[1 + \left(\frac{d}{r}\right)^2 + \frac{2dz}{r^2}\right]^{-1/2} \tag{A2.2.4}$$

At this stage, an important approximation is made, namely, that the distance $2d$ between the charges in the dipole is negligible compared with r. In other words, the approximation is made that

$$1 + \left(\frac{d}{r}\right)^2 + \frac{2dz}{r^2} \approx 1 + \frac{2dz}{r^2} \tag{A2.2.5}$$

It is clear that the validity of the approximation decreases, the closer the ion comes toward the dipole, i.e., as r decreases.

Making the above approximation, one has [see Eq. (A2.2.4)]

$$\frac{1}{r_1} \approx \frac{1}{r}\left(1 + \frac{2dz}{r^2}\right)^{-1/2} \tag{A2.2.6}$$

which, by the binomial expansion taken to two terms, gives

$$\frac{1}{r_1} = \frac{1}{r}\left(1 - \frac{dz}{r^2}\right) \tag{A2.2.7}$$

By similar reasoning,

$$\frac{1}{r_2} = \frac{1}{r}\left(1 + \frac{dz}{r^2}\right) \tag{A2.2.8}$$

By using Eqs. (A2.2.7) and (A2.2.8), Eq. (A2.2.2) becomes

$$\psi_r = -\frac{2dq}{r^2}\frac{z}{r} \tag{A2.2.9}$$

Since $z/r = \cos\theta$ and $2dq$ is the dipole moment μ,

$$\psi_r = -\frac{\mu\cos\theta}{r^2} \tag{A2.2.10}$$

or the ion–dipole interaction energy is given by

$$U_{I-D} = \frac{-z_i e_0 \mu \cos\theta}{r^2} \tag{A2.2.11}$$

APPENDIX 2.3. INTERACTION BETWEEN AN ION AND A WATER QUADRUPOLE

Instead of presenting a sophisticated general treatment for ion–quadrupole interactions, a particular case of these interactions will be worked out. The special case to be worked out is that corresponding to the water molecule being oriented with respect to a positive ion so that the interaction energy is a minimum.

In this orientation (see Fig. A2.3.1), the oxygen atom and a positive ion are on the y axis, which bisects the H–O–H angle. Further, the positive ion, the oxygen atom, and the two hydrogen atoms are all considered in the xy plane. The origin of the xy coordinate system is located at the point Q, which is the center of the water molecule. The ion is at a distance r from the origin.

The ion–quadrupole interaction energy U_{I-Q} is simply given by the charge on the ion times the potential ψ_r at the site of the ion due to the charges of the quadrupole,

$$U_{I-Q} = z_i e_0 \psi_r \tag{A2.3.1}$$

The potential ψ_r is the sum of the potentials due to the four charges q_1, q_2, q_3, and q_4 in the quadrupole (1 and 2 are the positive charges at the hydrogen, and 3 and 4 are the negative charges at the oxygen). That is,

$$\psi_r = \psi_1 + \psi_2 + \psi_3 + \psi_4 \tag{A2.3.2}$$

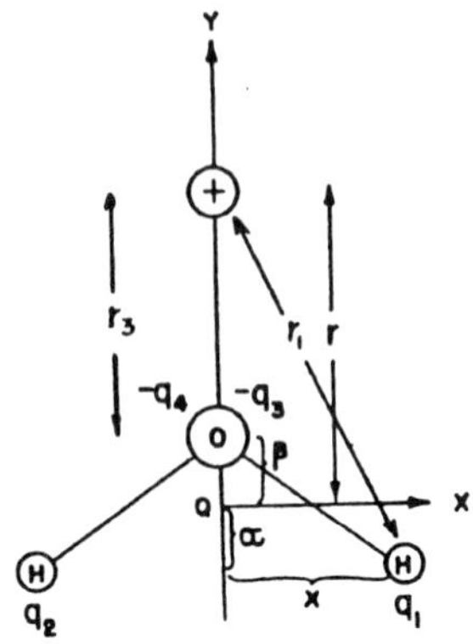

Fig. A2.3.1

Each of these potentials is given by the usual Coulombic expression for the potential

$$\psi_r = \frac{q_1}{r_1} + \frac{q_2}{r_2} - \frac{q_3}{r_3} - \frac{q_4}{r_4} \tag{A2.3.3}$$

where the minus sign appears before the third and fourth terms because q_3 and q_4 are negative charges. Further, the *magnitudes* of all the charges are equal

$$|q_1| = |q_2| = |q_3| = |q_4| = q \tag{A2.3.4}$$

and, because of symmetrical disposition of the water molecule,

$$r_2 = r_1 \quad \text{and} \quad r_4 = r_3 \tag{A2.3.5}$$

Hence, from Eqs. (A2.3.3), (A2.3.4), and (A2.3.5),

$$\psi_r = 2q\left(\frac{1}{r_1} - \frac{1}{r_3}\right) \tag{A2.3.6}$$

It is obvious (see Fig. A2.3.1) that

$$r_1^2 = (r + \alpha)^2 + x^2 \tag{A2.3.7}$$

$$= r^2\left(1 + \frac{\alpha^2 + x^2}{r^2} + \frac{2\alpha}{r}\right) \tag{A2.3.8}$$

and

$$r_3 = r - \beta \tag{A2.3.9}$$

$$= r\left(1 - \frac{\beta}{r}\right) \tag{A2.3.10}$$

Thus,

$$\frac{1}{r_1} = \frac{1}{r}\left(1 + \frac{\alpha^2 + x^2}{r^2} + \frac{2\alpha}{r}\right)^{-1/2} \tag{A2.3.11}$$

and

$$\frac{1}{r_3} = \frac{1}{r}\left(1 - \frac{\beta}{r}\right)^{-1} \tag{A2.3.12}$$

One can now use the binomial expansion, i.e.,

$$(1 \pm m)^{-n} = 1 \mp nm + \frac{n(n+1)}{2}m^2 \mp \cdots \tag{A2.3.13}$$

and drop off all terms higher than the *third*.[53] Thus,

$$\frac{1}{r_1} \approx \left\{\frac{1}{r} - \frac{1}{2}\frac{\alpha^2 + x^2}{r^3} - \frac{\alpha}{r^2} + \frac{3}{8}\left[\frac{(\alpha^2 + x^2)^2}{r^5} + \frac{4\alpha^2}{r^3} + \frac{4\alpha(\alpha^2 + x^2)}{r^4}\right]\right\} \tag{A2.3.14}$$

and, omitting all terms with powers r greater than 3, one has

$$\frac{1}{r_1} \approx \frac{1}{r} - \frac{\alpha}{r^2} + \frac{1}{2r^3}(2\alpha^2 - x^2) \tag{A2.3.15}$$

Further,

$$\frac{1}{r_3} \approx \frac{1}{r}\left(1 + \frac{\beta}{r} + \frac{\beta^2}{r^2}\right)$$

$$= \frac{1}{r} + \frac{\beta}{r^2} + \frac{\beta^2}{r^3} \tag{A2.3.16}$$

Subtracting Eq. (A2.3.16) from Eq. (A2.3.15), one has

---

[53] It is at this stage that the treatment of ion–quadrupole interactions diverges from that of ion–dipole interactions (*cf.* Appendix 2.2). In the latter, the binomial expansion was terminated after the second term.

$$\frac{1}{r_1} - \frac{1}{r_3} = -\frac{(\alpha+\beta)}{r^2} + \frac{1}{2r^3}(2\alpha^2 - x^2 - 2\beta^2) \tag{A2.3.17}$$

and therefore by substitution of Eq. (A2.3.17) in Eq. (A2.3.6),

$$\psi_r = -\frac{2q(\alpha+\beta)}{r^2} + \frac{1}{2r^3}[2(2q\alpha^2 - 2q\beta^2) - 2qx^2] \tag{A2.3.18}$$

The first term on the right-hand side of Eq. (A2.3.18) can be rearranged as follows:

$$\begin{aligned} 2q(\alpha+\beta) &= 2q\alpha + 2q\beta \\ &= \sum (2q)d \quad \text{where } d = \alpha \text{ or } \beta \end{aligned} \tag{A2.3.19}$$

Thus, as a first approximation, the water molecule can be represented as a dipolar charge distribution in which there is a positive charge of $+2q$ (due to the H atoms) at a distance α from the origin on the bisector of the H–O–H angle and a charge of $-2q$ (due to the lone electron pair) at a distance $-\beta$ from the origin; it follows that

$$\begin{aligned} \sum (2q)d = \sum &\text{ magnitude of each charge of dipole} \\ &\times \text{distance of the charge from origin} \end{aligned} \tag{A2.3.20}$$

The right-hand side of this expression is the *general* expression for the dipole moment μ, as is seen by considering the situation when $\alpha = \beta$, i.e., $\Sigma(2q)d = (2q)2d$, where $2d$ is the distance between the charges of the dipole, in which case one obtains the familiar expression for the dipole moment μ,

$$\mu = 2q2d \tag{A2.3.21}$$

Thus, the first term on the right-hand side of Eq. (A2.3.18) is

$$-\frac{2q(\alpha+\beta)}{r^2} = -\frac{\mu}{r^2} \tag{A2.3.22}$$

The second term can be interpreted as follows: Consider $(2q\alpha^2 - 2q\beta^2)$. It can be written thus

$$\begin{aligned} 2q\alpha^2 - 2q\beta^2 &= q\alpha^2 + q\alpha^2 + (-q)\beta^2 + (-q)\beta^2 \\ &= \sum qd_y^2 \quad \text{where } d_y = \alpha \text{ or } \beta \end{aligned} \tag{A2.3.23}$$

But the general definition of a quadrupole moment is the magnitude of each charge of quadrupole times the *square* of the distance of the charge from the origin. Thus, $\Sigma q d_y^2$ is the y component p_{yy} of the quadrupole moment for the particular coordinate system that has been chosen. Similarly, $\Sigma 2qx^2$ is the x component p_{xx} of the quadrupole moment. Thus,

$$2(2q\alpha^2 - 2q\beta^2) - 2qx^2 = 2p_{yy} - p_{xx} \tag{A2.3.24}$$

One can combine $2p_{yy} - p_{xx}$ into a single symbol and talk of the quadrupole moment p_W of the water molecule in the particular orientation of Fig. A2.3.1.

Hence,

$$2(2q\alpha^2 - 2q\beta^2) - 2qx^2 = p_W \tag{A2.3.25}$$

and therefore by substituting Eqs. (A2.3.22) and (A2.3.25) in Eq. (A2.3.18),

$$\psi_r = -\frac{\mu}{r^2} + \frac{p_W}{2r^3} \tag{A2.3.26}$$

The ion–quadrupole interaction energy [*cf.* Eq. (A2.3.1)] thus becomes

$$U_{I-Q} = -\frac{z_i e_0 \mu}{r^2} + \frac{z_i e_0 p_W}{2r^3} \tag{A2.3.27}$$

When a negative ion is considered, the water molecule turns around through π, and one obtains by an argument similar to that for positive ions

$$U_{I-Q} = -\frac{z_i e_0 \mu}{r^2} - \frac{z_i e_0 p_W}{2r^3} \tag{A2.3.28}$$

EXERCISES

1. The heat of evaporation of water is 2240 J g^{-1}. Calculate the total H-bond energy for 1 mole of water, and compare the two values.

2. The Cl^- ion has a radius of 180 pm. Find how much bigger the ion–dipole term is compared with the Born term. The dielectric constant in the Born equation is to be taken as 80.

3. The energy of the I^- ion–water dipole interaction is –171 kJ mol^{-1} at 298 K. Calculate the radius of a water molecule. Consider $r_{I^-} = 216$ pm and $\mu_W = 1.8$ D. (Contractor)

4. The heat of the chloride ion interaction with water is measured to be –347.3 kJ mol^{-1}, in which the contribution of the Born charging process is –152.5 kJ mol^{-1}. If the dielectric constant of water at 298 K is 78.3, estimate its rate of change with temperature at this temperature. Also calculate the percent error introduced in the Born charging term if the dielectric constant is assumed to be independent of temperature. Consider $r_{Cl^-} = 181$ pm and $r_w = 138$ pm. (Contractor)

5. When an F^- comes in contact with water molecules, its ion–quadrupole interaction energy is –394.6 kJ mol^{-1}. Calculate the quadrupole moment of water ($r_w = 138$ pm, $r_{F^-} = 136$ pm, $\mu_w = 1.8$ D). (Contractor)

6. If the total primary hydration number of NaCl in a 1 M solution is 6, make a rough calculation of the dielectric constant of the solution by assuming that the dielectric constant of pure water is 80 and that of the water molecule in the primary hydration sheath is 6.

7. For an NaCl solution, calculate the concentration at which the so-called "Gurney co-sphere" is reached. (Hint: For 1:1 electrolyte, the average separation 1(in Å) $= 9.40c^{-1/3}$, where c is the concentration in mol dm^{-3}. Assume the Gurney co-sphere is two water molecules beyond the ions periphery.) (Xu)

8. Suppose the results from Exercise 7 are true and calculate the solvation number for NaCl. Comment on the reliability of the result. (Xu)

9. (a) A Raman spectrum shows that in a 4.0 *M* NaCl solution, about 40% of the waters are in the primary sheath. Estimate the solvation number n_h. (b) If the SB region consists of only one layer of water molecules, is there any bulk water left in this solution? (Xu)

10. Solvation numbers for Na^+ and Cl^- have been measured at about 5 and 1, respectively. With this information, and recalling that the number of moles per liter of pure water is 55, calculate the dielectric constant of a 5 *M* solution of NaCl. (The dielectric constant of pure water is to be taken as 80 near room temperature; when the water molecules are held immobile in respect to the variations of an applied field, *M* drops to 6.)

11. Calculate the sum of the heats of hydration of K^+ and F^-. The lattice energy is 814.6 kJ mol^{-1}. The heat of the solution is –17.15 kJ mol^{-1}.

12. Using the Born equation as representing a part of the free energy of hydration of ions, derive an expression for the entropy of Born hydration. According to this, what would be the entropy of Born solvation of the I^- ion with a radius of 200 pm?

13. Calculate the ion–solvent interaction free energy for K^+, Ca^{2+}, F^-, and Cl^- in water. The ionic radii are 133, 99, 136, and 181 pm, respectively, and the dielectric constant for water is 78.3 at 25°C. (Kim)

14. Calculate the entropy change due to ion–solvent interaction for the ions in Exercise 13 using the relation $\partial\varepsilon/\partial T = 0.356/\text{K}$ for water. (Kim)

15. Evaluate the heat of solvation for the ions in Exercise 13 in terms of the ion–dipole approach ($r_w = 138$ pm, $\mu_w = 1.8$ D). (Kim)

16. Calculate the difference of the heat of hydration for the tons in Exercise 13 between the ion–quadrupole model and the ion–dipole model ($p_w = 3.9 \times 10^{-8}$ D cm). (Kim)

17. Calculate the absolute heats of hydration of Na^+ and Cl^- using the absolute heat of hydration of H^+ of -1113.0 kJ mol^{-1}. The heat of interaction between HCl and water is -1454.0 kJ mol^{-1}; the heat of solution of NaOH is $+3.8$ kJ mol^{-1}; and the heat of sublimation of NaCl is $+772.8$ kJ mol^{-1}. (Kim)

18. The adiabatic compressibilities of water and 0.1 *M* NaI solution at 298 K are 4.473×10^{-12} Pa^{-1} and 4.428×10^{-13} Pa^{-1}, respectively. Calculate the hydration number of the NaI solution when the density of the solution is 1.0086 cm^{-3}. (Kim)

19. The adiabatic compressibility of water at 25 °C is 4.473×10^{-12} Pa^{-1}. Calculate the adiabatic compressibility of a 0.101 *M* solution of $CaCl_2$ (density at 25 °C = 1.0059 g cm^{-3}) if the hydration number of the electrolyte is 12. (Contractor)

20. Sound velocity in water is measured to be 1496.95 m s^{-1} at 25 °C. Calculate the adiabatic compressibility of water in Pa^{-1}. (Xu)

21. In the text are data on compressibilities as a function of concentration. Use the Passynski equation to calculate the total solvation number of NaBr at infinite dilution.

22. The definition of compressibility is $\beta = -(1/V)(\partial V/\partial P)_T$. On the (often made but erroneous) assumption that β is constant with pressure, find V as a function of P. Why must your equation be applicable only over a limited range of pressures?

23. Calculate the hydration number of Na^+ when the mobility of the ion in water is 44×10^{-5} cm^2 s^{-1} V^{-1} and the viscosity of the solution is 0.01 poise, $r_{Na^+} = 95$ pm and $r_w = 138$ pm. (Kim)

24. Use the infinite-dilution equality between the accelerative force for ions under an applied electric field and the viscous drag to calculate the hydration number of Cl^- in HCl aqueous solution, using the result that the transport number of the cation is 0.83, while the equivalent conductivity at infinite dilution is 304 S cm^2 mol^{-1} (25 °C). Take the radius of water as 170 pm and the corresponding viscosity of water as 0.01 poise.

25. Calculate the change of solubility of 2,4-dinitrophenol in water due to primary solvation when 0.1 M NaCl is added to the solution and the solvation number of NaCl is assumed to be 5. (Kim)

26. Calculate the change of the solubility of ethyl ether (dielectric constant, 4,3; density, 0.7138 g cm^{-3}) in water due to the secondary solvation shell when 1 M of NaCl is added to the solution ($r_{Na^+} = 180$ pm, $r_{Cl^-} = 191$ pm, $r_w = 138$ pm). (Kim)

27. Show that the electrostatic interaction potential energy between two charges Q_1 and Q_2 can be represented by the nonconventional although widely used equation

$$E = \frac{132}{\varepsilon} \frac{Q_1 * Q_2}{R_{12}}$$

where Q_1, Q_2 are expressed in electronic charge units, R_{12} is the distance between the charges expressed in Å, ε is the relative dielectric constant of the medium, and E is given in units of kcal/mol. What is the Coulombic interaction between a proton and electron by 1 Å in vacuum? (Contractor)

28. (a) Li^+ has an ionic radius of 60 pm. Calculate the work of charging the lithium ion in vacuum. (b) Repeat the calculation for Cl^{-1}, whose ionic radius is 181 pm. (c) Calculate the work of charging Li^+ surrounded by *bulk* water with a relative dielectric constant of 80 at 293 K. (Contractor)

29. A positive charge of $+e_0$, and a negative charge of $-e_0$ are separated by 53 pm. Calculate their dipole moment. (Contractor)

30. Calculate the effective moment that a gas dipole of water exhibits in the direction of an external field of 10^3 V cm^{-1} when subject to electrical orienting and thermal randomizing forces at 25 °C. The dipole moment of water $p_w = 1.87 \times 10^{-15}$ D. (Contractor)

31. Estimate the average moment of a dipole cluster of water subject to an external electric field of 2.7×10^8 V cm^{-1}. Assume the average of the cosine of the angles between the dipole moment of the central water molecule and those of its bonded neighbors is 1/3 at 298 K. (Contractor)

32. For liquid water, the relative dielectric constant at 25 °C is 78. Calculate its deformation polarizability at this temperature if there are 10^{-3} moles of deformable molecules per unit volume.* Use the average of the cosines of the angles

*The deformable molecules are taken, in this problem to be less than the total number.

between the dipole moment of the central molecule and that of its bonded neighbors as 1/3. (Contractor)

33. Water often comes under extremely strong electric fields as, e.g., in the solvation of ions. By considering the total H-bond energy of water, calculate the electric field strength that will break up H-bonded water. At what field strength would liquid water dissociate, and to what?

34. An IR spectrum has a band at a wavenumber of 1.561×10^3 cm^{-1}. What are the wavelength and frequency of the corresponding bond? If the spectrum originates from water, calculate the force constant between O and H.

35. From the data in the text, the self-diffusion coefficients of certain ions (e.g., Li^+ and I^-) are known. The diffusion coefficient is related to the rate constant for diffusion by the equation $\frac{1}{2} l^2 = \frac{D}{k}$. What kind of value for l (jump distance) would you think reasonable? With the known values for D (see text), calculate the time the anion resides in one place.

36. The effect of electrolytes on the solubility of nonelectrolytes is generally to decrease the solubility of the nonelectrolyte (salting-out). Taking O_2 as the nonelectrolyte and the relevant solubility data from the text, obtain Setchenow's constant for HCl and $Ac_2(SO_4)_3$. Comment on the great difference.

PROBLEMS

1. Use the data of Table 2.8 to calculate the mean activity coefficient of a 5 *M* NaCl solution, assuming the total hydration number at this high concentration is <3. Values for *A* and *B* of the Debye–Hückel equation can be recovered from the text.

2. Define wavenumber and explain why ions of a small radius tend to give higher librative frequencies in their IR spectra in solution. Why do hydration numbers obtained from spectroscopic data tend to be lower than those from nonspectroscopic methods? Write down the expression for the coordination number of an ion in terms of the distribution function for the O in the first shell with respect to an Na^+ ion. Draw typical plots of g_r as a function of distance from an ion for (a) a strongly hydrated and (b) a weakly hydrated ion, respectively.

3. Explain what is meant by the statement: "The *relative* heat of hydration of I^- is -1385.7 kJ mol^{-1}." Halliwell and Nyburg's value for the heat of hydration of protons was -1087.8 kJ mol^{-1}. Use this value to establish the absolute heat of hydration of I^-.

4. Calculate the librative and vibrational contributions for the entropy of an Na^+ ion in dilute solutions assuming 6 water molecules in the first shell. Why is it

usual to neglect the translational entropy? (Give a numerical answer involving the free volume of Na^+ in water to this part of the question.)

5. Calculate the heats of interaction between individual ions and water: (a) $\Delta H_{Cl^--H_2O}$, (b) $\Delta H_{Na^+-H_2O}$, and (c) $\Delta H_{Br^--H_2O}$ using the experimental values of the heats of interaction between a salt and water, ΔH_{S-H_2O} at 25 °C (the Born model is considered valid). (Constantinescu)

Salt	ΔH_{S-H_2O} / kJ mol^{-1}		Ionic Radius / / pm
KF	−827.6	Na^+	95
KCl	−685.3	K^+	133
NaF	−911.2	F^-	136
NaBr	−791.8	Cl^-	181
		Br^-	195

6. Calculate the ion–water interaction for Cl^- using the ion-dipole model at 298 K. Use four coordination. (Contractor)

7. Calculate ΔH_{soln} and ΔH_h for AgCl, AgBr, and AgI and rationalize their low solubility (using the data in Table P.1). (Xu)

8. Calculate $\Delta H_{Cl^--H_2O}$ at 298 K. The ionic radius of Cl^- is 1.81 Å. The dielectric constant of water at three temperatures is ε (20°C) = 80.1, ε (25°C) = 78.3, ε (30°C) = 76.54. Use the Born model. (Contractor)

9. Estimate the error introduced by ignoring the size of the solvent molecules in calculating the heat of the Born–charging process of a Cs^+ ion interaction with water–water ($r_{Cs^+} = 169$ pm). (Contractor)

10. Estimate the error introduced in calculating the heat of an NaBr interaction with water by ignoring the distortability of the water molecules $r_{Bi} = 1.95$ Å, $r_{Na^+} = 0.95$ Å, $p_w = 3.9 \times 10^8$ D cm, $\alpha_w = 1.46 \times 10^{-26}$ cm^3, $\varepsilon_w(298\ \text{K}) = 78.3$, $\partial\varepsilon_w/\partial T = -0.356$ K^{-1}).

11. (a) Using data for solution enthalpy (ΔH_{soln}) and lattice enthalpy ($\Delta H_{lattice}$) (see Table P.2), calculate the hydration heat for various alkali halides (ΔH_h). Comment on the possible major error source. (b) Using the cation radii data, explain

TABLE P.1

	AgCl	AgBr	AgI
Equilibrium constant, K	1.77×10^{-10}		7.7×10^{-13}
$-S_{soln}$ (kJ K^{-1}mol^{-1})		32.98	51
$-H_{lattice}$ (kJ mol^{-1})			−91

TABLE P.2

	LiCl	LiBr	NaCl	NaBr	KCl	KBr
ΔH_{soln} (kJ/mol)	−37	−48.8	+3.89	−0.6	17.22	19.9
ΔS_{soln} (J/mol)	48	43	110	120	135	140
$\Delta H_{lattice}$ (kJ/mol)	−815	−787	−752	−717	−689	−653
r (cation) (Å)	0.59	0.59	1.02	1.02	1.38	1.38

the trend and the sign of ΔH_{soln}. What is the driving force of the solution process in case $\Delta H_{soln} > 0$? (Xu)

12. Chloride is surrounded by 4 water molecules. Calculate the ion–dipole interaction work ($r_{Cl^-} = 181$ pm). (Contractor)

13. Anomalous salting in is said to occur when the dipole moment of water is greater than that of the organic molecule concerned, but the solubility of the latter *increases* when ions are added to the solution. What kind of model could explain the observation that anomalous salting in occurs when the electrolyte consists of large ions?

    From the model you derive, calculate the minimum radius of ions required to salt-in benzoic acid in a 1 *M* solution of electrolyte. (Take the polarizability as r_i^3, where r_i is the radius of any entity involved.)

14. An ion of charge ze_0 and radius r is transferred from a solvent of dielectric constant ε_i to a solvent of dielectric constant $\varepsilon_{i'}$. Derive an expression for the free-energy change associated with this transfer using the Born model. (Contractor)

15. What is the free-energy change involved in transferring Cl^- from water to a nonpolar medium like carbon tetrachloride with a dielectric constant of 2.23 at 298 K? Is this an energetically favorable process? (Make use of Problem 3.) Water has a dielectric constant of 78.54 at 298 K. (Contractor)

16. Living cells are surrounded by membranes and on either side of the membrane an aqueous environment is present. The interior of the membrane is highly nonpolar. Based on the result of Problem 15, can you explain why it is difficult to transport or move charges across a membrane in a living cell? (Contractor)

17. For a given water molecule, what is the *maximum* number of hydrogen bonds that can be formed with other neighboring water molecules? Are these hydrogen bonds identical in bonding nature for this particular water molecule? Why? If there is a difference in bonding nature for these hydrogen bonds, how would you

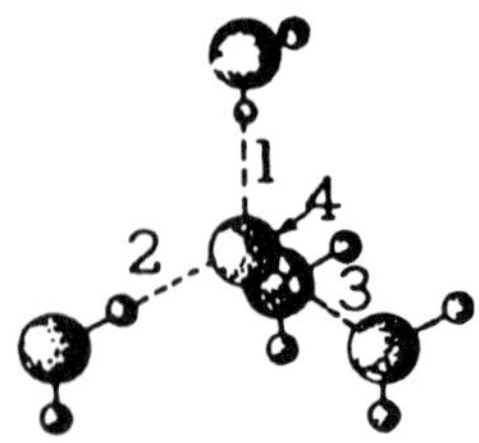

Fig. P2.1

differentiate these experimentally? (*Maximum* is emphasized because in a liquid state the actual number of hydrogen bonds per molecule is less than the maximum possible. Even in crystal ice II there are dangling hydrogen atoms, which do not participate in hydrogen bonding.) (Xu)

18. In an ion–solvent interaction model (Fig. P2.1), "solvated coordinated water" has two sites capable of forming hydrogen bonds with water molecules in the SB region. Are these two sites identical in bonding? For "nonsolvated coordinated water," there are three sites for hydrogen bonds. Are these three identical? Why? (Xu)

19. (a) Calculate the volume of bare and solvated ions sheathed by water, using the following ionic and water radii: $r_{Li^+} = 54$ pm, $r_{Na^+} = 102$ pm, $r_{K^+} = 138$ pm, and $r_w = 158$ pm. (b) Suppose the structure-breaking region consists of two layers of water molecules. Calculate the volume of water that has been affected by a single ion. (Xu)

20. Using the above results, calculate the percentage of bulk water in a 0.05 *M* NaCl solution. What if the concentration is 0.5 *M*? Comment on the significance of the result (assume the anion has the same SB structure as for cations ($r_{Cl^-} = 181$ pm). (Xu)

TABLE P.3

c_2, mol dm^{-3}	β (LiCl)	β (NaCl)	β (KCl)	β ($MgCl_2$)
0.05	44.49	44.45	44.46	44.21
0.09				43.88
0.10	44.30	44.20		
1.00	40.60	40.04	39.84	
1.05				35.94
2.00	37.15	36.64		
4.10			29.88	
5.00	29.93	26.75		

TABLE P.4

	LiCl	NaCl	KCl
ε at 25 °C	64.9	66.7	68.1

21. (a) Justify that in a dilute solution solvation number $n_s = [1 - (\beta/\beta_0)] \times (55.56/c_2)$, where c_2 is the electrolyte concentration in mol dm^{-3}. (b) The compressibility of LiCl, NaCl, KCl, and $MgCl_2$ are measured at 25 °C (Table P.3 in 10^{-6} bar^{-1}). Calculate the total solvation number of these electrolytes. (Xu)

22. Table P.4 lists measured dielectric constants at 25 °C for 1.0 *M* LiCl, NaCl, and KCl solutions, respectively. Calculate the percentage of water in the primary sheath and the total solvation number. Compare the results with those of the compressibility method (see problem 21) and comment on their reliability.

23. The densities of aqueous NaCl solutions at 25°C are given as a function of NaCl molality in Table P.5. (a) Obtain a graph for the partial molar volumes of both water V_1, and NaCl V_2, as a function of NaCl molality. Compare the limiting cases V_1 ($m \to 0$), V_2 ($m \to$ sat) with those of pure water and pure NaCl molar volumes, V_1° and V_2°, respectively. (b) Calculate V_2 for m = 0.5 and m = 2. (Mussini)

TABLE P.5

m NaCl / mol kg^{-1}	ρ / kg dm^{-3}
0	0.99707
0.11094	1.00158
0.23631	1.00663
0.56874	1.01970
0.85382	1.03071
1.47458	1.05353
2.51393	1.08963
3.09392	1.10849
3.9873	1.13600
5.24324	1.17290
5.4952	1.18030
5.8023	1.18880
5.82267	1.18888

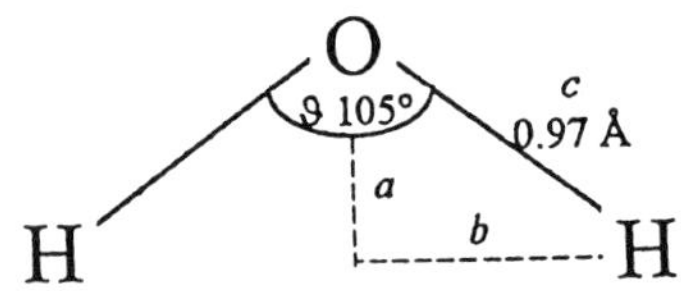

Fig. P2.2

24. (a) Calculate the dipole moment associated with water in the hypothesis of fully ionic OH bondings (i.e., considering a $-2e$ charge on the O atom and a + le charge on each H atom; e = elementary charge taken without sign). Draw the associated vector (refer to Fig. P2.2). (b) Knowing the dipole moment of water, write down an equation expressing it in terms of f the fractional charge on either end of the dipole and x the distance between the center of charge and the atomic nucleus. (c) Determine f and x in the hypothesis of a tetrahedrical angle HOH and draw the corresponding vector T. (Mussini)

25. (a) Write an expression for the potential in a point P of the electric field created by a dipole. (b) Then write an expression for the potential energy of the ion–dipole interaction. (c) Calculate the potential energy, E_p, of an ion-dipole interaction (in kJ mol^{-1}) between water (μ_w = 1.86 D) and a z-valent cation, as a function of the distance r, the ion charge z_i, the angle I and the relative permittivity ε_r. (d) Perform a complete calculation for the limiting cases $z_i = 1$, $r = 2$ Å, $\varepsilon_r = 4.5$ and $z_i = 1$, $r = 6$ Å, $\varepsilon_r = 80$. Assume that the relative positions are as in Fig. P.2.3 and the negative end of the water dipole faces the positive ion. (e) Assuming the intermediate values ε_r to increase exponentially, draw a complete E_p versus r characteristic in the interval $2 < r < 6$, and mark the region

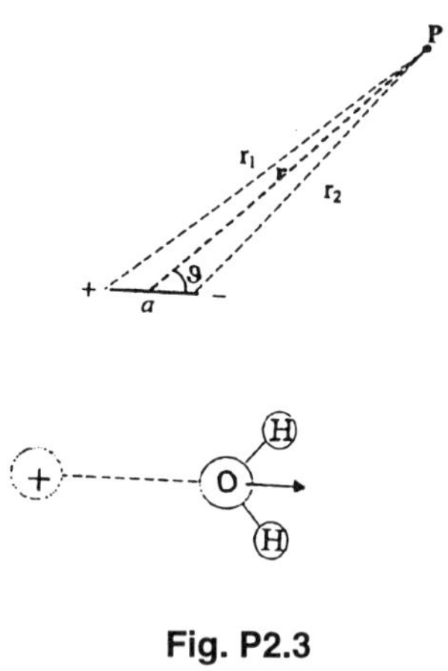

Fig. P2.3

in which the thermal energy, ~ RT, is competitive with the electric interaction. (Mussini)

26. The heat of hydration of an electrolyte is obtained by measuring the heat of dissolution at various concentrations and extrapolating its values to $c_{\text{electrolyte}} = 0$. This value is then used in conjunction with the lattice energy to yield the desired heat. Explain, then, how the free energy (hence the entropy) of the hydration of an electrolyte is obtained. Apart from the heat of solution, what other essential measurements would be necessary?

27. Table P.6 gives the density of a number of solutions of AgCl. Find the apparent molar volume of the electrolyte at each concentration. Then, by using the method described in the text, find the partial molar volume of the electrolyte at infinite dissolution.

28. The (idealized) radius of a polystyrene styrene sulfonate may be several hundred angstroms. Were you to measure the self-diffusion coefficient, what equation would you use to obtain a measure of the ion's size? Explain the principles (showing appropriate equations) of obtaining an individual ionic entropy and an individual ionic entropy of hydration.

TABLE P.6

c/M	$\rho/\text{g cm}^{-3}$
0.0010294	1.000170
0.0051484	1.000862
0.010904	1.001823
0.018974	1.003175
0.045680	1.00763
0.080311	1.01339
0.11868	1.01970
0.23881	1.03944
0.47773	1.0786
0.84786	1.1396
1.2005	1.1973
1.4717	1.2420
1.7321	1.2846
2.0451	1.3358
2.4640	1.4017

MICRO RESEARCH PROBLEMS

1. Define electrostriction.

   In the functioning of drugs, the physiological effects are dependent on the precise structure of the molecule. Among other reasons, this is because of the need to "fit" onto the structure of a relevant enzyme. In work on theoretical drug designs, some computations are made on models of the organic molecules constituting the drug in a force-free field, i.e., in a hypothetical vacuum.

   Consider an $(-NH_3)^+$ group associated with a designer drug. What happens to the group when it is introduced into a saline solution 0.4 *M* in NaCl? Examine the likely hydration in respect to energy and structure. Calculate the change in volume of the $(-NH_3)^+$ group due to electrostriction. Examine carboxypeptides (Fig. 2.4) and make computations that would lead to an estimate of the stabilization energy due to hydration.

2. (a) The origin of the hydrogen bond is the intermolecular dipole interaction caused by the polarized covalent bond. The existence of this additional intermolecular force accounts for the abnormally high boiling point for water. Using the electronegativity (χ) data provided, try to rationalize semiquantitatively the elevation of boiling point in terms of bond polarization for a series of hydrides (*Fig. 1*). (χ: H 2.2; N 3.0; O 3.4; F 4.0; Zn 1.6. Hint: The total intermolecular force caused by the H bond is proportional to the total number of H bonds and dipole moment of an individual H bond, which in turn vary approximately with X. (b) The elevation of boiling point is by no means unique to hydrogen. Explain the abnormal boiling point of ZnF_2 in the light of the "zinc bond." Compare the "zinc bond" with the hydrogen bond in (a) and account for the especially strong effect of the "zinc bond" on boiling point. (Assume that each ZnF_2 molecule can form 6 Zn bonds.)

CHAPTER 3

ION–ION INTERACTIONS

3.1. INTRODUCTION

A model has been given for the breaking up of an ionic crystal into free ions which stabilize themselves in solution with solvent sheaths. One central theme guided the account, the interaction of an ion with its neighboring water molecules.

However, ion–solvent interactions are only part of the story relating an ion to its environment. When an ion looks out upon its surroundings, it sees not only solvent dipoles but also other ions. The mutual interaction between these ions constitutes an essential part of the picture of an electrolytic solution.

Why are ion–ion interactions important? Because, as will be shown, they affect the equilibrium properties of ionic solutions, and also because they interfere with the drift of ions, for instance, under an externally applied electric field (Chapter 4).

Now, the degree to which these interactions affect the properties of solutions will depend on the mean distance apart of the ions, i.e., on how densely the solution is populated with ions, because the interionic fields are distance dependent. This ionic population density will in turn depend on the nature of the electrolyte, i.e., on the extent to which the electrolyte gives rise to ions in solution.

3.2. TRUE AND POTENTIAL ELECTROLYTES

3.2.1. Ionic Crystals Form True Electrolytes

An important point to recall regarding the dissolution of an ionic crystal (Chapter 2) is that ionic lattices consist of ions *even before they come in contact with a solvent.* In fact, all that a polar solvent does is to use ion–dipole (or ion–quadrupole) forces to disengage the ions from the lattice sites, solvate them, and disperse them in solution.

Such ionic crystals are known as true electrolytes or *ionophores* (the Greek suffix *phore* means "bearer of"; thus, an ionophore is a "substance that bears ions"). When a true electrolyte is melted, its ionic lattice is dismantled and the pure liquid true electrolyte shows considerable ionic conduction (Chapter 2). Thus, the characteristic of a true electrolyte is that in the pure liquid form it is an ionic conductor. All salts belong to this class. Sodium chloride therefore is a typical true electrolyte.

3.2.2. Potential Electrolytes: Nonionic Substances that React with the Solvent to Yield Ions

A large number of substances (e.g., organic acids) show little conductivity in the pure liquid state. Evidently there must be some fundamental difference in structure between organic acids and inorganic salts, and this difference is responsible for the fact that one pure liquid (the true electrolyte) is an ionic conductor and the other is not.

What is this difference between, say, sodium chloride and acetic acid? Electron diffraction studies furnish an answer. They show that gaseous acetic acid consists of separate, neutral molecules and the bonding of the atoms inside these molecules is essentially nonionic. These neutral molecules retain their identity and separate existence when the gas condenses to give liquid acetic acid. Hence, there are hardly any ions in liquid acetic acid and therefore little conductivity.

Now, the first requirement of an electrolyte is that it should give rise to a conducting solution. From this point of view, it appears that acetic acid will never answer the requirements of an electrolyte; it is nonionic. When, however, acetic acid is dissolved in water, an interesting phenomenon occurs: ions are produced, and therefore the solutions conduct electricity. Thus, acetic acid, too, is a type of electrolyte; it is not a true electrolyte, but a potential one ("one which can, but has not yet, become"). Potential electrolytes are also called *ionogens*, i.e., "ion producers."

How does acetic acid, which does not consist of ions in the pure liquid state, generate ions when dissolved in water? In short, how do potential electrolytes work? Obviously, there must be some reaction between neutral acetic acid molecules and water, and this reaction must lead to the splitting of the acetic acid molecules into charged fragments, or ions.

A simple picture is as follows. Suppose that an acetic acid molecule collides with a water molecule and in the process the H of the acetic acid OH group is transferred from the oxygen atom of the OH to the oxygen atom of the H_2O. A proton has been transferred from CH_3COOH to H_2O.

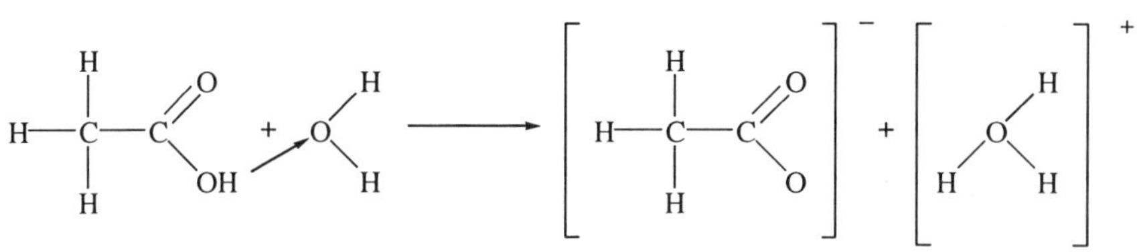

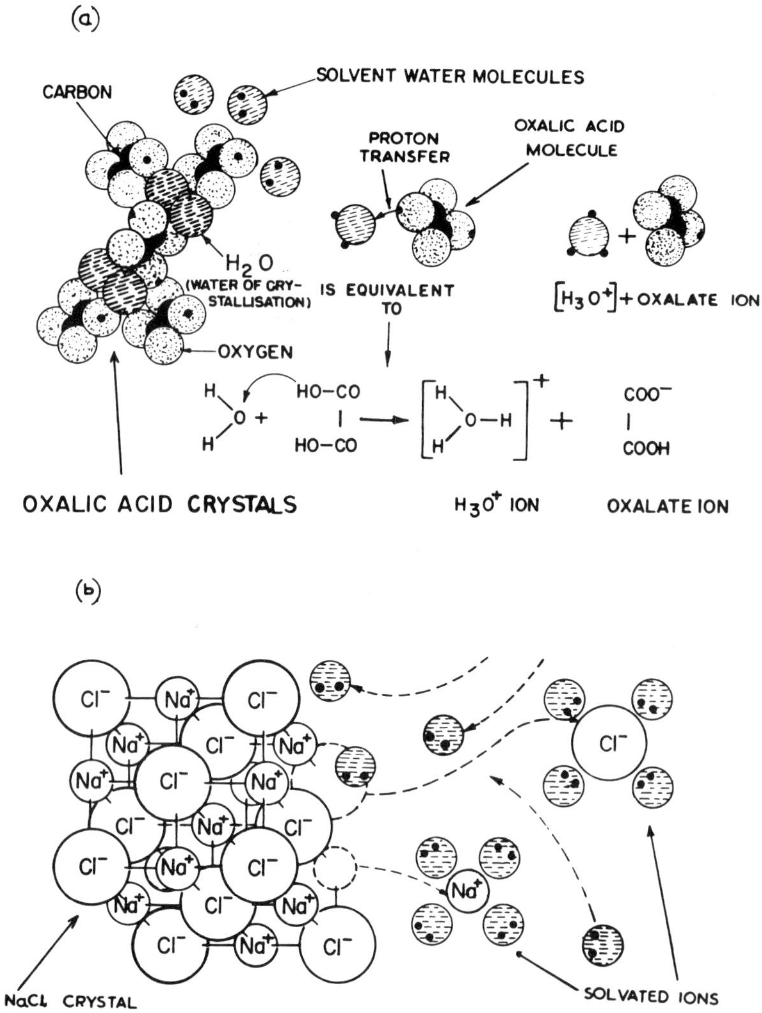

Fig. 3.1. Schematic diagram to illustrate the difference in the way potential electrolytes and true electrolytes dissolve to give ionic solutions: (a) Oxalic acid (a potential electrolyte) undergoes a proton-transfer chemical reaction with water to give rise to hydrogen ions and oxalate ions. (b) Sodium chloride (a true electrolyte) dissolves by the solvation of the Na^+ and Cl^+ ions in the crystal.

The result of the proton transfer is that two ions have been produced: (1) an acetate ion and (2) a hydrated proton. Thus, potential electrolytes (organic acids and most bases) dissociate into ions by ionogenic, or ion-forming, chemical reactions with solvent molecules, in contrast to true electrolytes, which often give rise to ionic solutions by physical interactions between ions present in the ionic crystal and solvent molecules (Fig. 3.1).

3.2.3. An Obsolete Classification: Strong and Weak Electrolytes

The classification into true and potential electrolytes is a modern one. It is based on a knowledge of the structure of the electrolyte: whether in the pure form it consists of an ionic lattice (true electrolytes) or neutral molecules (potential electrolytes) (Fig. 3.2). It is not based on the behavior of the solute in any particular solvent.

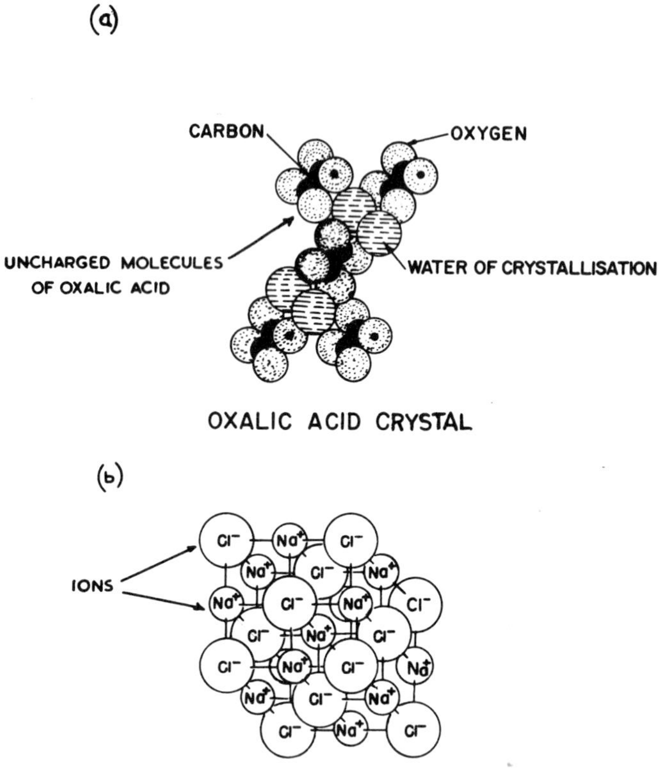

Fig. 3.2. Electrolytes can be classified as (a) potential electrolytes (e.g., oxalic acid), which in the pure state consist of uncharged molecules, and (b) true electrolytes (e.g., sodium chloride), which in the pure state consist of ions.

TABLE 3.1
Conductance Behavior of Substances in Different Media

	Equivalent Conductance	
	Water	Liquid Ammonia
NaCl	106.7	284.0
Acetic acid	4.7	216.6

Historically, however, the classification of electrolytes was made on the basis of their behavior in one particular solvent, i.e., water. *Weak* electrolytes were those that yielded relatively poorly conducting solutions when dissolved in water, and *strong* electrolytes were those that gave highly conducting solutions when dissolved in water.

The disadvantage of this classification into strong and weak electrolytes lies in the following fact: As soon as a different solvent (i.e., a nonaqueous solvent) is chosen, what was a strong electrolyte in water may behave as a weak electrolyte in the nonaqueous solvent. For example, sodium chloride behaves like a strong electrolyte (i.e., yields highly conducting solutions) in water and acetic acid behaves like a weak electrolyte. In liquid ammonia, however, the conductance behavior of acetic acid is similar to that of sodium chloride in water, i.e., the solutions are highly conducting (Table 3.1). This is an embarrassing situation. Can one say: Acetic acid is weak in water and strong in liquid ammonia? What is wanted is a classification of electrolytes that is independent of the solvent concerned. The classification into true and potential electrolytes is such a classification. It does not depend on the solvent, but rather upon the degree of ionicity of the substance constituting the solid lattice.

3.2.4. The Nature of the Electrolyte and the Relevance of Ion–Ion Interactions

Solutions of most potential electrolytes in water generally contain only small concentrations of ions, and therefore ion–ion interactions in these solutions are negligible; the ions are on the average too far apart. The behavior of such solutions is governed predominantly by the position of the equilibrium in the proton-transfer reaction between the potential electrolyte and water.

In contrast, true electrolytes are completely dissociated into ions when the parent salts are dissolved in water. The resulting solutions generally consist only of solvated ions and solvent molecules. The dependence of many of their properties on concentration (and therefore mean distance apart of the ions in the solution) is determined by the interactions between ions. To understand these properties, one must understand ion–ion interactions.

3.3. THE DEBYE–HÜCKEL (OR ION-CLOUD) THEORY OF ION–ION INTERACTIONS

3.3.1. A Strategy for a Quantitative Understanding of Ion–Ion Interactions

The first task in thinking in detail about ion–ion interactions is to evolve a quantitative measure of these interactions.[1] One approach is to follow a procedure similar to that used in the discussion of ion–solvent interactions (Section 2.4). Thus, one can consider an initial state in which ion–ion interactions do not exist (are "switched off") and a final state in which the interactions are in play (are "switched on"). Then, the free-energy change in going from the initial state to the final state can be considered the free energy ΔG_{I-I} of ion–ion interactions (Fig. 3.3).

The final state is obvious; it is ions in solution. The initial state is not so straightforward; one cannot take ions in vacuum, because then there will be ion–solvent interactions when these ions enter the solvent. The following approach is therefore adopted. One conceives of a hypothetical situation in which the ions are there in solution but are nevertheless not interacting. Now, if ion–ion interactions are assumed to be electrostatic in origin, then the imaginary initial state of noninteracting ions implies an assembly of discharged ions.

Thus, the process of going from an initial state of noninteracting ions to a final state of ion–ion interactions is equivalent to taking an assembly of discharged ions, charging them up, and setting the electrostatic charging work equal to the free energy ΔG_{I-I} of ion–ion interactions (Fig. 3.4).

One point about the above procedure should be borne in mind. Since, in the charging process, both the positively charged and negatively charged ionic species are charged up, one obtains a free-energy change that involves *all* the ionic species

[1]The question of how one obtains an experimental measure of ion–ion interactions is discussed in Section 3.4.

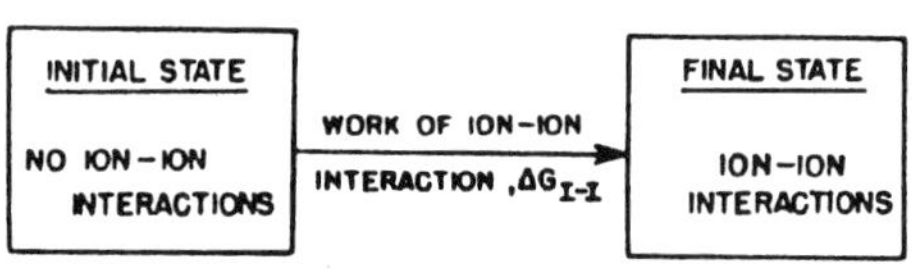

Fig. 3.3. The free energy ΔG_{I-I} of ion–ion interactions is the free-energy change in going from a hypothetical electrolytic solution, in which ion-ion interactions do not operate, to a real solution, in which these interactions do operate.

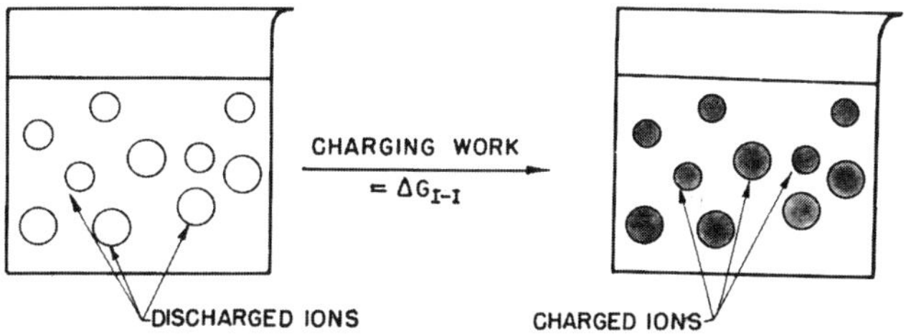

Fig. 3.4. The free energy ΔG_{I-I} of ion–ion interactions is the electrostatic work of taking an imaginary assembly of discharged ions and charging them up to obtain a solution of charged ions.

constituting the electrolyte. Generally, however, the desire is to isolate the contribution to the free energy of ion–ion interactions arising from one ionic species i only. This partial free-energy change is by definition the *chemical-potential change* $\Delta\mu_{i-I}$ arising from the interactions of one ionic species with the ionic assembly.

To compute this chemical-potential change $\Delta\mu_{i-I}$, rather than the free-energy change ΔG_{I-I}, one must adopt an approach similar to that used in the Born theory of solvation. One thinks of an ion of species i and imagines that this reference ion alone of all the ions in solution is in a state of zero charge (Fig. 3.5). If one computes the work of charging up the reference ion (of radius r_i) from a state of zero charge to its final charge of z_ie_0, then the charging work W times the Avogadro number N_A is equal to the partial molar free energy of ion–ion interactions, i.e., to the chemical potential of ion–ion interactions:

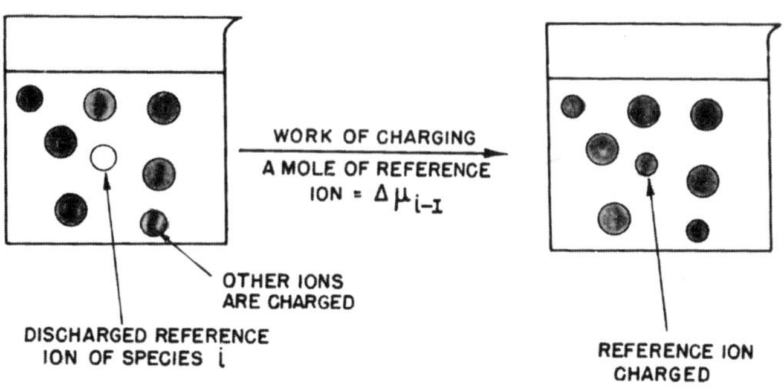

Fig. 3.5. The chemical potential $\Delta\mu_i$ arising from the interactions of an ionic species i with the electrolytic solution is equal to the Avogadro number times the electrostatic work of taking an imaginary solution in which one reference ion alone is discharged and charging this reference ion up to its normal charge.

$$\Delta\mu_{i-I} = N_A W \tag{3.1}$$

Then as the work of charging an electrical conductor is proved to be

$$W = 1/2\ [(\text{charge on the conductor}) \times (\text{conductor's electrostatic potential})] \tag{3.2}$$

one obtains

$$\Delta\mu_{i-I} = N_A W = \frac{N_A z_i e_0}{2}\,\psi \tag{3.3}$$

where ψ is the electrostatic potential of the ion due to the influence on it by the electrostatic interactions of the surrounding field.

The essence of the task therefore in computing the chemical-potential change due to the interactions of the ionic species i with the ionic solution is the calculation of the electrostatic potential produced at a reference ion by the rest of the ions in solution. Theory must aim at this quantity.

If one knew the *time-averaged spatial distribution* of the ions, then one could find out how all the other charges are distributed as a function of distance from the reference ion. At that stage, one of the fundamental laws of electrostatics could be used, namely, the law of the superposition of potentials, according to which the potential at a point due to an assembly of charges is the sum of the potentials due to each of the charges in the assembly.

Thus, the problem of calculating the chemical-potential change $\Delta\mu_{i-I}$ due to the interactions between one ionic species and the assembly of all the other ions has been reduced to the following problem: On a time average, how are the ions distributed around any specified ion? If that distribution became known, it would then be easy to calculate the electrostatic potential of the specified ion due to the other ions and then, by Eq. (3.3), the energy of that interaction. Thus, the task is to develop a model that describes the equilibrium spatial distribution of ions inside an electrolytic solution and then to describe that model mathematically.

3.3.2. A Prelude to the Ionic-Cloud Theory

A spectacular advance in the understanding of the distribution of charges around an ion in solution was achieved in 1923 by Debye and Hückel. It is as significant in the understanding of ionic solutions as the Maxwell theory of the distribution of velocities is in the understanding of gases.

Before going into the details of their theory, a moment's reflection on the magnitude of the problem will promote appreciation of their achievement. Consider, for example, a 10^{-3} M (mol dm^{-3}) aqueous solution of sodium chloride. There will be $10^{-6} \times 6.023 \times 10^{23}$ sodium ions per cubic centimeter of solution and the same number of chloride ions, together, of course, with the corresponding number of water molecules. Nature takes these $2 \times 6.023 \times 10^{17}$ ions cm^{-3} and arranges them so that there is a particular time-averaged[2] spatial distribution of the ions. The number of particles involved is enormous, and the situation appears far too complex for mathematical treatment.

However, there exist conceptual techniques for tackling complex situations. One of them is model building. This involves conceiving a model that contains only the essential features of the real situation. All the thinking and mathematical analysis is done on the (relatively simple) model and then the theoretical predictions are compared with the experimental behavior of the real system. A good model simulates nature. If the model yields wrong answers, then one tries again by changing the imagined model until one arrives at a model, the theoretical predictions of which agree well with experimental observations.

The genius of Debye and Hückel lay in their formulation of a very simple but powerful model for the time-averaged distribution of ions in very dilute solutions of electrolytes. From this distribution they were able to obtain the electrostatic potential contributed by the surrounding ions to the total electrostatic potential at the reference ion and hence the chemical-potential change arising from ion–ion interactions [Eq.(3.3)]. Attention will now be focused on their approach.

The electrolytic solution consists of solvated ions and water molecules. The first step in the Debye–Hückel approach is to select arbitrarily any one ion out of the assembly and call it a *reference ion* or *central ion*. Only the reference ion is given the individuality of a discrete charge. What is done with the water molecules and the remaining ions? The water molecules are looked upon as a continuous dielectric medium. The remaining ions of the solution (i.e., all ions except the central ion) lapse into anonymity, their charges being "smeared out" into a continuous spatial distribution of charge (Fig. 3.6). Whenever the concentration of ions of one sign exceeds that of the opposite sign, there will arise a net or excess charge in the particular region under consideration. Obviously, the total charge in the atmosphere must be of opposite sign and exactly equal to the charge on the reference ion.

Thus, the electrolytic solution is considered to consist of a central ion standing alone in a continuum. Thanks to the water molecules, this continuum acquires a dielectric constant (taken to be the value for bulk water). The charges of the discrete ions that populate the environment of the central ion are thought of as smoothed out and contribute to the continuum dielectric a net charge density (excess charge per unit volume). In this way, water enters the analysis in the guise of a dielectric constant ε; and the ions, except the specific one chosen as the central ion, in the form of an excess charge density ϱ (Fig. 3.7).

[2]Using an imaginary camera (with exposure time of $\sim 10^{-12}$ s), suppose that it were possible to take snapshots of the ions in an electrolytic solution. Different snapshots would show the ions distributed differently in the space containing the solution, but the scrutiny of a large enough number of snapshots (say, $\sim 10^{12}$) would permit one to recognize a certain average distribution characterized by average positions of the ions; this is the time-averaged spatial distribution of the ions.

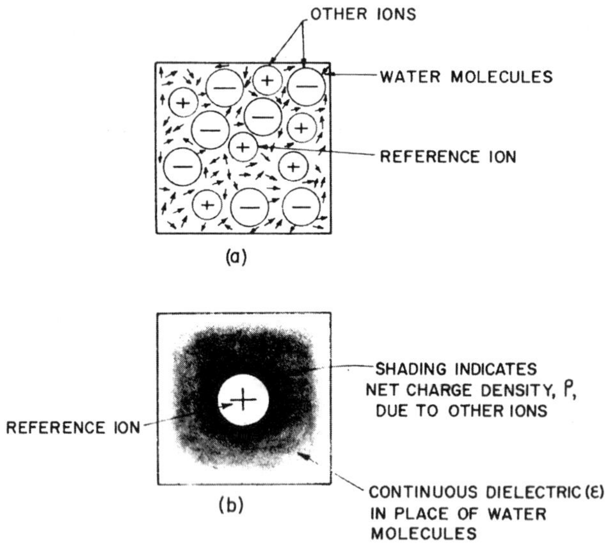

Fig. 3.6. A schematic comparison of (a) the assembly of ions and solvent molecules that constitute a real electrolytic solution and (b) the Debye–Hückel picture in which a reference ion is surrounded by net charge density ϱ due to the surrounding ions and a dielectric continuum of the same dielectric constant ε as the bulk solvent.

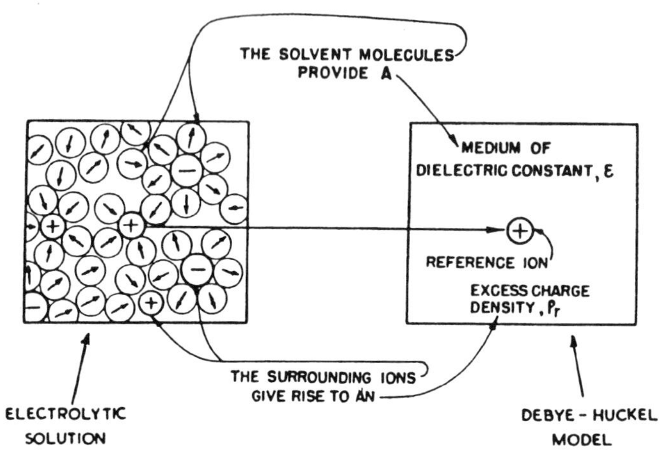

Fig. 3.7. The Debye–Hückel model is based upon selecting one ion as a reference ion, replacing the solvent molecules by a continuous medium of dielectric constant ε and the remaining ions by an excess charge density ϱ_r (the shading usually used in this book to represent the charge density is not indicated in this figure).

Hence, the complicated problem of the time-averaged distribution of ions inside an electrolytic solution reduces, in the Debye–Hückel model, to the mathematically simpler problem of finding out how the excess charge density ϱ varies with distance r from the central ion.

An objection may be raised at this point. The electrolytic solution as a whole is electroneutral, i.e., the net charge density ϱ is zero. Then why is not $\varrho = 0$ everywhere?

So as not to anticipate the detailed discussion, an intuitive answer will first be given. If the central ion is, for example, positive, it will exert an attraction for negative ions; hence, there should be a greater aggregation of negative ions than of positive ions in the neighborhood of the central positive ion, i.e., $\varrho \neq 0$. An analogous situation, but with a change in sign, obtains near a central negative ion. At the same time, the thermal forces are knocking the ions about in all directions and trying to establish electroneutrality, i.e., the thermal motions try to smooth everything to $\varrho = 0$. Thus, the time average of the electrostatic forces of ordering and the thermal forces of disordering is a local excess of negative charge near a positive ion and an excess of positive charge near a negative ion. Of course, the excess positive charge near a negative ion compensates for the excess negative charge near a positive ion, and the overall effect is electroneutrality, i.e., a ϱ of zero for the whole solution.

3.3.3. Charge Density near the Central Ion Is Determined by Electrostatics: Poisson's Equation

Consider an infinitesimally small volume element dV situated at a distance r from the arbitrarily selected central ion, upon which attention is to be fixed during the discussion (Fig. 3.8), and let the net charge density inside the volume element be ϱ_r. Further, let the average[3] electrostatic potential in the volume element be ψ_r. The question is: What is the relation between the excess density ϱ_r in the volume element and the time-averaged electrostatic potential ψ_r?

One relation between ϱ_r and ψ_r is given by Poisson's equation (Appendix 3.1). There is no reason to doubt that there is spherically symmetrical distribution of positive and negative charge and therefore excess charge density around a given central ion. Hence, Poisson's equation can be written as

$$\frac{1}{r^2}\frac{d}{dr}\left(r^2\frac{d\psi_r}{dr}\right) = -\frac{4\pi}{\varepsilon}\varrho_r \tag{3.4}$$

where ε is the dielectric constant of the medium and is taken to be that of bulk water, an acceptable approximation for dilute solutions.

[3]Actually, there are discrete charges in the neighborhood of the central ion and therefore discontinuous variations in the potential. But because in the Debye–Hückel model the charges are smoothed out, the potential is averaged out.

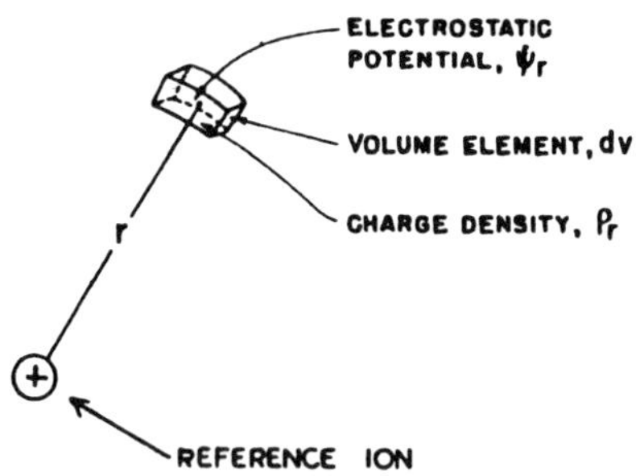

Fig. 3.8. At a distance r from the reference ion, the excess charge density and electrostatic potential in an infinitesimal volume element dV are ϱ_r and ψ_r, respectively.

3.3.4. Excess Charge Density near the Central Ion Is Given by a Classical Law for the Distribution of Point Charges in a Coulombic Field

The excess charge density in the volume element dV is equal to the total ion density (total number of ions per unit volume) times the charge on these ions. Let there be, per unit volume, n_1 ions of type 1, each bearing charge z_1e_0, n_2 of type 2 with charge z_2e_0, and n_i of type i with charge z_ie_0, where z_i is the valence of the ion and e_0 is the electronic charge. Then, the excess charge density ϱ_r in the volume element dV is given by

$$\varrho_r = n_1z_1e_0 + n_2z_2e_0 + \ldots + n_iz_ie_0 \tag{3.5}$$

$$= \sum n_iz_ie_0 \tag{3.6}$$

In order to proceed further, one must link up the unknown quantities $n_1, n_2, \ldots, n_i, \ldots$ to known quantities. The link is made on the basis of the Boltzmann distribution law of classical statistical mechanics. Thus, one writes

$$n_i = n_i^0 e^{-U/(kT)} \tag{3.7}$$

where U can be described either as the change in potential energy of the i particles when their concentration in the volume element dV is changed from the bulk value n_i^0 to n_i or as the work that must be done by a hypothetical external agency against the time average of the electrical and other forces between ions in producing the above concentration change. Since the potential energy U relates to the time average of the forces between ions rather than to the actual forces for a given distribution, it is also known as the *potential of average force.*

If there are no ion–ion interactional forces, $U = 0$; then, $n_i = n_i^0$, which means that the local concentration would be equal to the bulk concentration. If the forces are attractive, then the potential-energy change U is negative (i.e., negative work is done by the hypothetical external agency) and $n_i > n_i^0$; there is a local accumulation of ions in excess of their bulk concentrations. If the forces are repulsive, the potential-energy change is positive (i.e., the work done by the external agency is positive) and $n_i < n_i^0$; there is local depletion of ions.

In the first instance, and as a first approximation valid for very dilute solutions, one may ignore all types of ion–ion interactions except those deriving from simple Coulombic[4] forces. Thus, short-range interactions (e.g., dispersion interactions) are excluded. This is a fundamental assumption of the Debye–Hückel theory. Then the potential of average force U simply becomes the *Coulombic* potential energy of an ion of charge $z_i e_0$ in the volume element dV, i.e., the charge $z_i e_0$ on the ion times the electrostatic potential ψ_r in the volume element dV. That is,

$$U = z_i e_0 \psi_r \tag{3.8}$$

The Boltzmann distribution law (3.7) thus assumes the form

$$n_i = n_i^0 e^{-z_i e_0 \psi_r/(kT)} \tag{3.9}$$

Now that n_i, the concentration of the ionic species i in the volume element dV, has been related to its bulk concentration n_i^0, the expression (3.6) for the excess charge density in the volume element dV becomes

$$\varrho_r = \sum_i n_i z_i e_0 = \sum_i n_i^0 z_i e_0 e^{-z_i e_0 \psi_r/(kT)} \tag{3.10}$$

3.3.5. A Vital Step in the Debye–Hückel Theory of the Charge Distribution around Ions: Linearization of the Boltzmann Equation

At this point of the theory, Debye and Hückel made a move that was not only mathematically expedient but also turned out to be wise. They decided to carry out the

---

[4]In this book, the term *Coulombic* is restricted to forces (with r^{-2} dependence on distance) which are based directly on Coulomb's law. More complex forces, e.g., those which vary as r^{-4} or r^{-7}, may occur as a net force from the combination of several different Coulombic interactions. Nevertheless, such more complex results of the interplay of several Coulombic forces will be called *non-Coulombic.*

analysis only for systems in which the average electrostatic potential ψ_r would be much smaller than the thermal energy kT. Then:

$$z_i e_0 \psi_r << kT \quad \text{or} \quad \frac{z_i e_0 \psi_r}{kT} << 1 \tag{3.11}$$

Based on this assumption, one can expand the exponential of Eq. (3.10) in a Taylor series, i.e.,

$$e^{-z_i e_0 \psi_r/(kT)} = 1 - \frac{z_i e_0 \psi_r}{kT} + \frac{1}{2}\left(\frac{z_i e_0 \psi_r}{kT}\right)^2 + \ldots \tag{3.12}$$

and neglect all except the first two terms. Thus, in Eq. (3.10),

$$\varrho_r = \sum_i n_i^0 z_i e_0 \left(1 - \frac{z_i e_0 \psi_r}{kT}\right) \tag{3.13}$$

$$= \sum_i n_i^0 z_i e_0 - \sum_i \frac{n_i^0 z_i^2 e_0^2 \psi_r}{kT} \tag{3.14}$$

The first term $\Sigma n_i^0 z_i e_0$ gives the charge on the electrolytic solution as a whole. This is zero because the solution as a whole must be electrically neutral. The local excess charge densities near ions cancel out because the excess positive charge density near a negative ion is compensated for by an excess negative charge density near a positive ion. Hence,

$$\sum_i n_i^0 z_i e_0 = 0 \tag{3.15}$$

and one is left with

$$\varrho_r = -\sum_i \frac{n_i^0 z_i^2 e_0^2 \psi_r}{kT} \tag{3.16}$$

3.3.6. The Linearized Poisson–Boltzmann Equation

The stage is now set for the calculation of the potential ψ_r and the charge density ϱ_r in terms of known parameters of the solution.

Notice that one has obtained two expressions for the charge density ϱ_r in the volume element dV at a distance r from the central ion. One has the Poisson equation [Eq. (3.4)]

$$\varrho_r = -\frac{\varepsilon}{4\pi}\left[\frac{1}{r^2}\frac{d}{dr}\left(r^2\frac{d\psi_r}{dr}\right)\right] \tag{3.17}$$

and one also has the "linearized" Boltzmann distribution

$$\varrho_r = -\sum_i \frac{n_i^0 z_i^2 e_0^2 \psi_r}{kT} \tag{3.18}$$

where Σ_i refers to the summation over all species of ions typified by i.

If one equates these two expressions, one can obtain the linearized Poisson–Boltzmann (P–B) expression

$$\frac{1}{r^2}\frac{d}{dr}\left(r^2\frac{d\psi_r}{dr}\right) = \left(\frac{4\pi}{\varepsilon kT}\sum_i n_i^0 z_i^2 e_0^2\right)\psi_r \tag{3.19}$$

The constants in the right-hand parentheses can all be lumped together and called a new constant κ^2, i.e.,

$$\kappa^2 = \frac{4\pi}{\varepsilon kT}\sum_i n_i^0 z_i^2 e_0^2 \tag{3.20}$$

At this point, the symbol κ is used only to reduce the tedium of writing. It turns out later, however, that κ is not only a shorthand symbol; it contains information concerning several fundamental aspects of the distribution of ions around an ion in solution. In Chapter 6 it will be shown that it also contains information concerning the distribution of charges near a metal surface in contact with an ionic solution. In terms of κ, the linearized P–B expression (3.19) is

$$\frac{1}{r^2}\frac{d}{dr}\left(r^2\frac{d\psi_r}{dr}\right) = \kappa^2\psi_r \tag{3.21}$$

3.3.7. Solution of the Linearized P–B Equation

The rather messy-looking linearized P–B equation (3.21) can be tidied up by a mathematical trick. Introducing a new variable μ defined by

$$\psi_r = \frac{\mu}{r} \tag{3.22}$$

one has

$$\frac{d\psi_r}{dr} = \frac{d}{dr}\frac{\mu}{r} = -\frac{\mu}{r^2} + \frac{1}{r}\frac{d\mu}{dr}$$

and therefore

$$\frac{1}{r^2}\frac{d}{dr}\left(r^2\frac{d\psi_r}{dr}\right) = \frac{1}{r^2}\frac{d}{dr}\left(-\mu + r\frac{d\mu}{dr}\right)$$

$$= \frac{1}{r^2}\left(-\frac{d\mu}{dr} + r\frac{d^2\mu}{dr^2} + \frac{d\mu}{dr}\right)$$

$$= \frac{1}{r}\frac{d^2\mu}{dr^2} \tag{3.23}$$

Hence, the differential equation (3.21) becomes

$$\frac{1}{r}\frac{d^2\mu}{dr^2} = \kappa^2\frac{\mu}{r} \tag{3.24}$$

or

$$\frac{d^2\mu}{dr^2} = \kappa^2\mu \tag{3.25}$$

To solve this differential equation, it is recalled that the differentiation of an exponential function results in the multiplication of that function by the constant in the exponent. For example,

$$\frac{d}{dr}e^{\pm\kappa r} = \pm\kappa e^{\pm\kappa r} \tag{3.26}$$

and

$$\frac{d^2}{dr^2}e^{\pm\kappa r} = \kappa^2 e^{\pm\kappa r}$$

Hence, if μ is an exponential function of r, one will obtain a differential equation of the form of Eq. (3.25). In other words, the "primitive" or "origin" of the differential equation must have had an exponential in κr.

Two possible exponential functions, however, would lead to the same final differential equation; one of them would have a positive exponent and the other a negative one [Eq. (3.26)]. The general solution of the linearized P–B equation can therefore be written as

$$\mu = Ae^{-\kappa r} + Be^{+\kappa r} \tag{3.27}$$

where A and B are constants to be evaluated. Or, from Eq. (3.22),

$$\psi_r = A\frac{e^{-\kappa r}}{r} + B\frac{e^{+\kappa r}}{r} \tag{3.28}$$

The constant B is evaluated by using the boundary condition that far enough from a central ion situated at $r = 0$, the thermal forces completely dominate the Coulombic forces, which decrease as r^2, and there is electroneutrality (i.e., the electrostatic potential ψ_r vanishes at distances sufficiently far from such an ion, $\psi_r \to 0$ as $r \to \infty$). This condition would be satisfied only if $B = 0$. Thus, if B had a finite value, Eq. (3.28) shows that the electrostatic potential would shoot up to infinity (i.e., $\psi_r \to \infty$ as $r \to \infty$), a physically unreasonable proposition. Hence,

$$\psi_r = A\frac{e^{-\kappa r}}{r} \tag{3.29}$$

To evaluate the integration constant A, a hypothetical condition will be considered in which the solution is so dilute and on the average the ions are so far apart that there is a negligible interionic field. Further, the central ion is assumed to be a point charge, i.e., to have a radius negligible compared with the distances otherwise to be considered. Hence, the potential near the central ion is, in this special case, simply that due to an isolated point charge of value $z_i e_0$.

This is given directly from Coulomb's law as

$$\psi_r = \frac{z_i e_0}{\varepsilon r} \tag{3.30}$$

At the same time, for this hypothetical solution in which the concentration tends to zero, i.e., $n_i^0 \to 0$, it is seen from Eq. (3.20) that $\kappa \to 0$. Thus, in Eq. (3.29), $e^{-\kappa r} \to 1$, and one has

$$\psi_r = \frac{A}{r} \tag{3.31}$$

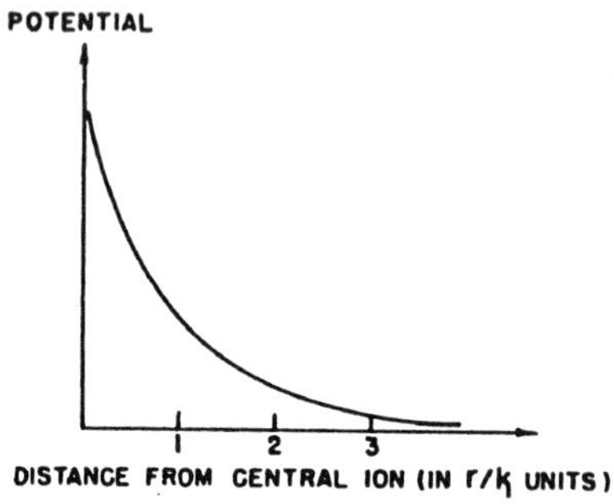

Fig. 3.9. The variation of the electrostatic potential ψ as a function of distance from the central ion expressed in units of r/κ.

Hence, by combining Eqs. (3.30) and (3.31),

$$A = \frac{z_i e_0}{\varepsilon} \tag{3.32}$$

By introducing this expression for A into Eq. (3.29), the result is

$$\psi_r = \frac{z_i e_0}{\varepsilon} \frac{e^{-\kappa r}}{r} \tag{3.33}$$

Here then is the appropriate solution of the *linearized* P–B equation (3.21). It shows how the electrostatic potential varies with distance r from an arbitrarily chosen reference ion (Fig. 3.9).

3.3.8. The Ionic Cloud around a Central Ion

In the imaginative Debye–Hückel model of a dilute electrolytic solution, a reference ion sitting at the origin of the spherical coordinate system is surrounded by the smoothed-out charge of the other ions. Further, because of the local inequalities in the concentrations of the positive and negative ions, the smoothed-out charge of one sign does not (locally) cancel out the smoothed-out charge of the opposite sign; there is a local excess charge density of one sign opposite to that of the central ion.

Now, as explained in Section 3.3.2, the principal objective of the Debye–Hückel theory is to calculate the time-averaged spatial distribution of the excess charge density around a reference ion. How is this objective attained?

The Poisson equation (3.4) relates the potential at r from the sample ion to the charge density at r, i.e.,

$$\frac{1}{r^2}\frac{d}{dr}\left(r^2\frac{d\psi_r}{dr}\right) = -\frac{4\pi}{\varepsilon}\varrho_r \tag{3.4}$$

Further, one has the linearized P–B equation

$$\frac{1}{r^2}\frac{d}{dr}\left(r^2\frac{d\psi_r}{dr}\right) = \kappa^2\psi_r \tag{3.21}$$

From these two equations, one has the linear relation between excess charge density and potential, i.e.,

$$\varrho_r = -\frac{\varepsilon}{4\pi}\kappa^2\,\psi_r \tag{3.34}$$

and by inserting the solution (3.33) for the linearized P–B equation, the result is

$$\varrho_r = -\frac{z_i e_0}{4\pi}\kappa^2\frac{e^{-\kappa r}}{r} \tag{3.35}$$

Here then is the desired expression for the spatial distribution of the charge density with distance r from the central ion (Fig. 3.10). Since the excess charge density results from an unequal distribution of positive and negative ions, Eq. (3.35) also describes the distribution of ions around a reference or sample ion.

To understand this distribution of ions, however, one must be sufficiently attuned to mathematical language to read the physical significance of Eq. (3.35). The physical ideas implicit in the distribution will therefore be stated in pictorial terms. One can say that the central reference ion is surrounded by a cloud, or atmosphere, of excess charge (Fig. 3.11). This ionic cloud extends into the solution (i.e., r increases), and the excess

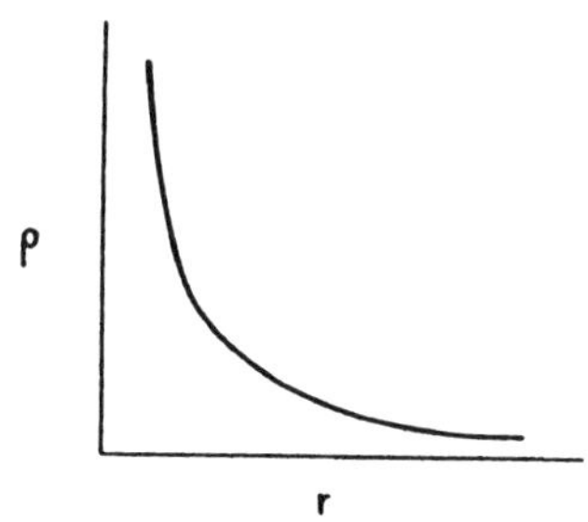

Fig. 3.10. The variation of the excess charge density ϱ as a function of distance from the central ion.

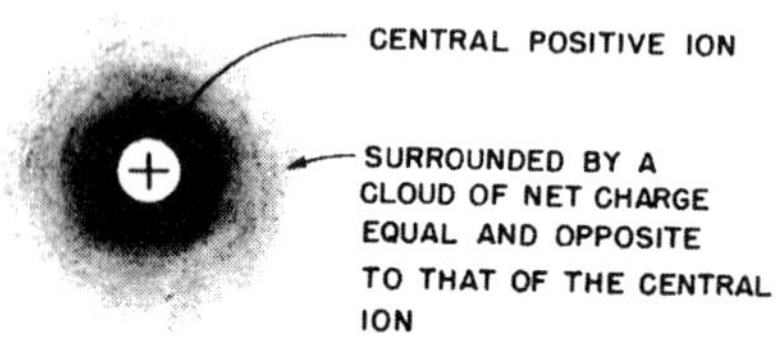

Fig. 3.11. The distribution of excess charge density around a central ion can be pictured as a cloud, or atmosphere, of net charge around the central ion.

charge density ϱ decays with distance r in an exponential way. The excess charge residing on the ion cloud is opposite in sign to that of the central ion. Thus, a positively charged reference ion has a negatively charged ion atmosphere, and vice versa (Fig. 3.12).

Up to now, the charge density at a given distance has been discussed. The total excess charge contained in the ionic atmosphere that surrounds the central ion can, however, easily be computed. Consider a spherical shell of thickness dr at a distance r from the origin, i.e., from the center of the reference ion (Fig. 3.13). The charge dq in this thin shell is equal to the charge density ϱ_r times the volume $4\pi r^2 dr$ of the shell, i.e.,

$$dq = \varrho_r 4\pi r^2 dr \tag{3.36}$$

The total charge q_{cloud} contained in the ion atmosphere is obtained by summing the charges dq contained in all the infinitesimally thick spherical shells. In other words, the total excess charge surrounding the reference ion is computed by integrating dq (which is a function of the distance r from the central ion) from a lower limit corresponding to the distance from the central ion at which the cloud is taken to commence to the point where the cloud ends. Now, the ion atmosphere begins at the surface of the ion, so the lower limit depends upon the model of the ion. The first model

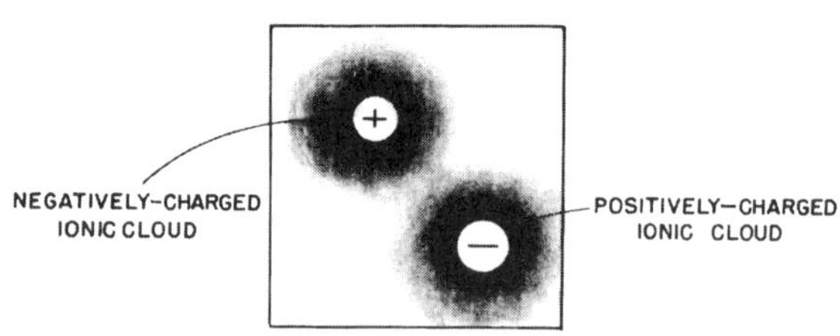

Fig. 3.12. A positively charged ion has a negatively charged ionic cloud, and vice versa.

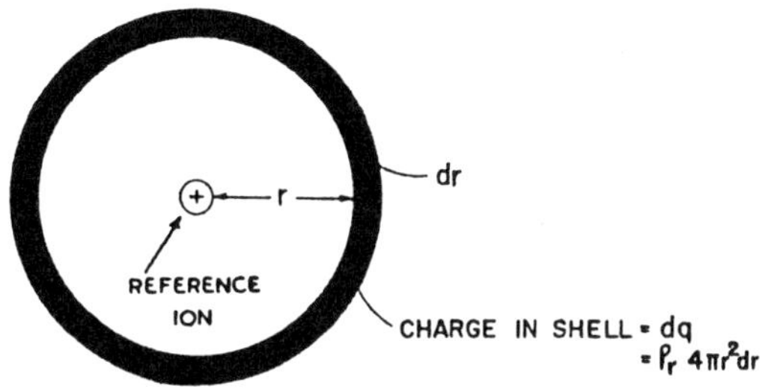

Fig. 3.13. A spherical shell of thickness *dr* at a distance *r* from the center of the reference ion.

chosen by Debye and Hückel was that of point-charge ions, in which case the lower limit is $r = 0$. The upper limit for the integration is $r \to \infty$ because the charge of the ionic cloud decays exponentially into the solution and becomes zero only in the limit $r \to \infty$.

Thus,

$$q_{\text{cloud}} = \int_{r=0}^{r\to\infty} dq = \int_{r=0}^{r\to\infty} \varrho_r 4\pi r^2 dr \tag{3.37}$$

and by substituting for ϱ_r from Eq. (3.35), the result is

$$q_{\text{cloud}} = -\int_{r=0}^{r\to\infty} \frac{z_i e_0}{4\pi} \kappa^2 \frac{e^{-\kappa r}}{r} 4\pi r^2 dr$$

$$= -z_i e_0 \int_{r=0}^{r\to\infty} e^{-\kappa r}(\kappa r)\, d(\kappa r) \tag{3.38}$$

The integration can be done by parts (Appendix 3.2), leading to the result

$$q_{\text{cloud}} = -z_i e_0 \tag{3.39}$$

which means that a central ion of charge $+z_i e_0$ is enveloped by a cloud containing a total charge of $-z_i e_0$ (Fig. 3.14). Thus, the total charge on the surrounding volume is just equal and opposite to that on the reference ion. This is of course precisely how things should be so that there can be electroneutrality for the ionic solution taken as a whole; a given ion, together with its cloud, has a zero net charge.

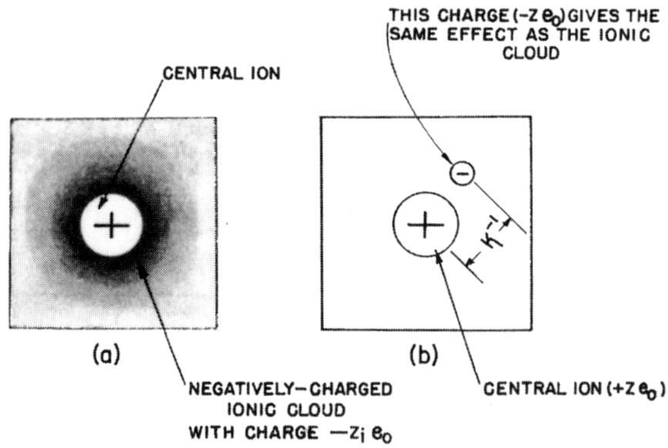

Fig. 3.14. The total charge $-z_i e_0$ on the ionic cloud is just equal and opposite to the charge $+z_i e_0$ on the central ion.

How is this equal and opposite charge of the ion atmosphere distributed in the space around the central ion? It is seen from Eqs. (3.35) and (3.36) that the net charge in a spherical shell of thickness dr and at a distance r from the origin is

$$dq = -z_i e_0 e^{-\kappa r} \kappa^2 r\, dr \tag{3.40}$$

Thus, the excess charge on a spherical shell varies with r and has a maximum value for a value of r given by

$$0 = \frac{dq}{dr}$$

$$= \frac{d}{dr}[-z_i e_0 \kappa^2 (e^{-\kappa r} r)]$$

$$= -z_i e_0 \kappa^2 \frac{d}{dr}(e^{-\kappa r} r)$$

$$= -z_i e_0 \kappa^2 (e^{-\kappa r} - r\kappa e^{-\kappa r}) \tag{3.41}$$

Since $(z_i e_0 \kappa^2)$ is finite, Eq. (3.41) can be true only when

$$0 = e^{-\kappa r} - r\kappa e^{-\kappa r}$$

or

$$r = \kappa^{-1} \tag{3.42}$$

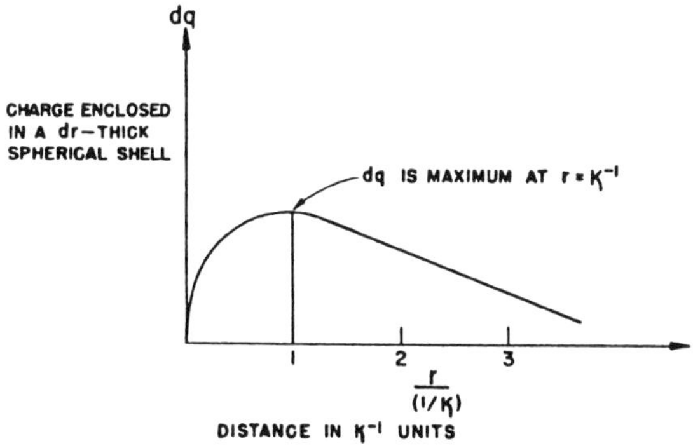

Fig. 3.15. The distance variation (in κ^{-1} units) of the charge *dq* enclosed in a *dr*-thick spherical shell, showing that *dq* is a maximum at $r = \kappa^{-1}$.

Hence, the maximum value of the charge contained in a spherical shell (of infinitesimal thickness *dr*) is attained when the spherical shell is at a distance $r = \kappa^{-1}$ from the reference ion (Fig. 3.15). For this reason (but see also Section 3.3.9), κ^{-1} is known as the *thickness*, or *radius*, of the ionic cloud that surrounds a reference ion. An elementary dimensional analysis [e.g., of Eq. (3.43)] will indeed reveal that κ^{-1} has the dimensions of length. Consequently, κ^{-1} is sometimes referred to as the *Debye–Hückel length.*

It may be recalled that κ^{-1} is given [from Eq. (3.20)] by

$$\kappa^{-1} = \left(\frac{\varepsilon kT}{4\pi} \frac{1}{\sum_i n_i^0 z_i^2 e_0^2} \right)^{1/2} \tag{3.43}$$

As the concentration tends toward zero, the cloud tends to spread out increasingly (Fig. 3.16). Values of the thickness of the ion atmosphere for various concentrations of the electrolyte are presented in Table 3.2.

3.3.9. Contribution of the Ionic Cloud to the Electrostatic Potential ψ_r at a Distance *r* from the Central Ion

An improved feel for the effects of ionic clouds emerges from considering the following interesting problem. Imagine, in a thought experiment, that the charge on the ionic cloud does not exist. There is only one charge now, that on the central ion. What is the potential at distance *r* from the central ion? It is simply given by the familiar formula for the potential at a distance *r* from a single charge, namely,

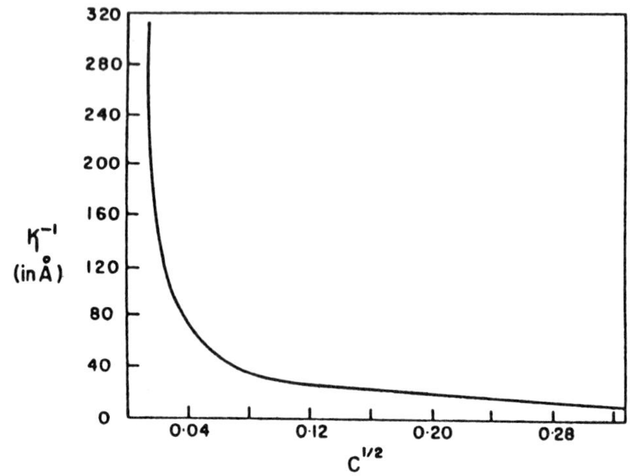

Fig. 3.16. The variation in the thickness κ^{-1} of the ionic cloud as a function of electrolyte concentration (1 Å = 0.1 nm).

$$\psi_r = \frac{z_i e_0}{\varepsilon r} \tag{3.44}$$

Then let the charge on the cloud be switched on. The potential ψ_r at the distance r from the central ion is no longer given by the central ion only. It is given by the law of superposition of potentials (Fig. 3.17); i.e., ψ_r is the sum of the potential due to the central ion and that due to the ionic cloud

$$\psi_r = \psi_{\text{ion}} + \psi_{\text{cloud}} \tag{3.45}$$

TABLE 3.2

Thickness of Ionic Atmosphere (nm) at Various Concentrations and for Various Types of Salts

Concentration (mol dm^{-3})	Type of Salt			
	1:1	1:2	2:2	1:3
10^{-4}	30.4	17.6	15.2	12.4
10^{-3}	9.6	5.55	4.81	3.93
10^{-2}	3.04	1.76	1.52	1.24
10^{-1}	0.96	0.55	0.48	0.39

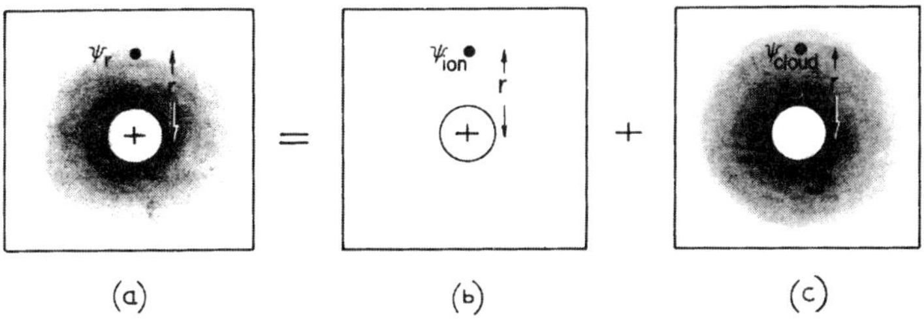

Fig. 3.17. The superposition of the potential ψ_{ion} due to the ion and the potential ψ_{cloud} due to the cloud yields the total potential at a distance r from the central ion.

The contribution ψ_{cloud} can thus be easily found. One rearranges Eq. (3.45) to read

$$\psi_{cloud} = \psi_r - \psi_{ion} \tag{3.46}$$

and substitutes for ψ_{ion} with Eq. (3.44) and for ψ_r with the Debye–Hückel expression [Eq. (3.33)]. Then,

$$\psi_{cloud} = \frac{z_i e_0}{\varepsilon}\frac{\varepsilon^{-\kappa r}}{r} - \frac{z_i e_0}{\varepsilon r}$$

$$= \frac{z_i e_0}{\varepsilon r}(e^{-\kappa r} - 1) \tag{3.47}$$

The value of κ [*cf.* Eq. (3.20)] is proportional to $\Sigma n_i^0 z_i^2 e_0^2$. In sufficiently dilute solutions, $\Sigma n_i^0 z_i^2 e_0^2$ can be taken as sufficiently small to make $\kappa r << 1$,

$$e^{-\kappa r} - 1 \approx 1 - \kappa r - 1 \approx -\kappa r \tag{3.48}$$

and based on this approximation,

$$\psi_{cloud} = -\frac{z_i e_0}{\varepsilon \kappa^{-1}} \tag{3.49}$$

By introducing the expressions (3.44) and (3.49) into the expression (3.45) for the total potential ψ_r at a distance r from the central ion, it follows that

$$\psi_r = \frac{z_i e_0}{\varepsilon r} - \frac{z_i e_0}{\varepsilon \kappa^{-1}} \tag{3.50}$$

The second term, which arises from the cloud, reduces the value of the potential to a value less than that if there were no cloud. This is consistent with the model; the cloud has a charge opposite to that on the central ion and must therefore alter the potential in a sense opposite to that due to the central ion.

The expression

$$\psi_{\text{cloud}} = -\frac{z_i e_0}{\varepsilon\kappa^{-1}} \tag{3.49}$$

leads to another, and helpful, way of looking at the quantity κ^{-1}. It is seen that ψ_{cloud} is independent of r, and therefore the contribution of the cloud to the potential *at the site of the point-charge central ion* can be considered to be given by Eq. (3.49). But, if the entire charge of the ionic atmosphere [which is $-z_i e_0$ as required by electroneutrality—Eq. (3.39)] were placed at a distance κ^{-1} from the central ion, then the potential produced at the reference ion would be $-z_i e_0/(\varepsilon\kappa^{-1})$. It is seen therefore from Eq. (3.49) that the effect of the ion cloud, namely, ψ_{cloud}, is equivalent to that of a single charge, equal in magnitude but opposite in sign to that of the central ion, placed at a distance κ^{-1} from the reference ion (Fig. 3.18). This is an added and more important reason that the quantity κ^{-1} is termed the effective thickness or radius of the ion atmosphere surrounding a central ion (see Section 3.3.8).

3.3.10. The ionic Cloud and the Chemical-Potential Change Arising from Ion–Ion Interactions

It will be recalled (see Section 3.3.1) that it was the potential at the surface of the reference ion which needed to be known in order to calculate the chemical-potential change $\Delta\mu_{i-I}$ arising from the interactions between a particular ionic species i and the rest of the ions of the solution, i.e., one needed to know ψ_{cloud} in Eq. (3.3),

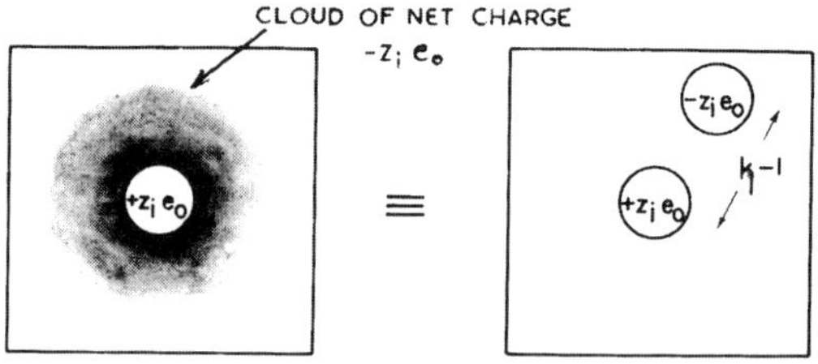

Fig. 3.18. The contribution ψ_{cloud} of the ionic cloud to the potential at the central ion is equivalent to the potential due to a single charge, equal in magnitude and opposite sign to that of the central ion, placed at a distance κ^{-1} from the central ion.

$$\Delta\mu_{i-I} = \frac{N_A z_i e_0}{2} \psi_{\text{cloud}} \tag{3.3}$$

It was to obtain this potential ψ that Debye and Hückel conceived their model of an ionic solution. The analysis presented the picture of an ion being enveloped in an ionic cloud. What is the origin of the ionic cloud? It is born of the interactions between the central ion and the ions of the environment. If there were no interactions (e.g., Coulombic forces between ions), thermal forces would prevail, distribute the ions randomly ($\varrho = 0$), and wash out the ionic atmosphere. It appears therefore that the simple ionic cloud picture has not only led to success in describing the distribution of ions but also given the electrostatic potential ψ_{cloud} at the surface of a reference ion due to the interactions between this reference ion and the rest of the ions in the solution (the quantity required for reasons explained in Section 3.3.1).

Thus, the expression (3.49) for ψ_{cloud} can be substituted for ψ in Eq. (3.3) with the result that

$$\Delta\mu_{i-I} = -\frac{N_A (z_i e_0)^2}{2\varepsilon\kappa^{-1}} \tag{3.51}$$

The Debye–Hückel ionic-cloud model for the distribution of ions in an electrolytic solution has permitted the theoretical calculation of the chemical-potential change arising from ion–ion interactions. How is this theoretical expression to be checked, i.e., connected with a measured quantity? It is to this testing of the Debye–Hückel theory that attention will now be turned.

3.4. ACTIVITY COEFFICIENTS AND ION–ION INTERACTIONS

3.4.1. Evolution of the Concept of an Activity Coefficient

The existence of ions in solution, of interactions between these ions, and of a chemical-potential change $\Delta\mu_{i-I}$ arising from ion–ion interactions have all been taken to be self-evident in the treatment hitherto presented here. This, however, is a modern point of view. The thinking about electrolytic solutions actually developed along a different path.

Ionic solutions were at first treated in the same way as nonelectrolytic solutions, though the latter do not contain (interacting) charged species. The starting point was the classical thermodynamic formula for the chemical potential μ_i of a nonelectrolyte solute

$$\mu_i = \mu_i^0 + RT \ln x_i \tag{3.52}$$

In this expression, x_i is the concentration[5] of the solute in mole fraction units, and μ_i^0 is its chemical potential in the standard state, i.e., when x_i assumes a standard or normalized value of unity

$$\mu_i = \mu_i^0 \quad \text{when} \quad x_i = 1 \tag{3.53}$$

Since the solute particles in a solution of a nonelectrolyte are uncharged, they do not engage in long-range Coulombic interactions. The short-range interactions arising from dipole–dipole or dispersion forces become significant only when the mean distance between the solute particles is small, i.e., when the concentration of the solute is high. Thus, one can to a good approximation say that there are no interactions between solute particles in dilute nonelectrolyte solutions. Hence, if Eq. (3.52) for the chemical potential of a solute in a nonelectrolyte solution (with noninteracting particles) is used for the chemical potential of an ionic species i in an electrolytic solution, then it is tantamount to ignoring the long-range Coulombic interactions between ions. In an actual electrolytic solution, however, ion–ion interactions operate whether one ignores them or not. It is obvious therefore that measurements of the chemical potential μ_i of an ionic species—or, rather, measurements of any property that depends on the chemical potential—would reveal the error in Eq. (3.52), which is blind to ion–ion interactions. In other words, experiments show that *even in dilute solutions*,

$$\mu_i - \mu_i^0 \neq RT \ln x_i$$

In this context, a frankly empirical approach was adopted by earlier workers not yet blessed by Debye and Hückel's light. Solutions that obeyed Eq. (3.52) were characterized as *ideal* solutions since this equation applies to systems of noninteracting solute particles, i.e., ideal particles. Electrolytic solutions that do not obey the equation were said to be *nonideal*. In order to use an equation of the form of Eq. (3.52) to treat nonideal electrolytic solutions, an empirical correction factor f_i was introduced by Lewis as a modifier of the concentration term[6]

$$\mu_i - \mu_i^0 = RT \ln x_i f_i \tag{3.54}$$

---

[5]The value of x_i in the case of an electrolyte derives from the number of moles of ions of species i actually present in solution. This number need not be equal to the number of moles of i expected of dissolved electrolyte; if, for instance, the electrolyte is a potential one, then only a fraction of the electrolyte may react with the solvent to form ions, i.e., the electrolyte may be incompletely dissociated.

[6]The standard chemical potential μ_i^0 has the same significance here as in Eq. (3.52) for ideal solutions. Thus, μ_i^0 can be defined either as the chemical potential of an ideal solution in its standard state of $x_i = 1$ or as the chemical potential of a solution in its state of $x_i = 1$ and $f_i = 1$, i.e., $a_i = 1$. No real solution can have $f_i = 1$ when $x_i = 1$; so the standard state pertains to the same hypothetical solution as the standard state of an ideal solution.

It was argued that, in nonideal solutions, it was not just the analytical concentration x_i of species i, but its effective concentration $x_i f_i$ which determined the chemical-potential change $\mu_i - \mu_i^0$. This effective concentration $x_i f_i$ was also known as the *activity* a_i of the species i, i.e.,

$$a_i = x_i f_i \tag{3.55}$$

and the correction factor f_i, as the *activity coefficient*. For ideal solutions, the activity coefficient is unity, and the activity a_i becomes identical to the concentration x_i, i.e.,

$$a_i = x_i \quad \text{when} \quad f_i = 1 \tag{3.56}$$

Thus, the chemical-potential change in going from the standard state to the final state can be written as

$$\mu_i - \mu_i^0 = RT \ln x_i + RT \ln f_i \tag{3.57}$$

Equation (3.57) summarizes the empirical or formal treatment of the behavior of electrolytic solutions. Such a treatment cannot furnish a theoretical expression for the activity coefficient f_i. It merely recognizes that expressions such as (3.52) must be modified if significant interaction forces exist between solute particles.

3.4.2. The Physical Significance of Activity Coefficients

For a hypothetical system of ideal (noninteracting) particles, the chemical potential has been stated to be given by

$$\mu_i\,(\text{ideal}) = \mu_i^0 + RT \ln x_i \tag{3.52}$$

For a real system of interacting particles, the chemical potential has been expressed in the form

$$\mu_i\,(\text{real}) = \mu_i^0 + RT \ln x_i + RT \ln f_i \tag{3.57}$$

Hence, to analyze the physical significance of the activity coefficient term in Eq. (3.57), it is necessary to compare this equation with Eq. (3.52). It is obvious that when Eq. (3 52) is subtracted from Eq. (3.57), the difference [i.e., μ_i (real) – μ_i (ideal)] is the chemical-potential change $\Delta\mu_{i-I}$ arising from interactions between the solute particles (ions in the case of electrolyte solutions). That is,

$$\mu_i(\text{real}) - \mu_i\,(\text{ideal}) = \Delta\mu_{i-I} \tag{3.58}$$

and therefore,

$$\Delta\mu_{i-I} = RT \ln f_i \tag{3.59}$$

Thus, the activity coefficient is a measure of the chemical-potential change arising from ion–ion interactions. There are several well-established methods of experimentally determining activity coefficients, and these methods are treated in adequate detail in standard treatises (see Further Reading at the end of this section).

Now, according to the Debye–Hückel theory, the chemical-potential change $\Delta\mu_{i-I}$ arising from ion–ion interactions has been shown to be given by

$$\Delta\mu_{i-I} = -\frac{N_A(z_i e_0)^2}{2\varepsilon\kappa^{-1}} \tag{3.51}$$

Hence, by combining Eqs. (3.51) and (3.58), the result is

$$RT \ln f_i = -\frac{N_A(z_i e_0)^2}{2\varepsilon\kappa^{-1}} \tag{3.60}$$

Thus, the Debye–Hückel ionic-cloud model for ion–ion interactions has permitted a theoretical calculation of activity coefficients resulting in Eq. (3.60).

The activity coefficient in Eq. (3.59) arises from the formula (3.57) for the chemical potential, in which the concentration of the species i is expressed in mole fraction units x_i. One can also express the concentration in moles per liter (1 liter = 1 dm^3) of solution (molarity) or in moles per kilogram of solvent (molality). Thus, alternative formulas for the chemical potential of a species i in an ideal solution read

$$\mu_i = \mu_i^0(c) + RT \ln c_i \tag{3.61}$$

and

$$\mu_i = \mu_i^0(m) + RT \ln m_i \tag{3.62}$$

where c_i and m_i are the molarity and molality of the species i, respectively, and $\mu_i^0(c)$ and $\mu_i^0(m)$ are the corresponding standard chemical potentials.

When the concentration of the ionic species in a real solution is expressed as a molarity c_i or a molality m_i, there are corresponding activity coefficients γ_c and γ_m and corresponding expressions for μ_i,

$$\mu_i = \mu_i^0(c) + RT \ln c_i + RT \ln \gamma_c \tag{3.63}$$

and

$$\mu_i = \mu_i^0(m) + RT \ln m_i + RT \ln \gamma_m \tag{3.64}$$

3.4.3. The Activity Coefficient of a Single Ionic Species Cannot Be Measured

Before the activity coefficients calculated on the basis of the Debye–Hückel model can be compared with experiment, there arises a problem similar to one faced in the discussion of ion–solvent interactions (Chapter 2). There, it was realized the heat of hydration of an individual ionic species could not be measured because such a measurement would involve the transfer of ions of only one species into a solvent instead of ions of two species with equal and opposite charges. Even if such a transfer were physically possible, it would result in a charged solution[7] and therefore an extra, undesired interaction between the ions and the electrified solution. The only way out was to transfer a neutral electrolyte (an equal number of positive and negative ions) into the solvent, but this meant that one could only measure the heat of interactions of a *salt* with the solvent and this experimental quantity could not be separated into the individual ionic heats of hydration.

Here, in the case of ion–ion interactions, the desired quantity is the activity coefficient f_i,[8] which depends through Eq. (3.57) on $\mu_i - \mu_i^0$. This means that one seeks the free-energy change of an ionic solution per mole of ions of a single species i. To measure this quantity, one would have a problem similar to that experienced with ion–solvent interactions, namely, the measurement of the change in free energy of a solution resulting from a change in the concentration of one ionic species only.

This change in free energy associated with the addition of one ionic species only would include an undesired work term representing the electrical work of interaction between the ionic species being added and the charged solution. To avoid free-energy changes associated with interacting with a solution, it is necessary that after a change in the concentration of the ionic species, the electrolytic solution should end up uncharged and electroneutral. This aim is easily accomplished by adding an electroneutral electrolyte containing the ionic species i. Thus, the concentration of sodium ions can be altered by adding sodium chloride. The solvent, water, maintains its electroneutrality when the uncharged ionic lattice (containing two ionic species of opposite charge) is dissolved in it.

When ionic lattices, i.e., salts, are dissolved instead of individual ionic species, one eliminates the problem of ending up with charged solutions but another problem emerges. If one increases the concentration of sodium ions by adding the salt sodium chloride, one has perforce to produce a simultaneous increase in the concentration of chloride ions. This means, however, that there are two contributions to the change in

---

[7]The solution may not be initially charged but will become so once an ionic species is added to it.

[8]The use of the symbol γ for the activity coefficients when the concentration is expressed in molarities and molalities should be noted. When the concentration is expressed as a mole fraction, f_i has been used here. For dilute solutions, the numerical values of activity coefficients for these different systems of units are almost the same.

free energy associated with a change in salt concentration: (1) the contribution of the positive ions and (2) the contribution of the negative ions.

Since neither the positive nor the negative ions can be added separately, the individual contributions of the ionic species to the free energy of the system are difficult to determine. Normally, one can only measure the activity coefficient of the net electrolyte, i.e., of at least two ionic species together. It is necessary therefore to establish a conceptual link between the activity coefficient of an electrolyte in solution (that quantity directly accessible to experiment) and that of only one of its ionic species [not accessible to experiment, but calculable theoretically from Eq. (3.60)].

3.4.4. The Mean Ionic Activity Coefficient

Consider a uni-univalent electrolyte MA (e.g., NaCl). The chemical potential of the M^+ ions is [Eq. (3.57)]

$$\mu_{M^+} = \mu^0_{M^+} + RT \ln x_{M^+} + RT \ln f_{M^+} \tag{3.65}$$

and the chemical potential of the A^- ions is

$$\mu_{A^-} = \mu^0_{A^-} + RT \ln x_{A^-} + RT \ln f_{A^-} \tag{3.66}$$

Adding the two expressions, one obtains

$$\mu_{M^+} + \mu_{A^-} = (\mu^0_{M^+} + \mu^0_{A^-}) + RT \ln (x_{M^+} x_{A^-}) + RT \ln (f_{M^+} f_{A^-}) \tag{3.67}$$

What has been obtained here is the change in the free energy of the system due to the addition of 2 moles of ions—1 mole of M^+ ions and 1 mole of A^- ions—which are contained in 1 mole of electroneutral salt MA.

Now, suppose that one is only interested in the *average* contribution to the free energy of the system from 1 mole of both M^+ and A^- ions. One has to divide Eq. (3.67) by 2

$$\frac{\mu_{M^+} + \mu_{A^-}}{2} = \frac{\mu^0_{M^+} + \mu^0_{A^-}}{2} + RT \ln (x_{M^+} x_{A^-})^{1/2} + RT \ln (f_{M^+} f_{A^-})^{1/2} \tag{3.68}$$

At this stage, one can define several new quantities

$$\mu_{\pm} = \frac{\mu_{M^+} + \mu_{A^-}}{2} \tag{3.69}$$

$$\mu^0_{\pm} = \frac{\mu^0_{M^+} + \mu^0_{A^-}}{2} \tag{3.70}$$

$$x_{\pm} = (x_{M^+} x_{A^-})^{1/2} \tag{3.71}$$

and

$$f_{\pm} = (f_{M^+} f_{A^-})^{1/2} \tag{3.72}$$

What is the significance of these quantities $\mu_{\pm}$, $\mu^0_{\pm}$, $x_{\pm}$, and $f_{\pm}$? It is obvious they are all average quantities—the mean chemical potential $\mu_{\pm}$, the mean standard chemical potential $\mu^0_{\pm}$, the mean ionic mole fraction $x_{\pm}$, and the mean ionic-activity coefficient $f_{\pm}$. In the case of $\mu_{\pm}$ and $\mu^0_{\pm}$, the arithmetic mean (half the sum) is taken because free energies are additive, but in the case of $x_{\pm}$ and $f_{\pm}$, the geometric mean (the square root of the product) is taken because the effects of mole fraction and activity coefficient on free energy are multiplicative.

In this notation, Eq. (3.68) for the average contribution of a mole of ions to the free energy of the system becomes

$$\mu_{\pm} = \mu^0_{\pm} + RT \ln x_{\pm} + RT \ln f_{\pm} \tag{3.73}$$

since a mole of ions is produced by the dissolution of half a mole of salt. In other words, $\mu_{\pm}$ is half the chemical potential μ_{MA} of the salt.[9]

$$\tfrac{1}{2}\mu_{MA} = \mu_{\pm} = \mu^0_{\pm} + RT \ln x_{\pm} + RT \ln f_{\pm} \tag{3.74}$$

Thus, a clear connection has been set up between observed free-energy changes μ_{MA} consequent upon the change from a state in which the two ionic species of a salt are infinitely far apart to a state corresponding to the given concentration and its mean ionic-activity coefficient $f_{\pm}$. Hence the value of $f_{\pm}$ is experimentally measurable. What can be obtained from $f_{\pm}$ is the product of the individual ionic-activity coefficients [Eq. (3.72)]. The theoretical approach must be to calculate the activity coefficients f_+ and f_- for the positive and negative ions [Eq. (3.60)] and combine them through Eq. (3.72) into a mean ionic-activity coefficient $f_{\pm}$ which can be compared with the easily experimentally derived mean ionic-activity coefficient.

3.4.5. Conversion of Theoretical Activity-Coefficient Expressions into a Testable Form

Individual ionic-activity coefficients are experimentally inaccessible (Section 3.4.3); hence, it is necessary to relate the theoretical individual activity coefficient f_i

---

[9]The symbol μ_{MA} should not be taken to mean that *molecules* of MA exist in the solution; μ_{MA} is the observed free-energy change of the system resulting from the dissolution of a mole of electrolyte.

[Eq. (3.60)] to the experimentally accessible *mean* ionic-activity coefficient $f_\pm$ so that the Debye–Hückel model can be tested.

The procedure is to make use of the relation (3.72)

$$f_\pm = (f_{M^+} f_{A^-})^{1/2} \tag{3.72}$$

of which the general form for an electrolyte that dissolves to give ν_+ z_+-valent positive ions and ν_- z_--valent negative ions can be shown to be (Appendix 3.3)

$$f_\pm = (f_+^{\nu_+} f_-^{\nu_-})^{1/\nu} \tag{3.75}$$

where f_+ and f_- are the activity coefficients of the positive and negative ions, and

$$\nu = \nu_+ + \nu_- \tag{3.76}$$

By taking logarithms of both sides of Eq. (3.75), the result is

$$\ln f_\pm = \frac{1}{\nu}(\nu_+ \ln f_+ + \nu_- \ln f_-) \tag{3.77}$$

At this stage, the Debye–Hückel expressions (3.60) for f_+ and f_- can be introduced into Eq. (3.77) to give

$$\ln f_\pm = -\frac{1}{\nu}\left[\frac{N_A e_0^2}{2\varepsilon RT}\kappa\,(\nu_+ z_+^2 + \nu_- z_-^2)\right] \tag{3.78}$$

Since the solution as a whole is electroneutral, $\nu_+ z_+$ must be equal to $\nu_- z_-$ and therefore

$$\begin{aligned} \nu_+ z_+^2 + \nu_- z_-^2 &= \nu_- z_- z_+ + \nu_+ z_+ z_- \\ &= z_+ z_-(\nu_+ + \nu_-) \\ &= z_+ z_- \nu \end{aligned} \tag{3.79}$$

Using this relation in Eq. (3.78), one obtains

$$\ln f_\pm = -\frac{N_A (z_+ z_-) e_0^2}{2\varepsilon RT}\kappa \tag{3.80}$$

Now, one can substitute for κ from Eq. (3.43)

$$\kappa = \left(\frac{4\pi}{\varepsilon kT} \sum n_i^0 z_i^2 e_0^2 \right)^{1/2} \tag{3.43}$$

but before this substitution is made, κ can be expressed in a different form. Since

$$n_i^0 = \frac{c_i N_A}{1000} \tag{3.81}$$

where c is the concentration in moles per liter, it follows that

$$\sum n_i^0 z_i^2 e_0^2 = \frac{N_A e_0^2}{1000} \sum c_i z_i^2 \tag{3.82}$$

Prior to the Debye–Hückel theory, $\frac{1}{2}\Sigma c_i z_i^2$ had been empirically introduced by Lewis as a quantity of importance in the treatment of ionic solutions. Since it quantifies the charge in an electrolytic solution, it was known as the *ionic strength* and given the symbol I

$$I = \tfrac{1}{2} \sum c_i z_i^2 \tag{3.83}$$

In terms of the ionic strength I, κ can be written as [Eqs. (3.43), (3.82), and (3.83)]

$$\kappa = \left(\frac{8\pi N_A e_0^2}{1000\varepsilon kT} \right)^{1/2} I^{1/2} \tag{3.84}$$

or as

$$\kappa = BI^{1/2} \tag{3.85}$$

where

$$B = \left(\frac{8\pi N_A e_0^2}{1000\varepsilon kT} \right)^{1/2} \tag{3.86}$$

Values of B for water at various temperatures are given in Table 3.3.

On the basis of the expression (3.85) for κ, Eq. (3.80) becomes

$$\ln f_\pm = -\frac{N_A(z_+z_-)e_0^2}{2\varepsilon RT} BI^{1/2} \tag{3.87}$$

or

TABLE 3.3
Values of the Parameter *B* for Water at Various Temperatures

Temperature (°C)	$10^{-8}B$
0	0.3248
10	0.3264
20	0.3282
25	0.3291
30	0.3301
35	0.3312
40	0.3323
50	0.3346
60	0.3371
80	0.3426
100	0.3488

$$\log f_{\pm} = -\frac{1}{2.303}\frac{N_A e_0^2}{2\varepsilon RT} B(z_+z_-)I^{1/2} \tag{3.88}$$

For greater compactness, one can define a constant A given by

$$A = \frac{1}{2.303}\frac{N_A e_0^2}{2\varepsilon RT} B \tag{3.89}$$

and write Eq. (3.88) in the form

$$\log f_{\pm} = -A(z_+z_-)I^{1/2} \tag{3.90}$$

For 1:1-valent electrolytes, $z_+ = z_- = 1$ and $I = c$, and therefore

$$\log f_{\pm} = -Ac^{1/2} \tag{3.91}$$

Values of the constant A for water at various temperatures are given in Table 3.4.

In Eqs. (3.90) and (3.91), the theoretical mean ionic-activity coefficients are in a form directly comparable with experiment. How are such experiments carried out?

3.4.6. Experimental Determination of Activity Coefficients

A reasonably informative account has now been given as to how—albeit in very low concentrations—theoretical developments concerning interionic interaction give rise to theoretical values of the quantity (activity coefficient) by which the real

TABLE 3.4
Values of Constant *A* for Water at Various Temperatures

Temperature (°C)	*A*
0	0.4918
10	0.4989
20	0.5070
25	0.5115
30	0.5161
40	0.5262
50	0.5373
60	0.5494
80	0.5767
100	0.6086

(observed) behavior of electrolytes in solution can be linked to the behavior expected if there were zero electrostatic interactions between ions in solution.

During this presentation, the experimental mean activity coefficients showing up in the deductions have been taken for granted—nothing has been said about how they have been obtained. Obviously, one must be sure when dealing with theory that the experimental values with which the theory is compared are soundly based.

It is time then that some account be given about the means by which experimental values of activity coefficients are known. Only two methods will be presented because the material contains no new ideas and is only presented so the reader is assured that the ground is firm.

3.4.7. How to Obtain *Solute* Activities from Data on *Solvent* Activities

A characteristic of an ionic solution is that any vapor pressure due to the dissolved electrolyte itself is effectively zero. The vapor pressure of the solvent in the solution therefore falls with increasing concentration of the electrolyte in the solution. Thus, the solvent vapor pressure in the solution will be less than the vapor pressure of the pure solvent because the nonvolatile ions block out part of the surface from which, in the pure solvent, solvent molecules would evaporate.

Now, there is nothing mysterious about a solvent activity. It is determined by

$$a_1 = \frac{P_1}{P_1^*} \tag{3.92}$$

where P_1^* is the vapor pressure of *pure solvent* and P_1 is the vapor pressure of the solvent when it is a component of a solution. The relation of P_1 to P_1^* in the ideal condition will be governed by Raoult's law, that is,

$$P_1 = x_1 P_1^* \tag{3.93}$$

where x_1 is the mole fraction of the solvent. One allows for the nonideal behavior of the solvent in respect to its vapor pressure by writing

$$a_1 = f_1 x_1 \tag{3.94}$$

As $x_1 \to 1, f_1 \to 1$. Thus, the "standard state" is the pure solvent.

At this point, one calls into play the Gibbs–Duhem equation of thermodynamics, according to which[10]

$$\sum n_i \, d\mu_i = 0 \tag{3.95}$$

Or, for a two-component system (solvent and electrolyte)

$$n_1 \, d\mu_1 + n_2 \, d\mu_2 = 0 \tag{3.96}$$

where the subscript 1 represents the solvent and the subscript 2 represents the electrolyte solute.

It follows from $\mu_i = \mu_i^0 + RT \ln a_i$, that

$$d \ln a_2 = -\frac{n_1}{n_2} d \ln a_1 \tag{3.97}$$

or

$$\ln \frac{a_2}{c_2} = -\int_1^{a_1} \frac{n_1}{n_2} \, d \ln a_1 \tag{3.98}$$

Thus, if one measures a number of values of the vapor pressure of the solvent P_1 at a corresponding number of solute concentrations, x_2 (to which there are matching solvent concentrations x_1), one can plot the ln a_1 values against the n_1/n_2 ratios. Then the area of that plot will give ln a_2, the a_2 being the solute activity corresponding to the limit of the integral at a_1 (this a_1 being the measured solvent activity for a solution containing a solute, the activity of which is a_2).

The left-hand side of Eq. (3.98) came from

---

[10]Any initial impression that there is something unreasonable about the Gibbs–Duhem equation should be instantly quelled. It merely tells one that (for a two-component system) when an increase in $n_1 \, d\mu_1$ occurs, it causes a decrease in $n_2 \, d\mu_2$ of equal magnitude, a typically powerful and general result of thermodynamic reasoning.

$$\int_{a_2 \to 0}^{a_2} d \ln a_2$$

and $a_2 \to 0$ is simply c_2 because when the solute concentration (hence also activity) tends to zero, its activity becomes equal to its concentration.

This solvent vapor pressure method for measuring the activity of electrolytes has the advantage that the actual experiments one has to do are simple.[11] The method can be applied to any concentration (e.g., a 15 *M* solution!). The difficulty comes at low concentrations when the difference of the vapor pressure between the solution and that of the solvent becomes limitingly small. A huge amount of data (see Table 3.5) have been determined by this method, particularly in the 1950s by a long-term Australian–New Zealand collaboration between professors Stokes and Robinson.

3.4.8. A Second Method by Which One May Obtain Solute Activities: From Data on Concentration Cells and Transport Numbers

Thermodynamics treats electrochemical cells in equilibrium and indeed such hoary material, going back to the work of the great German physical chemist Nernst,[12] is a part of classical electrochemistry that is still being taught in universities to students as if it were representative of modern electrochemistry! Consider then the chemical potential of a metal as μ_M in the solid electrode.

The chemical potential of the ion is that of a solute in a solution and hence is given by

$$\mu_{M^+} = \mu^0_{M^+} + kT \ln a_{M^+} \tag{3.99}$$

[11]There is often no need for an absolute determination of vapor pressure. The solvent vapor pressure can be determined simply by setting up a closed system that contains a solution of large volume having an already known solvent activity. The unknown solution will change its concentrations (and hence its weight) until its solvent activity is the same as that of the reference system, which is known. Great accuracy in the weighing is essential and one should use platinum vessels to minimize possible dissolution.

[12]Walter Nernst was professor of chemistry in Berlin in the early years of the twentieth century. He epitomized the professor as a "Great Man." Among his many achievements was the work that led (via the Nernst heat theorem) to the third law of thermodynamics. He was active not only in chemistry but also made significant contributions to the theory of the expanding universe. Nernst was famous not only for his real (many and great) contributions to physical electrochemistry but also for the cold and rigid discipline he demanded from those who aspired to be his collaborators. Were one of these to arrive at his workplace after the scheduled hour of 7:00 a.m., he might find a note from the professor reminding him of the number of applicants who were waiting to occupy it.

One such collaborator (later himself a famous physical chemist) is known to have remarked that, in making the mixture for Nernst, the Herr Creator had put in an extra dose of the intellectual but left out the humanity.

TABLE 3.5

Experimental Activity Coefficients of NaCl at 298 K

$I(=c)$	$-\log f_{\pm}$ (Experimental Values)
0.000997	0.0155
0.001994	0.0214
0.004985	0.0327
0.009969	0.0446
0.01993	0.0599
0.049891	0.0859
0.09953	0.1072
0.1987	0.1308
0.4940	0.1593
0.9788	0.1671
1.921	0.1453
3.696	0.0477
5.305	−0.0789

Source: Reprinted from R. A. Robinson and R. H. Stokes, *Electrolyte Solutions*, 2nd ed. rev., Butterworth, London, 1968.

where a_{M^+} is the metal ion's activity in solution. Now the reaction that goes on at an electrode when an ion in solution exchanges electrons with the electrode can be written as

$$M^+ + e_0^- \rightleftharpoons M \tag{3.100}$$

Clearly, one has to make an allowance for the electron in the Nernst-type theories—which are thermodynamically valid (model free) and therefore can be used today. This was done by a term Ee_0 (potential × the electronic charge = energy in electrostatics). The E was thought of by Nernst as the potential between the metal and the solution.

Now, the thermodynamic equilibrium for Eq. (3.100)[13]

$$\mu_{M^+} - Ee_0 = \mu_M^0 - \mu_e^0 \tag{3.101}$$

Substituting μ_{M^+} from Eq. (3.99),

$$\mu_{M^+}^0 + kT \ln a_{M^+} - Ee_0 = \mu_M^0 - \mu_e^0 \tag{3.102}$$

or

[13]The negative sign arises because e_0^-, the magnitude of the charge on the electron, bears a negative sign.

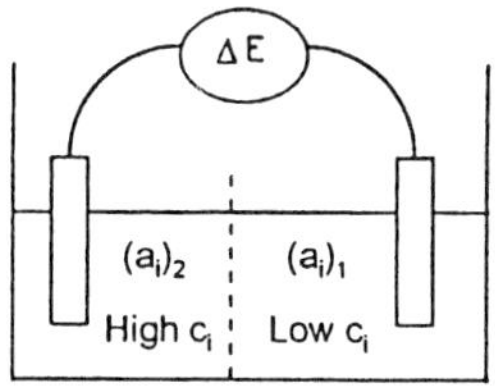

Fig. 3.19. Determination of individual ionic activity from the cell potential.

$$E = \frac{\Delta\mu^0}{e_0} + \frac{kT}{e_0} \ln a_{M^+} = E^0 + \frac{RT}{F} \ln a_{M^+} \tag{3.103}$$

where E^0 is the standard potential of the electrode reaction, at 25 °C and a unit activity of the ion i in solution; $N_A k = R$; $N_A e_0 = F$, the Faraday, or charge on 1 gram ion.

It follows that if one has two electrodes, each in contact with the same ions but at different activities, and the reactions are in thermodynamic equilibrium, then neglecting for a moment any potential that might exist at the contact between the two solutions, i.e., any *liquid-junction* potential, from Eq. (3.103),

$$E_{\text{soln2}} - E_{\text{soln1}} = E_{\text{cell}} = \frac{RT}{F} \ln \frac{a_{M^+,\text{soln2}}}{a_{M^+,\text{soln1}}} \tag{3.104}$$

and a schematic representation of this idea is shown in Fig. 3.19.

Thus, one could argue as follows: if one of the solutions in the cell has a sufficiently low concentration, e.g., $< 5 \times 10^{-3}$ mol dm^{-3} for 1:1 electrolyte, then the Debye–Hückel limiting law applies excellently. Hence, if one of the solutions has a concentration of $< 5 \times 10$ mol dm^{-3}, we know $a_{i,1}$, the activity of the ion i in that cell, so that Eq. (3.104) would give at once $a_{i,2}$ the activity of species i in the more concentrated cell.

Moreover—and still keeping the question of the liquid junction rigidly suppressed in one's mind—the answer would have one big advantage, it would give an *individual* ionic activity coefficient, f_i.

A method of such virtue must indeed have a compensating complication, and the truth is that the neglected liquid junction potential (LJP) may not be negligible at all.[14]

[14]It turns out that $E_{LJP}/E_{\text{cell}} = (t_+ - t_-)$ where t_+ and t_- are, respectively, the cationic and anionic transport numbers. There are cases (e.g., for junctions of solutions of KCl) where t_+ and t_- are almost the same and hence the $E_{LJP} = 0$ and the correction due to the liquid junction is negligible. In some cases the difference $t_+ - t_-$ may be quite considerable.

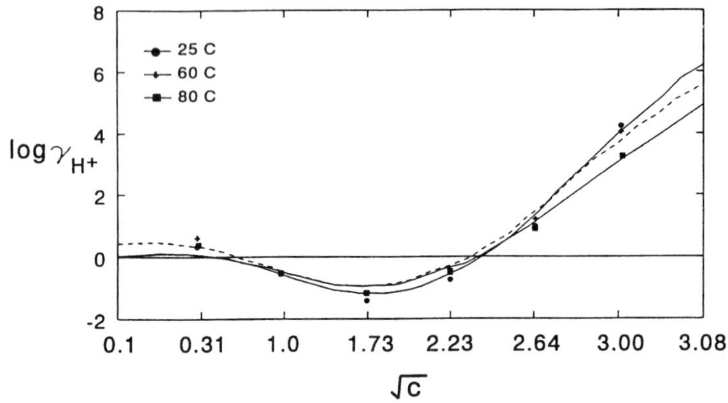

Fig. 3.20. Plot of log γ_{H^+} vs. $\sqrt{c}$ at different temperatures for CF_3SO_3H. (Reprinted from R. C. Bhardwaj, M. A. Enayetullah, and J. O'M. Bockris, *J. Electrochem. Soc.* **137**: 2070, 1990.)

Finding an equation for E_{LJP} is surprisingly difficult, but it is finally shown that [*cf.* the deduction of the Planck–Henderson equation, Eq. (4.291)]

$$E_{LJP} = (2t_+ - 1)\frac{RT}{F}\ln\frac{a_{i,\text{soln2}}}{a_{i,\text{soln1}}} \tag{4.291}$$

Using Eqs. (3.104) and (4.291), the cell potential *with* the liquid-junction potential is

$$E = \frac{RT}{F}\ln\frac{a_{M^+,\text{soln2}}}{a_{M^+,\text{soln1}}} + (2t_+ - 1)\frac{RT}{F}\ln\frac{a_{M^+,\text{soln2}}}{a_{M^+,\text{soln1}}} = 2t_+\frac{RT}{F}\ln\frac{a_{M^+,\text{soln2}}}{a_{M^+,\text{soln1}}} \tag{3.105}$$

If one knows t_+ at the concentration at which it is desired to know $a_{M^+,\text{soln2}}$, one can find the latter value by using the limiting law for $a_{M^+,\text{soln1}}$. Compared with the first method (Section 3.4.7), the advantage of being able to aim for an individual ion's activity coefficient is significant.

However, there are penalties to pay if one uses the electrochemical cell method. First, there is the question of the value of t_+—it should be known as a function of concentration, and such values are often not available and imply the need for a separate determination. Further, there is a nasty experimental point. One talks of "the liquid-junction potential" as though it were a clear and definite entity. The thermodynamic equation [Eq. (4.291)] *assumes that there is a sharp boundary with linear change of concentration across a small distance* (see Section 4.5.9). These conditions assumed in the deduction only last for a short time after the two solutions have been brought

TABLE 3.6

Experimental Values of Activity Coefficients of Various Electrolytes at Different Concentrations at 298 K

1:1 electrolyte, HCl					
Concentration, molal	0.0001	0.0002	0.0005	0.001	0.002
Mean activity coefficient	0.9891	0.9842	0.9752	0.9656	0.9521
2:1 electrolyte, $CaCl_2$					
Concentration (moles dm^{-3})	0.0018	0.0061	0.0095		
Mean activity coefficient	0.8588	0.7745	0.7361		
2:2 electrolyte, $CdSO_4$					
Concentration, molal	0.0005	0.001	0.005		
Mean activity coefficient	0.774	0.697	0.476		

into contact, even when the boundary is made very carefully.[15] Hence, the conditions under which Eq. (3.104) is valid are subtle and difficult to achieve. Nevertheless, the method has been used—with guarded admissions about the dangers of using it—when information on single ionic activities is desirable, e.g., in the study of O_2 reduction ($O_2 + 4\ H^+ + 4\ e \rightarrow 2\ H_2O$). Some results of the latter study— at very high concentrations—are given in Fig. 3.20. Note the extremely high values of the proton measured in the concentrated acid.

Further Reading

Seminal

1. P. Debye and E. Hückel, "The Interionic Attraction Theory of Deviations from Ideal Behavior in Solution," *Z. Phys.* **24**: 120 (1923).
2. H. S. Harned and B. B. Owen, *The Physical Chemistry of Electrolytic Solutions*, 3rd ed., Reinhold Publishing, New York (1958).
3. H. L. Friedman, "Electrolytic Solutions," in *Modern Aspects of Electrochemistry, No. 8*, J. O'M. Bockris and B. E. Conway, eds., Plenum, New York (1971).

Reviews

1. H. L. Friedman, "Theory of Ionic Solutions in Equilibrium," in *Physical Chemistry of Aqueous Ionic Solutions,* M.-C. Bellisent-Fund and G. W. Nielson, eds., *NATO ASI Series C* **205**: 61 (1986).

[15]There are several techniques for making an undisturbed boundary near in reality to that implicitly assumed in the theory (Section 5.6.7.2). In the one most usually used, two solutions of different concentrations are held apart by a glass slide that is slowly removed, allowing contact between the solutions with minimal coerciveness. In another, the two solutions are held apart in a tube, the one on top being restrained against gravity by means of reduced pressure. Very slow release of this pressure allows the gradual descent of the top solution to make a gentle junction with the lower one. The aim of each method is to avoid disturbance of the assumed ideal exact boundary.

2. J. C. Rasaiah, "Theories of Electrolyte Solutions," in *The Liquid State and Its Electrical Properties*, E. E. Kunhardt, L. G. Christophorou, and L. H. Luessen, eds., *NATO ASI Series B* **193**: 135 (1987).

Papers

1. C. F. Baes, Jr., E. J. Reardon, and B. A. Bloyer, *J. Phys. Chem.* **97**: 12343 (1993).
2. H. P. Diogo, M. E. Minas da Piedade, and J. J. Moura Ramos, *J. Chem. Ed.* **70**: A227 (1993).
3. H. Schönert, *J. Phys. Chem.* **98**: 643 (1994).
4. H. Schönert, *J. Phys. Chem.* **98**: 654 (1994).
5. B. Honig and A. Micholls, *Science* **268**: 1144 (1995).
6. B. B. Laird and A. D. J. Haymet, *J. Chem. Phys.* **100**: 3775 (1996).

3.5. THE TRIUMPHS AND LIMITATIONS OF THE DEBYE–HÜCKEL THEORY OF ACTIVITY COEFFICIENTS

3.5.1. How Well Does the Debye–Hückel Theoretical Expression for Activity Coefficients Predict Experimental Values?

The approximate theoretical equation

$$\log f_{\pm} = -A(z_+z_-)I^{1/2} \tag{3.90}$$

indicates that the logarithm of the activity coefficient must decrease linearly with the square root of the ionic strength or, in the case of 1:1-valent electrolytes,[16] with $c^{1/2}$. Further, the slope of the $\log f_{\pm}$ versus $I^{1/2}$ straight line can be unambiguously evaluated from fundamental physical constants and from (z_+z_-). Finally, the slope does not depend on the particular electrolyte (i.e., whether it is NaCl or KBr, etc.) but only on its valence type, i.e., on the charges borne by the ions of the electrolyte, whether it is a 1:1-valent or 2:2-valent electrolyte, etc. These are clear-cut predictions.

Even before any detailed comparison with experiment, one can use an elementary spot check: At infinite dilution, where the interionic forces are negligible, does the theory yield the activity coefficient that one would expect from experiment, i.e., unity? At infinite dilution, c or $I \rightarrow 0$, which means that $\log f_{\pm} \rightarrow 0$ or $f_{\pm} \rightarrow 1$. The properties of an extremely dilute solution of ions should be the same as those of a solution containing nonelectrolyte particles. Thus, the Debye–Hückel theory emerges successfully from the infinite dilution test.

Furthermore, if one takes the experimental values of the activity coefficient (Table 3.6) at extremely low electrolyte concentration and plots $\log f_{\pm}$ versus $I^{1/2}$ curves, it

---

[16]That is, $I = \frac{1}{2}\Sigma c_i z_i^2$. For a 1:1 electrolyte, $I = \frac{1}{2}(c_i 1^2 + c_j 1^2)$. As $c_i = c_j = c$, $I = c$.

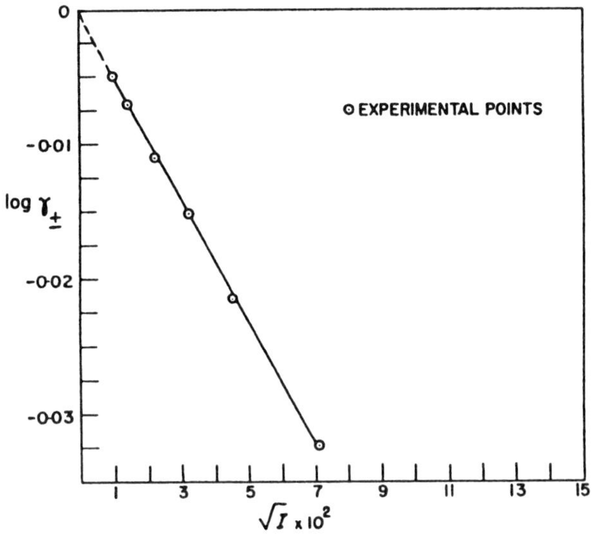

Fig. 3.21. The logarithm of the experimental mean activity coefficient of HCl varies linearly with the square root of the ionic strength.

is seen that: (1) They are linear (Fig. 3.21), and (2) they are grouped according to the valence type of the electrolyte (Fig. 3.22). Finally, when one compares the calculated and observed slopes, it becomes clear that there is excellent agreement to an error of ±0.5% (Table 3.7 and Fig. 3.23) between the results of experiment and the conclusions

TABLE 3.7

Experimental and Calculated Values of the Slope of log $f_\pm - \sqrt{I}$ for Alcohol–Water Mixtures at 298 K

Solvent Mole Fraction Water	Dielectric Constant	Slope	
		Observed	Calculated
	1:1 *type of salt, Croceo tetranitro diamino cobaltiate*		
1.00	78.8	0.50	0.50
0.80	54.0	0.89	0.89
	1:2 *type of salt, Croceo sulfate*		
1.00	78.8	1.10	1.08
0.80	54.0	1.74	1.76
	3:1 *type of salt, Luteo iodate*		
1.00	78.8	1.52	1.51

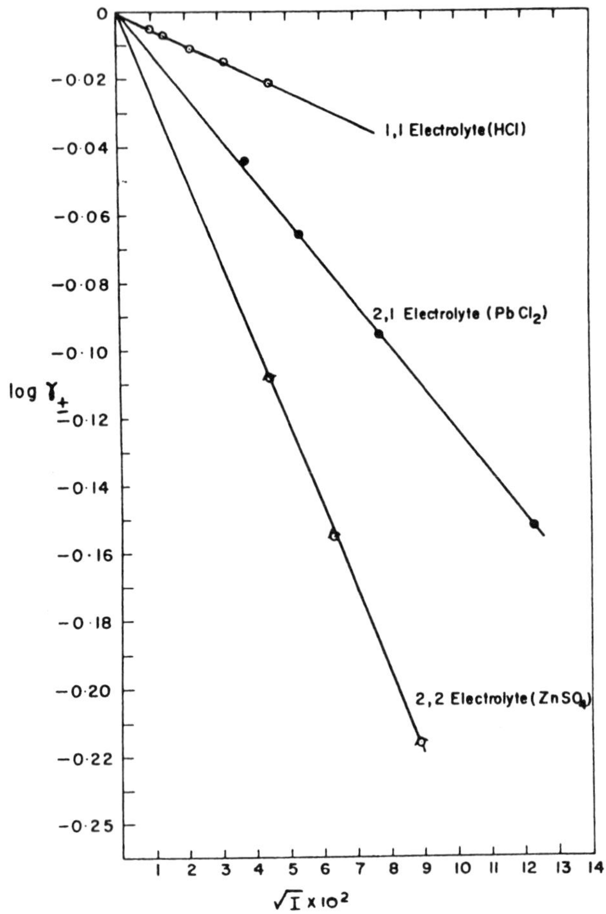

Fig. 3.22. The experimental log $f_{\pm}$ versus $I^{1/2}$ straight-line plots for different electrolytes can be grouped according to valence type.

emerging from an analysis of the ionic-cloud model of the distribution of ions in an electrolyte. Since Eq. (3.90) has been found to be valid at limiting low electrolyte concentrations, it is generally referred to as the *Debye–Hückel limiting law*.

The success of the Debye–Hückel limiting law is no mean achievement. One has only to think of the complex nature of the real system, of the presence of the solvent, which has been recognized only through a dielectric constant, of the simplicity of the Coulomb force law used, and, finally, of the fact that the ions are not point charges, to realize (Table 3.7) that the simple ionic cloud model has been brilliantly successful—almost unexpectedly so. It has grasped the essential truth about electrolytic

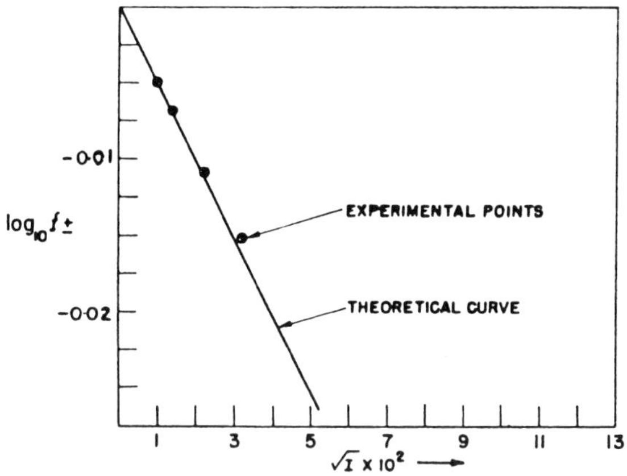

Fig. 3.23. The comparison of the experimentally observed mean activity coefficients of HCl and those that are calculated from the Debye–Hückel limiting law.

solutions, albeit about solutions of extreme dilution. The success of the model is so remarkable and the implications are so wide (see Section 3.5.6) that the Debye–Hückel approach is to be regarded as one of the most significant pieces of theory in the ionics part of electrochemistry. It even rates among the leading pieces of physical chemistry of the first half of the twentieth century.

The Debye–Hückel approach is an excellent example of electrochemical theory. Electrostatics is introduced into the problem in the form of Poisson's equation, and the chemistry is contained in the Boltzmann distribution law and the concept of true electrolytes (Section 3.2). The union of the electrostatic and chemical modes of

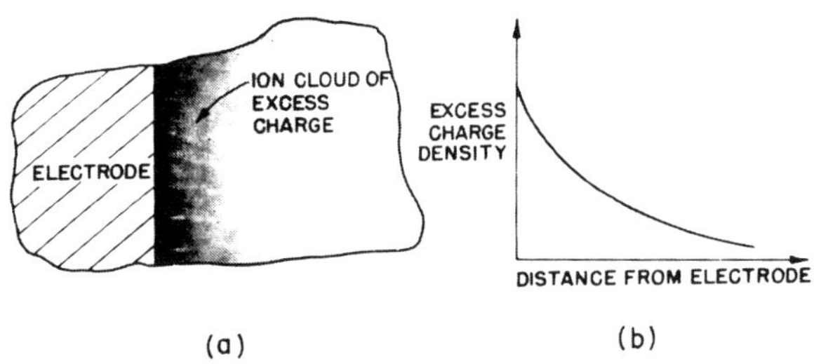

Fig. 3.24. An electrode immersed in an ionic solution is often enveloped by an ionic cloud [see Fig. 3.11] in which the excess charge density varies with distance as shown in (b).

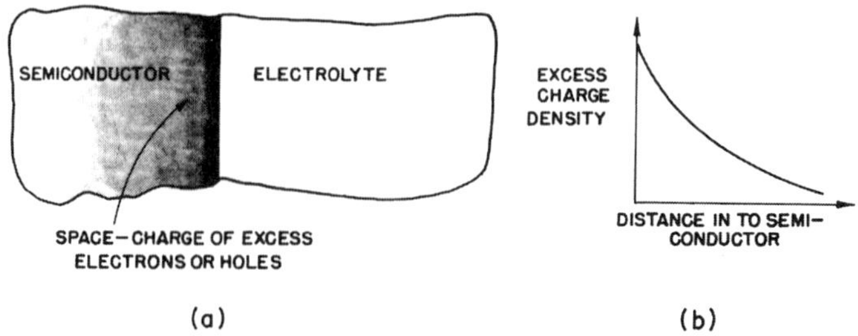

Fig. 3.25. (a) A space charge produced by excess electrons or holes often exists inside the semiconductor. (b) The space charge density varies with distance from the semiconductor–electrolyte interface.

description to give the linearized Poisson–Boltzmann equation illustrates therefore a characteristic development of electrochemical thinking.

It is not surprising that the Poisson–Boltzmann approach has been used frequently in computing interactions between charged entities. Mention may be made of the Gouy theory (Fig. 3.24) of the interaction between a charged electrode and the ions in a solution (see Chapter 6). Other examples are the distribution (Fig. 3.25) of electrons or holes inside a semiconductor in the vicinity of the semiconductor–electrolyte interface (see Chapter 6) and the distribution (Fig. 3.26) of charges near a polyelectrolyte molecule or a colloidal particle (see Chapter 6).

However, one must not overstress the triumphs of the Debye–Hückel limiting law [Eq. (3.90)]. Models are always simplifications of reality. They never treat all its complexities and thus there can never be a perfect fit between experiment and the predictions based on a model.

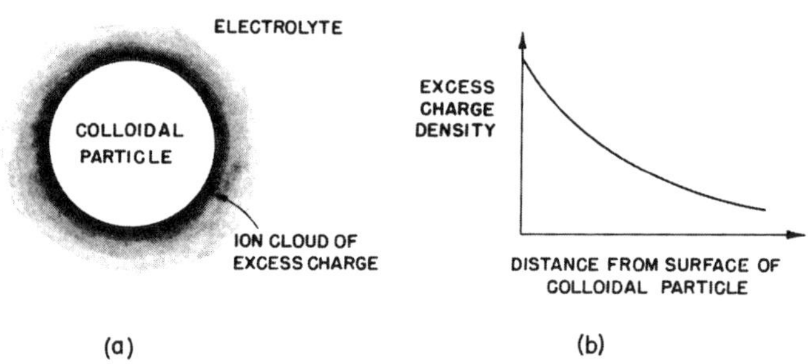

Fig. 3.26. (a) A colloidal particle is surrounded by an ionic cloud of excess charge density, which (b) varies with distance from the surface of the colloidal particle.

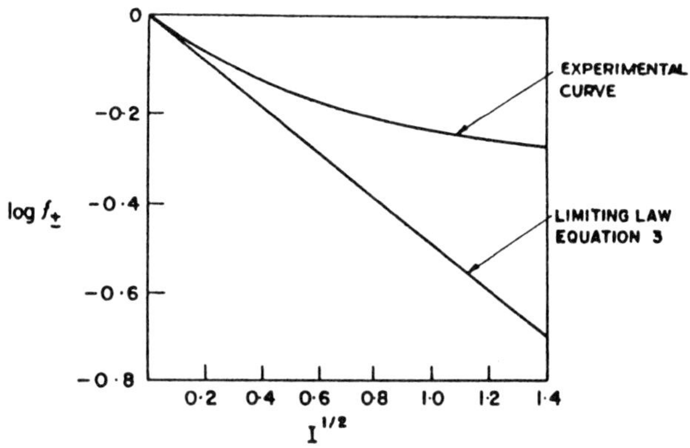

Fig. 3.27. The experimental log $f_\pm$ versus $I^{1/2}$ curve is a straight line only at extremely low concentrations.

What then are the inadequacies of the Debye–Hückel limiting law? One does not have to look far. If one examines the experimental $\log f_\pm$ versus $I^{1/2}$ curve, not just in the extreme dilution regions, but at higher concentrations, it turns out that the simple Debye–Hückel limiting law falters. The plot of $\log f_\pm$ versus $I^{1/2}$ is a curve (Fig. 3.27 and Table 3.8) and not a straight line as promised by Eq. (3.90). Further, the curves depend not only on valence type (e.g., 1:1 or 2:2) but also (Fig. 3.28) on the particular electrolyte (e.g., NaCl or KCl).

It appears that the Debye–Hückel law is the law for the tangent to the $\log f_\pm$ versus $I^{1/2}$ curve at very low concentrations, say, up to 0.01 N for 1:1 electrolytes in aqueous solutions. At higher concentrations, the model must be improved. What refinements can be made?

3.5.2. Ions Are of Finite Size, They Are Not Point Charges

One of the general procedures for refining a model that has been successful in an extreme situation is to liberate the theory from its approximations. So one has to recall

TABLE 3.8

Comparison of Calculated [Eq. (3.90)] and Experimental Values of log $f_\pm$ for NaCl at 298 K

Concentration (molal)	$-\log f_\pm$ Experimental	$-\log f_\pm$ Calculated
0.001	0.0155	0.0162
0.002	0.0214	0.0229
0.005	0.0327	0.0361
0.01	0.0446	0.0510
0.02	0.0599	0.0722

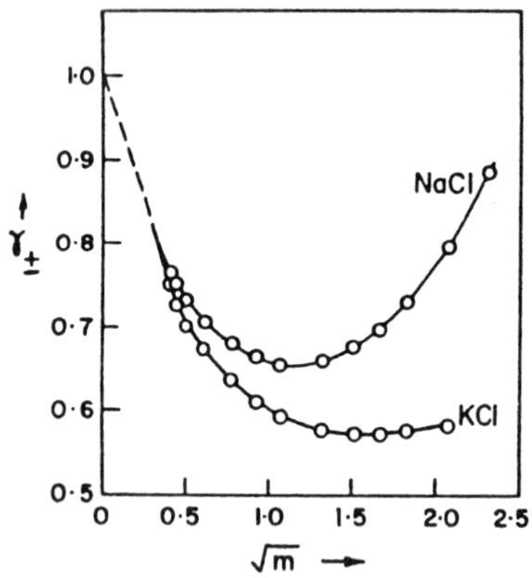

Fig. 3.28. Even though NaCl and KCl are 1:1 electrolytes, their activity coefficients vary indifferent ways with concentration as soon as one examines higher concentrations.

what approximations have been used to derive the Debye–Hückel limiting law. The first one that comes to mind is the point-charge approximation.[17] One now asks: Is it reasonable to consider ions as point charges?

It has been shown (Section 3.3.8) that the mean thickness κ^{-1} of the ionic cloud depends on the concentration. As the concentration of a 1:1 electrolyte increases from 0.001 N to 0.01 N to 0.1 N, κ^{-1} decreases from about 10 to 3 to about 1 nm. This means that the relative dimensions of the ion cloud and of the ion change with concentration. Whereas the radius of the cloud is 100 times the radius of an ion at 0.001 N, it is only about 10 times the dimensions of an ion at 0.1 N. Obviously, under these latter circumstances, an ion cannot be considered a geometrical point charge in comparison with a dimension only 10 times its size (Fig. 3.29). The more concentrated the solution (i.e., the smaller the size κ^{-1} of the ion cloud; Section 3.3.8), the less valid is the point-charge approximation. If therefore one wants the theory to be applicable to 0.1 N solutions or to solutions of even higher concentration, the finite size of the ions must be introduced into the mathematical formulation.

To remove the assumption that ions can be treated as point charges, it is necessary at first to recall at what stage in the derivation of the theory the assumption was

[17]Another approximation in the Debye–Hückel model involves the use of Poisson's equation, which is based on the smearing out of the charges into a continuously varying charge density. At high concentrations, the mean distance between charges is low and the ions see each other as discrete point charges, not as smoothed-out charges. Thus, the use of Poisson's equation becomes less and less justified as the solution becomes more and more concentrated.

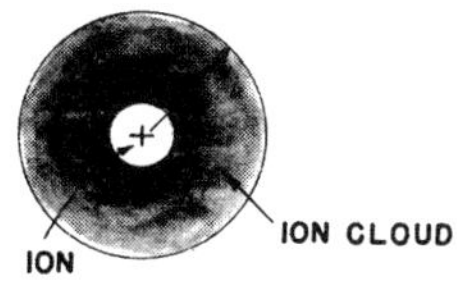

Fig. 3.29. At 0.1 *N*, the thickness of the ion cloud is only 10 times the radius of the central ion.

invoked. The linearized P–B equation involved neither the point-charge approximation nor any considerations of the dimensions of the ions. Hence, the basic differential equation

$$\frac{1}{r^2}\frac{d}{dr}\left(r^2\frac{d\psi_r}{dr}\right)=\kappa^2\psi_r \tag{3.21}$$

and its general solution, i.e.,

$$\psi_r = A\frac{e^{-\kappa r}}{r} + B\frac{e^{+\kappa r}}{r} \tag{3.28}$$

can be taken as the basis for the generalization of the theory for finite-sized ions.

As before (Section 3.3.7), the integration constant B must be zero because otherwise one cannot satisfy the requirement of physical sense that, as $r \to \infty$, $\psi \to 0$. Hence, Eq. (3.28) reduces to

$$\psi_r = A\,\frac{e^{-\kappa r}}{r} \tag{3.29}$$

In evaluating the constant A, a procedure different from that used after (3.29) is adopted. The charge dq in any particular spherical shell (of thickness dr) situated at a distance r from the origin is, as argued earlier,

$$dq = \varrho_r 4\pi r^2 dr \tag{3.36}$$

The charge density ϱ_r is obtained thus

$$\varrho_r = -\frac{\varepsilon}{4\pi}\left[\frac{1}{r^2}\frac{d}{dr}\left(r^2\frac{d\psi_r}{dr}\right)\right] = -\frac{\varepsilon}{4\pi}\kappa^2\psi_r \tag{3.34}$$

and inserting the expression for ψ_r from Eq. (3.29), one obtains

$$\varrho_r = -\frac{\varepsilon}{4\pi}\kappa^2 A \frac{e^{-\kappa r}}{r} \tag{3.106}$$

Thus, by combining Eqs. (3.36) and (3.106)

$$dq = -A\kappa^2\varepsilon(e^{-\kappa r} r\, dr) \tag{3.107}$$

The total charge in the ion cloud q_{cloud} is, on the one hand, equal to $-z_i e_0$ [Eq. (3.39)] as required by the electroneutrality condition and, on the other hand, the result of integrating dq. Thus,

$$q_{\text{cloud}} = -z_i e_0 = \int_{?}^{\infty} dq\, dr = -A\kappa^2\varepsilon \int_{?}^{\infty} e^{-\kappa r} r\, dr \tag{3.108}$$

What lower limit should be used for the integration? In the point-charge model, one used a lower limit of zero, meaning that the ion cloud commences from zero (i.e., from the surface of a zero-radius ion) and extends to infinity. However, now the ions are taken to be of finite size, and a lower limit of zero is obviously wrong. The lower limit should be a distance corresponding to the distance from the ion center at which the ionic atmosphere starts (Fig. 3.30).

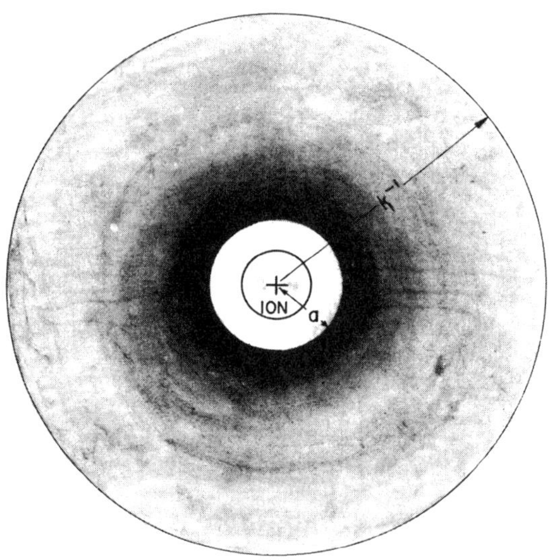

Fig. 3.30. For a finite-sized ion, the ion atmosphere starts at a distance *a* from the center of the reference ion.

As a first step, one can use for the lower limit of the integration a distance parameter that is greater than zero. Then one can go through the mathematics and later worry about the physical implications of the ion size parameter. Let this procedure be adopted and symbol a be used for the ion size parameter.

One has then

$$\int_a^\infty dq\, dr = -A\kappa^2\varepsilon \int_a^\infty e^{-\kappa r} r\, dr$$
$$= -A\varepsilon \int_a^\infty \kappa r e^{-\kappa r}\, d(\kappa r) \tag{3.109}$$

As before (Appendix 3.2), one can integrate by parts; thus,

$$\int_a^\infty \kappa r e^{-\kappa r}\, d\kappa r = -[\kappa r e^{-\kappa r}]_a^\infty + \int_a^\infty e^{-\kappa r}\, d(\kappa r)$$
$$= \kappa a e^{-\kappa a} - [e^{-\kappa r}]_a^\infty \tag{3.110}$$

Inserting Eq. (3.110) in Eq. (3.109), one obtains

$$\int_a^\infty dq\, dr = -A\varepsilon e^{-\kappa a}(1 + \kappa a) = -z_i e_0 \tag{3.111}$$

from which

$$A = \frac{z_i e_0}{\varepsilon} \frac{e^{\kappa a}}{1 + \kappa a} \tag{3.112}$$

Using this value of A in Eq. (3.29), one obtains a new and less approximate expression for the potential ψ_r at a distance r from a finite-sized central ion,

$$\psi_r = \frac{z_i e_0}{\varepsilon} \frac{e^{\kappa a}}{1 + \kappa a} \frac{e^{-\kappa r}}{r} \tag{3.113}$$

3.5.3. The Theoretical Mean Ionic-Activity Coefficient in the Case of Ionic Clouds with Finite-Sized Ions

Once again (see Section 3.3.9), one can use the law of superposition of potentials to obtain the ionic-atmosphere contribution ψ_{cloud} to the potential ψ_r at a distance r from the central ion. From Eq. (3.46), i.e.,

$$\psi_{\text{cloud}} = \psi_r - \psi_{\text{ion}} \tag{3.46}$$

it follows by substitution of the expression (3.113) for ψ_r and Eq. (3.44) for ψ_{ion} that

$$\psi_{\text{cloud}} = \frac{z_i e_0}{\varepsilon r} \frac{e^{\kappa(a-r)}}{1+\kappa a} - \frac{z_i e_0}{\varepsilon r}$$

$$= \frac{z_i e_0}{\varepsilon r}\left[\frac{e^{\kappa(a-r)}}{1+\kappa a} - 1\right] \tag{3.114}$$

It will be recalled, however, that, in order to calculate the activity coefficient from the expressions

$$RT \ln f_i = \Delta\mu_{i-I} \tag{3.59}$$

and

$$\Delta\mu_{i-I} = \frac{N_A z_i e_0}{2}\psi \tag{3.3}$$

i.e., from

$$\ln f_i = \frac{N_A z_i e_0}{2RT}\psi \tag{3.115}$$

it is necessary to know ψ, which is the potential at the surface of the ion due to the surrounding ions, i.e., due to the cloud. Since, in the finite-ion-size model, the ion is taken to have a size a, it means that ψ is the value of ψ_{cloud} at $r = a$,

$$\psi = \psi_{\text{cloud}} \quad r = a \tag{3.116}$$

The value of ψ_{cloud} at $r = a$ is obtained by setting $r = a$ in Eq. (3.114). Hence,

$$\psi = \psi_{\text{cloud}(r=a)} = -\frac{z_i e_0}{\varepsilon\kappa^{-1}} \frac{1}{1+\kappa a} \tag{3.117}$$

By substitution of the expression (3.117) for $\psi = \psi_{\text{cloud}(r=a)}$ in Eq. (3.115), one obtains

$$\ln f_i = -\frac{N_A (z_i e_0)^2}{2\varepsilon RT\kappa^{-1}} \frac{1}{1+\kappa a} \tag{3.118}$$

This individual ionic-activity coefficient can be transformed into a mean ionic-activity coefficient by the same procedure as for the Debye–Hückel limiting law (see Section 3.4.4). On going through the algebra, one finds that the expression for $\log f_{\pm}$ in the finite-ion-size model is

$$\log f_{\pm} = -\frac{A(z_{+}z_{-})I^{1/2}}{1+\kappa a} \tag{3.119}$$

It will be recalled, however, that the thickness κ^{-1} of the ionic cloud can be written as [Eq. (3.85)]

$$\kappa = BI^{1/2} \tag{3.85}$$

Using this notation, one ends up with the final expression

$$\log f_{\pm} = -\frac{A(z_{+}z_{-})I^{1/2}}{1+BaI^{1/2}} \tag{3.120}$$

If one compares Eq. (3.119) of the finite-ion-size model with Eq. (3.90) of the point-charge approximation, it is clear that the only difference between the two expressions is that the former contains a term $1/(1+\kappa a)$ in the denominator. Now, one of the tests of a more general version of a theory is the *correspondence principle*; i.e., the general version of a theory must reduce to the approximate version under the conditions of applicability of the latter. Does Eq. (3.119) from the finite-ion-size model reduce to Eq. (3.90) from the point-charge model?

Rewrite Eq. (3.119) in the form

$$\log f_{\pm} = -A(z_{+}z_{-})I^{1/2}\frac{1}{1+a/\kappa^{-1}} \tag{3.121}$$

and consider the term a/κ^{-1}. As the solution becomes increasingly dilute, the radius κ^{-1} of the ionic cloud becomes increasingly large compared with the ion size, and simultaneously a/κ^{-1} becomes increasingly small compared with unity, or

$$\frac{1}{1+a/\kappa^{-1}} \approx 1 \tag{3.122}$$

Thus, when the solution is sufficiently dilute to make $a \ll \kappa^{-1}$, i.e., to make the ion size insignificant in comparison with the radius of the ion atmosphere,the finite-ion-size model Eq (3.119) reduces to the corresponding Eq. (3.90) of the point-charge model because the extra term $1/(1+a/\kappa^{-1})$ tends to unity

$$-\left[\frac{A(z_+z_-)I^{1/2}}{1+\kappa a}\right]_{a<<\kappa^{-1}} = -A(z_+z_-)I^{1/2} \tag{3.123}$$

The physical significance of $a/\kappa^{-1} << 1$ is that at very low concentrations the ion atmosphere has such a large radius compared with that of the ion that one need not consider the ion as having a finite size a. Considering $a/\kappa^{-1} << 1$ is tantamount to reverting to the point-charge model.

One can now proceed rapidly to compare this theoretical expression for $\log f_\pm$ with experiment; but what value of the ion size parameter should be used? The time has come to worry about the precise physical meaning of the parameter a that was introduced to allow for the finite size of ions.

3.5.4. The Ion Size Parameter *a*

One can at first try to speculate on what value of the ion size parameter is appropriate. A lower limit is the sum of the *crystallographic* radii of the positive and negative ions present in solution; ions cannot come closer than this distance [Fig. 3.31(a)]. But in a solution the ions are generally solvated (Chapter 2). So perhaps the sum of the solvated radii should be used [Fig. 3.31(b)]. However when two solvated ions collide, is it not likely [Fig. 3.31(c)] that their hydration shells are crushed to some extent? This means that the ion size parameter a should be greater than the sum of the crystallographic radii and perhaps less than the sum of the solvated radii. It should best be called the *mean distance of closest approach*, but beneath the apparent wisdom of this term there lies a measure of ignorance. For example, an attempted calculation of just how crushed together two solvated ions are would involve many difficulties.

To circumvent the uncertainty in the quantitative definition of a, it is best to regard it as a parameter in Eq. (3.120), i.e., a quantity the numerical value of which is left to be calibrated or adjusted on the basis of experiment. The procedure (Fig. 3.32) is to assume that the expression for $\log f_\pm$ [Eq. (3.120)] is correct at one concentration. then to equate this theoretical expression to the experimental value of $\log f_\pm$ corresponding to that concentration and to solve the resulting equation for a. Once the ion size parameter, or mean distance of closest approach, is thus obtained at one concentration, the value can be used to calculate values of the activity coefficient over a range of other and higher concentrations. Then the situation is regarded as satisfactory if the value of a obtained from experiments at one concentration can be used in Eq. (3.120) to reproduce the results of experiments over a range of concentrations.

3.5.5. Comparison of the Finite-Ion-Size Model with Experiment

After taking into account the fact that ions have finite dimensions and cannot therefore be treated as point charges, the following expression has been derived for the logarithm of the activity coefficient:

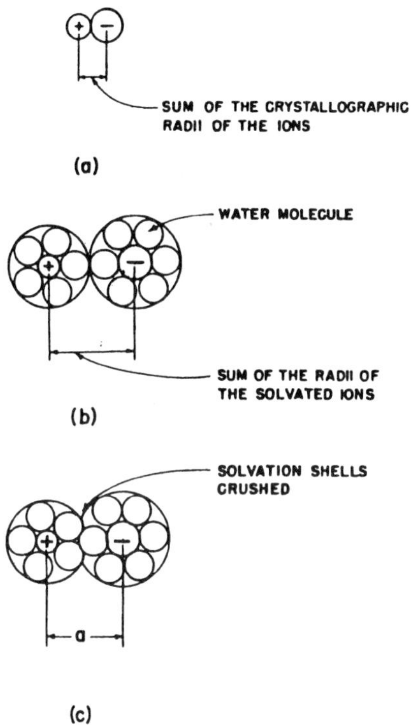

Fig. 3.31. The ion size parameter cannot be (a) less than the sum of the crystallographic radii of the ions or (b) more than the sum of the radii of the solvated ions and is most probably (c) less than the sum of the radii of the solvated ions because the solvation shells may be crushed.

$$\log f_{\pm} = -\frac{A(z_+z_-)I^{1/2}}{1 + BaI^{1/2}} \tag{3.120}$$

How does the general form of this expression compare with the Debye–Hückel limiting law as far as agreement with experiment is concerned? To see what the extra term $(1 + BaI^{1/2})^{-1}$ does to the shape of the $\log f_{\pm}$ versus $I^{1/2}$ curve, one can expand it in the form of a binomial series

$$\frac{1}{1+x} = (1+x)^{-1} = 1 - x + \frac{x^2}{2!} - \ldots \tag{3.124}$$

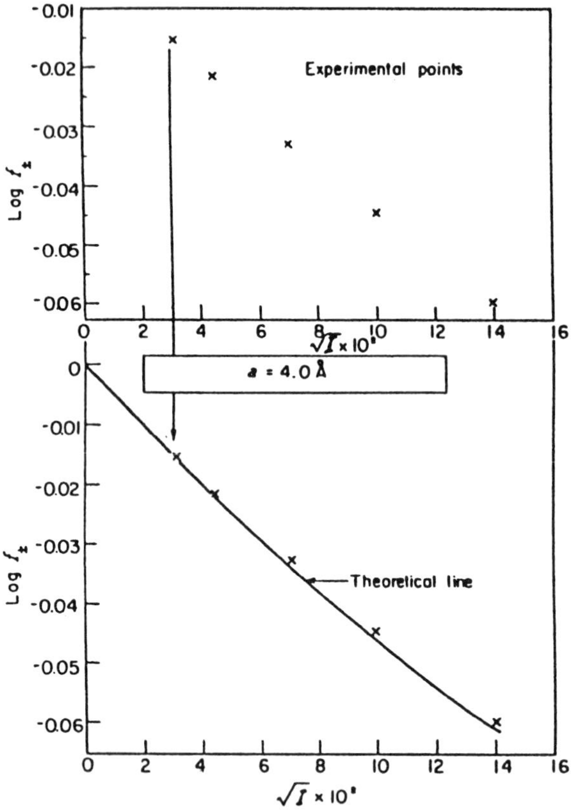

Fig. 3.32. Procedure for recovering the ion size parameter from experiment and then using it to produce a theoretical log $f_\pm$ versus $I^{1/2}$ curve that can be compared with an experimental curve.

and use only the first two terms. Thus,

$$\frac{1}{1 + BaI^{1/2}} = 1 - BaI^{1/2} \tag{3.125}$$

and therefore

$$\log f_\pm \approx -A(z_+z_-)I^{1/2}(1 - BaI^{1/2}) \tag{3.126}$$

$$\approx -A(z_+z_-)I^{1/2} + \text{constant}(I^{1/2})^2 \tag{3.127}$$

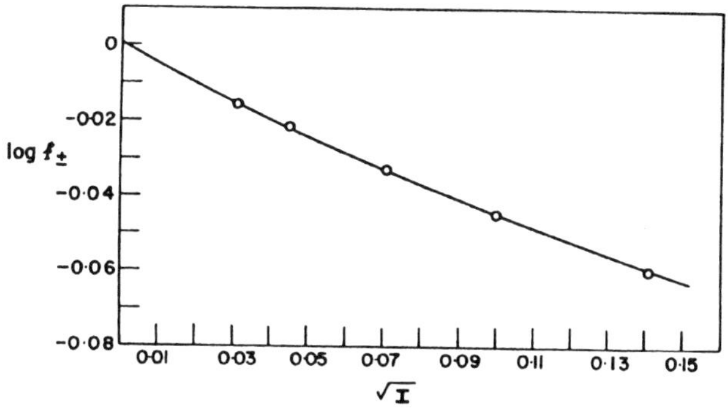

Fig. 3.33. Comparison of the experimental mean activity coefficients with theory for Eq. (3.126).

This result is encouraging. It shows that the $\log f_{\pm}$ versus $I^{1/2}$ curve gives values of $\log f_{\pm}$ higher than those given by the limiting law, the deviation increasing with concentration. In fact, the general shape of the predicted curve (Fig. 3.33) is very much on the right lines.

The values of the ion size parameter, or distance of closest approach, which are recovered from experiment are physically reasonable for many electrolytes. They lie around 0.3 to 0.5 nm, which is greater than the sum of the crystallographic radii of the positive and negative ions and pertains more to the solvated ion (Table 3.9).

By choosing a reasonable value of the ion size parameter a, independent of concentration, it is found that in many cases Eq. (3.126) gives a very good fit with experiment, often for ionic strengths up to 0.1. For example, on the basis of $a = 0.4$ nm, Eq. (3.126) gives an almost exact agreement up to 0.02 M in the case of sodium chloride (Fig. 3.34 and Table 3.10).

The ion size parameter a has done part of the job of extending the range of concentration in which the Debye–Hückel theory of ionic clouds agrees with experiment. Has it done the whole job? One must start looking for discrepancies between theory and fact and for the less satisfactory features of the model.

TABLE 3.9

Values of Ion Size Parameter for a Few Electrolytes

Salt	a (nm)
HCl	0.45
HBr	0.52
LiCl	0.43
NaCl	0.40
KCl	0.36

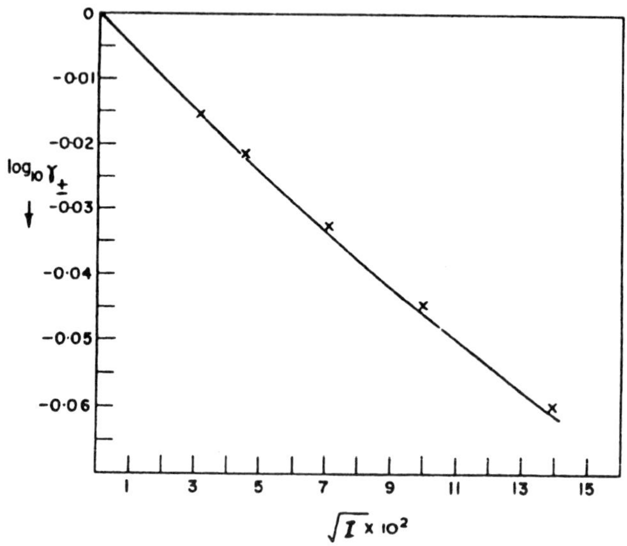

Fig. 3.34. Comparison of the experimental mean activity coefficients for sodium chloride with the theoretical $\log f_\pm$ versus $I^{1/2}$ curve based on Eq. (3.126) with $a = 0.4$ nm.

The most obvious drawback of the finite-ion-size version of the Debye–Hückel theory lies in the fact that a is an *adjustable parameter*. When parameters that have to be taken from experiment enter a theory, they imply that the physical situation has been incompletely comprehended or is too complex to be mathematically analyzed. In contrast, the constants of the limiting law were calculated without recourse to experiment.

TABLE 3.10

Experimental Mean Activity Coefficients and Those Calculated from Eq. (3.126) with $a = 0.4$ nm at 25 °C at Various Concentrations of NaCl

Molality	Experimental Mean Activity Coefficient, $-\log f_\pm$	Calculated
0.001	0.0155	0.0155
0.002	0.0214	0.0216
0.005	0.0327	0.0330
0.01	0.0446	0.0451
0.02	0.0599	0.0609

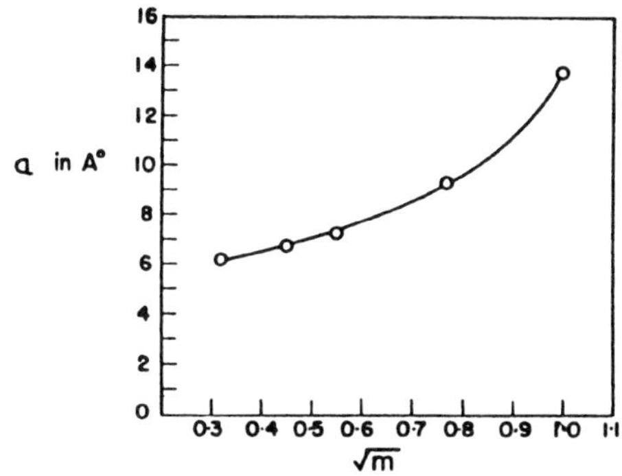

Fig. 3.35. The variation of the ion size parameter with concentration of NaCl.

The best illustration of the fact that *a* has to be adjusted is its concentration dependence. As the concentration changes, the ion size parameter has to be modified (Fig. 3.35). Further, for some electrolytes at higher concentrations, *a* has to assume quite impossible (i.e., large negative, irregular) values to fit the theory to experiment (Table 3.11).

Evidently there are factors at work in an electrolytic solution that have not yet been reckoned with, and the ion size parameter is being asked to include the effects of all these factors simultaneously, even though these other factors probably have little to do with the size of the ions and may vary with concentration. If this were so, the ion size parameter *a*, calculated back from experiment, would indeed have to vary with concentration. The problem therefore is: What factors, forces, and interactions were neglected in the Debye–Hückel theory of ionic clouds?

TABLE 3.11

Values of Parameter *a* at Higher Concentrations

Concentration (molality)	Value of *a* for HCl (nm)	Concentration (molality)	Value of *a* for LiCl (nm)
1	1.38	2	4.13
1.4	2.45		
1.8	8.50	2.5	−14.19
2	−41.12		
2.5	−2.79	3	−2.64
3	−1.48		

3.5.6. The Debye–Hückel Theory of Ionic Solutions: An Assessment

It is appropriate at this stage to recount the achievements in the theory of ionic solutions described thus far. Starting with the point of view that ion–ion interactions are bound to operate in an electrolytic solution, in going from a hypothetical state of noninteracting ions to a state in which the ions of species i interact with the ionic solution, the chemical-potential change $\Delta\mu_{i-I}$ was considered a quantitative measure of these interactions. As a first approximation, the ion–ion interactions were assumed to be purely Coulombic in origin. Hence, the chemical-potential change arising from the interactions of species i with the electrolytic solution is given by the Avogadro number times the electrostatic work W resulting from taking a discharged reference ion and charging it up in the solution to its final charge In other words, the charging work is given by the same formula as that used in the Born theory of solvation, i.e.,

$$W = \frac{z_i e_0}{2} \psi \tag{3.3}$$

where ψ is the electrostatic potential at the surface of the reference ion that is contributed by the other ions in the ionic solution. The problem therefore was to obtain a theoretical expression for the potential ψ. This involved an understanding of the distribution of ions around a given reference ion.

It was in tackling this apparently complicated task that appeal was made to the Debye–Hückel simplifying model for the distribution of ions in an ionic solution. This model treats only one ion—the central ion—as a discrete charge, the charge of the other ions being smoothed out to give a continuous charge density. Because of the tendency of negative charge to accumulate near a positive ion, and vice versa, the smoothed-out positive and negative charge densities do not cancel out; rather, their imbalance gives rise to an excess local charge density ϱ_r, which of course dies away toward zero as the distance from the central ion is increased. Thus, the calculation of the distribution of ions in an electrolytic solution reduces to the calculation of the variation of excess charge density ϱ_r with distance r from the central ion.

The excess charge density ϱ_r was taken to be given, on the one hand, by Poisson's equation of electrostatics

$$\varrho_r = -\frac{\varepsilon}{4\pi}\left[\frac{1}{r^2}\frac{d}{dr}\left(r^2\frac{d\psi_r}{dr}\right)\right] \tag{3.17}$$

and on the other, by the linearized Boltzmann distribution law

$$\varrho_r = -\sum \frac{n_i^0 z_i^2 e_0^2 \psi_r}{kT} \tag{3.18}$$

Poisson Equation

$$\varrho_r = -\frac{\varepsilon}{4\pi}\left[\frac{1}{r^2}\frac{d}{dr}\left(r^2\frac{d\psi_r}{dr}\right)\right]$$

Boltzmann Equation

$$\varrho_r = \sum n_i^0 z_i e_0 \exp\left(-\frac{z_i e_0 \psi_r}{kT}\right)$$

Linearized Boltzmann Equation

$$\varrho_r = \sum n_i^0 z_i e_0 - \sum \frac{n_i^0 z_i^2 e_0^2 \psi_r}{kT}$$

Electroneutrality

$$\sum n_i^0 z_i e_0 = 0$$

$$\varrho_r = -\sum \frac{n_i z_i^2 e_0^2 \psi_r}{kT}$$

Linearized Poisson–Boltzmann Equation

$$\frac{1}{r^2}\frac{d}{dr}\left(r^2\frac{d\psi_r}{dr}\right) = \left[\frac{4\pi}{\varepsilon kT}\sum n_i^0 z_i^2 e_0^2\right]\psi_r$$

$$\varkappa^2 = \frac{4\pi}{\varepsilon kT}\sum n_i^0 z_i^2 e_0^2$$

$$\frac{1}{r^2}\frac{d}{dr}\left(r^2\frac{d\psi_r}{dr}\right) = \varkappa^2\psi_r$$

Fig. 3.36. Steps in the derivation of the linearized Poisson–Boltzmann equation.

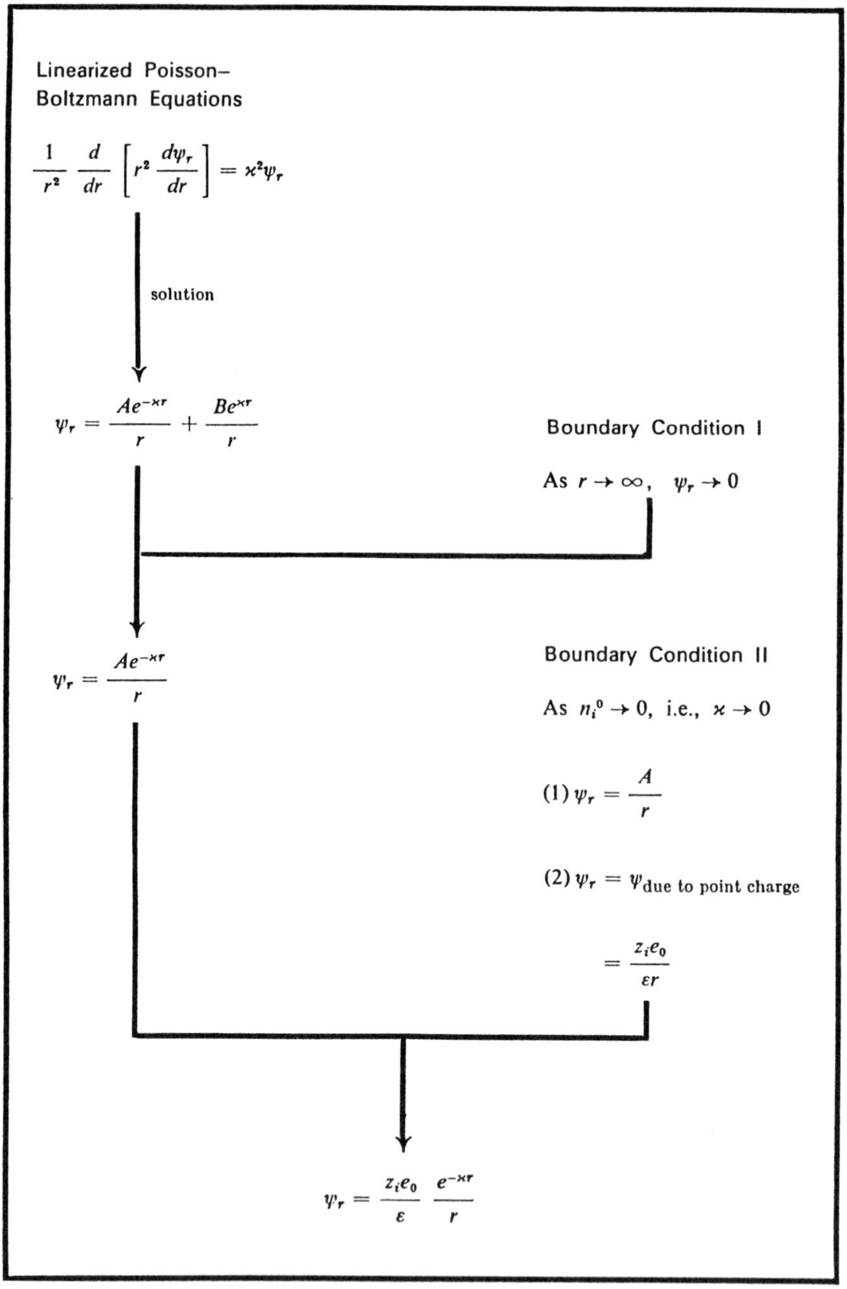

Fig. 3.37. Steps in the solution of the linearized Poisson–Boltzmann equation for point-charge ions.

The result of equating these two expressions for the excess charge density is the fundamental partial differential equation of the Debye–Hückel model, the linearized P–B equation (Fig. 3.36),

$$\frac{1}{r^2}\frac{d}{dr}\left(r^2\frac{d\psi_r}{dr}\right)=\kappa^2\psi_r \tag{3.21}$$

where

$$\kappa^2=\frac{4\pi}{\varepsilon kT}\sum n_i^0 z_i^2 e_0^2 \tag{3.20}$$

By assuming that ions can be regarded as point charges, the solution of the linearized P–B equation turns out to be (Fig. 3.37)

$$\psi_r=\frac{z_i e_0}{\varepsilon}\frac{e^{-\kappa r}}{r} \tag{3.33}$$

Such a variation of potential with distance from a typical (central or reference) ion corresponds to a charge distribution that can be expressed as a function of distance r from the central ion by

$$\varrho_r=-\frac{z_i e_0}{4\pi}\kappa^2\frac{e^{-\kappa r}}{r} \tag{3.35}$$

This variation of the excess charge density with distance around the central or typical ion yielded a simple physical picture. A reference positive ion can be thought of as being surrounded by a cloud of negative charge of radius κ^{-1}. The charge density in this ionic atmosphere, or ionic cloud, decays in the manner indicated by Eq. (3.35). Thus, the interactions between a reference ion and the surrounding ions of the solution are equivalent to the interactions between the reference ion and the ionic cloud, which in the point-charge model set up at the central ion a potential ψ_{cloud} given by

$$\psi_{\text{cloud}}=-\frac{z_i e_0}{\varepsilon\kappa^{-1}} \tag{3.49}$$

The magnitude of central ion–ionic-cloud interactions is given by introducing the expression for ψ_{cloud} into the expression (3.3) for the work of creating the ionic cloud, i.e., setting up the ionic interaction situation. Thus, one obtains for the energy of such interactions

$$\Delta\mu_{i-I}=-\frac{N_A(z_i e_0)^2}{2\varepsilon\kappa^{-1}} \tag{3.51}$$

In order to test these predictions, attention was drawn to an empirical treatment of ionic solutions. For solutions of noninteracting particles, the chemical-potential change in going from a solution of unit concentration to one of concentration x_i is described by the equation

$$\mu_i - \mu_i^0 = RT\ln x_i \tag{3.52}$$

However, in the case of an electrolytic solution in which there are ion–ion interactions, it is experimentally observed that

$$\mu_i - \mu_i^0 \neq RT\ln x_i$$

If one is unaware of the nature of these interactions, one can write an empirical equation to compensate for one's ignorance

$$RT\ln f_i = (\mu_i - \mu_i^0) - RT\ln x_i \tag{3.57}$$

and say that solutions behave ideally if the so-called activity coefficient f_i is unity, i.e., $RT\ln f_i = 0$, and, in real solutions, $f_i \neq 1$. It is clear that f_i corresponds to a coefficient to account for the behavior of ionic solutions, which differs from that of solutions in which there are no charges. Thus, f_i accounts for the interactions of the charges, so that

$$RT\ln f_i = \Delta\mu_{i-I} = -\frac{N_A(z_i e_0)^2}{2\varepsilon\kappa^{-1}} \tag{3.60}$$

Thus arose the Debye–Hückel expression for the experimentally inaccessible individual ionic-activity coefficient. This expression could be transformed into the Debye–Hückel limiting law for the experimentally measurable mean ionic-activity coefficient

$$\log f_{\pm} = -A(z_+z_-)I^{1/2} \tag{3.90}$$

which would indicate that the logarithm of the mean activity coefficient falls linearly with the square root of the ionic strength $I(=\frac{1}{2}\Sigma c_i z_i^2)$, which is a measure of the total number of electric charges in the solution.

The agreement of the Debye–Hückel limiting law with experiment improved with decreasing electrolyte concentration and became excellent for the limiting tangent to the $\log f_{\pm}$ versus $I^{1/2}$ curve. With increasing concentration, however, experiment deviated more and more from theory and, at concentrations above 1 N, even showed an *increase* in $f_{\pm}$ with an increase in concentration, whereas theory indicated a continued decrease.

An obvious improvement of the theory consisted in removing the assumption of point-charge ions and taking into account their finite size. With the use of an ion size parameter a, the expression for the mean ionic-activity coefficient became

$$\log f_{\pm} = -\frac{A(z_+z_-)I^{1/2}}{1+\kappa a} \tag{3.119}$$

However, the value of the ion size parameter a could not be theoretically evaluated. Hence, an experimentally calibrated value was used. With this calibrated value for a, the values of $f_{\pm}$ at other concentrations [calculated from Eq. (3.119)] were compared with experiment.

The finite-ion-size model yielded agreement with experiment at concentrations up to 0.1 N. It also introduced through the value of a, the ion size parameter, a specificity to the electrolyte (making NaCl different from KCl), whereas the point-charge model yielded activity coefficients that depended only upon the valence type of electrolyte. Thus, while the limiting law sees only the charges on the ions, it is blind to the specific characteristics that an ionic species may have, and this defect is overcome by the finite-ion-size model.

Unfortunately, the value of a obtained from experiment by Eq. (3.119) varies with concentration (as it would not if it represented simply the collisional diameters), and as the concentration increases beyond about 0.1 mol dm^{-3}, a sometimes has to assume physically impossible (e.g., negative) values. Evidently these changes demanded by experiment not only reflect real changes in the sizes of ions but represent other effects neglected in the simplifying Debye–Hückel model. Hence, the basic postulates of the Debye–Hückel model must be scrutinized.

The basic postulates can be put down as follows: (1) The central ion sees the surrounding ions in the form of a smoothed-out charge density and not as discrete charges. (2) All the ions in the electrolytic solution are free to contribute to the charge density and there is, for instance, no pairing up of positive and negative ions to form any electrically neutral couples. (3) Only long-range Coulombic forces are relevant to ion–ion interactions; short-range non-Coulombic forces, such as dispersion forces, play a negligible role. (4) The solution is sufficiently dilute to make ψ_r [which depends on concentration through κ—*cf.* Eq. (3.20)] small enough to warrant the linearization of the Boltzmann equation (3.10). (5) The only role of the solvent is to provide a dielectric medium for the operation of interionic forces; i.e., the removal of a number of ions from the solvent to cling more or less permanently to ions other than the central ion is neglected.

It is because it is implicitly attempting to represent all these various aspects of the real situation inside an ionic solution that the experimentally calibrated ion size parameter varies with concentration. Of course, a certain amount of concentration variation of the ion size parameter is understandable because the parameter depends upon the radius of solvated ions and this time-averaged radius might be expected to decrease with an increase of concentration. One must try to isolate that part of the changes in the ion size parameter that does not reflect real changes in the sizes of ions but represents the impact of, for instance, ionic solvation upon activity coefficients. This question of the influence of ion–solvent interactions (Chapter 2) upon the ion–ion interactions will be considered in Section 3.6.

3.5.7. Parentage of the Theory of Ion–Ion Interactions

Stress has been laid on the contribution of Debye and Hückel (1923) to the development of the theory of ion–ion interactions. It was Debye and Hückel who ushered in the electrostatic theory of ionic solutions and worked out predictions that precisely fitted experiments for sufficiently low concentrations of ions. It is not often realized, however, that the credit due to Debye and Hückel as the parents of the theory of ionic solutions is the credit that is quite justifiably accorded to foster parents. The true parents were Milner and Gouy. These authors made important contributions very early in the growth of the theory of ion–ion interactions.

Milner's contribution (1912) was direct. He attempted to find out the virial[18] equation for a mixture of ions. However. Milner's statistical mechanical approach lacked the mathematical simplicity of the ionic-cloud model of Debye and Hückel and proved too unwieldy to yield a general solution testable by experiment. Nevertheless, his contribution was a seminal one in that for the first time the behavior of an ionic solution had been linked mathematically to the interionic forces.

The contribution of Gouy (1910) was indirect.[19] Milner's treatment was not sufficiently fruitful because he did not formulate a mathematically treatable model. Gouy developed such a model in his treatment of the distribution of the excess charge density in the solution near an electrode. Whereas Milner sought to describe the interactions between series of discrete ions, it was Gouy who suggested the smoothing out of the ionic charges into a continuous distribution of charge and took the vital step of using Poisson's equation to relate the electrostatic potential and the charge density in the continuum. Thus, Gouy was the first to evolve the ionic-atmosphere model.

It was with an awareness of the work of Milner and Gouy that Debye and Hückel attacked the problem. Their contributions, however, were vital ones. By choosing one ion out of the ionic solution and making an analogy between this charged reference ion and the charged electrode of Gouy, by using the Gouy type of approach to obtain the variation of charge density and potential with distance from the central ion and

---

[18]*Virial* is derived from the Latin word for *force*, and the virial equation of state is a relationship between pressure, volume, and temperature of the form

$$\frac{PV}{RT} = 1 + \frac{K_2}{V} + \frac{K_3}{V^2} + \ldots$$

where $K_2, K_3, \ldots$, the virial coefficients, represent interactions between constituent particles.

[19]Chapman made an independent contribution in 1913 on the same lines as that of Gouy.

thus to get the contribution to the potential arising from interionic forces, and, finally, by evolving a charging process to get the chemical-potential change due to ion–ion interactions, they were able to link the chemical-potential change caused by interionic forces to the experimentally measurable activity coefficient. Without these essential contributions of Debye and Hückel, a viable theory of ionic solutions would not have emerged.

Further Reading

Seminal

1. G. Gouy, "About the Electric Charge on the Surface of an Electrolyte," *J. Phys.* **9**: 457 (1910).
2. S. R. Milner, "The Virial of a Mixture of Ions," *Phil. Mag.* **6**: 551 (1912).
3. P. Debye and E. Hückel, "The Interionic Attraction Theory of Deviations from Ideal Behavior in Solution," *Z. Phys.* **24**: 185 (1923).

Review

1. K. S. Pitzer, "Activity Coefficients in Electrolyte Solutions," in *Activity Coefficients in Electrolyte Systems*, K. S. Pitzer, ed., 2nd ed., CRC Press, Boca Raton, FL (1991).

Papers

1. C. F. Baes, Jr., E. J. Reardon, and B. A. Bloyer, *J. Phys. Chem.* **97**: 12343 (1993).
2. D. Dolar and M. Bester, *J. Phys. Chem.* **99**: 4763 (1995).
3. G. M. Kontogeorgis, A. Saraiva, A. Fredenslund, and D. P. Tassios, *Ind. Eng. Chem. Res.* **34**: 1823 (1995).
4. A. H. Meniai and D. M. T. Newsham, *Chem. Eng. Res. Des.* **73**: 842 (1995).
5. M. K. Khoshkbarchi and J. H. Vera, *AIChE J.* **42**: 249 (1996).

3.6. ION–SOLVENT INTERACTIONS AND THE ACTIVITY COEFFICIENT

3.6.1. Effect of Water Bound to Ions on the Theory of Deviations from Ideality

The theory of behavior in ionic solutions arising from ion–ion interactions has been seen (Section 3.5) to give rise to expressions in which as the ionic concentration increases, the activity coefficient decreases. In spite of the excellent numerical agreement between the predictions of the interionic attraction theory and experimental values of activity coefficients at sufficiently low concentrations (e.g., $< 3 \times 10^{-3}$ *N*), there is a most sharp disagreement at concentrations above about 1 *N*, when the activity coefficient begins to increase back toward the values it had in limitingly dilute solutions. In fact, at sufficiently high concentrations (one might have argued, when the ionic interactions are greatest), the activity coefficient, instead of continuing to decrease, begins to exceed the value of unity characteristic of the reference state of noninteraction, i.e., of infinite dilution.

A qualitative picture of the events leading to these apparently anomalous happenings has already been given in Chapter 2. There it has been argued that ions exist in solution in various states of interaction with solvent particles. There is a consequence that must therefore follow for the effectiveness of some of these water molecules in counting as part of the solvent. Those that are tightly bound to certain ions cannot be effective in dissolving further ions added (Fig. 3.38). As the concentration of electrolyte increases, therefore the amount of *effective* or free solvent decreases. In this way the apparently anomalous increase in the activity coefficient occurs. The activity coefficient is in effect that factor which multiplies the simple, apparent ionic concentration and makes it the *effective* concentration, i.e., the activity. If the hydration of the ions reduces the amount of free solvent from that present for a given stoichiometric concentration, then the effective concentration increases and the activity coefficient must increase so that its multiplying effect on the simple stoichiometric concentration is such as to increase it to take into account the reduction of the effective solvent. Experiment shows that sometimes these increases more than compensate for the decrease due to interionic forces, and it is thus not unreasonable that the activity coefficient should rise above unity.

Some glimmering of the quantitative side of this can be seen by taking the number of waters in the primary hydration sheath of the ions as those that are no longer effective solvent particles. For NaCl, for example, Table 2.18 indicates that this number is about 7. If the salt concentration is, e.g., 10^{-2} N, the moles of water per liter withdrawn from effect as free solvent would be 0.07. Since the number of moles of water per liter is $1000/18 = 55.5$, the number of moles of free water is 55.43 and the effects arising from such a small change are not observable. Now consider a 1 N solution of NaCl. The water withdrawn is 7 mol dm^{-3}, and the change in the number of moles of free water is from 55.5 in the infinitely dilute situation to 48.5, a significant change. At 5 N NaCl,

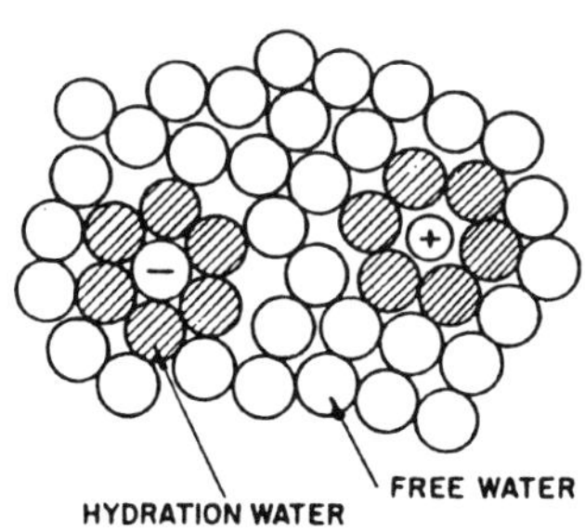

Fig. 3.38. The distinction between free water and hydration water that is locked up in the solvent sheaths of ions.

more than half of the water in the solution is associated with the ions, and a sharp increase of activity coefficient, somewhat of a doubling in fact, would be expected to express the increase in effective concentration of the ions. To what extent can this rough sketch be turned into a quantitative model?

3.6.2. Quantitative Theory of the Activity of an Electrolyte as a Function of the Hydration Number

The basic thought here has to be similar to that which lay beneath the theory of electrostatic interactions: to calculate the work done in going from a state in which the ions are too far apart to feel any interionic attraction to the state at a finite concentration c at which part of the ions' behavior is due to this. This work was then [Eq. (3.59)] placed equal to $RT \ln f$, where f is the activity coefficient, which was thereby calculated.

When one realizes that a reversal of the direction in which the activity coefficient varies with concentration has to be explained by taking into account the removal of some of the solvent from effectively partaking in the ionic solution's activity, the philosophy behind the calculation of this effect on the activity coefficient becomes clear. One must calculate the work done in the changes caused by solvent removal and add this to the work done in building up the ionic atmosphere. What, then, is the contribution due to these water-removing processes? [Note that ions are hydrated at all times in which they are in the solution. One is not going to calculate the heat of hydration; that was done in Chapter 2. Here, the task is to calculate the work done as a consequence of the fact that when water molecules enter the solvation sheath, they are, so to speak, no longer operating as far as the solvent is concerned. It is to be an $RT \ln(c_1/c_2)$ type of calculation.]

As a device for the calculation, let it be assumed that the interionic attraction is switched off. (Reasons for employing this artifice will be given.) Thus, there are two kinds of work that must be taken into account.

1. The $RT \ln(c_1/c_2)$ kind of work done when there is a change in concentration of the free solvent *water* caused by introducing ions of a certain concentration into the solution.
2. The $RT \ln(c_1/c_2)$ kind of work done when there is a corresponding change in concentration of the *ions* due to the removal of the water to their sheaths.

The work done for process 1 is easy to calculate. Before the ions have been added, the concentration of the water is unaffected by anything; it is the concentration of pure water and its activity, the activity of a pure substance, can be regarded as unity. After the ions are there, the activity of the water is, say, a_w. Then, the work when the activity of the water goes from 1 to a_w is $RT \ln (a_w/1)$.

However, one wishes to know the change of activity caused in the *electrolyte* by this change of activity of the water. Furthermore, the calculation must be reduced to that for 1 mole of electrolyte. Let the sum of the moles of water in the primary sheath per liter of solution for both ions of the imagined 1:1 electrolyte be n_h (for 1-molar solutions, this is the hydration number). Then, if there are n moles of electrolyte in the water, the change in free energy due to the removal of the water to the ions' sheaths is $-(n_h/n)RT\ln a_w$ per mole of electrolyte.

One now comes to the second kind of work and realizes why the calculation is best done as a thought process in which the interionic attraction is shut off while the work is calculated. One wants to be able to use the ideal-solution (no interaction) equation for the work done, $RT\ln(c_1/c_2)$, and not $RT\ln(a_1/a_2)$. Thus, using the latter expression would be awkward; it needs a knowledge of the activities themselves and that is what one is trying to calculate.

Now, the change in free energy change due to the change in the concentration of the ions after the removal of the effective solvent molecule is

$$RT\ln\left[\frac{x_{\text{after water removal from free to solvated state}}}{x_{\text{before water removal from free to solvated state}}}\right]$$

where x is the mole fraction of the electrolyte in the solution.

Before the water is removed,

$$x_{\text{before}}=\frac{n}{n_w+n} \tag{3.128}$$

where n is the number of moles of electrolyte present in n_w moles of water. Then after the water is removed to the sheaths,

$$x_{\text{after}}=\frac{n}{n_w-n_h+n} \tag{3.129}$$

The change in free energy is

$$RT\ln\frac{n_w+n}{n_w+n-n_h}$$

Hence, the total free-energy change in the solution, calculated per mole of the electrolyte present, is

$$-\frac{n_h}{n}RT\ln a_w+RT\ln\frac{n_w+n}{n_w+n-n_h}$$

Now, one has to switch back to the Coulombic interactions. If the expression for the work done in building up an ionic atmosphere [e.g., Eq. (3.120)] were still valid in the region of relatively high concentrations in which the effect of change of concentration is occurring, then,[20]

[20]Here $\sqrt{c}$ has been written instead of the $I^{1/2}$ of Eq. (3.119). For 1:1 electrolytes, c and I are identical.

$$RT \log f_{\pm(\exp)} = -\frac{A\sqrt{c}}{1 + Ba\sqrt{c}} - 2.303\, RT\frac{n_h}{n} \log a_w$$

$$+ 2.303\, RT \log \frac{n_w + n}{n_w + n - n_h} \quad (3.130)$$

One sees at once that there is a possibility of a change in direction for the change in $\log f_{\pm}$ with an increase in concentration in the solution. If the last term predominates, $RT \log f_{\pm}$ may increase with concentration.

The situation here does have a fairly large shadow on it because of the use of the expression (3.120) in $\sqrt{c}$. It will be seen (Section 3.14) that, at concentrations as high as 1 N, there are some fundamental difficulties for the ionic-cloud model on which this $\sqrt{c}$ expression of Eq. (3.120) was based (the ionic atmosphere can no longer be considered a continuum of smoothed-out charge). It is clear that when the necessary mathematics can be done, there will be an improvement on the $\sqrt{c}$ expression, and one will hope to get it more correct than it now is. Because of this shadow, a comparison of Eq. (3.130) with experiment to test the validity of the model for removing solvent molecules to the ions' sheathes should be done a little with tongue in cheek.

3.6.3. The Water Removal Theory of Activity Coefficients and Its Apparent Consistency with Experiment at High Electrolytic Concentrations

If one examines the ion–solvent terms in Eq. (3.130), one sees that since $a_w \leq 1$ and in general $n_h > n$ (more than one hydration water per ion), both the terms are positive. Hence, one can conclude that the Debye–Hückel treatment, which ignores the withdrawal of solvent from solution, gives values of activity coefficients that are smaller than those which take these effects into account. Furthermore, the difference arises from the ion–solvent terms, i.e.,

$$-2.303\, RT\frac{n_h}{n} \log a_w + 2.303\, RT \log \frac{n_w + n}{n_w + n - n_h}$$

As the electrolyte concentration increases, a_w decreases and n_h increases; hence both ion–solvent terms increase the value of $\log f$. Furthermore, the numerical evaluation shows that the above ion–solvent term can equal and become larger than the Debye–Hückel ($\sqrt{c}$) Coulombic term. This means that the $\log f_{\pm}$ versus $I^{1/2}$ curve can pass through a minimum and then start rising, which is precisely what is observed (Fig. 3.39, where an activity coefficient is plotted against the corresponding molality).

On the other hand, with increasing dilution, $n_w + n >> n_h$, or $n_w + n - n_h \approx n_w + n$ and $a_w \rightarrow 1$, and hence the terms vanish, which indicates that ion–solvent interactions (which are of short range) are significant for the theory of activity

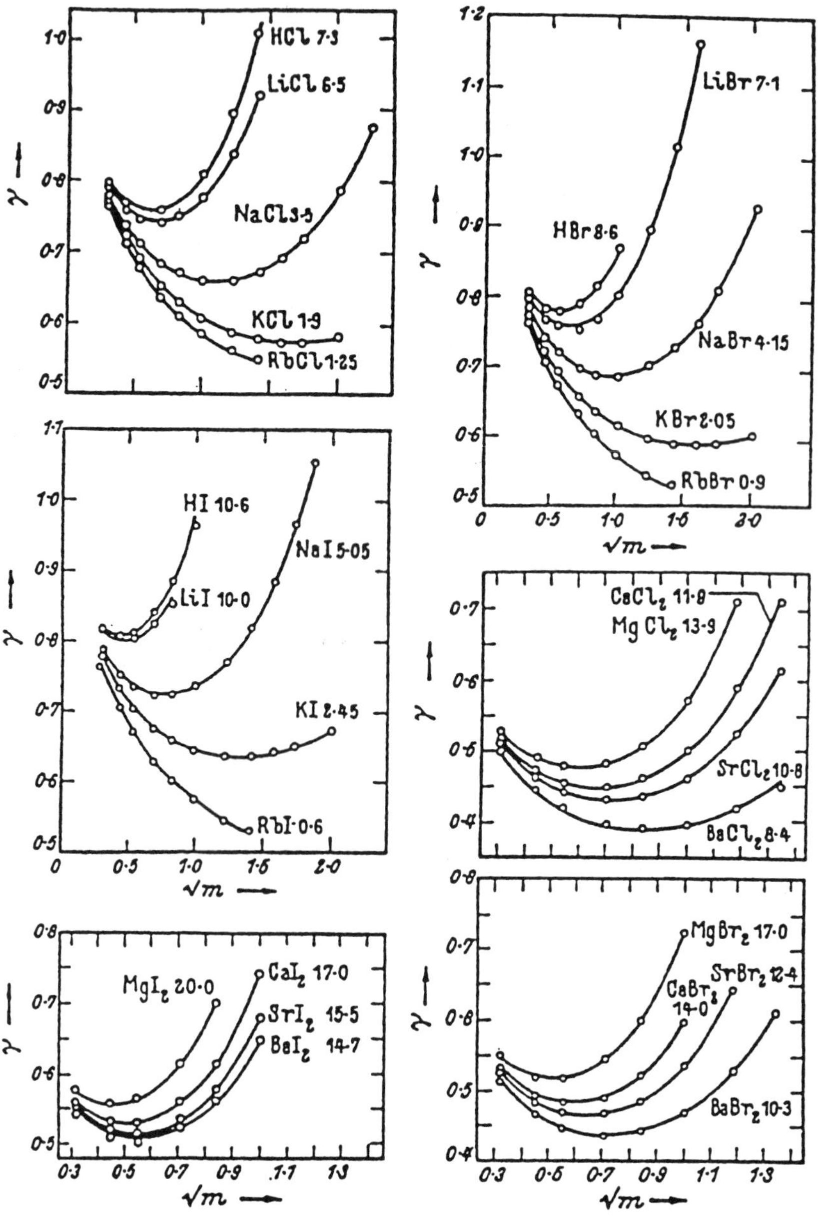

Fig. 3.39. Comparison of experimental activity coefficients with those predicted by the one-parameter equation as shown. (Reprinted from R. H. Stokes and R. A. Robinson, *J. Am. Chem. Soc.* **70**: 1870, 1948.)

TABLE 3.12
Water Activities in Sodium Chloride Solutions

M	a_w	M	a_w
0.1	0.99665	3	0.8932
0.5	0.98355	3.4	0.8769
1	0.96686	3.8	0.8600
1.4	0.9532	4.2	0.8428
1.8	0.9389	4.6	0.8250
2.2	0.9242	5	0.8068
2.6	0.9089	5.4	0.7883

coefficients only in concentrated solutions. At extreme dilutions, only the ion–ion long-range Coulombic interactions are important.

In order to test Eq. (3.130), which is a quantitative statement of the influence of ionic hydration on activity coefficients, it is necessary to know the quantity n_h and the activity a_w of water, it being assumed that an experimentally calibrated value of the ion size parameter is available. The activity of water can be obtained from independent experiments (Table 3.12). The quantity n_h can be used as a parameter. If Eq. (3.130) is tested as a two-parameter equation (n_h and a being the two parameters; Table 3.13), it is found that theory is in excellent accord with experiment. For instance, in the case of NaCl, the calculated activity coefficient agrees with the experimental value for solutions as concentrated as 5 mol dm^{-3} (Fig. 3.39).

Since the quantity n_h is the number of moles of water used up in solvating $n = n_+ + n_-$ moles of ions, it can be split up into two terms: $(n_h)^+$ moles required to hydrate n_+ moles of cations, and $(n_h)^-$ moles required to hydrate n_- moles of anions. It follows that $[(n_h)^+/n_+]$ and $[(n_h)^-/n_-]$ are the hydration numbers (see Chapter 2) of

TABLE 3.13
Values of n_h of a 1 *M* Solution and *a*

Salt	n_h	a (pm)
HCl	8.0	447
HBr	8.6	518
NaCl	3.5	397
NaBr	4.2	424
KCl	1.9	363
$MgCl_2$	13.7	502
$MgBr_2$	17.0	546

TABLE 3.14

Nearest-Integer Hydration Numbers of Electrolytes from Eq. (3.130) and the Most Probable Value from Independent Experiments

Salt	Hydration Number from Eq. (3.130) (nearest integer)	Hydration Number from Other Experimental Methods
LiCl	7	7 ± 1
LiBr	8	7 ± 1
NaCl	4	6 ± 2
KCl	2	5 ± 2
KI	3	4 ± 2

the positive and negative ions and $[(n_h^+)/n_+] + [(n_h^-)/n_-]$ is the hydration number of the electrolyte.

It has been found that, in the case of several electrolytes, the values of the hydration numbers obtained by fitting the theory [Eq. (3.130)] to experiment are in reasonable agreement with hydration numbers determined by independent methods (Table 3.14). Alternatively, one can say that, when independently obtained hydration numbers are substituted in Eq. (3.130), the resulting values of $\log f_\pm$ show fair agreement with experiment.

In conclusion, therefore, it may be said that the treatment of the influence of ion–solvent interactions on ion–ion interactions has extended the range of concentration of an ionic solution which is accessible to theory. Whereas the finite-ion-size version of the Debye–Hückel theory did not permit theory to deal with solutions in a range of concentrations corresponding to those of real life, Eq. (3.130) advances theory into the range of practical concentrations. Apart from this numerical agreement with experiment, Eq. (3.130) unites two basic aspects of the situation inside an electrolytic solution, namely, ion–solvent interactions and ion–ion interactions.

3.7. THE SO-CALLED "RIGOROUS" SOLUTIONS OF THE POISSON–BOLTZMANN EQUATION

One approach to understanding the discrepancies between the experimental values of the activity coefficient and the predictions of the Debye–Hückel model has just been described (Section 3.6); it involved a consideration of the influence of solvation.

An alternative approach is based on the view that the failure of the Debye–Hückel theory at high concentrations stems from the fact that the development of the theory involved the linearization of the Boltzmann equation (see Section 3.3.5). If such a view is taken, there is an obvious solution to the problem: instead of linearizing the

Boltzmann equation, one can take the higher terms. Thus, one obtains the unlinearized P–B equation

$$\frac{1}{r^2}\frac{d}{dr}\left(r^2\frac{d\psi_r}{dr}\right) = -\frac{4\pi}{\varepsilon}\,\varrho_r = -\frac{4\pi}{\varepsilon}\sum z_i e_0 n_i^0 e^{-z_i e_0 \psi_r/kT} \tag{3.131}$$

In the special case of a symmetrical electrolyte ($z_+ = -z_- = z$) with equal concentrations of positive and negative ions, i e., $n_+^0 = n_-^0 = n^0$, one gets

$$\begin{aligned}\sum z_i e_0 n_i^0 e^{-z_i e_0 \psi_r/kT} &= n^0 z_+ e_0 e^{-z_+ e_0 \psi_r/kT} - n^0 z_- e_0 e^{+z_- e_0 \psi_r/kT}\\ &= n^0 z e_0 (e^{-z e_0 \psi_r/kT} - e^{+z e_0 \psi_r/kT})\end{aligned} \tag{3.132}$$

But

$$e^{+x} - e^{-x} = 2\ \sinh x$$

and therefore

$$\varrho_r = -2n^0 z e_0 \sinh\frac{z e_0 \psi_r}{kT} \tag{3.133}$$

or

$$\frac{1}{r^2}\frac{d}{dr}\left(r^2\frac{d\psi}{dr}\right) = \frac{8\pi z e_0 n^0}{\varepsilon}\sinh\frac{z e_0 \psi}{kT} \tag{3.134}$$

By utilizing a suitable software, one could obtain from Eq. (3.134) so-called rigorous solutions.

Before proceeding further, however, it is appropriate to stress a logical inconsistency in working with the unlinearized P–B equation (3.131). The unlinearized Boltzmann equation (3.10) implies a *nonlinear* relationship between charge density and potential. In contrast, the *linearized* Boltzmann equation (3.16) implies a *linear* relationship of ϱ_r to ψ_r.

Now, a linear charge density–potential relation is consistent with the law of superposition of potentials, which states that the electrostatic potential at a point due to an assembly of charges is the sum of the potentials due to the individual charges. Thus, when one uses an unlinearized P–B equation, one is assuming the validity of the law of superposition of potentials in the Poisson equation and its invalidity in the Boltzmann equation. This is a basic logical inconsistency which must reveal itself in the predictions that emerge from the so-called rigorous solutions. This is indeed the case, as will be shown below.

Recall that, after the contribution of the ionic atmosphere to the potential at the central ion was obtained, the Coulombic interaction between the central ions and the cloud was calculated by an imaginary charging process, generally known as the *Guntelberg charging process* in recognition of its originator.

In the Guntelberg charging process, the central ion i is assumed to be in a hypothetical condition of zero charge. The rest of the ions, fully charged, are in the positions that they would hypothetically have were the central ion charged to its normal value z_ie_0; i.e., the other ions constitute an ionic atmosphere enveloping the central ion (Fig. 3.40). The ionic cloud sets up a potential $\psi_{\text{cloud}} = -(z_ie_0/\varepsilon\kappa^{-1})$ at the site of the central ion. Now, the charge of the central ion is built up (Fig. 3.40) from zero to its final value z_ie_0, and the work done in this process is calculated by the usual formula for the electrostatic work of charging a sphere (see Section 3.3.1), i.e.,

$$W = \frac{z_ie_0}{2}\psi \tag{3.3}$$

Since during the charging only ions of the ith type are considered, the Guntelberg charging process gives that part of the chemical potential due to electrostatic interactions.

Now, the Guntelberg charging process was suggested several years after Debye and Hückel made their theoretical calculation of the activity coefficient. These authors

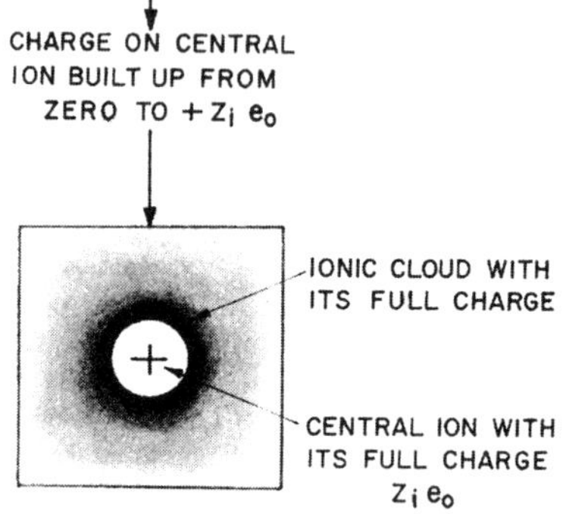

Fig. 3.40. The Guntelberg charging process.

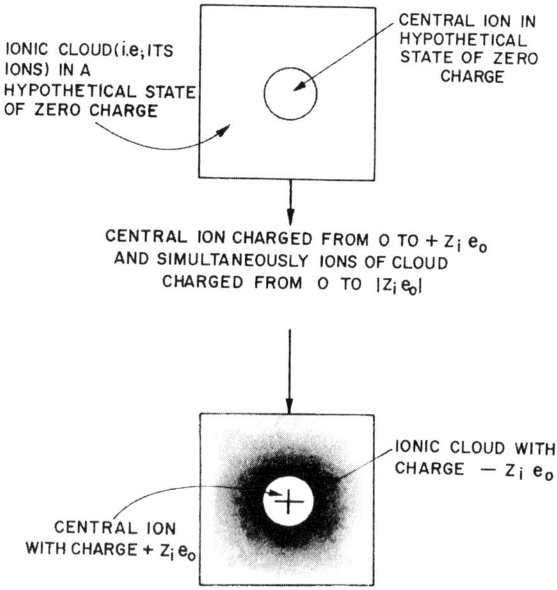

Fig. 3.41. The Debye charging process.

carried out another charging process, the *Debye*[†] *obtaining charging process*. All the ions are assumed to be in their equilibrium, or time-average positions in the ionic atmosphere (Fig. 3.41), but the central ion and the cloud ions are *all* considered in a hypothetical condition of zero charge. All the ions of the assembly are then simultaneously brought to their final values of charge by an imaginary charging process in which there are small additions of charges to each. Since ions of all types (not only of the *i*th type) are considered, the work done in this charging process yields the free-energy change arising from the electrostatic interactions in solution. Differentiation of the free energy with respect to the number of moles of the *i*th species gives the chemical potential

[†]Peter Debye is known not only for the seminal theory of the ionic atmosphere: he is the originator of the theory of dielectric constants in polar gases; and of ionic vibration potentials in electrolytes. During the '40's and '50's, his name was perhaps the most well known in the world of physical chemistry. He had a most fertile mind and it was humiliating to bring to him weighty problems which had puzzled you and your colleagues for months, for he generally had the solution in a few minutes and asked "Now what else shall we talk about?" He retained affection for his famous theory of the ionic atmosphere but when asked about it in later years would say: "You know, it applies much better than it should."

Debye was also a much appreciated lecturer at Cornell University in the '50's—particularly when he illustrated the random nature of diffusion movements by doing his "drunkard's walk" in front of the class. However, his eagerness to be an effective administrator was not so clearly manifest and after a year as Head of the Chemistry Department, he returned back to full-time research and teaching.

$$\frac{\partial \Delta G_{I-I}}{\partial n_i} = \Delta \mu_{i-I} \qquad (3.135)$$

Using the rigorous solutions of the unlinearized P–B equation, one gets the cloud contribution to the electrostatic potential at the central ion and when this value of the electrostatic potential is used in the two charging processes to get the chemical-potential change $\Delta\mu_{i-I}$ arising from ion–ion interactions, it is found that the Guntelberg and Debye charging processes give discordant results. As shown by Onsager, this discrepancy is not due to the invalidity of either of the two charging processes; it is a symptom of the logical inconsistency intrinsic in the unlinearized P–B equation.

This discussion of rigorous solutions has thus brought out an important point: The disagreement between the chemical-potential change $\Delta\mu_{i-I}$ calculated by the Debye and Guntelberg charging processes cuts off one approach to an improved theory of higher concentrations for it prevents our using the unlinearized P–B equation, which is needed when the concentration is too high for the use of $z_i e_0 \psi / kT << 1$.

3.8. TEMPORARY ION ASSOCIATION IN AN ELECTROLYTIC SOLUTION: FORMATION OF PAIRS, TRIPLETS

3.8.1. Positive and Negative ions Can Stick Together: Ion-Pair Formation

The Debye–Hückel model assumed the ions to be in almost random thermal motion and therefore in almost random positions. The slight deviation from randomness was pictured as giving rise to an ionic cloud around a given ion, a positive ion (of charge $+ze_0$) being surrounded by a cloud of excess negative charge ($-ze_0$). However, the possibility was not considered that some negative ions in the cloud would get sufficiently close to the central positive ion in the course of their quasi-random solution movements so that their thermal translational energy would not be sufficient for them to continue their independent movements in the solution. Bjerrum suggested that a pair of oppositely charged ions may get trapped in each other's Coulombic field. An *ion pair* may be formed.

The ions of the pair together form an ionic dipole on which the net charge is zero. Within the ionic cloud, the locations of such uncharged ion pairs are completely random, since, being uncharged, they are not acted upon by the Coulombic field of the central ion. Furthermore, on the average, a certain fraction of the ions in the electrolytic solution will be stuck together in the form of ion pairs. This fraction will now be evaluated.

3.8.2. Probability of Finding Oppositely Charged Ions near Each Other

Consider a spherical shell of thickness *dr* and of radius *r* from a reference positive ion (Fig. 3.42). The probability P_r that a negative ion is in the spherical shell is

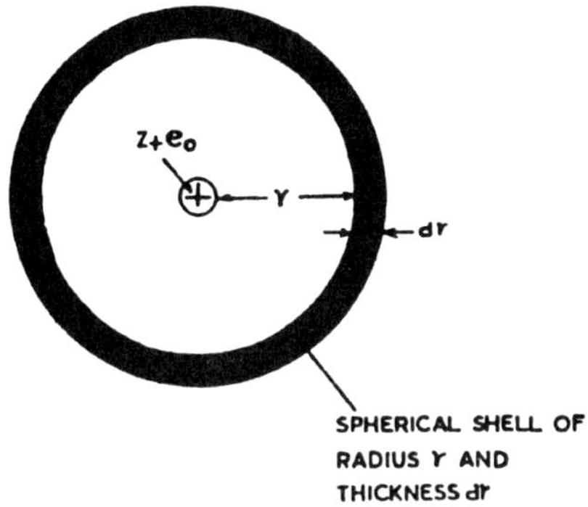

Fig. 3.42. The probability P_r of finding an ion of charge z_-e_0 in a dr-thick spherical shell of radius r around a reference ion of charge z_+e_0.

proportional, first, to the ratio of the volume $4\pi r^2\,dr$ of the shell to the total volume V of the solution; second, to the total number N_- of negative ions present; and third, to the Boltzmann factor $\exp(-U/kT)$, where U is the potential energy of a negative ion at a distance r from a cation, i.e.,

$$P_r = 4\pi r^2\,dr\frac{N_-}{V}\,e^{-U/kT} \tag{3.136}$$

Since N_-/V is the concentration n_-^0 of negative ions in the solution and

$$U = \frac{-z_-z_+e_0^2}{\varepsilon r} \tag{3.137}$$

it is clear that

$$P_r = (4\pi n_-^0)r^2 e^{z_-z_+e_0^2/\varepsilon rkT}\,dr \tag{3.138}$$

or, writing

$$\lambda = \frac{z_-z_+e_0^2}{\varepsilon kT} \tag{3.139}$$

one has

$$P_r = (4\pi n_-^0)e^{\lambda/r}r^2\,dr \tag{3.140}$$

A similar equation is valid for the probability of finding a positive ion in a dr-thick shell at a radius r from a reference negative ion. Hence, in general, one may write for

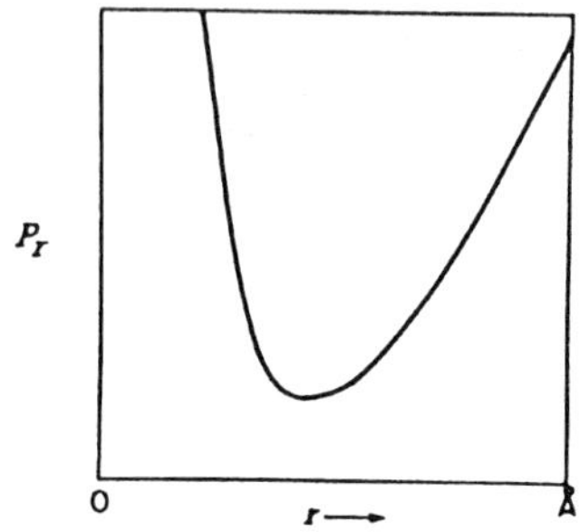

Fig. 3.43. The probability P_r of finding an ion of one type of charge as a function of distance.

the probability of finding an i type of ion in a dr-thick spherical shell at a radius r from a reference ion k of opposite charge

$$P_r = (4\pi n_i^0) e^{\lambda/r} r^2 \, dr \tag{3.141}$$

where

$$\lambda = \frac{z_i z_k e_0^2}{\varepsilon k T} \tag{3.142}$$

This probability of finding an ion of one type of charge near an ion of the opposite charge varies in an interesting way with distance (Fig. 3.43). For small values of r, the function P_r is dominated by $e^{\lambda/r}$ rather than by r^2, and under these conditions P_r *increases with decreasing* r; for large values of r, $e^{\lambda/r} \to 1$ and P_r *increases with*

TABLE 3.15

Number of Ions in Spherical Shells at Various Distances

	Number of Ions in Shell $\times 10^{22}$	
r (pm)	Of Opposite Charge	Of Like Charge
200	$1.77n_i$	$0.001n_j$
250	$1.37n_i$	$0.005n_j$
300	$1.22n_i$	$0.01n_j$
357	$1.18n_i$	$0.02n_j$
400	$1.20n_i$	$0.03n_j$
500	$1.31n_i$	$0.08n_j$

increasing r because the volume $4\pi r^2\,dr$ of the spherical shell increases as r^2. It follows from these considerations that P_r goes through a minimum for a particular, critical value of r. This conclusion may also be reached by computing the number of ions in a series of shells, each of an arbitrarily selected thickness of 0.01 nm (Table 3.15).

3.8.3. The Fraction of Ion Pairs, According to Bjerrum

If one integrates P_r between a lower and an upper limit, one gets the probability P_r of finding a negative ion within a distance from the reference positive ion, defined by the limits. Now, for two oppositely charged ions to stick together to form an ion pair, it is necessary that they should be close enough for the Coulombic attractive energy to overcome the thermal energy that scatters them apart. Let this "close-enough" distance be q. Then one can say that an ion pair will form when the distance r between a positive and a negative ion becomes less than q. Thus, the probability of ion-pair formation is given by the integral of P_r between a lower limit of a, the distance of closest approach of ions, and an upper limit of q.

Now, the probability of any particular event is the number of times that the particular event is expected to be observed divided by the total number of observations. Hence, the probability of ion-pair formation is the number of ions of species i that are associated into ion pairs divided by the total number of i ions; i.e., the probability of ion-pair formation is the fraction θ of ions that are associated into ion pairs. Thus,

$$\theta = \int_a^q P_r\,dr = \int_a^q 4\pi n_i^0 e^{\lambda/r} r^2\,dr \tag{3.143}$$

It is seen from Fig. 3.44 that the integral in Eq. (3.143) is the area under the curve between the limits $r = a$ and $r = q$. It is obvious that as r increases past the minimum, the integral becomes greater than unity. Since, however, θ is a fraction, this means that the integral diverges.

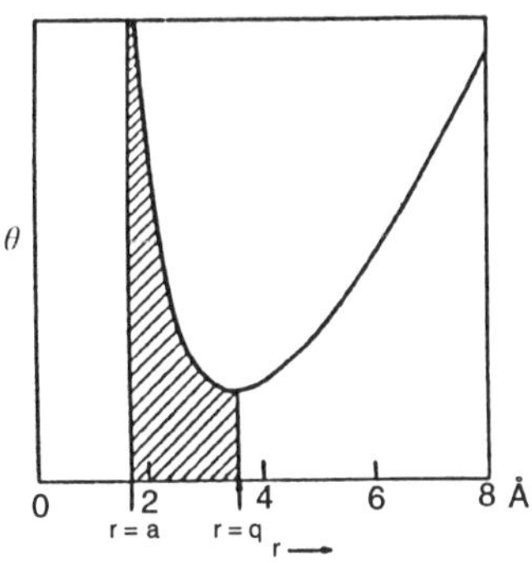

Fig. 3.44. Graphical representation of the integral in Eq. (3.143), between the limits a and q.

In this context, Bjerrum took an arbitrary step and cut off the integral at the value of $r = q$ corresponding to the minimum of the P_r vs. r curve. This minimum probability can easily be shown (Appendix 3.4) to occur at

$$q = \frac{z_+ z_- e_0^2}{2\varepsilon kT} = \frac{\lambda}{2} \tag{3.144}$$

Bjerrum argued that it is only short-range Coulombic interactions that lead to ion-pair formation and, further, when a pair of oppositely charged ions are situated at a distance apart of $r > q$, it is more appropriate to consider them free ions.

Bjerrum concluded therefore that ion-pair formation occurs when an ion of one type of charge (e.g., a negative ion) enters a sphere of radius q drawn around a reference ion of the opposite charge (e.g., a positive ion). However, it is the ion size parameter that defines the distance of closest approach of a pair of ions. The Bjerrum hypothesis can therefore be stated as follows: If $a < q$, then ion-pair formation can occur; if $a > q$, the ions remain free (Fig. 3.45).

Now that the upper limit of the integral in Eq. (3.143) has been taken to be $q = \lambda/2$, the fraction of ion pairs is given by carrying out the integration. It is

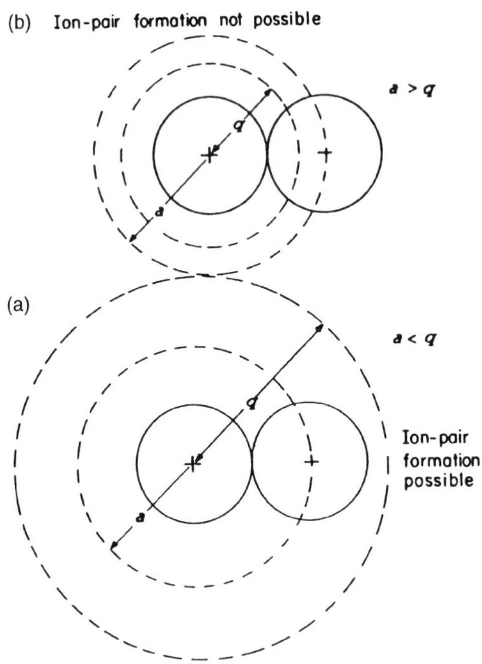

Fig. 3.45. (a) Ion-pair formation occurs if $a < q$. (b) Ion-pair formation does not occur if $a > q$.

TABLE 3.16

Values of the Integral $\int_2^b e^y y^{-4}\, dy$

b	$\int_2^b e^y y^{-4}\, dy$	b	$\int_2^b e^y y^{-4}\, dy$	b	$\int_2^b e^y y^{-4}\, dy$
2.0	0	3	0.326	10	4.63
2.1	0.0440	3.5	0.442	15	93.0
2.2	0.0843	4	0.550		
2.4	0.156	5	0.771		
2.6	0.218	6	1.041		
2.8	0.274				

$$\theta = 4\pi n_i^0 \int_a^{q=\lambda/2} e^{\lambda/r} r^2\, dr \tag{3.145}$$

For mathematical convenience, a new variable y is defined as

$$y = \frac{\lambda}{r} = \frac{2q}{r} \tag{3.146}$$

Hence, in terms of the new variable y, Eq. (3.145) becomes (Appendix 3.5)

$$\theta = 4\pi n_i^0 \left(\frac{z_+ z_- e_0^2}{\varepsilon kT}\right)^3 \int_2^b e^y y^{-4} dy \tag{3.147}$$

where

$$b = \frac{\lambda}{a} = \frac{z_+ z_- e_0^2}{\varepsilon akT} = \frac{2q}{a} \tag{3.148}$$

Bjerrum has tabulated the integral $\int_2^b e^y y^{-4} dy$ for various values of b (Table 3.16). This means that, by reading off the value of the corresponding integral and substituting for thc various other terms in Eq. (3.147), the degree of association of an electrolyte may be computed if the ion sizes, the dielectric constant, and the concentrations are known (Table 3.17).

3.8.4. The Ion-Association Constant K_A of Bjerrum

The quantity θ yields a definite idea of the fraction of ions that are associated in ion pairs in a particular electrolytic solution at a given concentration. The ions that get associated are those that get sufficiently close to an ion of opposite sign so that the energy of Coulombic attraction is greater than the thermal energy of the pair.

TABLE 3.17
Fraction of Association, θ, of Univalent Ions in Water at 291 K

$a \times 10^8$ cm:	2.82	2.35	1.76	1.01	0.70	0.47
q/a:	2.5	3	4	7	10	15
c, moles dm^{-3} [a]						
0.0001	—	—	—	—	0.001	0.027
0.0002	—	—	—	—	0.002	0.049
0.0005	—	—	—	0.002	0.006	0.106
0.001	—	0.001	0.001	0.004	0.011	0.177
0.002	0.002	0.002	0.003	0.007	0.021	0.274
0.005	0.002	0.004	0.007	0.016	0.048	0.418
0.01	0.005	0.008	0.012	0.030	0.030	0.529
0.02	0.008	0.013	0.022	0.053	0.137	0.632
0.05	0.017	0.028	0.046	0.105	0.240	0.741
0.1	0.029	0.048	0.072	0.163	0.336	0.804
0.2	0.048	0.079	0.121	0.240	0.437	0.854

[a] c in moles dm^{-3} = $1000n_i^0/N_A$.

It would, however, be advantageous if each electrolyte [e.g., NaCl, $BaSO_4$, and $La(NO_3)_3$] were assigned a particular number that would reveal, without going through the calculation of θ, the extent to which the ions of that electrolyte associate in ion pairs. The quantitative measure chosen to represent the tendency for ion-pair formation was guided by historical considerations.

Arrhenius in 1887 had suggested that many properties of electrolytes could be explained by a *dissociation* hypothesis: The neutral molecules AB of the electrolyte dissociate to form ions A^+ and B^-, and this dissociation is governed by an equilibrium

$$AB \rightleftharpoons A^+ + B^- \tag{3.149}$$

Applying the law of mass action to this equilibrium, one can define a dissociation constant

$$K = \frac{a_{A^+}a_{B^-}}{a_{AB}} \tag{3.150}$$

By analogy,[21] one can define an association constant K_A for ion-pair formation. Thus, one can consider an equilibrium between free ions (the positive M^+ ions and the negative A^- ions) and the associated ion pairs (symbolized IP)

[21]The analogy must not be carried too far because it is only a formal analogy. Arrhenius's hypothesis can now be seen to be valid for ionogens (i.e., potential electrolytes), in which case the neutral ionogenic molecules (e.g., acetic acid) consist of aggregates of atoms held together by covalent bonds. What is under discussion here is ion association, or ion-pair formation, of ionophores (i.e., true electrolytes). In these ion pairs, the positive and negative ions retain their identity as ions and are held together by electrostatic attraction.

$$M^+ + A^- \rightleftharpoons IP \tag{3.151}$$

The equilibrium sanctions the use of the law of mass action

$$K_A = \frac{a_{IP}}{a_{M^+}a_{A^-}} \tag{3.152}$$

where the a's are the activities of the relevant species. From (3.152) it is seen that K_A is the reciprocal of the ion pair's dissociation constant.

Since θ is the fraction of ions in the form of ion pairs, θc is the *concentration* of ion pairs, and $(1 - \theta)c$ is the concentration of free ions. If the activity coefficients of the positive and negative free ions are f_+ and f_-, respectively, and that of the ion pairs is f_{IP}, one can write

$$K_A = \frac{\theta c f_{IP}}{(1 - \theta)cf_+(1 - \theta cf_-}$$

$$= \frac{\theta}{(1 - \theta)^2} \frac{1}{c} \frac{f_{IP}}{f_+f_-} \tag{3.153}$$

or, using the definition of the mean ionic-activity coefficient [Eq. (3.72)],

$$K_A = \frac{\theta}{(1 - \theta)^2} \frac{1}{c} \frac{f_{IP}}{f_\pm^2} \tag{3.154}$$

Some simplifications can now be introduced. The ion-pair activity coefficient f_{IP} is assumed to be unity because deviations of activity coefficients from unity are ascribed in the Debye–Hückel theory to electrostatic interactions. But ion pairs are not involved in such interactions owing to their zero charge and hence they behave ideally like uncharged particles, i.e., $f_{IP} = 1$.

Furthermore, in very dilute solutions: (1) The ions rarely come close enough together (i.e., to within a distance q) to form ion pairs, and one can consider $\theta << 1$ or $1 - \theta \approx 1$; (2) activity coefficients tend to unity, i.e., f_i or $f_\pm \rightarrow 1$.

Hence, under these conditions of very dilute solutions, Eq. (3.154) becomes

$$K_A \approx \frac{\theta}{c} \tag{3.155}$$

and substituting for θ from Eq. (3.147), one has

$$K_A = \frac{4\pi n_i^0}{c} \left(\frac{z_+z_-e_0^2}{\varepsilon kT} \right)^3 \int_2^b e^y y^{-4} dy \tag{3.156}$$

TABLE 3.18

Ion-Association Constant K_A: Extent to Which Ion-Pair Formation Occurs

Salt	Solvent	Temperature (K)	ε	K_A
KBr	Acetic acid	303	6.20	9.09×10^6
KBr	Ammonia	239	22	5.29×10^2
CsCl	Ethanol	298	24.30	1.51×10^2
KI	Acetone	298	20.70	1.25×10^2
KI	Pyridine	298	12.0	4.76×10^3

But

$$n_i^0 = \frac{cN_A}{1000} \tag{3.157}$$

and therefore,

$$K_A = \frac{4\pi N_A}{1000}\left(\frac{z_+z_-e_0^2}{\varepsilon kT}\right)^3 \int_2^b e^y y^{-1}\,dy \tag{3.158}$$

The value of the association constant provides an indication of whether ion-pair formation is significant. The higher the value of K_A, the more extensive is the ion-pair formation (Table 3.18).

What are the factors that increase K_A and therefore increase the degree of ion-pair formation? From Eq. (3.158), it can be seen that the factors that increase K_A are (1) low dielectric constant ε; (2) small ionic radii, which lead to a small value of a and hence [Eq. (3.148)] to a large value of the upper limit b of the integral in Eq. (3.158); and (3) large z_+ and z_-.

These ideas based on Bjerrum's picture of ion-pair formation have received considerable experimental support. Thus, in Fig. 3.46, the association constant is seen to increase markedly with decrease of dielectric constant.[22] The dependence of ion-pair formation on the distance of closest approach is seen in Fig. 3.47.

When numerical calculations are carried out with these equations, the essential conclusion that emerges is that in aqueous media, ion association in pairs scarcely occurs for 1:1-valent electrolytes but can be important for 2:2-valent electrolytes. (However, see the results of post-Bjerrum calculations in 3.13.) The reason is that K_A depends on z_+z_- through Eq. (3.156). In nonaqueous solutions, most of which have dielectric constants much less than that of water ($\varepsilon = 80$), ion association is extremely important.

[22]The critical dielectric constant above which there is no more ion-pair formation (as indicated by Fig. 3.46) is really a result of the arbitrary cutting off of ion-pair formation at the distance q.

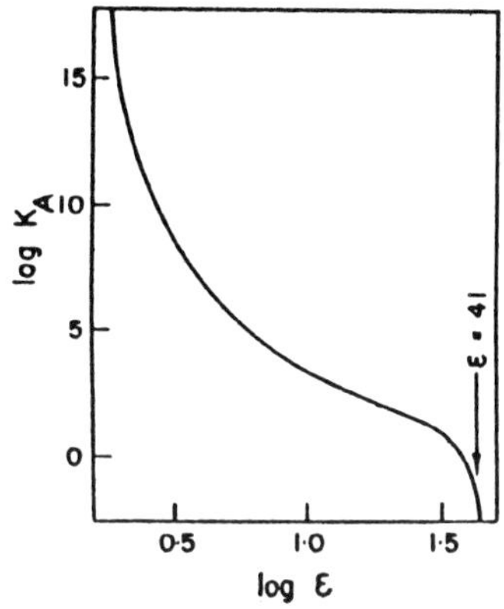

Fig. 3.46. Variation of the association constant K_A with the dielectric constant for 1:1 salts.

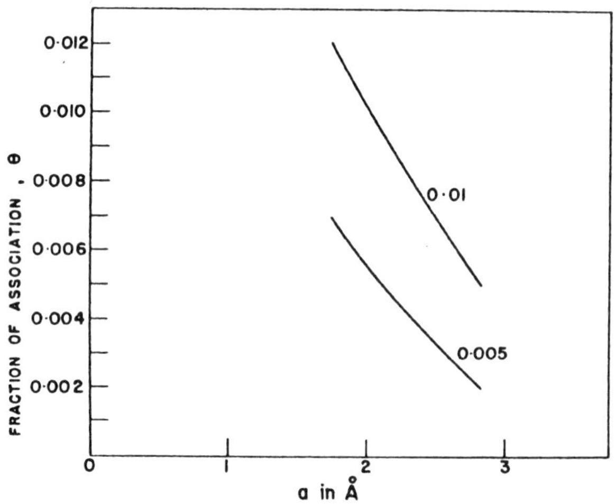

Fig. 3.47. Variation of θ, the fraction of associated ions, with a, the distance of closest approach (i.e., the ion size parameter).

3.8.5. Activity Coefficients, Bjerrum's Ion Pairs, and Debye's Free Ions

What direct role do the ion pairs have in the Debye–Hückel electrostatic theory of activity coefficients? The answer is simply: None. Since ion pairs carry no net charge,[23] they are ineligible for membership in the ion cloud, where the essential qualification is *charge*. Hence, ion pairs are dismissed from a direct consideration in the Debye–Hückel theory.

This does not mean that the Debye–Hückel theory gives the right answer when there is ion-pair formation. The extent of ion-pair formation decides the value of the concentration to be used in the ionic-cloud model. By removing a fraction θ of the total number of ions, only a fraction $1 - \theta$ of the ions remain for the Debye–Hückel treatment, which interests itself only in the free charges. Thus, the Debye–Hückel expression for the activity coefficient [Eq. (3.120)] is valid for the free ions, with two important modifications: (1) Instead of there being a concentration c of ions, there is only $(1 - \theta)c$; the remainder θc is not reckoned with owing to association. (2) The distance of closest approach of free ions is q and not a. These modifications yield

$$\log f_{\pm} = -\frac{A(z_+z_-)\sqrt{(1-\theta)c}}{1 + Bq\sqrt{(1-\theta)c}} \tag{3.159}$$

This calculated mean activity coefficient is related to the measured mean activity coefficient of the electrolyte $(f_{\pm})_{\text{obs}}$ by the relation (for the derivation, see Appendix 3.6)

$$(f_{\pm})_{\text{obs}} = (1-\theta)f_{\pm} \tag{3.160}$$

or

$$\begin{aligned}\log (f_{\pm})_{\text{obs}} &= \log f_{\pm} + \log(1-\theta)\\ &= -\frac{A(z_+z_-)\sqrt{(1-\theta)c}}{1 + Bq\sqrt{(1-\theta)c}} + \log(1-\theta)\end{aligned} \tag{3.161}$$

This equation indicates how the activity coefficient depends on the extent of ion association. In fact, this equation constitutes the bridge between the treatment of solutions of true electrolytes and that of solutions of potential electrolytes.

3.8.6. From Ion Pairs to Triple Ions to Clusters of Ions

The Coulombic attractive forces given by $z_+z_-e_0^2/\varepsilon r^2$ are large when the dielectric constant is small. When nonaqueous solvents of low dielectric constant are used, the

[23]Remember that the equations for the Bjerrum theory as presented here are correct only for electrolytes yielding ions of the same valence z, i.e., only for symmetrical 1:1- or 2:2-valent electrolytes.

values of dielectric constant are small. In such solutions of electrolytes therefore it has already been stated that ion-pair formation is favored.

Suppose that the electrostatic forces are sufficiently strong then it may well happen that the ion-pair "dipoles" may attract ions and *triple ions* be formed; thus

$$M^+ + (A^-M^+)_{\text{ion pair}} = [M^+A^-M^+]_{\text{triple ion}} \tag{3.162}$$

or

$$A^- + (M^+A^-)_{\text{ion pair}} = [A^-M^+A^-]_{\text{triple ion}} \tag{3.163}$$

Charged triple ions have been formed from uncharged ion pairs. These charged triple ions play a role in determining activity coefficients. Triple-ion formation has been suggested in solvents for which $\varepsilon < 15$. The question of triple-ion formation can be treated on the same lines as those for ion-pair formation.

A further decrease of dielectric constant below a value of about 10 may make possible the formation of still larger clusters of four, five, or more ions. In fact, there is some evidence for the clustering of ions into groups containing four ions in solvents of low dielectric constant.

3.9. THE VIRIAL COEFFICIENT APPROACH TO DEALING WITH SOLUTIONS

The material so far presented has shown that a model taking into account the size of the ions, ion-pair formation, and the idea that some of the water in the solution was not to be counted in estimating the "effective" concentration—the activity—has allowed a fair accounting for the main arbiter of the interionic attraction energy, the activity coefficient, even at concentrations up to nearly 5 mol dm^{-3} (see, e.g., Fig. 3.39).

This position of ionic solution theory has, however, a challenger,[24] and, during the 1970s and 1980s, it was this radically different approach to ionic solution theory

---

[24]There might be some who would actually say that there was "something wrong" with the theory of Debye and Hückel, but this claim depends on which version of the theory one means. The limiting-law equation certainly is inconsistent with experiments above ~ 5×10^{-3} mol dm^{-3} for a 1:1 electrolyte and even smaller concentrations for electrolytes possessing ions with a valence above unity. The later developments of the theory, which take into account the space occupied by the ions, do much better and taking the effect of solvation into account gives agreement with experiment to concentrations up to 10 mol dm^{-3} (Fig. 3.39). One question relates to whether cations and anions should have the same activity coefficient (the simple, original Debye and Hückel theory predicts this), but if one extends the model to account for "dead water" around ions, it turns out that there is more of this with small cations (they cling to water tighter) than with big anions, where the ion–water electric field is less and hence adherent dead waters are less in number. This solvational difference would imply a higher activity coefficient at high concentrations for cations than for anions, for which there is some evidence.

that absorbed most of the creative energy of physical electrochemists interested in this field. Suppose, says this '70s and '80s view, that one manages to remove from one's mind the evidence that some ions stick together in pairs, that (at least with smaller ions) they stick to water, and that the word "solvent" applies only to the free water. Then one tries to replicate experimental data with bare ions (always present in the stoichiometric amount and always free; there are no associated pairs) and free water alone.

Suppose also one regarded the basic idea of an "atmosphere" of ions (see Fig. 3.11) as a bit too theoretical for comfort and wanted to try an approach that looked more directly at the ionic interaction of individual ions. Then one might easily think of an approach made earlier in the century to deal with intermolecular force interactions and size effects in real gases: van der Waals's theory of imperfect gases. McMillan and Mayer devised a theory for liquids of this kind (although it was based on the theory of imperfect gases) in 1945 but tried to do much more than van der Waals had done. They used some highly generalized statistical mechanical arguments to deduce equations for pressure in terms of a number of virial (= force) coefficients

$$\frac{PV}{RT} = 1 + \frac{A}{V} + \frac{B}{V^2} + \frac{C}{V^3} + \ldots \tag{3.164}$$

where P, V, and RT are as usually understood in physical chemistry; A, B, C, etc., are the so-called *virial coefficients* theorized to take into account the intermolecular forces.

Applied to ionic solutions by Mayer in 1950, this theory aimed to express the *activity coefficient* and the *osmotic pressure* of ionic solutions while neglecting the effects of solvation that had been introduced by Bjerrum and developed by Stokes and Robinson to give experimental consistency of such impressive power at high concentrations.

However, Mayer found it to be more difficult to obtain a total of the interionic forces than that of forces between uncharged molecules. The reason is that the latter decline very rapidly, their dependence on distance being proportional to r^{-7}. Hence, accounting for nearest-neighbor interactions only becomes an acceptable proposition.

Consider now the interaction between an ion and those in a surrounding spherical section of ions chosen here—in this simple example of the kind of difficulty Mayer encountered—to be all of opposite sign to that of the central ion.

The energy of attraction of the central positive ion and *one* negative ion is $-e_0^2/\varepsilon r$, where r is the distance apart of two ions of opposite sign and ε is the dielectric constant. If the ionic concentration is N_i ions cm^{-3}, the number of ions in the spherical shell of thickness dr_i is $4\pi r^2 N_i\, dr$; therefore the energy of interaction of the central ion with all the negative ones becomes

$$W = \frac{4\pi N_i e_0^2}{\varepsilon} \int_{r_i}^{\infty} r\, dr \to \infty \tag{3.165}$$

This is a difficulty that occurs with many problems involving Coulombic interaction.

Mayer then conceived a most helpful stratagem. Instead of taking the potential of an ion at a distance r as $\psi = e_0/\varepsilon r$, he took it as $\psi = e_0 e^{-\kappa r}/\varepsilon r$ [where κ is the same κ of Debye–Hückel; Eq. (3.20)]. With this approach he found that the integrals in his theory, which diverged earlier [Eq. (3.165)] now *converged.* Hence, the calculation of the interionic interaction energy—the interaction of a representative ion with both negative and positive ions surrounding it—could yield manageable results.

There are many writers who would continue here with an account of the degree to which experiments agree with Mayer's theory and pay scant attention to the justification of the move that underlies it.[25] However, it *is* possible to give some basis to Mayer's equation (Friedman, 1989) because in a mixture of ions, interaction at a distance occurs through many other ions. Such intervening ions might be perceived as screening the interaction of one ion from its distant sister, and one feels intuitively that this screening might well be modeled by multiplying the simple $e_0/\varepsilon r$ by $e^{-\kappa r}$, for the new term decreases the interaction at a given distance and avoids the catastrophe of Eq. (3.165).

Does Mayer's theory of calculating the virial coefficients in equations such as Eq. (3.165) (which gives rise directly to the expression for the osmotic pressure of an ionic solution and less directly to those for activity coefficients) really improve on the second and third generations of the Debye–Hückel theory—those involving, respectively, an accounting for ion size and for the water removed into long-lived hydration shells?

Figure 3.48 shows two ways of expressing the results of Mayer's virial coefficient approach using the osmotic pressure[26] of an ionic solution as the test quantity. Two versions of the Mayer theory are indicated. In the one marked DHLL + B_2, the authors have taken the Debye–Hückel limiting-law theory, redone for osmotic pressure instead of activity coefficient, and then added to it the results of Mayer's calculation of the second virial coefficient, B. In the upper curve of Fig. 3.48, the approximation within the Mayer theory used in summing integrals (the one called hypernetted chain or HNC) is indicated. The former replicates experiment better than the latter. The two approxi-

[25]One of the reasons for passing over the physical basis of the modified equation for the potential due to an ion in Mayer's view is that several mathematical techniques are still needed to obtain final answers in Mayer's evaluation of an activity coefficient. (To replace the ionic cloud, he calculates the distribution of ions around each other and from this the sum of their interactions.) Among these occur equations that are approximation procedures for solving sums of integrals. To a degree, the mathematical struggle seems to have taken attention away from the validity of the modified equation for the potential due to an ion at distance r. These useful approximations consist of complex mathematical series (which is too much detail for us here) but it may be worthwhile noting their names (which are frequently mentioned in the relevant literature) for the reader sufficiently motivated to delve deeply into calculations using them. They are, in the order in which they were first published, the Ornstein–Zernicke equation, the Percus–Yevich equation, and the "hypernetted chain" approach.

[26]Since Mayer's theory originated in a theory for imperfect *gases*, it naturally tends to calculate the nearest analogue of gas pressure that an ionic solution exhibits—osmotic pressure.

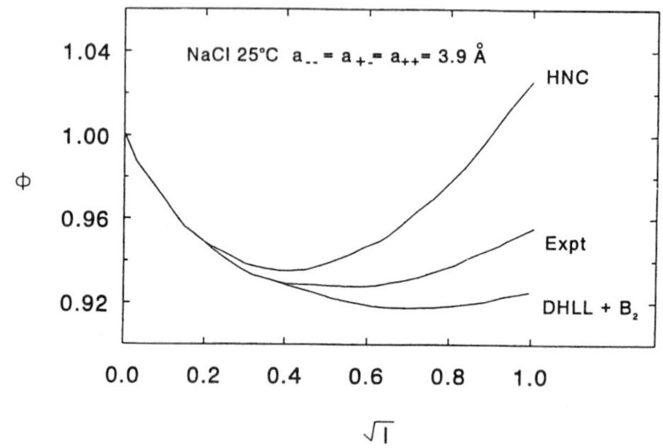

Fig. 3.48. Osmotic coefficients for the primitive model electrolyte compared with the experimental results for NaCl in aqueous solutions at 298 K. The $a_{\pm}$ parameters in the HNC and DHLL + B_2, approximations have been chosen to fit the data below 0.05 mol dm^{-3}. I is the ionic strength. (Reprinted from J. C. Rasaiah, *J. Chem. Phys.* **52:** 704, 1970.)

mations differ by ~12%. To get results which agree really well (±1%) with experiment up to, say, 1 mol dm^{-3}, the original theory tested here is not enough. One has to develop the interaction energy more realistically than by the use of Mayer's perceptive $e_0^2 e^{-\kappa r}/\varepsilon r$, using the addition of a number of "extra potentials," representing, for example, short-range repulsion energies on contact (Section 3.10.1). It is these closer approaches to physical reality rather than the solutions to difficult mathematical equations representing an overly simple model that now mark the way for electrolyte theory.

Further Reading

Seminal

1. N. Bjerrum, "Ionic Association," *Kgl. Danske Videnskab. Selskab.* **7**: 9 (1926).
2. W. G. McMillan and J. E. Mayer, "A Virial Coefficient Approach to the Theory of Fluids," *J. Chem. Phys.* **13**: 276 (1945).
3. J. E. Mayer, "A Virial Theory of Ionic Solutions," *J. Chem. Phys.* **18**: 1426 (1950).
4. R. M. Fuoss and L. Onsager, "Ionic Association in Electrolyte Theory," *Proc. Natl. Acad. Sci. U.S.A.* **41**: 274, 1010 (1955).
5. T. L. Hill, "A Simplified Version of the McMillan-Meyer Theory," *J. Chem. Phys.* **30**: 93 (1959).

Review

1. H. L. Friedman, E. O. Raineri, and O. M. D. Wodd, "Ion-Ion Interactions," *Chemica Scripta* **29A**: 49 (1989).
2. J. C. Rasaiah, "A Model for Weak Electrolytes," *Int. J. Thermodyn.* **11**: 1 (1990).
3. L. M. Schwartz, "Ion Pair Complexation in Moderately Strong Aqueous Acids," *J. Chem. Ed.* **72**: 823 (1995).

Papers

1. J.-L. Dandurand and J. Schott, *J. Phys. Chem.* **96**: 7770 (1992).
2. E. H. Oelkers and H. C. Helgeson, *Science* **261**: 888 (1993).
3. J. Gao, *J. Phys. Chem.* **98**: 6049 (1994).
4. J. Wang and A. J. Haymet, *J. Chem. Phys.* **100**: 3767 (1994).
5. M. Madhusoodana and B. L. Tembe, *J. Phys. Chem.* **99**: 44 (1995).
6. G. Sese, E. Guardia, and J. A. Padro, *J. Phys. Chem.* **99**: 12647 (1995).
7. M. Ue and S. Mori, *J. Electrochem. Soc.* **142**: 2577 (1995).

3.10. COMPUTER SIMULATION IN THE THEORY OF IONIC SOLUTIONS

All parts of the physical sciences are now served by calculation techniques that would not have been possible without the speed of electronic computers. Such approaches are creative in the sense that, given the law of the energy of interaction between the particles, the software allows one to predict experimental quantities. If agreement with experiment is obtained, it tells us that the energy of interaction law assumed is correct. Sometimes this approach can be used to calculate properties that are difficult to determine experimentally. Such calculations may allow increased insight into what is really happening in the system concerned or they may be used simply as rapid methods of obtaining the numerical value of a quantity.[27] There are two main computational approaches and these will be discussed next.

3.10.1. The Monte Carlo Approach

The Monte Carlo approach was invented by Metropolis (Metropolis, 1953). The system concerned is considered in terms of a small number of particles—a few hundred. The basic decision that has to be made is: What law are the particles i and j going to follow in expressing the interaction energy between them as a function of their distance apart? For example, a useful law might be the well-known Lennard-Jones equation:

---

[27]The major attraction of such computer simulation approaches is that they often result in lower costs. However, a prerequisite to their use is an experimental value on some related system, so that the *A*s and *B*s of equations such as Eq. (3.166) can be calibrated.

$$U_{ij} = -\frac{A}{r^6} + \frac{B}{r^{12}} \tag{3.166}$$

where r is the interparticle distance and A and B are constants, the values of which are not usually found from independent determinations but by assuming the law of interaction calculation procedure to be correct and finding the A's and B's that have then to be used to fit the experimental quantities.

What is the procedure? The N particles are started off in any configuration, e.g., in that of a regular lattice. Then each particle is moved randomly (hence the title of the procedure). A single but vital question is asked about each move: does it increase the energy of the particle (make its potential energy move positive) or decrease it (potential energy more negative)? If the former, the move is not counted as contributing to the final equilibrium stage of the system. If the latter, it is counted.

Such a calculation is then carried out successively on each particle many times. Since each can move to any point within the allotted space, a large enough number of moves allows one to reach the equilibrium state of the system while calculating a targeted quantity, e.g., the pressure of an imperfect gas. The result is compared with that known experimentally, thus confirming or denying the force law assumed (and other assumptions implicit in the calculation).

Card and Valleau (1970) were the first to apply the Metropolis Monte Carlo method to an electrolytic solution. Their basic assumption was that if

$$r < a_{ij} \quad U_{ij} = \infty$$

$$r > a_{ij} \quad U_{ij} = \frac{z_i z_j e_0^2}{\varepsilon} \tag{3.167}$$

In spite of the long-range nature of the interionic forces, they assumed that the yes/no answers obtained on the basis of the nearest-neighbor interactions could be relied upon. Using this nearest-neighbor-only approximation, and neglecting ion association and the effects of hydration in removing some of the water from circulation, their calculation replicated the experimentally observed minimum in the log $\gamma_{\pm}$ vs. $\sqrt{c}$ plot up to 1.4 mol dm^{-3}, a point supporting the approach.

3.10.2. Molecular Dynamic Simulations

Another and now more widely used computational approach to predicting the properties of ions in solution follows from the Monte Carlo method. Thus, in the latter, the particle is made mentally to move randomly in each "experiment" but only one question is posed: Does the random move cause an increase or decrease of energy? In molecular dynamic (MD) simulations, much more is asked and calculated. In fact, a random micromovement is subjected to all the questions that classical dynamics can ask and answer. By repeating calculations of momentum and energy exchanges

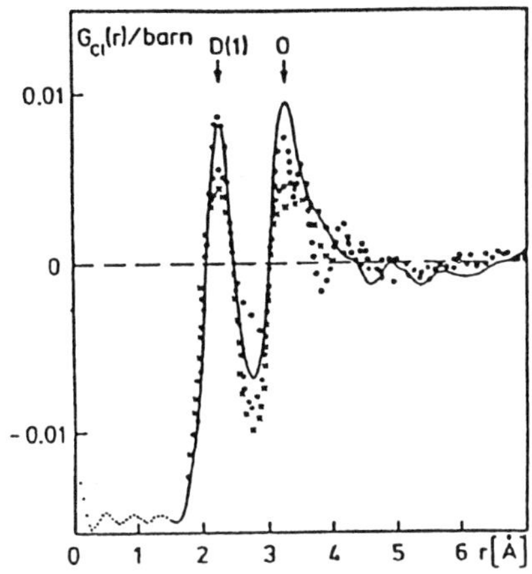

Fig. 3.49. Comparison of the weighted Cl^--water radial distribution function from an MD simulation of a 1.1 *M* $MgCl_2$ solution (solid line) with results from neutron diffraction studies of a 5.32 *M* NaCl (o), a 3 *M* $NiCl_2$ (x), and a 9.95 *M* LiCl (•) solution (1 Å = 0.1 nm). (Reprinted from P. Bopp, *NATO ASI Series* **206**: 237, 1987.)

between particles as they approach each other at intervals of time of the order of a few femtoseconds, and then working out the dynamic consequences for each encounter for the few hundred particles considered, it is possible to calculate the distribution functions of the particles in respect to a central particle, $g(r)$. Knowing this calculated distribution function, many properties of an ionic system can be computed. Of course, as with the Monte Carlo approach, the MD calculations pertain to reality insofar as the two-body energy–distance law between the particles has been made to fit the system by using some experimental results to calculate the parameters in such equations as (3.166). The procedure is thus designed to give an experiment-consistent answer.

P. Bopp has carried out many MD simulations for aqueous ionic solutions, following the introduction of the method by Adler and Wainwright (1959). The type of agreement he can obtain between theory and experiment is illustrated in Fig. 3.49.

3.10.3. The Pair-Potential Interaction

Computer simulation studies are based on the need for an experimental value to start with! The A and B of Eq. (3.166) are not obtained from independent methods. Rather, they are adjusted so that even with a simple pair potential, the calculation is correct. Thus in mechanics it is fundamentally not possible to calculate the final result of the interaction of more than two particles! Hence, the implicit assumption is made that the properties of a body depend on *pair*-potential (i.e., *two*-body) interactions only. Of course, this is not the case, and indeed the influence of particles apart from nearest neighbors is particularly strong for Coulombic interactions because of their long range. Thus A and B are not the real values for the interaction of a pair of particles in Eq. (3.166). They are values that make the simplified equations (two-body-interaction-only assumption) correct. The computation then becomes possible. The ability of the computation to get the right answer is bred into the calculation at an early stage.

This is a burden that MD (and Monte Carlo) procedures carry with them. Since the parameters are built in to give the right results on approximate model assumptions (no ion association, for example), the procedure is better suited for calculating quantities correctly than for finding out what is going on. Thus in electrolytic solution theory, the mean spherical approximation (MSA) and other procedures produce experimentally consistent values and properties, such as osmotic and activity coefficients up to a high concentration of 2 mol dm^{-3} (Section 3.16). However, in getting the right answer, they neglect the effects of hydration on the availability of water to act as part of the solution, although a substantial fraction of the ions are known to be associated and all the free ions are hydrated. The right answer is produced on the wrong model. Thus, computer-simulated solutions are good "answer-getters," but one had better look to spectroscopy to find out what is really going on.

3.10.4. Experiments and Monte Carlo and MD Techniques

In the brief description of computational approaches to the properties of solutions presented in this chapter, only the barest bones have been indicated (see also Chapters 2 and 5). The techniques actually used to move from the application of mechanics to the movement of particles to the computation of observable properties are beyond the scope of this book.

However, it is right to face an interesting question. The computational techniques described are widely used in research. They led to the rapid calculation of many properties of systems formerly laboriously determined by large numbers of white-coated experimentalists. They even allow calculation of some properties for which experimental determinations are nearly impossible.[28] Are the days of the white coats numbered?

There seems to be much going for this proposition. Before acceding to it completely, it is necessary to recall what has been written earlier about the use of pair potentials in place of multiparticle interactions. The quantities A and B in the Lennard-

---

[28]Excellent examples are the properties of the first one to two layers of water near an ion in a dilute solution or the properties of silicates at great pressures deep in the interior of the planet.

Jones equation are calculated from *experimentally obtained values*. Furthermore, it should be an experiment that duplicates as closely as possible the phenomena for which answers are being sought by means of calculation.

There is another reservation that should be mentioned for those who see Monte Carlo and MD techniques as all-conquering: they use classical mechanics. This is all very well for some movements, e.g., translation in imperfect gases. But what of the quantized vibration? Or what of quantal aspects in rate calculations or the tunneling of protons in the conduction of aqueous acid solutions (see Section 4.11.5), for example?

So, manufacturers of white coats may still have a market for a generation or more. But it will be a diminishing market! There is no doubt that computer-simulation techniques are making inroads into experimental data collection and that they will be increasingly used because they save time and cost.

Further Reading

Seminal

1. N. Metropolis, *J. Chem. Phys.* **21**: 1087 (1953).
2. B. J. Adler and T. E. Wainwright, *J. Chem. Phys.* **27**: 1208 (1959).
3. B. J. Adler and T. E. Wainwright, "Computer Simulation in Electrolytes," *J. Chem. Phys.* **33**: 1439 (1960).
4. H. L. Friedman, *Ionic Solution*, Interscience, New York (1962).
5. S. G. Brush, H. L. Saklin, and E. Teller, "A Monte Carlo Calculation of a One Component Plasma," *J. Chem. Phys.* **45**: 2102 (1965).
6. J. C. Rasaiah and H. L. Friedman, "Integral Equation Methods for Computations about Ionic Solutions," *J. Chem. Phys.* **48**: 2742 (1965).
7. G. Snell and J. L. Lebowitz, "Computational Techniques in Electrolytic Calculations," *J. Chem. Phys.* **48**: 3706 (1968).
8. P. S. Ramanathen and H. L. Friedman, "Study of a Refined Model for Electrolytes," *J. Chem. Phys.* **54**: 1080 (1971).
9. J. C. Rasaiah, "Ionic Solutions in Equilibrium," *J. Chem. Phys.* **56**: 3071 (1972).
10. L. Blum, "Mean Spherical Model for a Mixture of Hard Spheres and Hard Dipoles," *Chem. Phys. Lett.* **26**: 200 (1976).

Review

1. P. Turq, "Computer Simulation of Electrolytic Solutions," in *The Physics and Chemistry of Aqueous Solutions*, M. C. Bellisent-Funel and G. W. Neilson, eds., *NATO ASI Series*, **205:** 409 (1987).

Papers

1. D. N. Card and J. P. Valleau, *J. Chem. Phys.* **52**: 6232 (1970).
2. D. Smith, Y. Kalyuzhnyi, and A. D. J. Haymet, *J. Chem. Phys.* **95**: 9165 (1991).

3. T. Cartailler, P. Turq, and L. Blum, *J. Phys. Chem.* **96**: 6766 (1992).
4. J.-L. Dandurand and J. Schott, *J. Phys. Chem.* **96**: 7770 (1992).
5. B. Jamnik and V. Vlachy, *J. Am. Chem. Soc.* **115**: 660 (1993).
6. J. Gao, *J. Phys. Chem.* **98**: 6049 (1994).
7. D. E. Smith and L. X. Dang, *J. Chem. Phys.* **100**: 3757 (1994).
8. J. Wang and A. D. J. Haymet, *J. Chem. Phys.* **100**: 3767 (1994).
9. C. E. Dykstra, *J. Phys. Chem.* **99**: 11680 (1995).
10. C.-Y. Shew and P. Mills, *J. Phys. Chem.* **99**: 12988 (1995).

3.11. THE CORRELATION FUNCTION APPROACH

3.11.1. Introduction

Debye and Hückel's theory of ionic atmospheres was the first to present an account of the activity of ions in solution. Mayer showed that a virial coefficient approach relating back to the treatment of the properties of real gases could be used to extend the range of the successful treatment of the "excess" properties of solutions from 10^{-3} to 1 mol dm^{-3}. Monte Carlo and molecular dynamics are two computational techniques for calculating many properties of liquids or solutions. There is one more approach, which is likely to be the last. Thus, as shown later, if one knows the correlation functions for the species in a solution, one can calculate its properties. Now, correlation functions can be obtained in two ways that complement each other. On the one hand, neutron diffraction measurements allow their experimental determination. On the other, Monte Carlo and molecular dynamics approaches can be used to compute them. This gives a pathway purely to calculate the properties of ionic solutions.

3.11.2. Obtaining Solution Properties from Correlation Functions

To understand this section, it is best to review briefly the idea of a correlation function (see Section 2.11.3). Consider a hypothetical photograph of a number of ions in a solution. One looks for one ionic species i in a small volume dV at a series of distances r from a molecule of species j. The chance of finding the ion i in dV is

$$c_i g_{ij}(r)\, dV \tag{3.168}$$

where $c_i = N_i/V$ is the number of i ions per unit volume. In Eq. (3.168), g_{ij} is called the *correlation function*. This is a measure of the effect of species j in increasing or decreasing the number of i ions in dV. One way of writing g_{ij} is

$$g_{ij}(r) = c^{-w_{ij}/kT} \tag{3.169}$$

where w_{ij} is the reversible work to bring i and j from infinitely far apart to the distance r.

There are two kinds of ways to obtain g values, i.e., g_{ij}, g_{ii}, and g_{jj}. One is a computational approach using the Monte Carlo method or MD. In Fig. 3.49 one has seen the example of P. Bopp's results of such a determination using MD.

On the other hand, one can experimentally determine the various g's. Such determinations are made by a combination of X-ray and neutron diffraction measurements (see Section 2.11.3 and 5.2.3).

One now has to use the correlation functions to calculate a known quantity. One way of doing this (Friedman, 1962) is to use the following equation:

$$\left(\frac{\partial c}{\partial P}\right)_{V,T} = \frac{kT}{1 + 4Pr\int_0^{N/V} [g(r) - 1]r^2 dr} \tag{3.170}$$

The excess free energy can be shown to be

$$A^{ex} = N\int_0^c (P - ckT)c^{-2} dc \tag{3.171}$$

The process of relating A^{ex} back to $g(r)$, obtained theoretically or experimentally, is done by first calculating the pressure by computing $g(r)$ at various values of the distance r to get $\partial P/\partial c$ at a series of c's down to $P \sim ckT$, the ideal value, after which one integrates the equation to get

$$P(c) = ckT + \int \left\{\left(\frac{\partial P}{\partial c}\right) - kT\right\} dc \tag{3.172}$$

Having thus obtained P on the basis of a knowledge of the g's, one can use the expression given to obtain A^{ex} as a function of P. The procedure is neither brief nor simple but broad and general. It indicates that if one knows the distribution functions, one can compute the energy excess over that for zero interaction (and by implication any excess property).[29]

Why should one go to all this trouble and do all these integrations if there are other, less complex methods available to theorize about ionic solutions? The reason is that the correlation function method is open-ended. The equations by which one goes from the gs to properties are not under suspicion. There are no model assumptions in the experimental determination of the g's. This contrasts with the Debye–Hückel theory (limited by the absence of repulsive forces), with Mayer's theory (no misty closure procedures), and even with MD (with its pair potential used as approximations to reality). The correlation function approach can be also used to test any theory in the future because all theories can be made to give $g(r)$ and thereafter, as shown, the properties of ionic solutions.

---

[29]An excess property is one that indicates the difference between the real values of the property and the ideal value of the property that would exist in the absence of interionic attraction.

3.12. HOW FAR HAS THE MEAN SPHERICAL APPROXIMATION GONE IN THE DEVELOPMENT OF ESTIMATION OF PROPERTIES FOR ELECTROLYTE SOLUTIONS?

The latest models propose to represent electrolyte solutions as a collections of hard spheres of equal size, ions, immersed in a dielectric continuum, the solvent. For such a system, what is called the Mean Spherical Approximation, MSA, has been successful in estimating osmotic and mean activity coefficients for aqueous 1:1 electrolyte solutions, and has provided a reasonable fit to experimental data for dilute solutions of concentrations up to ~0.3 mol dm^{-3}. The advantage in this approach is that only one adjustable parameter, namely the single effective ionic radius, has to be considered.[30]

The abbreviation MSA refers here to the model implied by the simple correlation function stated and with the development which follows. More generally, the term refers to one or several integral equations used in the theory of liquids. It is appropriate to what is discussed here (charged hard spheres) because MSA equations have an analytical solution for that case.

What about the difficult problem of modeling electrolyte solutions at higher concentrations? In the MSA approach, attempts were made to extend the treatment by allowing the ion size parameter to be a function of ionic strength. Unfortunately, such an approach became unrealistic because of the sharp reduction of effective cation size with increasing electrolyte concentration.

In another attempt (Fawcett and Tikenen, 1996), the introduction of a changing dielectric constant of the solvent (although taken from experimental data) as a function of concentration has been used to estimate activity coefficients of simple 1:1 electrolyte solutions for concentrations up to 2.5 mol dm^{-3}.

In the mean spherical approximation, the interaction energy between any two particles is given by their pairwise interaction energy, independent of direction from one of the particles. This interaction is then used to define the direct correlation function. Under this approach,

$$g_{ij}(r) = 0 \quad \text{when} \quad r < \sigma$$

where σ is the diameter of the reference ion. In this hard sphere version, there is no steep tuning of the repulsive potential, but suddenly, at $r = \sigma$, $U_{ij} \rightarrow +\infty$.[31] However, when $r > \sigma$, the equation becomes complex to solve.

---

[30]Let's not forget that a model is only an attempt to represent the much more complex reality. Using one of different approaches—MSA, Monte Carlo, MD, etc.,—the scientist proposes a model starting with simple but reasonable assumptions that compares it with experimentally determined parameters. If the agreement is reasonable, then a complexity is added to the model, and the understanding of the system—of the reality—advances. Of course, all this is subjected to the limitations of the approach chosen, e.g., long computer times and complex integrals to solve.

[31]This is in contrast to the "soft-sphere approximation" where some distance-dependent function such as the second term in the Lennard-Jones equation (3.166) is used.

The simplest pairwise interaction equation for the electrostatic attraction of a positive and a negative ion to each other is

$$U_{ij} = \frac{z_i z_j e_0^2}{\varepsilon r} \tag{3.173}$$

Mayer achieved closure in integrals involved in his application of the McMillan–Mayer virial approach to ionic solutions by multiplying this equation by the factor $e^{-\kappa r}$ where κ^{-1} is the Debye length. Something similar is done in equations for the electrostatic attraction part of the MSA theory, but in this case the Ornstein–Zernicke integral equation is introduced.[32] Instead of κr, the term used is $2\Gamma r$ where

$$\Gamma = \frac{(1 + 2\sigma\kappa)^{1/2} - 1}{2\sigma} \tag{3.174}$$

If one takes the pairwise interaction equation (3.173) and modifies it by multiplying by $e^{-2\Gamma r}$, then one finds the correlation function becomes

$$g_{ij}(r) = \frac{z_i z_j e_0^2}{\varepsilon r k T} e^{-2\Gamma r} \tag{3.175}$$

In the primitive Debye–Hückel theory—one that did not allow for the size of ions—the value for the activity coefficient is given by Eq. (3.60). The corresponding equation in the MSA is

$$\ln (\gamma_i)_{es} = -\frac{z_i^2 e_0^2 \Gamma}{kT\varepsilon(1 + \Gamma\sigma)} \tag{3.176}$$

However, this is only the electrostatic attraction part of the theory. There is also the effect of the hard-sphere part of the theory, that upon contact ions immediately repel with infinite energy. This part of the activity coefficient is found to be given by

$$\ln (\gamma_i)_{hs} = \frac{6\eta}{1 - \eta} + \frac{3\eta^2}{(1 - \eta)^2} + \frac{2\eta}{(1 - \eta)^3} \tag{3.177}$$

Here the term η represents a *packing factor* and is given by

$$\eta = \frac{\pi\sigma^3}{6} \sum_i c_i \tag{3.178}$$

where c_i is the ionic concentration.

[32]This equation was first used to deal with problems in the theory of liquids. See, e.g., D.A. McQuarie, *Statistical Mechanics*, p. 269, Harper Collins, New York (1976).

Finally,

$$\ln \gamma_i = \ln(\gamma_i)_{es} + \ln(\gamma_i)_{hs} \tag{3.179}$$

It is easy then to develop equations for actual observables such as the mean activity coefficients (Section 3.4.4).

These equations all contain the value of ε, the dielectric constant of the solution. Many workers have approximated it by using $\varepsilon = 78$, the value for water. However, Blum (1977) was the first to point out that better agreement with experiment is achieved by using instead the actual dielectric constant of the solution. Fawcett and Tikanen (1996) took this into account, and by using a fit to results obtained much earlier by Hasted, found

$$\varepsilon = \varepsilon_0 - \delta_i c_i + b c_i^{3/2} \tag{3.180}$$

The σ values that were found by Fawcett and Tikanen to fit best are shown in Table 3.19 and the results for NaBr are shown in Fig. 3.50.

It is interesting to note that the $\gamma - c^{1/2}$ plot bends upward at higher concentrations. It is easy to see why. As the concentration increases, the solution, as it were, gets filled up with ions and there tends to be no more room. This is equivalent to too many ions per solvent and the *effective* ionic concentration becomes higher than would be

TABLE 3.19

Values of the Average Ionic Diameter σ Obtained from the Best Fit of the Model to Experiment[a]

	σ (pm)			
	F^-	Cl^-	Br^-	I^-
Li^+	—	435 (1.6 *M*)	448 (1.5 *M*)	489 (0.5 *M*)
Na^+	328 (1 *M*)[b]	388 (3 *M*)	407 (2.5 *M*)	427 (2.5 *M*)
K^+	365 (1.4 *M*)	362 (2.5 *M*)	376 (3 *M*)	394 (2.5 *M*)
Rb^+	389 (0.9 *M*)	349 (2.5 *M*)	349 (1.2 *M*)	351 (0.8 M)
Cs^+	408 (0.5 *M*)	317 (1 *M*)	318 (0.9 *M*)	311 (0.6 *M*)

Source: Reprinted from W. R. Fawcett and A. C. Tikanen, *J. Phys. Chem.* **100**: 4251, 1996.

[a]The concentration in parentheses gives the value over which a successful fit between theory and experiment was obtained.

[b]Experimental data available only up to 1 mol dm^{-3}.

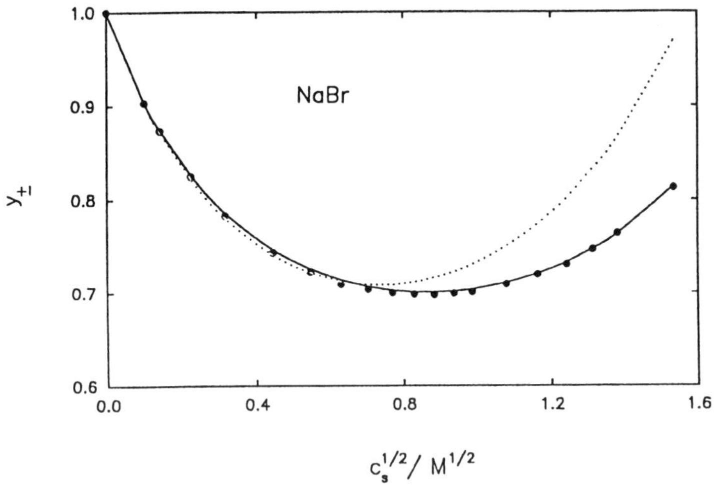

Fig. 3.50. Plot of the mean ionic activity coefficient for aqueous NaBr solutions against the square root of molarity. The solid line through the experimental points shows the MSA estimates with an optimized value of 407 pm for σ and a varying solvent ε. The dotted durve was calculated with an ε equal to that of the pure solvent and an optimized value of 366 pm for σ. (Reprinted from W. R. Fawcett and A. C. Tikanen, *J. Phys. Chem.* **100**: 4251, 1996.)

calculated on stoichiometric grounds, overcoming the tendency of γ to fall with increasing concentration due to increasing interionic attraction.[33]

The degree of fit in Fig. 3.50 looks excellent. Detailed considerations of results for all the alkali halides, however, show that there are discrepancies and these are most likely to be due to the neglected ion pairing.

3.13. COMPUTATIONS OF DIMER AND TRIMER FORMATION IN IONIC SOLUTION

In forming post-2000 theories of solutions, it is important to know what fraction of the ions in a solution are associated to form dimers and trimers and perhaps higher aggregates. Dimers are not charged and will interact with the surrounding ions as

[33]A similar result is obtained in the Stokes and Robinson application of the idea, i.e., that there is a removal of effective free solvent into hydration shells around ions (Section 2.4.1). Both ideas are similar to the effect of the $V - b$ term in van der Waals's equation of state for gases. If the a/V^2 attraction term is neglected, $P = kT/(V - b)$. As V is reduced to be comparable in value to b, P (which is analogous to the ionic activity) increases above that for the simple $PV = kT$ equation.

dipoles. Bjerrum was the first, in 1926, to give an estimate of dimer concentration (Section 3.8). In this theory, it was simply assumed that if the ions approach each other so that their mutual Coulombic attraction exceeds their total thermal energy ($2kT$), they will form a charge-free pair.

Experimental methods for determining dimer formation were published in an early book (Davies, 1962). The author interpreted conductance data in terms of degrees of association. His book had less impact than might have been thought reasonable because of ambiguities arising from allowance of the parallel effects of interionic attraction, as well as ion association to dipoles, in lowering equivalent conductance as the concentration increased.

One might suspect that Raman spectra would give an indication of ion pairs, but this is only true for pairs (such as those involving NO_3^-) where the ion pairing affects a molecular orbital (such as that in N–O). In a pair such as Na^+Cl^-, there are *no chemical bonds*—and it is these that produce spectra—so that ion pairs may exist without a spectral signature.

It follows that experimental determination of ion pairing has been difficult to carry out. Some theoretical works by Wertheim (1984) have contributed a correlation function approach to ion pairing that improves on the original work by Bjerrum (1926). The major difference is the form of the interaction energy between the ions, $U_{A^+B^-}$. In

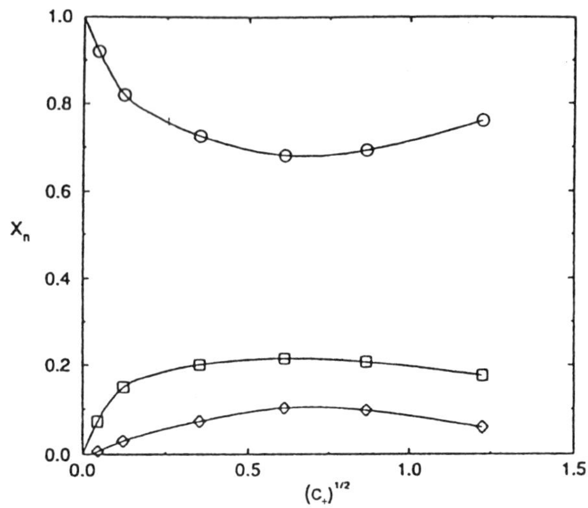

Fig. 3.51. Mole fraction of monomers, dimers, and + – + trimers as a function of the square root of the concentration $c_+^{1/2}$: monomer fraction (circles), dimer fraction (squares), and the fraction of + – + trimers (diamonds). (The lines are to guide the eye only.) (Reprinted from J. Wang and D. J. Haymet, *J. Chem. Phys.* **100**: 3767, 1994.)

Bjerrum's work it was simply Coulombic attraction, but in Wertheim the governing equation involves not only the attraction but also the repulsive forces, somewhat as in the Lennard-Jones equation. The actual equation used originated in work by Rossky and Friedman (1980).

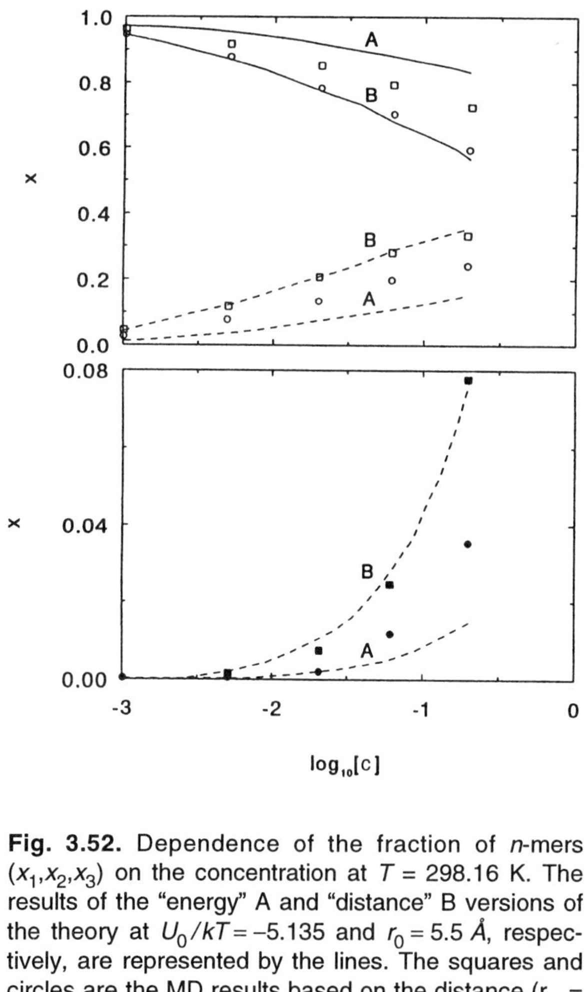

Fig. 3.52. Dependence of the fraction of *n*-mers (x_1,x_2,x_3) on the concentration at T = 298.16 K. The results of the "energy" A and "distance" B versions of the theory at $U_0/kT = -5.135$ and $r_0 = 5.5$ Å, respectively, are represented by the lines. The squares and circles are the MD results based on the distance (r_{in} = 3.191 Å, r_{out} = 5.5 Å) and energy [$U_0^{(MD)}/kT = -5.135$] criteria, respectively. The open symbols and solid lines represent the results for monomers, half-filled symbols and dashed lines represent the results for dimers, and filled symbols and dotted lines are the results for trimers and higher *n*-mers. (Reprinted from Y. V. Kalyuzhnyi et al. *J. Chem. Phys.* **95**: 9151, 1991.)

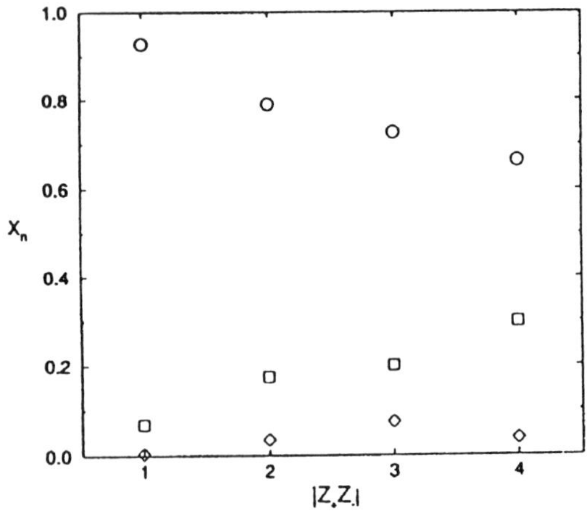

Fig. 3.53. The mole fraction of monomers, dimers, and total trimers for four model electrolytes at Debye length κ^{-1} = 0.61 nm, as a function of $|z_+z_-|$: monomer fraction (circles), dimer fraction (squares), and total trimer fraction (diamonds). (Reprinted from J. Wang and A. D. J. Haymet, *J. Chem. Phys.* **100**: 3767, 1994.)

This formalism for calculating dimer formation begun by Wertheim has been applied extensively by Haymet (1991 and onward) to electrolytes. For example, 1:3 electrolytes (e.g., Na_3PO_4) have been examined following Wertheim's interaction equation. Some of the results are shown in Fig. 3.51. The calculations show not only that there is significant dimer formation at 0.1 *M* but that trimers of the type + – + are present at higher concentrations than those that are – + –.

Haymet also studied 2:2 electrolytes. Here, even at 0.1 mol dm^{-3}, the association with dimers is 10–20% (Fig. 3.52). The dependence of dimer and trimer formation on z_+z_- in Fig. 3.53 is for a constant value of the Debye–Hückel length, $1/\kappa$.

What is the significance of these results on dimer and trimer formation for ionic solution theory? In the post-Debye and Hückel world, particularly between about 1950 and 1980 (applications of the Mayer theory), some theorists made calculations in which it was assumed that all electrolytes were completely dissociated at least up to 3 mol dm^{-3}. The present work shows that the degree of association, even for 1:1 salts, is ~10% at only 0.1 mol dm^{-3}. One sees that these results are higher than those of the primitive Bjerrum theory.

There is much work to do to recover from the misestimate that entered into calculations made after Debye and Hückel by neglecting ion pairing and "dead water" (Section 3.8).

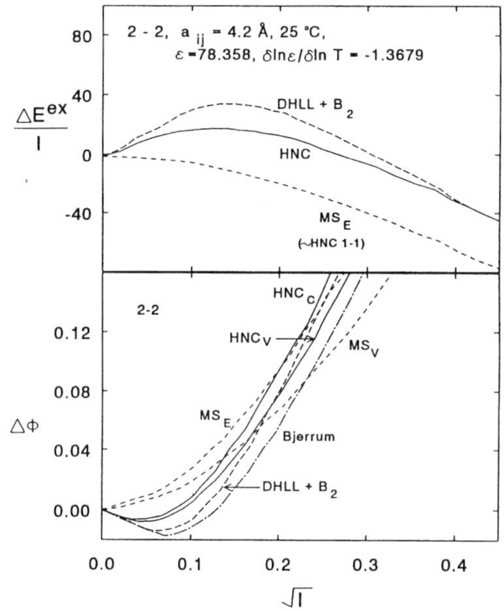

Fig. 3.54. Deviations from the Debye–Hückel limiting law (DHLL) for E^{ex} and ϕ of a 2:2 electrolyte for several theories. The ion-pairing cutoff distance d for the Bjerrum curve is 1.43 nm. I is the ionic strength. (Reprinted from J. C. Rasaiah, *J. Chem. Phys.* **56**: 3071, 1972.)

3.14. MORE DETAILED MODELS

By now this chapter has presented several modeling pathways that give rise to extensions of the Debye–Hückel limiting law of 1923. The results obtained from a few models developed in the half century that followed are illustrated in Fig. 3.54. The lines denoted "HNC" refer to versions of the Mayer theory utilizing the hypernetted chain approximation for evaluation of the integrals and the designation "MS" corresponds to theories using the mean spherical approximation.

No great superiority of one approximation over the other is apparent. However, there is something a little inconsistent, even dubious, about the comparisons. The approximation marked "Bjerrum" allows for ion association, and this is a fact known to be present at higher concentrations, as independently shown by Raman spectra of electrolytes containing, e.g., NO_3^- ions. The other models assume, however, that all the ions are free and unassociated, even in concentrations in which experiment shows strong ion-pair formation. Also, where is the allowance for Bjerrum's "dead water," i.e., water fixed in hydration shells that cannot be counted in calculating the ionic concentration?

In the post-Debye and Hückel developments surveyed above (virial coefficient theory, computer simulation, and distribution coefficients), it is frequently stated that the quantity that has the greatest influence on the result of an ionic solution is $U_{A^+B^-}$, the interaction potential. In discussing the validity of Monte Carlo and molecular dynamics calculations, the difficulty of expressing the potential of an ion in terms only of the nearest-neighbor interaction (when long-range Coulomb forces indicate that other interactions should be accounted for) was pointed out. Because computations that go further are too lengthy, the simple pair potential is the one generally used, but the previous generation's papers have progressively improved the potential.

Thus, theories using an U_{ij} that is simply

$$U_{ij} = \frac{e_i e_j}{\varepsilon r} \tag{3.181}$$

are said (Friedman, 1987) to be using the *primitive* model.

The simplest improvement is the mean spherical approximation model (Section 3.12), but a somewhat better version of this is what can be pictorially called the "mound model," because instead of having an abrupt change from simple Coulomb attraction to total repulsion between ions of opposite sign when they meet, this model (Rasaiah and Friedman, 1968) allows for a softer collision before the plus infinity of the hard wall is met (Fig. 3.55).

The fullest development along these lines (Ramanathan et al., 1987) is given by the following series:

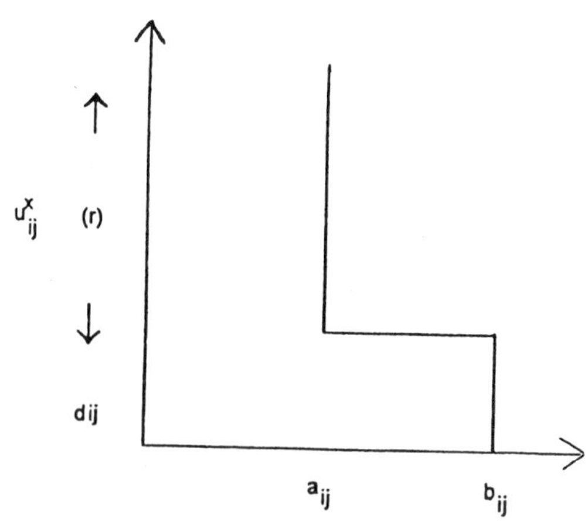

Fig. 3.55. The "mound model."

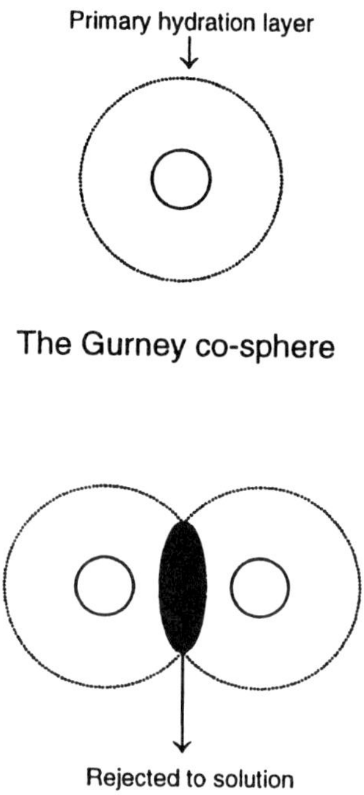

Fig. 3.56. Excluded solvent formed upon contact between two ions.

$$U_{ij}(r) = \frac{e_i e_j}{\varepsilon r} + COR_{ij} + CAV_{ij} + GUR_{ij} \tag{3.182}$$

Here COR_{ij} represents a short-range repulsion, the second term in the Lennard-Jones equation, or it is sometimes substituted for by a term becoming exponentially more positive as the interionic distance diminishes. CAV_{ij} is a term that accounts for the fact that the ions themselves are dielectric cavities in the solution that have a dielectric constant (<5) differing dramatically from that of the solution (~80). Thus, owing to this, one acts on the other in a way not allowed for in the first term in Eq. (3.182). (On the whole, this term is less important than the others.) Finally, the GUR_{ij} term gets its name from Gurney (Gurney, 1953), who stressed the existence of a primary solvation layer (Bockris, 1949). In effect, as far as interactions are concerned, this is like a hard rubber tire around the ion. Upon collision (*cf.* the mound model), two ions squelch into each other, knocking out a water molecule or two and making the distance of closest approach less than the sum of the hydrated radii. This hydrated layer around the ion, called by Bockris the "primary hydration layer," is called by Friedman the "Gurney co-sphere" (Fig. 3.56).

Figure 3.57 (originally drawn by Friedman) shows clearly how these various contributions to U_{ij} vary with distance as two ions approach each other. The diagram indicates

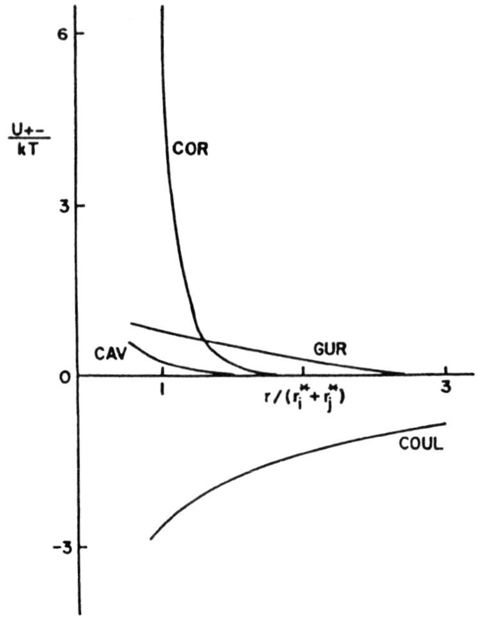

Fig. 3.57. Terms of the model potential for a (+ −) pair in units of kT for an aqueous 1:1 electrolyte with $r_{+}^{*} + r_{-}^{*} = 0.28$ nm, $\varepsilon_{CAV} = 2$, and $(A_g)_{+-} = 100$ cal/mol. (Reprinted from H. L. Friedman, *Pure Appl. Chem.* **59**: 1003, 1987.)

the complexity of ionic solution theory when the concentration is so high that most of the energies involved are effective the whole time and not only on closest approach.

Further Reading

Seminal

1. R. Gurney, *Ions in Solution*, Cambridge University Press, Cambridge (1940).
2. H. L. Friedman, *Ionic Solution*, Interscience, New York (1962).
3. C. W. Davies, *Ionic Association*, Butterworth, London (1962).
4. D. A. McQuarie, *Statistical Mechanics*, pp. 274–277, 355–356, Harper Collins, New York (1969).
5. H. L. Friedman, Computations of ionic properties, *Modern Aspects of Electrochemistry*, Vol. 6, J. O'M Bockris and B. E. Conway, eds., Plenum, New York (1971).
6. J. C. Rasaiah, "Comparison of Models for Ions in Solution," *J. Chem. Phys.* **56**: 5061 (1972).
7. L. Blum, *Mol. Phys.* **30**: 1529 (1975).

Reviews

1. K. S. Pitzer, "Activity Coefficients in Electrolyte Solutions," in *Activity Coefficients in Electrolyte Systems*, K. Pitzer, ed., 2nd ed., CRC Press, Boca Raton, FL (1991).
2. J. E. Enderby, *Chem. Soc. Review* **24**: 159 (1995).
3. G. W. Neilson and J. E. Enderby, "Aqueous Solutions," *J. Phys. Chem.* **100**: 1317 (1996).

Papers

1. J. O'M. Bockris, *Quant. Rev. Chem. Soc., London* **111**: 173 (1949).
2. R. Gurney, *J. Chem. Phys.* **19**: 1499 (1953).
3. J. C. Rasaiah and H. L. Friedman, *J. Chem. Phys.* **48**: 2742 (1968).
4. E. L. Blum, *Inorg. Chem.* **24**: 139 (1977).
5. P. J. Rossky and H. L. Friedman, *J. Chem. Phys.* **72**: 5694 (1980).
6. G. K. Wertheim, *Phys. Rev.* **30**: 4343 (1984).
7. H. L. Friedman, *NATO ASI Series C* **205**: 61 (1987).
8. K. V. Ramanathan and R. Pinto, *J. Electrochem. Soc.* **134**: 165 (1987).
9. J. Barthel and R. Buckner, *Pure Appl. Chem.* **63**: 1473 (1991).
10. Y. Wei, P. Chiang, and S. Sridhar, *J. Chem. Phys.* **96**: 4569 (1992).
11. J. F. Lu, Y. Yu, and Y. G. Li, *Fluid Phase Equilibria*, **85**: 81 (1993).
12. D. H. Powell, G. W. Nielson, and J. E. Enderby, *J. Phys. Condens. Matter* **5**: 5723 (1993).
13. R. H. Tromp, G. W. Nielson, and M. L. Bellisent-Fund, *J. Phys. Chem.* **98**: 13195 (1994).
14. S. Angell, R. H. Tromp and G. W. Nielson, *J. Phys. Condens. Matter* **7**: 1513 (1995).
15. J. Barthel, H. J. Gores, and L. Kraml, *J. Phys. Chem.* **100**: 1283 (1996).
16. W. R. Fawcett and A. C. Tikanen, *J. Phys. Chem.* **100**: 4251 (1996).

3.15. SPECTROSCOPIC APPROACHES TO THE CONSTITUTION OF ELECTROLYTIC SOLUTIONS

An understanding of the concentration dependence of activity coefficients required the postulation of the concepts of ion-pair formation and complex formation. Certain structural questions, however, could not be answered unequivocally by these considerations alone. For instance, it was not possible to decide whether pure Coulombic or chemical forces were involved in the process of ion association, i.e., whether the associated entities were ion pairs or complexes. The approach has been to postulate one of these types of association, then to work out the effect of such an association on the value of the activity coefficient, and finally to compare the observed and calculated values. Proceeding on this basis, it is inevitable that the postulate will always stand in need of confirmation because the path from postulate to fact is indirect.

It is fortunately possible to gain direct information on the constitution and concentrations of the various species in an electrolytic solution by "seeing" the species. These methods of seeing are based upon shining "light"[34] into the electrolytic solution

[34] The word *light* is used here to represent not only the visible range of the electromagnetic spectrum but all the other ranges for which analytical methods have been developed.

and studying the light that is transmitted, scattered, or refracted. The light emerging from the electrolytic solution in the ways just mentioned is altered or modulated as a result of the interaction between the free ionic or associated species and the incident light (*cf.* the corresponding use of light as a tool in Section 2.3.3).

There are many types of interactions between electromagnetic radiation and matter. A species as a whole can change its rotational state; the bonds (if any) within a species can bend, stretch, or twist and thus alter its vibrational state; the electrons in the species may undergo transitions between various energy states; and finally the atomic nuclei of the species can absorb energy from the incident radiation by making transitions between the different orientations to an externally imposed magnetic field.

All these responses of a species to the stimulus of the incident electromagnetic radiation involve energy exchange. According to the quantum laws, this energy exchange must occur only in finite jumps of energy. Hence, the light that has been modulated by these interactions contains information about the energy that has been exchanged between the incident light and the species present in the electrolytic solution. If the free, unassociated ions and the associated aggregates (ion pairs, triplets, complex ions, etc.) interact differently with the incident radiation, then information concerning the structures will be observed in the radiation emerging from the electrolytic solution.

There are various kinds of spectroscopy: visible and ultraviolet (UV) absorption spectroscopy, Raman and infrared spectroscopy, nuclear magnetic resonance spectroscopy, and electron-spin resonance (ESR) spectroscopy. A brief description of the principles of these techniques and their application to the study of ions in solution follows (see also Section 2.11).

3.15.1 Visible and Ultraviolet Absorption Spectroscopy

Atoms, neutral molecules, and ions (simple and associated) can exist in several possible electronic states; this is the basis of visible and UV absorption spectroscopy. Transitions between these energy states occur by the absorption of discrete energy quanta ΔE which are related to the frequency ν of the light absorbed by the well-known relation

$$\Delta E = h\nu$$

where h is Planck's constant. When light is passed through the electrolytic solution, there is absorption at the characteristic frequencies corresponding to the electronic transitions of the species present in the solution. Owing to the absorption, the intensity of transmitted light of the absorption frequency is less than that of the incident light. The falloff in intensity at a particular wavelength is given by an exponential relation known as the Beer–Lambert law

$$I = I^0 e^{-\varepsilon c l}$$

where I^0 and I are the incident and transmitted intensities, ε is a characteristic of the absorbing species and is known as *molar absorptivity*, l is the length of the material (e.g., electrolytic solution) through which the light passes, and c is the concentration of the absorbing molecules.

A historic use of the Beer–Lambert law was by Bjerrum, who studied the absorption spectra of dilute copper sulfate solutions and found that the molar absorptivity was independent of the concentration. Bjerrum concluded that the only species present in dilute copper sulfate solutions are free, unassociated copper and sulfate ions and not, as was thought at the time, undissociated copper sulfate molecules that dissociate into ions to an extent that depends on the concentration. For if any undissociated molecules were present, then the molar absorptivity of the copper sulfate solution would have been dependent on the concentration.

In recent years, it has been found that the molar absorptivity of *concentrated* copper sulfate solutions does show a slight concentration dependence. This concentration dependence has been attributed to ion-pair formation occurring through the operation of Coulombic forces between the copper and sulfate ions. This is perhaps ironic because Bjerrum's concept of ion pairs is being used to contradict his conclusion that there are only free ions in copper sulfate solutions. Nevertheless, there is a fundamental difference between the erroneous idea that a copper sulfate crystal dissolves to give copper sulfate molecules, which then dissociate into free ions, and the modern point of view that the ions of an ionic crystal pass into solution as free solvated ions which, under certain conditions, associate into ion pairs.

The method of visible and UV absorption spectroscopy is at its best when the absorption spectra of the free ions and the associated ions are quite different and known. When the associated ions cannot be chemically isolated and their spectra studied, the type of absorption by the associated ions has to be attributed to electronic transitions known from other well-studied systems. For example, there can be an electron transfer to the ion from its immediate environment (charge-transfer spectra), i.e., from the entities associated with the ion; or transitions between new electronic levels produced in the ion under the influence of the electrostatic field of the species associated with the ion (crystal-field splitting). Thus, there is an influence of the environment on the absorption characteristics of a species, and this influence reduces the clarity with which spectra are characteristic of species rather than of their environment. Herein lies what may be considered a disadvantage of visible and UV absorption spectroscopy.

3.15.2. Raman Spectroscopy

Visible and UV absorption spectroscopy are based on studying that part of the incident light transmitted (after absorption) through an electrolytic solution in the same direction as the original beam. However, a certain amount of light is scattered in other directions.

A simple view of the origin of the Raman effect (see also Section 2.11.5) is as follows: *Rayleigh scattering* is produced because the electric field of the incident light induces a dipole moment in the scattering species and since the incident field is oscillating, the induced dipole moment also varies periodically. Such an oscillating dipole acts as an antenna and radiates light of the same frequency as the incident light.

It is the deformation polarizability α_{deform} that determines the magnitude of a dipole moment induced by a particular field. If this polarizability changes from its time-averaged value, then the induced-dipole antenna will be radiating at a new frequency that is different from the incident frequency; in other words, there is a Raman shift. The changes in polarizability of the scattering species can be correlated with their rotations and vibrations and also their symmetry characteristics. Hence, the Raman shifts are characteristic of the rotational and vibrational energy levels of the scattering species and provide direct information about these levels. Though the presence of electrostatic bonding produces second-order perturbations in the Raman lines, it is the species with covalent bonds to which Raman spectroscopy is sensitive, and not purely electrostatic ion pairs. Another feature that makes Raman spectra useful is the fact that the integrated intensity of the Raman line is proportional to the concentration of the scattering species, i.e., allows the amount present to be determined.

The problem with Raman spectroscopy is the low intensity of the Raman lines, which permits easy detection of species only in concentrated solutions. Fortunately, the availability of lasers, which are intense sources of monochromatic light, is stimulating further applications of this powerful technique, which is noted for the lack of ambiguity with which it can report on the species in an electrolytic solution. Devices that distinguish a low-intensity signal from noise also help.

3.15.3. Infrared Spectroscopy

Infrared spectroscopy resembles Raman spectroscopy in that it provides information on the vibrational and rotational energy levels of a species, but it differs from the latter technique in that it is based on studying the light transmitted *through* a medium after absorption and not that *scattered* by it (see Section 2.11.2).

The techniques of Raman and IR spectroscopy are generally considered complementary in the gas and solid phases because some of the species under study may reveal themselves in only one of the techniques. Nevertheless, it must be stressed that Raman scattering is not affected by an aqueous medium, whereas the strong absorption in the infrared shown by water proves to be a troublesome interfering factor in the study of aqueous solutions by the IR method.

3.15.4. Nuclear Magnetic Resonance Spectroscopy

The nuclei of atoms can be likened in some respects to elementary magnets. In a strong magnetic field, the different orientations that the elementary magnets assume correspond to different energies. Thus, transitions of the nuclear magnets between these different energy levels correspond to different frequencies of radiation in the short-wave, radio-frequency range. Hence, if an electrolytic solution is placed in a strong magnetic field and an oscillating electromagnetic field is applied, the nuclear magnets exchange energy (exhibit resonant absorption) when the incident frequency equals that for the transitions of nuclei between various levels (see Section 2.11.6).

Were this NMR to depend only on the nuclei of the species present in the solution, the technique could not be used to identify ion pairs. However, the nuclei sense the applied field as modified by the environment of the nuclei. The modification is almost exclusively due to the nuclei and electrons in the neighborhood of the sensing nucleus, i.e., the adjacent atoms and bonds. Thus, NMR studies can be used to provide information on the type of association between an ion and its environmental particles, e.g., in ion association.

Further Reading

Seminal Papers

1. M. Falk and P. A. Giguere, "I.R. Spectra in Solution," *Can. J. Chem.* **35**: 1195 (1957).
2. H. G. Hertz, "N.M.R. Spectra of Solutions," *Ber. Bunsen. Ges.* **67**: 311 (1963).
3. G. E. Walrafen, "Raman Spectra in Solution," *J. Chem. Phys.* **40**: 3249 (1964).

Review

1. P. L. Goggin and C. Carr, "Infrared Spectroscopy in Aqueous Solution," in *Water and Aqueous Solutions*, G. W. Nielson and J. E. Enderby, eds., Adam Hilger, Bristol, U.K. (1986).

Papers

1. H. J. Reich, J. P. Borst, R. R. Dykstra, and D. P. Green, *J. Am. Chem. Soc.* **115**: 8728 (1993).
2. E. Vauthey, A. W. Parker, B. Nohova, and D. Phillips, *J. Am. Chem. Soc.* **116**: 9182 (1994).
3. Y. Wang and Y. Tominaga, *J. Chem. Phys.* **101**: 3453 (1994).
4. F. Mafune, Y. Hashimoto, M. Hashimoto, and T. Kondow, *J. Phys. Chem.* **99**: 13814 (1995).
5. D. C. Duffy, P. B. Davies, and A. M. Creeth, *Langmuir* **11**: 2931 (1995).
6. P. W. Faguy, N. S. Marinkovic, and R. R. Adzic, *J. Electroanal. Chem.* **407**: 209 (1996).
7. V. Razumas, K. Larsson, Y. Miezis, and T. Nylander, *J. Phys. Chem.* **100**: 11766 (1996).
8. X. M. Ren and P. G. Pickup, *Electrochim. Acta* **41**: 1877 (1996).

3.16. IONIC SOLUTION THEORY IN THE TWENTY-FIRST CENTURY

Looking first back to the publication date of the famous theory of Debye and Hückel (1923), there is no doubt that their ionic-atmosphere calculation (Section 3.3) of the activity coefficient at very low concentrations is still the dominating peak of this century in the field of ionic solutions. Not only is the theory effective in giving the theoretical means to calculate ionic activity coefficients at low concentrations, but its success also had ripple effects on other parts of the physical chemistry of solutions.

Bjerrum published his work on the calculations of the association of ions formed from salts (Section 3.8) in 1926 and then nothing much happened until the Mayer theory of 1950. The development of this exceedingly complex piece of statistical mechanics, largely by Friedman and Rasaiah, produced works of ever-increasing mathematical difficulty, but today these look much more like exercises in very difficult integrations rather than insights into what is really happening. The enormous improvement over the Debye and Hückel equations with which the virial approach was credited was based on a comparison with the most simple ("limiting law") equation of Debye and Hückel, rather than later versions of their theory. These later versions take into account the volume occupied by ions, ion pairing, and the removal of solvent in the solution from its natural role as a dissolver to spend significant time clamped tightly to ions, unable to move and act as a solvent to newly introduced ions. On the basis of this model, Stokes and Robinson were able to calculate good numerical values for activity coefficients in solutions of concentrations up to 10 mol dm^{-3}.

By the 1980s, it had to be admitted (Rasaiah, 1987) that several versions of the post-1950 theories (mostly developments based on Mayer's theory) could not be distinguished as to virtue when ranked by their ability to replicate experiments, nor were they markedly better than those obtainable from later developments of Debye and Hückel's ionic atmosphere theory.

However, there is no doubt that from the 1980s on, a very hopeful type of development has been taking place in ionic solution theories. It is the correlation function approach, not a theory or a model, but an open-ended way to obtain a realistic idea of how an ionic solution works (Fig. 3.58). In this approach, pair correlation functions that are experimentally determined from neutron diffraction measurements represent "the truth," without the obstructions sometimes introduced by a model. From a knowledge of the pair correlation function, it is possible to calculate properties (osmotic pressure, activities). The pair correlation function acts as an ever-ready test for new models, for the models no longer have to be asked to re-replicate specific properties of solutions, but can be asked to what degree they can replicate the known pair correlation functions.

It is rather easy to make a list of milestones in ionic solution theory:

- The peaklike beginning in 1923.
- The MM stage,[35] the theory of ionic solution based on concepts used to interpret the behavior of imperfect gases (1950–1980). During this long stage (a

[35]MM refers to the McMillan-Mayer theory of 1945. This was a general theory of interaction in gases which was applied to interpret the behaviors of solutions. Mayer's theory was produced in 1950 [J.E. Mayer, *J. Chem. Phys.* **18**: 1426 (1950)] and applied the MM theory particularly to ionic solutions where the dominance of long range interactions causes special mathematical difficulties.

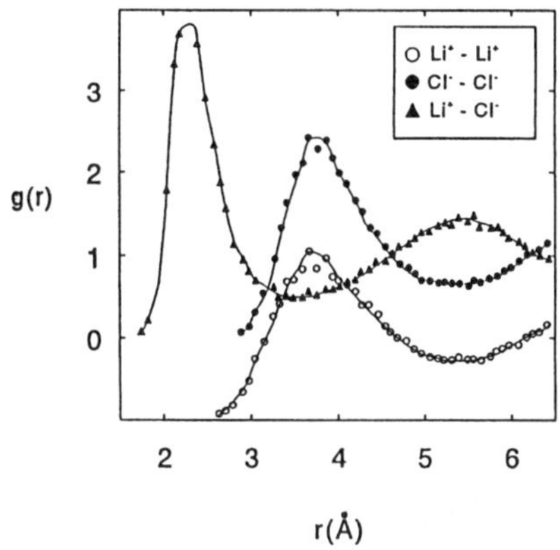

Fig. 3.58. Pair correlation functions from Monte Carlo simulation for charged soft sphere model for LiCl liquid at 883 K and 28.3 $cm^3 mol^{-1}$. (Reprinted from I. R. MacDonald and K. Singer, *Chemistry in Britain* **9**: 54, 1973.)

generation!), there developed an increasingly complex series of additions to the interaction energy U_{AB} which had first begun at the simplest level of $z_1z_2e_0^2/\varepsilon r$ but grew to encompass several extra terms, some of which had only a minimal effect on results.

- Pair correlation functions were discussed from the 1960s on, but not until the 1980s with the neutron diffraction contribution of Enderby and Neilson was it possible to make such determination with relative ease and certainty.
- Computer simulations began nearly half a century ago, but they did not really come into wide use for ionic solutions until the late 1980s (Heinziger, Bopp, Haymet). The convenience of being able to calculate the consequences of assuming a certain pair interaction law, to simulate an effect (the influence of high temperature and pressure, for example), and then find via the pair correlation functions the corresponding properties is a major advance because it brings a great increase in speed of obtaining new knowledge and a decrease in its expense.

There is yet a further stage for ionic solution theory that will not be presented here because it is still fragmentary in achievement and densely complex in the physical theory that underlies it. This "final" stage is the quantum mechanical one, in which an attempt is made to describe a solution at the Schrödinger equation level. Such work has been pioneered by Clementi and his colleagues since the mid-1980s.

APPENDIX 3.1. POISSON'S EQUATION FOR A SPHERICALLY SYMMETRICAL CHARGE DISTRIBUTION

A starting point for the derivation of Poisson's equation is Gauss's law, which can be stated as follows: Consider a sphere of radius r. The electric field X_r due to charge in the sphere will be normal to the surface of the sphere and equal everywhere on this surface. The total field over the surface of the sphere (i.e., the surface integral of the normal component of the field) will be equal to X_r times the area of the surface, i.e., $X_r 4\pi r^2$. According to Gauss's law, this surface integral of the normal component of the field is equal to $4\pi/\varepsilon$ times the total charge contained in the sphere. If ϱ_r is the charge density at a distance r from the center of the sphere, then $4\pi r^2 \varrho_r\, dr$ is the charge contained in a dr-thick shell at a distance r from the center and $\int_0^r 4\pi r^2 \varrho_r\, dr$ is the total charge contained in the sphere. Thus, Gauss's law states that

$$X_r 4\pi r^2 = \frac{4\pi}{\varepsilon} \int_0^r 4\pi r^2 \varrho_r\, dr$$

i.e.,

$$r^2 X_r = \frac{4\pi}{\varepsilon} \int_0^r r^2 \varrho_r\, dr \tag{A3.1.1}$$

Now, according to the definition of electrostatic potential ψ_r, at a point r,

$$\psi_r = -\int_\infty^r X_r\, dr \tag{A3.1.2}$$

or

$$X_r = -\frac{d\psi_r}{dr} \tag{A3.1.3}$$

Substituting Eq. (A3.1.2) in Eq. (A3.1.1), one gets

$$r^2 \frac{d\psi_r}{dr} = -\frac{4\pi}{\varepsilon} \int_0^r r^2 \varrho_r\, dr \tag{A3.1.4}$$

Differentiating both sides with respect to r,

$$\frac{d}{dr}\left(r^2 \frac{d\psi_r}{dr}\right) = -\frac{4\pi}{\varepsilon} r^2 \varrho_r$$

i.e.,

$$\frac{1}{r^2}\frac{d}{dr}\left(r^2\frac{d\psi_r}{dr}\right)=-\frac{4\pi}{\varepsilon}\varrho_r \tag{A3.1.5}$$

which is Poisson's equation for a spherically symmetrical charge distribution.

APPENDIX 3.2. EVALUATION OF THE INTEGRAL $\int_{r=0}^{r\to\infty} e^{-\kappa r}(\kappa r)\, d(\kappa r)$

According to the rule for integration by parts,

$$\int v\, du = uv - \int u\, dv$$

Thus,

$$\int_{r=0}^{r\to\infty}(\kappa r)e^{-\kappa r}\, d(\kappa r)=\left[-\kappa re^{-\kappa r}-\int\frac{e^{-\kappa r}}{(-1)}d(\kappa r)\right]_{r=0}^{r\to\infty}$$

$$=\left[-\kappa re^{-\kappa r}-e^{-\kappa r}\right]_{r=0}^{r\to\infty}$$

$$=\left[-e^{-\kappa r}(\kappa r+1)\right]_{r=0}^{r\to\infty}$$

$$=+1 \tag{A3.2.1}$$

APPENDIX 3.3. DERIVATION OF THE RESULT $f_{\pm}=(f_+^{\nu_+}f_-^{\nu_-})^{1/\nu}$

Suppose that on dissolution 1 mole of salt gives rise to ν_+ moles of positive ions and ν_- moles of negative ions. Then, instead of Eqs. (3.65) and (3.66), one has

$$\nu_+\mu_+=\nu_+\mu_+^0+\nu_+RT\ln x_++\nu_+RT\ln f_+ \tag{A.3.3.1}$$

and

$$\nu_-\mu_-=\nu_-\mu_-^0+\nu_-RT\ln x_-+\nu_-RT\ln f_- \tag{A.3.3.2}$$

Upon adding the two expressions (A3.3.1) and (A3.3.2) and dividing by $\nu=\nu_++\nu_-$ to get the average contribution to the free-energy change of the system per mole of both positive and negative ions, the result is

$$\mu_{\pm} = \frac{\nu_+\mu_+ + \nu_-\mu_-}{\nu} = \frac{\nu_+\mu_+^0 + \nu_-\mu_-^0}{\nu} + RT\ln(x_+^{\nu_+}x_-^{\nu_-})^{1/\nu} + RT\ln(f_+^{\nu_+}f_-^{\nu_-})^{1/\nu} \quad \text{(A3.3.3)}$$

which may be written as

$$\mu_{\pm} = \mu_{\pm}^0 + RT\ln x_{\pm} + RT\ln f_{\pm} \quad \text{(A3.3.4)}$$

where

$$\mu_{\pm}^0 = \frac{\nu_+\mu_+^0 + \nu_-\mu_-^0}{\nu} \quad \text{(A3.3.5)}$$

$$x_{\pm} = (x_+^{\nu_+}x_-^{\nu_-})^{1/\nu} \quad \text{(A3.3.6)}$$

and

$$f_{\pm} = (f_+^{\nu_+}f_-^{\nu_-})^{1/\nu} \quad \text{(A3.3.7)}$$

APPENDIX 3.4. TO SHOW THAT THE MINIMUM IN THE P_r VERSUS r CURVE OCCURS AT $r = \lambda/2$

From

$$P_r = 4\pi n_i^0 e^{\lambda/r} r^2 \quad \text{(A3.4.1)}$$

one finds the minimum in the P_r versus r curve by differentiating the expression for P_r with respect to r and setting the result equal to zero. That is,

$$\frac{dP_r}{dr} = 4\pi n_i^0 e^{\lambda/r} 2r - 4\pi n_i^0 r^2 e^{\lambda/r} \frac{\lambda}{r^2} = 0 \quad \text{(A3.4.2)}$$

i.e.,

$$2r_{\min} - \lambda = 0$$

or

$$r_{\min} = q = \frac{\lambda}{2} \quad \text{(A3.4.3)}$$

APPENDIX 3.5. TRANSFORMATION FROM THE VARIABLE r TO THE VARIABLE $y = \lambda/r$

From $y = \lambda/r$, it is obvious that

$$r^2 = \frac{\lambda^2}{y^2} \tag{A3.5.1}$$

$$dr = -\frac{\lambda}{y^2}\,dy \tag{A3.5.2}$$

$$e^{\lambda/r} = e^y \tag{A3.5.3}$$

Further, when $r = q = \lambda/2$, it is clear that

$$y = 2 \tag{A3.5.4}$$

and, when $r = a$, it follows that

$$y = \frac{\lambda}{a} = b \tag{A3.5.5}$$

By introducing the substitutions (A3.5.1), (A3.5.2), (A3.5.3), (A3.5.4), and (A3.5.5) into Eq. (3.145), the result is

$$\theta = 4\pi n_i^0 \lambda^3 \int_{y=2}^{y=b} e^y y^{-4}\,dy$$

or

$$\theta = 4\pi n_i^0 \left(\frac{z_+ z_- e_0^2}{\varepsilon \kappa T}\right)^3 \int_{y=2}^{y=b} e^y y^{-4}\,dy \tag{A3.5.6}$$

APPENDIX 3.6. RELATION BETWEEN CALCULATED AND OBSERVED ACTIVITY COEFFICIENTS

Consider a solution consisting of 1 kg of water in which is dissolved m moles of 1:1-valent salt. Ignoring the question of ion-pair formation, there are m moles of positive ions and m moles of negative ions in the solution. Hence, the total free energy G of the solution is given by

$$G = \left(\frac{1000}{M_{H_2O}}\right)\mu_{H_2O} + m\mu_+ + m\mu_- \tag{A3.6.1}$$

where M_{H_2O} is the molecular weight of the water and μ_{H_2O}, μ_+, and μ_- are the chemical potentials of water, positive ions, and negative ions, respectively.

Now, if a fraction θ of the positive and negative ions form ion pairs designated $(+\,-)$, then the free energy of the solution assuming ion-pair formation is

$$G = \left(\frac{1000}{M_{H_2O}}\right)\mu_{H_2O} + (1-\theta)m\mu'_+ + (1-\theta)m\mu'_- + \theta m\mu'_{+-} \tag{A3.6.2}$$

where the primed chemical potentials are based on ion association.

Since, however, the free energy of the solution must be independent of whether ion association is considered, Eqs. (A3.6.1) and (A3.6.2) can be equated. Hence

$$\mu_+ + \mu_- = (1-\theta)(\mu'_+ + \mu'_-) + \theta\mu'_{+-} \tag{A3.6.3}$$

But ions are in equilibrium with ion pairs, and therefore

$$\mu_{+-} = \mu'_+ + \mu'_- \tag{A3.6 4}$$

Combining Eqs. (A3.6.4) and (A3.6.3), one has

$$\mu_+ + \mu_- = \mu'_+ + \mu'_- \tag{A3.6.5}$$

or

$$\mu_+^0 + RT\ln\gamma_+ m + \mu_-^0 + RT\ln\gamma_- m = \mu_+^{0\prime} + RT\ln\gamma'_+(1-\theta)m + \mu_-^{0\prime}$$
$$+ RT\ln\gamma'_-(1-\theta)m \tag{A3.6.6}$$

It is clear, however, that $\mu_+^0 = \mu_+^{0\prime}$ and $\mu_-^0 = \mu_-^{0\prime}$ since, whether ion association is considered or not, the standard chemical potentials of the positive and negative ions are the chemical potentials corresponding to ideal solutions of unit molality of these ions. Hence

$$\gamma_+\gamma_- = (1-\theta)^2\gamma'_+\gamma'_- \tag{A3.6.7}$$

or

$$\gamma_{\pm} = (1 - \theta)\gamma'_{\pm} \tag{A3.6.8}$$

But $\gamma_{\pm}$ is the observed or stoichiometric activity coefficient and therefore

$$(\gamma_{\pm})_{\text{obs}} = (1 - \theta)\gamma'_{\pm} \tag{A3.6.9}$$

Similarly,

$$(f_{\pm})_{\text{obs}} = (1 - \theta)f_{\pm} \tag{3.160}$$

EXERCISES

1. Calculate the potentials due to the ionic cloud around cations in the following solutions: (a) 10^{-3} *M* NaCl, (b) 0.1 *M* NaCl, (c) 10^{-3} *M* $CaCl_2$, and (d) 10^{-3} *M* $CaSO_4$. (Kim)

2. For 0.001 *M* aqueous KCl solution, calculate the total potential in the ionic atmosphere and then that part of the total potential due to the ionic cloud.

3. Calculate the mean activity coefficient of a 0.002 *M* $LaCl_3$ solution, assuming that at this low concentration there is a negligible association between cations and anions.

4. What is the electrical work done in charging a body up to a charge q and a potential ψ? Prove the expression.

5. Compare the molarities and ionic strengths of 1:1, 2:1, 2:2, and 3:1 valent electrolytes in a solution of molarity c. (Constantinescu)

6. Calculate the ionic strength of the following solutions: (a) 0.04 *M* KBr, (b) 0.35 *M* $BaCl_2$, and (c) 0.02 *M* Na_2SO_4 + 0.004 *M* Na_3PO_4 + 0.01 *M* $AlCl_3$. (Constantinescu)

7. A 0.001 *M* $BaCl_2$ solution is mixed with an 0.001 *M* Na_2SO_4 solution. Calculate the ionic strength.

8. Assuming that a 10% error can be tolerated, calculate from the Debye–Hückel limiting law the highest concentration at 25°C at which activity can be replaced by concentration in the cases of (a) NaCl and (b) $CaSO_4$ solutions. (Constantinescu)

9. Calculate the activity coefficients in the following solution at 25°C by the Debye–Hückel limiting law: (a) 10^{-3} *M* NaCl, (b) 0.1 *M* NaCl, (c) 10^{-3} *M* $CaCl_2$, and (d) 10^{-3} *M* $CaSO_4$. (Kim)

TABLE P.1

KCl mol kg^{-1}	0.025	0.050	0.10	0.20
TlCl mol kg^{-1}	8.69×10^{-3}	5.90×10^{-3}	3.96×10^{-3}	2.68×10^{-3}

10. Calculate the activity coefficient of a 10^{-2} *M* NaCl solution by the extended Debye–Hückel limiting law with an ion-size parameter of 400 pm. (Kim)

11. Calculate the effect of water molecules on the activity coefficient in 1 *M* NaCl. The water activity in the solution is 0.96686 and the hydration number of the electrolyte is 3.5. The density of the solution is 1.02 g cm^{-3}. (Kim)

12. Calculate the mean activity coefficient of thallous chloride, the solubility of which has been measured in water and in the presence of various concentrations of potassium chloride solutions at 25 °C, as given in Table P.1. The solubility of this salt in pure water is 1.607×10^{-2} mol kg^{-1}. (Constantinescu)

13. Calculate the solubility product of $C_2O_4Ag_2$ if its solubility in pure water at 25 °C is 8×10^{-5} mol dm^{-3}. (Constantinescu)

14. Calculate the mean activity coefficient at 25 °C of 0.01 *N* solutions of (a) HCl, (b) $ZnCl_2$, and (c) $ZnSO_4$. (Constantinescu)

15. Utilize the known values of the Debye–Hückel constants *A* and *B* for water to calculate the mean activity coefficients for 1:1, 1:2, and 2:2-valent electrolytes in water at the ionic strengths 0.1 and 0.01 at 25 °C. The mean distance of closest approach of the ions *a* may be taken as 300 pm in each case. (Constantinescu)

16. Draw a diagram representing the Gurney co-sphere. When two ions collide, how does this sphere influence the parameters that go into the calculation of the activity coefficient?

17. In the text, an explanation is given which shows how the mean activity coefficient of an electrolyte can be given if the activity of water in the solution is known. Suppose you know the vapor pressure of water over the solution of an electrolyte. Could you still get the activity coefficient of the electrolyte in solution? How? Illustrate your answer by using the data in Table 3.5.

18. Calculate the chemical-potential change of the cations in the following solutions by the Debye–Hückel theory: (a) 10^{-3} *M* NaCl, (b) 0.1 *M* NaCl, (c) 10^{-3} *M* $CaCl_2$, and (d) 10^{-3} *M* $CaSO_4$. (Kim)

19. Use the values of the Debye–Hückel constants *A* and *B* at 25 °C to calculate $-\log f_{\pm}$ for a 1:1-valent electrolyte for ionic strengths 0.01, 0.1, 0.5, and 1.0,

assuming in turn that the mean distance of closest approach of the ions, a, is either zero or 100, 200, and 400 pm. (Constantinescu)

20. Electrolytes for which the concentration is less than 10^{-3} M can usually be dealt with by the Debye–Hückel limiting law. Utilize the Debye–Hückel theory extended by allowance for ion size and also for removal of some of the active solvent into the ion's primary solvation shell to calculate the activity coefficient of 5 M NaCl and 1 M $LaCl_3$ solutions (neglecting ion association or complexing). Take the total hydration number at the 5 M solution as 3 and at the 1 M solution as 5. Take r_i as 320 pm.

21. Use the Debye–Hückel limiting law to derive an expression for the solvent activity in dilute solutions. (Xu)

22. Use the above results to calculate the activity coefficients for water (Table 2.8). (Xu)

23. Calculate the Debye–Hückel reciprocal lengths for the following solutions: (a) 10^{-3} M NaCl, (b) 0.1 M NaCl, (c) 10^{-3} M $CaCl_2$, and (d) 10^{-3} M $CaSO_4$. (Kim)

24. Calculate the thickness of the ionic atmosphere in 0.1 N solutions of a uni-univalent electrolyte in the following solvents: (a) nitrobenzene ($\varepsilon = 34.8$), (b) ethyl alcohol ($\varepsilon = 24.3$), and (c) ethylene dichloride ($\varepsilon = 10.4$) at 25 °C. (Constantinescu)

25. Calculate the minimum distance at which the attractive electrostatic energy of the ions having charges z_+ and z_- in an electrolyte is greater than the thermal energy of an interacting cation and anion.

26. Determine the mean distance between ions in a solution of concentration c for a 1:1 electrolyte. Calculate $1/\kappa$ for a solution of 0.01 mol dm^{-3}.

27. Explain why κ^{-1} is called "the Debye length." Draw a figure that shows what this term means in terms of the concept of an ionic atmosphere.

28. Assume as a rough approximation that the distance of closest approach between two ions a is the sum of the ionic radii of cation and anion plus the diameter of water. Calculate κa for concentrations of CsCl of 10^{-4}, 5×10^{-4}, 10^{-3}, 5×10^{-3}, 10^{-2}, 5×10^{-2}, and 10^{-1} M at 25 °C and find the maximum concentration at which $\kappa a < 0.1$, i.e., the limiting law is applicable.

29. Prove that the ionic strength is actually a parameter reflecting the electrostatic field in the solution. (Hint: Show the effect of ionic valence charge on I.) (Xu)

30. Compute the ionic activity of 0.0001 M KCl. What change will be caused to the activity of KCl if 0.01 mole of $ZnCl_2$ is added to 1000 ml of the above solution? With the added salt, will the Debye–Hückel reciprocal length κ^{-1} increase or decrease? Why? (Xu)

31. Calculate the ion association constant K_A of CsCl in ethanol ($\varepsilon = 24.3$) according to Bjerrum's concept. $r(Cs^+) = 1.69$ Å and $r(Cl^-) = 1.81$ Å. (Kim)

32. Write down the full Taylor–MacClaurin expansion of e^x and e^{-x}. Find the difference between e^x and the expansion for $x = 0.1$, 0.5, and 0.9.

33. Derive the linearized Poisson–Boltzmann equation.

34. Write down a typical equation for the sum of the attractive and repulsive energies of an ion in an electrolyte. What is meant by the MSA approximation? Represent it diagrammatically and with equations.

35. Describe the basic principles of the Monte Carlo and molecular dynamic simulation methods applied to electrolytes. How are the adjustable parameters determined?

36. Explain in about 250 words the essential approach of the Mayer theory of ionic solutions and how it differs from the ionic-atmosphere view. The parent of Mayer's theory was the McMillan–Mayer theory of 1950. With what classical equation for imperfect gases might it be likened?

37. Among the outstanding contributors to the theory of ionic solutions after 1950 were Mayer, Friedman, Davies, Rasaiah, Blum, and Haymet. Write a few lines on the contributions due to each.

38. A number of acronyms are used in work on the theory of ion–ion interactions. Some are DHLL, HNC, MSA, COUL, and GUR. Give the full meaning of each and explain the significance of the concept it represents in the field of ion–ion interactions.

39. Spectroscopic methods are used to find out about the structure of an ionic solution. Write a few lines on what particular aspects of the solution structure are revealed by the use of the following kinds of spectra: (a) UV-visible, (b) Raman, (c) infrared, and (d) nuclear magnetic resonance.

PROBLEMS

1. Find the maximum potential of the ionic atmosphere at which the Poisson–Boltzmann equation can be linearized in a 1:1 electrolyte at 25 °C. After linearization, the charge density in the atmosphere is proportional to the potential. Can you see any fundamental objection to having ρ not proportional to ψ.

2. When the Debye–Hückel model was developed, an important hypothesis was made for mathematical convenience, i.e., $z_i e_0 \psi_r < kT$. Now, using an aqueous 1:1 electrolyte solution at 25 °C as an example, reassess the validity of the hypothesis. What is the physical nature of the hypothesis? (Xu)

TABLE P.2

Lowering of Vapor Pressure by Salts in Aqueous Solutions

$\frac{\text{g-mole salt}}{\text{liter of water}}$	KNO_3	KNO_2	LiBr	NaBr	NaI
0.5	10.3	11.1	12.2	12.6	12.1
1.0	21.1	22.8	26.2	24.9	25.6
2.0	40.1	44.8	60.0	57.0	60.2
3.0	57.6	67.0	97.0	89.2	99.5
4.0	74.5	90.0	140.0	124.2	136.7
5.0	88.2	110.0	186.3	159.5	177.3
6.0	102.1	130.7	241.5	197.5	221.0
8.0	126.3	167.0	341.5	268.0	301.5
10.0	148.0	198.8	438.0	—	370.0

Source: *CRC Handbook of Chemistry and Physics*, 58th ed., CRC, Boca Raton, FL

3. Explain from the Gibbs–Duhem equation how the determination of the activity of water can give rise to the activity of the electrolyte dissolved in it. Table P.2 shows data from the lowering of the vapor pressure of water by different salts at 100 °C at which $P^*_{H_2O} = 760$ torr. Plot $\log f_\pm$ against the ionic strength for each of the different salts. Discuss your results in the light of the data in Table P4.

4. Describe the difference between the Guntelberg and Debye charging processes. Which process should be more effective in deriving an equation for the activity coefficient?

5. Show that within the concentration range of the limiting law, the slope of the $\log f_\pm$ vs. $-\sqrt{I}$ line is 0.509 for aqueous solutions at 25 °C.

6. (a) The whitecoats built the cell below and managed to measure its potential at 25 °C as 0.35 V:

$$\text{Ag} \mid \text{AgCl(s)} \mid \text{AgCl(aq, saturated)} \mid$$
$$\mid \text{AgCl(aq, saturated in 0.05 } M \text{ CaCl}_2) \mid \text{AgCl(s)} \mid \text{Ag}$$

The solubility product of AgCl is known to be 1.77×10^{-10} at room temperature. Calculate the individual activity coefficient for Ag^+ in the $CaCl_2$ solution.
(b) Is there a way to obtain the individual activity coefficient for Cl^- in the same solution? (Xu)

7. The excess charge on an ionic atmosphere varies with distance out from the central ion. Show that the net change in the charge of a spherical shell of thickness dr is

$$dq = -z_i e_0 \, e^{-\kappa r} \kappa^2 r \, dr$$

Hence at a certain distance from the central ion, there will be a ring with a maximum charge. Find this distance.

8. Plot the values for the activity coefficient of the electrolyte as calculated in Problem 3 against the ionic strength. Then see what degree of match you can obtain from the Debye–Hückel law (one-parameter equation). Do a similar calculation with the equation in the text which brings in the distance of closest approach (a) and allows for the removal of water from the solution (two-parameter equation). Describe which values of a_i and n_h (the hydration number) fit best. Discuss the degree to which the values you had to use were physically sensible.

9. The radius of the ionic atmosphere is $1/\kappa$ where κ is defined in the text. Work out the average distance between ions (d) in terms of the concentration c_0, in mol dm^{-3}. If $d > 1/\kappa$, then one is confronting a situation in which the radius of the atmosphere is less than the average distance between ions. Describe what this means. Derive a general expression for c_0 at which this problem (coarse grainedness) occurs for a 2:2 electrolyte. Do you think an ionic atmosphere model applies when $d > 1/\kappa$?

10. Suppose the finite-sized center ion has a size parameter a; what will be the radial distribution of the total excessive charge $q(r)$ and $dq(r)$? Use the "correspondence principle" to confirm the validity of this expression. In the above case will the Debye–Hückel reciprocal length κ^{-1} depend on the ion size parameter a? (Xu)

11. Bjerrum's theory of ionic association gives rise to an expression for the fraction of ions in an ionic solution which are associated. Use the theory to calculate the degree of association of a 0.01 M $MgCl_2$ solution in ethanol ($\varepsilon = 32$).

12. Use the equations in the text that concern ion association. Find the concentration in an ethanol solution at which KCl manifests a 10% association. Compare the value you find with Haymet's calculations.

13. (a) As alternative to Bjerrum's model of ion association, you can use the results of the Debye–Hückel model to develop a qualitative but facile method for predicting ion-pair formation. (Hint: Compare the Debye–Hückel reciprocal length κ^{-1} and the effective distance for ion formation q.)
(b) In a lithium battery, usually a nonaqueous solution of lithium salts in organic solvents is used as an electrolyte. While the dielectric constants of such solvents range between 5 and 10, estimate the degree of ion-pair formation. (Xu)

14. The McMillan–Mayer theory is an alternative to the Debye–Hückel theory. It is called the virial coefficient approach and its equations bear some conceptual resemblance to the virial equation of state for gases. The key contribution in

applying the theory to electrolytes was made by Mayer, when he finally solved the equations by introducing a screening factor $e^{-\kappa r}$ when $1/\kappa$ is the Debye length. Plot $e_0/\varepsilon r$ and $e_0 e^{-\kappa r}/\varepsilon r$ against $1/r$ in the range $0.15 < r < 2.5$ nm for RbCl at 0.1 *M*. Determine the numerical effect of Mayer's screening factor.

15. Calculate $\log f_{\pm}$ from the Debye–Hückel equation with allowance for both the ion size and hydration terms for $AgNO_3$ solutions from 10^{-3} to 2 mol dm^{-3} and compare the result with those of the MSA equation:

$$\log f_{\pm} = \frac{e_0^2 z_i^2 \Gamma}{kT\varepsilon(1 + \Gamma\sigma)}$$

The term σ can be taken as the same as a_i in the Debye–Hückel expressions. Γ is defined in the text.

Compare each with the experimental values of Table P.3. Which is the more experimentally consistent approach?

16. Haymet and co-workers have calculated the mole fraction of dimers (associated ions) in electrolytic solutions, and some of their results are shown in Fig. 3.51. Use the equations of the Bjerrum theory applied to Na_3PO_4 and compare the results with those of the correlation function approach used by Haymet et al. The essential difference between the Haymet approach and that of Bjerrum is that

TABLE P.3

Molal Activity Coefficients of $AgNO_3$ Solutions at 25 °C

m	γ	m	γ
0.001	0.964	0.800	0.465
0.002	0.950	0.900	0.447
0.005	0.924	1.000	0.430
0.010	0.896	1.200	0.400
0.020	0.859	1.400	0.375
0.050	0.794	1.600	0.353
0.100	0.732	1.800	0.334
0.200	0.656	2.000	0.316
0.300	0.606	2.500	0.280
0.400	0.567	3.000	0.252
0.500	0.536	3.500	0.229
0.600	0.509	4.000	0.210
0.700	0.486	4.500	0.194

Source: Reprinted from W. Hamer and Y.-C. Wu, *J. Phys. Chem. Ref Data,* **1**: 1047 (1972).

the former more realistically involves repulsive as well as attractive forces. What is the percent difference in the two answers at 1 mol dm^{-3}?

17. What is the total charge on an ionic atmosphere around an anion of valence z? From the data in the text, examine $\log f_{\pm}$ vs. $\sqrt{m}$, where m is the molality of the solution, from 0 to 1 mol dm^{-3}. The plots always pass through a minimum. Use the fully extended Debye–Hückel theory, including the Bjerrum–Stokes and Robinson terms, to find the significance of the minimum at which the electrolyte concentration increases with the increase of the cation radius.

18. In most of the modern versions of the Debye–Hückel theory of 1923, it is still assumed that the dielectric constant to be used is that of water. The dielectric constant of solutions decreases linearly with an increase in the concentration of the electrolyte. Using data in the chapter, calculate the mean activity coefficient for NaCl from 0.1 *M* to 2 *M* solutions, using the full equation with correction for the space taken up by the ions and the water removed by hydration. Compare the new calculation with those of Stokes and Robinson. Discuss the change in "*a*" you had to assume.

19. Consider KCl and take "*a*" to be the sum of the ionic radii. Use data from tables to get these. Thus, one can calculate "*b*" of the Bjerrum theory over a reasonable concentration range and, using appropriate tables, obtain the value of the fraction of associated ions. Now recalculate the values of $\log f_{\pm}$ for KCl for 0.1 to 2 *M* solutions from the full Debye–Hückel theory involving allowance for ion size and hydration — but now also taking into account θ. In this approach, $c_0(1 - \theta)$ is the concentration of the ions that count in the expressions. (See Appendix 3.6.) Does this accounting for θ improve the fit?

20. Using the "point-charge version" of the Debye–Hückel model, derive the radial distribution of total excessive charge $q(r)$ from the central ion. Comment on the difference between $q(r)$ and the excessive charge in a dr-thickness shell $dq(r)$. (Xu)

21. (a) Using the above result, calculate the total excessive charge within the sphere of the radius of the Debye–Hückel reciprocal length κ^{-1}. How much of the overall excessive charge has been accounted for within the sphere of radius κ^{-1}? (b) Plot $q(r)$ vs. r for an aqueous solution of 10^{-3} *M* of a 1:1 electrolyte at 25 °C. (Xu)

22. Evaluate the Debye–Hückel constants A and B for ethyl alcohol and use the values to calculate the mean activity coefficients for 1:1, 1:2, and 2:2-valent electrolytes in ethyl alcohol at ionic strengths 0.1 and 0.01 at 25 °C. The mean distance of closest approach of the ions a may be taken as 300 pm in each case. Dielectric constant $\varepsilon = 24.3$. (Constantinescu)

23. Describe the correlation function as applied to ionic solutions. Sketch out a schematic of such a quantity for a hypothetical solution of $FeCl_2$. There are two entirely different types of methods of obtaining $g(r)$; state them. Finally, describe the "point" of knowing this quantity and comment on the meaning of the statement: "From a knowledge of the correlation function, it is possible to calculate solution properties (which may, in turn, be compared with the results of experiment). Hence the calculation of the correlation function is the aim of all new theoretical work on solutions."

24. The text gives the results of molecular dynamic calculations of the fraction of ion pairs as a function of concentration for univalent ions. Compare these values with those calculated by the Bjerrum theory.

25. Early work on solvation was dominated by the work of Born, Bernal and Fowler, Gurney and Frank. Which among these authors made the most lasting contributions? Why?

MICRO RESEARCH

1. One could use the Debye–Hückel ionic-atmosphere model to study how ions of opposite charges attract each other. (a) Derive the radial distribution of cation (n_+) and anion (n_-) concentration, respectively, around a central positive ion in a dilute aqueous solution of 1:1 electrolyte. (b) Plot these distributions and compare this model with Bjerrum's model of ion association. Comment on the applicability of this model in the study of ion association behavior. (c) Using the data in Table 3.2, compute the cation/anion concentrations at Debye–Hückel reciprocal lengths for NaCl concentrations of 10^{-4} and 10^{-3} mol dm^{-3}, respectively. Explain the applicability of the expressions derived. (Xu)

2. In the development of the theory of Debye and Hückel, it was assumed that two main changes had to be made: in the ion size (represented by the distance of closest approach a) and the diminution of the "available" waters due to hydration. Ion association was taken into account also.

   In a parallel series of developments, starting with the Mayer theory and continuing with the so-called mean spherical approximation, the effects of hydration and ion association were arbitrarily removed from consideration, in spite of their undeniable presence in nature.

   Utilizing equations of the text and values in Table P.4, follow through the calculations made to fit experiments in both approaches. Discover what adjustments are made in the MSA approach so that fair agreement (as in the work of Fawcett and Blum) with experiments can be obtained up to 2 mol dm^{-3}, using model assumptions that neglect association and hydration.

TABLE P.4

Recommended Values for the Mean Activity Coefficient γ and Water Activity of $CaCl_2$ in H_2O at 25 °C

m/mol kg^{-1}	γ	a_w
0.001	0.8885	0.999948
0.002	0.8508	0.999897
0.003	0.8245	0.999848
0.004	0.8039	0.999798
0.005	0.7869	0.999749
0.006	0.7724	0.999701
0.007	0.7596	0.999653
0.008	0.7483	0.999605
0.009	0.7380	0.999557
0.010	0.7287	0.999510
0.020	0.6644	0.999042
0.030	0.6256	0.998583
0.040	0.5982	0.998127
0.050	0.5773	0.997674
0.060	0.5607	0.997221
0.070	0.5470	0.996769
0.080	0.5355	0.996316
0.090	0.5256	0.995863
0.100	0.5171	0.995408
0.200	0.4692	0.990782
0.300	0.4508	0.985960
0.400	0.4442	0.980912
0.500	0.4442	0.975621
0.600	0.4486	0.970072
0.700	0.4564	0.964256
0.800	0.4670	0.958163
0.900	0.4801	0.951785
1.000	0.4956	0.945117
1.250	0.5440	0.927142
1.500	0.6070	0.907271
1.750	0.6861	0.885497
2.000	0.7842	0.861853
2.250	0.9049	0.836413
2.500	1.0529	0.809293
2.750	1.2339	0.780655
3.000	1.4550	0.750702

Source: Reprinted from R. N. Goldberg and R. L. Nuttall, *J. Phys. Chem. Ref. Data* **7**: 263 (1978).

3. The MSA equation used by Fawcett, Blum, and their colleagues shows good results in the calculation of activity coefficients up to a 2 *M* solution. Such calculations are described in the text. In this approach to electrolyte behavior, there are two properties of solutions that are neglected: (1) The degree of association, θ. How to calculate it from Bjerrum's point of view and the results of calculating it from MD, are described in the text, together with how to take θ into account in activity coefficient calculations. (2) The effect on activity calculations of hydration, also described in the text. What effect would it have on the Blum–Fawcett MSA calculations if these two neglected phenomena were also accounted for?

4. This text contains descriptions of numerous methods for obtaining the solvation number of ions (corresponding tables of data are given). By and large, the methods can be divided into two classes: solution properties (mobility, entropy, compressibility, partial molar volume, dielectric constant, Debye potential) and spectroscopic methods (neutron diffraction, Raman, and NMR). The values that arise from the solution methods hang together fairly well, though it is clear that the methods are not very precise. The results of the spectroscopic methods are similar to each other. However, it has to be admitted that for a given ion, the spectroscopic values are much lower (about one-third to one-half lower) than the self-consistent values determined from the properties of solutions. By examining the literature quoted in the text (and other literature), validate the statements made above for the following ions: Na^+, Ca^+, and Cr^{3+}. Then examine the evidence in favor of and against the following two hypotheses concerning the discrepancy noted:

   (a) There is no basic discrepancy. The two approaches are measuring the same quantity. However, solvation numbers are dependent on concentration; they get larger as the dilution increases. In concentrations above about 1 *M* (as shown by the molar volume–Debye potential approach), the solvation number declines considerably. However, it is only at the higher concentrations that most of the spectroscopic measurements have been made. Conversely, the solution properties have tended to be measured in the 10^{-3} to 10^{-1} *M* range, where the solvation number is indeed larger.
   (b) There is a basic discrepancy. The spectroscopic methods are more precise and with some of them one can obtain the time of residence of the water molecule and be quite precise about waters that stay with the ion, for how long, etc. The solution properties methods are a heterogeneous group. It is remarkable that they lead to a fairly consistent sense of numbers. However, the molar volume–Debye potential method should be precise.

   Analyze all this and produce a clear and reasoned judgment, buttressed by reasonings in a multipage analysis. Include in your discussion a ranking of methods for primary solvation numbers.

CHAPTER 4

ION TRANSPORT IN SOLUTIONS

4.1. INTRODUCTION

The interaction of an ion in solution with its environment of solvent molecules and other ions has been the subject of the previous two chapters. Now, attention will be focused on the motion of ions through their environment. The treatment is restricted to solutions of true electrolytes.

There are two aspects to these ionic motions. First, there is the *individual* aspect. This concerns the dynamic behavior of ions as individuals—the trajectories they trace out in the electrolyte, and the speeds with which they dart around. These ionic movements are basically random in direction and speed. Second, ionic motions have a *group* aspect that is of particular significance when more ions move in certain directions than in others and produce a drift, or flux,[1] of ions. This drift has important consequences because an ion has a mass and bears a charge. Consequently, the flux of ions in a preferred direction results in the transport of matter and a flow of charge.

If the directional drift of ions did not occur, the interfaces between the electrodes and electrolyte of an electrochemical system would run out of ions to fuel the charge-transfer reactions that occur at such interfaces. Hence, the movements and drift of ions is of vital significance to the continued functioning of an electrochemical system.

A flux of ions can come about in three ways. If there is a difference in the concentration of ions in different regions of the electrolyte, the resulting concentration gradient produces a flow of ions. This phenomenon is termed *diffusion* (Fig. 4.1). If there are differences in electrostatic potential at various points in the electrolyte, then

---

[1]The word *flux* occurs frequently in the treatment of transport phenomena. The flux of any species i is the number of moles of that species crossing a unit area of a reference plane in 1 s; hence, *flux is the rate of transport*.

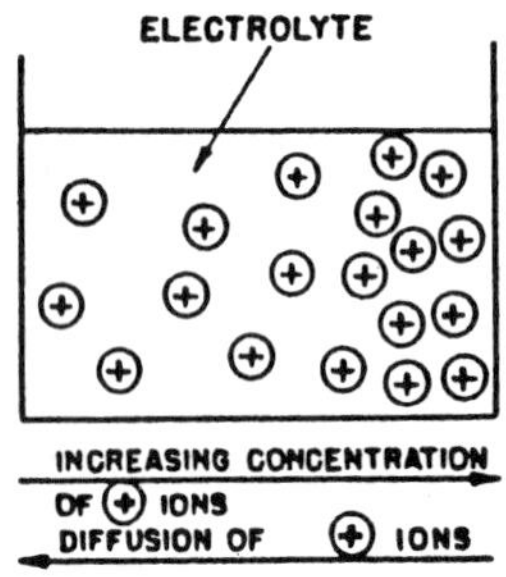

Fig. 4.1. The diffusion of positive ions resulting from a concentration gradient of these ions in an electrolytic solution. The directions of increasing ionic concentration and of ionic diffusion are shown below the diagram.

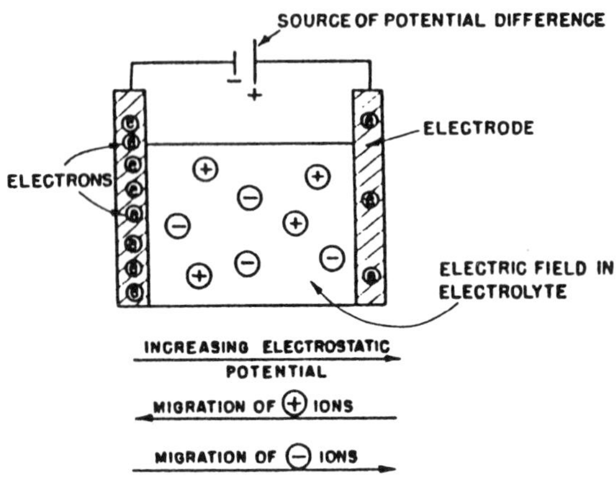

Fig. 4.2. The migration of ions resulting from a gradient of electrostatic potential (i.e., an electric field) in an electrolyte. The electric field is produced by the application of a potential difference between two electrodes immersed in the electrolyte. The directions of increasing electrostatic potentials and of ionic migration are shown below the diagram.

the resulting electric field produces a flow of charge in the direction of the field. This is termed *migration* or *conduction* (Fig. 4.2). Finally, if a difference in pressure or density or temperature exists in various parts of the electrolyte, then the liquid begins to move as a whole or parts of it move relative to other parts. This is *hydrodynamic* flow.

It is intended to restrict the present discussion to the transport processes of diffusion and conduction and their interconnection. (The laws of hydrodynamic flow will not be described, mainly because they are not particular to the flow of electrolytes; they are characteristic of the flow of all gases and liquids, i.e., of fluids.) The initial treatment of diffusion and conduction will be in phenomenological terms; then the molecular events underlying these transport processes will be explored.

In looking at ion–solvent and ion–ion interactions, it has been possible to present the phenomenological or nonstructural treatment in the framework of equilibrium thermodynamics, which excludes time and therefore fluxes, from its analyses. Such a straightforward application of thermodynamics cannot be made, however, to transport processes. The drift of ions occurs precisely because the system is not at equilibrium; rather, the system is seeking to attain equilibrium. In other words, the system undergoes change (there cannot be transport without temporal change!) because the free energy is not uniform and tends to reach a minimum. It is the existence of such gradients of free energy that sets up the process of ionic drift and makes the system strive to attain equilibrium by the dissipation of free energy.

4.2. IONIC DRIFT UNDER A CHEMICAL-POTENTIAL GRADIENT: DIFFUSION

4.2.1. The Driving Force for Diffusion

It has been remarked in the previous section that diffusion occurs when a concentration gradient exists. The theoretical basis of this observation will now be examined.

Consider that in an electrolytic solution, the concentration of an ionic species i varies in the x direction but is constant in the y and z directions. If desired, one can map equiconcentration surfaces (they will be parallel to the yz plane) (Fig. 4.3).

The situation pictured in Fig. 4.3 can also be considered in terms of the partial molar free energy, or chemical potential, of the particular species i. This is achieved through the use of the defining equation for the chemical potential [Eq. (3.61)]

$$\mu_i = \mu_i^0 + RT \ln c_i$$

(The use of concentration rather than activity implies that the solution is assumed to behave ideally.) Since c_i is a function of x, the chemical potential also is a function of x. Thus, the chemical potential varies along the x coordinate, and, if desired, equi-μ surfaces can be drawn. Once again, these surfaces will be parallel to the yz plane.

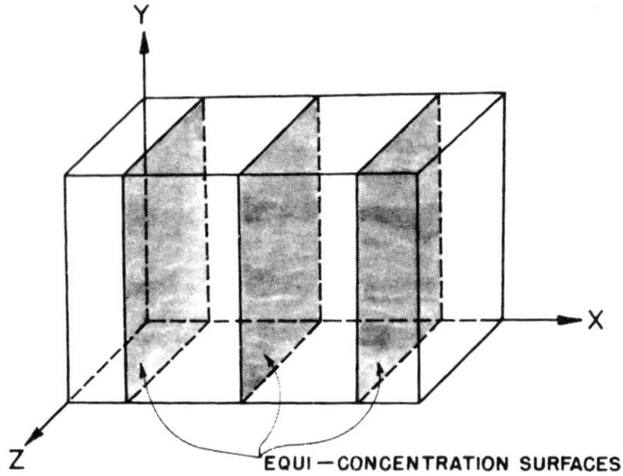

Fig. 4.3. A schematic representation of a slice of electrolytic solution in which the concentration of a species *i* is constant on the shaded equiconcentration surfaces parallel to the *yz* plane.

Now, if one transfers a mole of the species i from an initial concentration c_I at x_I to a final concentration c_F at x_F, then the change in free energy, or chemical potential, of the system is (Fig. 4.4):

$$\Delta\mu = \mu_F - \mu_I = RT \ln \frac{c_F}{c_I} \tag{4.1}$$

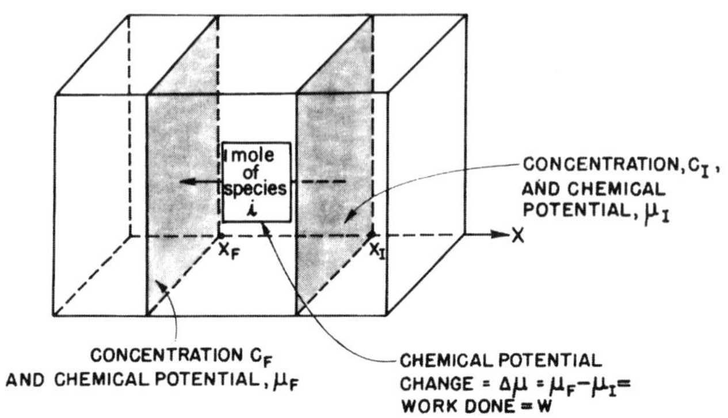

Fig. 4.4. A schematic representation of the work *W* done in transporting a mole of species *i* from an equiconcentration surface where its concentration and chemical potential are c_I and μ_I to a surface where its concentration and chemical potential are c_F and μ_F.

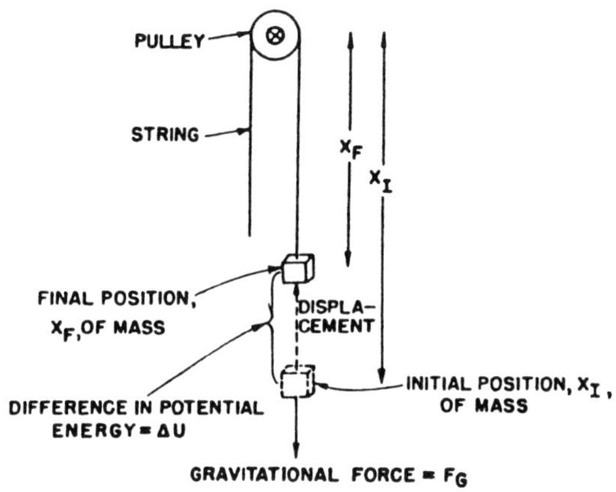

Fig. 4.5. Schematic diagram to illustrate that the mechanical work W done in lifting a mass from an initial height x_I to a final height x_F is $W = \Delta U = -F_G(x_I - x_F)$.

However, the change in free energy is equal to the net work done *on* the system in an isothermal, constant-pressure reversible process. Thus, the work done to transport a mole of species i from x_I to x_F is

$$W = \Delta\mu \tag{4.2}$$

Think of the analogous situation in mechanics. The work done to lift a mass from an initial height x_I to a final height x_F is equal to the difference in gravitational potential (energies) ΔU at the two positions (Fig. 4.5):

$$W = \Delta U \tag{4.3}$$

One may go further and say that this work has to be done because a gravitational force F_G acts on the body and that[2]

$$W = -F_G(x_I - x_F) = -F_G\Delta x = \Delta U \tag{4.4}$$

[2]The minus sign arises from the following argument: The displacement $x_I - x_F$ of the mass is *upward* and the force acts *downward*; hence, the product of the displacement and force vectors is negative. If a minus sign is not introduced, the work done W will turn out to be negative. It is desirable to have W as a positive quantity because of the convention that work done on a system is taken to be positive; hence, a minus sign must be inserted.

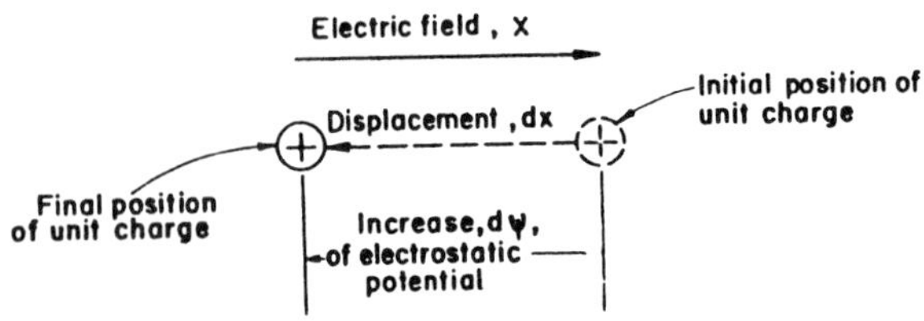

Fig. 4.6. Schematic diagram to illustrate the electrostatic work $W = d\psi = -X\,dx$ done in moving a unit positive charge through a distance *dx* against an electric field *X*.

In other words, the gravitational force can be defined thus:

$$F_G = -\frac{\Delta U}{\Delta x} \tag{4.5}$$

The potential energy, however, may not vary linearly with distance, and thus the ratio $\Delta U/\Delta x$ may not be a constant. So it is better to consider infinitesimal changes in energy and distance and write

$$F_G = -\frac{dU}{dx} \tag{4.6}$$

Thus, the gravitational force is given by the *gradient* of the gravitational potential energy, and the region of space in which it operates is said to be a *gravitational field.*

A similar situation exists in electrostatics. The electrostatic work done in moving a unit charge from x to $x + dx$ defines the difference $d\psi$ in electrostatic potential between the two points (Fig. 4.6)

$$W = d\psi \tag{4.7}$$

Further, the electrostatic work is the product of the electric field, or force per unit charge, X and the distance dx

$$-X\,dx = d\psi \tag{4.8}$$

or

$$X = -\frac{d\psi}{dx} \tag{4.9}$$

The electric force per unit charge is therefore given by the negative of the gradient of the electrostatic potentials, and the region of space in which the force operates is known as the *electric field.*

TABLE 4.1

Certain Forces in the Phenomenological Treatment of Transport

Force	Acting On	Results In
$-\frac{dU}{dx}$	Mass	Movement of mass
$-\frac{d\psi}{dx}$	Charge	Movement of charge (current)
$-\frac{d\mu}{dx}$	Species *i*	Movement of species *i*, i.e., diffusional flux of species *i*

Since the negative of the gradient of gravitational potential energy defines the gravitational force, and the negative of the gradient of electrostatic potential defines the electric force, one would expect that the negative of the gradient of the chemical potential would act formally like a force. Furthermore, just as the gravitational force results in the motion of a mass and the electric force results in the motion of a charge, the chemical-potential gradient results in the net motion, or transfer, of the species *i* from a region of high chemical potential to a region of low chemical potential. This net flow of the species *i* down the chemical-potential gradient is diffusion, and therefore the gradient of chemical potential may be looked upon[3] as the diffusional force F_D. Thus, one can write

$$F_D = -\frac{d\mu_i}{dx} \tag{4.10}$$

by analogy with the gravitational and electric forces [Eqs. (4.6) and (4.9)] and consider that the diffusional force produces a diffusional flux *J*, the number of moles of species *i* crossing per second per unit area of a plane normal to the flow direction (Table 4.1).

4.2.2. The "Deduction" of an Empirical Law: Fick's First Law of Steady-State Diffusion

Qualitatively speaking, the macroscopic description of the transport process of diffusion is simple. The gradient of chemical potential resulting from a nonuniform concentration is equivalent to a driving force for diffusion and produces a diffusion

---

[3]It will be shown later on (Section 4.2.6) that for the phenomenon of diffusion to occur, all that is necessary is an inequality of the net number of diffusing particles in different regions; there is, in fact, no directed force on the individual particles. Thus, $-d\mu/dx$ is only a *pseudo*force like the centrifugal force; it is *formally* equivalent to a force.

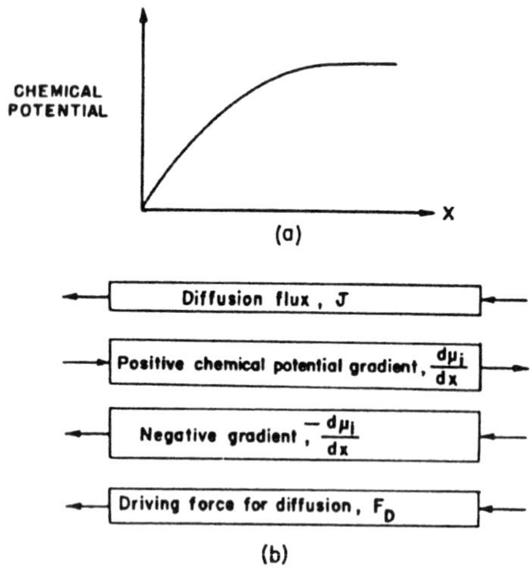

Fig. 4.7. Schematic diagram to show (a) the distance variation of the chemical potential of a species *i* and (b) the relative directions of the diffusion flux, driving force for diffusion, etc.

flux (Fig. 4.7). What is the quantitative cause-and-effect relation between the driving force $d\mu_i/dx$ and the flux J? This question must now be considered.

Suppose that when diffusion is occurring, the driving force F_D and the flux $\boldsymbol{J}$ reach values that do not change with time. The system can be said to have attained a *steady state*. Then the as-yet-unknown relation between the diffusion flux $\boldsymbol{J}$ and the diffusional force F_D can be represented quite generally by a power series

$$J = A + BF_D + CF_D^2 + DF_D^3 + \ldots \tag{4.11}$$

where A, B, C, etc., are constants. If, however, F_D is less than unity and sufficiently small,[4] the terms containing the powers (of F_D) greater than unity can be neglected.

Thus, one is left with

$$J = A + BF_D \tag{4.12}$$

but the constant A must be equal to zero; otherwise, it will mean that one would have the impossible situation of having diffusion even though there is no driving force for diffusion.

---

[4]Caution should be exercised in applying the criterion. The value of F_D that will give rise to unity will depend on the units chosen to express F_D. Thus, the extent to which F_D is less than unity will depend on the units, but one can always restrict F_D to an appropriately small value.

Hence, the assumption of a sufficiently small driving force leads to the result

$$J = BF_D \tag{4.13}$$

i.e., the flux is linearly related to the driving force. The value of $F_D = 0$ (zero driving force) corresponds to an equilibrium situation; therefore, the assumption of a small value of F_D required to ensure the linear relation (4.13) between flux and force is tantamount to saying that the system is *near equilibrium, but not at equilibrium.*

The driving force on 1 mole of ions has been stated to be $-d\mu_i/dx$ [Eq. (4.10)]. If, therefore, the concentration of the diffusing species adjacent to the transit plane (Fig. 4.8), across which the flux is reckoned, is c_i moles per unit volume, the driving force F_D at this plane is $-c_i(d\mu_i/dx)$. Thus, from relation (4.13), one obtains

$$J_i = -Bc_i \frac{d\mu_i}{dx} \tag{4.14}$$

Writing

$$\mu = \mu^0 + RT \ln c_i$$

which is tantamount to assuming ideal behavior, Eq. (4.14) becomes

$$J_i = -Bc_i \frac{RT}{c_i} \frac{dc_i}{dx} = -BRT \frac{dc_i}{dx} \tag{4.15}$$

Thus, *the steady-state diffusion flux has been theoretically shown to be proportional to the gradient of concentration.* That such a proportionality existed has been

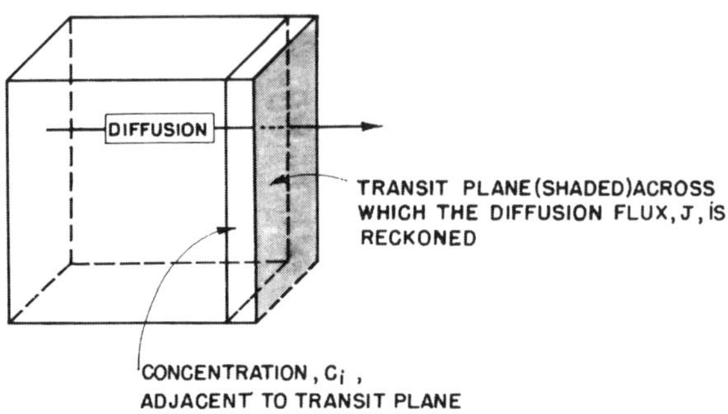

Fig. 4.8. Diagram for the derivation of the linear relation between the diffusion flux J_i and the concentration gradient dc_i/dx.

TABLE 4.2
Diffusion Coefficient *D* of Ions in Aqueous Solutions

Ion	Diffusion Coefficient, ($cm^2 s^{-1}$)
Li^+	1.028×10^{-5}
Na^+	1.334×10^{-5}
K^+	1.569×10^{-5}
Cl^-	2.032×10^{-5}
Br^-	2.080×10^{-5}

known *empirically* since 1855 through the statement of Fick's first law of steady-state diffusion, which reads

$$J_i = -D\frac{dc_i}{dx} \tag{4.16}$$

where D is termed the *diffusion coefficient* (Table 4.2).

4.2.3. The Diffusion Coefficient *D*

It is important to stress that, in the empirical Fick's first law, the concentration c is expressed in moles per cubic centimeter, and not in moles per liter. The flux is expressed in moles of diffusing material crossing a unit area of a transit plane per unit of time, i.e., in moles per square centimeter per second, and therefore the diffusion coefficient D has the dimensions of centimeters squared per second. The negative sign is usually inserted in the right-hand side of the empirical Fick's law for the following reason: The flux J_i and the concentration gradient dc_i/dx are vectors, or quantities which have both magnitude and direction. However, the vector $\mathbf{J}$ is in an opposite sense to the vector representing a positive gradient dc_i/dx. Matter flows downhill (Fig. 4.9). Hence, if J_i is taken as positive, dc_i/dx must be negative, and, if there is no negative sign in Fick's first law, the diffusion coefficient will appear as a negative quantity—perhaps an undesirable state of affairs. Hence, to make D come out a positive quantity, a negative sign is added to the right-hand side of the equation that states the empirical law of Fick.

Equating the coefficients of dc_i/dx in the phenomenological equation (4.15) with that in Fick's law [Eq. (4.16)], is seen that

$$BRT = D \tag{4.17}$$

Now, is the diffusion coefficient a concentration-independent constant? A naive answer would run thus: B is a constant and therefore it *appears* that D also is a constant.

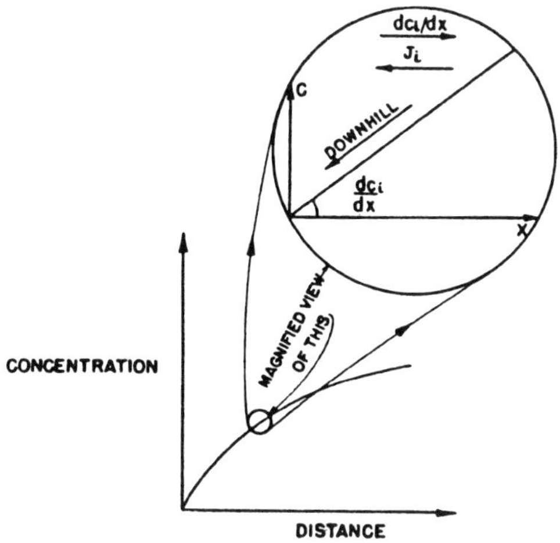

Fig. 4.9. Diagram to show that diffusion flow J_i is in a direction opposite to the direction of positive concentration gradient dc_i/dx. Matter flows downhill, i.e., down the concentration gradient.

However, expression (4.17) was obtained only because an ideal solution was considered, and activity coefficients f_i were ignored in Eq. (3.61). Activity coefficients, however, are concentration dependent. So, if the solution does not behave ideally, one has, starting from Eq. (4.14), and using Eq. (3.63),

$$
\begin{aligned}
\mathbf{J}_i &= -Bc_i \frac{d\mu_i}{dx} \\
&= -Bc_i \frac{d}{dx}(\mu_i^0 + RT\ln f_i c_i) \\
&= -Bc_i \frac{RT}{f_i c_i}\frac{d}{dx} f_i c_i \\
&= -BRT\frac{dc_i}{dx} - \frac{BRTc_i}{f_i}\frac{df_i}{dx} \\
&= -BRT\frac{dc_i}{dx} - \frac{BRTc_i}{f_i}\frac{df_i}{dc_i}\frac{dc_i}{dx} \\
&= -BRT\frac{dc_i}{dx}\left(1 + \frac{c_i}{f_i}\frac{df_i}{dc_i}\right)
\end{aligned}
$$

TABLE 4.3
Variation of the Diffusion Coefficient *D* with Concentration

	Diffusion Coefficient *D* in Units of 10^{-5} cm^2 s^{-1} at Concentration (in molarity)			
Electrolyte	0.05	0.1	0.2	0.5
HCl	3.07	3.05	3.06	3.18
LiCl	1.28	1.27	1.27	1.28
NaCl	1.51	1.48	1.48	1.47

$$= - BRT \frac{dc_i}{dx}\left(1 + \frac{d \ln f_i}{d \ln c_i}\right) \tag{4.18}$$

and therefore

$$D = BRT\left(1 + \frac{d \ln f_i}{d \ln c_i}\right) \tag{4.19}$$

Rigorously speaking, the diffusion coefficient is not a constant (Table 4.3). If, however, the variation of the activity coefficient is not significant over *the concentration difference that produces diffusion*, then $(c_i/f_i)(\partial f_i/\partial c_i) \ll 1$ and for all practical purposes D is a constant.[5] *This effective constancy of D with concentration will be assumed in most of the discussions presented here.*

The treatment so far has been phenomenological and therefore the dependence of the diffusion coefficient on factors such as temperature and type of ion can be theoretically understood only by an atomistic analysis. The quantity D can be understood in a fundamental way only by probing into the ionic movements, the results of which show up in the macroscopic world as the phenomenon of diffusion. What are these ionic movements, and how do they produce diffusion? The answering of these two questions will constitute the next topic.

4.2.4. Ionic Movements: A Case of the Random Walk

Long before the movements of ions in solution were analyzed, the kinetic theory of gases was developed and it involved the movements of gas molecules. The *overall*

[5]For example, in diffusion between solutions that have a large concentration difference, such as 0.1 to 0.01 mol dm^{-3}, a rough calculation suggests that the activity-coefficient correction is on the order of a few percent.

pattern of ionic movements is quite similar to that of gas molecules and therefore the latter will be recalled first.

Imagine a hypothetical situation in which all the gas molecules except one are at rest. According to Newton's first law of motion, the moving molecule will travel with a uniform velocity until it collides with a stationary molecule. During the collision, there is a transfer of momentum (mass m times velocity v). So, the moving molecule loses some speed in the collision, but the stationary molecule is set in motion. Now both molecules are moving, and they will undergo further collisions. The number of collisions will increase with time, and soon all the molecules of the gas will be continually moving, colliding, and changing their directions of motion and their velocities—a scene of hectic activity.

It would be of interest to have an idea of the path of such a gas molecule in the course of time. One might think that the detailed paths of all the particles could be predicted by applying Newton's laws to the motions of molecules. The problem, however, is obviously too complex for a practical solution. To use the laws of motion requires a knowledge of the position and velocity of each particle and even in 1 mole there are 6.023×10^{23} (the Avogadro number) particles.

One can, however, try another approach. Is the ceaseless jostling of molecules manifested in any gross (macroscopic) phenomenon? Consider a frictionless piston in mechanical equilibrium with a mass of gas enclosed in a cylinder. Owing to its weight, the piston exerts a force on the gas. What force balances the piston's weight? One says that the gas exerts a pressure (force per unit area) on the piston owing to the continual buffeting that the piston receives from the gas molecules. Despite this fact, the bombardment by the gas molecules does not produce any *visible* motion of the piston. Evidently the mass of the piston is so large compared with that of the gas molecules that the movements of the piston are too small to be detected.

Now let the mass of the piston be reduced. Then the jiggling of the extremely light piston as a result of being struck by gas molecules should make itself apparent to an observer. This is what happens if one tries to make a mirror galvanometer more and more sensitive. The essential part of this instrument is a thin quartz fiber that supports a light coil of wire seated in a magnetic field (Fig. 4.10). The deflections of the coil are made visible by fixing a mirror onto the quartz fiber and bouncing a beam of light off the mirror onto a scale. To increase the sensitivity of the instrument, one tries lighter coils, lighter mirrors, and thinner fibers. There comes a stage, however, when the "kicks" which the fiber-mirror-coil assembly receives from the air molecules are sufficient to make the assembly jiggle about. The reflected light beam then jumps about on the scale (Fig. 4.11). The movements of the spot about a mean position on the scale represent *noise*. (It is as if each collision produced a sound, in which case the irregular bombardment of the mirror assembly would result in a nonstop noise.) Signals (coil deflections) that are of this same order of magnitude obviously cannot be separated from the noise.

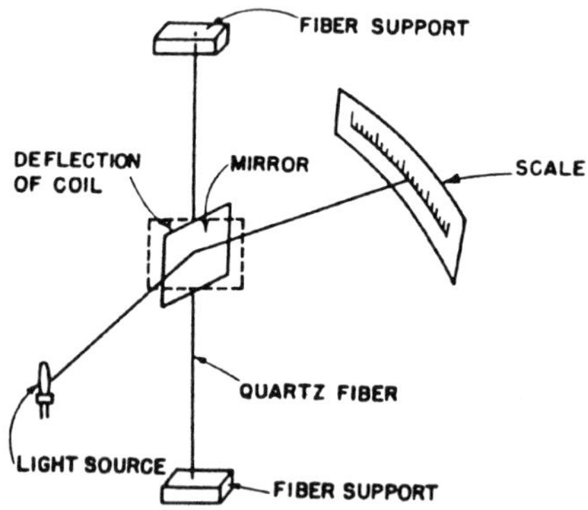

Fig. 4.10. Schematic representation of the essential parts of a mirror galvanometer, used for detecting Brownian motion.

Instead of mirrors, one could equally consider pistons of a microscopic size, large enough to be seen with the aid of a microscope but small enough to display motions due to collisions with molecules. Such small "pistons" are present in nature. A colloidal particle in a liquid medium behaves as such a piston if it is observed in a microscope. It shows a haphazard, zigzag motion as shown in Fig. 4.12. The irregular path of the particle must be a slow-motion version of the *random-walk* motion of the molecules in the liquid.

One has therefore a picture of the solvated ions (in an electrolytic solution) in ceaseless motion, perpetually colliding, changing direction, staggering hither and thither from site to site. This is the qualitative picture of ionic *movements*.

4.2.5. The Mean Square Distance Traveled in a Time *t* by a Random-Walking Particle

The movements executed by an ion in solution are a three-dimensional affair because the ion has a three-dimensional space available for roaming around. So one

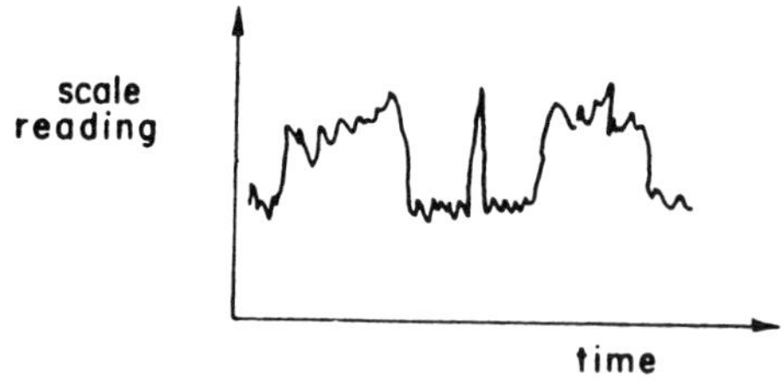

Fig. 4.11. The time variation of the reading on the scale of a mirror galvanometer.

Fig. 4.12. The haphazard zigzag motion of a colloidal particle.

must really call the movements a "random flight" because the ion *flies* in three dimensions and *walks* in two dimensions. This fine difference, however, will be ignored and the term *random walk* will be retained.

The aim now is to seek a quantitative description of ionic random-walk movements. There are many exotic ways of stating the random-walk problem. It is said, for example, that a drunken sailor emerges from a bar. He intends to get back to his ship, but he is in no state to control the direction in which he takes a step. In other words, the direction of each step is completely random, all directions being equally likely. The question is: On the average, how far does the drunken sailor progress in a time t?

For the sake of simplicity, the special case of a one-dimensional random walk will be considered. The sailor starts off from $x = 0$ on the x axis. He tosses a coin: heads—he moves forward in the positive x direction, tails—he moves backward. Since, for an "honest" coin, heads are as likely as tails, the sailor is equally as likely to take a forward step as a backward step. Of course, each step is decided on a fresh toss and is uninfluenced by the results of the previous tosses. After allowing him N steps, the distance x from the origin is noted[6] (Fig. 4.13). Then the sailor is brought back to the bar ($x = 0$) and started off on another try of N steps.

The process of starting the sailor off from $x = 0$, allowing him N steps, and measuring the distance x traversed is repeated many times, and it is found that the distances traversed from the origin are $x(1)$, $x(2)$, $x(3)$, . . . , $x(i)$, where $x(i)$ is the distance from the origin traversed in the ith try. The *average distance* $<x>$ from the origin is

$$<x> = \frac{\sum_i x(i)}{\sum i} = \frac{\text{Sum of distancesfrom origin}}{\text{Number of tries}} \tag{4.20}$$

[6]It will be zero only if an equal number $N/2$ of heads and tails turns up. For a small number of trials, this will not happen every time. So, after each small number of tosses, the sailor is not certain to be back where he started (i.e., in the bar).

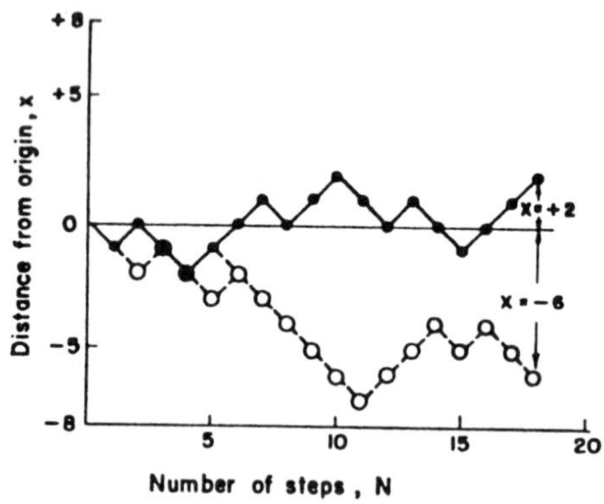

Fig. 4.13. The distance x (from the origin) traversed by the drunken sailor in two tries, each of $N = 18$ steps.

Since the distance x traversed by the sailor in N steps is as likely for a large number of trials to be in the plus-x direction as in the minus-x direction, it is obvious from the canceling out of the positive and negative values of x that the *mean* progress from the origin is given by

$$\langle x \rangle = 0 \tag{4.21}$$

Hence, it is not very fruitful to compute the mean distance $\langle x \rangle$ traversed by the sailor in N steps. To avoid such an unenlightening result, which arises because $x(i)$ can take either positive or negative values, it is best to consider the square of $x(i)$, which is always a positive quantity whether $x(i)$ itself is negative or positive. Hence, if $x(1)$, $x(2)$, $x(3)$, . . . , $x(i)$ are all squared and the mean of these quantities is taken, then one can obtain the *mean square distance* $\langle x^2 \rangle$, i.e.,

$$\langle x^2 \rangle = \frac{\sum_i [x(i)]^2}{\sum i} = \frac{\text{Sum of distances } x \text{ squared}}{\text{Number of tries}} \tag{4.22}$$

Since the square of $x(i)$, i.e., $[x(i)]^2$, is always a positive quantity, the mean square distance traversed by the sailor is always a positive nonzero quantity (Table 4.4).[7]

[7]That $\Sigma x/n = 0$ but $\Sigma x^2/n$ is finite at first seems difficult to comprehend. One can get to it by recalling that the drunken sailor may not make *net* progress ($\Sigma x/n = 0$), but the range of his lurching to the right or to the left is also interesting and this is obtained if one eliminates the sign of x (which causes the mean of the sum of the xs to be zero) and deals in $\sqrt{\langle x^2 \rangle}$, the square root of the mean of the sum of x^2. Then this root-mean-square value of x indicates the range of the drunk's wandering, no matter in what direction (Section 4.2.14).

TABLE 4.4
One-Dimensional Random Walk of the Sailor[a]

Trial	N_F [b]	N_B [c]	x [d]	x^2
1	11	19	−8	64
2	16	14	+2	4
3	17	13	+4	16
4	15	15	0	0
5	17	13	+4	16
6	16	14	+2	4
7	19	11	+8	64
8	18	12	+6	36
9	15	15	0	0
10	13	17	−4	16
11	11	19	−8	64
12	17	13	+4	16
13	17	13	+4	16
14	12	18	−6	36
15	20	10	+10	100
16	23	7	+16	256
17	11	19	−8	64
18	16	14	+2	4
19	17	13	+4	16
20	14	16	−2	4
21	16	14	+2	4
22	12	18	−6	36
23	15	15	0	0
24	10	20	−10	100
25	7	23	−16	256
26	15	15	0	0
	$\langle x \rangle = 0$	$\langle x^2 \rangle = 47.67$	$\sqrt{\langle x^2 \rangle} = 6.90$	

[a] Total number of steps in each trial = $N = 30$.
[b] Number of forward steps = N_F.
[c] Number of backwards steps = N_B.
[d] Distance from the origin = x.

Furthermore, it can easily be shown (Appendix 4.1) that the magnitude of $\langle x^2 \rangle$ is proportional to N, the number of steps and since N itself increases linearly with time, it follows that the mean square distance traversed by the random-walking sailor is proportional to time

$$\langle x^2 \rangle \propto t \tag{4.23}$$

It is to be noted that it is the *mean square distance*—and not the *mean distance*—that is proportional to time. If the mean distance were proportional to time, then the drunken sailor (or the ion) would be proceeding at a uniform velocity. This is not the case because the mean distance $\langle x \rangle$ traveled is zero. The only type of progress that the ion is making from the origin is such that the mean *square* distance is proportional to time. This is the characteristic of a random walk.

4.2.6. Random-Walking Ions and Diffusion: The Einstein–Smoluchowski Equation

Consider a situation in an electrolytic solution where the concentration of the ionic species of interest is constant in the yz plane but varies in the x direction. To analyze the diffusion of ions, imagine a *unit area* of a reference plane normal to the x direction. This reference plane will be termed the *transit plane* of unit area (Fig. 4.14). There is a random walk of ions across this plane both from left to right and from right to left. On either side of the transit plane, one can imagine two planes L and R that are parallel to the transit plane and situated at a distance $\sqrt{\langle x^2 \rangle}$ from it. In other words, the region under consideration has been divided into left and right compartments in which the concentrations of ions are different and designated by c_L and c_R, respectively.

In a time of t s, a random-walking ion covers a mean square distance of $\langle x^2 \rangle$, or a mean distance of $\sqrt{\langle x^2 \rangle}$. Thus, by choosing the plane L to be at a distance $\sqrt{\langle x^2 \rangle}$ from the transit plane, one has ensured that all the ions in the left compartment will cross the transit plane in a time t *provided* they are moving in a left-to-right direction.

The number of moles of ions in the left compartment is equal to the volume $\sqrt{\langle x^2 \rangle}$ of this compartment times the concentration c_L of ions. It follows that the number of moles of ions that make the $L \rightarrow T$ crossing in t s is $\sqrt{\langle x^2 \rangle}\, c_L$ times the fraction of ions making left-to-right movements. Since the ions are random-walking,

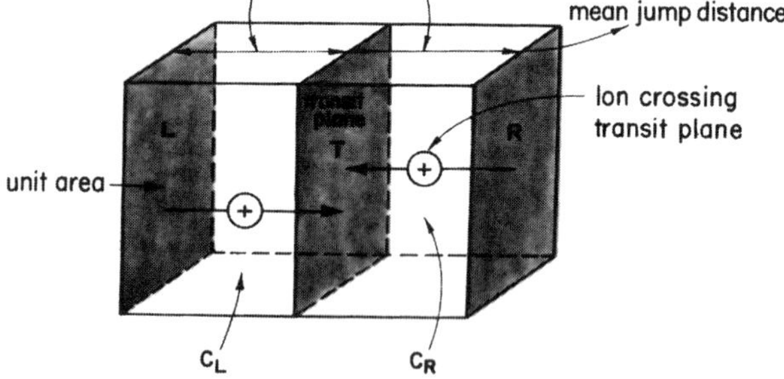

Fig. 4.14. Schematic diagram for the derivation of the Einstein–Smoluchowski relation, showing the transit plane T in between and at a distance $\sqrt{\langle x^2 \rangle}$ from the left L and right R planes. The concentrations in the left and right compartments are c_L and c_R, respectively.

right-to-left movements are as likely as left-to-right movements, i.e., only half the ions in the left compartment are moving toward the right compartment. Thus, in t s the number of moles of ions making the $L \rightarrow T$ crossing is $1/2\sqrt{<x^2>}c_L$ and therefore the number of moles of ions making the $L \rightarrow T$ crossing in 1 s is $1/2(\sqrt{<x^2>}/t)c_L$. Similarly, the number of moles of ions making the $R \rightarrow T$ crossing in 1 s is $1/2(\sqrt{<x^2>}/t)c_R$.

Hence, the diffusion flux of ions across the transit plane (i.e., the net number of moles of ions crossing unit area of the transit plane per second from left to right) is given by

$$J = \frac{1}{2}\frac{\sqrt{<x^2>}}{t}(c_L - c_R) \tag{4.24}$$

This equation reveals that all that is required to have diffusion is *a difference in the numbers* per unit volume of particles in two regions. The important point is that no special diffusive force acts *on the particles* in the direction of the flux.

If no forces are pushing particles in the direction of the flow, then what about the driving force for diffusion, i.e., the gradient of chemical potential (Section 4.2.1)? The latter is only *formally equivalent* to a force in a macroscopic treatment; it is a sort of pseudoforce like a centrifugal force. The chemical-potential gradient is not a true force that acts on the individual diffusing particles and from this point of view is quite unlike, for example, the Coulombic force, which acts on individual charges.

Now, the concentration gradient dc/dx in the left-to-right direction can be written

$$\frac{dc}{dx} = \frac{c_R - c_L}{\sqrt{<x^2>}} = -\frac{c_L - c_R}{\sqrt{<x^2>}}$$

or

$$c_L - c_R = -\sqrt{<x^2>}\frac{dc}{dx} \tag{4.25}$$

This result for $c_L - c_R$ can be substituted in Eq. (4.24) to give

$$J = -\frac{1}{2}\frac{<x^2>}{t}\frac{dc}{dx} \tag{4.26}$$

and, by equating the coefficients of this equation with that of Fick's first law [Eq. (4.16)], one has

$$\frac{<x^2>}{2t} = D$$

or

$$<x^2> = 2Dt \tag{4.27}$$

This is the Einstein–Smoluchowski equation; it provides a bridge between the microscopic view of random-walking ions and the coefficient D of the macroscopic Fick's law.

A coefficient of 2 is intimately connected with the approximate nature of the derivation, i.e., a *one-dimensional* random walk with ions being permitted to jump forward and backward only. More rigorous arguments may yield other values for the numerical coefficient, e.g., 6.

The characteristic of a random walk in the Einstein–Smoluchowski equation is the appearance of the mean square distance (i.e., square centimeters), and since this mean square distance is proportional to time (seconds), the proportionality constant D in Eq. (4.27) must have the dimensions of centimeters squared per second. It must not be taken to mean that every ion that starts off on a random walk travels in time t a mean square distance $<x^2>$ given by the Einstein-Smoluchowski relation (4.27). If a certain number of ions are, in an imaginary or thought experiment, suddenly introduced on the yz plane at $x = 0$, then, in t s, some ions would progress a distance x_1; others, x_2; still others, x_3; etc. The Einstein–Smoluchowski relation only says that [*cf.* Eqs. (4.22) and (4.27)]

$$\frac{x_1^2 + x_2^2 + \cdots + x_n^2}{n} = <x^2> = 2Dt \tag{4.28}$$

How many ions travel a distance x_1; how many, x_2; etc.? In other words, how are the ions *spatially distributed* after a time t, and how does the spatial distribution vary with time? This spatial distribution of ions will be analyzed, but only after a phenomenological treatment of *non*steady-state diffusion is presented.

4.2.7. The Gross View of Nonsteady-State Diffusion

What has been done so far is to consider steady-state diffusion in which neither the flux nor the concentration of diffusing particles in various regions changes with time. In other words, the whole transport process is time independent. What happens if a concentration gradient is suddenly produced in an electrolyte initially in a time-invariant equilibrium condition? Diffusion starts of course, but it will not immediately reach a steady state that does not change with time. For example, the distance variation of concentration, which is zero at equilibrium, will not instantaneously hit the final steady-state pattern. How does the concentration vary with time?

Consider a parallelepiped (Fig. 4.15) of unit area and length dx. Ions are diffusing in through the left face of the parallelepiped and *out* through the right face. Let the concentration of the diffusing ions be a continuous function of x. If c is the concentration of ions at the left face, the concentration at the right force is

$$c + \frac{dc}{dx}\,dx$$

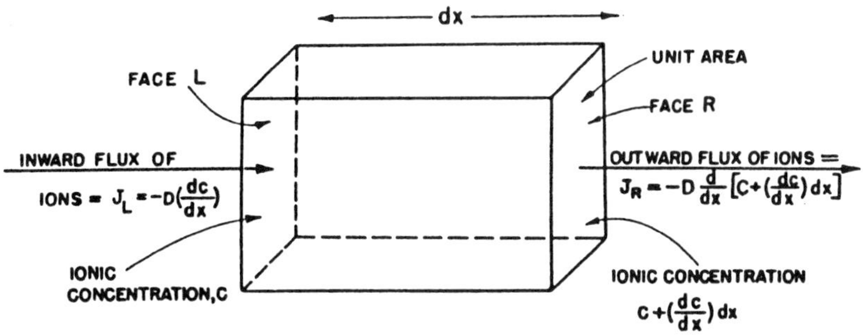

Fig. 4.15. The parallelepiped of electrolyte used in the derivation of Fick's second law.

Fick's law [Eq. (4.16)] is used to express the flux into and out of the parallelepiped. Thus the flux into the left face J_L is

$$J_L = -D\frac{dc}{dx} \tag{4.29}$$

and the flux out of the right face is

$$\begin{aligned} J_R &= -D\frac{d}{dx}\left(c + \frac{dc}{dx}dx\right) \\ &= -D\frac{dc}{dx} - D\frac{d^2c}{dx^2}dx \end{aligned} \tag{4.30}$$

The net *out*flow of material from the parallelepiped of volume dx is

$$J_L - J_R = D\frac{d^2c}{dx^2}dx \tag{4.31}$$

Hence, the net outflow of ions *per unit volume* per unit time is $D\,(d^2c/dx^2)$. But this net outflow of ions per unit volume per unit time from the parallepiped is in fact the sought-for variation of concentration with time, i.e., dc/dt. One obtains partial differentials because the concentration depends both on time and distance, but the subscripts x and t are generally omitted because it is, for example, obvious that the time variation is at a fixed region of space, i.e., constant x. Hence,

$$\frac{\partial c}{\partial t} = D\frac{\partial^2 c}{\partial x^2} \tag{4.32}$$

This partial differential equation is known as *Fick's second law*. It is the basis for the treatment of most time-dependent diffusion problems in electrochemistry.

That Fick's second law is in the form of a differential equation implies that it describes what is common to *all* diffusion problems and it has "squeezed out" what is characteristic of any *particular* diffusion problem.[8] Thus, one always has to calculate the precise functional relationship

$$c = f(x, t)$$

for a particular situation. The process of calculating the functional relationship consists in solving the partial differential equation, which is Fick's second law, i.e., Eq. (4.32).

4.2.8. An Often-Used Device for Solving Electrochemical Diffusion Problems: The Laplace Transformation

Partial differential equations, such as Fick's second law (in which the concentration is a function of both time and space), are generally more difficult to solve than *total* differential equations, in which the dependent variable is a function of only one independent variable. An example of a total differential equation (of second order[9]) is the linearized Poisson–Boltzmann equation

$$\frac{1}{r^2}\frac{d}{dr}\left(r^2\frac{d\psi}{dr}\right) = \kappa^2\psi$$

It has been shown (Section 3.3.7) that the solution of this equation (with ψ dependent on r only) was easily accomplished.

One may conclude therefore that the solution of Fick's second law (a partial differential equation) would proceed smoothly if some mathematical device could be utilized to convert it into the form of a total differential equation. The Laplace transformation method is often used as such a device.

Since the method is based on the operation[10] of the Laplace transformation, a digression on the nature of this operation is given before using it to solve the partial differential equation involved in nonsteady-state electrochemical diffusion problems, namely, Fick's second law.

Consider a function y of the variable z, i.e., $y = f(z)$, represented by the plot of y against z. The familiar operation of differentiation performed on the function y consists in finding the slope of the curve representing $y = f(z)$ for various values of z, i.e., the

---

[8]This point is dealt with at greater length in Section 4.2.9.

[9]The order of a differential equation is the order of its highest derivatives, which in the example quoted is a second-order derivative, $d^2\psi/dr^2$.

[10]A mathematical operation is a rule for converting one function into another.

differentiation operation consists in evaluating dy/dz. The integration operation consists in finding the area under the curve, i.e., it consists in evaluating $\int_{z_1}^{z_2} f(z)\, dz$.

The operation of Laplace transformation performed on the function $y = f(z)$ consists of two steps:

1. Multiplying $y = f(z)$ by e^{-pz}, where p is a positive quantity that is independent of z
2. Integrating the resulting product ye^{-pz} [$= f(z)e^{-pz}$] with respect to z between the limits $z = 0$ and $z = \infty$

In short, the Laplace transform $y = f(z)$ is

$$\int_0^\infty e^{-pz} y\, dz \quad \text{or} \quad \int_0^\infty e^{-pz} f(z)\, dz$$

Just as one often symbolizes the result of the differentiation of y by y', the result of the operation of Laplace transformation performed on y is often represented by a symbol $\bar{y}$. Thus,

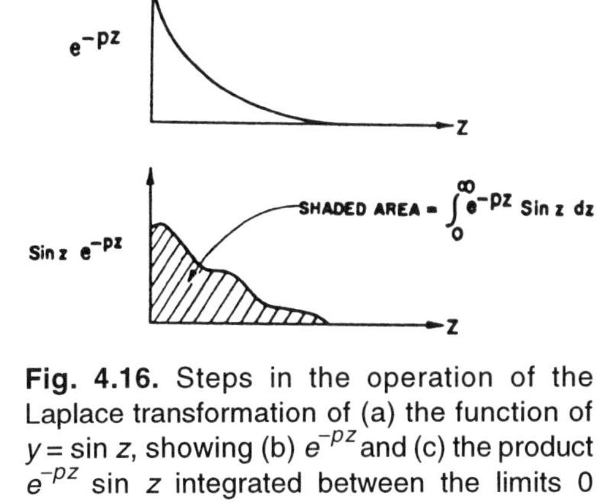

Fig. 4.16. Steps in the operation of the Laplace transformation of (a) the function of $y = \sin z$, showing (b) e^{-pz} and (c) the product $e^{-pz} \sin z$ integrated between the limits 0 and ∞.

TABLE 4.5
Laplace Transforms

Function	Transform
$f(t)$	$\bar{f} = \int_0^\infty e^{-pt} f(t)\, dt$
1	$\frac{1}{p}$
λ (a constant)	$\frac{\lambda}{p}$
$\frac{1}{\sqrt{\pi t}}$	$\frac{1}{\sqrt{p}}$
$2\sqrt{\frac{t}{\pi}}$	$p^{-3/2}$
$e^{\omega t}$	$\frac{1}{p-\omega}$
$\frac{1}{\sqrt{\pi t}} \exp\left(-\frac{k^2}{4t}\right)$	$\frac{1}{\sqrt{p}} e^{-k\sqrt{p}}$ $[k \geq 0]$
$2\sqrt{\frac{t}{\pi}} \exp\left(-\frac{k^2}{4t}\right) - k\,\mathrm{erfc}\left(\frac{k}{2\sqrt{t}}\right)$	$p^{-3/2} e^{-k\sqrt{p}}$ $[k \geq 0]$
$\cos \omega t$	$\frac{p}{p^2 + \omega^2}$

$$\bar{y} = \text{Laplace transform of } y = \int_0^\infty e^{-pz} y\, dz \tag{4.33}$$

What happens during Laplace transformation can be easily visualized by choosing a function, say, $y = \sin z$, and representing the operation in a figure (Fig. 4.16). It can be seen[11] that the operation consists in finding the area under the curve ye^{-pz} between

[11]From Fig. 4.16, it can also be seen that apart from having to make the integral converge, the exact value of p is not significant because p disappears after the operation of inverse transformation (see Section 4.2.11).

the limits $z = 0$ and $z = \infty$. The Laplace transforms of some functions encountered in diffusion problems are collected in Table 4.5.

4.2.9. Laplace Transformation Converts the Partial Differential Equation into a Total Differential Equation

It will now be shown that by using the operation of Laplace transformation, Fick's second law—a partial differential equation—is converted into a total differential equation that can be readily solved. Since whatever operation is carried out on the left-hand side of an equation must be repeated on the right-hand side, both sides of Fick's second law will be subject to the operation of Laplace transformation (*cf.* Eq. (4.33)]

$$\int_0^\infty e^{-pt}\frac{\partial c}{\partial t}\,dt = \int_0^\infty e^{-pt}\,D\frac{\partial^2 c}{\partial x^2}\,dt \tag{4.34}$$

which, by using the symbol for a Laplace-transformed function, can be written

$$\overline{\frac{\partial c}{\partial t}} = D\,\overline{\frac{\partial^2 c}{\partial x^2}} \tag{4.35}$$

To proceed further, one must evaluate the integrals of Eq. (4.34). Consider the Laplace transform

$$\overline{\frac{\partial c}{\partial t}} = \int_0^\infty e^{-pt}\frac{\partial c}{\partial t}\,dt \tag{4.36}$$

The integral can be evaluated by the rule for integration by parts as follows:

$$\int_0^\infty \underset{u}{e^{-pt}}\,\underset{dv}{\frac{\partial c}{\partial t}\,dt} = \underset{u}{e^{-pt}}\underset{v}{\int_0^\infty \partial c} - \int_0^\infty\left[\underset{v}{\int_0^\infty \partial c}\right]\underset{du}{de^{-pt}}$$

$$= \left[e^{-pt}c\right]_0^\infty + p\int_0^\infty e^{-pt}\,c\,dt \tag{4.37}$$

Since $\int_0^\infty e^{-pt}c\,dt$ is in fact the Laplace transform of c [*cf.* the defining equation (4.33)], and for conciseness is represented by the symbol $\overline{c}$, and since e^{-pt} is zero when $t \to \infty$ and unity when $t = 0$, Eq. (4.36) reduces to

$$\overline{\frac{\partial c}{\partial t}} = \int_0^\infty e^{-pt} \frac{\partial c}{\partial t}\, dt = -\, c[t=0] + p\bar{c} \tag{4.38}$$

where $c[t = 0]$ is the value of the concentration c at $t = 0$.

Next, one must evaluate the integral on the right-hand side of Eq. (4.34), i.e.,

$$D\,\overline{\frac{\partial^2 c}{\partial x^2}} = \int_0^\infty e^{-pt}\, D\,\frac{\partial^2 c}{\partial x^2}\, dt \tag{4.39}$$

Since the integration is with respect to the variable t and the differentiation is with respect to x, their order can be interchanged. Furthermore, one can move the constant D outside the integral sign. Hence, one can write

$$D\,\overline{\frac{\partial^2 c}{\partial x^2}} = D\,\frac{\partial^2}{\partial x^2} \int_0^\infty e^{-pt} c\, dt \tag{4.40}$$

Once again, it is clear from Eq. (4.33) that $\int_0^\infty e^{-pt} c\, dt$ is the Laplace transform of c, i.e., $\bar{c}$, and therefore

$$D\,\overline{\frac{\partial^2 c}{\partial x^2}} = D\,\frac{\partial^2}{\partial x^2}\,\bar{c} \tag{4.41}$$

From Eqs. (4.32), (4.38), and (4.41), it follows that after Laplace transformation, Fick's second law takes the form

$$p\bar{c} - c[t=0] = D\,\frac{d^2\bar{c}}{dx^2} \tag{4.42}$$

This, however, is a total differential equation because it contains only the variable x. Thus, by using the operation of Laplace transformation, Fick's second law has been converted into a more easily solvable total differential equation involving $\bar{c}$, the Laplace transform of the concentration.

4.2.10. Initial and Boundary Conditions for the Diffusion Process Stimulated by a Constant Current (or Flux)

A differential equation can be arrived at by differentiating an original equation, or *primitive*, as it is called. In the case of Fick's second law, the primitive is the equation that gives the precise nature of the functional dependence of concentration on space and time; i.e., the primitive is an elaboration on

$$c = f(x, t)$$

Since, in the process of differentiation, constants are eliminated and since three differentiations (two with respect to x and one with respect to time) are necessary to arrive at Fick's second law, three constants have been eliminated in the process of going from the precise concentration dependence that characterizes a particular problem to the general relation between the time and space derivatives of concentration that describes any nonstationary diffusion situation.

The three characteristics, or *conditions*, as they are called, of a particular diffusion process cannot be rediscovered by mathematical argument applied to the differential equation. To get at the three conditions, one has to resort to a physical understanding of the diffusion process. Only then can one proceed with the solution of the (now) total differential equation (4.42) and get the precise functional relationship between concentration, distance, and time.

Instead of attempting a general discussion of the three conditions characterizing a particular diffusion problem, it is best to treat a typical electrochemical diffusion problem. Consider that in an electrochemical system a constant current is switched on at a time arbitrarily designated $t = 0$ (Fig. 4.17). The current is due to charge-transfer reactions at the electrode–solution interfaces, and these reactions consume a species. Since the concentration of this species at the interface falls below the bulk concentration, a concentration gradient for the species is set up and it diffuses toward the interface. Thus, the externally controlled current sets up[12] a diffusion flux within the solution.

The diffusion is described by Fick's second law

$$\frac{\partial c}{\partial t} = D \frac{\partial^2 c}{\partial x^2} \tag{4.32}$$

or, after Laplace transformation, by

$$p\bar{c} - c[t = 0] = D \frac{d^2 \bar{c}}{dx^2} \tag{4.42}$$

To analyze the diffusion problem, one must solve the differential equation, i.e., describe how the concentration of the diffusing species varies with distance x from the electrode and with the time that has elapsed since the constant current was switched

[12]When the externally imposed current sets up charge-transfer reactions that provoke the diffusion of ions, there is a very simple relation between the current density and the diffusion flux. The diffusion flux is a mole flux (number of moles crossing 1 cm^2 in 1 s), and the current density is a charge flux (Table 4.1). Hence, the current density j, or charge flux, is equal to the charge zF per mole of ions (z is the valence of the diffusing ion and F is the Faraday constant) times the diffusion flux J, i.e., $j = zFJ$.

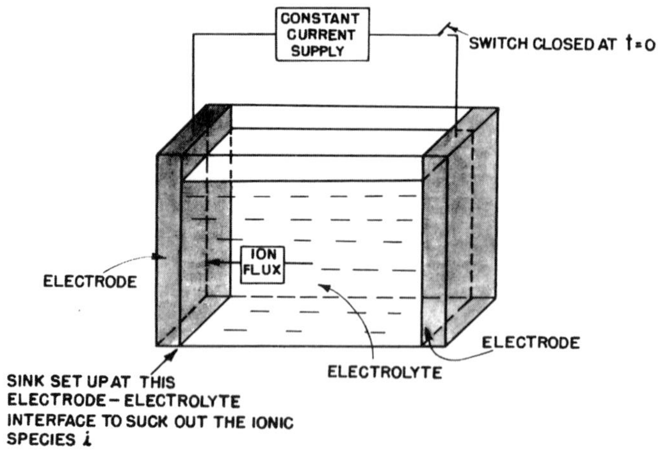

Fig. 4.17. Schematic representation of an electrochemical system connected to a constant current supply that is switched on at $t = 0$. The current promotes a charge-transfer reaction at the electrode–electrolyte interfaces, which results in the diffusion flux of the species i toward the interface.

on. First one must think out the three characteristics, or conditions, of the diffusion process described above.

The nature of one of the conditions becomes clear from the term $c[t = 0]$ in the Laplace-transformed version [see Eq. (4.42)] of Fick's second law. The term $c[t = 0]$ refers to the concentration before the start of diffusion; i.e., it describes the initial condition of the electrolytic solution in which diffusion is made to occur by the passage of a constant current. Since before the constant current is switched on and diffusion starts, one has an unperturbed system, the concentration c of the species that subsequently diffuses must be the same throughout the system and equal to the bulk concentration c^0. Thus, the initial condition of the electrolytic solution is

$$c[t = 0] = c^0 \tag{4.43}$$

The other two conditions pertain to the situation after the diffusion begins, e.g., after the diffusion-causing current is switched on. Since these two conditions often pertain to what is happening to the boundaries of the system (in which diffusion is occurring), they are usually known as *boundary conditions.*

The first boundary condition is the expression of an obvious point, namely, that very far from the boundary at which the diffusion source or sink is set up, the concentration of the diffusing species is unperturbed and remains the same as in the initial condition

$$c[x \to \infty] = c[t = 0] = c^0 \tag{4.44}$$

Thus, the concentration of the diffusing species has the same value c^0 at any x at $t = 0$ or for any $t > 0$ at $x \to \infty$. This is true for almost all electrochemical diffusion problems in which one switches on (at $t = 0$) the appropriate current or potential difference across the interface and thus sets up interfacial charge-transfer reactions which, by consuming or producing a species, provoke a diffusion flux of that species.

What is characteristic of one particular electrochemical diffusion process and distinguishes it from all others is the nature of the diffusion flux that is started off at $t = 0$. Thus, the essential characteristic of the diffusion problem under discussion is the switching on of the constant current, which means that the diffusing species is consumed at a constant rate at the interface and the species diffuses across the interface at a constant rate. In other words, the flux of the diffusing species at the $x = 0$ boundary of the solution is a constant.

It is convenient from many points of view to assume that the constant value of the flux is unity, i.e., 1 mole of the diffusing species crossing 1 cm^2 of the electrode–solution interface per second. This unit flux corresponds to a constant current density of 1 A cm^{-2}. This normalization of the flux scarcely affects the generality of the treatment because it will later be seen that the concentration response to an arbitrary flux can easily be obtained from the concentration response to a unit flux.

If one looks at the time variation of current or the flux across the solution boundary, it is seen that for $t < 0$, $J = 0$ and for $t > 0$, there is a constant flux $J = 1$ (Fig. 4.18) corresponding to the constant current switched on at $t = 0$. In other words, the time variation of the flux is like a step; that is why the flux produced in this setup is often known as a *step function* (of time).

At any instant of time, the constant flux across the boundary is related to the concentration gradient there through Fick's first law, i.e.,

$$J_{x=0} = 1 = -D\left(\frac{\partial c}{\partial x}\right)_{x=0} \tag{4.45}$$

The above initial and boundary conditions can be summarized thus:

$$c[t = 0] = c^0 \tag{4.43}$$

$$c[x \to \infty] = c^0 \tag{4.44}$$

$$\left(\frac{\partial c}{\partial x}\right)_{x=0} = -\frac{1}{D} \tag{4.45}$$

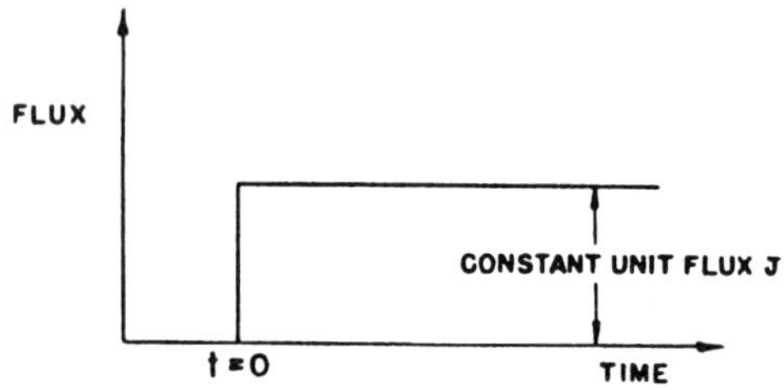

Fig. 4.18. When a constant unit flux of 1 mol $\text{cm}^{-2}\ \text{s}^{-1}$ is switched on at $t = 0$, the variation of flux with time resembles a step. (It is only an ideal switch that makes the current and therefore the flux instantaneously rise from zero to its constant value; this problem of technique is ignored in the diagram.)

The three conditions just listed describe the special features of the constant (unit)-flux diffusion problem. They will now be used to solve Fick's second law.

4.2.11. Concentration Response to a Constant Flux Switched On at $t = 0$

It has been shown (Section 4.2.9) that after Laplace transformation, Fick's second law takes the form

$$p\bar{c} - c[t = 0] = D\frac{d^2\bar{c}}{dx^2} \tag{4.42}$$

The solution of an equation of this type is facilitated if the second term is zero. This objective can be attained by introducing a new variable c_1 defined as

$$c_1 = c^0 - c \tag{4.46}$$

The variable c_1 can be recognized as the departure $c^0 - c$ of the concentration from its initial value c^0. In other words, c_1 represents the *perturbation* from the initial concentration (Fig. 4.19).

The partial differential equation [Eq. (4.32)] and the initial and boundary conditions now have to be restated in terms of the new variable c_1. This is easily done by using Eq. (4.46) in Eqs. (4.32), (4.43), (4.44), and (4.45). One obtains

$$\frac{\partial c_1}{\partial t} = D\frac{\partial^2 c_1}{\partial x^2} \tag{4.47}$$

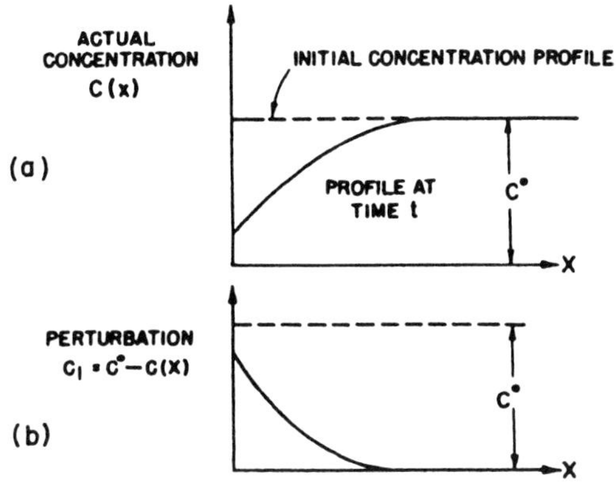

Fig. 4.19. Schematic representation of (a) the variation of concentration with distance x from the electrode at $t = 0$ and $t = t$ and (b) the variation of the perturbation $c_1 = c^0 - c_x$ in concentrations.

$$c_1[t = 0] = 0 \tag{4.48}$$

$$c_1[x \to \infty] = 0 \tag{4.49}$$

$$\left(\frac{\partial c_1}{\partial x}\right)_{x=0} = -\frac{1}{D} \tag{4.50}$$

After Laplace transformation of Eq. (4.47), the differential equation becomes

$$p\bar{c}_1 - c_1[t = 0] = D\frac{d^2\bar{c}_1}{dx^2} \tag{4.51}$$

Since, however, $c_1[t = 0] = 0$ [*cf.* Eq. (4.48)], it is clear that

$$\frac{d^2\bar{c}_1}{dx^2} = \frac{p}{D}\bar{c}_1 \tag{4.52}$$

This equation is identical in form to the linearized P–B equation [*cf.* Eq. (3.21)] and therefore must have the same general solution, i.e.,

$$\bar{c}_1 = Ae^{-(p/D)^{1/2}x} + Be^{(p/D)^{1/2}x} \tag{4.53}$$

where A and B are the arbitrary integration constants to be evaluated by the use of the boundary conditions. If the Laplace transformation method had not been used, the solution of Eq. (4.47) would not have been so simple.

The constant B must be zero by virtue of the following argument. From the boundary condition $c_1[x \to \infty] = 0$, i.e., Eq. (4.49), it is clear that after Laplace transformation,

$$\bar{c}_1[x \to \infty] = 0 \tag{4.54}$$

Hence, as $x \to \infty$, $\bar{c}_1 \to 0$, but this will be true only if $B = 0$ because otherwise $\bar{c}_1$ in Eq. (4.53) will go to infinity instead of zero.

One is left with

$$\bar{c}_1 = A e^{-(p/D)^{1/2} x} \tag{4.55}$$

Differentiating this equation with respect to x, one obtains

$$\frac{d\bar{c}_1}{dx} = -\sqrt{\frac{p}{D}} A e^{-(p/D)^{1/2} x} \tag{4.56}$$

which at $x = 0$ leads to

$$\left(\frac{d\bar{c}_1}{dx}\right)_{x=0} = -\sqrt{\frac{p}{D}} A \tag{4.57}$$

Another expression for $(d\bar{c}_1/dx)_{x=0}$ can be obtained by applying the operation of Laplace transformation to the constant-flux boundary condition (4.50). Laplace transformation on the left-hand side of the boundary condition leads to $(d\bar{c}_1/dx)_{x=0}$; and the same operation performed on the right-hand side, to $-1/Dp$ (Appendix 4.2). Thus, from the boundary condition (4.50) one gets

$$\left(\frac{d\bar{c}_1}{dx}\right)_{x=0} = -\frac{1}{Dp} \tag{4.58}$$

Hence, from Eqs. (4.57) and (4.58), it is found that

$$A = \frac{1}{p^{3/2} D^{1/2}} \tag{4.59}$$

Upon inserting this expression for A into Eq. (4.55), it follows that

$$\bar{c}_1 = \frac{1}{D^{1/2} p^{3/2}} e^{-(p/D)^{1/2} x} \tag{4.60}$$

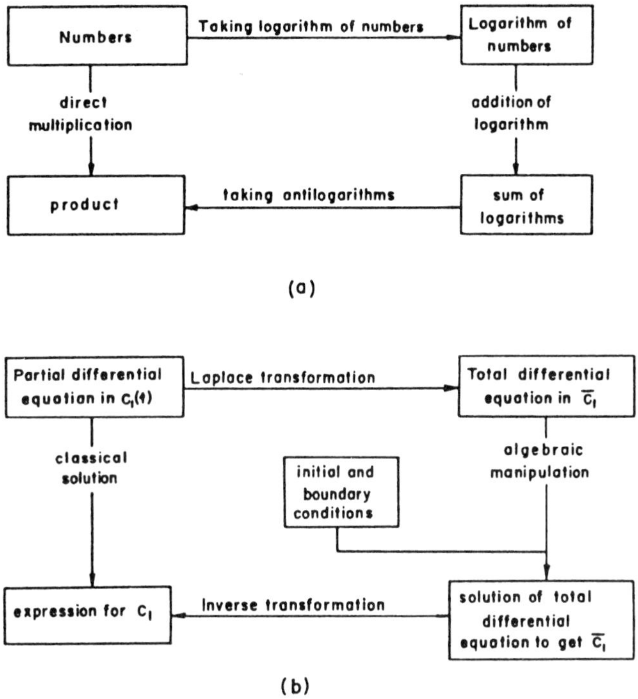

Fig. 4.20. Comparison of the use of (a) logarithms and (b) Laplace transformation.

The ultimate aim, however, is not to get an expression for $\bar{c}_1$, the Laplace transform of c_1, but to get an expression for c_1 (or c) as a function of distance x and time t. The expression $\bar{c}_1$ has been obtained by a Laplace transformation of c_1; hence, to go from $\bar{c}_1$ to c_1, one must do an inverse transformation. The situation is analogous to using logarithms to facilitate the working-out of a problem (Fig. 4.20). In order to get c_1 from $\bar{c}_1$, one asks the question: Under Laplace transformation, what function c_1 would give the Laplace transform $\bar{c}_1$ of Eq. (4.60)? In other words, one has to find c_1 in the equation

$$\int_0^\infty e^{-pt} c_1 \, dt = \bar{c}_1 = \frac{1}{D^{1/2} p^{3/2}} e^{-(p/D)^{1/2} x} \tag{4.61}$$

A mathematician would find the function c_1 (i.e., do the inverse transformation) by making use of the theory of functions of variables which are complex. Since, however, there are extensive tables of functions y and their transforms $\bar{y}$, it is only necessary to look up the column of Laplace transforms in the tables (Table 4.5). It is seen that corresponding to the transform of the equation arising from Eq. (4.60)

$$p^{-3/2}\exp\left(-\frac{x}{\sqrt{D}}p^{1/2}\right)$$

is the function

$$2\pi^{-1/2}t^{1/2}\exp\left(-\frac{x^2}{4Dt}\right)-xD^{-1/2}\operatorname{erfc}\left(\frac{x^2}{4Dt}\right)^{1/2}$$

where erfc is the *error function complement* defined thus:

$$\operatorname{erfc}(y)=1-\operatorname{erf}(y) \tag{4.62}$$

erf(y) being the *error function* given by (Fig. 4.21)

$$\operatorname{erf}(y)=\frac{2}{\sqrt{\pi}}\int_0^y e^{-u^2}\,du \tag{4.63}$$

Hence, the expression for the concentration perturbation c_1 in Eq. (4.61) must be

$$c_1=\frac{1}{D^{1/2}}\left[\frac{2t^{1/2}}{\pi^{1/2}}\exp\left(-\frac{x^2}{4Dt}\right)-xD^{-1/2}\operatorname{erfc}\left(\frac{x^2}{4Dt}\right)^{1/2}\right] \tag{4.64}$$

If one is interested in the true concentration c, rather than the deviation c_1 in the concentration from the initial value c^0, one must use the defining equation for c_1

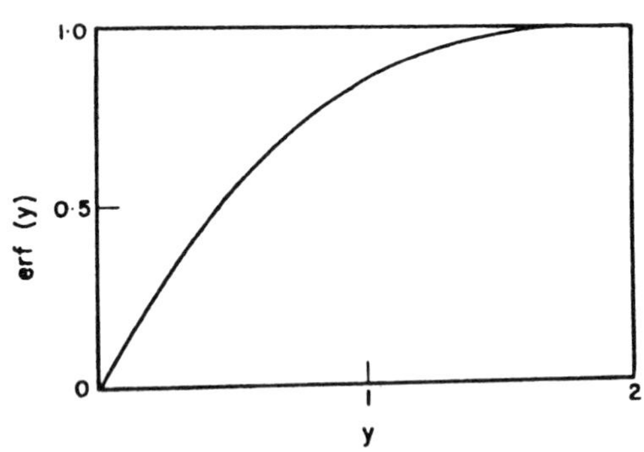

Fig. 4.21. Variation of the error function erf(y) with argument y.

$$c_1 = c^0 - c \tag{4.46}$$

The result is

$$c = c^0 - \frac{1}{D^{1/2}}\left[\frac{2t^{1/2}}{\pi^{1/2}}\exp\left(-\frac{x^2}{4Dt}\right) - \frac{x}{D^{1/2}}\operatorname{erfc}\left(\frac{x^2}{4Dt}\right)^{1/2}\right] \tag{4.65}$$

This then is the fundamental equation showing how the concentration of the diffusing species varies with distance x from the electrode–solution interface and with

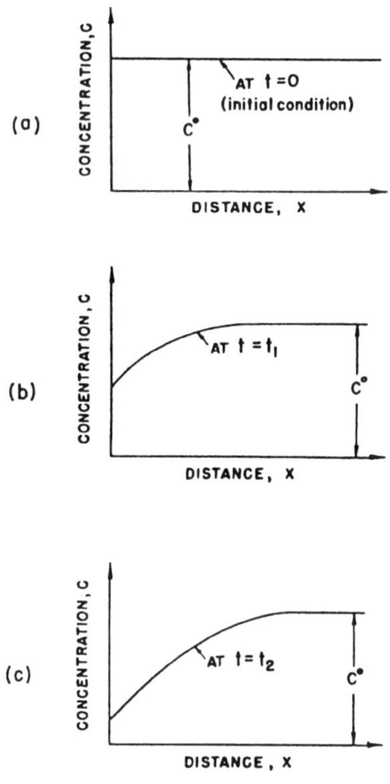

Fig. 4.22. Graphical representation of the variation of the concentration c with distance x from the electrode or diffusion sink. (a) The initial condition at $t = 0$; (b) and (c) the conditions at t_1 and t_2, where $t_2 > t_1 > t = 0$. Note that, at $t > 0$, $(dc/dx)_{x=0}$ is a constant, as it should be in the constant-flux diffusion problem.

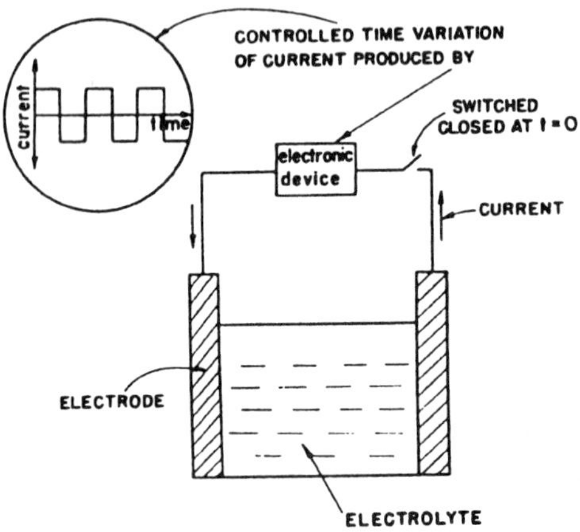

Fig. 4.23. Use of an electronic device to vary the current passing through an electrochemical system in a controlled way.

the time t that has elapsed since a constant unit flux was switched on. In other words, Eq. (4.65) describes the diffusional response of an electrolytic solution to the stimulus of a flux which is in the form of a step function of time. The nature of the response is best appreciated by seeing how the concentration profile of the species diffusing varies as a function of time [Fig. 4.22(a), (b), and (c)]. Equation (4.65) is also of fundamental importance in describing the response of an electrochemical system to a current density that varies as a step function, i.e., to a constant current density switched on at $t = 0$.

4.2.12. How the Solution of the Constant-Flux Diffusion Problem Leads to the Solution of Other Problems

The space and time variation of a concentration in response to the switching on of a constant flux has been analyzed. Suppose, however, that, instead of a constant flux, one switches on a sinusoidally varying flux.[13] What is the resultant space and time variation of the concentration of the diffusing species?

One approach to this question is to set up the new diffusion problem with the initial and boundary conditions characteristic of the sinusoidally varying flux and to obtain a solution. There is, however, a simpler approach. Using the property of Laplace transforms, one can use the solution (4.65) of the constant-flux diffusion problem to generate solutions for other problems.

[13]If one feels that *current* is a more familiar word than flux, one can substitute the word current because these diffusion fluxes are often, but not always, provoked by controlling the current across an electrode-solution interface.

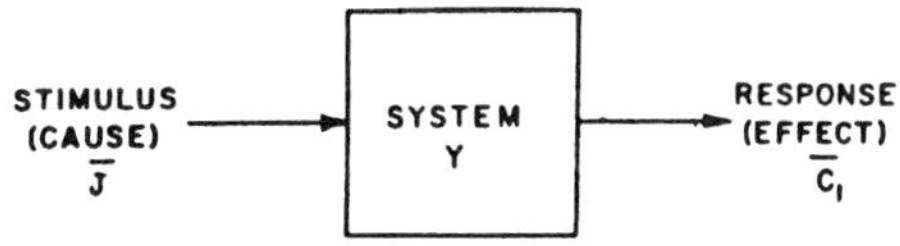

Fig. 4.24. Schematic representation of the response $\bar{c}_1$ of a system Y to a stimulus $\bar{J}$.

An electronic device is connected to an electrochemical system so that it switches on a current (Fig. 4.23) that is made to vary with time in a controllable way. The current provokes charge-transfer reactions that lead to a diffusion flux of the species involved in the reactions. This diffusion flux varies with time in the same way as the current. The time variation of the flux will be represented thus

$$\boldsymbol{J} = g(t) \tag{4.66}$$

The imposition of this time-varying flux stimulates the electrolytic solution to respond with a space and time variation in the concentration c or a perturbation in concentration, c_1. The response depends on the stimulus, and the mathematical relationship between the cause $\boldsymbol{J}(t)$ and the effect c_1 can be represented (Fig. 4.24) quite generally thus[14]:

$$\bar{c}_1 = y\bar{J} \tag{4.67}$$

where $\bar{c}_1$ and $\bar{J}$ are the Laplace transforms of the perturbation in concentration and the flux, and y is the to-be-determined function that links the cause and effect and is characteristic of the system.

The relationship (4.67) has been defined for a flux that has an arbitrary variation with time; hence, it must also be true for the constant unit flux described in Section 4.2.10. The Laplace transform of this constant unit flux $\boldsymbol{J} = 1$ is $1/p$ according to Appendix 4.2; and the Laplace transform of the concentration response to the constant unit flux is given by Eq. (4.60), i.e.,

$$\bar{c}_1 = D^{-1/2} p^{-3/2} \exp\left(-\sqrt{\frac{p}{D}}x\right) \tag{4.60}$$

Hence, by substituting $1/p$ for $\bar{J}$ and $D^{-1/2}p^{-3/2}e^{-(p/D)^{1/2}x}$ for $\bar{c}_1$ in Eq. (4.67), one has

[14]It will be seen further on that one uses a relationship between the Laplace transforms of the concentration perturbation and the flux rather than the quantities c_1 and J themselves, because the treatment in Laplace transforms is not only elegant but fruitful.

$$y = D^{-1/2} p^{-1/2} \exp\left(-\sqrt{\frac{p}{D}}\, x\right) \tag{4.68}$$

On introducing this expression for y into the general relationship (4.67), the result is

$$\bar{c}_1 = \left[D^{-1/2} p^{-1/2} \exp\left(-\sqrt{\frac{p}{D}}\, x\right)\right]\bar{J} \tag{4.69}$$

This is an important result: Through the evaluation of y, it contains the concentration response to a constant unit flux switched on at $t = 0$. In addition, it shows how to obtain the concentration response to a flux $\boldsymbol{J}(t)$ that is varying in a known way. All one has to do is to take the Laplace transform $\bar{J}$ of this flux $\boldsymbol{J}(t)$ switched on at time $t = 0$, substitute this $\bar{J}$ in Eq. (4.69), and get $\bar{c}_1$. If one inverse-transforms the resulting expression for $\bar{c}_1$, one will obtain c_1, the perturbation in concentration as a function of x and t.

Consider a few examples. Suppose that instead of switching on a constant unit flux at $t = 0$ (see Section 4.2.10), one imposes a flux that is a constant but now has a magnitude of λ mol cm^{-2} s^{-1}, i.e., $\boldsymbol{J} = \lambda$. Since the transform of a constant is $1/p$ times the constant (Appendix 4.2), one obtains

$$\bar{J} = \frac{\lambda}{p} \tag{4.70}$$

which when introduced into Eq. (4.69) gives

$$\bar{c}_1 = \frac{\lambda}{D^{1/2} p^{3/2}}\, e^{-(x/D^{1/2})p^{1/2}} \tag{4.71}$$

The inverse transform of the right-hand side of Eq. (4.71) is identical to that for the unit step function [*cf.* Eq. (4.60)] except that it is multiplied by λ. That is,

$$c_1 = \frac{\lambda}{D^{1/2}}\left[\frac{2t^{1/2}}{\pi^{1/2}} \exp\left(-\frac{x^2}{4Dt}\right) - \frac{x}{D^{1/2}}\, \text{erfc}\left(\frac{x^2}{4Dt}\right)^{1/2}\right] \tag{4.72}$$

In other words, the concentration response of the system to a $J = \lambda$ flux is a magnified-λ-times version of the response to a constant unit flux.

One can also understand what happens if instead of sucking ions out of the system, the flux acts as a source and pumps in ions. This condition can be brought about by changing the direction of the constant current going through the interface and thus changing the direction of the charge-transfer reactions so that the diffusing species is produced rather than consumed. Thus, diffusion from the interface into the solution occurs. Because the direction of the flux vector is reversed, one has

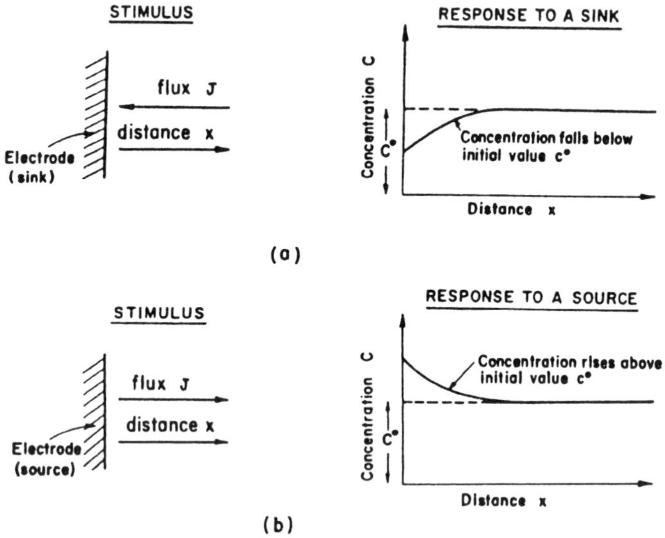

Fig. 4.25. The difference in the concentration response to the stimulus of (a) a sink and (b) a source at the electrode–electrolyte interface.

$$\boldsymbol{J} = -\lambda \quad \text{or} \quad \bar{J} = -\frac{\lambda}{p} \tag{4.73}$$

and thus

$$c_1 = \frac{-\lambda}{D^{1/2}}\left[\frac{2t^{1/2}}{\pi^{1/2}}\exp\left(-\frac{x^2}{4Dt}\right) - \frac{x}{D^{1/2}}\operatorname{erfc}\left(\frac{x^2}{4Dt}\right)^{1/2}\right] \tag{4.74}$$

or, in view of Eq. (4.46),

$$c = c^0 + \frac{\lambda}{D^{1/2}}\left[\frac{2t^{1/2}}{\pi^{1/2}}\exp\left(-\frac{x^2}{4Dt}\right) - \frac{x}{D^{1/2}}\operatorname{erfc}\left(\frac{x^2}{4Dt}\right)^{1/2}\right] \tag{4.75}$$

Note the plus sign; it indicates that the concentration c rises above the initial value c^0 (Fig. 4.25).

Consider now a more interesting type of stimulus involving a periodically varying flux (Fig. 4.26). After representing the imposed flux by a cosine function

$$\boldsymbol{J} = \boldsymbol{J}_{\max}\cos\omega t \tag{4.76}$$

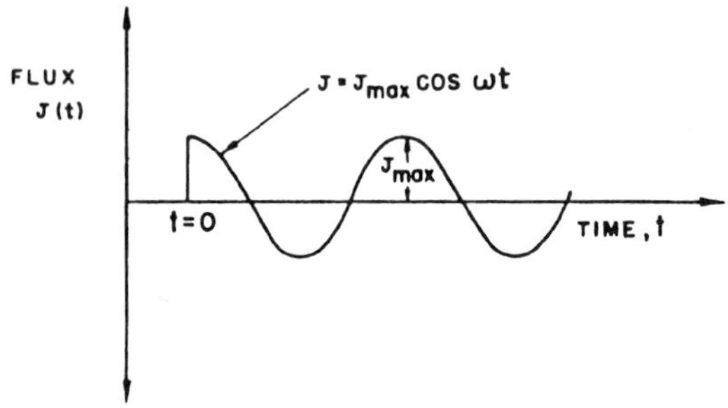

Fig. 4.26. A sinusoidally varying flux at an electrode–electrolyte interface, produced by passing a sinusoidally varying flux through the electrochemical system.

its Laplace transform is (see Table 4.5)

$$\bar{J} = J_{\max} \frac{p}{p^2 + \omega^2} \tag{4.77}$$

When this is combined with Eq. (4.69), one gets

$$\bar{c}_1 = \frac{p}{p^2 + \omega^2} \left[\frac{1}{D^{1/2} p^{1/2}} e^{-(x/D^{1/2}) p^{1/2}} \right] J_{\max} \tag{4.78}$$

To simplify matters, the response of the system will only be considered at the boundary, i.e., at $x = 0$. Hence, one can set $x = 0$ in Eq. (4.78), in which case,

$$\bar{c}_1[x = 0] = \frac{p}{p^2 + \omega^2} \frac{1}{D^{1/2} p^{1/2}} J_{\max} \tag{4.79}$$

The inverse transform reads

$$c_1[x = 0] = \frac{J_{\max}}{(D\omega)^{1/2}} \cos\left(\omega t - \frac{\pi}{4}\right) \tag{4.80}$$

which shows that, corresponding to a periodically varying flux (or current), the concentration perturbation also varies periodically, but there is a $\pi/4$ phase difference between the flux and the concentration response (Fig. 4.27).

This is an extremely important result because an alternating flux can be produced by an alternating current density at the electrode–electrolyte interface, and in the case of sufficiently fast charge-transfer reactions, the concentration at the boundary is related to the potential difference across the interface. Thus, the current density and

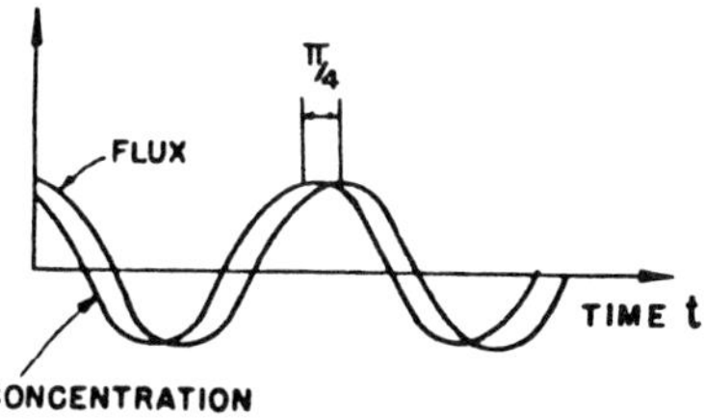

Fig. 4.27. When a sinusoidally varying diffusion flux is produced in an electrochemical system by passing a sinusoidally varying current through it, the perturbation in the concentration of the diffusing species also varies sinusoidally with time, but with a phase difference of $\pi/4$.

the potential difference both vary periodically with time, and it turns out that the phase relationship between them provides information on the rate of the charge-transfer reaction.

4.2.13. Diffusion Resulting from an Instantaneous Current Pulse

There is another important diffusion problem, the solution of which can be generated from the concentration response to a constant current (or a flux). Consider that in an electrochemical system there is a plane electrode at the boundary of the electrolyte. Now, suppose that with the aid of an electronic pulse generator, an extremely short time current pulse is sent through the system (Fig. 4.28). The current is directed so as to dissolve the metal of the electrode; hence, the effect of the pulse is to produce a burst of metal dissolution in which a layer of metal ions is piled up at the interface (Fig. 4.29).

Because the concentration of metal ions at the interface is far in excess of that in the bulk of the solution, diffusion into the solution begins. Since the source of the diffusing ions is an ion layer parallel to the plane electrode, it is known as a *plane source*; and since the diffusing ions are produced in an instantaneous pulse, a fuller description of the source is contained in the term *instantaneous plane source*.

As the ions from the instantaneous plane source diffuse into the solution, their concentration at various distances will change with time. The problem is to calculate the distance and time variation of this concentration.

The starting point for this calculation is the general relation between the Laplace transforms of the concentration perturbation c_1 and the time-varying flux $\boldsymbol{J}(t)$

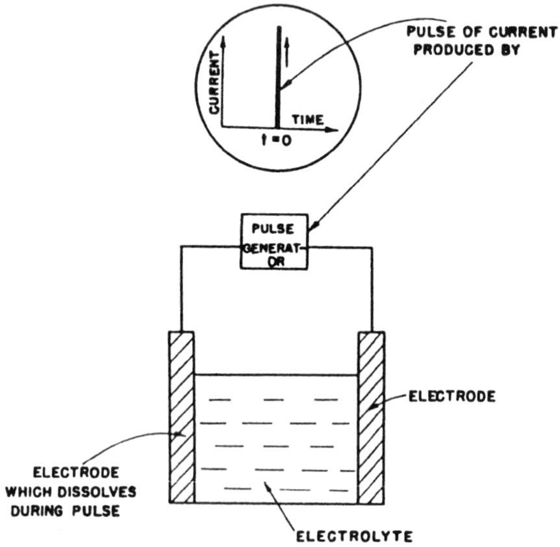

Fig. 4.28. The use of an electronic pulse generator to send an extremely short time current pulse through an electrochemical system so that there is dissolution at one electrode during the pulse.

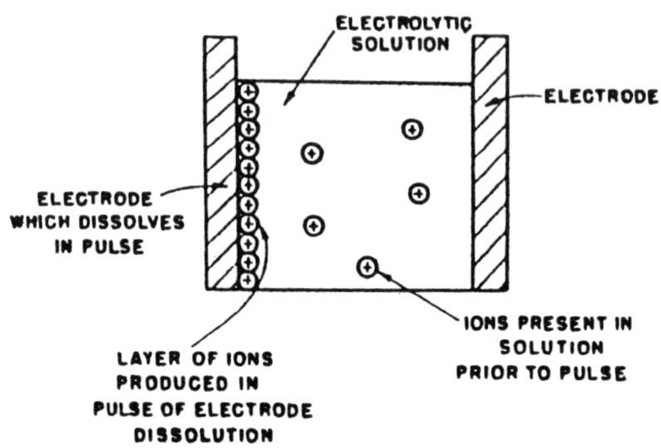

Fig. 4.29. The burst of electrode dissolution during the current pulse produces a layer of ions adjacent to the dissolving electrode (negative ions are not shown in the diagram).

$$\bar{c}_1 = D^{-1/2} p^{-1/2} \exp\left(-\sqrt{\frac{p}{D}}x\right)\bar{J} \tag{4.69}$$

One has to substitute for $\bar{J}$ the Laplace transform of a flux that is an instantaneous pulse. This is done with the help of the following interesting observation.

If one takes any quantity that varies with time as a step, then the differential of that quantity with respect to time varies with time as an instantaneous pulse (Fig. 4.30). In other words, the time derivative of a step function is an instantaneous pulse. Suppose therefore one considers a constant flux (or current) switched on at $t = 0$ (i.e., the flux is a step function of time and will be designated $\boldsymbol{J}_{\text{step}}$); then the time derivative of that constant flux is a pulse of flux (or current) at $t = 0$, referred to by the symbol $\boldsymbol{J}_{\text{pulse}}$, i.e.,

$$\boldsymbol{J}_{\text{pulse}} = \frac{d}{dt}\boldsymbol{J}_{\text{step}} \tag{4.81}$$

If, now, one takes Laplace transforms of both sides and uses Eq. (4.38) to evaluate the right-hand side, one has

$$\bar{J}_{\text{pulse}} = p\bar{J}_{\text{step}} - \bar{J}_{\text{step}}\,[t = 0] \tag{4.82}$$

But at $t = 0$, the magnitude of a flux that is a step function of time is zero. Hence,

$$\bar{J}_{\text{pulse}} = p\bar{J}_{\text{step}} \tag{4.83}$$

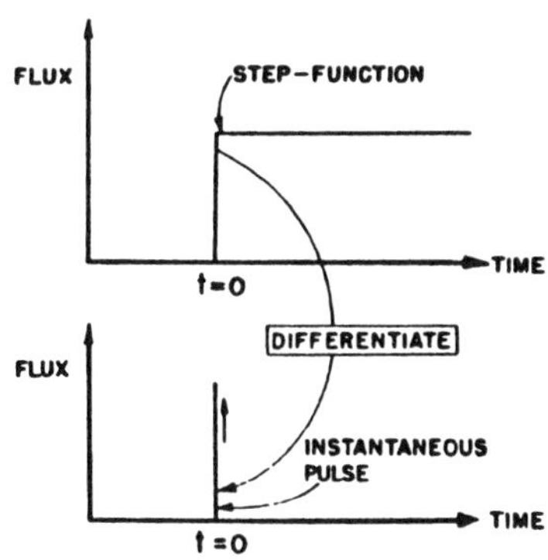

Fig. 4.30. The time derivative of a flux that is a step function of time is an instantaneous pulse of flux.

If the pumping of diffusing particles into the system by the step-function flux consists in switching on, at $t = 0$, a flux of $-\lambda$ mol cm^{-2} s^{-1}, then

$$\boldsymbol{J}_{\text{step}} = -\lambda \tag{4.84}$$

and

$$\bar{J}_{\text{step}} = -\frac{\lambda}{p} \tag{4.85}$$

Using this relation in Eq. (4.83), one has

$$\bar{J}_{\text{pulse}} = -\lambda \tag{4.86}$$

and, by substitution in Eq. (4.69),

$$\bar{c}_1 = -\lambda D^{-1/2} p^{-1/2} \exp\left(-\sqrt{\frac{p}{D}}\, x\right) \tag{4.87}$$

By inverse transformation (Table 4.5),

$$c_1 = -\lambda(\pi Dt)^{-1/2} \exp\left(-\frac{x^2}{4Dt}\right) \tag{4.88}$$

or by referring to the actual concentration c instead of the perturbation c_1 in concentration, the result is

$$c = c^0 + \lambda(\pi Dt)^{-1/2} \exp\left(-\frac{x^2}{4Dt}\right) \tag{4.89}$$

If, prior to the current pulse, there is a zero concentration of the species produced by metal dissolution, i.e.,

$$c[t = 0] = c^0 = 0 \tag{4.90}$$

then Eq. (4.89) reduces to

$$c = \frac{\lambda}{(\pi Dt)^{1/2}}\, e^{-x^2/4Dt} \tag{4.91}$$

It can be seen from Eq. (4.86) that λ is the Laplace transform of the pulse of flux. However, a Laplace transform is an integral with respect to time. Hence, λ, which is a flux (of moles per square centimeter per second) in the constant-flux problem (see Section 4.2.12), is in fact the total concentration (moles per square centimeter) of the

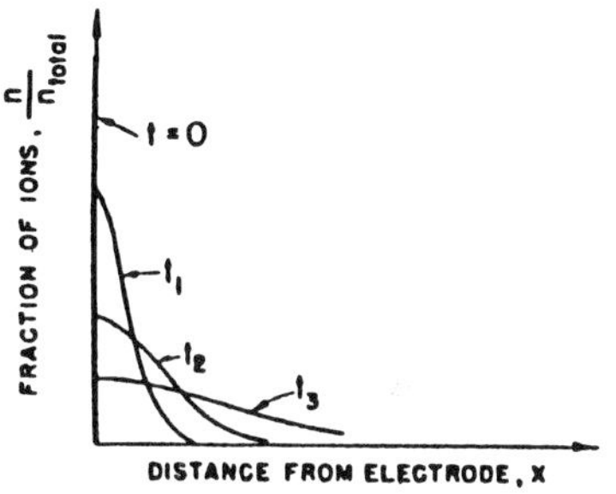

Fig. 4.31. Plots of the fraction n/n_{total} of ions (produced in the pulse of electrode dissolution) against the distance x from the electrode. At $t = 0$, all the ions are on the $x = 0$ plane, and at $t > 0$, they are distributed in the solution as a result of dissolution and diffusion. In the diagram, $t_3 > t_2 > t_1 > 0$, and the distribution curve becomes flatter and flatter.

diffusing ions produced on the $x = 0$ plane in the burst of metal dissolution. If, instead of dealing with concentrations, one deals with *numbers* of ions, the result is

$$n = \frac{n_{total}}{(\pi Dt)^{1/2}} e^{-x^2/4Dt} \tag{4.92}$$

where n is the number of ions at a distance x and a time t, and n_{total} is the number of ions set up on the $x = 0$ plane at $t = 0$; i.e., n_{total} is the total number of diffusing ions.

This is the solution to the instantaneous-plane-source problem. When n/n_{total} is plotted against x for various times, one obtains curves (Fig. 4.31) that show how the ions injected into the $x = 0$ plane at $t = 0$ (e.g., ions produced at the electrode in an impulse of metal dissolution) are distributed in space at various times. At any particular time t, a semi-bell-shaped distribution curve is obtained that shows that the ions are mainly clustered near the $x = 0$ plane, but there is a "spread." With increasing time, the spread of ions increases. This is the result of diffusion, and after an infinitely long time, there are equal numbers of ions at any distance.

4.2.14. Fraction of Ions Traveling the Mean Square Distance $\langle x^2 \rangle$ in the Einstein–Smoluchowski Equation

In the previous section, an experiment was described in which a pulse of current dissolves out of the electrode a certain number n_{total} of ions, which then start diffusing

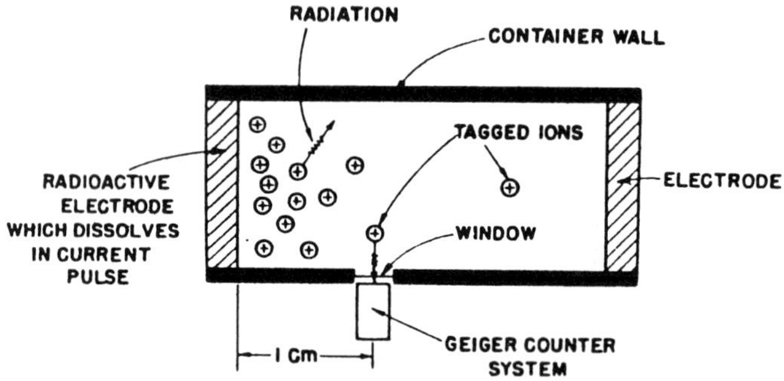

Fig. 4.32. Schematic of experiment to note the time interval between the pulse of electrode dissolution (at $t = 0$) and the arrival of radioactive ions at the window where they are registered in the Geiger-counter system.

into the solution. Now suppose that the electrode material is made radioactive so that the ions produced by dissolution are detectable by a counter (Fig. 4.32). The counter head is then placed near a window in the cell at a distance of 1 cm from the dissolving electrode, so that as soon as the tagged ions pass the window, they are registered by the counter. How long after the current pulse at $t = 0$ does the counter note the arrival of the ions?

It is experimentally observed that the counter begins to register within a few seconds of the termination of the instantaneous current pulse. Suppose, however, that one attempted a theoretical calculation based on the Einstein–Smoluchowski equation (4.27), i.e.,

$$\langle x^2 \rangle = 2Dt \tag{4.27}$$

using, for the diffusion coefficient of ions, the experimental value of 10^{-5} cm^2 s^{-1}. Then, the estimated time for the radioactive ions to reach the counter is

$$t = \frac{1}{2 \times 10^{-5} \times 60} \text{ min}$$

or

$$t \approx 10^3 \text{ min}$$

This is several orders of magnitude larger than is indicated by experience.

The dilemma may be resolved as follows. If $\langle x^2 \rangle$ in the Einstein–Smoluchowski relation pertains to the mean square distance traversed by a majority of the radioactive particles and if Geiger counters can—as is the case—detect a very small number of

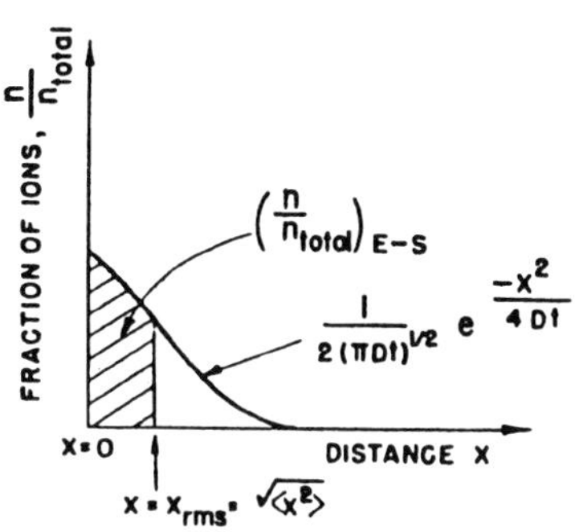

Fig. 4.33. The Einstein–Smoluchowski equation, $<x^2> = 2\,Dt$, pertains to the fraction $(n/n_{total})_{ES}$, of ions between $x = 0$ and $x = x_{rms} = \sqrt{<x^2>}$. This fraction is found by integrating the distribution curve between $x = 0$ and $x = x_{rms}$.

particles, then one can qualitatively see that there is no contradiction between the observed time and that estimated from Eq. (4.27). The time of 10^3 min estimated by the Einstein–Smoluchowski equation is far too large because it pertains to a number of radioactive ions far greater than the number needed to register in the counting apparatus. The way in which the diffusing particles spread out with time, i.e., the distribution curve for the diffusing species (Fig. 4.31), shows that even after very short times, some particles have diffused to very large distances, and these are the particles registered by the counter in a time far less than that predicted by the Einstein–Smoluchowski equation.

The qualitative argument just presented can now be quantified. The central question is: To what fraction n/n_{total} of the ions (released at the instantaneous-plane source) does $<x^2> = 2Dt$ apply?[15] This question can be answered easily by integrating the n versus x distribution curve (Fig. 4.33) between the lower limit $x = 0$ (the location of the plane source) and the value of x corresponding to the square root of $<x^2>$. This upper limit of $\sqrt{<x^2>}$—the root-mean-square distance—is, for conciseness, represented by the symbol x_{rms}, i.e.,

$$x_{rms} = \sqrt{<x^2>} \tag{4.93}$$

Thus, the Einstein–Smoluchowski fraction $(n/n_{total})_{ES}$ is given by [*cf.* Eq. (4.92)]

[15]This fraction will be termed the *Einstein–Smoluchowski fraction*.

$$\left(\frac{n}{n_{\text{total}}}\right)_{\text{ES}} = \int_0^{x_{\text{rms}}} \frac{1}{(\pi D t)^{1/2}} e^{-x^2/4Dt}\, dx$$

$$= \frac{\sqrt{2}}{\pi^{1/2}(2Dt)^{1/2}} \int_0^{x_{\text{rms}}} e^{-x^2/2(2Dt)}\, dx \tag{4.94}$$

According to the Einstein–Smoluchowski relation,

$$2Dt = \langle x^2 \rangle$$

Hence,

$$(2Dt)^{1/2} = \sqrt{\langle x^2 \rangle} = x_{\text{rms}} \tag{4.95}$$

By using this relation, Eq. (4.94) becomes

$$\left(\frac{n}{n_{\text{total}}}\right)_{\text{ES}} = \frac{\sqrt{2}}{\pi^{1/2} x_{\text{rms}}} \int_0^{x_{\text{rms}}} e^{-\frac{1}{2}(x/x_{\text{rms}})^2} dx \tag{4.96}$$

To facilitate the integration, substitute

$$\frac{x}{x_{\text{rms}}} = \sqrt{2}\, u \tag{4.97}$$

in which case several relations follow

$$\left(\frac{x}{x_{\text{rms}}}\right)^2 = 2u^2 \tag{4.98}$$

$$dx = \sqrt{2}\, x_{\text{rms}}\, du \tag{4.99}$$

$$u = \frac{1}{\sqrt{2}} \quad \text{when} \quad x = x_{\text{rms}} \tag{4.100}$$

and

$$u = 0 \quad \text{when} \quad x = 0 \tag{4.101}$$

With the use of these relations, Eq. (4.96) becomes

$$\left(\frac{n}{n_{\text{total}}}\right)_{\text{ES}} = \frac{2}{\pi^{1/2}} \int_0^{1/\sqrt{2}} e^{-u^2}\, du \tag{4.102}$$

The integral on the right-hand side is the error function of $1/\sqrt{2}$ [*cf.* Eq. (4.63)].

Values of the error function have been tabulated in detail (Table 4.6). The value of the error function of $1/\sqrt{2}$, i.e.,

$$\frac{2}{\pi^{1/2}} \int_0^{1/\sqrt{2}} e^{-u^2} du$$

is 0.68. Hence,

$$\left(\frac{n}{n_{\text{total}}}\right)_{\text{ES}} = 0.68 \tag{4.103}$$

and therefore about two-thirds (68%) of the diffusing species are within the region from $x = 0$ to $x_{\text{rms}} = \sqrt{2Dt}$. This means, however, that the remaining fraction, namely one-third, have crossed beyond this distance. Of course, the radioactive ions that are sensed by the counter almost immediately after the pulse of metal dissolution belong to this one-third group (Fig. 4.34).

TABLE 4.6
The Value of the Integral $\frac{2}{\sqrt{\pi}} \int_0^{1/\sqrt{2}} e^{-u^2} du$

y	Value
0.00	0.00000
0.01	0.01128
0.02	0.02256
0.10	0.11246
0.20	0.22270
0.30	0.32863
0.40	0.42839
0.50	0.52050
0.60	0.60386
0.70	0.67780
0.80	0.74210
0.90	0.79691
1.00	0.84270
2.00	1.00000

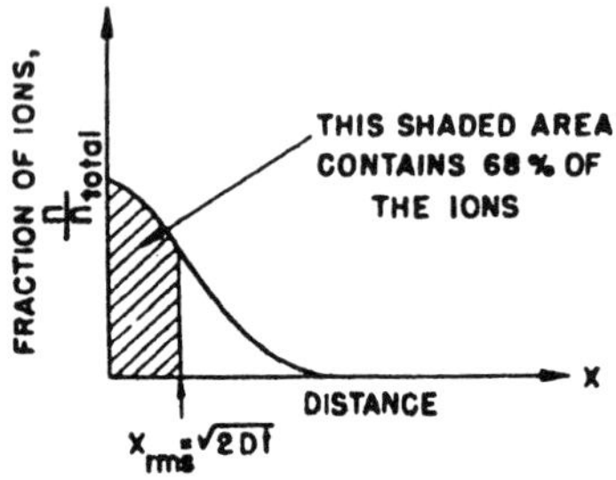

Fig. 4.34. When diffusion occurs from an instantaneous plane source (set up, e.g., by a pulse of electrode dissolution), then 68% of the ions produced in the pulse lie between $x = 0$ and $x = x_{rms}$, after the time t.

In the above experiment, diffusion toward the $x \to -\infty$ direction is prevented by the presence of a physical boundary (i.e., the electrode). If no such boundary exists and diffusion in both the $+x$ and $-x$ directions is possible, then 68% of the particles will distribute themselves in the region from $x = -x_{rms} = -\sqrt{2Dt}$ to $x = +x_{rms} = +\sqrt{2Dt}$. From symmetry considerations, one would expect 34% to be within $x = 0$ and $x = +x_{rms}$ and an equal amount to be on the other side (Fig. 4.35).

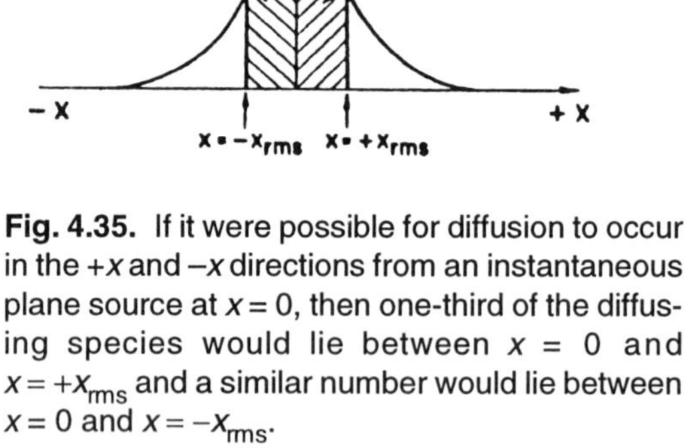

Fig. 4.35. If it were possible for diffusion to occur in the $+x$ and $-x$ directions from an instantaneous plane source at $x = 0$, then one-third of the diffusing species would lie between $x = 0$ and $x = +x_{rms}$ and a similar number would lie between $x = 0$ and $x = -x_{rms}$.

From the above discussion, the advantages and limitations of using the Einstein–Smoluchowski relation become clear. If one is considering phenomena involving a few particles, then one can be misled by making Einstein–Smoluchowski calculations. If, however, one wants to know about the diffusion of a sizable fraction of the total number of particles, then the relation provides easily obtained, although rough, answers without having to go through the labor of obtaining the exact solution for the diffusion problem (see, e.g., Section 4.6.8).

4.2.15. How Can the Diffusion Coefficient Be Related to Molecular Quantities?

The diffusion coefficient D has appeared in both the macroscopic (Section 4.2.2) and the atomistic (Section 4.2.6) views of diffusion. How does the diffusion coefficient depend on the structure of the medium and the interatomic forces that operate? To answer this question, one should have a deeper understanding of this coefficient than that provided by the empirical first law of Fick, in which D appeared simply as the proportionality constant relating the flux J and the concentration gradient dc/dx. Even the random-walk interpretation of the diffusion coefficient as embodied in the Einstein–Smoluchowski equation (4.27) is not fundamental enough because it is based on the mean square distance traversed by the ion after N steps taken in a time t and does not probe into the laws governing each step taken by the random-walking ion.

This search for the atomistic basis of the diffusion coefficient will commence from the picture of random-walking ions (see Sections 4.2.4 to 4.2.6). It will be recalled that a net diffusive transport of ions occurs in spite of the completely random zigzag dance of individual ions, because of unequal numbers of ions in different regions.

Consider one of these random-walking ions. It can be proved (see Appendix 4.1) that the mean square distance $<x^2>$ traveled by an ion depends on the number N of jumps the ion takes and the mean jump distance l in the following manner (Fig. 4.36):

$$<x^2> = Nl^2 \tag{4.104}$$

It has further been shown (Section 4.2.6) that in the case of a one-dimensional random walk, $<x^2>$ depends on time according to the Einstein–Smoluchowski equation

$$<x^2> = 2Dt \tag{4.27}$$

By combining Eqs. (4.104) and (4.27), one obtains the equation

$$Nl^2 = 2Dt \tag{4.105}$$

which relates the number of jumps and the time. If now only one jump of the ion is considered, i.e., $N = 1$, Eq. (4.105) reduces to

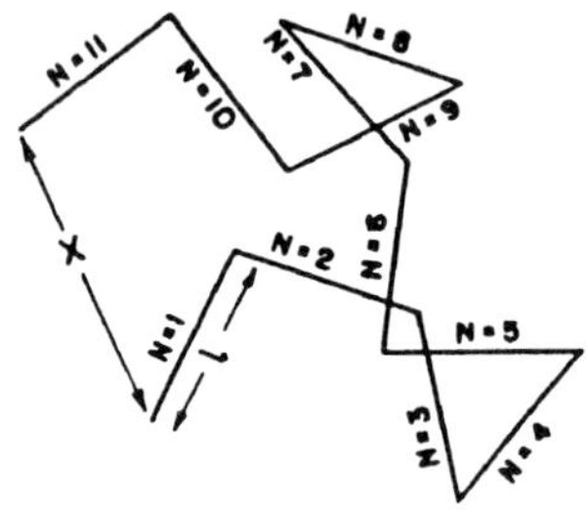

Fig. 4.36. Schematic representation of $N = 11$ steps (each of mean length l) in the random walk of an ion. After 11 steps, the ion is at a distance x from the starting point.

$$D=\frac{1}{2}\frac{l^2}{\tau} \tag{4.106}$$

where τ is the *mean jump time* to cover the mean jump distance l. This mean jump time is the number of seconds per jump,[16] and therefore $1/\tau$ is the jump frequency, i.e., the number of jumps per second. Putting

$$k=\frac{1}{\tau} \tag{4.107}$$

one can write Eq. (4.106) thus:

$$D=\frac{1}{2}l^2k \tag{4.108}$$

Equation (4.108) shows that the diffusion depends on how far, on average, an ion jumps and how frequently these jumps occur.

4.2.16. The Mean Jump Distance l, a Structural Question

To go further than Eq. (4.108), one has to examine the factors that govern the mean jump distance l and the jump frequency k. For this, the picture of a liquid (in which diffusion is occurring) as a structureless continuum is inadequate. In reality, the liquid has a structure—ions and molecules in definite arrangements at any one instant

[16]This mean jump time will include any waiting time between two successive jumps.

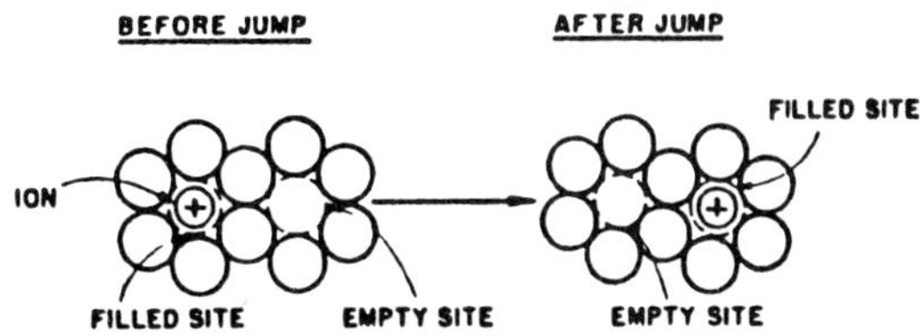

Fig. 4.37. Diagram to illustrate one step in the random walk of an ion.

of time. This arrangement in a liquid (unlike that in a solid) is local in extent, transitory in time, and mobile in space. The details of the structure are not necessary to continue the present discussion. What counts is that ions zigzag in a random walk, and for any particular step, the ion has to *maneuver out of* one site in the liquid structure into another site (Fig. 4.37). This maneuvering process can be symbolically represented thus:

$$\boxed{\text{Ion}} + \boxed{} \rightarrow \boxed{} + \boxed{\text{Ion}} \tag{4.109}$$

where $\boxed{\text{Ion}}$ is a site occupied by an ion and $\boxed{}$ is an empty acceptor site waiting to receive a jumping ion. The mean jump distance l is seen to be the mean distance between sites, and its numerical value depends upon the details of the structure of the liquid, i.e., upon the instantaneous and local atomic arrangement.

4.2.17. The Jump Frequency, a Rate-Process Question

The process of diffusion always occurs at a finite rate; it is a *rate process.* Chemical reactions, e.g., three-atom reactions of the type

$$AB + C \rightarrow A + BC \tag{4.110}$$

are also rate processes. Further, a three-atom reaction can be formally described as the jump of the particle B from a site in A to a site in C (Fig. 4.38). With this description, it can be seen that the notation (4.109) used to represent the jump of an ion has in fact established an analogy between the two rate processes, i.e., diffusion and chemical reaction. Thus, the basic theory of rate processes should be applicable to the processes of both diffusion and chemical reactions.

The basis of this theory is that the potential energy (and standard free energy) of the system of particles involved in the rate processes varies as the particles move to accomplish the process. Very often, the movements crucial to the process are those of a single particle, as is the case with the diffusive jump of an ion from site to site. If the free energy of the system is plotted as a function of the position of the crucial particle,

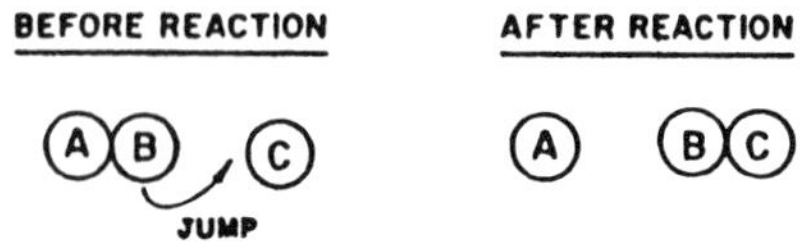

Fig. 4.38. A three-atom chemical rection $AB + C \rightarrow A + BC$ viewed as the jump of atom B from atom A to atom C.

e.g., the jumping ion, then the standard free energy of the system has to attain a critical value (Fig. 4.39)—the activation free energy ΔG°—for the process to be accomplished. One says that the system has to cross an *energy barrier* for the rate process to occur. The number of times per second that the rate process occurs, k, i.e., the jump frequency in the case of diffusion, can be shown to be given by[17]

$$\vec{k} = \frac{kT}{h} e^{-\Delta G^{\circ\ddagger}/RT} \tag{4.111}$$

4.2.18. The Rate-Process Expression for the Diffusion Coefficient

To obtain the diffusion coefficient in terms of atomistic quantities, one has to insert the expression for the jump frequency (4.111) into that for the diffusion coefficient [Eq. (4.108)]. The result is

$$D = \frac{1}{2} l^2 \vec{k}$$

$$= \frac{1}{2} l^2 \frac{kT}{h} e^{-\Delta G^{\circ\ddagger}/RT} \tag{4.112}$$

The numerical coefficient $\frac{1}{2}$ has entered here only because the Einstein–Smoluchowski equation $\langle x^2 \rangle = 2Dt$ for a *one*-dimensional random walk was considered. In general, it is related to the probability of the ion's jumping in various directions, not just forward and backward. For convenience, therefore, the coefficient will be taken to be unity, in which case

$$D \approx l^2 \vec{k} \tag{4.113}$$

$$\approx l^2 \frac{kT}{h} e^{-\Delta G^{\circ\ddagger}/RT} \tag{4.114}$$

---

[17]The $\vec{k}$ on the left-hand side is the jump frequency; the k in the term kT/h is the Boltzmann constant.

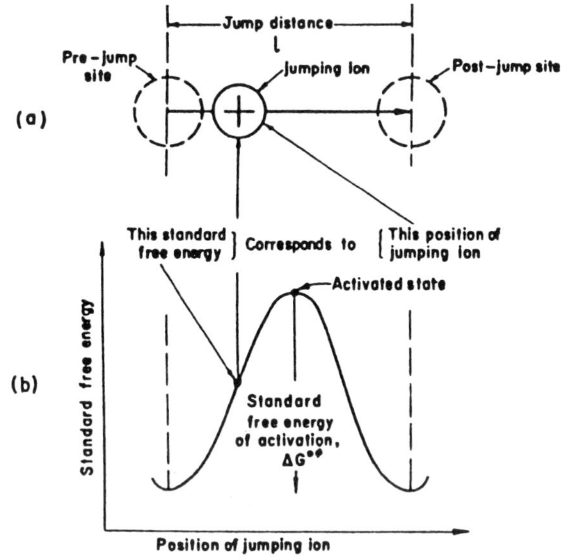

Fig. 4.39. (a) The jump of an ion from a prejump site to a postjump site through a jump distance l. (b) Corresponding to each position of the jumping ion, the system (sites + jumping ion) has a standard free energy. Thus, the standard free-energy changes corresponding to the ionic jump can be represented by the passage of a point (representing the standard free energy of the system) over the barrier by the acquisition of the standard free energy of activation.

As to the value of l, this depends on the model process seen as the mechanism for diffusion. This is discussed for liquid electrolytes rather fully in Section 5.7. In one mechanism ("shuffling along"), the diffusing particles are seen to be analogues of persons pushing through a crowd. Each microstep is less than the distance between two sites. In another, the ion moves by jumps, taking the opportunity when a void or vacancy turns up beside it to move in to fill the void. Then the value of l would be equal to the intersite distance. Thus, in the shuffle-along model, the range of ls would be as little as 0.01 nm, and in the jump-into-a-hole model, perhaps 0.2–0.3 nm.

4.2.19. Ions and Autocorrelation Functions

Autocorrelation functions involve concepts that one sometimes reads about in analyses of transport in liquids and are therefore concepts with which the student should be familiar. These are mathematical devices used in some discussions of the theory of transport in liquids, e.g., in treatment of viscous flow and diffusion. It is possible, for example, to represent the diffusion coefficient D in terms of the autocorrelation function of a particle's velocity. When a particle in solution collides with another (say, an ion colliding with a solvent molecule), the collision is tantamount to a force operating for a very short time on the particle. This time is much shorter than the time needed for the normal so-called "Stokes' resistance" (see Section 4.4.7), which slows down the particle that has "received a knock" to the average velocity (e.g., in flow). Now consider dealing with diffusion coefficients in terms of molecular dynamics (see a description of this technique in Section 2.3.2). This method is used to determine how the dynamics of motion of a given particle will be affected by its collision with other particles. Thus, it is important to know the relation between the velocity of a particle at the beginning ($t = 0$), that is, just after collision, and that at some later time t.

This is where one brings in the concept of the *velocity autocorrelation function*. This indeed concerns the velocity at $t = 0$, v_0, and the velocity at some subsequent time, v_t. To what extent does the subsequent velocity depend upon the velocity at time $t = 0$? That is the sort of information given by the autocorrelation function.

There are several technical details in a rigorous definition of the autocorrelation function for velocity. First, one has to remember the vectorial character of velocity, because clearly the direction in which the particle is knocked is important to its subsequent dynamic history. Then, according to the way it is defined, one has to take the product of the velocity at $t = 0$, v_0, and that at the later chosen time, v_t. However, it is not as simple as just multiplying together the two vectors, v_0 and v_t. One has to allow for the distribution of positions and momenta of the particle in the system at the beginning, that is, at $t = 0$. To allow for this, one can introduce symbolically a probability distribution coefficient, g_0. Therefore, the expression for the autocorrelation function will involve the product $g_0 v_0 v_t$.

Thereafter, there is only one step left, but it is a vital one. One has to carry out an averaging process for the entire liquid (or solution) concerned. Such averaging processes can be carried out in more than one way. One of these involves an integration with respect to time. One ends up by writing down the full-blown expression for the autocorrelation function as a function of an expression dependent on time, $A(t)$. Then, in a general way, an autocorrelation function would run

$$C(\tau) = \langle A(t)A(t+\tau) \rangle \tag{4.115}$$

where the brackets represent "the average value of," as defined in Eq. (4.20).

How can this concept be used to calculate diffusion coefficients in ionic solutions? First one has to remember that for diffusion in one direction,

$$\langle x^2 \rangle = 2Dt \tag{4.116}$$

However, the displacement of the particle x is in reality a function of time and therefore can also be expressed in terms of an autocorrelation function similar to that presented in Eq. (4.115). One advantage of this procedure is that the autocorrelation function will depend only on a time interval, $\tau - t$, and not on the time itself. Through the use of Eq. (4.116) and some mechanics produced earlier in the century by Langevin, one finds that the diffusion coefficient D can be expressed by the *time integral of the velocity autocorrelation function* and eventually obtains the useful equation

$$D = \int_0^\infty \langle v(0)v(\tau)\rangle \, d\tau = \frac{kT}{f_c} \tag{4.117}$$

where f_c is the frictional coefficient.

Does this concern ions in solution and electrochemistry? It does indeed concern some approaches to diffusion and hence the related properties of conduction and viscous flow. It has been found that the autocorrelation function for the velocity of an ion diffusing in solution decays to zero very quickly, i.e., in about the same time as that of the random force due to collisions between the ion and the solvent. This is awkward because it is not consistent with one of the approximations used to derive analytical expressions for the autocorrelation function.[18] The result of this is that instead of an analytical expression, one has to deal with molecular dynamics simulations.

One of the simplest examples of this type of calculation involves the study of a system of rare-gas atoms, as in, e.g., calculations carried out on liquid argon. The relaxation time after a collision was found to be on the order of 10^{-13} s, which is about the same time as that for rather large ions (e.g., of 500 pm). Thus, much of what one learns from the MD study of molecular motion in liquid argon should be applicable to ionic diffusion.

Figure 4.40 shows the velocity autocorrelation function for liquid argon as calculated by Levesque et al. Looking now at this figure, one can see at first the fast exponential decay of the autocorrelation function. The function then becomes negative (reversal of velocity), indicating a scattering collision with another molecule. At longer times it trails off to zero, as expected, for eventually the argon atom's motion becomes unconnected to the original collision. The time for this to happen is relatively long, about 10^{-12} s.

In summary, then, autocorrelation functions are useful mathematical devices which, when applied to velocities, tell us to what degree the motion of a particle at a given instant is related to the impelling force of the last collision. Their usefulness is mainly in molecular dynamics, the principal computer-oriented method by which systems are increasingly being analyzed (Section 2.17).

---

[18]Here it is assumed that the influence of the collision lasts a lot longer than the force due to the collision itself.

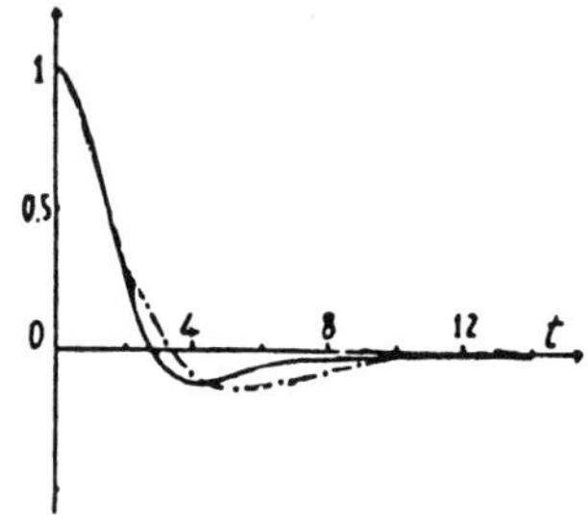

Fig. 4.40. Autocorrelation function of the velocity (time units 10^{-13} s) obtained by Levesque et al. (dash-dot curve). The solid curve is the molecular dynamics results (D. Levesque, L. Verlet, and J. Kurkijarvi, *Phys. Rev. A* **7**: 1690, 1973.)

4.2.20. Diffusion: An Overall View

An electrochemical system runs on the basis of charge-transfer reactions at the electrode–electrolyte interfaces. These reactions involve ions or molecules that are constituents of the electrolyte. Thus, the transport of particles to or away from the interface becomes an essential condition for the continued electrochemical transformation of reactants at the interface.

One of the basic mechanisms of ionic transport is diffusion. This type of transport occurs when the concentration of the diffusing species is different in different parts of the electrolyte. What makes diffusion occur? This question can be answered on a macroscopic and on a microscopic level.

The macroscopic view is based on the fact that when the concentration varies with distance, so does the chemical potential. However, a nonuniformity of chemical potential implies that the free energy is not the same everywhere and therefore the system is not at equilibrium. How can the system attain equilibrium? By equalizing the chemical potential, i.e., by transferring the species from the high-concentration regions to the low-concentration regions. Thus, the negative gradient of the chemical potential, $d\mu/dx$, behaves like a driving force (see Section 4.2.1) that produces a net flow, or flux, of the species.

When the driving force is small, it may be taken to be linearly related to the flux. On this basis, an equation can be derived for the rate of steady-state diffusion, which is identical in form to the empirical first law of Fick,

$$J = -D\frac{dc}{dx}$$

The microscopic view of diffusion starts with the movements of individual ions. Ions dart about haphazardly, executing a random walk. By an analysis of one-dimensional random walk, a simple law can be derived (see Section 4.2.6) for the mean square distance $<x^2>$ traversed by an ion in a time t. This is the Einstein–Smoluchowski equation

$$<x^2> = 2Dt \tag{4.27}$$

It also turns out that the random walk of individual ions is able to give rise to a flux, or flow, on the level of the group. Diffusion is simply the result of there being more random-walking particles in one region than in another (see Section 4.2.6). The gradient of chemical potential is therefore only a pseudoforce that can be regarded as operating on a society of ions but not on individual ions.

The first law of Fick tells one how the concentration gradient is related to the flux under steady-state conditions; it says nothing about how the system goes from nonequilibrium to steady state when a diffusion source or sink is set up inside or at the boundary of the system. Thus, it says nothing about how the concentration changes with time at different distances from the source or sink. In other words, Fick's first law is inapplicable to nonsteady-state diffusion. For this, one has to go to Fick's second law

$$\frac{\partial c}{\partial t} = D\frac{\partial^2 c}{\partial x^2} \tag{4.32}$$

which relates the time and space variations of the concentration during diffusion.

Fick's second law is a partial differential equation. Thus, it describes the general characteristics of all diffusion problems but not the details of any one particular diffusion process. Hence, the second law must be solved with the aid of the initial and boundary conditions that characterize the particular problem.

The solution of Fick's second law is facilitated by the use of Laplace transforms, which convert the partial differential equation into an easily integrable total differential equation. By utilizing Laplace transforms, the concentration of diffusing species as a function of time and distance from the diffusion sink when a constant normalized current, or flux, is switched on at $t = 0$ was shown to be

$$c = c^0 - \frac{1}{D^{1/2}}\left[\frac{2t^{1/2}}{\pi^{1/2}}\exp\left(-\frac{x^2}{4Dt}\right) - \frac{x}{D^{1/2}}\operatorname{erfc}\left(\frac{x^2}{4Dt}\right)^{1/2}\right] \tag{4.65}$$

With the solution of this problem (in which the flux varies as a unit-step function with time), one can easily generate the solution of other problems in which the current, or flux, varies with time in other ways, e.g., as a periodic function or as a single pulse.

When the current, or flux, is a single impulse, an instantaneous-plane source for diffusion is set up and the concentration variation is given by

$$c(x, t) = \frac{\lambda}{(\pi Dt)^{1/2}} e^{-x^2/4Dt} \tag{4.91}$$

in the presence of a boundary. From this expression, it turns out that in a time t, only a certain fraction (two-thirds) of the particles travel the distance given by the Einstein–Smoluchowski equation. Actually, the spatial distribution of the particles at a given time is given by a semi-bell-shaped distribution curve.

The final step involves the relation of the diffusion coefficient to the structure of the medium and the forces operating there. It is all a matter of the mean distance l through which ions jump during the course of their random walk and of the mean jump frequency k. The latter can be expressed in terms of the theory of rate processes, so that one ends up with an expression for the rate of diffusion that is in principle derivable from the local structure of the medium.

There is more than one way of calculating diffusion coefficients, and a method being used increasingly involves molecular dynamics. Some description of this technique is given in Sections 2.3.2 and 2.17. One aspect of it is the velocity autocorrelation function as explained in Section 4.2.19.

This then is an elementary picture of diffusion. The next task is to consider the phenomenon of conduction, i.e., the migration of ions in an electric field.

Further Reading

Seminal

1. M. Smoluchowski, "Diffusive Movements in Liquids," *Ann. Physik (Paris)* **25**: 205 (1908).
2. A. Einstein, *Investigations on the Theory of Brownian Movement*, Methuen & Co., London (1926).
3. Lin Yong and M. T. Simned, "Measurement of Diffusivity in Liquid Systems," in *Physicochemical Measurements at High Temperatures*, J. O'M. Bockris, J. L. White, and J. W. Tomlinson, eds., Butterworth, London (1959).

Review

1. J. G. Wijmans and R. W. Baker, "The Solution-Diffusion Model: A Review," *J. Membr. Sci.* **107**: 1 (1995).

Papers

1. T. Munakata and Y. Kaneko, *Phys. Rev. E.: Stat. Phys. Plasmas, Fluids, Relat. Interdiscip. Top.* **47**: 4076 (1993).
2. J. P. Simmon, P. Turq, and A. Calado, *J. Phys. Chem.* **97**: 5019 (1993).
3. S. Rondot, J. Cazaux, and O. Aaboubi, *Science* **263**: 1739 (1994).
4. E. Hawlichka, *Chem. Soc. Rev.* **24**: 367 (1994).
5. Z. A. Metaxiotou and S. G. Nychas, *AIChE J* **41**: 812 (1995).
6. R. Biswas, S. Roy, and B. Bagchi, *Phys. Rev. Lett.* **75**: 1098 (1995).

4.3. IONIC DRIFT UNDER AN ELECTRIC FIELD: CONDUCTION

4.3.1. Creation of an Electric Field in an Electrolyte

Assume that two plane-parallel electrodes are introduced into an electrolytic solution so as to cover the end walls of the rectangular insulating container (Fig. 4.41). With the aid of an external source, let a potential difference be applied across the electrodes. How does this applied potential difference affect the ions in the solution?

The potential in the solution has to vary from the value at one electrode, ψ_{I}, to that at the other electrode, ψ_{II}. The major portion of this potential drop $\psi_{\mathrm{I}} - \psi_{\mathrm{II}}$ occurs across the two electrode–solution interfaces (see Chapter 6); i.e., if the potentials on the solution side of the two interfaces are ψ'_{I} and ψ'_{II}, then the interfacial potential differences are $\psi_{\mathrm{I}} - \psi'_{\mathrm{I}}$ and $\psi'_{\mathrm{II}} - \psi_{\mathrm{II}}$ (Fig. 4.42). The remaining potential drop, $\psi'_{\mathrm{I}} - \psi'_{\mathrm{II}}$, occurs in the electrolytic solution. The electrolytic solution is therefore a region of space in which the potential at a point is a function of the distance of that point from the electrodes.

Let the test ion in the solution be at the point x_1, where the potential is ψ_1 (Fig. 4.43). This potential is by definition the work done to bring a unit of positive charge from infinity up to the particular point. [In the course of this journey of the test charge from infinity to the particular point, it may have to cross phase boundaries, for example, the electrolyte-air boundary, and thereby do extra work (see Chapter 6). Such surface work terms cancel out, however, in discussions of the differences in potential between two points in the same medium.] If another point x_2 is chosen on the normal from x to the electrodes, then the potential at x_2 is different from that at x_1 because of the variation

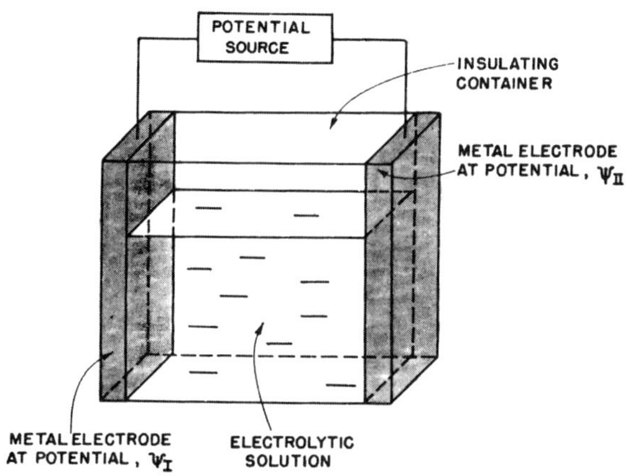

Fig. 4.41. An electrochemical system consisting of two plane-parallel electrodes immersed in an electrolytic solution is connected to a source of potential difference.

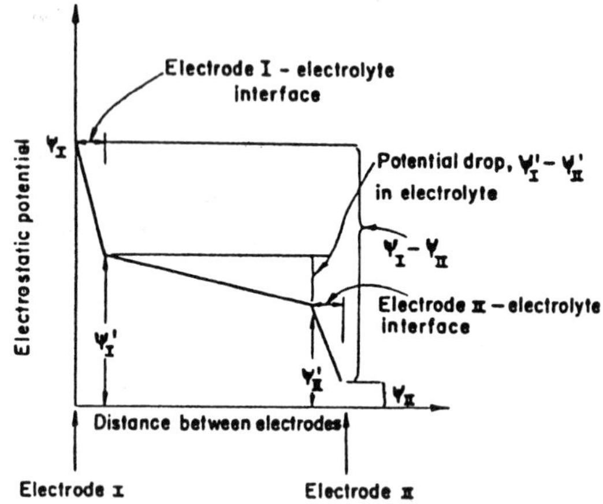

Fig. 4.42. Diagram to illustrate how the total potential difference $\psi_I - \psi_{II}$ is distributed in the region between the two electrodes.

of the potential along the distance coordinate between the electrodes. Let the potential at x_2 be ψ_2. Then $\psi_1 - \psi_2$, the difference in potential between the two points, is the work done to take a unit of charge from x_1 to x_2.

When this work $\psi_1 - \psi_2$ is divided by the distance over which the test charge is transferred, i.e., $x_1 - x_2$, one obtains the *force per unit charge*, or the electric field $\boldsymbol{X}$

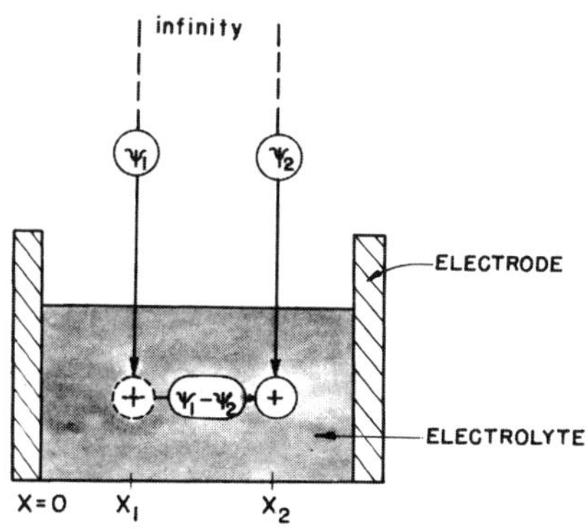

Fig. 4.43. The work done to transport a unit of positive charge from x_1 to x_2 in the solution is equal to the difference $\psi_1 - \psi_2$ in electrostatic potential at the two points.

$$X = -\frac{\psi_1 - \psi_2}{x_1 - x_2} \tag{4.118}$$

where the minus sign indicates that the force acts on a positive charge in a direction opposite to the direction of the positive gradient of the potential. In the particular case under discussion, i.e., parallel electrodes covering the end walls of a rectangular container, the potential drop in the electrolyte is linear (as in the case of a parallel-plate condenser), and one can write

$$X = -\frac{\text{Potential difference across the solution}}{\text{Distance across the solution}} \tag{4.119}$$

In general, however, it is best to be in a position to treat nonlinear potential drops. This is done by writing (Fig. 4.44)

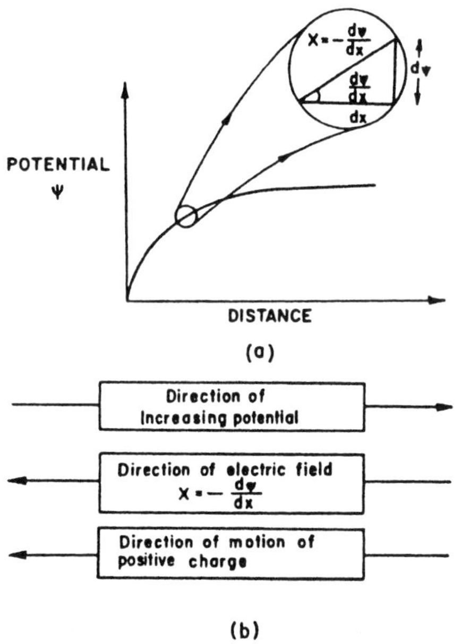

Fig. 4.44. (a) In the case of a nonlinear potential variation in the solution, the electric field at a point is the negative gradient of the electrostatic potential at that point. (b) The relative directions of increasing potential, field, and motion of a positive charge.

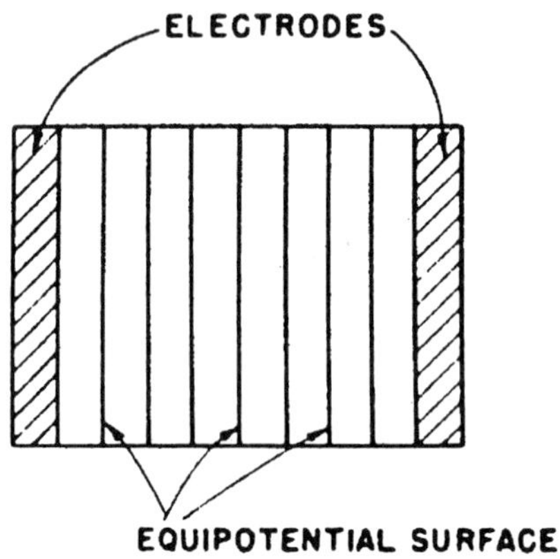

Fig. 4.45. A geometric representation of the electric field in an electrochemical system in which plane-parallel electrodes are immersed in an electrolyte so that they extend up to the walls of the rectangular insulating container. The equipotential surfaces are parallel to the electrodes.

$$X = -\frac{d\psi}{dx} \tag{4.120}$$

where now the electric field may be a function of x.

The imposition of a potential difference between two electrodes thus makes an electrolytic solution the scene of operation of an electric field (i.e., an electric force) acting upon the charges present. This field can be mapped by drawing equipotential surfaces (all points associated with the same potential lie on the same surface). The potential map yields a geometric representation of the field. In the case of plane-parallel electrodes extending to the walls of a rectangular cell, the equipotential surfaces are parallel to the electrodes (Fig. 4.45).

4.3.2. How Do Ions Respond to the Electric Field?

In the absence of an electric field, ions are in ceaseless random motion. This random walk of ions has been shown to have an important characteristic: The mean distance traversed by the ions as a whole is zero because while some are displaced in one direction, an equal number are displaced in the opposite direction. From a phenomenological view, therefore, the random walk of ions can be ignored because it does not lead to any net transport of matter (as long as there is no difference in

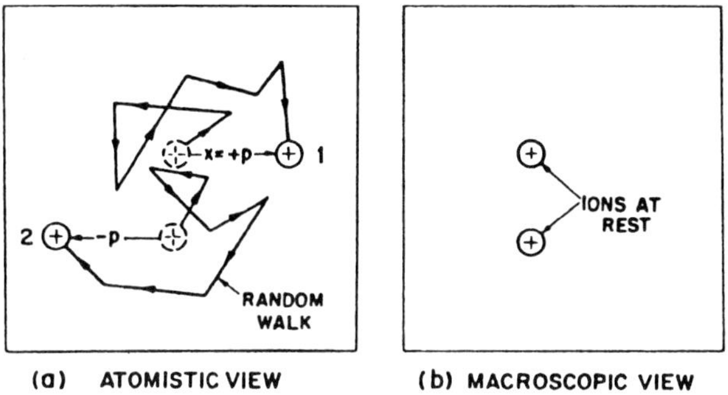

Fig. 4.46. (a) Schematic representation of the random walk of two ions showing that ion 1 is displaced a distance $x = +p$ and ion 2, a distance $x = -p$ and hence the mean distance traversed is zero. (b) Since the mean distance traversed by the ions is zero, there is no net flux of the ions and from a macroscopic point of view, they can legitimately be considered at rest.

concentration in various parts of the solution so that net diffusion down the concentration gradient occurs). The net result is as if the ions were at rest (Fig. 4.46).

Under the influence of an electric field, however, the net result of the zigzag jumping of ions is not zero. Ions feel the electric field; i.e., they experience a force directing them toward the electrode that is charged oppositely to the charge on the ion.

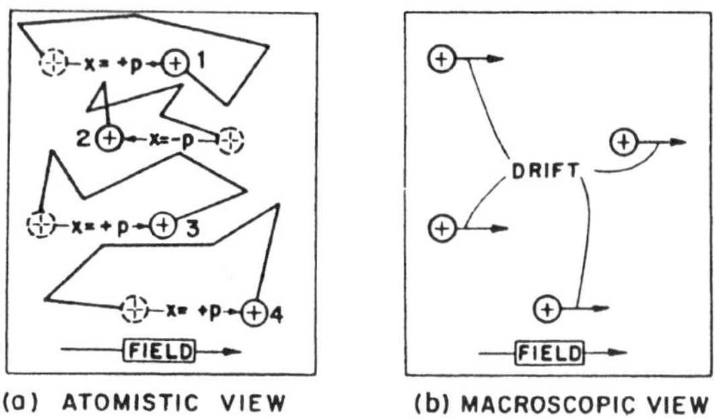

Fig. 4.47. (a) Schematic representation of the movements of four ions which random walk in the presence of a field. Their displacements are $+p$, $-p$, $+p$, and $+p$, i.e., the mean displacement is finite. (b) From a macroscopic point of view, one can ignore the random walk and consider that each ion drifts in the direction of the field.

This directed force is equal to the charge on the ion, $z_i e_0$, times the field at the point where the ion is situated. The driving force of the electric field produces in all ions of a particular species a velocity component in the direction p of the potential gradient. Thus, the establishment of a potential difference between the electrodes produces a drift, or flux, of ions (Fig. 4.47). This drift is the migration (or conduction) of ions in response to an electric field.

As in diffusion, the relationship between the steady-state flux $\boldsymbol{J}$ of ions and the driving force of the electric field will be represented by the expression

$$J = A + BX + CX^2 + \ldots \tag{4.121}$$

For small fields, the terms higher than BX will tend to zero. Further, the constant A must be zero because the flux of ions must vanish when the field is zero. Hence, for small fields, the flux of ions is proportional to the field (see Section 4.2.2)

$$J = BX \tag{4.122}$$

4.3.3. The Tendency for a Conflict between Electroneutrality and Conduction

When a potential gradient, i.e., electric field, exists in an electrolytic solution, the positive ions drift toward the negative electrode and the negative ions drift in the opposite direction. What is the effect of this ionic drift on the state of charge of an electrolytic solution?

Prior to the application of an external field, there is a time-averaged electroneutrality in the electrolyte over a distance that is large compared with κ^{-1} (see Section 3.3.8); i.e., the net charge in any macroscopic volume of solution is zero because the total charge due to the positive ions is equal to the total charge due to the negative ions. Owing to the electric field, however, ionic drift tends to produce a spatial separation of charge. Positive ions will try to segregate near the negatively charged electrode, and negative ions near the positively charged electrode.

This tendency for gross charge separation has an important implication: electroneutrality tends to be upset. Furthermore, the separated charge causing the lack of electroneutrality tends to set up its own field, which would run counter to the externally applied field. If the two fields were to become equal in magnitude, the net field in the solution would become zero. (Thus, the driving force on an ion would vanish and ion migration would stop.)

It appears from this argument that an electrolytic solution would sustain only a transient migration of ions and then the tendency to conform to the principle of electroneutrality would result in a halt in the drift of ions after a short time. A persistent flow of charge, an electric current, appears to be impossible. In practice, however, an electrolytic solution can act as a conductor of electricity and is able to pass a current, i.e., maintain a continuous flow of ions. Is there a paradox here?

4.3.4. Resolution of the Electroneutrality-versus-Conduction Dilemma: Electron-Transfer Reactions

The solution to the dilemma just posed can be found by comparing an electrolytic solution with a metallic conductor. In a metallic conductor, there is a lattice of positive ions that hold their equilibrium positions during the conduction process. In addition, there are the free conduction electrons which assume responsibility for the transport of charge. Contact is made to and from the metallic conductor by means of other metallic conductors [Fig. 4.48(a)]. Hence, electrons act as charge carriers *throughout the entire circuit.*

In the case of an electrolytic conductor, however, it is necessary to make electrical contact to and from the electrolyte by metallic conductors (wires). Thus, here one has the interesting situation in which electrons transport charge in the external circuit and

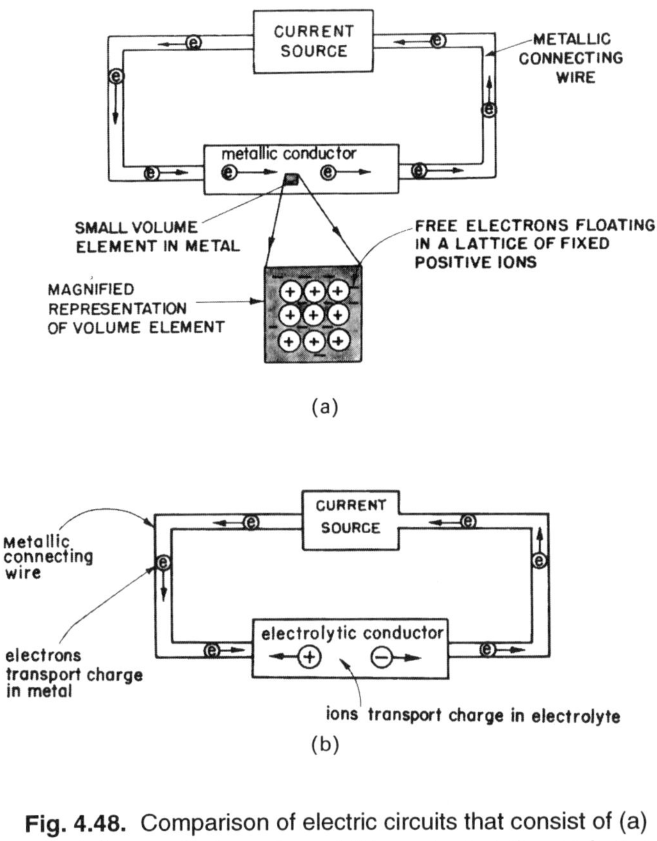

Fig. 4.48. Comparison of electric circuits that consist of (a) a metallic conductor only and (b) an electrolytic conductor as well as a metallic one.

ions carry the charge in the electrolytic solution [Fig. 4.48(b)]. Obviously, one can maintain a steady flow of charge (current) in the entire circuit only if there is a change of charge carrier at the electrode–electrolyte interface. In other words, for a current to flow in the circuit, ions have to hand over or take electrons from the electrodes.

Such electron transfers between ions and electrodes result in chemical changes (changes in the valence or oxidation state of the ions), i.e., in electrodic reactions. When ions receive electrons from the electrode, they are said to be "electronated," or *to undergo reduction*; when ions donate electrons to the electrodes, they are said to be "deelectronated," or *to undergo oxidation*.

The occurrence of a reaction at each electrode is tantamount to removal of equal amounts of positive and negative charge from the solution. Hence, when electron-transfer reactions occur at the electrodes, ionic drift does *not* lead to segregation of charges and the building up of an electroneutrality field (opposite to the applied field). Thus, the flow of charge can continue; i.e., the solution conducts. It is an ionic conductor.

4.3.5. Quantitative Link between Electron Flow in the Electrodes and Ion Flow in the Electrolyte: Faraday's Law

Charge transfer is the essence of an electrodic reaction. It constitutes the bridge between the current I_e of electrons in the electrode part of the electrical circuit and the current I_i of ions in the electrolytic part of the circuit (Fig 4.49). When a steady-state

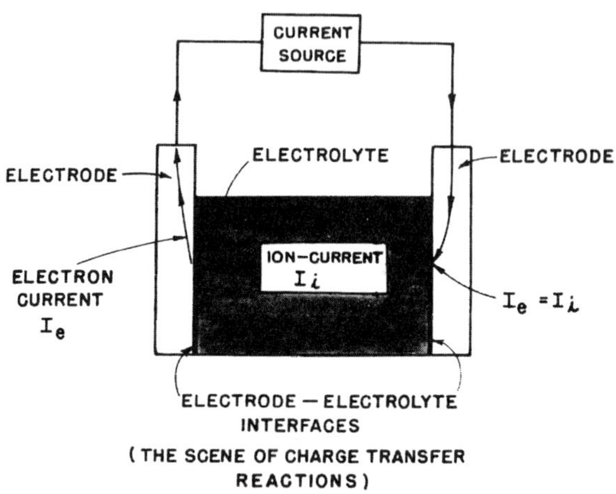

Fig. 4.49. Diagram for the derivation of Faraday's laws. The electron current I_e in the metallic part of the circuit must be equal to the ion current I_i in the electrolytic part of the circuit.

current is passing through the circuit, there must be a continuity in the currents at the electrode–electrolyte interfaces, i.e.,

$$I_e = I_i \tag{4.123}$$

(This is in fact an example of Kirchhoff's law, which says that the *algebraic* sum of the currents at any junction must be zero.) Further, if one multiplies both sides of Eq. (4.123) by the time *t*, one obtains

$$I_e t = I_i t \tag{4.124}$$

which indicates[19] that the quantity Q_e of electricity carried by the electrons is equal to that carried by the ions

$$Q_e = Q_i \tag{4.125}$$

Let the quantity of electricity due to electron flow be the charge borne by an Avogadro number of electrons, i.e., $Q_e = N_A e_0 = F$. If the charge on each ion participating in the electrodic reaction is $z_i e_0$, it is easily seen that the number of ions required to preserve equality of currents [Eq. (4.123)] and equality of charge transported across the interface in time *t* [Eq. (4.125)] is

$$\frac{Q_e}{z_i e_0} = \frac{N_A e_0}{z_i e_0} = \frac{N_A}{z_i} \text{ ions} = \frac{1}{z_i} \text{ mole of ions}$$

$$= 1 \text{ g-eq} \tag{4.126}$$

Thus, the requirement of steady-state continuity of current at the interface leads to the following law: The passage of 1 faraday (F) of charge results in the electrodic reaction of one equivalent ($1/z_i$ moles) of ions, each of charge $z_i e_0$. This is Faraday's law.[20] Conversely, if $1/z_i$ moles of ions undergo charge transfer, then 1 F of electricity passes through the circuit, or *zF* faradays per mole of ions transformed.

4.3.6. The Proportionality Constant Relating Electric Field and Current Density: Specific Conductivity

In the case of small fields, the steady-state flux of ions can be considered proportional to the driving force of an electric field (see Section 4.3.2), i.e.,

---

[19]The product of the current and time is the quantity of electricity.

[20]Alternatively, Faraday's law states that if a current of ***I*** amp passes for a time *t* s, then *It/zF* moles of reactants in the electronic reaction are produced or consumed.

$$J = BX \tag{4.122}$$

The quantity J is the number of moles of ions crossing a unit area per second. When J is multiplied by the charge borne by 1 mole of ions zF, one obtains the current density i, or *charge flux*, i.e., the quantity of charge crossing a unit area per second. Because i has direction, it will be written as a vector quantity, $\vec{j}$.

$$\vec{j} = JzF = zFBX \tag{4.127}$$

The constant zFB can be set equal to a new constant σ, which is known as the *specific conductivity* (Table 4.7). The relation between the current density i and the electric field X is therefore

$$X = \frac{1}{\sigma} j \tag{4.128}$$

The electric field is very simply related (Fig. 4.50) to the potential difference across the electrolyte, $\psi'_I - \psi'_{II}$ [see Eq. (4.119)],

$$X = \frac{\Delta\psi}{l} \tag{4.129}$$

where l is the distance across the electrolyte. Furthermore, the total current $\boldsymbol{I}$ is equal to the area A of the electrodes times the current density i

$$I = jA \tag{4.130}$$

Substituting these relations [Eqs. (4.129) and (4.130)] in the field-current-density relation [Eq. (4.128)], one has

TABLE 4.7

Representative Values of Specific Conductivity

Substance	Type of Conductor	Specific Conductivity ($S\ cm^{-1}$)	T (K)
Copper	Metallic	5.8×10^5	293
Lead	Metallic	4.9×10^5	273
Iron	Metallic	1.1×10^5	273
4 M H_2SO_4	Electrolytic	7.5×10^{-1}	291
0.1 M KCl	Electrolytic	1.3×10^{-2}	298
Xylene	Nonelectrolyte	1×10^{-19}	298
Water	Nonelectrolyte	4×10^{-8}	291

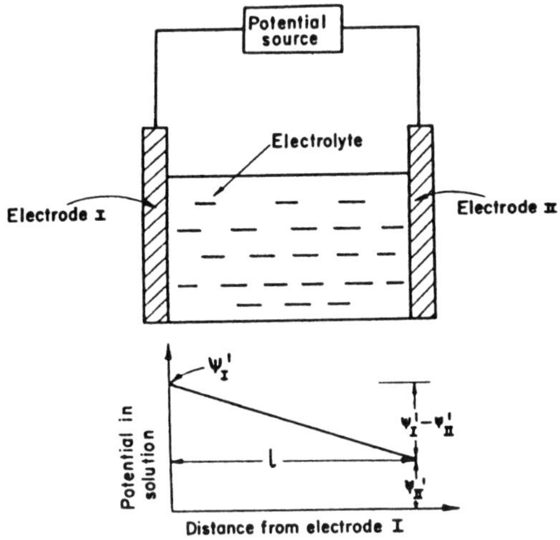

Fig. 4.50. Schematic representation of the variation of the potential in the electrolytic conductor of length *l*.

$$\frac{\Delta\psi}{l} = \frac{1}{\sigma}\frac{\boldsymbol{I}}{A}$$

or

$$\Delta\psi = \frac{l}{\sigma A}\boldsymbol{I} \tag{4.131}$$

The constants σ, l, and A determine the resistance R of the solution

$$R = \frac{l}{\sigma A} \tag{4.132}$$

and therefore one has the equation

$$\Delta\psi = R\boldsymbol{I} \tag{4.133}$$

which reexpresses in the conventional Ohm's law form the assumption Eq. of (4.122) concerning flux and driving force.

Thus, an electrolytic conductor obeys Ohm's law for all except very high fields and, *under steady-state conditions*, it can be represented in an electrical circuit (in which there is only a dc source) by a resistor. (An analogue must obey the same equation as the system it represents or simulates.)

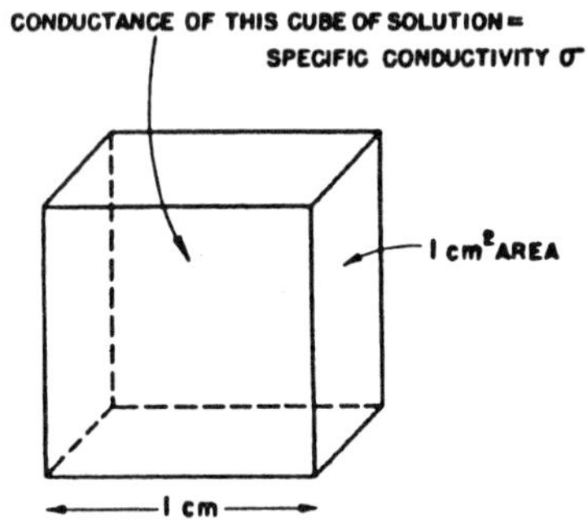

Fig. 4.51. Diagram to illustrate the meaning of the specific conductivity of an electrolyte.

As in the case of a resistor, the dc resistance of an electrolytic cell increases with the length of the conductor (distance between the electrodes) and decreases with the area [*cf.* Eq. (4.132)]. It can also be seen by rearranging this equation into the form

$$\sigma = \frac{1}{R}\frac{l}{A} \tag{4.134}$$

that the specific conductivity σ is the conductance $1/R$ of a cube of electrolytic solution 1 cm long and 1 cm^2 in area (Fig. 4.51).

4.3.7. Molar Conductivity and Equivalent Conductivity

In the case of metallic conductors, once the specific conductivity is defined, the macroscopic description of the conductor is complete. In the case of electrolytic

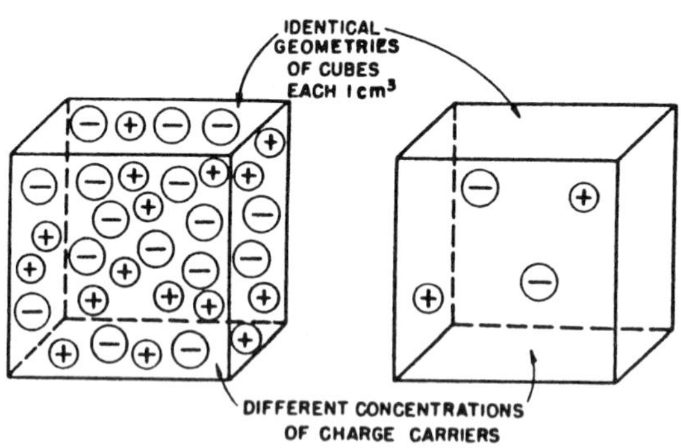

Fig. 4.52. A schematic explanation of the variation of the specific conductivity with electrolyte concentration.

TABLE 4.8
Specific Conductivities of KCl Solutions

	Specific Conductivity (S cm^{-1})		
KCl (g kg^{-1} of solution)	273 K	281 K	298 K
1.0	0.065144	0.097790	0.11187
0.1	0.0071344	0.0111612	0.012896
0.01	0.00077326	0.00121992	0.001427

conductors, further characterization is imperative because not only can the concentration of charge carriers vary but also the charge per charge carrier.

Thus, even though two electrolytic conductors have the same geometry, they need not necessarily have the same specific conductivity (Fig. 4.52 and Table 4.8); the number of charge carriers in that normalized geometry may be different, in which case their fluxes under an applied electric field will be different. Since *the specific conductivity of an electrolytic solution varies as the concentration*, one can write

$$\sigma = f(c) \tag{4.135}$$

where c is the concentration of the solution in gram-moles of solute dissolved in 1 cm^3 of solution.[21] The specific conductivities of two solutions can be compared only if they contain the same concentration of ions. The conclusion is that in order to compare the conductances of electrolytic conductors, one has to normalize (set the variable quantities equal to unity) not only the *geometry* but also the *concentration* of ions.

The normalization of the geometry (taking electrodes of 1 cm^2 in area and 1 cm apart) defines the specific conductivity; the additional normalization of the concentration (taking 1 mole of ions) defines a new quantity, the molar conductivity (Table 4.9),

$$\Lambda_m = \frac{\sigma}{c} = \sigma V \tag{4.136}$$

where V is the volume of solution containing 1 g-mole of solute (Fig. 4.53). Defined thus, it can be seen that the molar conductivity is the specific conductivity of a solution times the volume of that solution in which is dissolved 1 g-mole of solute; the molar conductivity is a kind of conductivity per particle.

[21] As in the case of diffusion fluxes, the concentrations used in the definition of conduction currents (or fluxes) and conductances are not in the usual moles per liter but in moles per cubic centimeter.

TABLE 4.9

Molar Conductivities of Electrolytes

Electrolyte	Molar Conductivity ($S\ cm^2\ mol^{-1}$ at 0.01 mol dm^{-3})
KCl	141.3
NaCl	118.5
$MgCl_2$	229.2
Na_2SO_4	224.8

One can usefully compare the molar conductivities of two electrolytic solutions only if the charges borne by the charge carriers in the two solutions are the same. If there are singly charged ions in one electrolyte (e.g., NaCl) and doubly charged ions in the other (e.g., $CuSO_4$), then the two solutions will contain different amounts of charge even though the same quantity of the two electrolytes is dissolved. In such a case, the specific conductivities of the two solutions can be compared only if they contain equivalent amounts of charge. This can be arranged by taking 1 mole of charge in each case, i.e., 1 mole of ions divided by z, or 1 g-eq of the substance. Thus, the *equivalent conductivity* Λ of a solution is the specific conductivity of a solution times the volume V of that solution containing 1 g-eq of solute dissolved in it (Fig. 4.54 and Table 4.10). Hence, the equivalent conductivity is given by[22]

$$\Lambda = \frac{\sigma}{cz} \tag{4.137}$$

where cz is the number of gram-equivalents per cubic centimeter of solution (see Fig. 4.55 for units of these quantities).

There is a simple relation between the molar and equivalent conductivities. It is [*cf.* Eqs. (4.136) and (4.137)]

$$\Lambda_m = z\Lambda \tag{4.138}$$

The equivalent conductivity Λ is in the region (±25%) of 100 $S\ cm^2\ mol^{-1}\ eq^{-1}$ for most dilute electrolytes of 1:1 salts.

4.3.8. Equivalent Conductivity Varies with Concentration

At first sight, the title of this section may appear surprising. The equivalent conductivity has been defined by normalizing the geometry of the system *and* the

[22]Since $1/z$ mole of ions is 1 g-eq, c moles is cz g-eq.

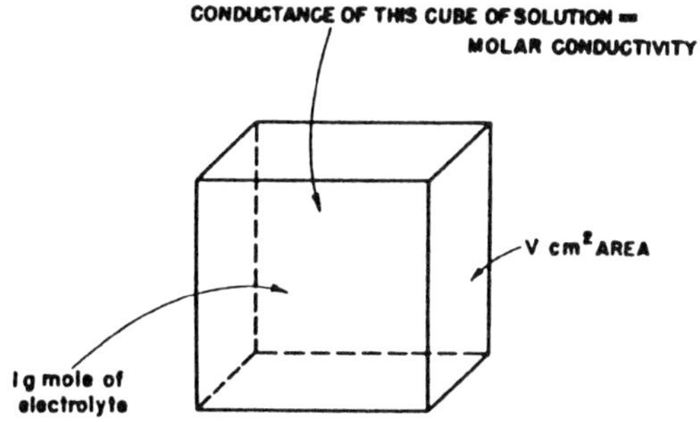

Fig. 4.53. Diagram to illustrate the meaning of the molar conductivity of an electrolyte.

charge of the ions; why then should it vary with concentration? Experiment, however, gives an unexpected answer. The equivalent conductivity varies significantly with the concentration of ions (Table 4.11). The direction of the variation may also surprise some, for the equivalent conductivity *increases* as the ionic concentration decreases (Fig. 4.56).

It would be awkward to have to refer to the concentration every time one wished to state the value of the equivalent conductivity of an electrolyte. One should be able to define some reference value for the equivalent conductivity. Here, the facts of the experimental variation of equivalent conductivity with concentration come to one's aid; as the electrolytic solution is made more dilute, the equivalent conductivity approaches a limiting value (Fig. 4.57). This limiting value should form an excellent basis for comparing the conducting powers of different electrolytes, for it is the only value in which the effects of ionic concentration are removed. The limiting value will be called the *equivalent conductivity at infinite dilution*, designated by the symbol Λ^0 (Table 4.12).

TABLE 4.10

Equivalent Conductivities of True Electrolytes in Dilute Aqueous Solutions at 298 K

Electrolyte	Equivalent Conductivity ($S\ cm^2\ mol^{-1}\ eq^{-1}$ at 0.005 g-eq dm^{-3})
KCl	143.55
NaOH	240.00
$AgNO_3$	127.20
$\frac{1}{2}$ $BaCl_2$	128.02
$\frac{1}{2}$ $NiSO_4$	93.20

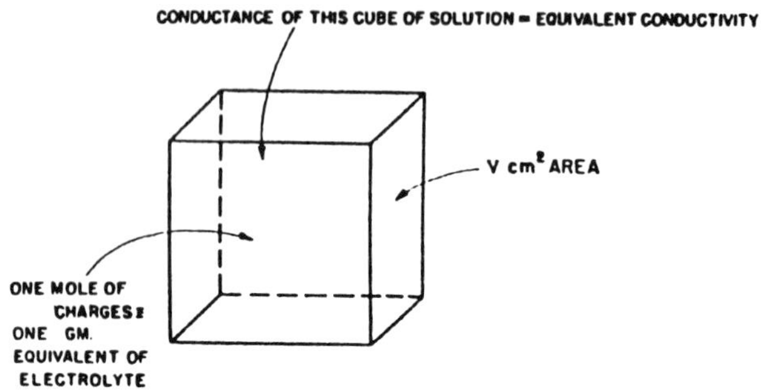

Fig. 4.54. Diagram to illustrate the meaning of the equivalent conductivity of an electrolyte.

It may be argued that if at infinite dilution there are no ions of the solute, how can the solution conduct? The procedure for determining the equivalent conductivity of an electrolyte at infinite dilution will clarify this problem. One takes solutions of a substance of various concentrations, determines the σ, and then normalizes each to the equivalent conductivity of particular solutions. If these values of Λ are then plotted against the logarithms of the concentration and this Λ versus log c curve is extrapo-

G	CONDUCTANCE	$\frac{1}{R}$	S or Ω^{-1}
ρ	RESISTIVITY	$\frac{RA}{\ell}$	S^{-1} cm
σ	SPECIFIC CONDUCTIVITY	$\frac{\ell}{RA}$	S cm^{-1}
Λ_m	MOLAR CONDUCTIVITY	$\frac{\sigma}{C}$	S cm^2 mol^{-1}
Λ	EQUIVALENT CONDUCTIVITY	$\frac{\sigma}{CZ}$	S cm^2 mol^{-1} eq^{-1}

Fig. 4.55. Definitions and units of the conductivity of electrolyte solutions.

TABLE 4.11

Equivalent Conductivity Varies with Concentration

Concentration (mol dm^{-3}) (KCl solutions)	Λ (S cm^2 mol^{-1} eq^{-1})
0.001	146.9
0.005	143.5
0.01	141.2
0.02	138.2
0.05	133.3
0.10	128.9

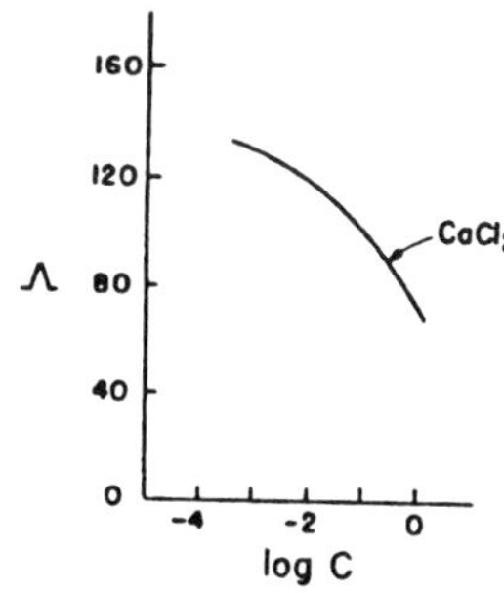

Fig. 4.56. The observed variation of the equivalent conductivity of $CaCl_2$ with concentration.

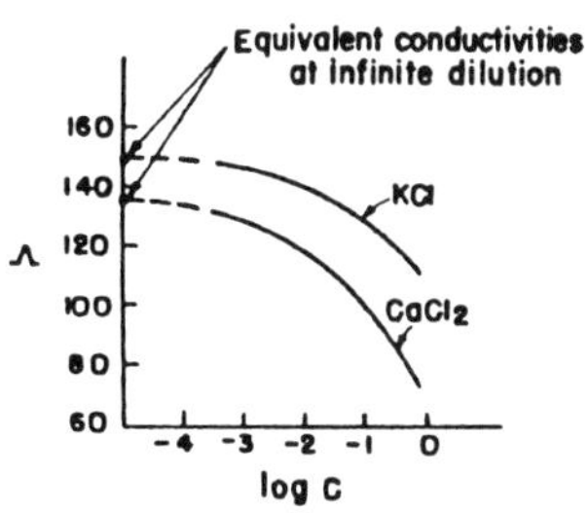

Fig. 4.57. The equivalent conductivity of an electrolyte at infinite dilution is obtained by extrapolating the Λ versus log *c* curve to zero concentration.

TABLE 4.12

Equivalent Conductivities at Infinite Dilution of Electrolytes in Aqueous Solution at 298 K

Electrolyte	Λ^0 (S cm^2 mol^{-1} eq^{-1})
HCl	426.0
NaOH	247.9
NaCl	126.4
KCl	149.8
$K_4Fe(CN)_6$	183.9
$CaCl_2$	135.7

lated, it approaches a limiting value (Fig. 4.57). It is this extrapolated value at zero concentration that is known as the equivalent conductivity at infinite dilution.

Anticipating the atomistic treatment of conduction that follows, it may be mentioned that at very low ionic concentrations, the ions are too far apart to exert appreciable interionic forces on each other. Only under these conditions does one obtain the pristine version of equivalent conductivity, i.e., values unperturbed by ion–ion interactions, which have been shown in Chapter 3 to be concentration dependent. The state of infinite dilution therefore is not only the reference state for the study of equilibrium properties (Section 3.3), it is also the reference state for the study of the nonequilibrium (irreversible) process, which is called *ionic conduction*, or *migration* (see Section 4.1).

4.3.9. How Equivalent Conductivity Changes with Concentration: Kohlrausch's Law

The experimental relationship between equivalent conductivity and the concentration of an electrolytic solution is found to be best brought out by plotting Λ against $c^{1/2}$. When this is done (Fig. 4.58), it can be seen that up to concentrations of about 0.01 N there is a linear relationship between Λ and $c^{1/2}$; thus,

$$\Lambda = \Lambda^0 - A c^{1/2} \tag{4.139}$$

where the intercept is the equivalent conductivity at infinite dilution Λ^0 and the slope of the straight line is a positive constant A.

This empirical relationship between the equivalent conductivity and the square root of concentration is a law named after Kohlrausch. His extremely careful measurements of the conductance of electrolytic solutions can be considered to have played a leading role in the initiation of ionics, the physical chemistry of ionic solutions.

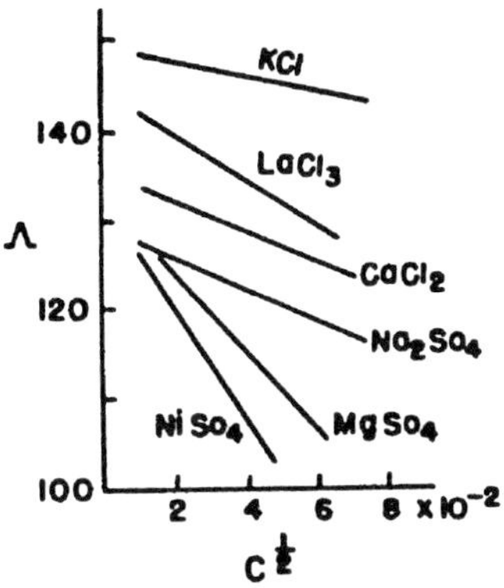

Fig. 4.58. The experimental basis for Kohlrausch's law: Λ versus $c^{1/2}$ plots consist of straight lines.

However, Kohlrausch's law [Eq. (4.139)] had to remain nearly 40 years without a theoretical basis.

The justification of Kohlrausch's law on theoretical grounds cannot be obtained within the framework of a macroscopic description of conduction. It requires an intimate view of ions in motion. A clue to the type of theory required emerges from the empirical findings by Kohlrausch: (1) the $c^{1/2}$ dependence and (2) the intercepts Λ^0 and slopes A of the Λ versus $c^{1/2}$ curves depend not so much on the particular electrolyte (whether it is KCl or NaCl) as on the type of electrolyte (whether it is a 1:1 or 2:2 electrolyte) (Fig. 4.59). All this is reminiscent of the dependence of the activity coefficient on $c^{1/2}$ (Chapter 3), to explain which the subtleties of ion–ion interactions had to be explored. Such interactions between positive and negative ions would determine to what extent they would influence each other when they move, and this would in turn bring about a fall in conductivity.

Kohlrausch's law will therefore be left now with only the sanction of experiment. Its incorporation into a theoretical scheme will be postponed until the section on the atomistic view of conduction is reached (see Section 4.6.12).

4.3.10. Vectorial Character of Current: Kohlrausch's Law of the Independent Migration of Ions

The driving force for ionic drift, i.e., the electric field X, not only has a particular magnitude, it also acts in a particular direction. It is a *vector*. Since the ionic current density j, i.e., the flow of electric charge, is proportional to the electric field operating in a solution [Eq. (4.128)],

$$j = \sigma X \tag{4.128}$$

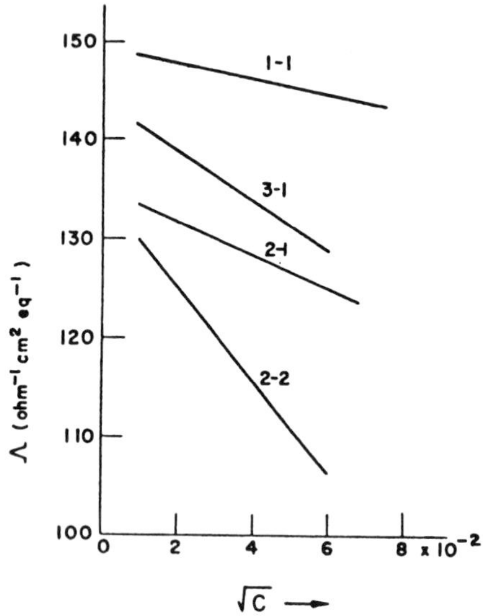

Fig. 4.59. The experimental Λ versus $c^{1/2}$ plots depend largely on the type of electrolyte.

the ionic current density must also be a vector. Vectorial quantities are often designated by arrows placed over the quantities (unless their directed character is obvious) or are indicated by bold type. Hence, Eq. (4.128) can be written

$$\vec{j} = \sigma \vec{X} \tag{4.140}$$

How is this current density constituted? What are its components? What is the structure of this ionic current density?

The imposition of an electric field on the electrolyte (Fig. 4.60) makes the positive ions drift toward the negative electrode and the negative ions drift in the opposite direction. The flux of positive ions $\vec{J}_+$ gives rise to a positive-ion current density $\vec{j}_+$, and the flux of negative ions in the opposite direction $*_-$ results in a negative-ion current density $\vec{j}_-$. By convention, the direction of current flow is taken to be either the direction in which positive charge flows or the direction opposite to that in which the negative charge flows. Hence, the positive-ion flux $\vec{J}_+$ corresponds to a current toward the negative electrode $\vec{j}_+$ and the negative-ion flux $*_-$ also corresponds to a current $\vec{j}_-$ *in the same direction* as that due to the positive ions.

It can be concluded therefore that the total current density $\vec{j}$ is made up of two contributions, one due to a flux of positive ions and the other due to a flux of negative ions. Furthermore, assuming for the moment that the drift of positive ions toward the

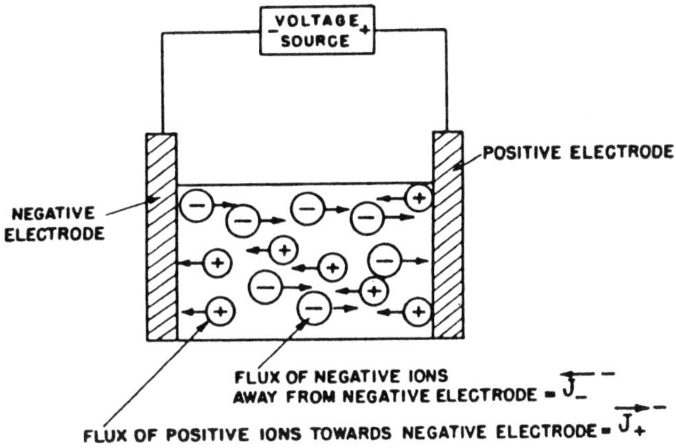

Fig. 4.60. Schematic representation of the direction of the drifts (and fluxes) of positive and negative ions acted on by an electric field.

negative electrode does not interfere with the drift of negative ions in the opposite direction, it follows that the component current densities are additive, i.e.,

$$\vec{j} = \vec{j}_+ + \vec{j}_- \tag{4.141}$$

Do ions migrate independently? Is the drift of the positive ions in one direction uninfluenced by the drift of the negative ions in the opposite direction? This is so if, and only if, the force fields of the ions do not overlap significantly, i.e., if there is negligible interaction or coupling between the ions. Coulombic ion–ion interactions usually establish such coupling. The only conditions under which the absence of ion–ion interactions can be assumed occur when the ions are infinitely far apart. Strictly speaking, therefore, ions migrate independently only at infinite dilution. Under these conditions, one can proceed from Eq. (4.141) to write

$$\frac{\vec{j}}{\vec{X}} = \frac{\vec{j}_+}{\vec{X}} + \frac{\vec{j}_-}{\vec{X}} \tag{4.142}$$

or [from Eq. (4.128)],

$$\sigma = \sigma_+ + \sigma_- \tag{4.143}$$

whence

$$\Lambda^0 = \lambda_+^0 + \lambda_-^0 \tag{4.144}$$

TABLE 4.13
Equivalent Conductances ($S\ cm^2\ mol^{-1}\ eq^{-1}$) of Individual Ions at Infinite Dilution at 298 K

Cation	λ_+^0	Anion	λ_-^0
H^+	349.82	OH^-	198.5
K^+	73.52	Br^-	78.4
Na^+	50.11	I^-	76.8
Li^+	38.69	Cl^-	76.34
$\frac{1}{2}Ba^{2+}$	63.64	$CH_3CO_2^-$	40.9

This is Kohlrausch's law of the independent migration of ions: The equivalent conductivity (at infinite dilution) of an electrolytic solution is the sum of the equivalent conductivities (at infinite dilution) of the ions constituting the electrolyte (Table 4.13).

At appreciable concentrations, the ions can be regarded as coupled or interacting with each other (see the ion-atmosphere model of Chapter 3). This results in the drift of positive ions toward the negative electrode, hindering the drift of negative ions toward the positive electrode; i.e., the interionic interaction results in the positive ions' equivalent conductivity reducing the magnitude of the negative ions' equivalent conductivity to below the infinite dilution value, and vice versa. To make quantitative estimates of these effects, however, one must calculate the influence of ionic-cloud effects on the phenomenon of conduction, a task that will be taken up further on.

4.4. A SIMPLE ATOMISTIC PICTURE OF IONIC MIGRATION

4.4.1. Ionic Movements under the Influence of an Applied Electric Field

In seeking an atomic view of the process of conduction, one approach is to begin with the picture of ionic movements as described in the treatment of diffusion (Section 4.2.4) and then to consider how these movements are perturbed by an electric field. In the treatment of ionic movements, it was stated that the ions in solution perform a random walk in which all possible directions are equally likely for any particular step. The analysis of such a random walk indicated that the mean displacement of ions is zero (Section 4.2.4), diffusion being the result of the statistical bias in the movement of ions, due to inequalities in their numbers in different regions.

When, however, the ions are situated in an electric field, their movements are affected by the fact that they are charged. Hence, the imposition of an electric field singles out one direction in space (the direction parallel to the field) for preferential ionic movement. Positively charged particles will prefer to move toward the negative electrode, and negatively charged particles, in the opposite direction. The walk is no longer quite random. The ions drift.

Another way of looking at ionic drift is to consider the fate of any particular ion under the field. The electric force field would impart to it an acceleration according to Newton's second law. Were the ion completely isolated (e.g., in vacuum), it would accelerate indefinitely until it collided with the electrode. In an electrolytic solution, however, the ion very soon collides with some other ion or solvent molecule that crosses its path. This collision introduces a discontinuity in its speed and direction. The motion of the ion is not smooth; it is as if the medium offers resistance to the motion of the ion. Thus, the ion stops and starts and zigzags. However, the applied electric field imparts to the ion a direction (that of the oppositely charged electrode), and the ion gradually works its way, though erratically, in the direction of this electrode. The ion drifts in a preferred direction.

4.4.2. Average Value of the Drift Velocity

Any particular ion starts off after a collision with a velocity that may be in any direction; this is the randomness in its walk. The initial velocity can be ignored precisely because it can take place in any direction and therefore does not contribute to the drift (preferred motion) of the ion. But the ion is all the time under the influence of the applied-force field.[23] This force imparts a component to the velocity of the ion, an extra velocity component in the same direction as the force vector $\vec{F}$. It is this additional velocity component due to the force $\vec{F}$ that is called the *drift velocity* v_d. What is its average value?

From Newton's second law, it is known that the force divided by the mass of the particle is equal to the acceleration. Thus,

$$\frac{\vec{F}}{m} = \frac{dv}{dt} \tag{4.145}$$

Now the time between collisions is a random quantity. Sometimes the collisions may occur in rapid succession; at others, there may be fairly long intervals. It is possible, however, to talk of a mean time between collisions, τ. In Section 4.2.5, it was shown that the number of collisions (steps) is proportional to the time. If N collisions occur in a time t, then the average time between collisions is t/N. Hence,

$$\tau = \frac{t}{N} \tag{4.146}$$

The average value of that component of the velocity of an ion picked up from the externally applied force is the product of the acceleration due to this force and the average time between collisions. Hence, the drift velocity v_d is given by

[23]The argument is developed in general for any force, not necessarily an electric force.

$$v_d = \frac{dv}{dt}\tau$$

$$= \frac{\vec{F}}{m}\tau \tag{4.147}$$

This is an important relation. It opens up many vistas. For example, through the mean time τ, one can relate the drift velocity to the details of ionic jumps between sites, as was done in the case of diffusion (Section 4.2.15).

Furthermore, the relation (4.147) shows that the drift velocity is proportional to the driving force of the electric field. The flux of ions will be shown (Section 4.4.4) to be related to the drift velocity[24] in the following way:

$$\text{Flux} = \text{Concentration of ions} \times \text{drift velocity} \tag{4.148}$$

Thus, if the $\vec{F}$ in Eq. (4.147) is the electric force that stimulates conduction, then this equation is the molecular basis of the fundamental relation used in the macroscopic view of conduction [see Eq. (4.122)], i.e.,

$$\text{flux} \propto \text{electric field}$$

The derivation of the basic relation (4.147) reveals the conditions under which the proportionality between drift velocity (or flux) and electric field breaks down. It is essential to the derivation that in a collision, an ion does not preserve any part of its extra velocity component arising from the force field. If it did, then the actual drift velocity would be greater than that calculated by Eq. (4.147) because there would be a cumulative carryover of the extra velocity from collision to collision. In other words, every collision must wipe out all traces of the force-derived extra velocity, and the ion must start afresh to acquire the additional velocity. This condition can be satisfied only if the drift velocity, and therefore the field, is small (see the autocorrelation function, Section 4.2.19).

4.4.3. Mobility of Ions

It has been shown that when random-walking ions are subjected to a directed force $\vec{F}$, they acquire a nonrandom, directed component of velocity—the drift velocity v_d. This drift velocity is in the direction of the force $\vec{F}$ and is proportional to it

$$v_d = \frac{\tau}{m}\vec{F} \tag{4.147}$$

[24]The dimensions of flux are moles per square centimeter per second, and they are equal to the product of the dimensions of concentration expressed in moles per cubic centimeter and velocity expressed in centimeters per second.

Since the proportionality constant τ/m is of considerable importance in discussions of ionic transport, it is useful to refer to it with a special name. It is called *the absolute mobility* because it is an index of how mobile the ions are. The absolute mobility, designated by the symbol $\bar{u}_{\text{abs}}$, is a measure of the drift velocity v_d acquired by an ionic species when it is subjected to a force $\vec{F}$, i.e.,

$$\bar{u}_{\text{abs}} = \frac{\tau}{m} = \frac{v_d}{\vec{F}} \tag{4.149}$$

which means that the absolute mobility is the drift velocity developed under unit applied force ($\vec{F} = 1$ dyne) and the units in which it is available in the literature are centimeters per second per dyne.

For example, one might have an electric field X of 0.05 V cm^{-1} in the electrolyte solution and observe a drift velocity of 2×10^{-5} cm s^{-1}. The electric force $\vec{F}$ operating on the ion is equal to the electric force per unit charge, i.e., the electric field X, times the charge $z_i e_0$ on each ion

$$\begin{aligned} \vec{F} &= z_i e_0 X \\ &= 1.60 \times 10^{-19} \times 0.05\ 10^2 \times 10^5 \\ &= 8 \times 10^{-14} \text{ dynes} \end{aligned} \tag{4.150}$$

(10^5 dynes = 1 Newton) for univalent ions. Hence, the absolute mobility is

$$\begin{aligned} \bar{u}_{\text{abs}} &= \frac{2 \times 10^{-5}}{8 \times 10^{-14}} \\ &= 2.5 \times 10^8 \text{ cm s}^{-1} \text{ dyn}^{-1} \end{aligned}$$

In electrochemical literature, however, mobilities of ions are not usually expressed in the absolute form defined in Eq. (4.149). Instead, they are more normally recorded as the drift velocities in *unit electric field* (1 V cm^{-1}) and will be referred to here as *conventional (electrochemical) mobilities* with the symbol u_{conv}.

The relation between the absolute and conventional mobilities follows from Eq. (4.149)

$$(V_{\text{Drift}})_{1 \text{ volt cm}^{-1}} = u_{\text{conv}} = \bar{u}_{\text{ABS}}\, z_i e_0 x \tag{4.151}$$

$$x = 1 \text{ volt}$$

i.e.,

$$u_{\text{conv}} = \bar{u}_{\text{abs}} z_i e_0 \tag{4.152}$$

Thus, the conventional and absolute mobilities are proportional to each other, the proportionality constant being an integral multiple z_i of the electronic charge. In the example cited earlier,

$$u_{\text{conv}} = \frac{2 \times 10^{-5}}{0.05}$$

$$= 4 \times 10^{-4}\ \text{cm}^2\ \text{V}^{-1}\ \text{s}^{-1}$$

Though the two types of mobilities are closely related, it must be stressed that the concept of absolute mobility is more general because it can be used for *any* force that determines the drift velocity of ions and not only the electric force used in the definition of conventional mobilities.

4.4.4. Current Density Associated with the Directed Movement of Ions in Solution, in Terms of Ionic Drift Velocities

It is the aim now to show how the concept of drift velocity can be used to obtain an expression for the ionic current density flowing through an electrolyte in response to an externally applied electric field. Consider a transit plane of unit area normal to the direction of drift (Fig. 4.61). Both the positive and the negative ions will drift across this plane. Consider the positive ions first, and let their drift velocity be $(v_d)_+$ or simply v_+. Then, in 1 s, all positive ions within a distance v_+ cm of the transit plane will cross it. The flux J_+ of positive ions (i.e., the number of moles of these ions arriving in 1 second at the plane of unit area) is equal to the number of moles of positive ions in a volume of 1 cm^2 in area and v cm in length (with $t = 1$ s). Hence, J_+ is equal to the volume v_+ in cubic centimeters times the concentration c_+ expressed in moles per cubic centimeter

$$\boldsymbol{J}_+ = c_+ v_+ \tag{4.153}$$

The flow of charge across the plane due to this flux of positive ions (i.e., the current density $\vec{j}_+$) is obtained by multiplying the flux $\boldsymbol{J}_+$ by the charge z_+F borne by 1 mole of ions

$$j_+ = z_+ F c_+ v_+ \tag{4.154}$$

This, however, is only the contribution of the positive ions. Other ionic species will make their own contributions of current density. In general, therefore, the current density due to the ith species will be

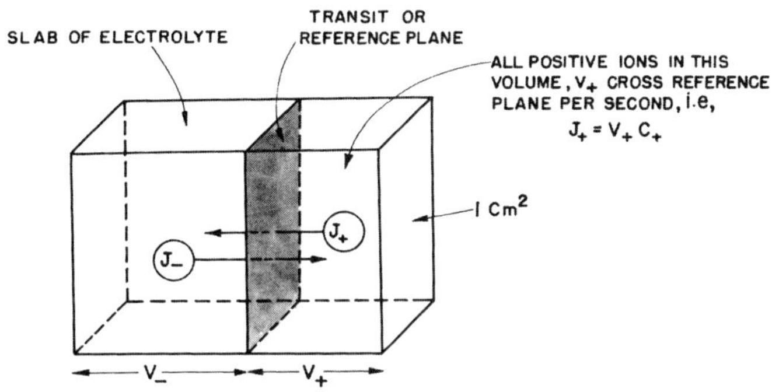

Fig. 4.61. Diagram for the derivation of a relation between the current density and the drift velocity.

$$\vec{j}_i = z_i F c_i v_i \tag{4.155}$$

The total current density due to the contribution of all the ionic species will therefore be

$$\vec{j} = \sum_i \vec{j}_i \tag{4.156}$$

$$= \sum_i z_i F c_i v_i \tag{4.157}$$

If a z:z-valent electrolyte is taken, then $z_+ = z_- = z$ and $c = c_+ = c_-$ and one has

$$\vec{j} = zFc(v_+ + v_-) \tag{4.158}$$

By recalling that the ionic drift velocities are related through the force operating on the ions to the ionic mobilities [Eq. (4.151)], it will be realized that Eq. (4.157) is the basic expression from which may be derived the expressions for conductance, equivalent conductivity, specific conductivity, etc.

4.4.5. Specific and Equivalent Conductivities in Terms of Ionic Mobilities

Let the fundamental expression for the drift velocity of ions [Eq. (4.151)] be substituted in Eq. (4.157) for current density. One obtains

$$\vec{j} = \sum z_i F c_i (u_{\text{conv}})_i X \tag{4.159}$$

or, from Eq. (4.128)

$$\sigma = \frac{\vec{j}}{X} = \Sigma z_i F c_i (u_{\text{conv}})_i \tag{4.160}$$

which reduces in the special case of a z:z-valent electrolyte to

$$\sigma = zFc[(u_{\text{conv}})_- + (u_{\text{conv}})_-] \tag{4.161}$$

Several conclusions follow from this atomistic expression for specific conductivity. First, it is obvious from this equation that the specific conductivity σ of an electrolyte cannot be a concentration-independent constant (as it is in the case of metals). It will vary because the number of moles of ions per unit volume c can be varied in an electrolytic solution.

Second, the specific conductivity can easily be related to the molar Λ_m and equivalent Λ conductivities. Take the case of a z:z-valent electrolyte. With Eqs. (4.161), (4.136), and (4.138), it is found that

$$\Lambda_m = \frac{\sigma}{c}$$

$$= zF[(u_{\text{conv}})_+ + (u_{\text{conv}})_-] \tag{4.162}$$

and

$$\Lambda = \frac{\Lambda_m}{z}$$

$$= F[(u_{\text{conv}})_+ + (u_{\text{conv}})_-] \tag{4.163}$$

What does Eq. (4.163) reveal? It shows that the equivalent conductivity will be a constant independent of concentration only if the electrical mobility does not vary with concentration. It will be seen, however, that ion–ion interactions (which have been shown in Section 3.3.8 to depend on concentration) prevent the electrical mobility from being a constant. Hence, the equivalent conductivity must be a function of concentration.

4.4.6. The Einstein Relation between the Absolute Mobility and the Diffusion Coefficient

The process of diffusion results from the random walk of ions; the process of migration (i.e., conduction) results from the drift velocity acquired by ions when they experience a force. The drift of ions does not obviate their random walk; in fact, it is superimposed on their random walk. Hence, the drift and the random walk must be intimately linked. Einstein† realized this and deduced a vital relation between the absolute mobility $\bar{u}_{abs}$, which is a quantitative characteristic of the drift, and the diffusion coefficient D, which is a quantitative characteristic of the random walk.

Both diffusion and conduction are nonequilibrium (irreversible) processes and are therefore not amenable to the methods of equilibrium thermodynamics or equilibrium statistical mechanics. In these latter disciplines, the concepts of time and change are absent. It is possible, however, to imagine a situation where the two processes oppose and balance each other and a "pseudoequilibrium" obtains. This is done as follows (Fig. 4.62).

Consider a solution of an electrolyte MX to which a certain amount of radioactive M^+ ions are added in the form of the salt MX. Further, suppose that the tracer ions are not dispersed uniformly throughout the solution; instead, let there be a concentration gradient of the tagged species so that its diffusion flux J_D is given by Fick's first law

$$J_D = -D\frac{dc}{dx} \tag{4.16}$$

Now let an electric field be applied. Each tagged ion feels the field, and the drift velocity is

$$v_d = \bar{u}_{abs}\vec{F} \tag{4.149}$$

This drift velocity produces a current density given by [*cf.* Eq. (4.154)][25]

$$\vec{j} = z_+ F c v_d \tag{4.164}$$

i.e., a conduction flux J_c that is arrived at by dividing the conduction current density $\vec{j}$ by the charge per mole of ions

†Albert Einstein's name remains firmly as that of the most well know scientist in the world, and his name gives rise to the image students have of what it is like to be a scientist. This is because he produced two theories which few other scientists understand but seem to show that, in extreme situations of mass and/or velocity, the world is not what it seems at all. But the fertility of Einstein's thought extended in several other directions and most of them are presented in this chapter. Thus, he produced a theory of Brownian motion (and related it to the net movement in one direction of a diffusing particle); he joined the ancient law of Stokes to diffusion and showed how knowledge of viscosity and the radius of a particle allowed one to know the corresponding diffusion coefficient; and he gave rise to a most unexpected connection of the coefficient of diffusion to the rate of the vectorial drift of ions under an electric field (later taken up by Nernst and connected to conductivity). If Relativity is Einstein's most famous work, that which has been most immediately useful lies in electrochemistry.

Many books have described Einstein's life, particularly after he came to Princeton. He was a stickler for time keeping and rode a bike to the office, 9:00 arrival, riding home again at 1:00 p.m. to work with his assistant in mathematics. But he sometimes caused embarrassment in the social scene—arriving, say, at a formal banquet in dinner jacketed and black tie, but wearing casual grey pants and sandals (no socks). To questions about this, his rejoinder was logical: "The invitation said "Dinner jacket and black tie."

[25]In Eq. (4.154), one will find v_+; the reason is that, in Section 4.4.4, the drift velocity of a positive ion, $(v_d)_+$, had been concisely written as v_+.

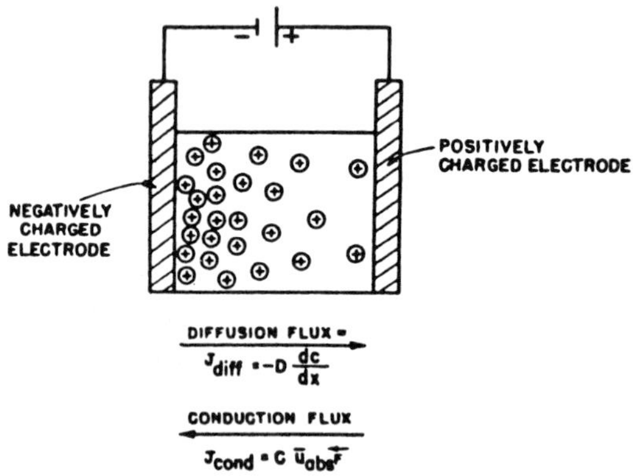

Fig. 4.62. An imaginary situation in which the applied field is adjusted so that the conduction flux of tagged ions (the only ones shown in the diagram) is exactly equal and opposite to the diffusion flux.

$$J_c = \frac{\vec{j}}{zF} = cv_d \tag{4.165}$$

By introducing the expression (4.149) for the drift velocity into (4.165), the conduction flux becomes

$$J_c = c\bar{u}_{\text{abs}}\vec{F}$$

The applied field is adjusted so that the conduction flux exactly compensates for the diffusion flux. In other words, if the tracer ions (which are positively charged) are diffusing toward the positive electrode, then the magnitude of the applied field is such that the positively charged electrode repels the positive tracer ions to an extent that their net flux is zero. Thus

$$J_D + J_c = 0 \tag{4.166}$$

or

$$-D\frac{dc}{dx} + c\bar{u}_{\text{abs}}\vec{F} = 0$$

i.e.,

$$\frac{dc}{dx} = \frac{c\bar{u}_{\text{abs}}\vec{F}}{D} \tag{4.167}$$

Under these "balanced" conditions, the situation may be regarded as tantamount to equilibrium because there is no net flux or transport of ions. Hence, the Boltzmann law can be used. The argument is that since the potential varies along the x direction, the concentration of ions at any distance x is given by

$$c = c^0 e^{-U/kT} \tag{4.168}$$

where U is the potential energy of an ion in the applied field and c^0 is the concentration in a region where the potential energy is zero. Differentiating this expression, one obtains

$$\begin{aligned}\frac{dc}{dx} &= -c^0 e^{-U/kT}\frac{1}{kT}\frac{dU}{dx}\\ &= -\frac{c}{kT}\frac{dU}{dx}\end{aligned} \tag{4.169}$$

But, by the definition of force,

$$\vec{F} = -\frac{dU}{dx} \tag{4.170}$$

Hence, from Eqs. (4.169) and (4.170), one obtains

$$\frac{dc}{dx} = \frac{c}{kT}\vec{F} \tag{4.171}$$

If, now, Eqs. (4.167) and (4.171) are compared, it is obvious that

$$\frac{\bar{u}_{\text{abs}}}{D} = \frac{1}{kT}$$

or

$$D = \bar{u}_{\text{abs}} kT \tag{4.172}$$

This is the Einstein relation. It is probably the most important relation in the theory of the movements and drift of ions, atoms, molecules, and other submicroscopic particles. It has been derived here in an atomistic way. It will be recalled that in the phenomenological treatment of the diffusion coefficient (Section 4.2.3), it was shown that

$$BRT = D \tag{4.17}$$

where B was an undetermined phenomenological coefficient. If one combines Eqs. (4.172) and (4.17), it is clear that

$$BRT = \bar{u}_{\text{abs}} kT$$

or

$$B = \frac{\overline{u}_{abs} kT}{RT} = \frac{\overline{u}_{abs}}{N_A} \tag{4.173}$$

Thus, one has provided a fundamental basis for the phenomenological coefficient B; it is the absolute mobility $\overline{u}_{abs}$ divided by the Avogadro number.

The Einstein relation also permits experiments on diffusion to be linked up with other phenomena involving the mobility of ions, i.e., phenomena in which there are forces that produce drift velocities. Two such forces are the force experienced by an ion when it overcomes the viscous drag of a solution and the force arising from an applied electric field. Thus, the diffusion coefficient may be linked up to the viscosity (the Stokes–Einstein relation) and to the equivalent conductivity (the Nernst–Einstein relation).

4.4.7. Drag (or Viscous) Force Acting on an Ion in Solution

Striking advances in science sometimes arise from seeing the common factors in two apparently dissimilar situations. One such advance was made by Einstein when he intuitively asserted the similarity between a macroscopic sphere moving in an incompressible fluid and a particle (e.g., an ion) moving in a solution (Fig. 4.63).

The macroscopic sphere experiences a viscous, or drag, force that opposes its motion. The value of the drag force depends on several factors—the velocity v and diameter d of the sphere and the viscosity η and density ρ of the medium. These factors can all be combined and used to define a dimensionless quantity known as the *Reynolds number* (Re) defined thus:

$$\mathrm{Re} = vd\frac{\rho}{\eta} \tag{4.174}$$

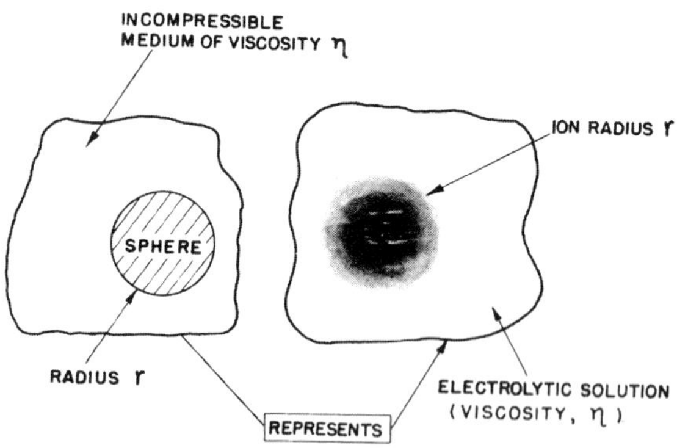

Fig. 4.63. An ion drifting in an electrolytic solution is like a sphere (of the same radius as the ion) moving in an incompressible medium (of the same viscosity as the electrolyte).

When the hydrodynamic conditions are such that this Reynolds number is much smaller than unity, Stokes showed that the drag force F opposing the sphere is given by the following relation

$$F = 6\pi r\eta v \tag{4.175}$$

where v is the velocity of the macroscopic body. The relation is known as Stokes' law. Its derivation is lengthy and awkward because the most convenient coordinates to describe the sphere and its environment are spherical coordinates and those to describe the flow are rectangular coordinates.

The real question, of course, centers around the applicability of Stokes' law to microscopic ions moving in a structured medium in which the surrounding particles are roughly the same size as the ions. Initially, one can easily check on whether the Reynolds number *is* smaller than unity for ions drifting through an electrolyte. With the use of the values $d_{\text{ion}} \approx 10^{-8}$ cm, $v_{\text{ion}} = v_d \approx 10^{-4}$ cm s^{-1}, $\eta \approx 10^{-2}$ poise, and $\rho \simeq 1$, it turns out that the Reynolds number for an ion moving through an electrolyte is about 10^{-10}. Thus, the hydrodynamic condition Re << 1 required for the validity of Stokes' law is easily satisfied by an ion in solution.

However, the hydrodynamic problem that Stokes solved to get $F = 6\pi r\eta v$ pertains to a sphere moving in an incompressible *continuum* fluid. This is a far cry indeed from the actuality of an ion drifting inside a discontinuous electrolyte containing particles (solvent molecules, other ions, etc.) of about the same size as the ion. Furthermore, the ions considered may not be spherical.

From this point of view, the use of Stokes' law for the viscous force experienced by ions is a bold step. Several attempts have been made over a long time to theorize about the viscous drag on tenths of nanometer-sized particles in terms of a more realistic model than that used by Stokes. It has been shown, for example, that if the moving particle is cylindrical and not spherical, the factor 6π should be replaced by 4π. While refraining from the none-too-easy analysis of the degree of applicability of Stokes' law to ions in electrolytes, one point must be stressed. For sufficiently small ions, Stokes' law does not have a numerical significance[26] greater than about ±50%. Attempts to tackle the problem of the flow of ions in solution without resorting to Stokes' law do not give much better results.

[26]Stokes' law is often used in electrochemical problems, but its approximate nature is not always brought out. Apart from the validity of extrapolating from the macroscopic–sphere-continuum-fluid model of Stokes to atomic near-spheres in a molecular liquid, another reason for the limited validity of Stokes' law arises from questions concerning the radii which should be substituted in any application of the law. These should not be the crystallographic radii, and an appraisal of the correct value implies a rather detailed knowledge of the structure of the solvation sheath (see Section 2.4). Furthermore, the viscosity used in Stokes' law is the bulk average viscosity of the whole solution, whereas it is the local viscosity in the neighborhood of the ion that should be taken. The two may not be the same, because the ion's field may affect the solvent structure and hence its viscosity.

4.4.8. The Stokes–Einstein Relation

During the course of diffusion, the individual particles are executing the complicated starts, accelerations, collisions, stops, and zigzags that have come to be known as random walk. When a particle is engaged in its random walk, it is of course subject to the viscous drag force exerted by its environment. The application of Stokes' law to these detailed random motions is no easy matter because of the haphazard variation in the speed and direction of the particles. Instead, one can apply Stokes' law to the diffusional movements of ions by adopting the following artifice suggested by Einstein.

When diffusion is occurring, it can be considered that there is a driving force $-d\mu/dx$ operating on the particles. This driving force produces a steady-state diffusion flux J, corresponding to which [*cf.* Eq. (4.14)] one can imagine a drift velocity v_d for the diffusing particles.[27] Since this velocity v_d is a steady-state velocity, the diffusional driving force $-d\mu/dx$ must be opposed by an equal resistive force, which will be taken to be the Stokes viscous force $6\pi r\eta v_d$. Hence,

$$-\frac{d\mu}{dx} = 6\pi r\eta v_d \tag{4.176}$$

The existence of a charge on a moving body has the following effect on a polar solvent: It tends to produce an orientation of solvent dipoles in the vicinity of the ion. Since, however, the charge is moving, once oriented, the dipoles take some finite relaxation time τ to disorient. During this relaxation time, a relaxation force operates on the ion; this relaxation force is equivalent to an additional frictional force on the ion and results in an expression for the drag force of the form

$$F = 6\pi\eta vr - 6\pi\eta v\frac{s}{\varepsilon}$$

where s is $(4/9)(\tau/6\pi\eta)e_0^2/r^3$ and ε is the dielectric constant of the medium. The correction may be as much as 25% but will be neglected here in the interest of deriving the classical Stokes–Einstein relation.

One can therefore define the absolute mobility $\bar{u}_{\text{abs}}$ for the diffusing particles by dividing the drift velocity by either the diffusional driving force or the equal and opposite Stokes viscous force

$$\bar{u}_{\text{abs}} = \frac{v_d}{-d\mu/dx} = \frac{v_d}{6\pi r\eta v_d} = \frac{1}{6\pi r\eta} \tag{4.177}$$

---

[27]The hypothetical nature of the argument lies in the fact that in diffusion, there is no actual force exerted on the particles. Consequently, there is not the actual force-derived component of the velocity; i.e., there is no actual drift velocity (see Section 4.2.1). Thus, the drift velocity enters the argument only as a device.

The fundamental expression (4.172) relating the diffusion coefficient and the absolute mobility can be written thus:

$$\bar{u}_{\text{abs}} = \frac{D}{kT} \tag{4.178}$$

By equating Eqs. (4.177) and (4.178), the Stokes–Einstein relation is obtained.

$$D = \frac{kT}{6\pi r\eta} \tag{4.179}$$

It links the processes of diffusion and viscous flow.

The Stokes–Einstein relation proved extremely useful in the classical work of Perrin. Using an ultramicroscope, he watched the random walk of a colloidal particle, and from the mean square distance $<x^2>$ traveled in a time t, he obtained the diffusion coefficient D from the relation (4.27)

$$D = \frac{<x^2>}{2t} \tag{4.27}$$

The weight of the colloidal particles and their density being known, their radius r was then obtained. Then the viscosity η of the medium could be used to obtain the Boltzmann constant

$$k = \frac{6\pi r\eta D}{T} \tag{4.180}$$

But

$$k = \frac{R}{N_A} \tag{4.181}$$

or

$$N_A = \frac{R}{k} \tag{4.182}$$

and thus the Avogadro number could be determined.

The use of Stokes' law also permits the derivation of a very simple relation between the viscosity of a medium and the conventional electrochemical mobility u_{conv}. Starting from the earlier derived equation (4.177)

$$\bar{u}_{\text{abs}} = \frac{1}{6\pi r\eta} \tag{4.177}$$

one substitutes for the absolute mobility the expression from Eq. (4.152)

$$u_{\text{conv}} = \bar{u}_{\text{abs}} z_i e_0 \tag{4.152}$$

and gets the result

$$u_{\text{conv}} = \frac{z_i e_0}{6\pi r\eta} \tag{4.183}$$

This relation shows that, owing to the Stokes viscous force, the conventional mobility of an ion depends on the charge and radius[28] of the solvated ion and the viscosity of the medium. The mobility given by Eq. (4.183) is often called the *Stokes mobility*. It will be seen later that the Stokes mobility is a highly simplified expression for mobility, and ion–ion interaction effects introduce a concentration dependence that is not seen in Eq. (4.183).

4.4.9. The Nernst–Einstein Equation

Now the Einstein relation (4.172) will be used to connect the transport processes of diffusion and conduction. The starting point is the basic equation relating the equivalent conductivity of a *z*:*z*-valent electrolyte to the conventional mobilities of the ions, i.e., to the drift velocities under a potential gradient of 1 V cm^{-1},

$$\Lambda = F[(u_{\text{conv}})_+ + (u_{\text{conv}})_-] \tag{4.163}$$

By using the relation between the conventional and absolute mobilities, Eq. (4.163) can be written

$$\Lambda = z_i e_0 F[(\bar{u}_{\text{abs}})_+ + (\bar{u}_{\text{abs}})_-] \tag{4.184}$$

With the aid of the Einstein relation (4.172),

$$(\bar{u}_{\text{abs}})_+ = \frac{D_+}{kT} \quad \text{and} \quad (\bar{u}_{\text{abs}})_- = \frac{D_-}{kT} \tag{4.185}$$

one can transform Eq. (4.184) into the form

$$\Lambda = \frac{z e_0 F}{kT}(D_+ + D_-) \tag{4.186}$$

[28]Earlier, the radius dependence of the conventional mobility was used to obtain information on the solvation number (see Section 2.8).

This is one form of the Nernst–Einstein equation; from a knowledge of the diffusion coefficients of the individual ions, it permits a calculation of the equivalent conductivity. A more usual form of the Nernst–Einstein equation is obtained by multiplying numerator and denominator by the Avogadro number, in which case it is obvious that

$$\Lambda = \frac{zF^2}{RT}(D_+ + D_-) \tag{4.187}$$

4.4.10. Some Limitations of the Nernst–Einstein Relation

There were several aspects of the Stokes–Einstein relation that reduced it to being only an approximate relation between the diffusion coefficient of an ionic species and the viscosity of the medium. In addition, there were fundamental questions regarding the extrapolation of a law derived for macroscopic spheres moving in an incompressible medium to a situation involving the movement of ions in an environment of solvent molecules and other ions. In the case of the Nernst–Einstein relation, the factors that limit its validity are more subtle.

An implicit but principal requirement for the Nernst–Einstein equation to hold is that the species involved in diffusion must also be the species responsible for conduction. Suppose now that the species M exists not only as ions M^{z+} but also as ion pairs $M^{z+}A^{z-}$ of the type described in Section 3.8.1.

The diffusive transport of M proceeds through both ions and ion pairs. In the conduction process, however, the situation is different (Fig. 4.64). The applied electric field exerts a driving force on only the charged particles. An ion pair as a whole is electrically neutral; it does not feel the electric field. Thus, ion pairs are not participants in the conduction process. This point is of considerable importance in conduction in nonaqueous media (see Section 4.7.12).

In systems where ion-pair formation is possible, the mobility calculated from the diffusion coefficient $\bar{u}_{abs} = D/kT$ is not equal to the mobility calculated from the equivalent conductivity $\bar{u}_{abs} = u_{conv}/z_i e_0 = (\Lambda/z_i e_0)F$ and therefore the Nernst–Einstein equation, which is based on equating these two mobilities, may not be completely valid. In practice, one finds a degree of nonapplicability of up to 25%.

Another important limitation on the Nernst–Einstein equation in electrolytic solutions may be approached through the following considerations. The diffusion coefficient is in general not a constant. This has been pointed out in Section 4.2.3, where the following expression was derived,

$$D = BRT\left(1 + \frac{d \ln f}{d \ln c}\right) \tag{4.19}$$

It is clear that BRT is the value of the diffusion coefficient when the solution behaves ideally, i.e., $f = 1$; this ideal value of the diffusion coefficient will be called D^0. Hence,

Fig. 4.64. The difference in the behavior of neutral ion pairs during diffusion and conduction.

$$D = D^0\left(1 + \frac{d\ln f}{d\ln c}\right)$$

$$= D^0\left(1 + c\,\frac{d\log f}{dc}\right) \tag{4.188}$$

and making use of the Debye–Hückel limiting law for the activity coefficient (see Section 3.5),

$$\log f = -Ac^{1/2}$$

one has

$$D = D^0(1 - \tfrac{1}{2}Ac^{1/2}) \tag{4.189}$$

an expression which shows how the diffusion coefficient varies with concentration. In addition, there is Kohlrausch's law

$$\Lambda = \Lambda^0 - Ac^{1/2} \tag{4.139}$$

where Λ^0 is the equivalent conductivity at infinite dilution, i.e., the ideal value.

From Eqs. (4.189) and (4.139), it is obvious that the diffusion coefficient D and the equivalent conductivity Λ have different dependencies on concentration (Fig. 4.65). This experimentally observed fact has an important implication as far as the

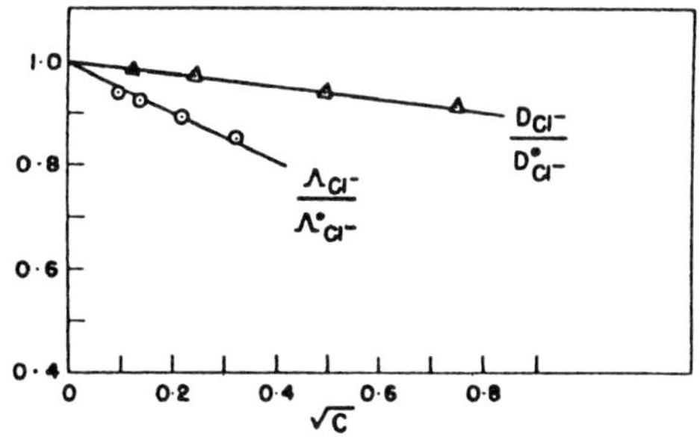

Fig. 4.65. The variation of the diffusion coefficient and the equivalent conductivity with concentration.

applicability of the Nernst–Einstein equation in electrolytic solutions is concerned. If the equation is true at one concentration, it cannot be true at another because the diffusion coefficient and the equivalent conductivity have varied to different extents in going from one concentration to the other.

The above argument brings out an important point about the limitations of the Nernst–Einstein equation. It does not matter whether the diffusion coefficient and the equivalent conductivity vary with concentration; to introduce deviations into the Nernst–Einstein equation, D and Λ must have different concentration dependencies. The concentration dependence of the diffusion coefficient has been shown to be due to nonideality ($f \neq 1$), i.e., due to ion–ion interactions, and it will be shown later that the concentration dependence of the equivalent conductivity is also due to ion–ion interactions. It is not the existence of interactions *per se* that underlies deviations from the Nernst–Einstein equation; otherwise, molten salts and ionic crystals, in which there are strong interionic forces, would show far more than the observed few percent deviation of experimental data from values calculated by the Nernst–Einstein equation. The essential point is that the interactions must affect the diffusion coefficient and the equivalent conductivity by *different mechanisms* and thus to different extents. How this comes about for diffusion and conduction in solution will be seen later.

In solutions of electrolytes, the $c^{1/2}$ terms in the expressions for D and Λ tend to zero as the concentration of the electrolytic solution decreases, and the differences in the concentration variation of D and Λ become more and more negligible; in other words, the Nernst–Einstein equation becomes increasingly valid for electrolytic solutions with increasing dilution.

4.4.11. The Apparent Ionic Charge

In Eqs. (4.172) and (4.186), two relations between the random movement of particles and directed drift (mobility under an electric field gradient) have been deduced. In Eq. (4.172), D is related to ionic mobility (Einstein equation) and in Eq. (4.186), it is related to the equivalent conductivity (Nernst–Einstein equation).

The relations derived were based upon very acceptable concepts, such as the random movement of particles in fluids, or their drift in the direction of an applied electric field. These ideas are so general that they cannot be doubted—they do not involve models later found to be wrong and to be displaced by other models, etc. For this reason, equations such as those named above are called *phenomenological*, meaning that they involve phenomena (drift and random movement) that cannot be doubted by the greatest skeptic. They happen.

One gets a mild shock therefore when one looks into the experimental tests of these equations, for only under extremely simple conditions (in fact, very dilute solutions) do they work out to be correct. Although the results never hint that the equations are wrong, there is sufficient discrepancy (e.g., between the diffusion coefficient calculated by using an experimental mobility substituted in the Einstein equation and that determined by direct experiment) for one to take notice and form some idea of a puzzle. How can simple mathematical reasoning based on the existence of movements undeniably present give rise to error?

There are two kinds of response to this challenge. In the first, one can invent special models to explain the observed discrepancy. One such model is displayed on several pages of Chapter 5—it deals with the deviations from the Einstein equations in high-temperature ionic liquids. It is suggested that the discrepancies arise because there are some kinds of molecular movements that contribute to diffusion but not to conduction. One kind of diffusive mode involves pairs of oppositely charged ions moving together. This would contribute to D, but because a pair of oppositely charged ions carries no net charge, its movement would not contribute to the mobility u_i or to the equivalent conductivity Λ, which reflects only the movement of individual charged particles moving under an applied field. This kind of movement—*individual* ionic movement—contributes to D_{exptl} also. Therefore, D_{exptl} receives two kinds of contributions (uncharged pairs and charged individuals), but the u_i receives only one (that of the individual charges). Clearly then, the D_{exptl} determined from radiotracer experiments would be greater than the D calculated from the Einstein or Nernst–Einstein equations because in their deduction only individual ions count—those that drift in one direction under the applied field, and not pairs of ions, which the field cannot affect.

This explanation is all very well for the liquid sodium chloride type of case, but deviations from the predictions of the Nernst–Einstein equation occur in dilute aqueous solutions also, and here the + and – ions are separated by stretches of water, and ion pairs do not form significantly until about 0.1 M.

Because deviations from the Nernst–Einstein equation are so widespread, and because the reasoning that gives rise to the equation is phenomenological, it is better to work out a general kind of noncommittal response—one that is free of a specific model such as that suggested in the molten salt case (see Section 5.2). The response is to suggest that it is useful to imagine that ions carry (effectively), not the charge we would normally associate with them (e.g., 2 for SO_4^{2-}), but a slightly different charge—one which, when used in the Nernst–Einstein equation in the form (4.186) will make the D to which it gives rise have the same value as the experimental one. It is clear that this way of describing a discrepancy from what is expected from reasoning that is evidently too simple bears some resemblance to the use of activity coefficients to describe why the use of concentrations (and not "activities") does not work well when describing the equilibrium properties of true electrolytes (Section 3.4.2).

The original Einstein equation is

$$D_{\text{calc}} = \bar{u}_{\text{abs}} kT \tag{4.190}$$

Now z is the formal charge on the ion (e.g., 3 for La^{3+}). One finds experimentally, however, that the D_{calc} obtained in this way is smaller than the one that is determined experimentally by means of radiotracers. It might be reasonable then to write

$$D_{\text{real}} = \frac{u_{\text{conv}}}{z_{\text{apparent}}} \frac{kT}{e_0} \tag{4.191}$$

The z_{apparent} is chosen so the equation comes out to fit the experiment. In this second response to the challenge of the deviant values from the Einstein equation, one draws picture or suggest a model for why z_{apparent} is not equal to z. The apparent charge is just a coverup for the deviation.

Now from Eq. (4.187), and for a symmetrical electrolyte

$$z_{\text{apparent}} = \frac{RT}{F^2} \frac{\Lambda}{(D_+ + D_-)} \tag{4.192}$$

At infinite dilution the ions are truly independent of each other (no interionic attraction) and there would be no reason for any difference between z_{apparent} and z.

In Section 5.6.6, one will see the details of the modeling treatment of the deviations for the Nernst–Einstein equation outlined here. Of course, a modeling explanation is more enlightening than a general-explanation type of approach. However, the difficulty is that the model of paired ions jumping together applies primarily to a pure liquid electrolyte, or alloy, where the existence of paired vacancies is a fact. Other models would have to be devised for other kinds of systems where deviations do occur (Fig. 4.66).

4.4.12. A Very Approximate Relation between Equivalent Conductivity and Viscosity: Walden's Rule

The Stokes–Einstein equation (4.179) connects the diffusion coefficient and the viscosity of the medium; the Nernst–Einstein equation (4.187) relates the diffusion coefficient to the equivalent conductivity. Hence, by eliminating the diffusion coeffi-

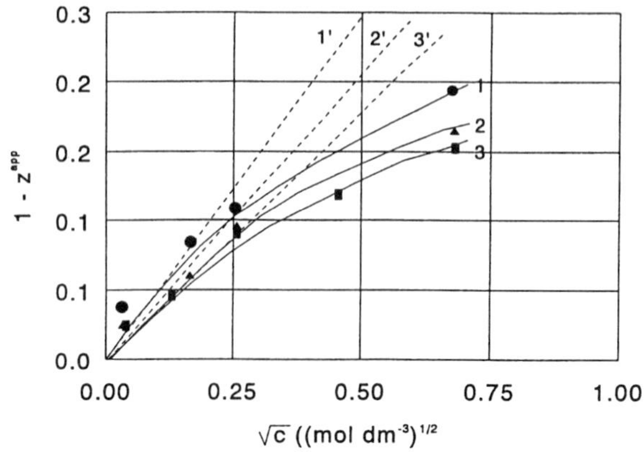

Fig. 4.66. Apparent charge of alkali halides as a function of concentration. Aqueous solutions of NaF (l), NaCl (2), and KCl (3) at 25 °C. Broken lines: limiting law of Debye and Hückel. (Reprinted from P. Turq, J. Barthel, and M. Chemla, in *Transport Relaxation and Kinetic Processes in Electrolyte Solutions*, Springer-Verlag, Berlin, 1992.)

cient in these two equations, it is possible to obtain a relation between the equivalent conductivity and the viscosity of the electrolyte. The algebra is as follows:

$$D = \frac{kT}{6\pi r\eta} = \frac{RT}{zF^2}\Lambda \tag{4.193}$$

and therefore

$$\Lambda = \frac{zF^2k}{6\pi R}\frac{1}{r\eta} \tag{4.194}$$

Since $F = N_A e_0$ and $k/R = 1/N_A$, one obtains

$$\Lambda = \frac{ze_0F}{6\pi}\frac{1}{r\eta}$$

$$\Lambda\eta = \frac{\text{constant}}{r} \tag{4.195}$$

Hence, if the radius of the moving (kinetic) entity in conduction, i.e., the solvated ion, can be considered the same in solvents of various viscosities, the following relation is obtained:

TABLE 4.14

Tests of Walden's Rule. The Product $\Lambda\eta$ for Potassium Iodide in Various Solvents at 298 K

Solvent	Λ	η	$\Lambda\eta$
Sulfur dioxide[a]	265	0.00394	1.044
Acetonitrile	198.2	0.00345	0.684
Acetone	185.5	0.00316	0.586
Nitromethane	124.0	0.00611	0.758
Methyl alcohol	114.8	0.00546	0.627
Pyridine[b]	71.3	0.00958	0.682
Ethyl alcohol	50.9	0.01096	0.560
Furfural	43.1	0.01490	0.642
Acetophenone	39.8	0.01620	0.644

[a]273 K.
[b]293 K.

$$\Lambda\eta = \text{constant} = \frac{ze_0F}{6\pi r} \tag{4.196}$$

This means that the product of the equivalent conductivity and the viscosity of the solvent for a particular electrolyte at a given temperature should be a constant (at one temperature). This is indeed what the empirical Walden's rule states.

Some experimental data on the product $\Lambda\eta$ are presented in Table 4.14 for solutions of potassium iodide in various solvents. Walden's rule has some rough applicability in organic solvents. When, however, the $\Lambda\eta$ products for a solute dissolved on the one hand in water and on the other in organic solvents are compared, it is found that there is considerable discrepancy (Table 4.15). This should hardly come as a surprise; one should expect differences in the solvation of ions in water and in organic solvents (Section 2.20) and the resulting differences in radii of the moving ions.

TABLE 4.15

The Product $\Lambda\eta$ for Sodium Chloride in Various Solvents at 298 K

Solvent	Λ	η	$\Lambda\eta$
Water	126.39	0.00895	1.131
Methyl alcohol	96.9	0.00546	0.529
Ethyl alcohol	42.5	0.01096	0.466

4.4.13. The Rate-Process Approach to Ionic Migration

The fundamental equation for the current density (flux of charge) as a function of the drift velocity has been shown to be

$$\vec{j} = zFcv_d \tag{4.164}$$

Hitherto, the drift velocity has been related to macroscopic forces (e.g., the Stokes viscous force $\vec{F} = 6\pi r\eta$ or the electric force $\vec{F} = ze_0X$) through the relation

$$v_d = \frac{\vec{F}}{m}\tau \tag{4.147}$$

Another approach to the drift velocity is by molecular models. The drift velocity v_d is considered the net velocity, i.e., the difference of the velocity $\vec{v}$ of ions in the direction of the force field and the velocity $\overleftarrow{v}$ of ions in a direction opposite to the field (Fig. 4.67). In symbols, one writes

$$v_d = \vec{v} - \overleftarrow{v} \tag{4.197}$$

Any velocity is given by the distance traveled divided by the time taken to travel that distance. In the present case, the distance is the jump distance l, i.e., the mean distance that an ion jumps in hopping from site to site in the course of its directed random walk, and the time is the mean time τ between successive jumps. This mean time includes the time the ion may wait in a "cell" of surrounding particles as well as the actual time involved in jumping. Thus,

$$\vec{v} = \frac{\text{Mean jump distance } l}{\text{Mean time between jumps } \tau} \tag{4.198}$$

The reciprocal of the mean time between jumps is the net jump frequency k, which is the number of jumps per unit of time. Hence,

$$\text{Velocity} = \text{Jump distance} \times \text{net jump frequency}$$

or

$$\vec{v} = l\vec{k} \tag{4.199}$$

and

$$\overleftarrow{v} = l\overleftarrow{k} \tag{4.200}$$

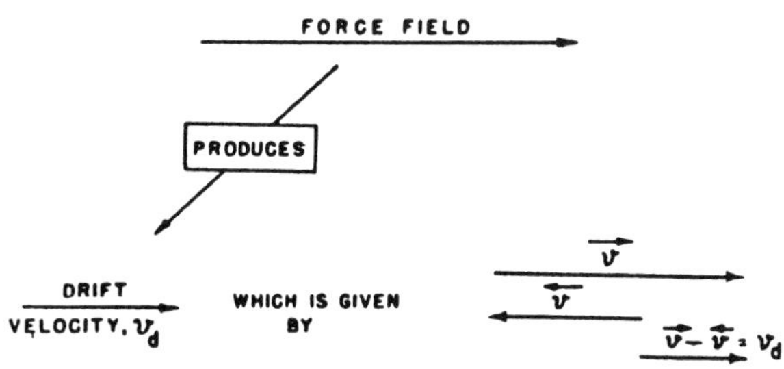

Fig. 4.67. The drift velocity v_d can be considered to be made up of a velocity $\vec{v}$ in the direction of the force field and a velocity * in a direction opposite to the force field.

For diffusion, the net jump frequency k was related to molecular quantities by viewing the ionic jumps as a rate process (Equation 4.111). In this view, for an ion to jump, it must possess a certain free energy of activation to surmount the free-energy barrier. It was shown that the net jump frequency is given by

$$\vec{k} = \frac{kT}{h} e^{-\Delta G^{\circ\ddagger}/RT} \tag{4.111}$$

To emphasize that this is the jump frequency for a pure diffusion process, in which case the ions are not subjected to an externally applied field, a subscript D will be appended to the net jump frequency and to the standard free energy of activation, i.e.,

$$\vec{k}_D = \frac{kT}{h} e^{-\Delta G_D^{\circ\ddagger}/RT} \tag{4.201}$$

Now, suppose an electric field is applied so that it *hinders* the movement of a positive ion from right to left. Then the work that is done on the ion in moving it from the equilibrium position to the top of the barrier (Fig. 4.68) is the product of the charge on the ion, z_+e_0, and the potential difference between the equilibrium position and the activated state, i.e., the position at the top of the barrier. Let this potential difference be a fraction β of the total potential difference (i.e., the applied electric field X times the distance l) between two equilibrium sites. Then the electrical work done on one positive ion in making it climb to the top of the barrier, i.e., in activating it, is equal to the charge on the ion z_+e_0 times the potential difference βXl through which it is transported. Thus the electrical work is $z_+e_0\beta Xl$ per ion, or $z_+F\beta Xl$ per mole of ions.

The electrical work of activation corresponds to a free-energy change. It appears therefore that there is a contribution to the total free energy of activation due to the electrical work done on the ion in making it climb the barrier. This electrical contribution to the free energy of activation is

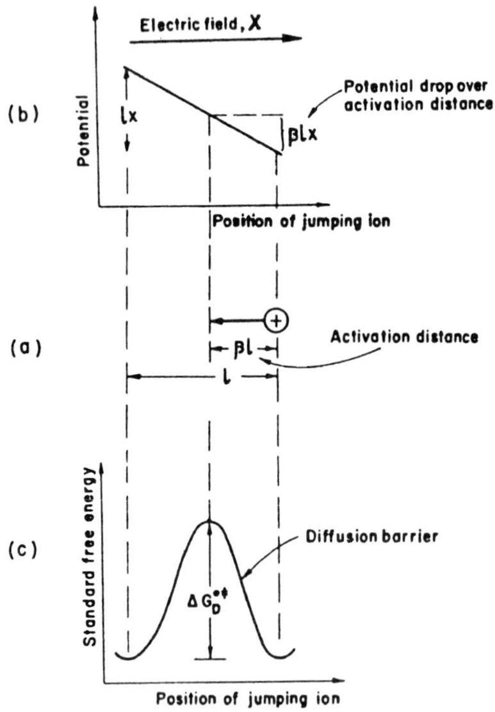

Fig. 4.68. (a) As the ion moves for the jump, it has to climb (b) the potential gradient arising from the electric field in the electrolyte, in addition to (c) the free-energy barrier for diffusion. To be activated, the ion has to climb the fraction βXl of the total potential difference Xl between the initial and final positions for the jump.

$$\Delta G_e^{\circ\ddagger} = z_+ F\beta Xl \tag{4.202}$$

Hence, the *total* free energy of activation (for positive ions moving from right to left) is

$$\Delta G_{\text{total}}^{\circ\ddagger} = \Delta G_D^{\circ\ddagger} + \Delta G_e^{\circ\ddagger} \tag{4.203}$$

Thus, in the presence of the field, the frequency of right $\rightarrow$ left jumps is

$$\overleftarrow{k} = \frac{kT}{h} e^{-(\Delta G_D^{\circ\ddagger} + z_+ F\beta Xl)/RT} \tag{4.204}$$

or

$$\overleftarrow{k} = k_D e^{-z_+ F\beta X l/RT} \tag{4.205}$$

By a similar argument, the left → right jump frequency $\overrightarrow{k}$, or the number of jumps per second from left to right, may be obtained. There are, however, two differences: When positive ions move from left to right, (1) they are moving with the field and therefore are *helped*, not *hindered*, by the field, and (2) they have to climb through only a fraction $1 - \beta$ of the barrier. Hence, the electrical work of activation is $-[z_+F(1-\beta)Xl]$, the minus sign indicating that the field assists the ion. Thus,

$$\overrightarrow{k} = k_D e^{z_+ F(1-\beta) X l/RT} \tag{4.206}$$

If the factor β is assumed to be $\frac{1}{2}$,[29] then $\beta = 1 - \beta = \frac{1}{2}$ and Eqs. (4.205) and (4.206) can be written

$$\overleftarrow{k} = k_D e^{-pX} \tag{4.207}$$

and

$$\overrightarrow{k} = k_D e^{+pX} \tag{4.208}$$

where for conciseness p is written instead of $z_+Fl/2RT$. It follows from these equations that $\overleftarrow{k} < k_D$ and $\overrightarrow{k} > k_D$, or $\overrightarrow{k} > \overleftarrow{k}$.

In the presence of the field therefore, the jumping frequency is anisotropic, i.e., it varies with direction. The jumping frequency $\overrightarrow{k}$ of an ion in the direction of the field is greater the jumping frequency that $\overleftarrow{k}$ against the field. When, however, there is no field, the jump frequency k_D is the same in all directions, and therefore jumps in all directions are equally likely. This is the characteristic of a random walk. The application of the field destroys the equivalence of all directions. The walk is not quite random. The field makes the ions more likely to move with it than against it. There is drift. In Eqs. (4.207) and (4.208), the k_D is a random-walk term, the exponential factors are the perturbations due to the field, and the result is a drift. The equations are therefore a quantitative expression of the qualitative statement made in Section 4.4.1.

$$\text{Drift due to field} = \text{Random walk in absence of field} \times \text{perturbation due to field} \tag{4.209}$$

4.4.14. The Rate-Process Expression for Equivalent Conductivity

Introducing the expressions (4.207) and (4.208) for $\overrightarrow{k}$ and $\overleftarrow{k}$ into the equations for the component forward and backward velocities $\overrightarrow{v}$ and $\overleftarrow{v}$ [i.e., into Eqs. (4.199) and (4.200)], one obtains

---

[29]This implies (Fig. 4.68) that the energy barrier is symmetrical.

$$\overleftarrow{v} = lk_D e^{-pX} \tag{4.210}$$

and

$$\overrightarrow{v} = lk_D e^{+pX} \tag{4.211}$$

where, as stated earlier,

$$p = \frac{z_+ Fl}{2RT}$$

The drift velocity v_d is obtained [*cf.* Eq. (4.197)] by subtracting Eq. (4.210) from Eq. (4.211), thus,

$$\begin{aligned} v_d = \overrightarrow{v} - \overleftarrow{v} &= lk_D e^{+pX} - lk_D e^{-pX} \\ &= lk_D(e^{+pX} - e^{-pX}) \\ &= 2lk_D \sinh pX \end{aligned} \tag{4.212}$$

The net charge transported per second across a unit area (i.e., the current density j) is given by Eq. (4.164),

$$\overrightarrow{j} = z_+ F c v_d \tag{4.164}$$

Upon inserting the expression (4.212) for the drift velocity into Eq. (4.164), it is clear that

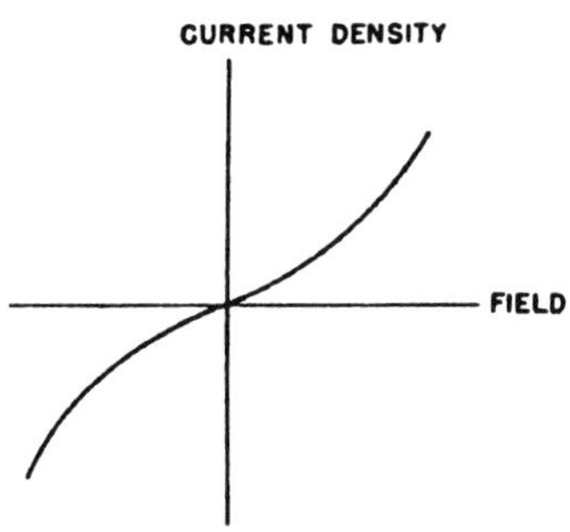

Fig. 4.69. The hyperbolic sine relation between the ionic current density and the electric field according to Eq. (4.213).

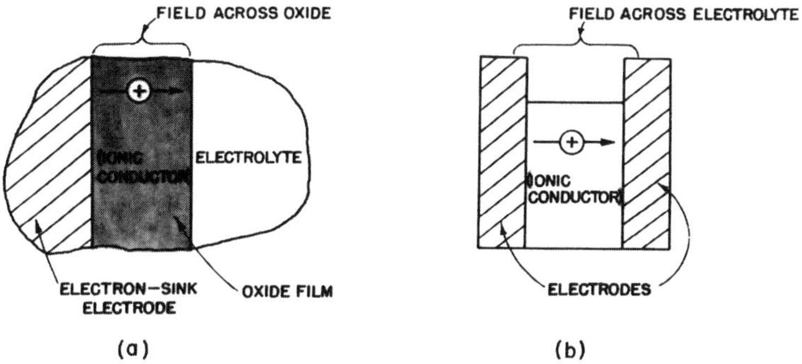

Fig. 4.70. The similarity between the ionic current flowing (a) through an oxide film between an electrode and an electrolyte and (b) through an electrolyte between two electrodes.

$$\vec{j} = z_+ Fc(2lk_D \sinh pX) \tag{4.213}$$

A picture of the hyperbolic sine relation between the ionic current density and the electric field that would result from Eq. (4.213) is shown in Fig. 4.69.

The fundamental thinking used in the derivation of Eq. (4.213) has wide applicability. Take the case of an oxide film that grows on an electron-sink electrode (anode). All one has to do is to consider an ionic crystal (the oxide) instead of an electrolytic solution, and all the arguments used to derive the hyperbolic sine relation (4.213) become immediately applicable to the ionic current flowing through the oxide in response to the potential gradient in the solid (Fig. 4.70). In fact, Eq. (4.213) is the basic equation describing the field-induced migration of ions in any ionic conductor. Equation (4.213) is also formally similar to the expression for the current density due to a charge-transfer electrodic reaction occurring under the electric field present at an electrode–electrolyte interface (Chapter 7).

In all these cases, two significant approximations can be made. One is the *high-field Tafel type* (see Chapter 7) of approximation, in which the absolute magnitude of the exponents $|pX|$ in Eq. (4.213) [i.e., the argument pX of the hyperbolic sine in Eq. (4.213)] is much greater than unity. Under this condition of $pX >> 1$, one obtains $\sinh pX \approx e^{pX}/2$ because one can neglect e^{-pX} in comparison with e^{pX}. Thus (Fig. 4.71)

$$j = z_+ Fclk_D e^{pX} \tag{4.214}$$

i.e., the current density bears an exponential relation to the field. Such an exponential dependence of current on field is commonly observed in oxide growth, at electrode–electrolyte interfaces, but not in electrolyte solutions.

In electrolytic solutions, however, the conditions for the high-field approximation are not often observed. The applied field X is generally relatively small, in which case $pX << 1$ and the following approximation can be used:

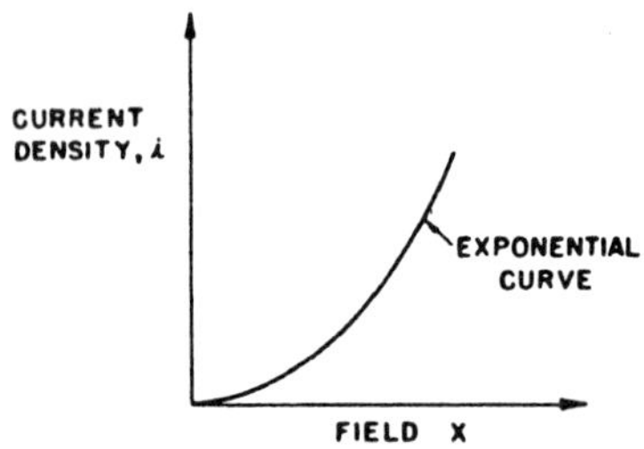

Fig. 4.71. Under high-field conditons, there is an exponential relation between ionic current density and the field across an oxide.

$$\sinh pX \approx pX = \frac{z_+ FlX}{2RT} \tag{4.215}$$

and the current density in Eq. (4.213) is approximately given (Fig. 4.72) by

$$\begin{aligned} \vec{j}_+ &= z_+ Fc2lk_D \frac{z_+ FlX}{2RT} \\ &= \left(z_+^2 F^2 c \frac{l^2 k_D}{RT} \right) X \end{aligned} \tag{4.216}$$

All the quantities within the parentheses are constants in a particular electrolyte and therefore

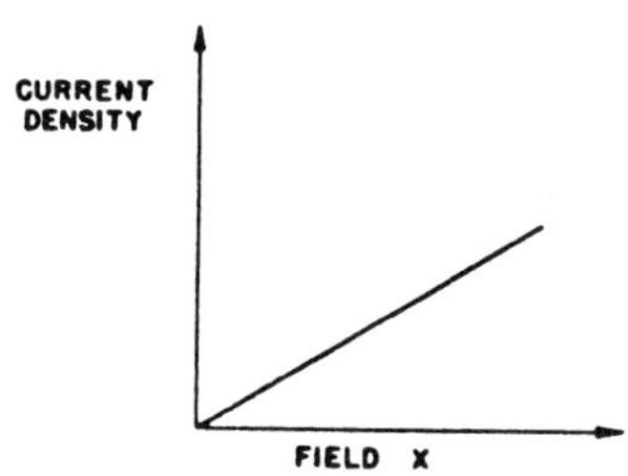

Fig. 4.72. Under low-field conditions, there is a linear relation between the ionic current density and the field in the electrolyte.

$$\vec{j}_+ = \text{constant} \times X \tag{4.217}$$

and according to Eq. (4.160), the value of the constant is equal to σ. This is the *low-field* approximation. It is in fact the rate-process version of Ohm's law. An important point, however, has emerged: *Ohm's law is valid only for sufficiently small fields*. Of course, this was accounted for in the phenomenological treatment of conduction where the general flux–force relation

$$J = A + BX + CX^2 + \ldots \tag{4.121}$$

reduced to the linear relation

$$J = BX \tag{4.122}$$

only for "small" fields.

However, words such as "small" and "large" are relative. If one proceeds to substitute numerical values in the conditions for a linear relation between current and field, one starts with $pX << 1$. For example, $zFlX/2RT < 0.1$; then, with $F/RT = 39.0$ V^{-1} at 25 °C, $z = 1$, and $l = 3 \times 10^{-8}$ cm, one obtains $X < 10^5$ V cm^{-1} as the condition for the linearization of Eq. (4.213). This condition also implies a proportionality of the current to the applied field in the electrolyte and hence [*cf.* Eq. (4.213)] independence of the conductivity on the potential applied to the cell.

Since one often works in the laboratory with $0 < X < 10$ V cm^{-1}, when one says that X has to be "small" for the conductivity to be independent of field strength, this becomes a very relative matter. It is better to say that the conductivity remains independent of the value of the applied field so long as it is not very high. Indeed, it has been found that the conductivity finally does increase with the applied field but only at 10^6 V cm^{-1}.

4.4.15. The Total Driving Force for Ionic Transport: The Gradient of the Electrochemical Potential

In the rate-process view of conduction that has just been presented, it has been assumed that the concentration is the same throughout the electrolyte. Suppose, however, that there is a concentration gradient of a particular ionic species, say, positively charged radiotracer ions. Further, let the concentration vary continuously in the x direction (see Fig. 4.73), so that if the concentration of positive ions at x on the left of the barrier is $(c_+)_x$, the concentration on the right (i.e., at $x + l$) is given by

$$\begin{matrix}\text{Concentration} \\ \text{on the right}\end{matrix} = \begin{matrix}\text{Concentration} \\ \text{on the left}\end{matrix} + \begin{matrix}\text{rate of change of} \\ c_+ \text{ with distance}\end{matrix} \times \text{distance}$$

i.e.,

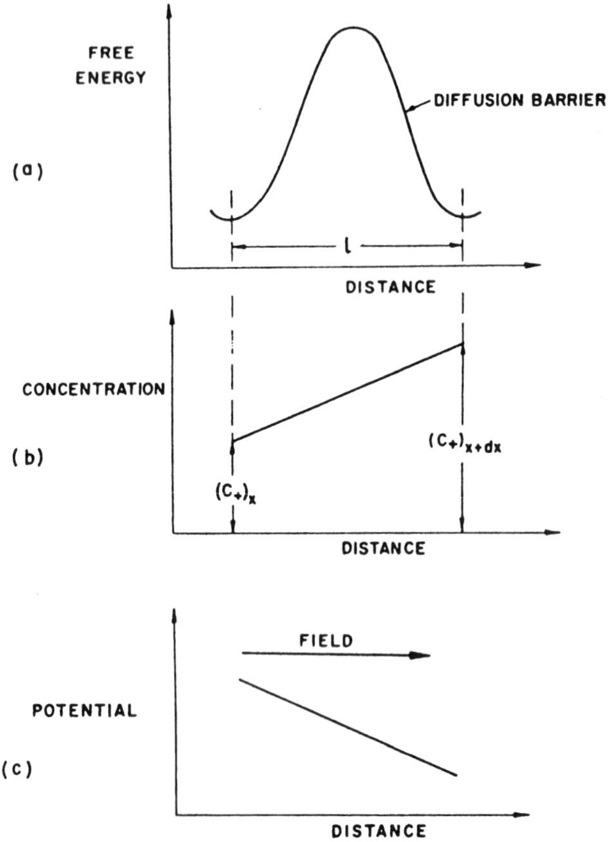

Fig. 4.73. Measurement of ions under both a concentration gradient and a potential gradient: (a) the free-energy barrier for the diffusive jump of an ion, (b) the concentration variation over the jump distance, and (c) the potential variation over the jump distance.

$$(c_+)_{x+l} = (c_+)_x + \frac{dc_+}{dx} l \tag{4.218}$$

In this case, there will be diffusion of the tracer ions and therefore the current density j_+ is not given by a conduction law, i.e., by Eq. (4.159), which governs the situation in the absence of a concentration gradient. Instead, the expression for the current density has to be written [the subscript x in $(c_+)_x$ has been dropped for the sake of convenience]

$$j_+ = z_+ F c_+ \overrightarrow{v} - z_+ F \left(c_+ + \frac{dc_+}{dx} l \right) \overleftarrow{v} \tag{4.219}$$

However, $\vec{v}$ and $\overleftarrow{v}$ have been evaluated as

$$\vec{v} = lk_D e^{pX} \tag{4.211}$$

and

$$\overleftarrow{v} = lk_D e^{-pX} \tag{4.210}$$

Under low-field conditions $pX << 1$, the exponentials can be expanded and linearized to give

$$\vec{v} \approx lk_D(1 + pX) \tag{4.220}$$

and

$$\overleftarrow{v} \approx lk_D(1 - pX) \tag{4.221}$$

Combining Eqs. (4.219), (4.220), and (4.221), one gets

$$\begin{aligned} j_+ &= z_+Fc_+\vec{v} - z_+Fc_+\overleftarrow{v} - z_+F\frac{dc_+}{dx}\,l\,\overleftarrow{v} \\ &= z_+Fc_+lk_D(1 + pX) - z_+Fc_+lk_D(1 - pX) \\ &\quad - z_+Fl^2k_D(1 - pX)\frac{dc_+}{dx} \\ &= 2z_+Fc_+lk_DpX - z_+Fl^2k_D(1 - pX)\frac{dc_+}{dx} \end{aligned} \tag{4.222}$$

This expression can be simplified further, first, by applying the low-field condition $pX << 1$. It becomes

$$j_+ = 2z_+Fc_+lk_D\,pX - z_+Fl^2k_D\frac{dc_+}{dx} \tag{4.223}$$

Second, by substituting $z_+Fl/(2RT)$ for p, one has

$$j_+ = z_+^2F^2c_+\frac{l^2k_D}{RT}X - z_+Fl^2k_D\frac{dc_+}{dx} \tag{4.224}$$

and finally by replacing l^2k_D with D_+ [*cf.* Eq. (4.113)], the result is

$$j_+ = \frac{z_+^2 F^2 c_+ D_+ X}{RT} - z_+ F D_+ \frac{dc_+}{dx} \tag{4.225}$$

To go from the current density j_+ to the flux J_+ of positive tracer ions is straightforward. Thus,

$$J_+ = \frac{j_+}{z_+ F} = \frac{D_+ c_+}{RT}(z_+ F X) - D_+ \frac{dc_+}{dx} \tag{4.226}$$

The second term on the right-hand side can be rewritten as

$$\begin{aligned} D_+ \frac{dc_+}{dx} &= \frac{D_+ c_+}{RT} \frac{RT}{c_+} \frac{dc_+}{dx} \\ &= \frac{D_+ c_+}{RT} \frac{d(RT \ln c_+)}{dx} \\ &= \frac{D_+ c_+}{RT} \frac{d(\mu_+^0 + RT \ln c_+)}{dx} \\ &= \frac{D_+ c_+}{RT} \frac{d\mu_+}{dx} \end{aligned} \tag{4.227}$$

since, according to the definition of the chemical potential for ideal solutions [*cf.* Eq. (3.54)],

$$\mu_+ = \mu_+^0 + RT \ln c_+$$

In addition, from Eq. (4.9), the electric field X is equal to minus the gradient of the electrostatic potential, i.e.,

$$X = -\frac{d\psi}{dx} \tag{4.9}$$

Hence,

$$\begin{aligned} J_+ = \frac{j_+}{z_+ F} &= -\frac{D_+ c_+}{RT}\left(z_+ F \frac{d\psi}{dx} + \frac{d\mu_+}{dx}\right) \\ &= -\frac{D_+ c_+}{RT} \frac{d}{dx}(z_+ F \psi + \mu_+) \end{aligned} \tag{4.228}$$

This is an interesting result. The negative gradient of the chemical potential $d\mu_+/dx$ is known to be the driving force for pure diffusion and $-z_+F(d\psi/dx) = z_+FX$, the driving force for pure conduction. However, when there is both a chemical potential (or concentration) gradient and an electric field $(-d\psi/dx)$, then the total driving force for ionic transport is the negative gradient of

$$\underset{\substack{\text{Electrostatic}\\ \text{potential}}}{z_+F\psi} \quad + \quad \underset{\substack{\text{Chemical}\\ \text{potential}}}{\mu_+}$$

This quantity $z_+F\psi + \mu_+$ could be called the *electrostatic-chemical* potential, or simply the *electrochemical potential*, of the positive ions and is denoted by the symbol $\overline{\mu}_+$. Thus,

$$\overline{\mu}_+ = \mu_+ + z_+F\psi \tag{4.229}$$

and the total driving force for the drift of ions is the gradient of the electrochemical potential. Thus, one can write the flux J_+ of Eq. (4.228) in the form

$$J_+ = \frac{j_+}{z_+F} = -\frac{D_+c_+}{RT}\frac{d\overline{\mu}_+}{dx} \tag{4.230}$$

Or, by making use of the Einstein relation,

$$D_+ = (\overline{u}_{\text{abs}})_+kT \tag{4.172}$$

and the relation between absolute and conventional mobilities, i.e.,

$$(u_{\text{conv}})_+ = (\overline{u}_{\text{abs}})_+z_+e_0 \tag{4.152}$$

one can rewrite Eq. (4.230) in the form

$$\begin{aligned} J_+ &= -\frac{D_+}{RT}c_+\frac{d\overline{\mu}_+}{dx} = -\frac{(\overline{u}_{\text{abs}})_+kT}{RT}c_+\frac{d\overline{\mu}_+}{dx} \\ &= -\frac{(u_{\text{conv}})_+}{z_+e_0}\frac{kT}{RT}c_+\frac{d\overline{\mu}_+}{dx} \\ &= -\frac{(u_{\text{conv}})_+}{z_+F}c_+\frac{d\overline{\mu}_+}{dx} \end{aligned} \tag{4.231}$$

Expression (4.231) is known as the *Nernst–Planck flux equation*. It is an important equation for the description of the flux or flow of a species under the total driving force of an electrochemical potential. The Nernst–Planck flux expression is useful in explaining, for example, the electrodeposition of silver from silver cyanide ions. In this process, the negatively charged $[Ag(CN)_2]^-$ ions travel to the negatively charged electron source or cathode, a fact that cannot be explained by considering that the only driving force on the $[Ag(CN)_2]^-$ ions is the electric field because the electric field drives these ions *away from* the negatively charged electrode. If, however, the concentration gradient of these ions in a direction normal to the electron source is such that in the expanded form of the Nernst–Planck equation, i.e.,

$$J = \frac{Dc}{RT} zFX - D\frac{dc}{dx} \tag{4.226}$$

the second term is larger than the first, then the flux of the $[Ag(CN)_2]^-$ ions is opposite to the direction of the electric field, i.e., toward the negatively charged electrode.

Further Reading

Seminal

1. A. Einstein, *Investigations on the Theory of Brownian Movement*, Methuen & Co., London (1926).
2. R. W. Gurney, *Ionic Processes in Solution*, Dover, New York (1953).
3. R. A. Robinson and R. H. Stokes, *Electrolyte Solutions*, Butterworth, London (1955).

Review

1. P. Turq, J. Barthel, and M. Chemla, *Transport, Relaxation and Kinetic Processes in Electrolyte Solutions*, Springer-Verlag, Berlin (1992).

Papers

1. J. Horno, J. Castilla, and C. F. Gonzalez-Fernandez, *J. Phys. Chem.* **96**: 854 (1992).
2. G. E. Spangler, *Anal. Chem.* **65**: 3010 (1993).
3. C. A. Lucy and T. L. McDonald, *Anal. Chem.* **67**: 1074 (1995).
4. R. Bausch, *J. Chem. Ed.* **72**: 713 (1995).
5. A. A. Moya, J. Castilla, and J. Horno, *J. Phys. Chem.* **99**: 1292 (1995).

4.5. THE INTERDEPENDENCE OF IONIC DRIFTS

4.5.1. The Drift of One Ionic Species May Influence the Drift of Another

The processes of diffusion and conduction have been treated so far with the assumption that each ionic species drifts independently of every other one. In general, however, the assumption is not realistic for electrolytic solutions because it presupposes the absence of ionic atmospheres resulting from ion–ion interactions. One has been talking therefore of ideal laws of ionic transport and expressed them in the Nernst–Planck equation for the independent flux of a species i.

$$J_i = -\frac{(u_{\text{conv}})_i}{z_i F} c_i \frac{d\bar{\mu}_i}{dx} \tag{4.232}$$

The time has come to free the treatment of ionic transport from the assumption of the independence of the various ionic fluxes and to consider some phenomena which depend on the fact that the drift of a species i is affected by the flows of other species present in the solution. It is the whole society of ions that displays a transport process, and each individual ionic species takes into account what all the other species are doing. Ions interact with each other through their Coulombic fields and thus it will be seen that the law of electroneutrality that seeks zero excess charge in any macroscopic volume element plays a fundamental role in phenomena where ionic flows influence each other.

A stimulating approach to the problem of the interdependence of ionic drifts can be developed as follows. Since different ions have different radii, their Stokes mobilities, given by

$$(u_{\text{conv}})_i = \frac{z_i e_0}{6\pi r\eta} \tag{4.183}$$

must be different. What are the consequences that result from the fact that different ionic species have unequal mobilities?

4.5.2. A Consequence of the Unequal Mobilities of Cations and Anions, the Transport Numbers

The current density $\vec{j}_i$ due to an ionic species i is related to the mobility $u_{\text{conv},i}$ in the following manner [*cf.* Eq. (4.159)]

$$\vec{j}_i = z_i F c_i (u_{\text{conv}})_i X \tag{4.159}$$

If therefore one considers a unit field $X = 1$ in an electrolyte solution containing a $z{:}z$-valent electrolyte (i.e., $z_+ = z_- = z$ and $c'_+ = c'_- = c$), then since $(u_{\text{conv}})_+ \neq (u_{\text{conv}})_-$, it follows that

$$j_+ \neq j_- \tag{4.233}$$

This is a thought-provoking result. It shows that although all ions feel the externally applied electric field to the extent of their charges, some respond by migrating more than others. It also shows that though the burden of carrying the current through the electrolytic solution falls on the whole community of ions, the burden is not shared equally among the various species of ions. Even if there are equal numbers

TABLE 4.16

Transport Numbers of Cations in Aqueous Solutions at 298 K in 0.1 *N* Solutions

Electrolyte	HCl	LiCl	NaCl	KCl	KNO_3	$AgNO_3$	$BaCl_2$
Transport number of cation, t_+	0.83	0.32	0.39	0.49	0.51	0.47	0.43

of the various ions, those that have higher mobility contribute more to the communal task of transporting the current through the electrolytic solution than the ions handicapped by lower mobilities.

It is logical under these circumstances to seek a quantitative measure of the extent to which each ionic species is taxed with the job of carrying current. This quantitative measure, known as the *transport number* (Table 4.16), should obviously be defined by the *fraction* of the total current carried by the particular ionic species, i.e.,

$$t_i = \frac{j_i}{j_T} = \frac{j_i}{\sum j_i} \tag{4.234}$$

This definition requires that the sum of the transport numbers of all the ionic species be unity for

$$\sum t_i = \sum \frac{j_i}{\sum j_i} = 1 \tag{4.235}$$

Thus, the conduction current carried by the species i (e.g., Na^+ ions in a solution containing NaCl and KCl) depends upon the current transported by all the other species. Here then is a clear and simple indication that the drift of the ith species depends on the drift of the other species.

For example, consider a 1:1-valent electrolyte (e.g., HCl) dissolved in water. The transport numbers will be given by[30]

$$t_{H^+} = \frac{j_{H^+}}{j_{H^+} + j_{Cl^-}} = \frac{z_{H^+} F c_{H^+} u_{H^+} X}{z_{H^+} F c_{H^+} u_{H^+} X + z_{Cl^-} F c_{Cl^-} u_{Cl^-} X}$$

However, $z_{H^+} = z_{Cl^-} = 1$ and $c_{H^+} = c_{Cl^-} = c$, and therefore

[30]To avoid cumbersome notation, the symbol u_{conv} for the conventional mobilities has been contracted to u. The absence of a bar above the u stresses that it is not the absolute mobility $\bar{u}_{abs}$.

$$t_{H^+} = \frac{u_{H^+}}{u_{H^+} + u_{Cl^-}} \tag{4.236}$$

Similarly,

$$t_{Cl^-} = \frac{u_{Cl^-}}{u_{H^+} + u_{Cl^-}} \tag{4.237}$$

The mobilities of the H^+ and Cl^- ions in 0.1 *N* HCl at 25 °C are 33.71×10^{-4} and 6.84×10^{-4} cm^2 s^{-1} V^{-1}, respectively, from which it turns out the transport numbers of the H^+ and Cl^- ions are 0.83 and 0.17, respectively. Thus, in this case, the positive ions carry a major fraction (~83%) of the current.

Now suppose that an excess of KCl is added to the HCI solution so that the concentration of H^+ is about 10^{-3} *M* in comparison with a K^+ concentration of 1 *M*. The transport numbers in the mixture of electrolytes will be

$$\frac{t_{K^+}}{t_{H^+}} = \frac{c_{K^+}u_{K^+}/\Sigma\, c_i u_i}{c_{H^+}u_{H^+}/\Sigma\, c_i u_i} \tag{4.238}$$

$$= \frac{c_{K^+}}{c_{H^+}} \frac{u_{K^+}}{u_{H^+}} \tag{4.239}$$

The ratio c_{K^+}/c_{H^+} is 10^3, and the ratio of mobilities is

$$\frac{u_{K^+}}{u_{H^+}} = \frac{6 \times 10^{-4}}{30 \times 10^{-4}} = \frac{1}{5}$$

Hence,

$$\frac{t_{K^+}}{t_{H^+}} = 200$$

which means that although the H^+ is about 5 times more mobile than the K^+ ion, it carries 200 times less current. Thus, the addition of the excess of KCl has reduced to a negligible value the fraction of the current carried by the H^+ ions.

In fact, the transport number of the H^+ ions under such circumstances is virtually zero, as shown from the following approximate calculation.

$$t_{H^+} = \frac{c_{H^+}u_{H^+}}{c_{H^+}u_{H^+} + c_{K^+}u_{K^+} + c_{Cl^-}u_{Cl^-}}$$

$$= \frac{10^{-6} u_{H^+}}{10^{-6} u_{H^+} + 10^{-3} u_{K^+} + 10^{-3} u_{Cl^-}} = 10^{-3} \tag{4.240}$$

Thus the conduction current carried by an ion depends very much on the concentration in which the other ions are present.

4.5.3. The Significance of a Transport Number of Zero

In the previous section, it was shown that the addition of an excess of KCl makes the fraction of the migration (i.e., conduction) current carried by the H^+ ions tend to zero. What happens if this mixture of HCl and KCl is placed between two electrodes and a potential difference is applied across the cell (Fig. 4.74)?

In response to the electric field developed in the electrolyte, a migration of ions occurs and there is a conduction current in the solution. Since this conduction current is almost completely borne by K^+ and Cl^- ions ($t_{H^+} \rightarrow 0$), there is a tendency for the

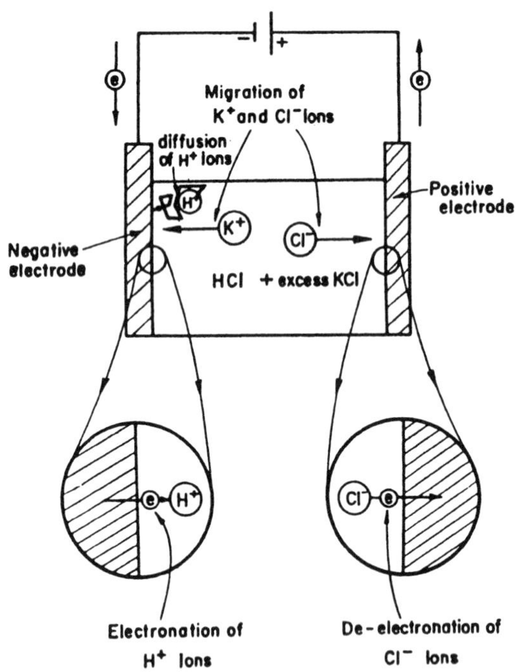

Fig. 4.74. A schematic diagram of the transport processes in an electrolyte (of HCl – KCl, with an excess of KCl) and of the reactions at the interfaces.

Cl^- ions to accumulate near the positive electrode, and the K^+ ions near the negative electrode. If the excess negative charge near the positive electrode and vice versa were to build up, then the resulting field due to lack of electroneutrality (see Section 4.3.3) would tend to bring the conduction current to a halt. It has been argued (Section 4.3.4), however, that conduction (i.e., migration) currents are sustained in an electrolyte because of charge-transfer reactions (at the electrode–electrolyte interfaces), which remove the excess charge that tends to build up near the electrodes.

In the case of the HCl + KCl electrolyte, the reaction at the positive electrode may be considered the deelectronation of the Cl^- ions. Furthermore, according to Faraday's law (see Section 4.3.5), 1 g-eq of Cl^- ions must be deelectronated at the positive electrode for the passage of 1 F of charge in the external circuit. This means, however, that at the other electrode, 1 g-eq of positive ions must be involved in a reaction. Thus, either the K^+ or the H^+ ions must react, but by keeping the potential difference within certain limits one can ensure that only the H^+ ions react.

There is no difficulty in effecting the reaction of the layer of H^+ near the negative electrode, but to keep the reaction going, there must be a flux of H^+ ions from the bulk of the solution toward the negative electrode. By what process does this flux occur? It cannot be by migration because the presence of the excess of K^+ ions makes the transport number of H^+ tend to zero. It is here that diffusion comes into the picture; the removal of H^+ ions by the charge-transfer reactions causes a depletion of these ions near the electrode, and the resulting concentration gradient provokes a diffusion of H^+ ions toward the electrode.

To provide a quantitative expression for the diffusion flux J_{H^+}, one cannot use the Nernst–Planck flux equation (4.231) because the latter describes the independent flow of one ionic species and in the case under discussion it has been shown that the migration current of the H^+ ions is profoundly affected by the concentration of the K^+ ions. A simple modification of the Nernst–Planck equation can be argued as follows.

Since conduction (i.e., migration) and diffusion are the two possible[31] modes of transport for an ionic species, the total flux J_i must be the sum of the conduction flux $(J_C)_i$ and the diffusion flux $(J_D)_i$. Thus,

$$J_i = (J_C)_i + (J_D)_i \tag{4.241}$$

The conduction flux is equal to $1/z_iF$ times the conduction current j_i borne by the particular species

$$(J_C)_i = \frac{j_i}{z_iF} \tag{4.242}$$

[31] Another possible mode of transport, hydrodynamic flow, is not considered in this chapter.

and the conduction current carried by the species i is related to the total conduction current $j_T = \sum j_i$ through the transport number of the species i [*cf.* Eq. (4.234)]

$$j_i = t_i \sum j_i = t_i j_T \tag{4.243}$$

Hence,

$$(J_C)_i = \frac{t_i j_T}{z_i F} \tag{4.244}$$

Furthermore, the diffusion flux $(J_D)_i$ is given by[32]

$$(J_D)_i = -Bc_i \frac{d\mu_i}{dx} \tag{4.14}$$

or approximately by Fick's first law

$$(J_D)_i = -D_i \frac{dc_i}{dx} \tag{4.16}$$

so that the total flux of species i is

$$J_i = \frac{t_i j_T}{z_i F} - Bc_i \frac{d\mu_i}{dx} \tag{4.245}$$

or approximately

$$J_i = \frac{t_i j_T}{z_i F} - D_i \frac{dc_i}{dx} \tag{4.246}$$

From these modified forms of the Nernst–Planck flux equation (4.231), one can see that even if $t_i \to 0$, it is still possible to have a flux of a species provided there is a concentration gradient, which is often brought into existence by interfacial charge-transfer reactions at the electrode–electrolyte interfaces consuming or generating the species.

From the modified Nernst–Planck flux equation (4.245), one can give a more precise definition of the transport number. If $d\mu_j/dx = 0$, in which case $dc_j/dx = 0$, then

---

[32]The constant B has been shown in Section 4.4.6 to be equal to $\bar{u}_{\text{abs}}/N_A$.

$$t_j = \left(\frac{z_j F J_j}{j_T}\right)_{d\mu_j/dx=0}$$

$$= \left(\frac{j_j}{j_T}\right)_{d\mu_j/dx=0} \tag{4.247}$$

It should be emphasized therefore that the transport number only pertains to the conduction flux (i.e., to that portion of the flux produced by an *electric field*) and any flux of an ionic species arising from a chemical potential gradient (i.e., any diffusion flux) is *not* counted in its transport number. From this definition, the transport number of a particular species can tend to zero, $t_i \to 0$, and at the same time its diffusion flux can be finite.

This is an important point in electroanalytical chemistry, where the general procedure is to arrange for the ions that are being analyzed to move to the electrode–electrolyte interface by diffusion only. Then if the experimental conditions correspond to clearly defined boundary conditions (e.g., constant flux), the partial differential equation (Fick's second law) can be solved exactly to give a theoretical expression for the bulk concentration of the substance to be analyzed. In other words, the transport number of the substance being analyzed must be made to tend to zero if the solution of Fick's second law is to be applicable. This is ensured by adding some other electrolyte in such excess that it takes on virtually the entire burden of the conduction current. The added electrolyte is known as the *indifferent* electrolyte. It is indifferent only to the electrodic reaction at the interface; it is far from indifferent to the conduction current.

4.5.4. The Diffusion Potential, Another Consequence of the Unequal Mobilities of Ions

Consider that a solution of a $z{:}z$-valent electrolyte (of concentration c moles dm^{-3}) is instantaneously brought into contact with water at the plane $x = 0$ (Fig. 4.75). A concentration gradient exists both for the positive ions and for the negative ions. They therefore start diffusing into the water.

Since, in general,[33] $\bar{u}_+ \neq \bar{u}_-$, let it be assumed that $\bar{u}_+ > \bar{u}_-$. With the use of the Einstein relation (4.172), it is clear that

$$D_+ = \bar{u}_+ kT \quad \text{and} \quad D_- = \bar{u}_- kT$$

or that

[33]Though the subscript "abs" has been dropped, it is clear from the presence of a bar over the *u*s that one is referring to absolute mobilities.

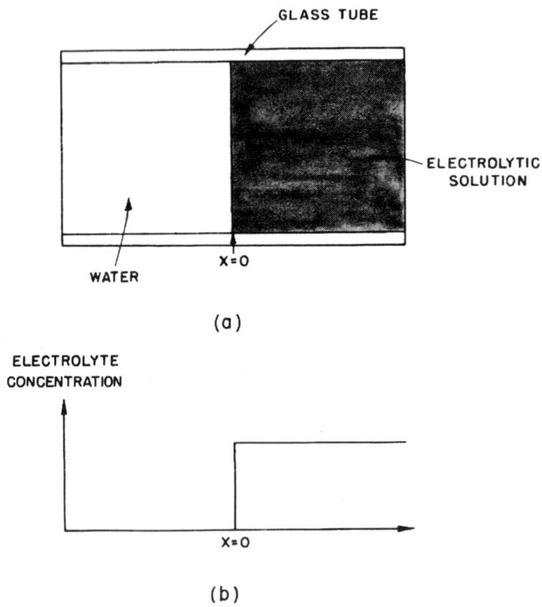

Fig. 4.75. (a) An electrolytic solution is instantaneously brought into contact with water at a plane $x = 0$; (b) the variation of the electrolyte concentration in the container at the instant of contact.

$$D_- < D_+$$

This means that the positive ions try to lead the negative ions in the diffusion into the water. But when an ionic species of one charge moves faster than a species of the opposite charge, any unit volume in the water phase will receive more ions of the faster-moving variety.

Compare two unit volumes (Fig. 4.76), one situated at x_2 and the other at x_1, where $x_2 > x_1$; i.e., x_2 is farther from the plane of contact ($x = 0$) of the two solutions. The positive ions are random-walking faster than the negative ions, and therefore the greater the value of x, the greater is the ratio c_{+x}/c_{-x}.

All this is another way of saying that the center of the positive charge tends to separate from the center of the negative charge (Fig. 4.76). Hence, there is a tendency for the segregation of charge and the breakdown of the law of electroneutrality.

When charges of opposite sign are spatially separated, a potential difference develops. This potential difference between two unit volumes at x_2 and x_1 *opposes* the attempt at charge segregation. The faster-moving positive ions face strong opposition from the electroneutrality field and they are slowed down. In contrast, the slower-

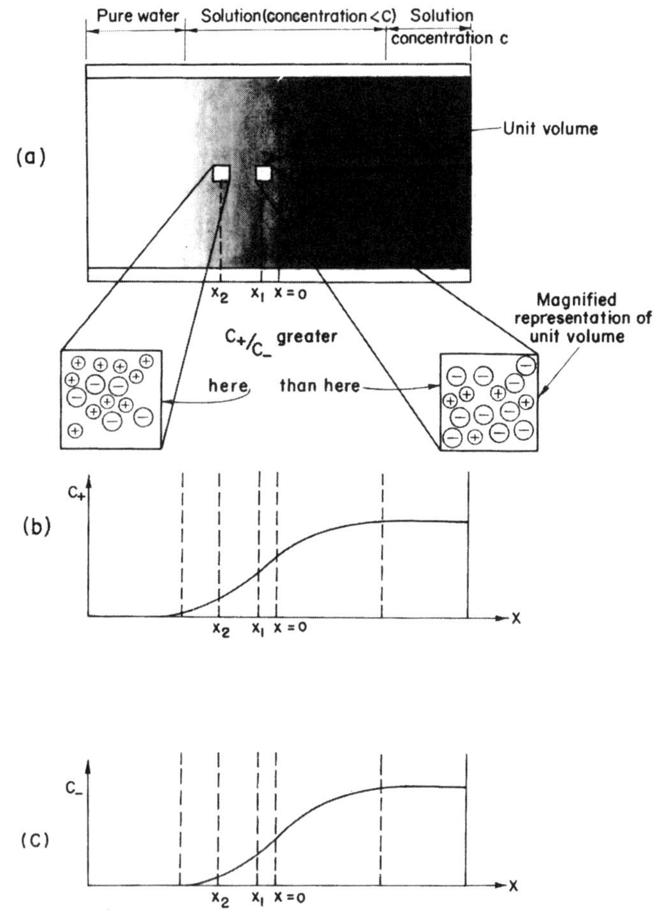

Fig. 4.76. (a) At a time $t > 0$ after the electrolyte and water are brought into contact, pure water and the electrolyte are separated by a region of mixing. In this mixing region, the c_+/c_- ratio increases from right to left because of the higher mobility of the positive ions. (b) and (c) The distance variations of the concentrations of positive and negative ions.

moving negative ions are assisted by the potential difference (arising from the incipient charge separation) and they are speeded up. When a steady state is reached, the acceleration of the slow negative ions and the deceleration of the initially fast, positive ions resulting from the electroneutrality field that develops exactly compensate for the inherent differences in mobilities. The electroneutrality field is the leveler of ionic mobilities, helping and retarding ions according to their need so as to keep the situation as electroneutral as possible.

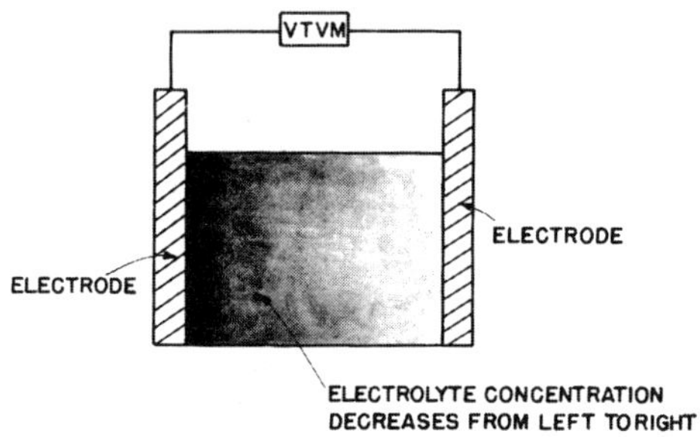

Fig. 4.77. A potential difference is registered by a vacuum-tube voltmeter (VTVM) connected to a concentration cell, i.e., to electrodes dipped in an electrolyte, the concentration of which varies from electrode to electrode.

The conclusion that may be drawn from this analysis has quite profound ramifications. The basic phenomenon is that whenever solutions of differing concentration are allowed to come into contact, diffusion occurs; there is a tendency for charge separation due to differences between ionic mobilities; and a potential difference develops across the interphase region in which there is a transition from the concentration of one solution to the concentration of the other.

This potential is known by the generic term *diffusion potential*. The precise name given to the potential varies with the situation, i.e., with the nature of the interphase region. If one ignores the interphase region and simply sticks two electrodes, one into each solution, in order to "tap" the potential difference, then the whole assembly is known as a *concentration cell* (Fig. 4.77). On the other hand, if one constrains or restricts the interphase region by interposing a sintered-glass disk or any uncharged membrane between the two solutions so that the concentrations of the two solutions are uniform up to the porous material, then one has a *liquid-junction potential*[34] (Fig. 4.78). A *membrane potential* is a more complicated affair for two main reasons: (1) There may be a pressure difference across the membrane, producing hydrodynamic flow of the solution, and (2) the membrane itself may consist of charged groups, some fixed and others exchangeable with the electrolytic solution, a situation equivalent to having sources of ions within the membrane.

[34]Now that the origin of a liquid-junction potential is understood, the method of minimizing it becomes clear. One chooses positive and negative ions with a negligible difference in mobilities; K^+ and Cl^- ions are the usual pair. This is the basis of the so-called "KCl salt bridge."

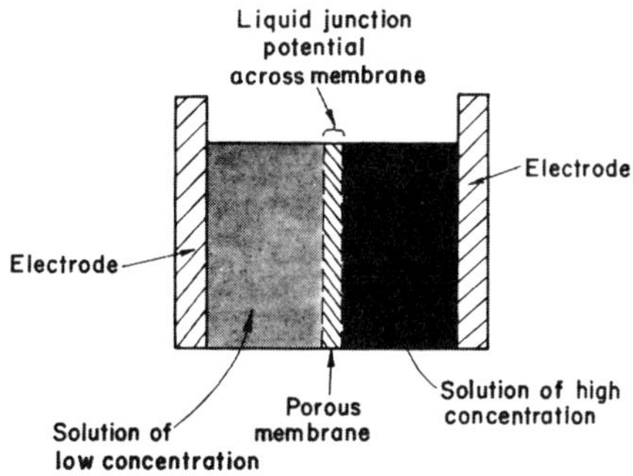

Fig. 4.78. A potential difference, the liquid-junction potential, is developed across a porous membrane introduced between two solutions of differing concentration.

4.5.5. Electroneutrality Coupling between the Drifts of Different Ionic Species

The picture of the development of the electroneutrality field raises a general question concerning the flow or drift of ions in an electrolytic solution. Is the flux of one ionic species dependent on the fluxes of the other species? In the diffusion experiment just discussed, is the diffusion of positive ions affected by the diffusion of negative ions? The answer to both these questions is in the affirmative.

Without doubt, the ionic flows start off as if they were completely independent, but it is this attempt to assert their freedom that leads to an incipient charge separation and the generation of an electroneutrality field. This field, which is dependent on the flows of all ionic species, curtails the independence of any one particular species. In this way, the flow of one ionic species is "coupled" to the flows of the other species.

In the absence of any interaction or coupling between flows, i.e., when the drift of any particular ionic species i is completely independent, the flow of that species i (i.e., the number of moles of i crossing per square centimeter per second) is described as follows.

The total driving force on an ionic species that is drifting independently of any other ionic species is the gradient of the electrochemical potential, $d\bar{\mu}_i/dx$. In terms of this total driving force, the expression for the total independent flux is given by the Nernst–Planck flux equation

$$J_i = \frac{j_i}{z_i F} = -\frac{u_i}{z_i F} c_i \frac{d\bar{\mu}_i}{dx} \tag{4.231}$$

When, however, the flux of the species i is affected by the flux of the species j through the electroneutrality field, then another modification (see Section 4.5.3) of the

Nernst–Planck flux equations has to be made. The modification that will now be described is a more detailed version which will lead us back to transport numbers.

4.5.6. How to Determine Transport Number

Before introducing the reader to the ways in which these transport numbers can be experimentally measured, it may be helpful to think again about just what they mean. The bare definition is given in Eq. (4.234) but this often only arouses puzzlement. For how, asks the thoughtful reader, can unequal amounts of negative and positive charges pass across a solution without doing injury to the electroneutrality condition that (understood over a time average) must always apply to the bulk of a solution?[35] The trouble seems at first a deep one because in some systems (an extreme case would be the liquid silicates) the alkali metal cation (a relatively small entity) has a t_+ near unity, whereas the anion (the giant silicate polyanion, in the case cited) hardly moves at all.

Looked at in terms of an analogy, the difficulty is seen to be less than real. Consider the ions as black (negative) and white (positive) balls. There are just two lines of black balls and white balls, an equal number of each, and both lines are at first stationary. Obviously, with an equal number of (oppositely charged) black and white balls, electroneutrality is preserved.

Now let the white balls all roll steadily off to the left while the black balls retain their stations. Let there be a machine that produces any required number of new white balls and inserts them into the beginning of this line on the right. Correspondingly, on the left, a consumer of white balls acts to bring about their disappearance. Consider then the middle part. The white balls move along while the black balls remain immovable. Then, framing a certain element of the bulk, the number of white and black balls remains always equal in number. As one white ball comes into the frame, one white ball also goes out of it. The blacks remain constant. Clearly, electroneutrality in the volume element considered is preserved. The same would be found for any piece of the bulk of the solution. The only reservation is the presence of the two machines, one of which produces the white balls on the right while another annihilates them on the left. In the real case, these functions would be supplied by the electrodes, the positive anode dissolving ions into the solution while the negative cathode destroys the ions by depositing them as atoms. So one can have t_+ and t_- of any value, with the proviso that for a 1:1 electrolyte, $t_+ + t_- = 1$. Now let three entirely different approaches for determining t's be described.

4.5.6.1. *Approximate Method for Sufficiently Dilute Solutions.*

In order to use the method, one has to use the Einstein equation

---

[35]To the bulk only? Yes. For near the interfaces there is an electrode reaction (an exchange of charge of ions with the electrode) and the positive ions (the cations) will tend to exceed the negative ones (the anions) near the negative electrode (and obviously vice versa at the positive electrode).

$$kT\bar{u}_i = D_i \tag{4.172}$$

From this equation and from the definition of transport number, i.e., Eq. (4.236),

$$t_i = \frac{D_i}{D_i + D_j} \tag{4.248}$$

and then t_i can be determined experimentally so long as the diffusion coefficients of the relevant ions can be measured (e.g., by means of radiotracers). However, there are a few reservations necessary before this method can be utilized.

1. The Einstein equation is exact only at very low concentrations in aqueous solution. As explained in a general way in Section 4.4.6, and in a detailed way for a given system in Section 5.6.6.2, there is usually some deviation—perhaps as much as 20%—between the results of Einstein's equation and experimental fact. Thus, outside very dilute solutions, using the Einstein equation to determine transport numbers is a rough-and-ready method and the results carry a burden of ±10%.

2. The method is clearly limited to ions that have suitable (β- or α-emitting) radiotracers. Those ions having radioactive isotopes that emit γ radiation are more difficult to measure because the long range and penetrating power of the γ radiation make it difficult precisely to determine the position of the radiotracer ions as they spread through a solution.

Another approximate approach to determining transport number is to use the zeroth approximation equation for ionic mobility, i.e., Eq. (4.183)

$$u_{\text{conv}} = \frac{z_i e_0}{6\pi r_S \eta} \tag{4.183}$$

where r_S is the Stokes radius of the ion concerned, η is the viscosity of the solution, and u_{conv} is the mobility or ionic velocity under a field gradient of 1 V cm^{-1}.

All one has to know here is the viscosity (for dilute solutions this is roughly equal to that of water) and the radius r_S of a hydrated ion. The principal approximation lies in the nature of the Stokes equation (4.175) (see Section 4.4.7). This may introduce an error of up to 25%.

4.5.6.2. *Hittorf's Method.* This method of determining transport numbers was devised as long ago as 1901 and has been described in innumerable papers and many books. Nevertheless, it is not all that simple to understand and contains a number of assumptions not always stated.

To start with, let an overall description of the method be given. The essentials of the apparatus (Fig. 4.79) are two clearly separated compartments joined by a substantial middle compartment. There is an aqueous electrolyte, say, silver nitrate, and if this

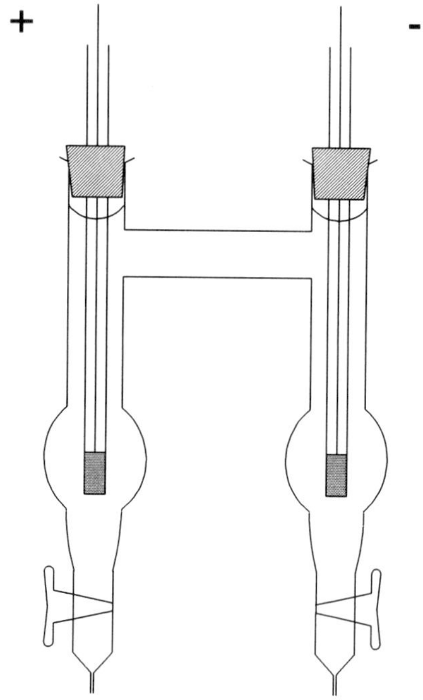

Fig. 4.79. Hittorf's cell. (After E. A. Muelwhyn Hughes, 1968)

is the case, the electrodes will each be made of silver. Before the experiment begins, the concentration of $AgNO_3$ is the same throughout the cell. The experiment involves passage of a direct electric current (from some power source not shown in the figure) through the cell.

At the left-hand electrode, Ag dissolves and increases the $AgNO_3$ concentration in this compartment. In the right-hand compartment, Ag^+ ions deposit so the $AgNO_3$ concentration decreases in the solution in the right-hand compartment. Measurement of the changes in concentration in each compartment after a 2–3 hr passage of current yields the transport number of the anion (since $t_+ + t_- = 1$, it also gives that of the cation). Now, let the analysis of what happens be written out.

The current gets passed for the requisite time. Thereafter, the anolyte (see Fig. 4.80) has an increased concentration c_1 and the catholyte a decreased concentration c_3. The middle compartment does not change its concentration of silver nitrate, which will be designated c_2.

After t seconds (s) at current I, the number of g-ions of Ag introduced into the anolyte is

$$N = \frac{It}{\mathrm{F}} \tag{4.249}$$

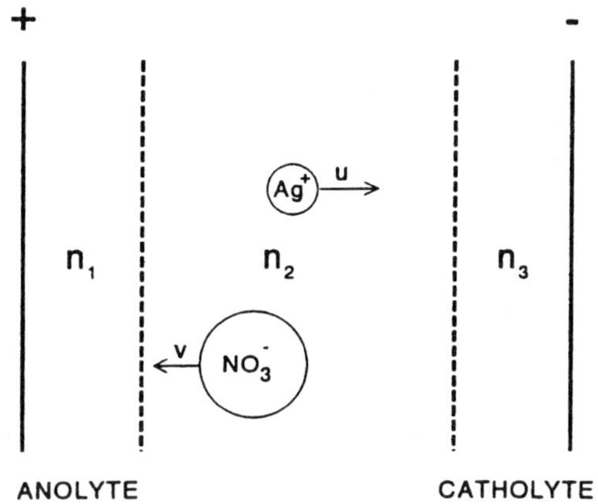

Fig. 4.80. The principle of Hittorf's experiment. (After E. A. Moelwhyn-Hughes, 1968)

where F is the faraday or electrical charge on one g-ion of Ag^+.

In the central compartment, in which the concentration is shown by experiment to remain constant, one can write for the current:

$$I = Ac_2\mathrm{F}(u_2 + v_2) \tag{4.250}$$

where A is the cross-sectional area of the central compartment and u_2 and v_2 are the ionic mobilities, respectively, of Ag^+ and NO_3^-, under a unit applied field. Therefore, from Eqs. (4.249) and (4.250)

$$N = Ac_2(u_2 + v_2)t \tag{4.251}$$

In the *left*-hand compartment, Ag^+ ions are not only produced, they also get moved out. Hence,

$$\frac{dN_1}{dt} = \frac{I}{\mathrm{F}} - Ac_1u_1 \tag{4.252}$$

The principle of electroneutrality demands that the concentration of both positive and negative ions in the left-hand compartment be the same. Therefore,

$$\frac{dN_1}{dt} = Ac_2v_2 \tag{4.253}$$

which represents the rate at which anions introduced by dissolution from the silver electrode move into the left-hand compartment to partner the cation. Integrating (4.253) gives

$$N_1 - N_1^0 = A\,c_2 v_2 t \tag{4.254}$$

where N_1^0 is the number of moles of $AgNO_3$ in the compartment before the current I was switched on.

Turning attention now to the catholyte, it follows that with the Ag^+ being removed by deposition and being augmented by transport from the middle compartment,

$$\frac{dN_3}{dt} = A\,c_2 u_2 - \frac{I}{\mathrm{F}} \tag{4.255}$$

However, diffusion of Ag^+ and its removal by deposition is not the only thing happening in the right-hand compartment. NO_3^- clearly moves out to allow electroneutrality to be maintained. It must move out at the same rate as Ag^+ disappears. Thus

$$\frac{dN_3}{dt} = -A\,c_3 v_3 \tag{4.256}$$

Integration of (4.256) gives

$$N_3^0 - N_3 = A c_3 v_3 t \tag{4.257}$$

It has been assumed all along (but it needs an experiment to verify it) that the central compartment keeps a constant concentration while the $AgNO_3$ is increasing on the anolyte and decreasing in the catholyte. Hence,

$$N_1 - N_1^0 = N_3^0 - N_3 \tag{4.258}$$

Now from Eqs. (4.251) and (4.254)

$$t_- = \frac{v_2}{u_2 + v_2} = \frac{N_1 - N_1^0}{N} = \frac{\text{Gain in weight in anolyte}}{\text{Loss of weight in anode}} \tag{4.259}$$

and from Eqs. (4.251) and (4.257),

$$\frac{N_3^0 - N_3}{N} = \frac{\text{Loss of weight of silver in catholyte}}{\text{Gain in weight of cathode}} \tag{4.260}$$

Is this all there is to be said about Dr. Hittorf's classic method? The reader may have noticed a weak point in the argument. Where does the middle compartment begin and end? This is not a silly question if by "the middle compartment" one does not mean that section of the apparatus shown in the figure to divide the two compartments, but that section of the electrolyte between the two compartments which maintains its concentration constant. Thus, the method does have an Achilles heel: one has to be

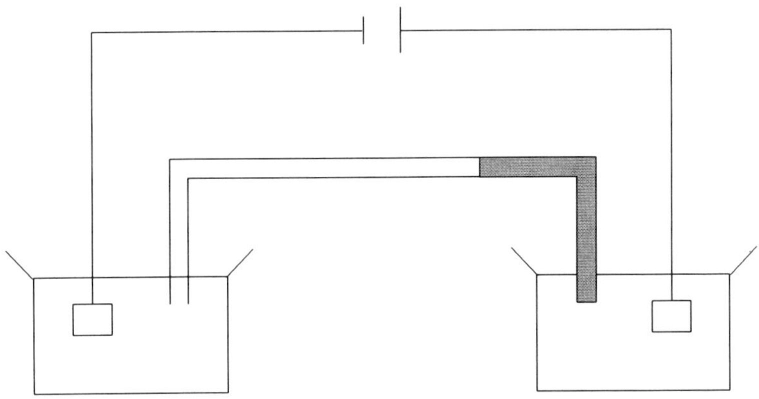

Fig. 4.81. Lodge's moving-boundary apparatus.

careful that the concentration changes that occur in anolyte and catholyte do not spread to the central compartment. This in turn means that the current must not flow for too long at a time and the downside of that is that the concentration changes in the cathode and anode compartments may not then be large enough to be accurately measurable.[36]

4.5.6.3. *Oliver Lodge's Experiment.* An experiment first done by the English physicist Oliver Lodge is the origin of a third method by which transport numbers can be obtained. Here also there is a limitation: one must be able to observe a boundary between two electrolytes, for knowledge of the boundary's movement is the observation upon which the method is based. This implies that the ions concerned must differ in color (not always an easy condition to fulfill) or at least in refractive index (but then the observation of the boundary may not always be easy).

To understand how this method works, let us have a look at Fig. 4.81. It involves a tube and in this tube there are two solutions and a boundary between them. Let the electrolyte in the upper compartment be named MR and be at a concentration c. The second (bottom) solution is $M'R$, containing the same anion R, but a different cation, M'.

Now a current is passed. The boundary moves and the velocity of that movement u is given by

$$u = \frac{x}{t} \tag{4.261}$$

with x being the distance the boundary moves in t s.

[36]However, this objection is much lessened by the improvement in differential analysis made in the last quarter-century.

Now from Eq. (4.250)

$$\mathrm{I} = AcF(u_+ + v_-) \tag{4.250}$$

where A is the cross section of the solution. Hence, from Eqs. (4.250) and (4.261)

$$t_+ = \frac{u_+}{u_+ + u_-} = \frac{xAcF}{It} \tag{4.262}$$

One can see that the method would work best with a colored cation, e.g., a cation that is chromophoric, such as Cr^{2+} ions.

4.5.7. The Onsager Phenomenological Equations

The development of an electroneutrality field introduces an interaction between flows and makes the flux of one species dependent on the fluxes of all the other species. To treat situations in which there is a coupling between the drift of one species and that of another, a general formalism will be developed. It is only when there is zero coupling or zero interaction that one can accurately write the Nernst–Planck flux equation

$$J_i = -\frac{u_i}{z_i F} c_i \frac{d\bar{\mu}_i}{dx} \tag{4.231}$$

Once the interaction (due to the electroneutrality field) develops, a correction term is required, i.e.,

$$J_i = -\frac{u_i}{z_i F} c_i \frac{d\bar{\mu}_i}{dx} + \text{Coupling correction} \tag{4.263}$$

It is in the treatment of such interacting transport processes, or coupled flows, that the methods of near-equilibrium thermodynamics yield a clear understanding of such phenomena, but only from a macroscopic or phenomenological point of view. These methods, as relevant to the present discussion, can be summarized with the following series of statements:

1. As long as the system remains close to equilibrium and the fluxes are independent, the fluxes are treated as proportional to the driving forces. Experience (Table 4.17) commends this view for diffusion [Fick's law, Eq. (4.16)], conduction [Ohm's law, Eq. (4.130)], and heat flow (Fourier's law). Thus, the *independent* flux of an ionic species 1 given by the Nernst–Planck equation (4.231) is written

$$J_1 = L_{11}\vec{F}_1 \tag{4.264}$$

where L_{11} is the proportionality or phenomenological constant and $\vec{F}_1$ is the driving force.

TABLE 4.17

Some Linear Flux–Force Laws

Phenomenon	Flux of	Driving Force	Law
Diffusion	Matter J	Concentration gradient dc/dx	Fick $J = -D\, dc/dx$
Migration (conduction)	Charge i	Potential gradient $X = -d\psi/dx$	Ohm $i = \sigma X$
Heat conduction	Heat J_{heat}	Temperature gradient	Fourier $J_{\text{heat}} = K\, dT/dx$

2. When there is coupling, the flux of one species (e.g., J_1) is not simply proportional to its own (or as it is called, *conjugate*) driving force (i.e., $\vec{F}_1$), but receives contributions from the driving forces on all the other particles. In symbols,

$$J_1 = L_{11}\vec{F}_1 + \text{Coupling correction}$$

$$= L_{11}\vec{F}_1 + \begin{array}{c}\text{Flux of 1 due}\\ \text{to driving force}\\ \text{on species 2}\end{array} + \begin{array}{c}\text{Flux of 1 due}\\ \text{to driving force}\\ \text{on species 3}\end{array} + \text{etc.} \tag{4.265}$$

3. The linearity or proportionality between fluxes and conjugate driving force is also valid for the contributions to the flux of one species from the forces on the *other* species. Hence, with this assumption, one can write Eq. (4.265) in the form

$$J_1 = L_{11}\vec{F}_1 + [L_{12}\vec{F}_2 + L_{13}\vec{F}_3 + \cdots] \tag{4.266}$$

4. Similar expressions are used for the fluxes of all the species in the system. If the system consists of an electrolyte dissolved in water, one has three species: positive ions, negative ions, and water. By using the symbol + for the positive ions, – for the negative ions, and 0 for the water, the fluxes are

$$J_+ = L_{++}\vec{F}_+ + L_{+-}\vec{F}_- + L_{+0}\vec{F}_0$$

$$J_- = L_{-+}\vec{F}_+ + L_{--}\vec{F}_- + L_{-0}\vec{F}_0$$

$$J_0 = L_{0+}\vec{F}_+ + L_{0-}\vec{F}_- + L_{00}\vec{F}_0 \tag{4.267}$$

These equations are known as the *Onsager phenomenological equations.* They represent a complete macroscopic description of the interacting flows when the system is near equilibrium.[37] It is clear that all the "straight" coefficients L_{ij}, where the indices

[37]If the system is not near equilibrium, the flows are no longer proportional to the driving forces [*cf.* Eq. (4.265)].

are equal, $i = j$, pertain to the independent, uncoupled fluxes. Thus $L_{++}X_+$ and $L_{--}X_-$ are the fluxes of the positive and negative ions, respectively, when there are no interactions. All other cross terms represent interactions between fluxes; e.g., $L_{+-}X_-$ represents the contribution to the flux of positive ions from the driving force on the negative ions.

5. What are the various coefficients L_{ij}? Onsager put forward the helpful *reciprocity relation*. According to this, all symmetrical coefficients are equal, i.e.,

$$L_{ij} = L_{ji} \tag{4.268}$$

This principle has the same status in nonequilibrium thermodynamics as the law of conservation of energy has in classical thermodynamics; it has not been disproved by experience.

4.5.8. An Expression for the Diffusion Potential

The expression for the diffusion potential can be obtained in a straightforward though hardly brief manner by using the Onsager phenomenological equations to describe the interaction flows. Consider an electrolytic solution consisting of the ionic species M^{z+} and A^{z-} and the solvent. When a transport process involves the ions in the system, there are two ionic fluxes, J_+ and J_-. Since, however, the ions are solvated, the solvent also participates in the motion of ions and hence there is also a solvent flux J_0.

If, however, the solvent is considered fixed, i.e., the solvent is taken as the coordinate system or the frame of reference,[38] then one can consider ionic fluxes *relative* to the solvent. Under this condition, $J_0 = 0$, and one has only two ionic fluxes. Thus, one can describe the interacting and independent ionic drifts by the following equations

$$J_+ = L_{++}\vec{F}_+ + L_{+-}\vec{F}_- \tag{4.269}$$

$$J_- = L_{-+}\vec{F}_+ + L_{--}\vec{F}_- \tag{4.270}$$

The straight coefficients L_{++} and L_{--} represent the independent flows, and the cross coefficients L_{+-} and L_{-+}, the coupling between the flows.

The important step in the derivation of the diffusion potential is the statement that under conditions of steady state, the electroneutrality field sees to it that the quantity of positive charge flowing into a volume element is equal in magnitude but opposite in sign to the quantity of negative charge flowing in (Fig. 4.82). That is,

---

[38] Coordinate systems are chosen for convenience.

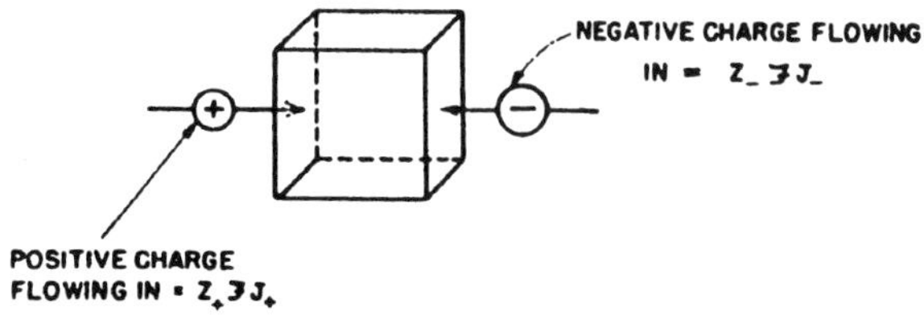

Fig. 4.82. According to the principle of electroneutrality as applied to the fluxes, the flux of positive ions into a volume element must be equal to the flux of negative ions into the volume element, so that the total negative charge is equal to the total positive charge.

$$z_+FJ_+ + z_-FJ_- = 0 \tag{4.271}$$

For convenience, z_+F and z_-F are written as q_+ and q_-, respectively. Now the expressions (4.269) and (4.270) for J_+ and J_- are substituted in Eq. (4.271)

$$q_+L_{++}\vec{F}_+ + q_+L_{+-}\vec{F}_- + q_-L_{-+}\vec{F}_+ + q_-L_{--}\vec{F}_- = 0$$

or

$$\vec{F}_+(q_+L_{++} + q_-L_{-+}) + \vec{F}_-(q_+L_{+-} + q_-L_{--}) = 0 \tag{4.272}$$

Using the symbols

$$p_+ = q_+L_{++} + q_-L_{-+} \tag{4.273}$$

and

$$p_- = q_+L_{+-} + q_-L_{--} \tag{4.274}$$

one has

$$p_+\vec{F}_+ + p_-\vec{F}_- = 0 \tag{4.275}$$

What are the driving forces $\vec{F}_+$ and $\vec{F}_-$ for the independent flows of the positive and negative ions? They are the gradients of electrochemical potential (see Section 4.4.15)

$$\vec{F}_+ = \frac{d\mu_+}{dx} + q_+\frac{d\varphi}{dx} \tag{4.276}$$

and

$$\vec{F}_- = \frac{d\mu_-}{dx} + q_- \frac{d\varphi}{dx} \tag{4.277}$$

With these expressions, Eq. (4.275) becomes

$$p_+ \frac{d\mu_+}{dx} + p_+ q_+ \frac{d\varphi}{dx} + p_- \frac{d\mu_-}{dx} + p_- q_- \frac{d\varphi}{dx} = 0$$

or

$$-\frac{d\psi}{dx} = \frac{p_+}{p_+ q_+ + p_- q_-} \frac{d\mu_+}{dx} + \frac{p_-}{p_+ q_+ + p_- q_-} \frac{d\mu_-}{dx} \tag{4.278}$$

It can be shown, however, that (Appendix 4.3)

$$\frac{p_+}{p_+ q_+ + p_- q_-} = \frac{t_+}{z_+ F} \tag{4.279}$$

and

$$\frac{p_-}{p_+ q_+ + p_- q_-} = \frac{t_-}{z_- F} \tag{4.280}$$

where t_+ and t_- are the transport numbers of the positive and negative ions. By making use of these relations, Eq (4.278) becomes

$$-\frac{d\psi}{dx} = \frac{t_+}{z_+ F} \frac{d\mu_+}{dx} + \frac{t_-}{z_- F} \frac{d\mu_-}{dx}$$

$$= \sum \frac{t_i}{z_i F} \frac{d\mu_i}{dx} \tag{4.281}$$

The negative sign before the electric field shows that it is opposite in direction to the chemical potential gradients of all the diffusing ions.

If one considers (Fig. 4.83) an infinitesimal length dx parallel to the direction of the electric- and chemical-potential fields, one can obtain the electric-potential difference $d\psi$ and the chemical-potential difference $d\mu$ across the length dx

$$-\frac{d\psi}{dx} dx = \sum \frac{t_i}{z_i F} \frac{d\mu_i}{dx} dx \tag{4.282}$$

or

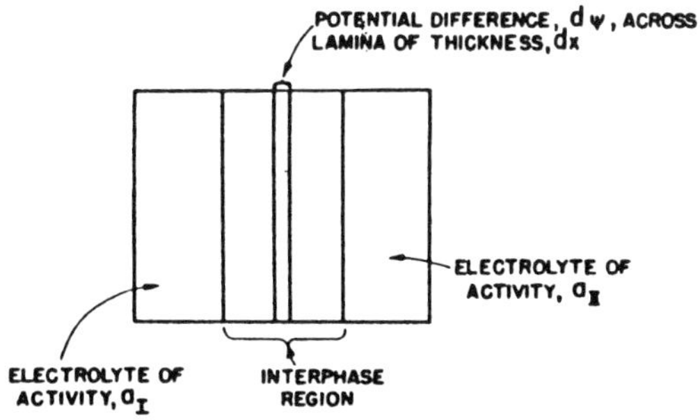

Fig. 4.83. Schematic representation of a *dx*-thick lamina in the interphase region between two electrolytes of activities a_{I} and a_{II}.

$$-d\psi = \frac{1}{F}\sum \frac{t_i}{z_i}\, d\mu_i \tag{4.283}$$

$$= \frac{RT}{F}\sum \frac{t_i}{z_i}\, d\ln a_i \tag{4.284}$$

This is the basic equation for the diffusion potential. It has been derived here on the basis of a realistic point of view, namely, that the diffusion potential arises from the *nonequilibrium* process of diffusion.

There is, however, another method[39] of deriving the diffusion potential. One takes note of the fact that when a steady-state electroneutrality field has developed, the system relevant to a study of the diffusion potential hangs together in a delicate balance. The diffusion flux is exactly balanced by the electric flux; the concentrations and the electrostatic potential throughout the interphase region *do not vary with time*. (Remember the derivation of the Einstein relation in Section 4.4.) In fact, one may turn a blind eye to the drift and pretend that the whole system is in equilibrium.

On this basis, one can equate to zero the sum of the electrical and diffusional work of transporting ions across a lamina *dx* of the interphase region (Fig. 4.84). If one equivalent of charge (both positive and negative ions) is taken across this lamina, the electrical work is $F\,d\psi$. But this one equivalent of charge consists of t_+/z_+ moles of positive ions and t_-/z_- moles of negative ions. Hence, the diffusional work is $d\mu_+$ per mole, or $(t_+/z_+)\,d\mu_+$ per t_+/z_+ moles, of positive ions and $(t_-/z_-)\,d\mu_-$ per t_-/z_- moles of negative ions. Thus,

---

[39]This method is based on Thomson's hypothesis, according to which it is legitimate to apply equilibrium thermodynamics to the reversible parts of a steady-state, nonequilibrium process.

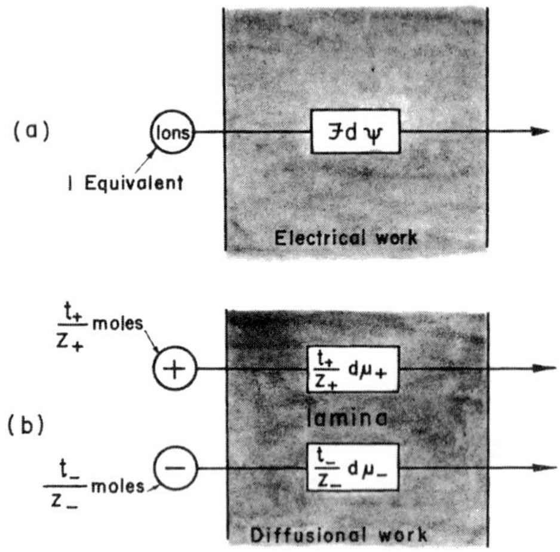

Fig. 4.84. The sum of (a) the electrical work $F\,d\psi$ and (b) the diffusional work $(t_+/z_+)d\mu_+ + (t_-/z_-)d\mu_-$ of transporting one equivalent of ions across a *dx*-thick lamina in the interphase region is equal to zero.

$$F\,d\psi + \frac{t_+}{z_+}d\mu_+ + \frac{t_-}{z_-}d\mu_- = 0 \tag{4.285}$$

or

$$-d\psi = \frac{1}{F}\sum \frac{t_i}{z_i}\,d\mu_i \tag{4.283}$$

$$= \frac{RT}{F}\sum \frac{t_i}{z_i}\,d\ln a_i \tag{4.284}$$

4.5.9. The Integration of the Differential Equation for Diffusion Potentials: The Planck–Henderson Equation

An equation has been derived for the diffusion potential [*cf.* Eq. (4.283)], but it is a *differential* equation relating the infinitesimal potential difference $d\psi$ developed across an infinitesimally thick lamina dx in the interphase region. What one measures experimentally, however, is the total potential difference $\Delta\psi = \psi^0 - \psi^l$ across a transition region extending from $x = 0$ to $x = l$ (Fig. 4.85). Hence, to theorize about the

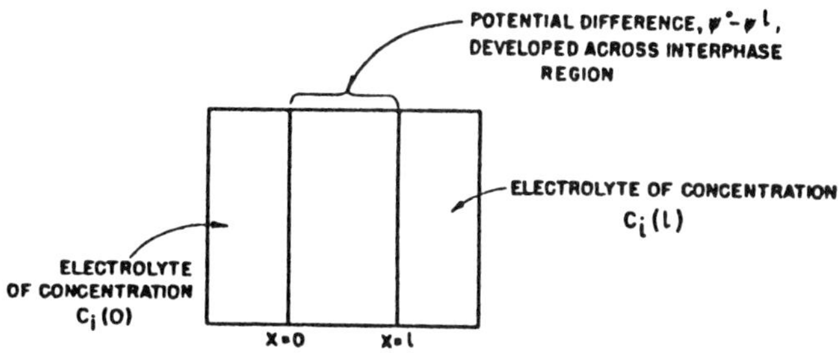

Fig. 4.85. The measured quantity is the total potential difference $\psi^0 - \psi^l$ across the whole interphase region between electrolytes of differing concentration $c_i(0)$ and $c_i(l)$.

measured potential differences, one has to integrate the differential equation (4.283); i.e.,

$$-\Delta\psi = \psi^0 - \psi^l = \frac{1}{F}\sum_i \int_{x=0}^{x=l} \frac{t_i}{z_i}\frac{d\mu_i}{dx}\,dx$$

$$= \frac{RT}{F}\sum_i \int_{x=0}^{x=l} \frac{t_i}{z_i}\frac{d\ln a_i}{dx}\,dx$$

$$= \frac{RT}{F}\sum_i \int_{x=0}^{x=l} \frac{t_i}{z_i}\frac{1}{f_i c_i}\frac{d(f_i c_i)}{dx}\,dx \qquad (4.286)$$

Here lies the problem. To carry out the integration, one must know:

1. How the concentrations of all the species vary in the transition region.
2. How the activity coefficients f_i vary with c_i.
3. How the transport number varies with c_i.

The general case is too difficult to solve analytically, but several special cases can be solved. For example (Fig. 4.86), the activity coefficients can be taken as unity, $f_i = 1$—ideal conditions; the transport numbers t_i can be assumed to be constant; and a linear variation of concentrations with distance can be assumed. The last assumption implies that the concentration $c_i(x)$ of the ith species at x is related to its concentration $c_i(0)$ at $x = 0$ in the following way

$$c_i(x) = c_i(0) + k_i x \qquad (4.287)$$

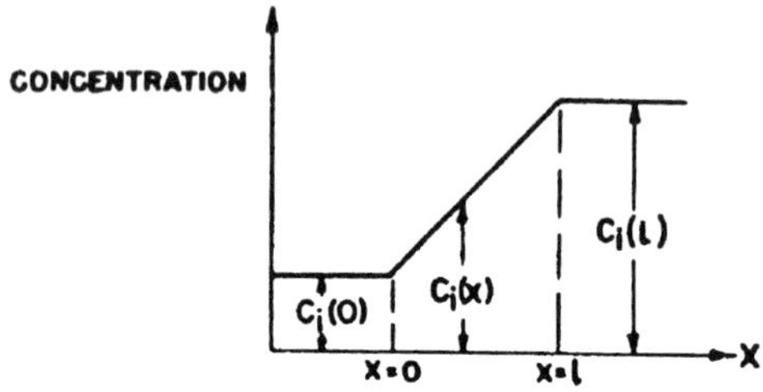

Fig. 4.86. In the derivation of the Planck–Henderson equation, a linear variation of concentration is assumed in the interphase region, which commences at $x = 0$ and ends at $x = l$.

and

$$\frac{dc_i}{dx} = \text{Constant} = k_i = \frac{c_i(l) - c_i(0)}{l} \tag{4.288}$$

With the aid of these assumptions, the integration becomes simple. Thus, with $t_i \neq f(x), f_i = 1$, and Eqs. (4.287) and (4.288), one has in (4.286)

$$-\Delta\psi = \frac{RT}{F}\sum_i \int_{x=0}^{x=l} \frac{t_i}{z_i} \frac{k_1\, dx}{c_i(0) + k_1 x}$$

$$= \frac{RT}{F}\sum_i \frac{t_i}{z_i} \int_{x=0}^{x=l} \frac{d[c_i(0) + k_1 x]}{c_i(0) + k_1 x}$$

$$= \frac{RT}{F}\sum_i \frac{t_i}{z_i} \left\{ \ln\left[c_i(0) + k_1 x\right]\right\}_{x=0}^{x=l}$$

$$= \frac{RT}{F}\sum_i \frac{t_i}{z_i} \ln \frac{c_i(l)}{c_i(0)} \tag{4.289}$$

This is known as the Planck–Henderson equation for diffusion or liquid-junction potentials.

In the special case of a $z{:}z$-valent electrolyte, $z_+ = z_- = z$ and $c_+ = c_- = c$, Eq. (4.289) reduces to

$$-\Delta\psi = \frac{RT}{zF}(t_+ - t_-) \ln \frac{c(l)}{c(0)} \tag{4.290}$$

and since $t_+ + t_- = 1$,

$$-\Delta\psi = (2t_+ - 1)\frac{RT}{zF}\ln\frac{c(l)}{c(0)} \tag{4.291}$$

In the highly simplified treatment of the diffusion potential that has just been presented, several drastic assumptions have been made. The one regarding the concentration variation within the transition region can be avoided. One may choose a more realistic concentration versus distance relationship either by thinking about it in more detail or by using experimental knowledge on the matter. Similarly, instead of assuming the activity coefficients to be unity, one can feed in the theoretical or experimental concentration dependence of the activity coefficients. Of course, the introduction of nonideality ($f_i \neq 1$) makes the mathematics awkward; in principle, however, the problem is understandable.

What about the assumption of the constancy of the transport number? Is this reasonable? In the case of a z:z-valent electrolyte, the transport number depends on the mobilities

$$t_+ = \frac{u_+}{u_+ + u_-} \quad \text{and} \quad t_- = \frac{u_-}{u_+ + u_-}$$

Thus, the constancy of the transport numbers with concentration depends on the degree to which the mobilities vary with concentration. That is something to be dealt with in the model-oriented arguments of the next section.

4.5.10. A Bird's Eye View of Ionic Transport

In the section on diffusion (Section 4.2), we were concerned with the random walk, but we learned that because this type of movement is purposeless in direction, it causes particles to spread out in all available directions. To bring about a diffusional flux in a given direction, all one has to do is to introduce a concentration gradient into the system and, although the movement of each particle is random, the fact that there are fewer ions in one direction than in others means that the random walk gradually raises the concentration in the dilute parts of the solution until all is uniform. In this sense, there is a directed diffusion flux down the concentration gradient.

Then we discussed the idea that this randomness can have superimposed upon it an electrical field that does indeed have a direction, so that there are still random motions by cations and the anions, but a little less randomness for the positive ion in the direction of the negative electrode and for the negative ion in the direction of the positive electrode. There thus arises a net drift of ions in the direction of the field.

In the phenomenological treatment of the directed drift that the field brings, we take the attitude that there is a stream of cations going toward the negative electrode and anions going toward the positive one. We now neglect the random diffusive movements; they do not contribute to the vectorial flow that produces an electrical current.

This discussion then follows on to give rise to the idea of a *transport number*. This is a term that describes the fact that when we talk about the drift velocity, cations and anions of equal but opposite charge do not have the same speed although the applied field is the same. On the whole, cations tend to be smaller than anions and we show that the drift velocity is inversely proportional to the radius of the solvated ion, so that transport numbers tend to be larger for the cation and smaller for the anion in an electrolyte.

This brings us to an apparent dilemma because, according to laws first laid down by Faraday himself, when one passes a certain amount of electricity through a solution for a certain time, the number of cations and anions carrying the same numerical charge put on the cathode and anode, respectively, is the same.

Because there is a difference in the cationic and anionic transport number and hence mobilities, this is at first difficult to understand. One would expect more of the ions with the higher transport number to be preferentially deposited.

The dilemma is solved by taking into account the fact that the lack of an equal supply rate for cations and anions carried toward the electrodes by the electric current will create a concentration gradient near the interface for the slower ions, and this concentration gradient will speed up the motion of the slower ions to compensate for their poorer performance. It is this diffusional component that makes Faraday's laws come true. The diffusional gradient pitches in to help the slower ions to the electrode at the same rate as the faster ones.

Several famous equations (Einstein, Stokes–Einstein, Nernst–Einstein, Nernst–Planck) are presented in this chapter. They derive from the heyday of phenomenological physical chemistry, when physical chemists were moving from the predominantly thermodynamic approach current at the end of the nineteenth century to the molecular approach that has characterized electrochemistry in this century. The equations were originated by Stokes and Nernst but the names of Einstein and Planck have been added, presumably because these scientists had examined and discussed the equations first suggested by the other men.

As stated earlier, these phenomenological relationships existed after physical chemistry had passed through its predominantly thermodynamic stage. However, Onsager produced a late bloom in the 1930s by applying *nonequilibrium thermodynamics* to transport in ionic solutions. In this subdiscipline, *transport coefficients* are described and their relation to ionic drift in terms of the interaction between the ions (sodium–sodium, chloride–chloride, and sodium–chloride) is represented mathematically. Accordingly, the potential difference between two liquid phases differing in ionic concentrations or even in ionic species is treated in terms of Onsager's theory, and the equation for the electrical potential across a liquid–liquid junction is derived and presented as the Planck–Henderson equation. All this then is a phenomenological prelude to applying ionic-atmosphere concepts (see Chapter 3) to ionic migration in order to provide the physical explanation of a famous empirical law due originally to

Kohlrausch—that equivalent conductivity decreases linearly with the increase of the square root of the concentration.

Further Reading

Seminal

1. M. Planck, "Diffusion and Potential at Liquid–Liquid Boundaries," *Ann. Physik* **40**: 561 (1890).
2. P. Henderson, "An Equation for the Calculation of Potential Difference at any Liquid Junction Boundary," *Z. Phys. Chem. (Leipzig)* **59**: 118 (1907).
3. C. A. Angell and J. O'M. Bockris, "The Measurement of Diffusion Coefficients," *J. Sci. Instrum.* **29**: 180 (1958).

Review

1. P. Turq, J. Barthel, and M. Chemla, *Transport, Relaxation and Kinetic Processes in Electrolyte Solutions*, Springer-Verlag, Berlin (1992).

Papers

1. O. J. Riveros, *J. Phys. Chem.* **96**: 6001 (1992).
2. T. F. Fuller and J. Newman, *J. Electrochem. Soc.* **139**: 1332 (1992).
3. P. Vanysek, *Adv. Chem. Ser.* **235**: 55 (1994).
4. M. Olejnik, A. Blahut, and A. B. Szymanski, *Appl. Opt.* **24**: 83 (1994).
5. M. Suzuki, *J. Electroanal. Chem.* **384**: 77 (1995).
6. K. Aoki, *J. Electroanal. Chem.* **386**: 17 (1995).

4.6. INFLUENCE OF IONIC ATMOSPHERES ON IONIC MIGRATION

4.6.1. Concentration Dependence of the Mobility of Ions

In the phenomenological treatment of conduction (Section 4.2.12), it was stated that the equivalent conductivity Λ varies with the concentration c of the electrolyte according to the empirical law of Kohlrausch [Eq. (4.139)]

$$\Lambda = \Lambda^0 - A c^{1/2} \tag{4.139}$$

where A is a constant and Λ^0 is the pristine or ungarbled value of equivalent conductivity, i.e., the value at infinite dilution.

The equivalent conductivity, however, has been related to the conventional electrochemical mobilities[40] u_+ and u_- of the current-carrying ions by the following expression

[40]To avoid cumbersome notation, the conventional mobilities are written in this section without the subscript "conv"; i.e., one writes u_+ instead of $(u_{\text{conv}})_+$.

$$\Lambda = F(u_+ + u_-) \tag{4.163}$$

from which it follows that

$$\Lambda^0 = F(u_+^0 + u_-^0) \tag{4.292}$$

where the u^0s are the conventional mobilities (i.e., drift velocities under a field of 1 V cm^{-1}) at infinite dilution. Thus, Eq. (4.163) can be written as

$$(u_+ + u_-) = (u_+^0 + u_-^0) - \left(\frac{A}{F}\frac{1}{2}\right)(c_+^{1/2} + c_-^{1/2}) \tag{4.293}$$

or it can be split up into two equations

$$u_+ = u_+^0 - A'c_+^{1/2} \tag{4.294}$$

and

$$u_- = u_-^0 - A'c_-^{1/2} \tag{4.295}$$

What is the origin of this experimentally observed dependence of ionic mobilities on concentration? Equations (4.294) and (4.295) indicate that the more ions there are per unit volume, the more they diminish each other's mobility. In other words, at appreciable concentrations, the movement of any particular ion does not seem to be independent of the existence and motions of the other ions, and there appear to be forces of interaction between ions. This coupling between the individual drifts of ions has already been recognized, but now the discussion is intended to be on an atomistic rather than phenomenological level. The interactions between ions can be succinctly expressed through the concept of the ionic cloud (Chapter 3). It is thus necessary to analyze and incorporate ion-atmosphere effects into the zero-approximation atomistic picture of conduction (Section 4.4) and in this way understand how the mobilities of ions depend on the concentration of the electrolyte.

Attention should be drawn to the fact that there has been a degree of inconsistency in the treatments of ionic clouds (Chapter 3) and the elementary theory of ionic drift (Section 4.4.2). When the ion atmosphere was described, the central ion was considered—from a time-averaged point of view—at rest. To the extent that one seeks to interpret the equilibrium properties of electrolytic solutions, this picture of a static central ion is quite reasonable. This is because in the absence of a spatially directed field acting on the ions, the only ionic motion to be considered is random walk, the characteristic of which is that the mean distance traveled by an ion (not the mean square distance; see Section 4.2.5) is zero. The central ion can therefore be considered to remain where it is, i.e., to be at rest.

When, however, the elementary picture of ionic drift (Section 4.4.2) was sketched, the ionic cloud around the central ion was ignored. This approximation is justified only when the ion atmosphere is so tenuous that its effects on the movement of ions can be neglected. This condition of extreme tenuosity (in which there is a negligible coupling between ions) obtains increasingly as the solution tends to infinite dilution. Hence, the simple, unclouded picture of conduction (Section 4.4) is valid only at infinite dilution.

To summarize the duality of the treatment so far: When the ion atmosphere was treated in Chapter 3, the motion of the central ion was ignored and only equilibrium properties fell within the scope of analysis; when the motion of the central ion under an applied electric field was considered, the ionic cloud (which is a convenient description of the interactions between an ion and its environment) was neglected and only the infinite-dilution conduction could be analyzed. Thus, a unified treatment of *ionic atmospheres around moving ions* is required. The central problem is: How does the interaction between an ion and its cloud affect the motion of the ion?

4.6.2. Ionic Clouds Attempt to Catch Up with Moving ions

In the absence of a driving force (e.g., an externally applied electric field), no direction in space from the central ion is privileged. The Coulombic field of the central ion has spherical symmetry and therefore the probability of finding, say, a negative ion at a distance r from the reference ion is the same irrespective of the direction in which the point r lies. On this basis, it was shown that the ionic cloud was spherically symmetrical (see Section 3.8.2).

When, however, the ions are subject to a driving force (be it an electric field, a velocity field due to the flow of an electrolyte, or a chemical-potential field producing diffusion), one direction in space becomes privileged. The distribution function (which is a measure of the probability of finding an ion of a certain charge in a particular volume element) has to be lopsided, or asymmetrical. The probability depends not only on the distance of the volume element from the central ion but also on the direction in which the volume element lies in relation to the direction of ionic motion. The procedure of Chapter 3 no longer applies. One cannot simply assume a Boltzmann distribution and, for the work done to bring an ion (of charge $z_i e_0$) to the volume element under consideration, use the electrostatic work $z_i e_0 \psi$ because the electrostatic potential ψ was, in the context of Chapter 3, a function of r only and one would then infer a symmetrical distribution function.

The rigorous but unfortunately mathematically difficult approach to the problem of ionic clouds around moving ions is to seek the asymmetrical distribution functions and then work out the implications of such functions for the electric fields developed among moving ions. A simpler approach will be followed here. This is the relaxation approach. The essence of relaxation analysis is to consider a system in one state, then perturb it slightly with a stimulus and analyze the *time dependence* of the system's response to the stimulus. (It will be seen later that relaxation techniques are much used in modern studies of the mechanism of electrode reactions.)

Consider therefore the spherically symmetrical ionic cloud around a stationary central ion. Now let the stimulus of a driving force displace the reference ion in the x direction. The erstwhile spherical symmetry of the ion atmosphere can be restored only if its contents (the ions and the solvent molecules) *immediately* readjust to the new position of the central ion. This is possible only if the movements involved in restoring spherical symmetry are instantaneous, i.e., if no frictional resistances are experienced in the course of these movements. But the readjustment of the ionic cloud involves ionic movements which are rate processes. Hence, a finite time is required to reestablish spherical symmetry.

Even if this time were available, spherical symmetry would obtain only if the central ion did not move still farther away while the ionic cloud was trying to readjust. Under the influence of the externally applied field, the central ion just keeps moving on, and its ionic atmosphere never quite catches up. It is as though the part of the cloud behind the central ion is left standing. This is because its reason for existence (the field of the central ion) has deserted it and thermal motions try to disperse this part of the ionic cloud. In front of the central ion, the cloud is being continually built up. When ions move therefore, one has a picture of the ions losing the part of the cloud behind them and building up the cloud in front of them.

4.6.3. An Egg-Shaped Ionic Cloud and the "Portable" Field on the Central Ion

The constant lead which the central ion has on its atmosphere means that the center of charge of the central ion is displaced from the center of charge of its cloud. The first implication of this argument is that the ionic cloud is no longer spherically symmetrical around the *moving* central ion. It is egg-shaped (see Fig. 4.87).

A more serious implication is that since the center of charge on a drifting central ion does not coincide with the center of charge of its oppositely charged (egg-shaped) ionic cloud, an electrical force develops between the ion and its cloud. The development of an electrical force between a moving ion and its lagging atmosphere means that the ion is then subject to an electric field. Since this field arises from the continual relaxation (or decay) of the cloud behind the ion and its buildup in front of the ion, it is known as a *relaxation field.* Notice, however, that the centers of charge of the ion and of the cloud lie on the path traced out by the moving ion (Fig. 4.88). Consequently, the relaxation field generated by this charge separation acts in a direction precisely opposite to that of the driving force on the ion (e.g., the externally applied field). Hence, a moving ion, by having an egg-shaped ionic cloud, carries along its own "portable" field of force, the relaxation field, which acts to retard the central ion and decrease its mobility compared with that which it would have were it only pulled on by the externally applied field and retarded by the Stokes force [the zeroth-order theory of conductance; see Eq. (4.183)].

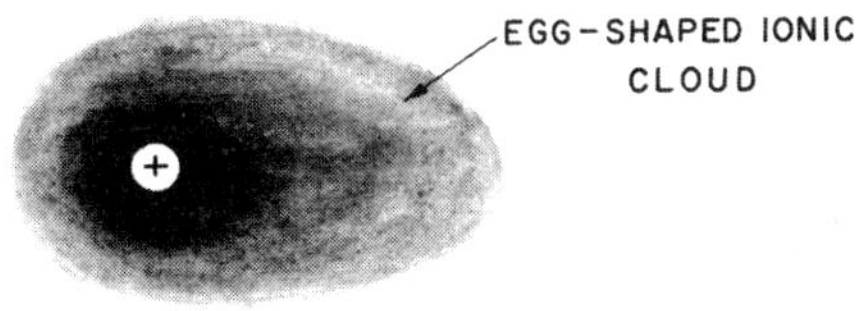

Fig. 4.87. The egg-shaped ionic cloud around a moving central ion.

4.6.4. A Second Braking Effect of the Ionic Cloud on the Central Ion: The Electrophoretic Effect

The externally applied electric field acts not only on the central ion but also on its oppositely charged cloud. Consequently, the ion and its atmosphere tend to move in *opposite directions*.

This poses an interesting problem. The ionic atmosphere can be considered a charge sphere of radius κ^{-1} (Fig. 4.89). The charged sphere moves under the action of an electric field. The thickness of the ionic cloud in a millimolar solution of a 1:1-valent electrolyte is about 10 nm (see Table 3.2). One is concerned therefore with the migration of a fairly large "particle" under the influence of the electric field. The term *electrophoresis* is generally used to describe the migration of particles of colloidal dimensions (1 to 1000 nm) in an electric field. It is appropriate therefore to describe the migration of the ionic cloud as an *electrophoretic effect*.

The interesting point is that when the ionic cloud moves, it tries to carry along its entire baggage: the ions and the solvent molecules constituting the cloud *plus* the central ion. Thus, not only does the moving central ion attract and try to keep its cloud (the *relaxation effect*), but the moving cloud also attracts and tries to keep its central ion by means of a force which is then termed the *electrophoretic force* $\vec{F}_E$.

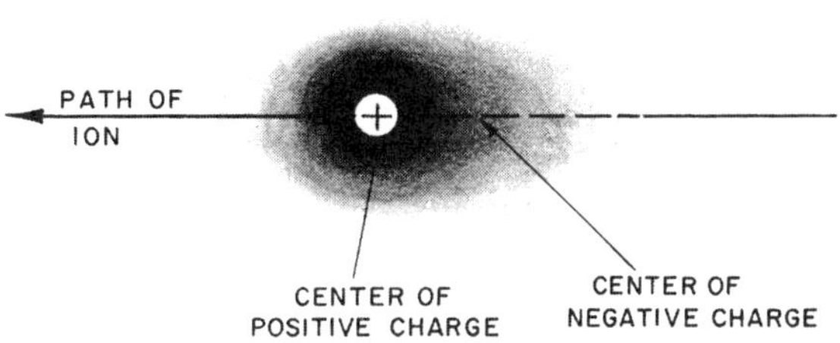

Fig. 4.88. The centers of charge of the ion and of the cloud lie on the path of the drifting ion.

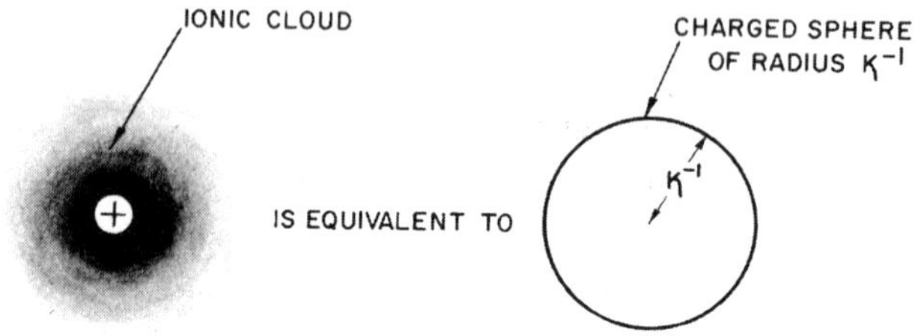

Fig. 4.89. The ionic atmosphere can be considered a charged sphere of radius κ^{-1}.

4.6.5. The Net Drift Velocity of an Ion Interacting with Its Atmosphere

In the elementary treatment of the migration of ions, it was assumed that the drift velocity of an ion was determined solely by the *electric force* $\vec{F}$ arising from the externally applied field. When, however, the mutual interactions between an ion and its cloud were considered, it turned out (Sections 4.6.3 and 4.6.4) that there were two other forces operating on an ion. These extra forces consisted of (1) the *relaxation force* $\vec{F}_R$ resulting from the distortion of the cloud around a moving ion and (2) the *electrophoretic force* $\vec{F}_E$ arising from the fact that the ion shares in the electrophoretic motion of its ionic cloud. Thus, in a rigorous treatment of the migrational drift velocity of ions, one must consider a *total force* $\vec{F}_{\text{total}}$, which is the resultant of force due to the applied electric field together with the relaxation and electrophoretic forces (Fig. 4.90)

$$\vec{F}_{\text{total}} = \vec{F} - (\vec{F}_E + \vec{F}_R) \tag{4.296}$$

The minus sign is used because both the electrophoretic and relaxation forces act in a direction opposite to that of the externally applied field.

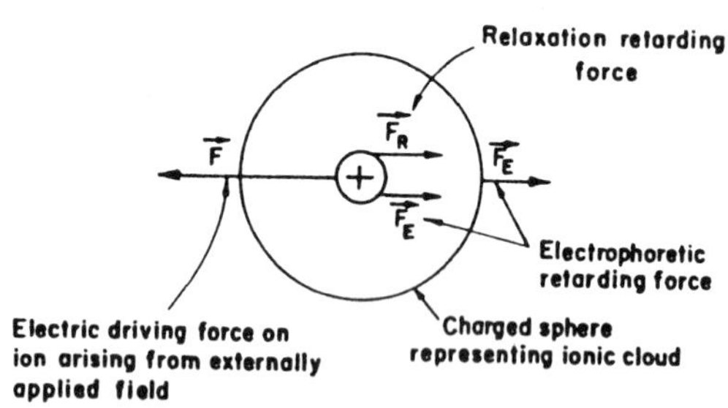

Fig. 4.90. The ion drift due to a net force which is a resultant of the electric driving force and two retarding forces, the relaxation and electrophoretic forces.

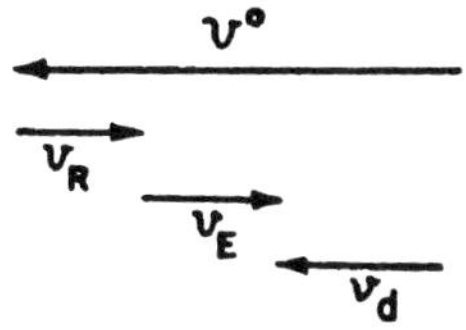

Fig. 4.91. The components of the overall drift volocity.

Since an ion is subject to a resultant or net force, its drift velocity also must be a *net* drift velocity resolvable into components. Furthermore, since each component force should produce a component of the overall drift velocity, there must be three components of the net drift velocity. The first component, which will be designated v^0, is the direct result of the externally applied field only and excludes the influence of interactions between the ion and the ionic cloud; the second is the electrophoretic component v_E and arises from the participation of the ion in the electrophoretic motion of its cloud; finally, the third component is the relaxation field component v_R originating from the relaxation force that retards the drift of the ion. Since the electrophoretic and relaxation forces act in a sense opposite to the externally applied electric field, it follows that the electrophoretic and relaxation components must diminish the overall drift velocity (Fig. 4.91), i.e.,

$$v_d = v^0 - (v_E + v_R) \tag{4.297}$$

The next step is to evaluate the electrophoretic and relaxation components of the net drift velocity of an ion.

4.6.6. Electrophoretic Component of the Drift Velocity

The electrophoretic component v_E of the drift velocity of an ion is equal to the electrophoretic velocity of its ionic cloud because the central ion shares in the motion of its cloud. If one ignores the asymmetry of the ionic cloud, a simple calculation of the electrophoretic velocity v_E can be made.

The ionic atmosphere is accelerated by the externally applied electric force ze_0X but is retarded by a Stokes viscous force. When the cloud attains a steady-state electrophoretic velocity v_E, then the viscous force is exactly equal and opposite to the electric force driving the cloud

$$ze_0X = \text{Stokes' force on cloud} \tag{4.298}$$

The general formula for Stokes' viscous force is $6\pi r\eta v$, where r and v are the radius and velocity of the moving sphere. In computing the viscous force on the cloud, one can substitute κ^{-1} for r and v_E for v in Stokes' formula. Thus,

$$ze_0X = 6\pi\kappa^{-1}\eta v_E \tag{4.299}$$

from which it follows that

$$v_E = \frac{ze_0}{6\pi\kappa^{-1}\eta} X \tag{4.300}$$

This is the expression for the electrophoretic contribution to the drift velocity of an ion.

4.6.7. Procedure for Calculating the Relaxation Component of the Drift Velocity

From the familiar relation [*cf.* Eq. (4.149)],

$$\text{Velocity} = \text{Absolute mobility} \times \text{force}$$

it is clear that the relaxation component v_R of the drift velocity of an ion can be obtained by substituting for the relaxation force $\vec{F}_R$ in

$$v_R = \bar{u}^0_{\text{abs}} \vec{F}_R \tag{4.301}$$

The problem therefore is to evaluate the relaxation force.

Since the latter arises from the distortion of the ionic cloud, one must derive a relation between the relaxation force and a quantity characterizing the distortion. It will be seen that the straightforward measure of the asymmetry of the cloud is the distance d through which the center of charge of the ion and the center of charge of the cloud are displaced.

However, the distortion d of the cloud itself depends on a relaxation process in which the part of the cloud in front of the moving ion is being built up and the part at the back is decaying. Hence, the distortion d and the relaxation force $\vec{F}_R$ must depend on the time taken by a cloud to relax, or decay.

Thus, it is necessary first to calculate how long an atmosphere would take to decay, then to compute the distortion parameter d, and finally to obtain an expression for the relaxation force $\vec{F}_R$. Once this force is evaluated, it can be introduced into Eq. (4.30) for the relaxation component v_R of the drift velocity.

4.6.8. Decay Time of an Ion Atmosphere

An idea of the time involved in the readjustment of the ionic cloud around the moving central ion can be obtained by a thought experiment suggested by Debye (Fig. 4.92). Consider a static central ion with an equilibrium, spherical ionic cloud around it. Let the central ion suddenly be discharged. This perturbation of the ion–ionic cloud system sets up a relaxation process. The ionic cloud is now at the mercy of the thermal

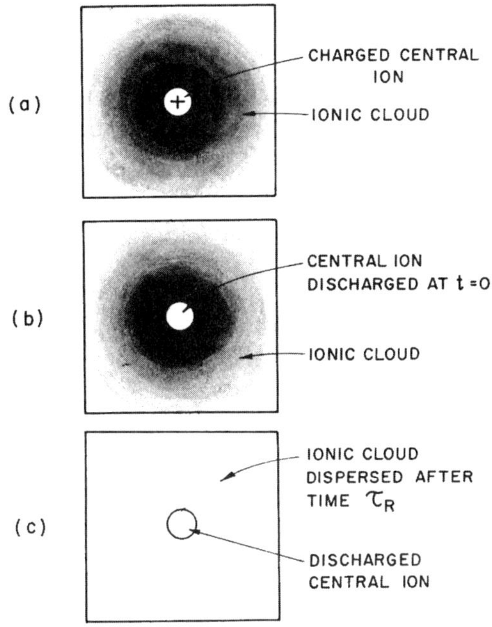

Fig. 4.92. Debye's thought experiment to calculate the time for the ion atmosphere to relax: (a) the ionic cloud around a central ion; (b) at $t = 0$, the central ion is discharged; and (c) after time τ_R, the ion atmosphere has relaxed or dispersed.

forces, which try to destroy the ordering effect previously maintained by the central ion and responsible for the creation of the cloud.

The actual mechanism by which the ions constituting the ionic atmosphere are dispersed is none other than the random-walk process described in Section 4.2. Hence, the time taken by the ionic cloud to relax or disperse may be estimated by the use of the Einstein–Smoluchowski relation (Section 4.2.6)

$$t = \frac{\langle x^2 \rangle}{2D} \tag{4.27}$$

What distance x is to be used? In other words, when can the ionic cloud be declared to have dispersed or relaxed? These questions may be answered by recalling the description of the ionic atmosphere where it was stated that the charge density in a dr-thick spherical shell in the cloud declines rapidly at distances greater than the Debye–Hückel length κ^{-1}. Hence, if the ions diffuse to a distance κ^{-1}, the central ion can be stated to have lost its cloud, and the time taken for this diffusion provides an estimate of the relaxation time τ_R. One has, by substituting κ^{-1} in the Einstein–Smoluchowski relation [Eq. (4.27)]

$$\tau_R = \frac{(\kappa^{-1})^2}{2D} \tag{4.302}$$

which, with the aid of the Einstein relation $D = \bar{u}_{\text{abs}}kT$ [Eq. (4.172)], can be transformed into the expression

$$\tau_R = \frac{(\kappa^{-1})^2}{2\bar{u}_{\text{abs}}kT} \tag{4.303}$$

4.6.9. The Quantitative Measure of the Asymmetry of the Ionic Cloud around a Moving Ion

To know how asymmetric the ionic cloud has become owing to the relaxation effect, one must calculate the distance d through which the central ion has moved in the relaxation time τ_R. This is easily done by multiplying the relaxation time τ_R by the velocity v^0 which the central ion acquires from the externally applied electric force, i.e.,

$$d = \tau_R v^0 \tag{4.304}$$

By substituting the expression (4.303) for the relaxation time τ_R, Eq. (4.304) becomes

$$d = \frac{v^0(\kappa^{-1})^2}{2\bar{u}^0_{\text{abs}}kT} \tag{4.305}$$

The center of charge of the relaxing ionic cloud coincides with the original location of the central ion; in the meantime, however, the central ion and its center of charge move through a distance d. The centers of charge of the central ion and its ionic cloud are displaced through the distance d, which is a quantitative measure of the egg-shapedness of the ion atmosphere around a moving ion.

4.6.10. Magnitude of the Relaxation Force and the Relaxation Component of the Drift Velocity

Consider first a *static* central ion. The ion may exert an electric force on the cloud and vice versa, but at first the net force is zero because of the spherical symmetry of the cloud around the static central ion.

When the central ion moves, it can be considered to be at a distance d from the center of its cloud. The net force due to the asymmetry of the cloud is nonzero. A rough calculation of the force can be made as follows.

The relaxation force is zero when the centers of charge of the ion and its cloud coincide, and it is nonzero when they are separated. So let it be assumed in this approximate treatment that the relaxation force is proportional to d, i.e., proportional to the distance through which the ion has moved from the original center of charge of the cloud. On this basis, the relaxation force $\vec{F}_R$ will be given by the maximum total force of the atmosphere on the central ion, i.e., $z^2e_0^2/[\varepsilon(\kappa^{-1})^2]$,[41] multiplied by the fraction of the radius of the cloud through which the central ion is displaced during its motion under the external field, i.e., d/κ^{-1}. Hence, the relaxation force is

$$\vec{F}_R = \frac{z^2e_0^2}{\varepsilon(\kappa^{-1})^2}\frac{d}{\kappa^{-1}} \tag{4.306}$$

and using Eq. (4.305) for d, one has

$$\vec{F}_R = \frac{z^2e_0^2}{\varepsilon(\kappa^{-1})^2}\frac{1}{\kappa^{-1}}\frac{v^0(\kappa^{-1})^2}{2\bar{u}^0_{\text{abs}}kT}$$

$$= \frac{z^2e_0^2\kappa}{2\varepsilon kT}\frac{v^0}{\bar{u}^0_{\text{abs}}} \tag{4.307}$$

Since, however, the velocity v^0 arises solely from the externally applied field and excludes the influence of ion–ion interactions, the ratio $v^0/\bar{u}^0_{\text{abs}}$ is equal to the applied electric force

$$\frac{v^0}{\bar{u}^0_{\text{abs}}} = \vec{F} = ze_0X \tag{4.308}$$

On inserting this into Eq. (4.307), it turns out that

$$\vec{F}_R = \frac{z^3e_0^3\kappa X}{2\varepsilon kT} \tag{4.309}$$

In the above treatment of the relaxation field, it has been assumed that the only motion of the central ion destroying the spherical symmetry of the ionic cloud is motion

[41]This total force is obtained by considering the ionic cloud equivalent to an equal and opposite charge placed at a distance κ^{-1} from the central ion. Then, Coulomb's law for the force between these two charges gives the result $z^2e_0^2/[\varepsilon(\kappa^{-1})^2]$.

in the direction of the applied external field. This latter directed motion is in fact a drift superimposed on a random walk. The random walk is itself a series of motions, and these motions are random in direction. Thus, the central ion exercises an erratic, rather than a consistent, leadership on its atmosphere.

Onsager considered the effect that this erratic character of the leadership would have on the time-averaged shape of the ionic cloud and therefore on the relaxation field. His final result differs from Eq. (4.309) in two respects: (1) Instead of the numerical factor $\frac{1}{2}$, there is a factor $\frac{1}{3}$; and (2) a correction factor $\omega/2z^2$ has to be introduced, the quantity ω being given by

$$\omega = z_+ z_- \frac{2q}{1+\sqrt{q}} \tag{4.310}$$

in which

$$q = \frac{z_+ z_-}{z_+ + z_-} \frac{\lambda_+ + \lambda_-}{z_+\lambda_+ + z_-\lambda_-} \tag{4.311}$$

where λ_+ and λ_- are related to mobilities of cation and anion, respectively. For symmetrical or z:z-valent electrolytes, the expression for q reduces to $\frac{1}{2}$, and that for ω becomes (Table 4.18)

$$\omega = z^2 \frac{1}{1+(1/\sqrt{2})} \tag{4.312}$$

Thus, a more rigorous expression for the relaxation force is

$$\vec{F}_R = \frac{z^3 e_0^3 \kappa}{3\varepsilon kT} \frac{\omega}{2z^2} X$$

$$= \frac{z e_0^3 \kappa \omega}{6\varepsilon kT} X \tag{4.313}$$

TABLE 4.18

Values of ω for Different Types of Electrolytes

Type of Electrolyte	ω
1:1	0.5859
2:2	2.3436

Substituting the expression (4.313) for the relaxation force in Eq. (4.301) for the relaxation component of the drift velocity, one gets

$$v_R = \overline{u}^0_{\text{abs}} \frac{ze_0^3 \kappa \omega}{6\varepsilon kT} X \tag{4.314}$$

Furthermore, from the definition (4.152) of the conventional mobility,

$$u^0 = u^0_{\text{conv}} = \overline{u}^0_{\text{abs}} ze_0$$

Eq. (4.314) becomes

$$v_R = \frac{u^0}{ze_0} \frac{ze_0^3 \kappa \omega}{6\varepsilon kT} X = \frac{u^0 e_0^2 \kappa \omega}{6\varepsilon kT} X \tag{4.315}$$

4.6.11. Net Drift Velocity and Mobility of an Ion Subject to Ion–Ion Interactions

Now that the electrophoretic and relaxation components in the drift velocity of an ion have been evaluated, they can be introduced into Eq. (4.297) to give

$$\begin{aligned} v_d &= v^0 - (v_E + v_R) \\ &= v^0 - \left(\frac{ze_0 \kappa}{6\pi\eta} X + \frac{u^0 e_0^2 \kappa \omega}{6\varepsilon kT} X \right) \\ &= v^0 - \left(\frac{ze_0}{6\pi\eta} + \frac{u^0 e_0^2 \omega}{6\varepsilon kT} \right) \kappa X \end{aligned}$$

If one divides throughout by X, then according to the definition of the conventional mobility $u_{\text{conv}} = u = v/X$, one has

$$u = u^0 - \left(\frac{ze_0}{6\pi\eta} + \frac{u^0 e_0^2 \omega}{6\varepsilon kT} \right) \kappa \tag{4.316}$$

An intelligent inspection of expression (4.316) shows that the mobility u of ions is not a constant independent of concentration. It depends on the Debye–Hückel reciprocal length κ. But this parameter κ is a function of concentration (see Eq. 3.84). Hence, Eq. (4.316) shows that the mobility of ions is a function of concentration, as was suspected (Section 4.6.1) on the basis of the empirical law of Kohlrausch.

As the concentration decreases, κ^{-1} increases and κ decreases, as can be seen from Eq. (3.84). In the limit of infinite dilution ($c \to 0$), $\kappa^{-1} \to \infty$ or $\kappa \to 0$. Under these conditions, the second and third terms in Eq. (4.316) drop out, which leaves

$$u_{\text{limit},c\to 0} = u^0 \tag{4.317}$$

The quantity u^0 is therefore the mobility at infinite dilution and can be considered to be given by the expression for the Stokes mobility (Section 4.4.8), i.e.,

$$u^0 = \frac{ze_0}{6\pi r\eta} \tag{4.183}$$

To go back to the question that concluded the previous section, it is clear that since transport numbers depend on ionic mobilities, which have been shown to vary with concentration, the transport number must itself be a concentration-dependent quantity (Table 4.19). However, it is seen that this variation is a small one.

4.6.12. The Debye–Hückel–Onsager Equation

The equivalent conductivity Λ of an electrolytic solution is simply related to the mobilities of the constituent ions [Eq. (4.163)]

$$\Lambda = F(u_+ + u_-)$$

Thus, to obtain the equivalent conductivity, one has only to write down the expression for the mobilities of the positive and negative ions, multiply both the expressions by the Faraday constant F, and then add up the two expressions. The result is

$$\Lambda = F(u_+ + u_-)$$

TABLE 4.19

Transport Numbers of Cations in Aqueous Solutions at 298 K

Concentration	HCl	LiCl	NaCl	KCl
0.01 N	0.8251	0.3289	0.3918	0.4902
0.02	0.8266	0.3261	0.3902	0.4901
0.05	0.8292	0.3211	0.3876	0.4899
0.1	0.8314	0.3168	0.3854	0.4898
0.2	0.8337	0.3112	0.3821	0.4894
0.5	—	0.300	—	0.4888
1.0	—	0.287	—	0.4882

$$= F\left[u_+^0 - \kappa\left(\frac{z_+e_0}{6\pi\eta} + \frac{e_0^2\omega}{6\varepsilon kT}u_+^0\right)\right] \tag{4.318}$$

$$+ F\left[u_-^0 - \kappa\left(\frac{z_-e_0}{6\pi\eta} + \frac{e_0^2\omega}{6\varepsilon kT}u_-^0\right)\right]$$

For a symmetrical electrolyte, $z_+ = z_- = z$ or $z_+ + z_- = 2z$, and Eq. (4.318) reduces to

$$\Lambda = F(u_+^0 + u_-^0) - \left[\frac{ze_0F\kappa}{3\pi\eta} + \frac{e_0^2\omega\kappa}{6\varepsilon kT}F(u_+^0 + u_-^0)\right] \tag{4.319}$$

However, according to Eq. (4.292)

$$\Lambda^0 = F(u_+^0 + u_-^0) \tag{4.292}$$

Hence,

$$\Lambda = \Lambda^0 - \left(\frac{ze_0F\kappa}{3\pi\eta} + \frac{e_0^2\omega\kappa}{6\varepsilon kT}\Lambda^0\right) \tag{4.320}$$

Replacing κ by the familiar expression (3.84), i.e.,

$$\kappa = \left(\frac{8\pi z^2e_0^2c}{\varepsilon kT}\right)^{1/2}\left(\frac{N_A}{1000}\right)^{1/2} \tag{3.84}$$

one has

$$\Lambda = \Lambda^0 - \left[\frac{ze_0F}{3\pi\eta}\left(\frac{8\pi z^2e_0^2N_A}{1000\varepsilon kT}\right)^{1/2} + \frac{e_0^2\omega}{6\varepsilon kT}\left(\frac{8\pi z^2e_0^2N_A}{1000\varepsilon kT}\right)^{1/2}\Lambda^0\right]c^{1/2} \tag{4.321}$$

This is the well-known Debye–Hückel–Onsager equation for a symmetrical electrolyte. By defining the following constants

$$A = \frac{ze_0F}{3\pi\eta}\left(\frac{8\pi z^2e_0^2N_A}{1000\varepsilon kT}\right)^{1/2} \quad \text{and} \quad B = \frac{e_0^2\omega}{6\varepsilon kT}\left(\frac{8\pi z^2e_0^2N_A}{1000\varepsilon kT}\right)^{1/2}$$

it can also be written

$$\Lambda = \Lambda^0 - (A + B\Lambda^0)c^{1/2}$$

$$= \Lambda^0 - \text{constant}\ c^{1/2} \tag{4.322}$$

Thus, the theory of ionic clouds has been able to give rise to an equation that has the same form as the empirical law of Kohlrausch (Section 4.3.9).

4.6.13. Theoretical Predictions of the Debye–Hückel–Onsager Equation versus the Observed Conductance Curves

The two constants

$$A = \frac{ze_0 F}{3\pi\eta}\left(\frac{8\pi z^2 e_0^2 N_A}{1000\varepsilon kT}\right)^{1/2} \tag{4.323}$$

and

$$B = \frac{e_0^2 \omega}{6\varepsilon kT}\left(\frac{8\pi z^2 e_0^2 N_A}{1000\varepsilon kT}\right)^{1/2} \tag{4.324}$$

in the Debye–Hückel–Onsager equation are completely determined (Table 4.20) by the valence type of the electrolyte z, the temperature T, the dielectric constant ε, and the viscosity η of the solution, and universal constants.

The Debye–Hückel–Onsager equation has been tested against a large body of accurate experimental data. A comparison of theory and experiment is shown in Fig. 4.93 and Table 4.21 for aqueous solutions of true electrolytes, i.e., substances that consisted of ions in their crystal lattices before they were dissolved in water. At very low concentrations ($< 0.003\ N$), the agreement between theory and experiment is very good. There is no doubt that the theoretical equation is a satisfactory expression for the *limiting tangent* to the experimentally obtained Λ versus $c^{1/2}$ curves.

One cannot, however, expect the Debye–Hückel–Onsager theory of the nonequilibrium conduction properties of ionic solutions to fare better at high concentration than the corresponding Debye–Hückel theory of the equilibrium properties (e.g.,

TABLE 4.20

Values of the Onsager Constants for Uni-Univalent Electrolytes at 298 K

Solvent	A	B
Water	60.20	0.229
Methyl alcohol	156.1	0.923
Ethyl alcohol	89.7	1.83
Acetone	32.8	1.63
Acetonitrile	22.9	0.716
Nitromethane	125.1	0.708
Nitrobenzene	44.2	0.776

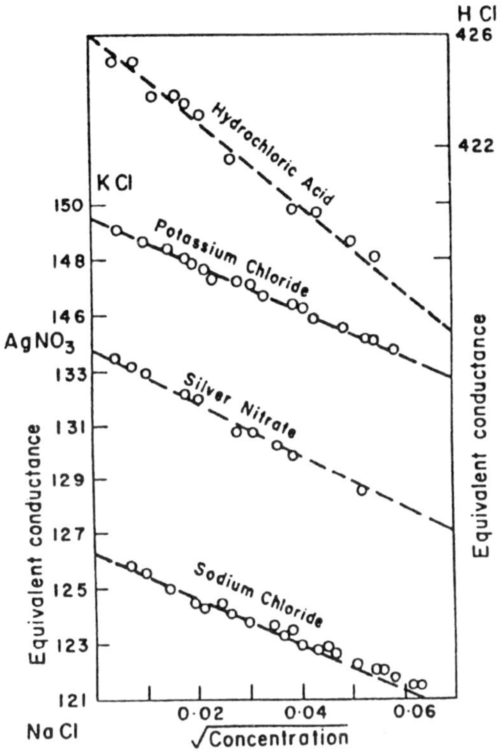

Fig. 4.93. Comparison of the equivalent conductivities of HCl and some salts predicted by the Debye–Hückel–Onsager equation (4.321) with those observed experimentally.

TABLE 4.21
Observed and Calculated Onsager Slopes in Aqueous Solutions at 298 K

Electrolyte	Observed Slope	Calculated Slope
LiCl	81.1	72.7
$NaNO_3$	82.4	74.3
KBr	87.9	80.2
KCNS	76.5	77.8
CsCl	76.0	80.5
$MgCl_2$	144.1	145.6
$Ba(NO_3)_2$	160.7	150.5
K_2SO_4	140.3	159.5

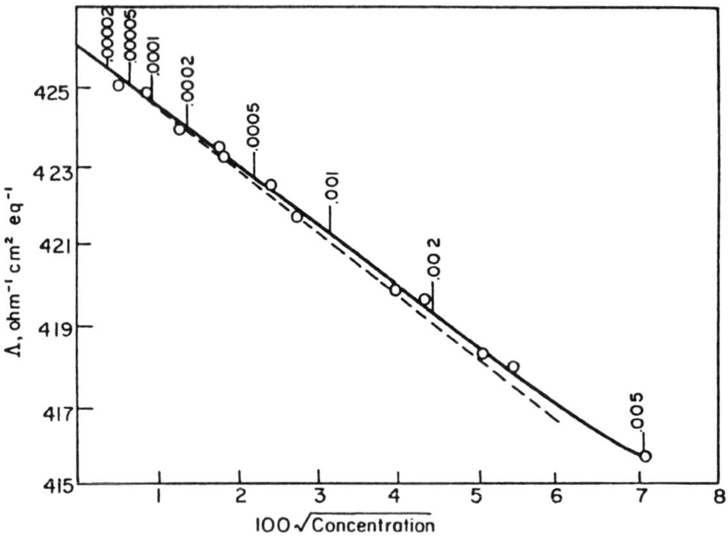

Fig. 4.94. Deviation of the predicted equivalent conductivities from those observed for HCl.

activity coefficients of electrolytic solutions); both theories are based on the ionic-cloud concept. In the case of the Debye–Hückel–Onsager equation, it is seen from Fig. 4.94 that as the concentration increases (particularly above 0.003 *N*), the disparity between the theoretical and experimental curves widens.

4.6.14. Changes to the Debye–Hückel–Onsager Theory of Conductance

The original Debye–Hückel–Onsager theory of conductance of ions in solution takes into account the two ionic atmosphere effects in reducing the mobility and therefore the lessening of the conductivity of the ions that occurs as their concentration is increased (see Section 4.6.12). The original theory formed a peak in the physical chemistry of the first half of the twentieth century, but it was first published some three-quarters of a century ago. Moreover, as indicated in Fig. 4.94, the original theory showed deviations from experiments even at really low aqueous concentrations. In nonaqueous solutions, where the dielectric constants are usually lower than the 80 one finds for water, the interionic attractions are higher (interionic forces are inversely proportional to the dielectric constant of the solution), and hence the deviations from the zeroeth approximation of no interionic interactions are greater. For this reason the deviation between theory and experiment begins at even lower concentrations than it does in aqueous solutions. Of course, if one removes some of the primitivities of the early theory (e.g., neglecting the fact that ions have finite size), then the concentration range for the applicability of the theory widens.

Apart from improvements made by taking into account the fact that ions do indeed take up some of the space in electrolytic solutions, one has to consider also that *ion association* occurs in true electrolytes,

$$Na^+ + Cl^- \rightleftharpoons [Na^+Cl^-]^0 \tag{4.325}$$

and the associated ionic molecule is a dipole, not an ion, and therefore is no longer in the running as far as contributing to the conductance is concerned. This was allowed for in an empirical equation due to Justice,

$$\frac{\Lambda_m}{\alpha} = \Lambda_m^0 - S\sqrt{\alpha c} + E\alpha c \ln(\alpha c) + J_1 \alpha c - J_2(\alpha c)^{3/2} \tag{4.326}$$

where α is the degree of association of the electrolyte, S is the Onsager limiting-law coefficient, c is the electrolyte concentration, and E, J_1, and J_2 are constants. The parameter α is related to an association constant K_A by

$$\frac{(1-\alpha)f_A}{\alpha^2 c f_\pm^2} = K_A \tag{4.327}$$

where f_A is the activity coefficient of the dipole. The $f_\pm$ is given by the relevant expression found in the activity coefficient of the Debye–Hückel theory (see Section 3.4.4).

What is the use of an empirical equation such as Eq. (4.326)? It acts as a hanger for the facts. One fits experimental data of Λ_m as a function of c to the equation and determines by a least-squares fitting procedure the values of Λ_m^0 and K_A.[42]

These empirical modifications of the Debye–Hückel–Onsager theory of electrolytes do not yet give much *physical* insight into what changes in the elderly (but still famous) theory might improve the theory of ionic conductance. A more relevant improvement can be attributed to Fuoss and to Lee and Wheaton. Instead of thinking about bare ions traveling in a structureless dielectric medium, these authors have taken the ion to have three regions, as shown in Fig. 4.95. In the first of these regions, that nearest to the ion, the water molecules are regarded as being "totally oriented" to the ion, so that their effective dielectric constant would be that of water dielectrically

[42]In his derivation, Justice suggested that the distance of closest approach a involved in the determination of $f_\pm$ [*cf.* Eq. (3.120)] be replaced by the Bjerrum parameter q. While q is the distance of closest approach of the unpaired ions that contribute to the conductance of the solution, ions separated by distances between q and a are ion-paired and do not contribute to the conductance. The advantage of this approximation is that a is not known *a priori*, whereas q is defined by Eq. (3.144).

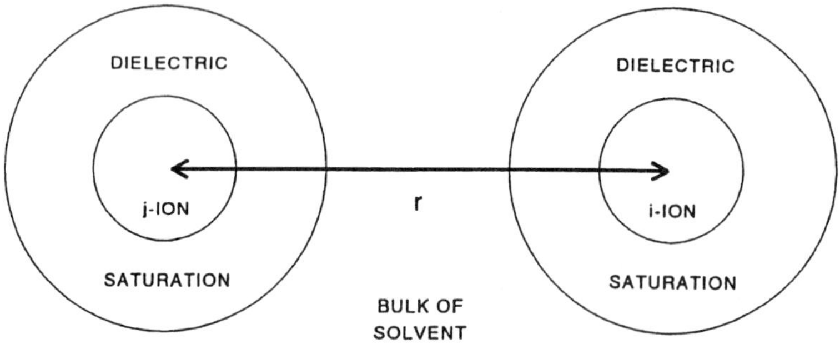

Fig. 4.95. Regions of dielectric saturation about *i*- and *j*-ions according to Lee and Wheaton (W. H. Lee and R. J. Wheaton, *J. Chem. Soc. Faraday Trans. II* **75**: 1128, 1979).

saturated, i.e., having all its orientation polarization used up by the electric field of the ion itself. In the second region, there is only partial dielectric saturation. This region is part of the secondary hydration. Outside these first two regions, the influence of the ion's field is taken to be only the long-range Coulomb forces generally used in arguments about the properties of ionic solutions.[43]

On the basis of this model, Lee and Wheaton arrived at an equation for Λ_m in terms of q, the Bjerrum distance. The equation is several lines long and clearly only fit for use in appropriate software. The application of experimental data of Λ_m to the equation allows one to find values of Λ_m^0, K_A, and the co-sphere radius R. These values are then taken as if they had been experimentally determined, an assumption that is true only in a secondary way because they depend on the validity of Lee and Wheaton's equation.

Another approach to the conductance of electrolytes, which is less complex than that of Lee and Wheaton, is due to Blum and his co-workers. This theory goes back to the original Debye–Hückel–Onsager concepts, for it does not embrace the ideas of Lee and Wheaton about the detailed structure around the ion. Instead, it uses the concept of mean spherical approximation of statistical mechanics. This is the rather portentous phrase used for a simple idea, which was fully described in Section 3.12. It is easy to see that this is an approximation because in reality an ionic collision with another ion will be softer than the brick-wall sort of idea used in an MSA approach. However, using MSA, the resulting mathematical treatment turns out to be relatively simple. The principal equation from the theory of Blum et al. is correspondingly simple and can be quoted. It runs

---

[43]These regions have been met before (Section 3.6.2) in discussions of recent models for finding activity and osmotic coefficients. They correspond to the 1949 models for primary and secondary hydration of Bockris [*cf.* the Gurney co-sphere (1971)].

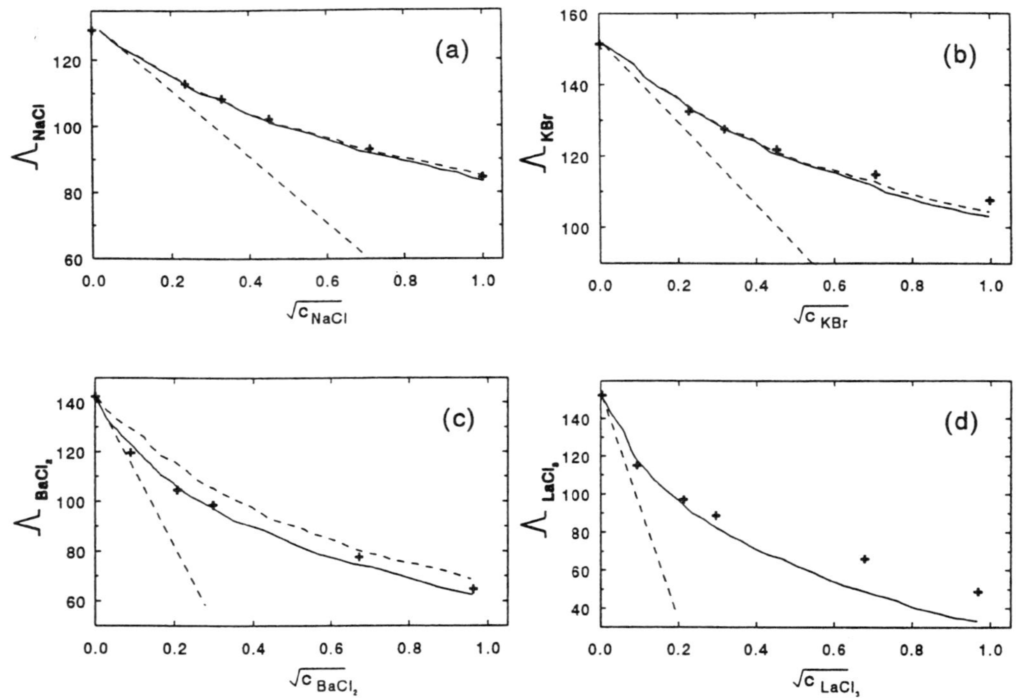

Fig. 4.96. Equivalent conductivity of aqueous electrolyte solutions as a function of the square root molarity: (+), experimental values; (—●—●—), Debye–Hückel–Onsager limiting law; (- - -) and (—), theoretical values predicted by Blum's approach. (a) Aqueous NaCl solutions. (b) Aqueous KBr solutions. (c) Aqueous $BaCl_2$ solutions. (d) Aqueous $LaCl_3$ solutions. (Reprinted from O. Bernard, W. Kunz, P. Turq, and L. Blum, *J. Phys. Chem.***96**: 3833, 1992.)

$$\Lambda_m = \sum_i \lambda_i^0 \left(1 + \frac{v_i^E}{v_i^0}\right)\left(1 + \frac{\Delta E}{E}\right) \tag{4.328}$$

with v_i^E, the drift velocity under a field E, defined as

$$v_i^E = \frac{z_i e_0 E D_i^0}{kT} \tag{4.329}$$

where E is the electric field applied across the solution, D_i^0 is the diffusion coefficient of the ion i, and λ_i^0 is the individual conductance of the ion i at infinite dilution.

The application of Blum's theory to experiment is unexpectedly impressive: it can even represent conductance up to 1 mol dm^{-3}. Figure 4.96 shows experimental data and both theories—Blum's theory and the Debye–Hückel–Onsager first approximation. What is so remarkable is that the Blum equations are able to show excellent agreement with experiment without taking into account the solvated state of the ion, as in Lee and Wheaton's model. However, it is noteworthy that Blum stops his comparison with experimental data at 1.0 *M*.

Blum's use of the MSA represents a significant advance, but it does not take into account either ionic association or Bjerrum's very reasonable idea (Section 3.8) about the removal of free water in the solution by means of hydration. Furthermore, Blum's equations do not explain the relation between conductance and concentration noted for many electrolytes, particularly at high concentrations, that is,

$$\Lambda_m = \Lambda_m^0 - Ac^{1/3} \tag{4.330}$$

There are some who see this equation as indicative that a whole different approach to conductance theory might be waiting in the wings, as it were. As the concentration increases, the idea of an ionic atmosphere becomes less useful and one might start at the other end, with ideas used to treat molten salts (Chapter 5), but in a diluted form. This would repeat the history of the theory of liquids which, in the early part of this century, was derived from the treatment of very compressed gases but later seemed to be more developable from modifications of how solids are treated.

4.7. DIVERSE RELAXATION PROCESSES IN ELECTROLYTIC SOLUTIONS

4.7.1. Definition of Relaxation Processes

The term *relaxation*, as applied in physical chemistry (*cf.* Section 4.6.8), refers to molecular processes occurring after the imposition of a stress on a system. Thus, one can have a system at equilibrium to which a new constraint is applied (e.g., an electric field switched on suddenly onto a dipole-containing liquid). The system is then constrained to a new position of equilibrium. The time it takes to change position is called the *relaxation time.*

To understand better what is meant by relaxation time, consider a system under the following equilibrium

$$A \underset{\overleftarrow{k}'}{\overset{\overrightarrow{k}'}{\rightleftharpoons}} B \tag{4.331}$$

The rate of change of A is given by

$$\frac{dc_A}{dt} = -\overrightarrow{k}'c_A' + \overleftarrow{k}'c_B' = 0 \tag{4.332}$$

Upon the imposition of a constraint, the system reaches a new equilibrium. Moreover, the constraint changes the rate constants that control the interconversion of states A and B. During the first equilibrium, these were $\overrightarrow{k}'$ and $\overleftarrow{k}'$, and they become $\overrightarrow{k}$ and $\overleftarrow{k}$ under the new conditions. While A is adjusting (relaxing) to its new equilibrium value, its concentration changes from c_A' to $c_A + x$ until it reaches its new equilibrium state, c_A. In the same way, under the constraint, the concentration of B changes to the new value c_B, but during the relaxation it is $c_B - x$. The concentration of A then changes as follows

$$\frac{dc_A}{dt} = -\overrightarrow{k}(c_A + x) + \overleftarrow{k}(c_B - x) \tag{4.333}$$

and after reaching the new equilibrium state,

$$\frac{dc_A}{dt} = -\overrightarrow{k}c_A + \overleftarrow{k}c_B = 0 \tag{4.334}$$

Therefore, Eq. (4.333) becomes

$$\frac{dc_A}{dt} = \frac{dx}{dt} = -(\overrightarrow{k} + \overleftarrow{k})x \tag{4.335}$$

Integrating from a time 0 when $x = x_0$, to a time t when $x = x_t$,

$$x_t = x_0 e^{-(\overrightarrow{k}+\overleftarrow{k})t} \tag{4.336}$$

The change of the concentration of A as it goes from the preconstraint equilibrium value x_0 to the after-constraint equilibrium value x_t follows the exponential change indicated by Eq. (4.336) and shown in Fig. (4.97).

What is called the relaxation time τ is a somewhat arbitrary quantity: it is taken to be the time that makes the exponential in Eq. (4.336) unity, i.e.,

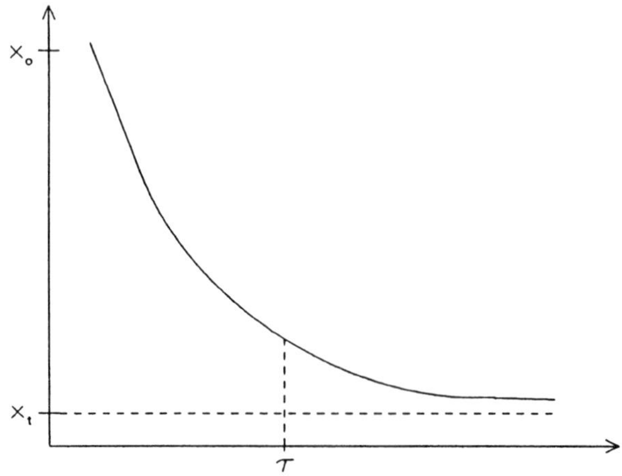

Fig. 4.97. Concentration change of species *A* from its preconstraint equilibrium value x_0 to the after-constraint equilibrium value x_t.

$$\tau = \frac{1}{\vec{k} + \overleftarrow{k}} \tag{4.337}$$

Hence, at the relaxation time,

$$\frac{x_\tau}{x_0} = e^{-1} = 0.368 \tag{4.338}$$

A trivial calculation shows that Eq. (4.338) implies that x at time τ changed some 63% of the way from the initial value of x_0 to the final value of x_t. Thus, τ is some *measure* of the time to complete the change caused by the constraint just as the half-life of a radioactive element is a measure of the lifetime of a radioactive element.[44] In the next subsections, the relaxation times of certain systems will be presented.

4.7.2. Dissymmetry of the Ionic Atmosphere

One kind of relaxation time has already been discussed in Section 4.6.8, namely, the time of adjustment to dissymmetry of the ionic atmosphere around an ion when an applied electric field is switched on. Its understanding is basic to our picture of ionic

[44]And for the same sort of reason. The change in both cases follows an exponential law and the rate of change slows down toward completion of the change so that formally the change is never quite finished.

solutions. Thus, in the absence of an electric field, the ions randomly jump about (or shuffle around) in all directions. When an electric field forces them to move preferentially in one direction, the ionic atmosphere around the ions becomes dissymmetric—egg-shaped—so that the longer part of the egg is behind the ion and, having more counter charge behind it than in front, retards the ion's forward motion.

An interesting effect in the conductivity of ions is related to the relaxation time of their ionic atmosphere. Imagine that the electrical conductance of ionic solutions is measured by using alternating currents of a certain frequency, let's say 1000 counts per second.[45] Imagine also the ions moving in the solution and following the dictates of the oscillating electric field. Since this is constantly altering its direction, for a millisecond it pulls the cations to the right and then for a millisecond pulls them to the left, with analogous movements but in opposite directions being forced upon the anions.

The frequency of 1000 s^{-1} has been mentioned because it is typical for measurement of electrical conductivity. To perform the measurement, the researcher varies the frequency (ν) and plots the corresponding measured resistance values against $1/\nu$, extrapolating the measured resistance in the ordinate to "infinite frequency."

Suppose, however, that instead of making conductance measurements at, say, 10, 100, 1000, and 10,000 s^{-1}, and performing the said extrapolation to infinite frequency, one goes on increasing the frequency past 100,000 s^{-1}. Think again now of a given ion and its atmosphere in the system. As the frequency increases, the ionic atmosphere has a harder time keeping the changes in its dissymmetry in tune with the changes in direction of drift of the ions—moving now to the right and now to the left. Below about 10^7 s^{-1}, the ionic atmosphere manages to adjust its shape every time the ion changes direction and present the appropriate asymmetric stance, slowing down the ionic movements.

Eventually there is a critical frequency above 10^7 s^{-1} at which the ionic cloud cannot adjust anymore to the ion's movements in the right way because there is too much inertia to execute the rapid changes required by the oscillating applied field. The reciprocal of this critical frequency is called the *relaxation time of the asymmetry of the ionic cloud*. As a consequence, an increase in conductivity occurs at this frequency because there is no longer more charge behind the ion than in front. This increase in conductance at the critical frequency is called the *Debye effect*. It is part of the evidence that shows that the ionic atmosphere is indeed present and functioning according to the way first calculated by Debye.

---

[45]The reason for this is that with a direct current, an unwanted ionic layer forms at the interfaces with the solution of the electrodes used to make contact with the outside power source. These nonequilibrium structures at the surface–solution boundaries create a new resistance that interferes with the solution resistance one is trying to measure. This is wiped out if an alternating current, which keeps on reversing the structure at the interface, is applied.

4.7.3. Dielectric Relaxation in Liquid Water

Water is not only the most abundant of all liquids on this planet but also one of the most complex (see Section 2.4). The dielectric constant (permittivity) at frequencies below about 10^{11} s^{-1} is about 78 at room temperature, and this is about one whole order of magnitude higher than the dielectric constant of simple liquids such as carbon tetrachloride. Kirkwood was the first to develop a model to explain why the dielectric constant of water is so high. He pictured groups of H_2O's coupled together by means of H bonding. His idea was that the dielectric constant of water consists of three parts.

First, there would be the part based on the distortion of the electronic shells of the atoms making up water molecules. Because their inertia is so small, electrons have no difficulty in keeping up with an applied field as its frequency increases. Such a contribution is part of the permittivity of any liquid.

The second part can be viewed as the distortion of the nuclei of the atoms making up water—how much the applied field disturbs the positions of the nuclei in the O and H atoms of molecular water. This part is also present in all liquids.

The third contribution involves the dipole moment of the individual molecules. In water and associated liquids, the dipoles should be taken in groups as a result of the intermolecular H bonding (Fig. 4.98). It is this coupling of the molecules that provides the huge permittivity of water.

Were water a simple unassociated dipolar liquid, the effect of an applied field would be simply to orient it, to inhibit its random libration and bend the average

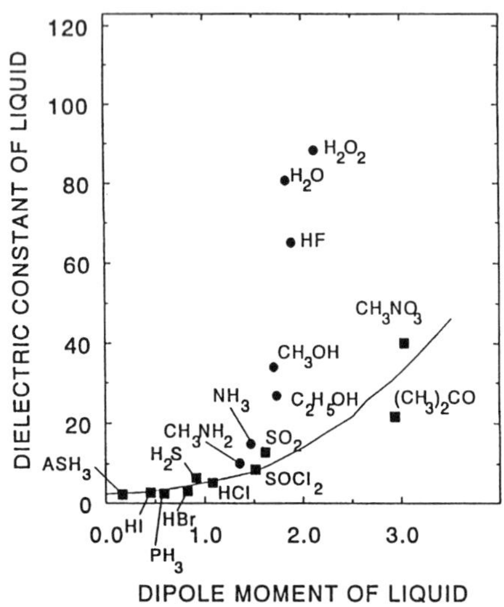

Fig. 4.98. The dielectric constants of liquids as a function of their dipole moments (■ = unassociated liquids, and ● = H-bonded liquids).

direction of the molecules to lie more along the lines of the field. With water, the molecular dipoles also feel the force of the field. The special feature is that each of the water molecules, being bound together with H bonds, pulls other waters with it as it bonds to the field. When one water molecule goes to obey the field and align with it, the ones to which it is bound get a pull from the waters to which they are bound and this helps them turn also. Thus, in an associated liquid one gets a double whammy from the oscillating applied field. It pulls each dipolar molecule around. This is an effect on the individual dipoles that one would find with any dipolar liquid, but there is also a correlated additive effect due to the orienting effect of one dipole on another.

Now, what of relaxation? According to the picture drawn here (Fig. 4.99), there should be two relaxation times. The first will be that corresponding to the state in which the dipoles can no longer react in consonance with the applied field. At some critical high frequency, the permittivity falls (loss of the biggest contributor—the netted dipolar groups), and the dielectric constant falls from 78 to ~5 at 298 K. The value of the associated time, that is, the reciprocal of the critical frequency of the applied ac field at which the permittivity falls, is 8.32×10^{-12} s, which corresponds to a frequency of 1.2×10^{11} s^{-1}.

At frequencies above 10^{11} s^{-1}, the next thing to be exceeded is the speed at which the nuclei in the molecules can react to changes in the direction of the field. The protons in the nuclei have an inertia approximately 2000 times greater than that of the electrons in the outer shell and accordingly a relaxation time much less than that of the electron shells. This value for water is 1.02×10^{-12} s, and the critical frequency is 9.8×10^{11} s^{-1}. The remaining permittivity at frequencies higher than this is due to distortion of the electron shell of the atoms. This last and most fundamental permittivity is often called the *optical* permittivity because it pertains to movements in the liquid (distortion of the electron shells), which occur near the speed of light).

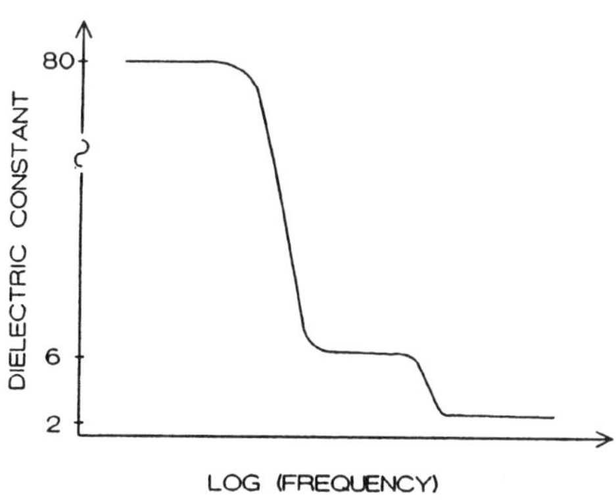

Fig. 4.99. Variation of the dielectric constant as a function of the frequency of an applied ac electric field.

TABLE 4.22

Dielectric Parameters of Several Polar Solvents Showing Discontinuous Relaxation Time Distributions

Solvent	ε_s	ε_∞	n_D^2	(τ_1/ps)	(τ_2/ps)	(τ_3/ps)
Water	78.36	4.49	1.7765	8.32		1.02
Methanol	32.63	2.79	1.7596	51.15	7.09	1.12
Ethanol	24.35	2.69	1.8473	163	8.97	1.81
Formamide	109.5	4.48	2.096	37.3		1.16
N-Methylformamide	185.98	3.20	2.0449	128	7.93	0.78
N,N-Dimethylformamide	37.24	2.94	2.0400	—	10.4	(0.76)
Acetonitrile	35.92	2.26	1.7800	—	3.21	—
Propylene carbonate	64.96	4.14	2.0153	—	43.1	—

Source: Reprinted from P. Turq, J. Barthel, and M. Chemla, *Transport, Relaxation and Kinetic Processes in Electrolyte Solutions*, Springer-Verlag, Heidelberg, 1992, p. 81.

Much information can be obtained from the study of dielectric relaxation times in liquids, for example, the extent to which the waters are netted together (Table 4.22).

4.7.4. Effects of Ions on the Relaxation Times of the Solvents in Their Solutions

There are three new "effects" related to the properties of relaxation time that arise when ions are added to water.

First, the solution's relaxation time appears to change. If solvent molecules are far away, say more than 1000 pm distant from an ion, the ion's effect on the relaxation time will be negligible. Conversely, water molecules bound to ions will be what is called *dielectrically saturated*; they will be so tightly held in the ion's local electric field that they will not be affected by the applied electric field used to measure the dielectric constant of the solution. The average relaxation time of all the waters will be increased, because the water molecules attached to the ions now have, in effect, an infinite relaxation time.

The second effect is related to the formation of ion pairs. If ion pairs or other ionic aggregates are present, they will introduce a new relaxation time above that exhibited by the pure solvent.

Figure 4.100 shows the Argand diagram† of water (curve 1) and the permittivity for 0.8 *M* KCl (curve 2) in water. The "structural" part of the spectrum is represented by curve 3. The difference of curves 2 and 3 is the result of electrolytic conductance.

†An Argand diagram (also called a Cole-Cole plot) is a diagram of the real ε' and imaginary ε'' components of the dielectric constant of the system.

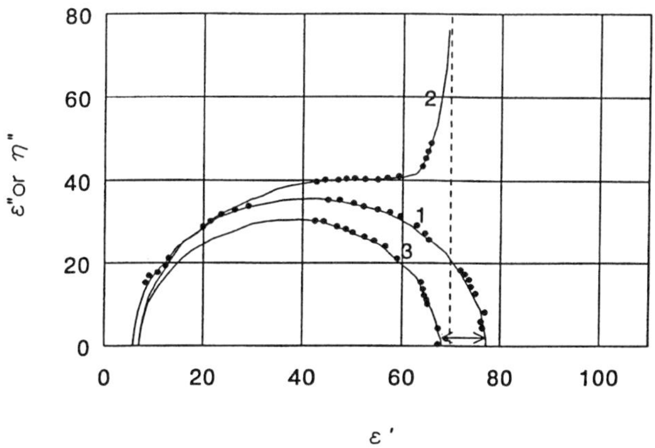

Fig. 4.100. Argand diagrams of a completely dissociated electrolyte and its pure solvent. Full circles: experimental data from frequency domain measurements on aqueous potassium chloride solutions at 25 °C. Curve 1: Argand diagram of pure water. Curve 2: Argand diagram, $\theta'' = f(\varepsilon')$, of an 0.8 *M* aqueous KCl solution, Curve 3: Argand diagram, $\varepsilon'' = f(\varepsilon')$, obtained from curve 2. (Reprinted from P. Turq, J. Barthel, and M. Chemla, in *Transport, Relaxation and Kinetic Processes in Electrolyte Solutions*, Springer-Verlag, Berlin, 1992, p. 78).

The permittivity of ionic solutions is less than that of the pure solvent and decreases linearly with an increase in concentration. The reason for this has already been discussed (Section 2.12.1): water dipoles held by the very strong local field of an ion cannot orient against the weak applied field used in measuring the dielectric constant. The average is therefore decreased.

The linear relation found between dielectric constant and concentration can be interpreted in a first approximation as the result of a number of "irrotationally bound" waters. Such waters would constitute the primary hydration water referred to in Section 2.4.

Further Reading

Seminal

1. P. Debye and E. Hückel, "The Interionic Attraction Theory of Deviations from Ideal Behavior in Solution," *Z. Phys.* **24**: 185 (1923).
2. L. Onsager, "Interionic Attraction Theory of Conductance," *Z. Phys.* **28**: 277 (1927).
3. R. M. Fuoss, "Conductance in Aqueous Solutions," *Proc. Natl. Acad. Sci. U.S.A.* **75**: 16 (1928).

4. R. M. Fuoss and C. A. Krauss, "Dimer and Trimer Formation in Ionic Solution," *J. Am. Chem. Soc.* **55**: 476 (1933).

Review

1. M. Chabanel, "Ionic Aggregates of 1-1 Salts in Non-aqueous Solutions: Structure, Thermodynamic and Solvations," *Pure Appl. Chem.* **62**: 35 (1990).

Papers

1. J. Barthel, H. Hetzenauer, and R. Buchner, *Ber. Bunsenges. Phys. Chem.* **96**: 988 (1992).
2. D. D. K. Chingakule, P. Gans, J. B. Gill, and P. L. Longdon, *Monatsh. Chem.* **123**: 521 (1992).
3. H. Ohshma, *J. Colloid Interface Sci.* **168**: 269 (1994).
4. M. Bachelin, P. Gans, and J. B. Gill, in *Chemistry of Nonaqueous Solutions: Current Progress*, G. Mamantov and A. I. Popov, eds., pp. 1–148, VCH Publishers, New York (1994).
5. M. R. Gittings and D. A. Saville, *Langmuir* **11**: 798 (1995).
6. A. R. Volkel and J. Noolandi, *J. Chem. Phys.* **102**: 5506 (1995).

4.8. NONAQUEOUS SOLUTIONS: A NEW FRONTIER IN IONICS

4.8.1. Water Is the Most Plentiful Solvent

Not only is water the most plentiful solvent, it is also a most successful and useful solvent. There are several facts that support this description. First, the dissolution of true electrolytes occurs by solvation (Chapter 2) and therefore depends on the free energy of solvation. A sizable fraction of this free energy depends on electrostatic forces. It follows that the greater the dielectric constant of the solvent, the greater is its ability to dissolve true electrolytes. Since water has a particularly high dielectric constant (Table 4.23), it is a successful solvent for true electrolytes.

A second advantage of water is that in addition to being able to dissolve electrolytes by the physical forces involved in solvation, it is also able to undergo *chemical* proton-transfer reactions with potential electrolytes and produce ionic solutions. Water is able to donate protons to, and to receive protons from, molecules of potential electrolytes. Thus, water can function as both a source and a sink for protons and consequently can enter into ion-forming reactions with a particularly large range of substances. This is why potential electrolytes often react best with water as a partner in the proton-transfer reactions. Finally, water is stable both chemically and physically at ambient temperature, unlike many organic solvents which tend to evaporate (Table 4.24) or decompose slowly with time.

On the whole, therefore, ionics is best practiced in water. Nevertheless, there are also good reasons why nonaqueous solutions of electrolytes are often of interest.

TABLE 4.23
Dielectric Constants of Some Solvents (Temperature 298 K unless Otherwise Noted)

Solvent	Dielectric Constant
Water	78.30
Acetone	20.70
Acetonitrile	36.70
Ammonia (239 K)	22.00
Benzene	2.27
Dimethylacetamide	37.78
Dimethyl sulfoxide	46.70
Dioxane	2.21
Ethanol	24.30
Ethylenediamine	12.90
Hydrogen cyanide (289 K)	118.30
Pyridine	12.00
Sulfuric acid	101.00

4.8.2. Water Is Often Not an Ideal Solvent

If water were the ideal solvent, there would be no need to consider other solvents. However, in many situations, water is hardly the ideal solvent. Take the electrolytic production of sodium metal, for example. If an aqueous solution of a sodium salt is taken in an electrolytic cell and a current is passed between two electrodes, then all that will happen at the cathode is the liberation of hydrogen gas; there will be no electrodeposition of sodium (see Chapter 7). Hence, sodium cannot be electrowon from aqueous solutions. This is why the electrolytic extraction of sodium has taken place from molten sodium hydroxide, i.e., from a medium free of hydrogen. This process requires the system to be kept molten (~ 600 °C) and therefore requires the

TABLE 4.24
Boiling Points of Some Solvents

Solvent	Boiling point (K)
Water	373.15
Acetone	329.35
Benzene	353.25
1,1-Dichloroethane	330.15
Methanol	338.11

use of high-temperature technology with its associated materials problems. It would be a boon to industry if one could use a low-temperature conducting solution having the capacity to maintain sodium ions in a nonaqueous solvent free of ionizing hydrogen. This argument is valid for many other metals which are extracted today by electrodic reactions in fused salts at high temperatures, with the attendant difficulties of corrosion and heat losses.

Another field attracting the development of nonaqueous electrochemistry is that of energy storage for automobiles. Many reasons (e.g., the growing danger of pollution from automobile exhausts; the increasing concentration of CO_2, with its consequences of planetary warming; and the accelerating consumption of oil reserves) make the search for an alternative to the internal combustion engine a necessity. Nuclear reactors with their attendant shielding problems will always be too heavy for the relatively small power needed in road vehicles. There would be attractive advantages to a zero-emission, vibration-free electric power source. However, the currently available cheap electrochemical storage device—the lead–acid battery—is too heavy for the electric energy it needs to store to offer a convenient distance range between recharging. The electrochemical energy storers available today that do have a sufficiently high energy capacity per unit weight are expensive. The highest energy density theoretically conceivable is in a storage device that utilizes the dissolution of lithium or beryllium. However, aqueous electrolytes are debarred because in them, these metals corrode wastefully rather than dissolving with useful power production. So one answer to the need for an electrochemically powered transport system is the development of a nonaqueous electrochemical energy storage system that incorporates alkali metal (in particular lithium) electrodes. Many other examples could be cited in which the use of water as a solvent is a nuisance. In all these cases, there may be an important future for applications using nonaqueous solutions.

A nonaqueous solution must be able to conduct electricity if it is going to be useful. What determines the conductivity of a nonaqueous solution? Here, the theoretical principles involved in the conductance behavior of *true* electrolytes in nonaqueous solvents will be sketched. However, before that, let the pluses and minuses of working with nonaqueous solutions (particularly those involving organic solvents) be laid out.

4.8.3. More Advantages and Disadvantages of Nonaqueous Electrolyte Solutions

First, compared with aqueous electrolytes, nonaqueous solutions generally are liquid over a larger temperature range; this may include temperatures below 273 K, which is useful for applications in cold climates. They allow the electroplating of substances that would be unstable in aqueous solutions, such as aluminum, beryllium, silicon, titanium, and tungsten. One can even plate high-temperature superconductors and oxidize and reduce many organic and inorganic materials in electrosyntheses as long as one uses organic solvents. (The reason for the preference for nonaqueous solutions here is the absence of protons and OH^- ions. These tend to undergo preferential electrode reactions, evolving, respectively, hydrogen and oxygen. Such reactions compete with the intended organic reaction and often dominate it. In a nonaqueous solutions, the "window of opportunity," the potential range in which something useful may be done, is as much as 6 V.) Nonaqueous solutions are useful in the electromachining of metals and, as mentioned earlier, in high-energy-density batteries, which eventually could allow the use of these batteries in automobiles.

On the other hand, the drawbacks of nonaqueous solutions include their lower conductivities and their toxicity and flammability. They need extreme purification and handling under a highly purified inert-gas atmosphere. They may not be exposed to the atmosphere because they will pick up water, which may give rise to the undesired co-deposition of hydrogen.

The most important concentration range of conductivity studies for these electrolytes is below 10^{-3} mol dm^{-3}. Their most determined enemy is water, which acts as a contaminant. If one considers that 20 ppm of water is equivalent to a 10^{-3} mol dm^{-3} solution of water in a nonaqueous solvent, it is no surprise that electrochemical quantities recorded in the literature are much less precise than those for aqueous solutions. Conductivities that are said to be as precise as $\pm 1\%$ are often $\pm 10\%$ in the nonaqueous literature. With materials that react with water (e.g., Li and Na) the water level has to be cut to less than 0.05 ppm and kept there; otherwise a monolayer of oxide forms on the metals' surfaces.

Drying of nonaqueous solutions can be carried out by various methods which include the addition of sodium, in the form of wires, or powdered solids such as barium oxide; alternatively, a high potential difference between auxciliary electrodes can be used to getter the H^+ and OH^- ions. The quality of the nonaqueous work in electrochemistry is increasing rapidly.

4.8.4. The Debye–Hückel–Onsager Theory for Nonaqueous Solutions

An examination of the Debye–Hückel–Onsager theory in Section 4.6.12 together with the recent developments described in Section 4.6.14 shows that the developments are in no way wedded to water as a solvent. Does experiment support the predicted Λ versus $c^{1/2}$ curve in nonaqueous solutions also?

Figure 4.101 shows the variation of the equivalent conductivity versus concentration for a number of alkali sulfocyanates in a methanol solvent. The agreement with the theoretical predictions demonstrates the applicability of the Debye–Hückel–Onsager equation up to at least 2×10^{-3} mol dm^{-3}.

When one switches from water to some nonaqueous solvent, the magnitudes of several quantities in the Debye–Hückel–Onsager equation alter, sometimes drastically, even if one considers the same true electrolyte in all these solvents. These quantities are the viscosity and the dielectric constant of the medium, and the distance of closest approach of the solvated ions (i.e., the sum of the radii of the solvated ions). As a result, the mobilities of the ions at infinite dilution, the slope of the Λ versus

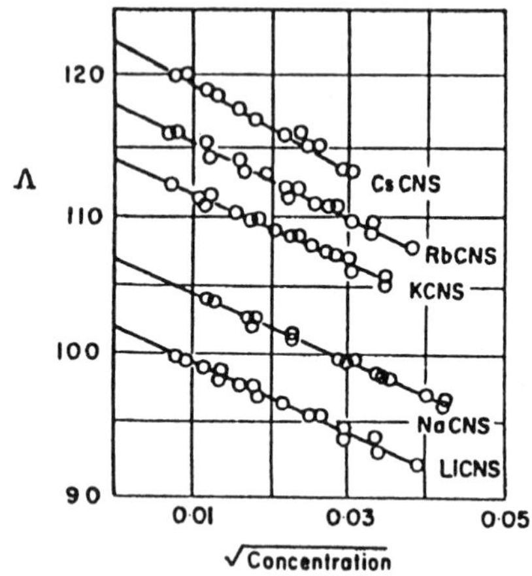

Fig. 4.101. Change in the equivalent conductivity of some alkali sulfocyanates with concentration in methyl alcohol.

$c^{1/2}$ curve, and finally, the concentration of *free* ions cause the conductance behavior of an electrolyte to vary when one goes from water to nonaqueous solvent. Before we get into equations which will deal with such effects to some degree, it is useful to look at the data available for nonaqueous solutions.

Among the goals of this brief survey is overcoming a common prejudice—that nonaqueous solutions are always less conductive than aqueous ones. A clear example is that of highly concentrated electrolyte solutions such as 5 *M* $LiBF_4$ in 1,2-dimethoxyethane (DME), which has a conductivity comparable to that in aqueous solutions at the same temperature.

4.8.5. What Type of Empirical Data Are Available for Nonaqueous Electrolytes?

4.8.5.1. *Effect of Electrolyte Concentration on Solution Conductivity.* It has been seen that reliable conductivity values are known only at low electrolyte concentrations. Under these conditions, even conductance equations for models such as the McMillan–Mayer theory (Sections 3.12 and 3.16) are known. However, the empirical extension of these equations to high concentration ranges has not been successful. One of the reasons is that conductivity measurements in nonaqueous solutions are still quite crude and literature values for a given system may vary by as much as 50% (doubtless due to purification problems).

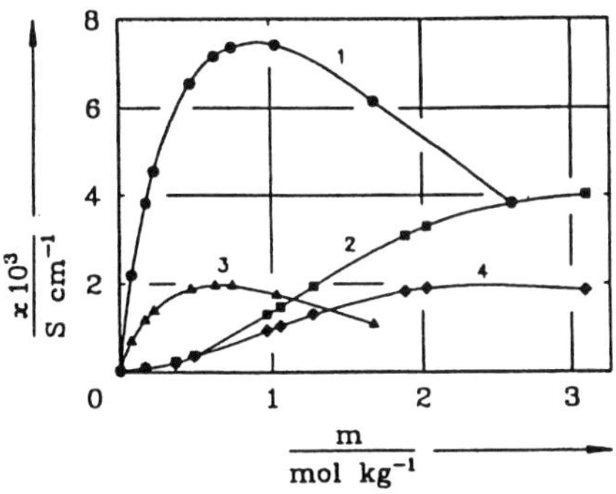

Fig. 4.102. Conductivity of $LiBF_4$ in γ-butyrolactone at (1) 298 K and (3) 238 K, and in tetrahydrofuran at (2) 298 K and (4) 238 K (M. Kindler, Dissertation, Regensburg, 1985).

One characteristic of conductivity curves when studied in a wide concentration range is the appearance of maxima, as shown in Fig. 4.102. These maxima are always observed when the solubility of the electrolyte in the given solvent is sufficiently high. They are the result of the competition between the increase of conductivity due to the

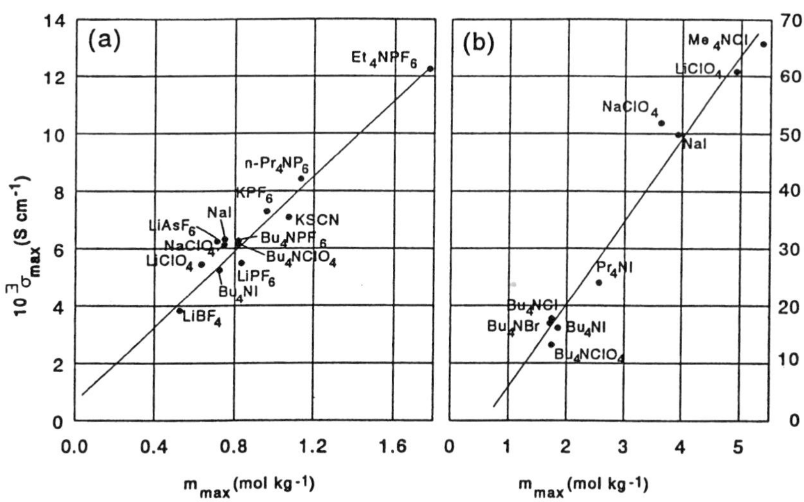

Fig. 4.103. Dependence of σ_{max} vs. c_{max} for various salts in (a) propylene carbonate and (b) methanol at 298 K (J. Barthel and H.-J. Gores, *Pure Appl. Chem.* **57**: 1071, 1985).

increase of free ions and the decrease of conductivity due to the decrease of ionic mobility as the electrolyte concentration increases. A linear relation has been found (Fig. 4.103) between the maxima plotted for specific conductivity and the concentration at which the maxima are observed.

Another type of behavior observed when plotting conductivities versus $c^{1/2}$ is the presence of temperature-dependent minima and maxima, as shown in Fig. 4.104. This unusual behavior has been attributed to triplet-ion formation (see Section 4.8.11). In this case Walden's rule (Section 4.4.12) has been used to calculate the values of equivalent conductivities at infinite dilution Λ^0.

4.8.5.2. *Ionic Equilibria and Their Effect on the Permittivity of Electrolyte Solutions.* Most of the commonly used solvents exhibit several relaxation processes that show up in the change of dielectric constant with frequency (see Section 2.12). These relaxation processes include rotation and libration of the molecules of the solvents, aggregates of ionic species, and H-bonding dynamics.

Relaxation times and dispersion amplitudes[46] change when ions are added. If ion pairs are formed, a new relaxation region appears on the solvent relaxation spectrum on the low-frequency side. Figure 4.105 shows the dielectric absorption spectrum of LiBr in acetonitrile, and how a maximum is developed in the low-frequency region as the concentration of solute increases and ion pairs are formed. Association constants can be determined from these data and contribute to the identification of the ion pair present.

4.8.5.3. *Ion–Ion Interactions in Nonaqueous Solutions Studied by Vibrational Spectroscopy.* Conventional methods to determine ion association measure a single property of the bulk solution, that is, an average of the interactions occurring over the time of the measurement. Microwave absorption studies exemplify such methods to determine solvation and ion association by studying, e.g., dielectric relaxation phenomena (see Section 2.12).

On the other hand, techniques that give information on the particular ion–ion and ion–solvent interactions would be of great help in the electrochemistry of nonaqueous solutions. Such help can be obtained from the various vibrational spectroscopic techniques, which are able to probe specific species in solution.

Raman spectra are related to the concentration of the species that give rise to them and offer a tool by which one may perform quantitative evaluations of ion-pair equilibria. For example, the ion association constant for ion pairing between Ag^+ and

---

[46]The dielectric constant (or relative permittivity) is an important property in the study of electrolyte conductivity of solutions and their solvents. However, the measurable quantity is the frequency-dependent complex permittivity, $\varepsilon(\omega)$. The static permittivity (dielectric constant) is obtained by extrapolation to zero frequency of the frequency-dependent complex permittivity $\varepsilon(\omega) = \varepsilon'(\omega) - \iota\varepsilon''(\omega)$, and both relaxation times and dispersion amplitudes can be obtained from these variables. The real part $\varepsilon'(\omega)$ yields the dispersion curve, and the imaginary part $\varepsilon''(\omega)$ the absorption curve of the dielectric spectrum.

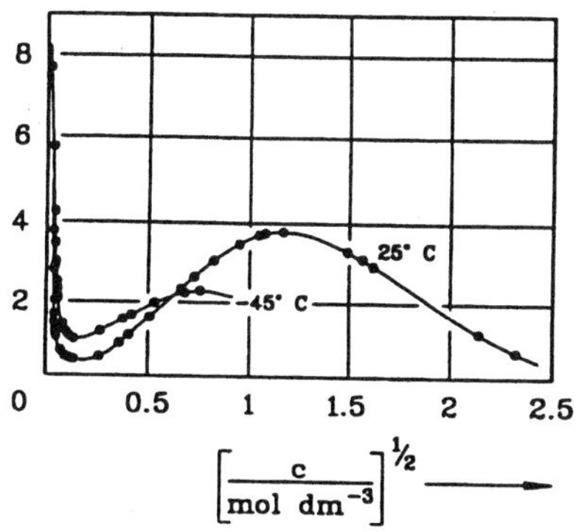

Fig. 4.104. Molar conductivity of $LiBF_4$ solutions in 1,2-dimethoxyethane at –45 and 25 °C, showing negative temperature coefficients at moderate concentrations (J. Barthel and H.-J. Gores, "Solution Chemistry: A Cutting Edge in Modern Electrochemical Technology," in *Chemistry of Nonaqueous Solutions: Current Progress*, G. Mamantov and A. I. Popov, eds., VCH Publishers, New York, 1994).

NO_3^- in acetonitrile has been obtained by Janz and Müller. Associated structures of ions have been studied in nonaqueous solvents over a wide range of dielectric constants. LiCNS in solvents of low dielectric constant, such as ethers and thioethers, gives rise to several different types of ion aggregates. Many different types of contact ion pairs or agglomerates have been identified, and the role the solvent has in this association—whether the solvent separates the ions or not—has been determined. The Bjerrum critical distance, that is, the distance at which the ion is able to interact with other ions to form ion-pair structures (see Section 4.8.8), is of great use in these types of studies. Table 4.25 shows some values for 1:1, 2:2, and 3:3 electrolytes in different solvents.

It might be expected that NMR would be the ultimate technique for the identification of ion pairs in nonaqueous solutions because of its specificity for differentiating resonances of nuclei in ions in different environments. Several studies that involve ^7Li, ^{13}C, and ^{17}O among others, have been examined satisfactorily. Nevertheless, NMR also has drawbacks, such as a lack of well-defined spectra, which make interpretation difficult. The main reason seems to be the short lifetimes of the complexes (the ion pair lasts probably less than the data acquisition time, which is on the order of 10^{-6} s), which allow only small chemical shifts to be detected.

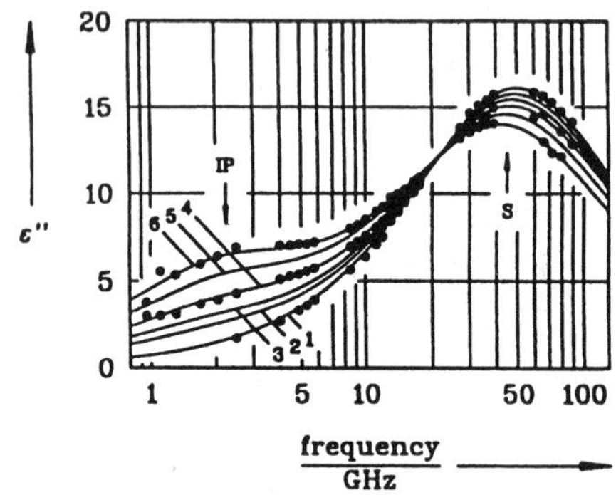

Fig. 4.105. Dielectric absorption spectrum (imaginary part of the complex permittivity, ε'') of LiBr solutions in acetonitrile at 25 °C. 1, Pure solvent; 2, 0.107 *M*; 3, 0.194 *M*; 4, 0.303 *M*; 5, 0.479 *M*; 6, 0.657 *M*. S and IP indicate the frequency regions of the relaxation processes of solvent and solute. For the sake of clarity, experimental data (•) are added only for curves 1, 4, and 6 (J. Barthel, H. Hetzenauer, and R. Buchner, *Ber. Bunsenges. Phys. Chem.* **96**: 988, 1992).

TABLE 4.25
Bjerrum Critical Distances for Three Solvents

	Temperature (K)	Permittivity (ε)	Bjerrum Critical Distances (pm) 1:1	2:2	3:3
Water	298	78.3	357	1428	3,213
Acetonitrile	298	36.7	3044	6849	
Liquid ammonia	203	26.3	1062	4248	9,558
	230	22.4	1248	4992	11,232
	298	16.9	1654	6616	14,886
	350	13.5	2071	8284	18,640

Source: Reprinted from J. B. Gill, "The Study of Ion–Ion Interactions in Electrolyte Solutions by Vibrational Spectroscopy," in *Chemistry of Nonaqueous Solutions: Current Progress*, G. Mamantov and A. I. Popov, eds., VCH Publishers, New York, 1994, p. 203.

It seems likely that improvements in the technology of the various spectroscopic techniques will make it possible to increase our knowledge of ion association in nonaqueous solutions. Knowledge of parameters such as $\Delta H_{\text{association}}$ and $\Delta S_{\text{association}}$ is becoming indispensable for understanding ionic association processes. For example, it is possible that the driving force to form ion pairs is a reorganizational one; that is, it is entropic and not enthalpic in nature, as might have been expected at first.

4.8.5.4. *Liquid Ammonia as a Preferred Nonaqueous Solvent.* Liquid ammonia has been widely used as a nonaqueous solvent in the study of ion-pair association. One of its advantages is the large range of dielectric constants—from 12 to 26—when the temperature is changed over 200 degrees. In contrast, the dielectric constant of water does not change throughout its normal liquid temperature range enough to move it out of the high dielectric constant range.

Furthermore, many salts are highly soluble in NH_3 because of the high solvation energies it makes possible. Finally, from the spectroscopic point of view, the small ammonia molecule has few molecular vibrations, which eases the interpretation of spectra observed in it—a characteristic difficult to find in other (more complex) solvents.

The first evidence of ion pairing in liquid ammonia came from a study of nitrate solutions by means of Raman spectroscopy. A number of bands larger than the ones expected for "free" nitrate ions were observed. A full understanding of these bands

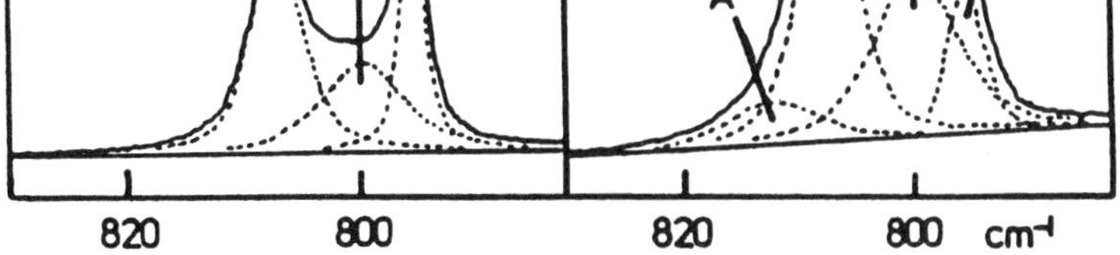

Fig. 4.106. The $\nu_2(NO_2)$ stretching region of the Raman spectra of sodium nitrite solutions at about 1.0 *M* in liquid ammonia at 293 K. The second spectrum is one of a solution in which $[NaClO_4]/[NaNO_2] = 4$. Band A, either ion triplet $[Na^+ \cdot NO_2^- \cdot Na^+]^+$ or contact ion pair containing bidentate NO_2^-; bands B and C, contact ion pairs $[Na^+ \cdot NO_2^-]^0$ and $[Na^+ \cdot ONO^-]^0$; band E, "free" solvated anion $[NO_2(NH_3)_x]^-$ (P. Gans and J. B. Gill, *J. Chem. Soc. Faraday Discuss.* **64**: 150, 1978).

has not been accomplished yet, but researchers agree that the observed changes in symmetry are due to ion association. Figure 4.106 illustrates these Raman spectra.

4.8.5.5. *Other Protonic Solvents and Ion Pairs.* Apart from information for the solvent liquid ammonia, little spectroscopic data are available on this topic. The reason seems to be the great complexity of spectra when solvents like methanol or ethanol are used.

4.8.6. The Solvent Effect on Mobility at Infinite Dilution

At infinite dilution, neither relaxation nor electrophoretic effects are operative on the drift of ions; both these effects depend for their existence on a finite-sized ionic cloud. Under these special conditions, the infinite-dilution mobility can be considered to be given by the Stokes mobility

$$u^0_{\text{conv}} = \frac{ze_0}{6\pi r\eta} \tag{4.183}$$

Considering the same ionic species in several solvents, one has

$$u^0_{\text{conv}} r\eta = \text{constant} \tag{4.339}$$

If the radius r of the solvated ion is independent of the solvent, then one can approximate Eq. (4.339) to[47]

$$u^0\eta = \text{constant} \tag{4.340}$$

Hence, an increase in the viscosity of the medium leads to a decrease in the infinite-dilution mobility and vice versa.

Tables 4.14 and 4.15 contains data on the equivalent conductivity, the viscosity, and the Walden constant $\Lambda^0\eta$ for two electrolytes and several solvents. It is seen that (1) Eq. (4.340) is a fair approximation for many solvents and (2) its validity is better for solvents other than water.

In fact, the radius of the kinetic entity may change in going from one solvent to another because of changes in the structure of the solvation sheath. Sometimes these solvation effects on the radius may be as much as 100%. Hence, it is only to a rough degree that one can use the approximate equation (4.340).

In some cases, the changes in the radii of the solvated ions are mainly due to the changes in the sizes of the solvent molecules in the solvation sheath. Thus, in the case of water, methanol, and ethanol, the size of the three solvent molecules increases in the order

---

[47]Actually, Eq. (4.340) containing the mobility is a form of Walden's rule [*cf.* Eq. (4.196)], which contains the equivalent conductivity. Since $\Lambda^0 = F[(u_{\text{conv}})^0_+ + (u_{\text{conv}})^0_-]$—*cf.* Eq. (4.292)—it is easy to transform Eq. (4.340) into the usual form of Walden's rule.

$$\text{Water} \rightarrow \text{methanol} \rightarrow \text{ethanol}$$

Since the radius of the solvated ions should also increase in the same order, it follows from Eq. (4.339) that the mobility or equivalent conductivity at infinite dilution should increase from ethanol to methanol to water. This is indeed what is observed (see Table 4.15).

One should be careful in using a simple Walden's rule $\Lambda\eta$ = constant, which assumes that the radii of the moving ions are independent of the solvent. Rather, one should use a generalized Walden's rule, namely,

$$u^0 r\eta = \text{constant} \tag{4.339}$$

where r is the radius of the ionic entity concerned in a given solvent.

4.8.7. Slope of the Λ versus $c^{1/2}$ Curve as a Function of the Solvent

If one takes the generalized Walden's rule (4.339) and calculates (from $\Lambda^0 = Fu^0$) the equivalent conductivity at *infinite dilution* for a number of nonaqueous solutions, it turns out that the values of Λ^0 in such solutions are relatively high. They are near those of water and are in some cases greater than those of water.

One might naively conclude from this fact that in using nonaqueous solutions instead of aqueous solutions in an electrochemical system, the conductivity presents no problem. Unfortunately, this is not the case. The crucial quantity that often determines the feasibility of using nonaqueous solutions in practical electrochemical systems is the *specific conductivity* σ *at a finite concentration*, not the equivalent conductivity Λ^0 at infinite dilution. The point is that it is the specific conductivity which, in conjunction with the electrode geometry, determines the electrolyte resistance R in an electrochemical system. This electrolyte resistance is an important factor in the operation of an electrochemical system because the extent to which useful power is diverted into the wasteful heating of the solution depends on I^2R, where I is the current passing through the electrolyte; hence, R must be reduced or the σ increased.

Now, σ is related to the equivalent conductivity Λ at the same concentration

$$\sigma = \Lambda zc \tag{4.341}$$

but Λ varies with concentration; this is what the Debye–Hückel–Onsager equation (4.321) was all about. To understand the specific conductivity σ at a concentration c, it is not enough to know the equivalent conductivity under a hypothetical condition of infinite dilution. One must be able to calculate,[48] for the nonaqueous solution, the

[48]Of course, the validity of the calculation depends upon whether the theoretical expression for the equivalent conductivity (e.g., the Debye–Hückel–Onsager equation) is valid in the given concentration range.

equivalent conductivity at finite concentrations, utilizing the Λ^0 value and the theoretical slope of the Λ versus $c^{1/2}$ curve. This will be possible if one knows the values of the constants A and B in the Debye–Hückel–Onsager equation

$$\Lambda = \Lambda^0 - (A + B\Lambda^0)c^{1/2}$$

where

$$A = \frac{ze_0F}{3\pi\eta}\left(\frac{8\pi z^2 e_0^2}{\varepsilon kT}\right)^{1/2}\left(\frac{N_A}{1000}\right)^{1/2}$$

and

$$B = \frac{e_0^2\omega}{6\varepsilon kT}\left(\frac{8\pi z^2 e_0^2}{\varepsilon kT}\right)^{1/2}\left(\frac{N_A}{1000}\right)^{1/2}$$

When one looks at the above expressions for A and B, it becomes obvious that as ε decreases, A and B increase; the result is that Λ and therefore σ decrease. Physically, this corresponds to the stronger interionic interactions arising as ε is reduced.

So the question of the specific conductivity of nonaqueous solutions *vis-à-vis* aqueous solutions hinges on whether the dielectric constant of nonaqueous solvents is lower or higher than that of water. Table 4.23 shows that many nonaqueous solvents have ε's considerably lower than that of water. There are some notable exceptions, namely, the hydrogen-bonded liquids.

Thus, because of the lower dielectric constant values, the effect of an increase of electrolyte concentration on lowering the equivalent conductance is much greater in nonaqueous than in aqueous solutions. The result is that the specific conductivity of nonaqueous solutions containing practical electrolyte concentrations is far less than the specific conductivity of aqueous solutions at the same electrolyte concentration (Table 4.26 and Fig. 4.107).

In summary, it is the lower dielectric constants of the typical nonaqueous solvent that cause a far greater decrease in equivalent conductivity with an increase of concentration than that which takes place in typical aqueous solutions over a similar concentration range. Even if the infinite-dilution value Λ^0 makes a nonaqueous electrochemical system look hopeful, the practically important values of the specific conductivity (i.e., the ones at real concentrations) are nearly always much less than those in the corresponding aqueous solution. That is another unfortunate aspect of nonaqueous solutions, to be added to the difficulty of keeping them free of water in ambient air.

TABLE 4.26

Specific Conductivities of Electrolytes in Aqueous and Nonaqueous Solvents at the Same Concentration

Electrolyte	Concentration (mol dm^{-3})	Specific Conductivity, ohm^{-1} cm^{-1} at 298 K		
		Water	Nonaqueous Solvent	
HCl	0.1	391.32×10^{-4}	(methanol)	122.5×10^{-4}
			(ethanol)	35.43×10^{-4}
NaCl	0.01	11.85×10^{-4}	(methanol)	7.671×10^{-4}
KCl	0.01	14.127×10^{-4}	(methanol)	8.232×10^{-4}

4.8.8. Effect of the Solvent on the Concentration of Free Ions: Ion Association

The concentration c that appears in the Debye–Hückel–Onsager equation pertains only to the free ions. This concentration becomes equal to the analytical concentration (which is designated here as c_a) only if every ion from the ionic lattice from which the electrolyte was produced is stabilized in solution as an independent mobile charge carrier; i.e., $c \neq c_a$ if there is ion-pair formation. Whether ion-pair formation occurs depends on the relative values of a, the distance of closest approach of oppositely charged ions, and the Bjerrum parameter $q = (z_+ z_- e_0^2/2kT)\, 1/\varepsilon$. When $a < q$, the condition for ion-pair formation is satisfied and when $a > q$, the ions remain free.

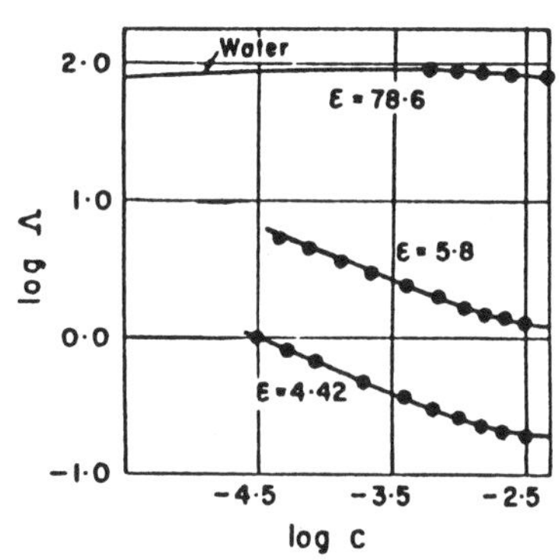

Fig. 4.107. Comparison of the concentration dependence of the equivalent conductivity of tetraisoamylammonium nitrate dissolved in water and in water–dioxane mixtures.

From the expression for q, it is clear that the lower the dielectric constant of the solvent, the larger is the magnitude of q. Hence, when one replaces water with a nonaqueous solvent, the likelihood of ion-pair formation increases because of the increasing q (assuming that a does not increase in proportion to q).

It has already been emphasized that, taken as a whole, an ion pair is electrically neutral and ceases to play its role in the ionic cloud (Section 3.8). For the same reason (i.e., that the ion pair is uncharged), the ion pair does not respond to an externally applied electric field. Hence, ion pairs do not participate in the conduction of current. A quantitative analysis of the extent to which ion-pair formation affects the conductivity of an electrolyte must now be considered.

4.8.9. Effect of Ion Association on Conductivity

In treating the thermodynamic consequences of ion-pair formation (Section 3.8.4), it was shown that the association constant K_A for the equilibrium between free ions and ion pairs is given by

$$K_A = \frac{\theta}{(1-\theta)^2} \frac{1}{c_a} \frac{f_{IP}}{f_{\pm}^2} \tag{3.154}$$

where θ is the fraction of ions that are associated, c_a is the analytical concentration of the electrolyte, $f_{\pm}$ is the mean activity coefficient, and f_{IP} is the activity coefficient for the ion pairs. Since neutral ion pairs are not involved in the ion–ion interactions responsible for activity coefficients deviating from unity, it is reasonable to assume that $f_{IP} \approx 1$, in which case,

$$K_A(1-\theta)^2 c_a f_{\pm}^2 = \theta \tag{4.342}$$

A relation between θ and the conductivity of the electrolyte will now be developed. The specific conductivity has been shown [*cf.* Eq. (4.161)] to be related to the concentration of mobile charge carriers (i.e., of free ions) in the following way:

$$\sigma = zF(u_+ + u_-)c_{\text{free ions}} \tag{4.343}$$

One can rewrite this equation in the form

$$\sigma = zF(u_+ + u_-)\frac{c_{\text{free ions}}}{c_a} c_a \tag{4.344}$$

Since $c_{\text{free ions}}/c_a$ is the fraction of ions that are *not* associated (i.e., are free), it is equal to unity minus the fraction of ions that *are* associated. Hence,

$$\frac{c_{\text{free ions}}}{c_a} = 1 - \theta \tag{4.345}$$

and using this result in Eq. (4.344)

$$\sigma = zF(u_+ + u_-)(1 - \theta)c_a \tag{4.346}$$

or, from the definition of equivalent conductivity, i.e.,

$$\Lambda = \frac{\sigma}{zc_a} \tag{4.341}$$

one can write

$$\Lambda = F(u_+ + u_-)(1 - \theta) \tag{4.347}$$

If there is no ion association, i.e., $\theta = 0$, then one can define a quantity $\Lambda_{\theta=0}$, which is given by [*cf.* Eq. (4.163)]

$$\Lambda_{\theta=0} = F(u_+ + u_-) \tag{4.348}$$

By dividing Eq. (4.347) by Eq. (4.348), the result is

$$1 - \theta = \frac{\Lambda}{\Lambda_{\theta=0}} \tag{4.349}$$

and

$$\theta = 1 - \frac{\Lambda}{\Lambda_{\theta=0}} \tag{4.350}$$

Introducing these expressions for θ and $1 - \theta$ into Eq. (4.342), one finds that

$$K_A \frac{\Lambda^2}{\Lambda_{\theta=0}^2} c_a f_\pm^2 = 1 - \frac{\Lambda}{\Lambda_{\theta=0}}$$

or

$$\frac{1}{\Lambda} = \frac{1}{\Lambda_{\theta=0}} + \frac{K_A f_\pm^2}{\Lambda_{\theta=0}^2} \Lambda c_a \tag{4.351}$$

Though Eq. (4.351) relates the equivalent conductivity to the electrolyte concentration, it contains the unevaluated quantity $\Lambda_{\theta=0}$. By combining Eqs. (4.346) and (4.348), one gets

$$\Lambda_{\theta=0} = F(u_+ + u_-) = \frac{\sigma}{z[(1-\theta)c_a]} \tag{4.352}$$

from which it is clear that $\Lambda_{\theta=0}$ is the equivalent conductivity of a solution in which there is no ion association but in which the concentration is $(1-\theta)c_a$. Thus, for small concentrations (see Section 4.6.12), one can express $\Lambda_{\theta=0}$ by the Debye–Hückel–Onsager equation (4.321), taking care to use the concentration $(1-\theta)c_a$. Thus,

$$\Lambda_{\theta=0} = \Lambda^0 - (A + B\Lambda^0)(1-\theta)^{1/2}c_a^{1/2} \tag{4.353}$$

which can be written in the form (see Appendix 4.4)

$$\Lambda_{\theta=0} = \Lambda^0 - (A + B\Lambda^0)(1-\theta)^{1/2}c_a^{1/2} = \Lambda^0 Z \tag{4.354}$$

where Z is the continued fraction

$$Z = 1 - z\{1 - z[1 - z(\cdots)^{-1/2}]^{-1/2}\}^{-1/2} \tag{4.355}$$

with

$$z = \frac{(A + B\Lambda^0)c_a^{1/2}\Lambda^{1/2}}{\Lambda^{0\,3/2}} \tag{4.356}$$

Introducing expression (4.354) for $\Lambda_{\theta=0}$ into Eq. (4.351), one has

$$\frac{1}{\Lambda} = \frac{1}{\Lambda^0 Z} + \frac{K_A f_\pm^2}{(\Lambda^0)^2 Z^2}\Lambda c_a$$

or

$$\frac{Z}{\Lambda} = \frac{1}{\Lambda^0} + \frac{K_A}{(\Lambda^0)^2}\frac{\Lambda c_a f_\pm^2}{Z} \tag{4.357}$$

This is an interesting result. It can be seen from Eq. (4.357) that the association of ions into ion pairs has entirely changed the form of the equivalent conductivity versus concentration curve. In the absence of significant association, Λ was linearly dependent on $c^{1/2}$, as empirically shown by Kohlrausch. When, however, there is

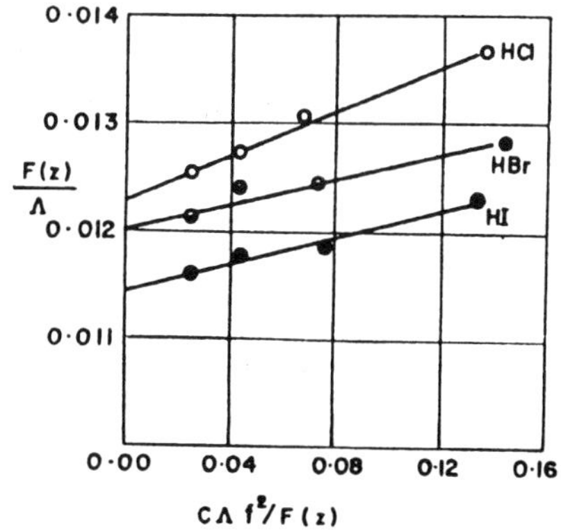

Fig. 4.108. Plots of Eq. (4.357) for the hydrogen halides in ethyl alcohol.

considerable ion-pair formation (as would be the case in nonaqueous solvents of low dielectric constant), instead of the Kohlrausch law, one finds that when Z/Λ is plotted against $\Lambda f_{\pm}^2 c_a/Z$, a straight line is obtained with slope $K_A/(\Lambda^0)^2$ and intercept $1/\Lambda^0$. Figure 4.108 shows the experimental demonstration of this conductance behavior.

4.8.10. Ion-Pair Formation and Non-Coulombic Forces

The theory of ion-pair formation in nonaqueous solutions has been substantially advanced by the work of Barthel, who demonstrated how important it is to take into account the non-Coulombic forces at small ionic distances in addition to the Coulombic ones used by Bjerrum. These non-Coulombic forces are represented by the mean force potential $W_{+-}^*(r)$ in the region $a \le r \le R$,[49]

$$K_A = 4\pi N_A \int_a^R r^2 \exp\left[\frac{2q'}{r} - \frac{W_{+-}^*}{kT}\right] dr \tag{4.358}$$

where q' is related to Bjerrum's parameter q (*cf.* Eq. 3.144).

Non-Coulombic forces are those that are responsible for different degrees of association of electrolytes in isodielectric solvents. For example, one can see the importance of non-Coulombic forces in respect to ion-pair formation when one compares the temperature dependence of the association constants of alkali metal salts and the tetraalkylammonium salts in protic solvents. The association constants of alkali metal halides show a monotonically increasing K_A when plotted against $(\varepsilon T)^{-1}$,

[49]In this region, the ion pair suppresses long-range interactions with other ions in solution.

TABLE 4.27

Association Constants, Gibbs Energies, Enthalpies, and Entropies of Ion-Pair Formation of Alkali Metal and Tetraalkylammonium Iodides in 1-Propanol from Temperature-Dependent Conductivity Measurements

Electrolyte	K_A (dm^3/mol)	ΔG_A (kJ/mol)	ΔH_A (kJ/mol)	ΔS_A (J K^{-1} mol^{-1})
NaI	205	−13.19	15.60	96.57
KI	336	−14.41	17.56	107.21
RbI	433	−15.04	16.28	105.03
Et_4NI	543	−15.60	5.29	70.05
Pr_4NI	515	−15.47	4.57	67.21
Bu_4NI	517	−15.48	4.33	66.44
$Pent_4NI$	537	−15.57	3.87	65.21

Source: Reprinted from J. Barthel, R. Wachter, G. Schmeer, and H. Hilbinger, *J. Sol. Chem.* **15:** 531, 1986.

whereas the tetraalkylammonium halide plot passes through a minimum at a temperature that is characteristic of the anion and solvent studied. Cation exchange has no effect on the position of this minimum.

In Table 4.27, one can see the K_A, free energies, enthalpies, and entropies of ion-pair formation. The enthalpies and entropies of the alkali metal salts are on the order of 16 kJ mol^{-1} and 100 J K^{-1} mol^{-1}, respectively. In contrast, the small enthalpies and entropies of the tetraalkylammonium ions reflect lesser solvation of the cations in the protic solvents.

4.8.11. Triple Ions and Higher Aggregates Formed in Nonaqueous Solutions

When the dielectric constant of the nonaqueous solvent goes below about 15, ions can associate not only in ion pairs but also in ion triplets. This comes about by one of the ions (e.g., M^+) of an ion pair $M^+\cdot\cdot A^-$ Coulombically attracting a free ion A^- strongly enough to overcome the thermal forces of dissociation

$$A^- + M^+ \cdot\cdot A^- \rightleftharpoons A^- \cdot\cdot M^+ \cdot\cdot A^-$$

From the conductance point of view, ion pairs and triple ions behave quite differently. The former, being uncharged, do not respond to an external field; the latter are charged and respond to the external field by drifting and contributing to the conductance.

The extent of ion-pair formation is governed by the equilibrium between free ions and ion pairs. In like fashion, the extent of triple-ion formation depends on the equilibrium between ion pairs and triple ions.

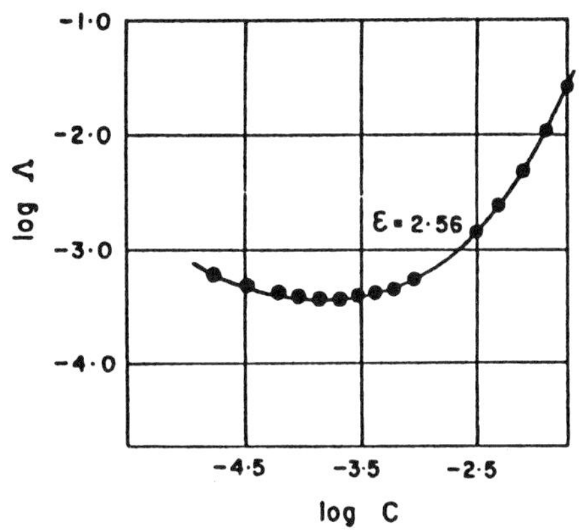

Fig. 4.109. Minimum in the curve for equivalent conductivity vs. concentration in the case of tetraisoamylammonium nitrate in a water–dioxane mixture of dielectric constant $\varepsilon = 2.56$.

$$M^+ + A^- \rightleftharpoons M^+ \cdot\cdot A^- \overset{A^-}{\rightleftharpoons} A^- \cdot\cdot M^+ \cdot\cdot A^-$$

Thus, the greater the stoichiometric concentration, the greater is the ion-pair formation and triple-ion formation.

With increasing concentration, therefore, ion-pair formation dominates the equivalent conductivity, which decreases with increasing concentration faster than if there had been no formation. At still higher concentrations, when triple-ion formation starts becoming significant, the equivalent conductivity starts increasing after passing through a minimum. This behavior has been experimentally demonstrated (Fig. 4.109).

Nevertheless, minima in plots of Λ vs. $\sqrt{c}$ do not unambiguously prove the existence of triple ions. Grigo showed that it is also possible to reproduce the Λ data for sodium iodide in butanol in terms of dielectric constant variation without the assumption of ion triplets. In summary, doubts still exist in relation to the formation of triple ions. Such doubts are most likely to be relieved by information that may become available through Raman and other kinds of spectroscopy.

4.8.12. Some Conclusions about the Conductance of Nonaqueous Solutions of True Electrolytes

The change from aqueous to nonaqueous solutions of *true* electrolytes results in characteristic effects on the conductance. The order of magnitude of the equivalent conductivity at infinite dilution is approximately the same in both types of solutions and is largely dependent on the viscosity of the solvent. However, the slope of the equivalent-conductivity versus concentration curve is considerably more negative in nonaqueous solutions than in the corresponding aqueous solutions. This means that the actual specific conductivity σ, which is the significant quantity as far as the conducting power of an actual solution is concerned, is much lower for nonaqueous solutions. Ion-pair formation worsens the conductance situation; triple-ion formation may be a slight help.

Thus, nonaqueous solutions of true electrolytes are not to be regarded with unrestrained optimism for applications in which there is a premium on high specific conductivity and minimum power losses through resistance heating. One may have to think of solutions of *potential* electrolytes where there is an ion-forming reaction between the electrolyte and the solvent (Section 2.4).

Further Reading

Seminal

1. R. M. Fuoss and F. Ascania, *Electrolytic Conductance*, Interscience, New York (1959).
2. H. J. Gores and J. M. C. Barthel, "Nonaqueous Solutions: New Materials for Devices and Processes Based on Recent Applied Research," *Pure Appl. Chem.* **67**: 919 (1995).

Review

1. J. Barthel and H. J. Gores, *Nonaqueous Solutions: Ionic Conductors with Widely Varying Properties*, in *Chemistry of Nonaqueous Solutions*, G. Mamantov and A. Popov, eds., p. 36, VCH Publishers, New York (1994).

Papers

1. A. N. Ogude and J. D. Bradley, *J. Chem. Ed.* **71**: 29 (1994).
2. Z. H. Wang and D. Scherson, *J. Electrochem. Soc.* **142**: 4225 (1995).
3. Y. Koga, V. J. Loo, and K. T. Puhacz, *Can. J. Chem.* **73**: 1294 (1995).
4. Y. C. Wu and P. A. Berezansky, *J. Res. Natl. Inst. Standards Technol.* **100**: 521 (1995).
5. A. A. Chialvo, P. T. Cummings, H. D. Cochran, J. M. Simonson, and R. E. Mesmer, *J. Chem. Phys.* **103**: 9379 (1995).
6. G. H. Zimmerman, M. S. Gruszkiewicz, and R. H. Wood, *J. Phys. Chem.* **99**: 11612 (1995).
7. M. G. Lee and J. Jorne, *J. Membr. Sci.* **99**: 39 (1995).
8. J. Barthel, H. J. Gores, and L. Kraml, *J. Phys. Chem.* **100**: 3671 (1996).

4.9. ELECTRONICALLY CONDUCTING ORGANIC COMPOUNDS IN ELECTROCHEMISTRY

4.9.1. Why Some Polymers Become Electronically Conducting

Although organic compounds by and large have very poor electronic conductivity with values of specific conductivity σ that are as much as 10^{12} times lower than those

TABLE 4.28

Conductivity Values for Conducting Polymers and Other Materials

	σ (S cm^{-1})
I-Polyacetylene[a]	1×10^4
Polysulfur nitride (film)	3×10^3
Polypyrrole[b]	40.7
Poly-3-methylthiophene (foil)[c]	16.7
PPy-p7MB/ClO_4[d]	13.74
Polythiophene (PTh) 95%[e]	0.6
P3TASH[f]	0.01
Cu	6×10^5
KCl (aq. 1 *M*)	0.1
CuO	1×10^{-5}
Glass	1×10^{-14}

[a] R. Menon, *Synthetic Metals* **80:** 223, 1996.
[b] 0.1 *M* $Fe_2(SO_4)_3$ oxidant; 0.022 sodium alkyl sufonate additive; Y. Kudoh, *Synthetic Metals* **79:** 17, 1996.
[c] U. Barsch and F. Beck, *Electrochim. Acta* **41:** 1761, 1996.
[d] PPy-p7MB/ClO_4 = Polypyrrole-poly(heptamethylene *p,p'*-bibenzoate); M. A. de la Plaza, M. J. Gonzalez-Tejera, E.S. de la Blanca, J. R. Jurado, and I. Hernandez-Fuentes, *Polymer Int.* **38:** 395, 1995.
[e] S. Yigit, J. Hacaloglu, U. Akbulut, and L. Toppare, *Synthetic Metals* **79:** 11, 1996.
[f] P3TASH = Poly[2-(3′-thienyl)ethanesulfonic acid]; S-A. Chen and M.-Y. Hua, *Macromolecules* **29:** 4919, 1996.

of a metal, certain organic polymers (conjugated compounds such as polyaniline and polypyrrole) conduct relatively well (see Table 4.28). The actual range of σ at room temperature observed among organic polymers is large, but some organic polymers have values of σ around unity. A few have specific conductivities that are even comparable with those of metals.

The degree of conductivity of organic polymers arises from their states of relative oxidation or reduction. In such states the polymer itself loses (for oxidation) or gains (for reduction) electrons in its structure. The number of monomer units that gain or lose an electron is variable but may be, e.g., 1 unit in 4. Once the polymer is electronically charged, counterions from solution enter the polymer fibrils to guard electrostatic neutrality. These neutralizing ions are often referred to as *dopants*. However, this is not doping in the sense of semiconductor doping (see Chapter 6), where the dopant itself ionizes to provide the charge carriers. In conducting polymers, the charge carriers are generated within the polymer chain. On the other hand, it is convenient to refer to the counterions in the charged polymers as dopants, so this term is widely used.

One model used to explain conductivity in polymeric structures is that of *polaron formation.*[50] Upon oxidation, double bonds along the chain are broken, leaving a

[50] A polaron is the term given to the quantum of electrostatic energy.

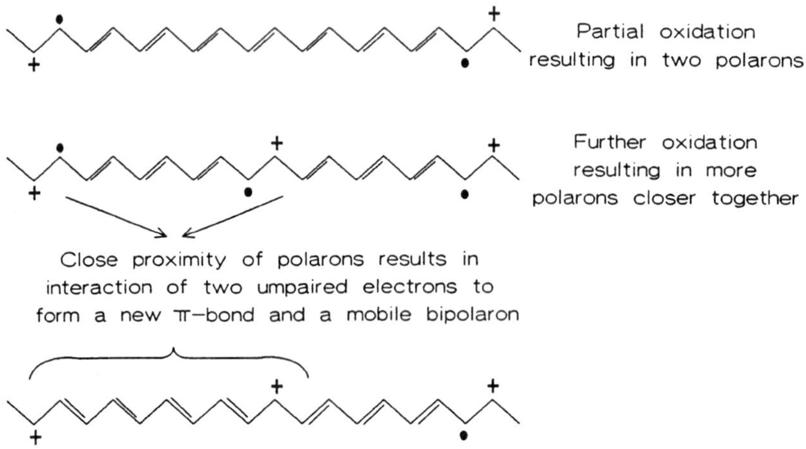

Fig. 4.110. Conductivity in polymeric structures by polaron formation. (After Lyons, 1996.)

radical and a positive charge unit on the polymer chain—and this positive charge is called a *polaron* (see Fig 4.110). After further oxidation, more polarons are formed along the chain. When the polaron concentration gets high enough, the radical cations spread out through adjacent π structures across approximately eight bond lengths. At this distance the polarons are able to "feel" each other, making contact between them. The combination of two radicals—one from each polaron—forms a new π bond. This π bond is more stable than the two radical-cation bonds; that is, the ΔG of the π bond is greater than the ΔG of dissociation of the two polarons. The result is a *bipolaron* that is more stable than two polarons the same distance apart. The bipolarons are then free to move along the polymer chain, which gives rise to electronic conductivity. It is at this point in the oxidation process that the conductivity undergoes a marked

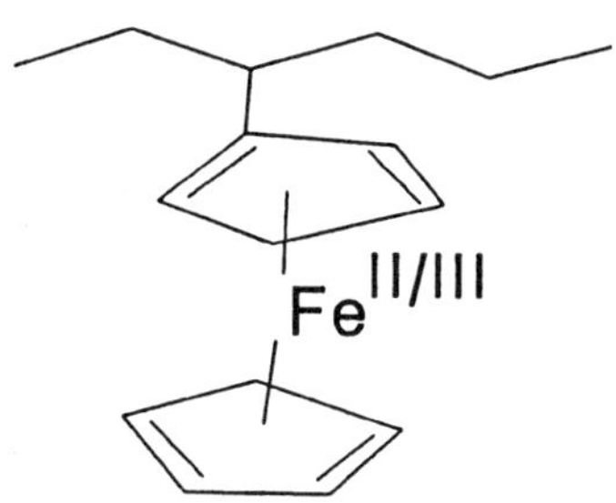

POLY(VINYLFERROCENE)

Fig. 4.111. An example of a redox polymer.

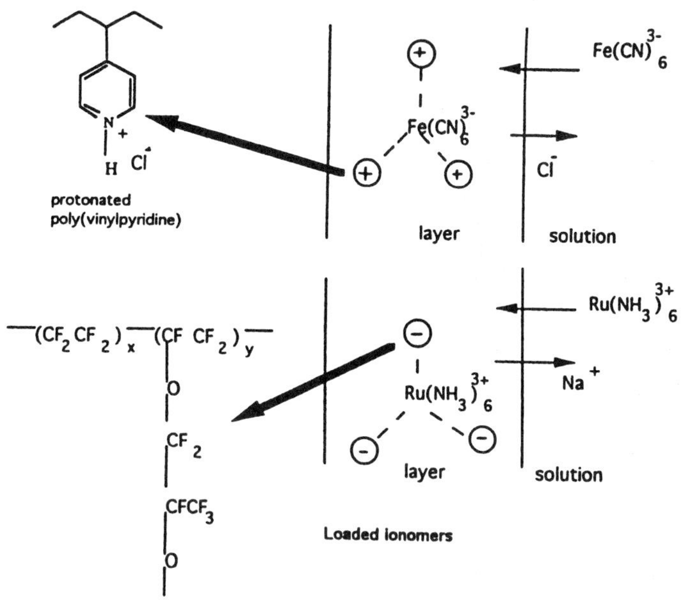

Fig. 4.112. Mechanism of conduction of loaded ionomers. (After Lyons, 1996.)

increase. Once the radical components of the polarons have combined to form π bonds, the remaining positive charges achieve high mobility along the chain.

Organic materials that conduct electricity are grouped into three classes. These are called, respectively (see Figs. 4.111–4.113),

1. redox polymers
2. loaded ionomer materials
3. electronically conducting polymers

The redox polymers contain, as the name indeed suggests, redox-active groups that are in turn bound to the polymer's spine, as shown in Fig. 4.111. Electrons travel macroscopic distances by hopping along using the redox groups attached to the spine at points between which the hops occur.

For electrodes at which redox processes occur (Chapter 7), the redox potential E is given by the expression

$$E = E^0 + \frac{RT}{nF} \ln \frac{c_{\text{ox}}}{c_{\text{red}}} \tag{4.359}$$

At the *standard redox potential* E^0 it is found that the conductivity passes through a maximum. This maximum of conduction occurs when $c_{\text{ox}} = c_{\text{red}}$. Hence, highly oxidized or highly reduced polymers do not conduct well at all.

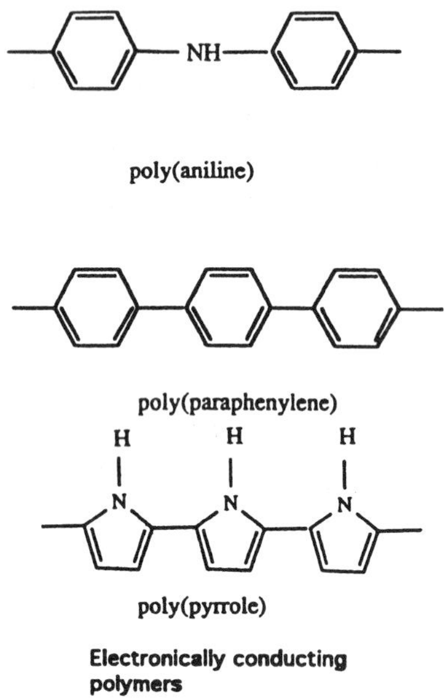

Fig. 4.113. Examples of electronically conducting polymers.

A second type of conducting electroactive organic substance is called a *loaded ionomer* (Fig. 4.112). Here the redox-active groups are mixed into the matrix of the ion-containing substance or *ionomer*. It can be seen in Fig. 4.112 what happens when an ionomer is placed in solution. As with the redox polymers, ionomers have charged sites attached to the polymer spine. Each of these charged sites has a partner or counterion of opposite sign. Once placed in a solution which itself contains ions that can take part in redox processes, the ionomer takes part in an ion-exchange process. The counterions originally belonging to the ionomer transport themselves into the surrounding solution, and the redox-active ions which were in the solution to begin with enter the ionomer and become electrostatically attached to the opposite charges in the polymer that makes up the ionomer material (usually a film). The conduction has the same mechanism as that for the redox polymer: electrons hop along between charged sites.

It is when one reaches the electronically conducting polymers that the really high conductances are found. There is a clear reason for this: for these substances (Fig. 4.113) the spine contains many conjugated structures and in these electrons are delocalized. Along such conjugated structures, charge transport is very fast and the rate-determining act is crossing from chain to chain.

Such materials also vary greatly in conductivity, depending on the state of oxidation. However, the conductance does not maximize at the standard redox potential—as with redox polymers—but when the polymer is in the highly oxidized state, so that many electrons are set free (i.e., $Fe^{2+} \rightarrow Fe^{3+} + e^{-}$). In the reduced state, electrons are withdrawn and the electronically conducting polymers become insulating polymers; the loss in conductivity is dramatically sudden and tends to disrupt the happiness of the experimenter. Thus, to keep the electronically conducting polymers conducting, they always have to be oxidized.[51]

4.9.2. Applications of Electronically Conducting Polymers in Electrochemical Science

4.9.2.1. *Electrocatalysis.* New materials that may act as electrocatalysts, that is, that may improve the desired path in a given reaction, are always needed if they are inexpensive and/or more easily handled than conventional materials. The development of conducting polymers to act as electrocatalysts is an area of research that began in the 1990s. It is relatively easy for organic compounds to be "adjusted," that is, the surface groups in the polymer can be varied to a great degree, which is just what is needed to give catalysts a variety and power corresponding to those of enzymes.

Corresponding to this is the idea of biosensors that could be implanted in the body for the electroanalysis of conceivably any chemical in the body. Thus, it may become possible to adsorb enzymes on the surface of electrodes and then tune these enzymes to react with appropriate biomolecules, as represented in Fig. 4.114. How would conducting polymers figure in such devices? They might be useful as the biosensor itself, since, being organic, they are more likely to interact positively with enzymes and biochemicals than metal electrodes would.

4.9.2.2. *Bioelectrochemistry.* A potential area of application of conducting polymers is in prosthetic devices. One of the difficulties in this area of research is blood clotting, which occurs when the electrical potential of the implanted device is negative insufficiently in potential. Blood clotting is common when metals are used as prosthetics, but given the great variety of materials that could become suitable through the use of electronically conducting organic polymers, the range of possible materials is much increased. The further possibility of organically conducting materials being able to be made into artificial organs[52]—replacing the difficult-to-find transplants—is an enticing one, particularly in the future development of electrochemistry in concepts such as the "cyborg" (Chapter 1).

---

[51]There are other factors that also affect the degree of conduction of electronically conducting polymers when they are in the oxidized state; one is alignment of the polymer chains. Thus, a rate-determining step in conduction may be the transfer of electrons from one unit in the spine to another: here linearity in the chain would help and junctions out of alignment would impede the continued passage of electrons along the chain.

[52] Although this goal is enticing, and perhaps not more than a decade away, a prerequisite to its achievement is much more knowledge of the surface properties of electronically conducting polymers.

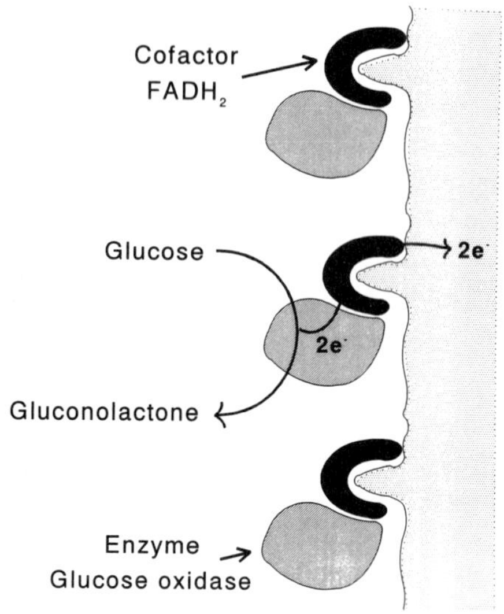

Fig. 4.114. Glucose sensor in which the enzyme (containing a reduced cofactor) is oxidized directly at an electrode.

4.9.2.3. *Batteries and Fuel Cells.* The work that directed commercial attention toward the use of electronically conducting polymers was initiated by Allen MacDiarmid at the University of Pennsylvania, who reported in 1980 on a battery using polyacetylene in propylene carbonate. The original battery is shown in Fig. 4.115. The conductivity of the polymer is adjusted by varying the content of lithium perchlorate, which acted as doping material in the polyacetylene. The maximum conductivity of the polyacetylene is on the order of 10^3 S cm^{-1}. The energy density obtained by this battery, which discharges with an initial current of 25 mA cm^{-2}, was about 170 W-hr per kilogram, compared with about 30 for the lead acid–battery. Since MacDiarmid's pioneer work, many other batteries involving conducting polymers have been studied. They are attractive, particularly for their potential as extremely cheap electricity storers, perhaps for widespread use in electrically powered bicycles.

An attraction of conducting organic materials as batteries is their low densities of about 1 g cm^{-3} compared with 10 g cm^{-3} for metals. Their use in aircraft might be cost effective. Such prospects must be balanced against the relatively small potential range in which the present electronically conducting polymers are stable.

4.9.2.4. *Other Applications of Electronically Conducting Polymers.* Future applications of electrochemistry in clean energy systems (based on solar light or chemically stimulated nuclear changes) seem possible. A major difficulty so far has been the expense of the materials. In this area, one of the initial studies involving

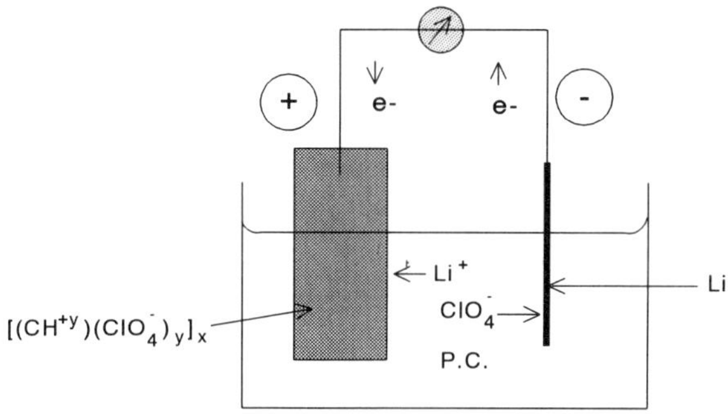

Fig. 4.115. Schematic representation of the discharge process in a $(CH)_x/LiClO_4/Li$ rechargeable storage battery cell (A. J. Heeger and A. G. MacDiarmid, in *The Physics and Chemistry of Low Dimensional Solids*, L. Alcacer, ed., D. Reidel, Boston, 1980, p. 45).

conducting polymers was carried out by Shirakawa et al. in the early 1990s. Their cell is shown in Fig. 4.116.

One of the fields in which conducting polymers may have a great potential for development is the one broadly classified as *molecular electronics*. The fabrication of reliable electronic devices based on organic molecules or biological macromolecules offers formidable challenges. Is there a possibility of utilizing conducting organic compounds in miniaturizing electronic devices so that eventually molecules take the place of wires, transistors, and memory devices? Electronically conducting polymers should be useful in such advanced developments.

Another area of potential applicability of conducting polymers is monitoring the composition of gaseous ambients. Although solid-state gas sensors are available, they present a disadvantage: the high temperatures needed for the sensor elements to operate. Here is where research may find conducting organic polymers useful. For example, it has been shown by Miasik et al. that the resistance of a polypyrrole deposited on a filter paper is sensitive to the presence of ammonia at room temperature. Thus, the resistance increases in the presence of a reducing gas, such as ammonia, and decreases in the presence of an oxidizing gas, such as nitrogen dioxide. The response of such an electrode is depicted in Fig. 4.117.

4.9.3. Summary

The major difficulty in the 1990s for the development of electronically conducting polymers lies in the limited understanding we have about the relation between the molecular structure of the organic material and the resulting electronic conductance.

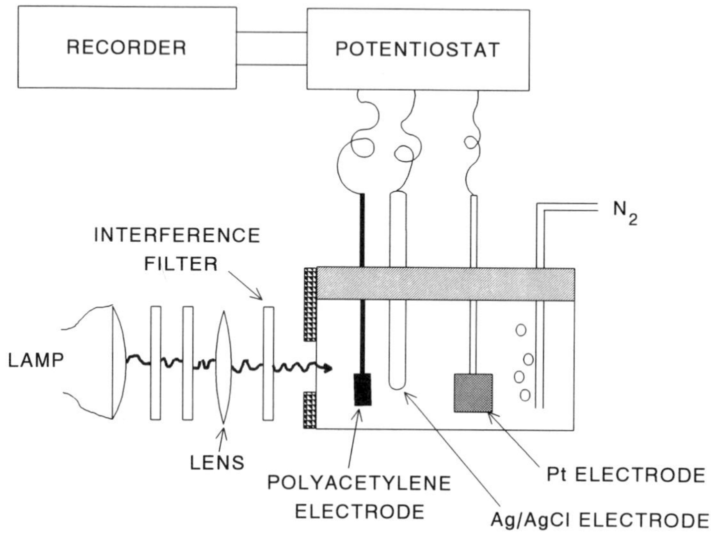

Fig. 4.116. Experimental apparatus used by Shirakawa et al. for photoelectrochemical measurements of a polyacetylene electrode (H. Shirakawa, S. Ikeda, M. Aizawa, J. Yoshitake, and

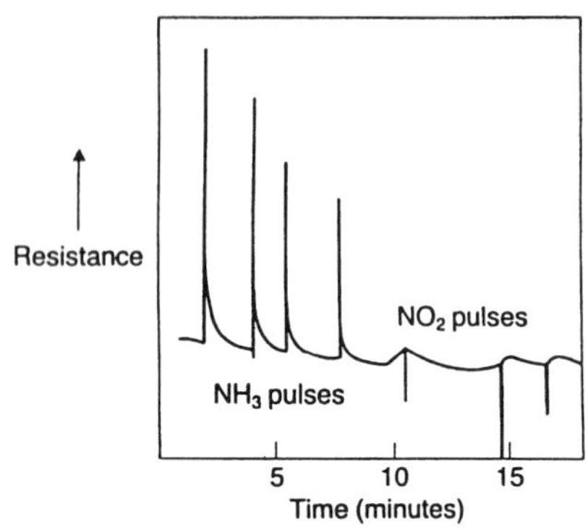

Fig. 4.117. Qualitative dc resistance response characteristic at room temperature of a conductive polypyrrole sensor to pulses of ammonia and nitrogen dioxide (J. J. Miasik, A. Hooper, P. T. Mosely, and B. C. Tofield, "Electronically Conducting Polymer Gas Sensors," in *Conducting Polymers*, Luis Alcacer, ed., Reidel, Dordrecht, The Netherlands, 1987).

However, the field is ripe for development, particularly in electrocatalysis for obtaining cheap, light, electrochemical storage and in molecular electronics coupled with prosthetics.

4.10. A BRIEF RERUN THROUGH THE CONDUCTION SECTIONS

We learned early on that equivalent conductivity and specific conductivity differed in that the former was not directly proportional to concentration but only secondarily so. However, it turned out that this secondary dependence was considerable and arose because the mobility of the ion itself decreased with an increase in concentration. Thus, as ions get near enough to feel each other through the interaction of their electric fields, they slow down.

At a molecular level, this slowdown is described in terms of two effects. One of these effects is called *electrophoretic* and the other, *relaxational.* The electrophoretic effect is easy to understand because it is a kind of electrical friction: as one ion passes the other within electrical hailing distance, both ions slow down in recognition of the electrical existence of the other.

The relaxation effect is a bit more difficult to explain. It has to do with the fact that when an ion moves in a given direction, inertia causes the ionic cloud around it to become egg shaped, and this dissymmetric ionic cloud has more counter charge toward its rear than toward its front. The dissymmetry of charges acts to counteract the effect of the directional electric field applied through the solution, and so this also slows the ion down. Both these effects combine to explain why mobility falls with increasing concentration, for the two effects increase in strength with the square root of the concentration.

Then, having broached the subject of the relaxation of the ion's atmosphere—its taking up a dissymmetric shape when the ion moves—we went on to tackle the subject of relaxation quite generally. For example, if an electric field is suddenly applied to a solution, it would orient the solvent dipoles therein. A new equilibrium would then be set up. The relaxation time is a measure of the time it takes to set up this new equilibrium. At first it seems peculiar that one should call it "a measure of" and not the time itself. However, the situation is similar to that of radioactive decay because in changing from state 1 to state 2, the concentration of a radioactive nucleus decreases exponentially with time, taking an infinite time to disappear completely. Since this is not a practical measure, we agree to use another measure of the rate of decay—the time to decline by 63%.

These ideas about relaxation times are applied to several phenomena, including the changes in asymmetry of the ionic atmosphere and the unusual behavior of the dielectric constant of water, which has three values according to the frequency with which it is measured. They are 78, 5, and 2.

Nonaqueous solutions are one of two final topics in this section. They form a subfield in which there has been immense progress since 1970. Nonaqueous solutions allow a much greater electrochemical window than do the aqueous solutions because of course in the latter, one is subject to hydrogen evolution if one goes too negative and evolution of oxygen if one goes too positive. In practice, in aqueous solutions, there is only about 1.5 V of practical working potential—a relatively short range when one recalls that the electrochemical series extends over about 4.5 V. This is one of the reasons why the study of nonaqueous solutions has increased in intensity so much in recent years. It opens up several new areas, and one of them concerns the strange new ions that are formed there. This is because the dielectric constant is so much lower in nonaqueous than in aqueous solutions and therefore the Coulombic attraction between ions of opposite sign is higher in the former solutions, so that there is a greater tendency to "stick together." Dimers, trimers, and even larger aggregates occur.

Last of all, we discussed the conductivity of certain polymers. This is a fairly new area and very promising for electrochemical devices, the breadth of application of which may be greatly extended. We started off by pointing out why it is that only certain types of organic materials produce relatively high concentrations of mobile electrons that can leave the tiny prescribed molecular area in which they usually have their being and become mobile along a whole polymer chain, by hopping along charged sites in the polymer spine. In this way organic materials (that normally would be considered insulators) may become significantly conducting (e.g., 10^3 S cm^{-1}).

Electronically conducting polymers form so new a field that it exists more in hope than in reality. As with the progress of much scientific research, it depends on the amount of funding available. Nonaqueous solutions offer tremendous scope because they allow one to introduce so much variety in properties. Among the applications already mentioned is the possibility of making cheap electricity energy storers (or batteries) utilizing, e.g., polyacetylene or polypyrrole. A really exciting suggestion for the future is the possibility that we might be able to manufacture artificial organs for the body by using conducting organics and hence avoid the wait for transplants.

Further Reading

Seminal

1. H. Kallmann and M. Pope, "Conductivity of Insulators in Contact with Ionic Solutions," *J. Chem. Phys.* **32**: 300 (1960).
2. H. Mehl, J. M. Hale, and F. Lohmaan, "Electrode Made out of Insulators," *J. Electrochem. Soc.* **113**: 1166 (1966).
3. R. Pethig and G. Szent-Györgyi, "Redox Properties in Biomolecules," *Proc. Natl. Acad. Sci. U.S.A.* **74**: 226 (1978).
4. J. Heeger and A. G. MacDiarmid, "Electronically Conducting Polyacetylene," in *The Physics and Chemistry of Low Dimensional Solids*, L. Alcacer, ed., p. 73, D. Reidel, Boston (1980).

Reviews

1. J. O'M. Bockris and D. Miller, "An Introduction to Conducting Polymers," in *Conducting Polymers*, Luis Alcacer, ed., Reidel, Dordrecht, The Netherlands (1987).

2. W. H. Smyrl and M. Lien, "Conductivity in Electronically Conducting Polymers," in *Applications of Electroactive Polymers*, B. Scrosati, ed., Chapman and Hall, London (1993).
3. M. E. G. Lyons, in *Electroactive Polymer Chemistry*, Part I. *Fundamentals*, M. E. G. Lyons, ed., Chapter I, Plenum Press, New York (1994).

Papers

1. D. L. Miller and J. O'M. Bockris, *J. Electrochem. Soc.* **139**: 967 (1992).
2. A. Szucs, G. D. Hitchens, and J. O'M. Bockris, *Electrochim. Acta* **37**: 277 (1992).
3. S. Li, C. W. Macosko, and H. S. White, *Science* **259**: 957 (1993).
4. G. Shi, S. Jin, and G. Xue, *Science* **267**: 994 (1995).
5. M. S. Walaszlowski, J. Orlikowski, and R. Juchniewicz, *Corros. Sci.* **37**: 645 (1995).
6. M. S. Freund and N. S. Lewis, *Proc. Natl. Acad. Sci. U.S.A.* **92**: 2652 (1995).
7. X. B. Chen, J. Devaux, J. P. Issi, and D. Billaud, *Polym. Eng. Sci.* **35**: 637 (1995).
8. B. Coffey, P. V. Madson, T. O. Poehler, and P. C. Searson, *J. Electrochem. Soc.* **142**: 321 (1995).

4.11. THE NONCONFORMING ION: THE PROTON

4.11.1. The Proton as a Different Sort of Ion

The reason for having a whole section in this book on proton conduction is that this particle exhibits characteristics quite different from those of other ions; moreover, the particle itself is *tiny* in dimension. Instead of carrying with it the electron shell normal to all other ions—hence having its radius expressed in nanometers—the proton is a nucleus only, and its radius is only about 1 fermi (F) (10^{-13} cm). This makes it the smallest ion and the lightest, giving it the property of being able to approach much closer to a neighboring ion or atom than any other ion or atom can. The tiny proton is, however, also a mighty proton: it attracts electrons much more powerfully than, say, Li. This particular property is shown by the magnitude of the ionization energy: 1309 kJ mol^{-1} for hydrogen vs. 519 kJ mol^{-1} for lithium. Apart from this display of power, protons have a unique characteristic—they form H bonds. Partly for this reason, the time a proton remains bare in water is small indeed; it spends less than 1% of its time alone. The remaining 99% of its time, a proton in solution is closely attached to an H_2O molecule, forming H_3O^+, a hydronium ion.

Being such a famous performer, the shape and size of the hydronium ion have been determined by NMR and other methods. The H_3O^+ ion has a rather flat trigonal-pyramidal structure with the hydrogen at the corners of the pyramid and the oxygen in the middle, as shown in Fig. 4.118. Its structure resembles that of the ammonia molecule.

A great deal can be learned about the proton simply by reviewing some energy quantities assigned to it. The energy change associated with the interaction between a

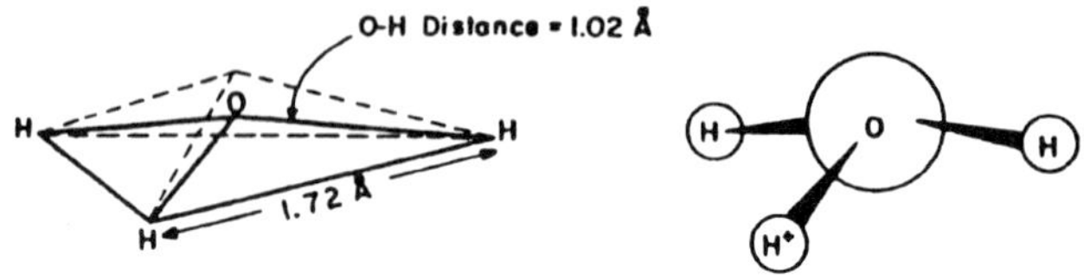

Fig. 4.118. The trigonal-pyramid structure of the hydronium ion H_3O^+.

proton and a single water molecule, the *proton affinity of water*, is –711 kJ mol^{-1}. This value corresponds to the interaction of H^+ and H_2O in the *gas phase*, not in solution. The corresponding value of the enthalpy or heat of hydration of the proton is –1112 kJ mol^{-1}. How are these two values related? If the *proton affinity* is subtracted from the *total hydration heat* of a proton, one gets then the *heat of hydration* of an H_3O^+ ion! This value is –401 kJ mol^{-1}, a reasonable value when one compares it with the heat of hydration of a K^+ ion (–344 kJ mol^{-1}), which has approximately the same radius as the H_3O^+ ion.

The value of –401 kJ mol^{-1} for the hydration energy of the hydronium ion indicates that it itself is hydrated. How many molecules of water hydrate it? A look at Fig. 4.119 shows that approximately three water molecules are associated to the hydronium ion, giving a structure of $(H_9O_4)^+$.[53]

How is the H^+ a nonconforming ion? For one thing, because of its special association with water; it is nearly always tightly bound to *one* water. This structure seems to exist as an entity which itself is being solvated by other waters.

This is not the only nor the main differentiating property of the nonconforming ion. Having seen that H^+ is usually to be found tightly attached to H_2O, one would expect that H_3O^+ would be the transporting entity. It is not! H_3O^+ movement contributes only about 10% of the transport of H^+ in aqueous solution, and the main mode of transport is, indeed, entirely different from that of other ions.

Why is it that one regards the proton as different from all other ions? There are three reasons, all connected with its tiny size and small mass: (1) The tiny size means that such an ion can go anywhere (e.g., diffusing in Pd). (2) Its small mass turns out to give rise to a mechanism of motion in solution quite different from that of any other ion (except its isotope, the deuteron). (3) In quantum mechanical tunneling (see also Chapter 9), low mass is a vital factor. The electron, the mass of which is nearly 2000 times less than that of a proton, can easily tunnel through barriers more than 2000 pm in thickness. The ability of the proton to tunnel is much less than that of the electron,

---

[53]In the coordination shell of an H_3O^+ ion, n water molecules are compressed owing to the electric field of the ion, and thus the molar volume is decreased compared to what one would expect from the volume of water. By measuring molar volumes or densities as a function of temperature and comparing them with curves calculated for an assumed value of n, the number of coordinating water molecules can be predicted.

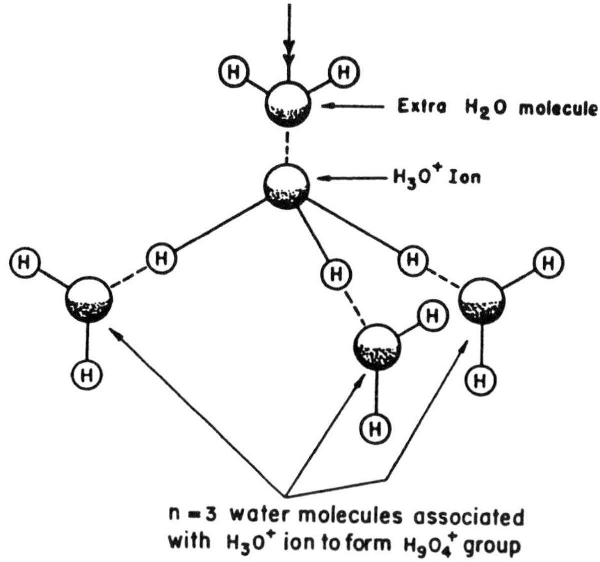

Fig. 4.119. Schematic configuration of $H_9O_4^+$ group shown with an extra H_2O molecule electrostatically bound.

but its quantum properties are still significant—and that is something that sets it apart from heavier ions, even He^+ and certainly Li^+. So this ion's properties deserve a special section of which this is the introduction.

4.11.2. Protons Transport Differently

The starting point to elucidate the way the proton moves in solution is to consider its movement through the solvent at a steady state—constant velocity—and at a concentration so low that there is no interionic interaction (zeroth approximation). This occurs when the electric driving force ze_0X balances the Stokes viscous force, $6\pi r\eta v$. Thus, the Stokes mobility is

$$u_0 = \frac{ze_0}{6\pi r\eta} \tag{4.183}$$

What radius should one use? Suppose one takes a rough-and-ready measure to consider the hydronium ion. Since it has approximately the same radius as that of a K^+ ion, one would expect its mobility to be approximately the same, i.e., about 5×10^{-4} cm^2 V^{-1} s^{-1}. It is here that one encounters the great anomaly—the nonconforming ion—for the mobility of the proton is in reality 36×10^{-4} cm^2 $V^{-1}s^{-1}$, a sevenfold excess.

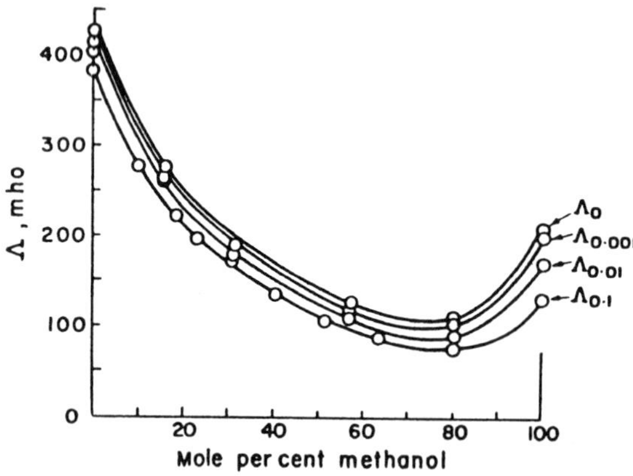

Fig. 4.120. The equivalent conductivity of HCl in methanol–water mixtures decreases with increasing alcohol content and reaches a minimum at about 80 mol % of alcohol. At this concentration, the abnormal mobility, i.e., the mobility of the H_3O^+ ion compared with that of the K^+ ion (which is of similar size), is reduced almost to zero.

One can pick up a clue as to the reason for this anomaly in mobility if one asks: What is the proton's mobility in other, related solvents? This rather vital question was addressed and solved in a Ph.D. thesis by an Austrian student, Hanna Rosenberg, more than 40 years ago. She found that if, for example, methanol was added to water, the anomalous mobility of the proton was decreased (Fig. 4.120). When methanol was replaced by other, larger alcohols (no water present), she was astounded to find that the anomalous mobility was greatly reduced until by the time *n*-propanol was reached, the difference between HCl and LiCl was greatly reduced (Table 4.29).

TABLE 4.29
Equivalent Conductivities (S cm^2 eq^{-1}) at Infinite Dilution of HCl and LiCl at 293 K

	Solvent			
	Water	Methanol	Ethanol	*n*-Propanol
Λ_{HCl}	426.2	192.0	84.3	22
Λ_{LiCl}	115.0	90.9	38.0	18
$\dfrac{\Lambda_{HCl}}{\Lambda_{LiCl}}$	3.71	2.11	2.22	1.22

Another fact came up in further work stimulated by Ms. Rosenberg: the ratio of the proton and deuterium mobilities in water had been found to be 1.4 at 25 °C, an unexpectedly large value. If the proton mobility was to be understood in terms of movements of H_3O^+ or D_3O^+ in a viscous fluid (i.e., if it were determined by Stokes' law), the two values should be about the same, because the actual difference in size of the two ions (H_3O^+ and D_3O^+) would be negligible. Finally, when one looks into the conductance of protons in water at various temperatures, one expects that as with the other ions, there will be a constant energy of activation over an appreciable range of temperatures. The truth is, the $\log \Lambda$ vs. $1/T$ plot shows curvature, suggesting that as the temperature increases, the entity contributing to the transport is changing.

The proton is indeed anomalous in its conductance and mobility. These properties do not vary with temperature in the expected, regular way. There is not the expected near-sameness for hydronium and deuterium ion mobilities. The conductance of protons in aqueous-non-aqueous media is wholly dependent on the mole-fraction of water present.

4.11.3. The Grotthuss Mechanism

The use of Stokes' law to calculate the mobility of an ion implies a mechanism for the way an ion goes through a solution. When Stokes derived his famous equation (the one that gives the resistive force to flow as $6\pi r\eta v$; (Section 4.4.7)), ions had not been thought of, and the movement that Stokes imagined was much more like a brass ball being pulled through molasses. As the decades passed, Stokes' law turned out to perform remarkably well, not only for brass balls in molasses, but also right down to particles of angstrom size, the size of normal, regular ions. So it is not unreasonable to conclude that ions, too, have a slow, viscous kind of movement through a solvent. The molecular-level picture is that of the ion—in the absence of an applied field—lurching hither and staggering thither, the direction of each lurch being randomly determined. When an electric field is applied, there is still this lurching all over the place, as in diffusion (Section 4.2), but now there is a slightly preferred component to the random movement—that of the ion's movement in the direction of the electric field. It is to this drift movement—movement in the direction of the electric field—that Stokes' law applies.

It was demonstrated earlier in several ways that although it may well apply to K^+ and Na^+, Stokes' law certainly *does not* apply to the proton, the nonconforming ion. There may be a slow, viscous drifting of H_3O^+ through the solvent. However, it does not explain the proton's movement.

A Swedish worker, Grotthus, had noticed that in a row of marbles in contact, the collision of the marble at one end of the row with a new marble caused a marble at the far end to detach itself and go off alone (Fig. 4.121). This sort of movement would provide a rapid way for an ion to appear to travel through a solution. There would then be no need for a whole H_3O^+ to lumber along, taking the proton with it. Could a proton

Fig. 4.121. Effect of marbles colliding in a row: the Grotthuss mechanism.

attach itself to one end of a chain of water molecules in solution and push one off at the other end of the chain?

Stokes and Grotthuss both worked during the Victorian age. It appeared that Stokes' drift of whole ions though solutions had won out as far as most ions are concerned, and Grotthuss's concept found no resonance until in 1933 Bernal and Fowler, in the first issue of the famous *Journal of Chemical Physics*, suggested a mechanism that borrowed some things from it. In Fig. 4.122 one sees a suggestion for the mechanism of a proton jumping from one water molecule to the next, which is vaguely what Grotthuss had suggested. Thus, when a proton arrives at one water, making it temporarily H_3O^+, another different proton from the same H_3O^+ detaches itself from the H_3O^+ for the next hop. It is at once clear that this "relay-like" mechanism provides an exciting possibility for more rapid transport than Stokes' law would allow. There is no need for the relatively heavy H_3O^+ to lumber along; the tiny proton itself hops from H_3O^+ to H_2O (making it H_3O^+), and new H_3O^+s are rapidly formed across the solution.

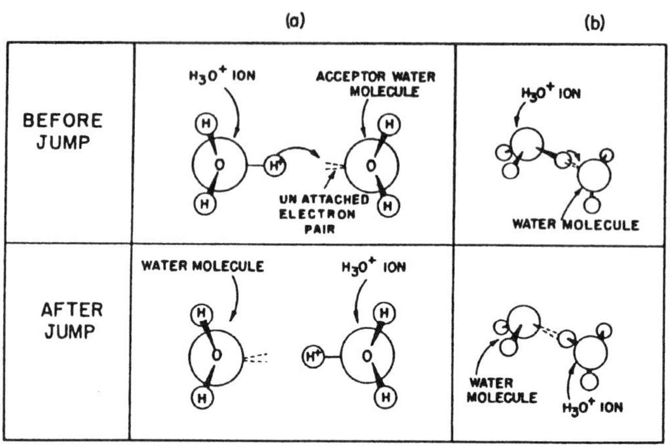

Fig. 4.122. Two schematic views (a) and (b) of a water molecule adjacent to an H_3O^+ ion. The free electron pair (orbital) of the O of the water molecule is oriented along the O (of H_3O^+)–H^+–O (of H_2O) line. The jump of the proton H^+ from the H_3O^+ ion to the water molecule converts the water molecule into an H_3O^+ ion and the H_3O^+ ion into a water molecule.

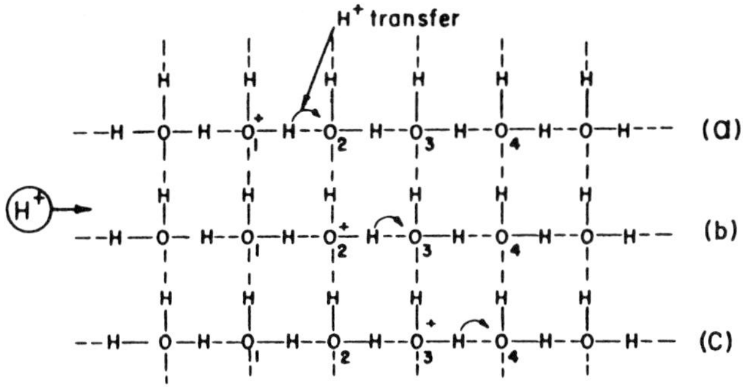

Fig. 4.123. If there are a series of proton jumps down a line of water molecules, the net result is equivalent to the migration of an H_3O^+ ion (indicated by a charge on the oxygen) along the line.

This latter impression is better seen by looking at Fig. 4.123 because there one sees that although in reality it is always a proton that does the jumping, the positions *a, b, c* represent an ensemble of waters netted together by means of H bonding, with a proton swinging away from one water to the next, making each *stationary* water[54] become momentarily a hydronium ion, and causing the casual onlooker to get the impression that H_3O^+ itself is moving through the solution.

4.11.4. The Machinery of Nonconformity: A Closer Look at How the Proton Moves

If one is going to consider protons actually jumping from one quasi-stationary water molecule to the next, the classical view would be to ask what fraction of them would be sufficiently activated to get over the top of the corresponding energy barrier. Another possibility was discussed not long after the introduction of quantum mechanics in 1928 by Bernal and Fowler. In a famous paper of 1933, they applied quantum mechanics to the possibility of tunneling through a barrier.

It is not reasonable just to say "the proton transfers to a water molecule," because that is a fairly vague statement. One starts asking questions like: Can it transfer to a water molecule when it arrives from *any* direction? Hardly, for it has to find an orbital on O to form a bond with—and orbitals have direction. Considering that water molecules librate and sometimes break H bonds and rotate, the next question which will have to be researched is: Does a water get into a proton-receiving posture all by

[54]Not quite stationary! Every time a water molecule is blessed by the arrival of a proton, it momentarily become an H_3O^+ and lumbers off a bit down the electric field gradient. But, as the proton detaches itself again, it jumps off to the next H_2O, and so each H_3O^+ undergoes a Stokeslike movement only to a small degree that corresponds to the short time an H^+ is attached to it.

itself, spontaneously, or does it have to be pushed by the powerful electric field of the approaching proton to turn around and offer an orbital to which the proton can jump?

To start finding out what happens to the energy of a proton when it leaps from one water molecule to the next, let the H_3O^+ system be simplified to be treated as though it were a diatomic entity, W–H^+, where W is the water molecule. Then the potential energy of any diatomic system can be represented accurately in the gas phase and roughly in solution by a Morse-type equation,

$$U_r = D_0\{1 - e^{[-a(r-r_{eq})]}\}^2 \tag{4.360}$$

where U_r is the energy of the diatomic system as a function of the actual distance r at any degree of stretching or compressing around the equilibrium distance r_{eq} of the two "atoms"; D_0 is the dissociation energy of W–H^+ and a is a constant as a function of the frequency of vibration. For D_0, the overall hydration energy of the proton is used, i.e., 1112 kJ mol^{-1}, and for r_{eq}, the distance is 98 pm.

Knowing these parameters, one can plot the U_r vs. r curve for the stretch of a proton along one of the p orbitals of oxygen. This plot is known as a *Morse curve* and is shown in Fig. 4.124 for our system. The energy trajectory under study here pertains to the reaction shown in Fig. 4.125. Because of the symmetry of the system, one has an identical curve for the approach of the proton to the second water molecule. Thus,

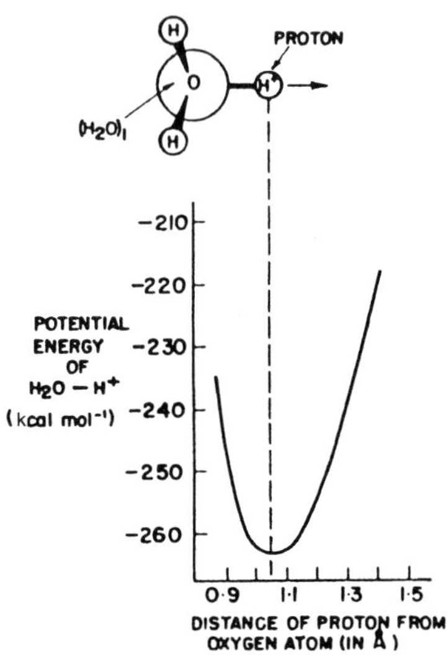

Fig. 4.124. Variation of the potential energy of the W_1–H^+ system with the stretching of the W_1–H^+ bond (Morse curve).

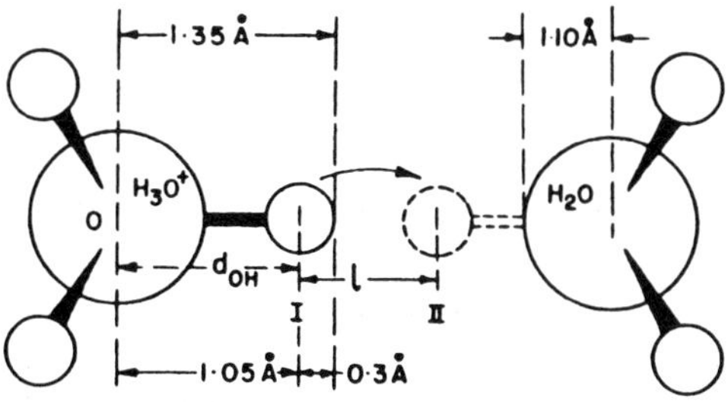

Fig. 4.125. Model for proton transfer between an H_3O^+ ion and a favorably oriented H_2O molecule.

the two Morse curves, one for W_1–H^+ and the other for H^+–W_2, are put together (Fig. 4.126) and form the potential-energy barrier for proton transfer.

This then is the energy barrier for the jump of a proton from one water molecule to another *over* a potential-energy barrier. From this barrier, one can calculate the energy of activation and thus the rate of proton transfer. How can one create and locate the curves with respect to each other?

Eyring and Stearn made the first study of the probability that the proton would have enough energy to climb over the potential-energy barrier. The basic application of the theory of absolute reaction rates made by Eyring begins with the equation for the frequency at which the proton crosses the barrier, i.e., the rate constant $\vec{k}$

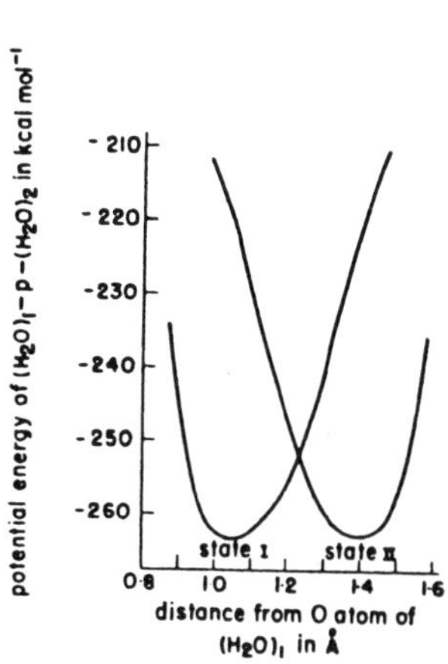

Fig. 4.126. Potential-energy barrier for proton transfer from H_3O^+ to H_2O.

$$\vec{k} = \frac{kT}{h} \exp\left[-\frac{\Delta G^{\neq} - XF\delta \cos\theta}{RT} \right] \tag{4.361}$$

where $\Delta G^{\neq}$ denotes the free energy of activation at zero local field X. Equation (4.361) is derived from the basic equation for all rate constants in Eyring's famous theory. However, the free energy of activation is modified (reduced from the proton's forward movement toward the negative pole) by the term $XF\delta \cos\theta$. Thus, X is the electric field applied to the solution at an angle θ to the proton's movement, F is the charge per g-ion on the proton, and δ is the activation or half-jump distance, i.e., half the distance the proton must travel in crossing a symmetrical barrier. This second electrical term in the exponent represents the change in energy which acts to reduce $\Delta G^{\neq}$ for a forward direction jump.

If we average over all angles θ of orientation of the jumping direction to the field, from 0 to π (180°), then the rate constant becomes

$$\vec{k} = \frac{kT}{h} \exp\left[-\frac{\Delta G^{\neq}}{RT} \right] \left[\frac{\int_0^{\pi} \exp\left(\frac{XF\delta \cos\theta}{RT} \right) \cos\theta \sin\theta \, d\theta}{\int_0^{\pi} \sin\theta \, d\theta} \right] \tag{4.362}$$

If one substitutes 1 V cm^{-1} for X, 3×10^{-8} cm for δ, and the usual values of F and R at room temperature, the exponential in the integral can become less than 10^{-6}, a rather small number, and the exponential can be linearized as approximately $1 + x$.

We can use this model to fix the relative positions of the minima of the Morse curves. When the two minima are fixed, we can see from Fig. 4.127 that the two Morse

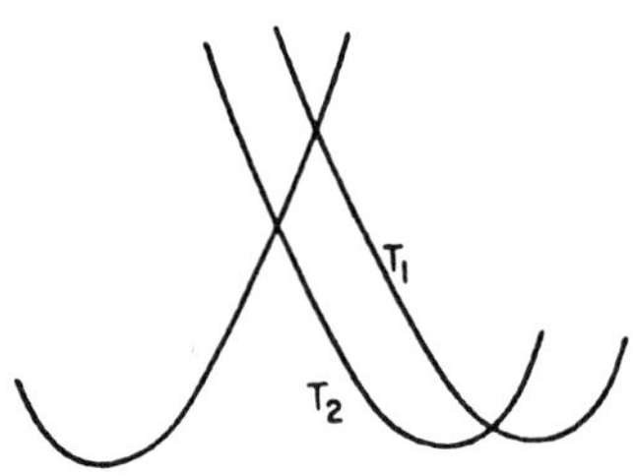

Fig. 4.127. Schematic diagram to show that the activation energy for proton transfer depends upon the distance between the minima of the Morse curves.

curves intersect and a potential-energy barrier is fabricated. To obtain the height of the energy barrier, the resonance energy of the activated complex is required. In an O–H···O hydrogen bond, resonance energy would account for half the hydrogen bond energy if the hydrogen were symmetrically distributed between the two oxygen atoms. The hydrogen bond energy has been established as –24.7 kJ mol^{-1}, and so the resonance energy of the activated state can be taken as –12.35 kJ mol^{-1}. This value must be subtracted from the energy of activation calculated from the point of intersection of the Morse curves. A microscopic look at Figure 4.126 shows that the intersection point occurs at about 28.45 kJ mol^{-1} above the zero-point energy for the proton, and so the energy of activation is about 16.1 kJ mol^{-1}. An important feature to note is that the barrier height is reasonably low and that the width (the transfer distance) is of atomic dimensions, about 35 pm.

Is Eyring's theory on proton mobility in water successful in predicting the experimental values of mobility? The unfortunate answer is that this classical calculation is not acceptable at all, partly because it gives mobilities that are much smaller than those observed and it does not fit the demanding criterion of $u_{H+}/u_{D+} = 1.41$. It is necessary to turn to another view.

4.11.5. Penetrating Energy Barriers by Means of Proton Tunneling

A third approach to understanding the anomalous behavior of the nonconforming ion was provided by Bernal and Fowler in the paper mentioned earlier. Their suggestion was that when the proton jumped from an

$$\begin{matrix}H\\H\end{matrix}\!\!>\!O{-}H^+ \text{ to an } O\!<\!\!\begin{matrix}H\\H\end{matrix}$$

converting the water on the right to an H_3O^+ and leaving a water behind, the crossing of the space between the two waters was accomplished, not by crossing above the energy barrier, but by going *through* it. This is quantum mechanical tunneling. The initial application of this type of novel idea—the earliest application of quantum mechanics to chemistry—had been made by Gurney and Condon to the escape of electrons (then called β-particles) from nuclei in radioactivity. To think of protons penetrating a barrier in this way was a big step, because a proton is 1840 times heavier than an electron. However, in the Gamow probability function for the penetration of barriers, the probability P_r is proportional to the exponential of the energies of the jumping particle, that is

$$P_r \propto \exp\left[-\frac{4\pi\ell}{h}\sqrt{m(E-U)}\right] \tag{4.363}$$

where m is the mass of the particle penetrating, ℓ is the length of the jump, and E and U are the total and potential energies of the jumping particle, respectively. There are three factors that could compensate for the great decrease of the tunneling probability obtained by increasing the mass of the particle in the expression. To begin with, the probability of tunneling found for electrons penetrating barriers of 1 to 10 eV (the magnitudes of barriers generally found in chemical problems) is approximately 1 up to a distance of 2000 pm. Nevertheless, the very large increase in mass of the proton compared with the electron—which would seem to greatly decrease the tunneling probability—is reduced in effect by being taken as $\sqrt{1840} = 42.8$. Furthermore, the energy quantities in penetrating the barrier between

$$\begin{matrix}\mathrm{H}\\ \mathrm{H}\end{matrix}\!\!>\!\mathrm{O{-}H^+} \quad \text{and} \quad \mathrm{O}\!<\!\!\begin{matrix}\mathrm{H}\\ \mathrm{H}\end{matrix}$$

are small—around 0.1 eV, compared with the 1 to 10 eV often observed in electron penetration problems. Putting this all together gives $P_{H^+}/P_{e^-} = 1/66$. This is the result of a very rough calculation, but it suggests that even if the path lengths ℓ for jumping were the same for an electron tunneling out of an electrode to a neighboring ion (see Chapter 7) and for a proton passing from one water to another, the proton tunneling probability would be at least 0.01. It is likely[55] to be much more.

This is somewhat the way Bernal and Fowler conjectured the proton to behave more than half a century ago in one of the earliest applications of quantum mechanics to chemistry. However, their conjecture was by no means satisfactory in a numerical sense. It was, as it were, too much of a good thing. The resulting values suggested a velocity for the proton *far greater* than the high rate observed, and also a quite wrong value for u_{H^+}/u_{D^+}. Something was still very much amiss.

4.11.6. One More Step in Understanding Proton Mobility: The Conway, Bockris, and Linton (CBL) Theory

Rosenberg, whose work on proton conduction in the alcohols led to insights into proton conduction, was also a coauthor of a paper that laid the foundation for the development of the theory of proton conduction in solutions. The theory utilized the idea of proton tunneling as outlined earlier, but added an essential limitation to its rate. Thus, in the Eyring theory, the only prerequisite for a proton to pass from an

$$\begin{matrix}\mathrm{H}\\ \mathrm{H}\end{matrix}\!\!>\!\mathrm{O{-}H^+} \quad \text{to an} \quad \mathrm{O}\!<\!\!\begin{matrix}\mathrm{H}\\ \mathrm{H}\end{matrix}$$

in water was a certain degree of activation from the ground state, which is governed by Maxwell's law. The answer was far too small. In the quantum mechanical theory of Bernal and Fowler, the fraction of proton neighbors of a given water which had

[55]The proton's passage turns out from model building to be about 30 pm whereas electrons jumping from an electrode to a neighboring ion tend to jump several thousand picometers.

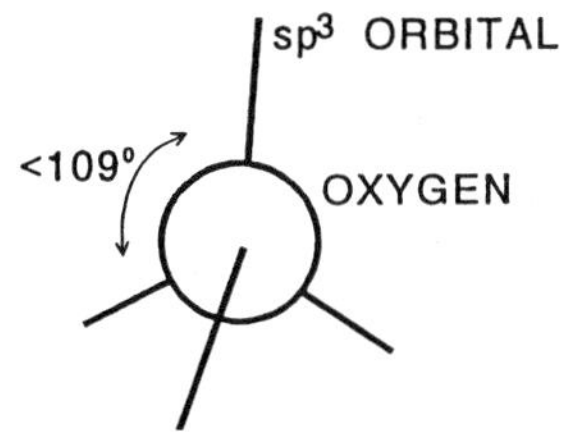

Fig. 4.128. The hybrid orbitals of an oxygen atom. (After Lyons, 1996.)

attained a certain activation energy was multiplied by the Gamow probability equation using the energy value corresponding to that of the activated proton [see Eq. (4.363)], and this product was taken as the probability of protons penetrating the energy barrier at a given height. The probabilities for all the heights were added to obtain the total barrier penetration probability. Now the answer was far too large.

A new group of researchers, Conway, Bockris, and Linton, however, asked and answered an important question: Does it not also matter whether the orbitals of the oxygen of the water molecule to which the proton is tunneling are correctly oriented to receive it? The oxygen atom has four orbitals (Fig. 4.128), two of them occupied by H atoms. It would be reasonable to consider it a nonevent if a proton attempted to tunnel to a water that was incorrectly oriented, as depicted in Fig. 4.129, and a successful event if the orientation was adequate, as in Fig. 4.130.

There are two processes that must cooperate for a successful proton transfer, the basis of proton mobility. The first is *water reorientation* and then the second is *proton tunneling*. Hence the rate of proton transfer will be limited by whichever of the two processes is slower. One must therefore suspect that the water reorientation is the *rate-determining step* in the process of proton transfer (because the tunneling through the barrier has already been shown to be too fast to be consistent with the mobility observed).

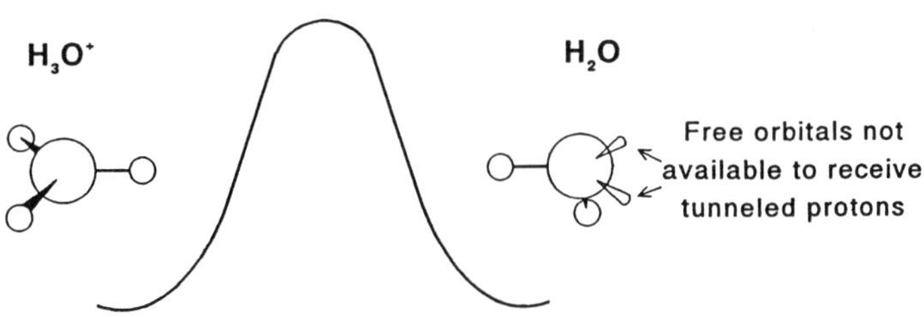

Fig. 4.129. None of the free orbitals of the water molecule are correctly aligned and therefore no proton transfer occurs. (After Lyons, 1996.)

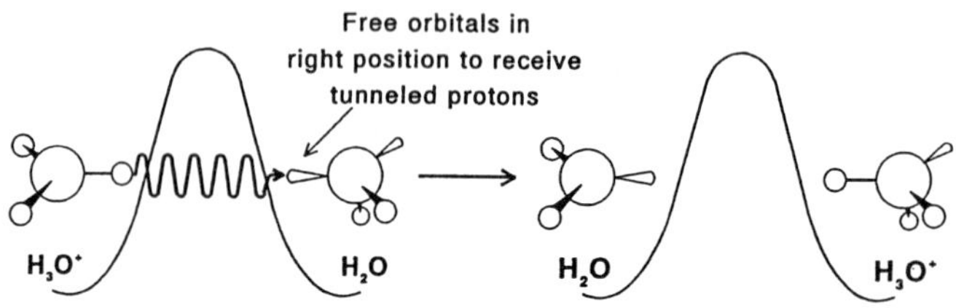

Fig. 4.130. One of the free orbitals of a water molecule is properly oriented and the proton is able to tunnel to the water molecule. (After Lyons, 1996.)

Thus, in the new theory by CBL, a model was formulated in which the librational properties of water played an important part. Even that was not the whole story. The calculation of the specific rate of water reorientation is a complex task. One cannot consider the reorienting water molecule as an isolated entity. If that were so, then one could work on the only basis of the rate of rotation of a *gas molecule* and calculate the rate. However, the water molecule is hydrogen bonded to other water molecules within the 3D lattice and therefore the reorientation involves the torsional stretching and breaking of the hydrogen bonds—an attempt that seldom succeeds.

To summarize, in order for the proton to tunnel successfully, it has to wait for the adjacent water molecules to turn so that they provide a properly oriented orbital. If the tunneling protons have to wait for the water molecules to turn around spontaneously and get into the right position every time a proton arrives on the scene,[56] the predicted mobility will be low compared with experiment. Figure 4.131 shows one of the CBL calculations for the variation of energy in this turning process. However, as is hinted by the figure, the energy of turning and finally breaking the H bonds that hold water down is sufficiently large that it will happen seldom and the excessive speedup due to tunneling will be slowed so much that the reduction in rate caused by the proton having to bide its time will lead to an overcorrection.

Therefore there must be another factor, one that helps turn the water molecule around in the "ready" position. The force to do this is readily found; it is provided by the field of the approaching proton itself.[57] The equation for the energy of a proton–water (ion–dipole) interaction is [see Eq. (A2.2.11)],

$$U_{H^+-W} = -\frac{z_{H^+} e_0 \mu_W \cos\theta}{\varepsilon r^2} \tag{4.364}$$

[56]This spontaneous turning of the water molecules by random motions corresponds to the acceptor water molecule's reorientation by the thermal motions without help from surrounding electric fields.

[57]This process is called the *field-induced reorientation* of the water molecules.

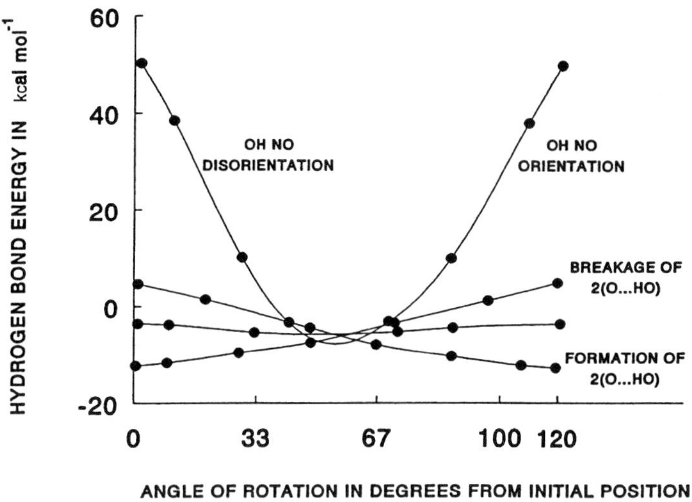

Fig. 4.131. Angular electrostatic potential-energy curves for rotation of H_2O near the H_3O^+ ion (ion dipole contribution not included) (B. E. Conway, J. O'M. Bockris and H. Linton, *J. Chem. Phys.* **24**: 834, 1956).

where θ is the angle between the water dipole and the line between the proton and the dipole center (Fig. 4.132), r is the ion–dipole distance, ε is the dielectric constant of the medium, and μ_W is the dipole moment of water. As θ is reduced to zero, U_{H^+-W}

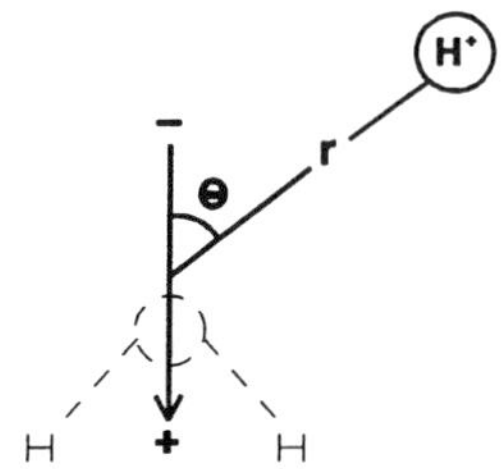

Fig. 4.132. Angle of interaction between a proton (positive charge) and a water molecule (dipole). Maximum interaction occurs when $\theta = 0$, and therefore $\cos\theta = 1$. (After Lyons, 1996.)

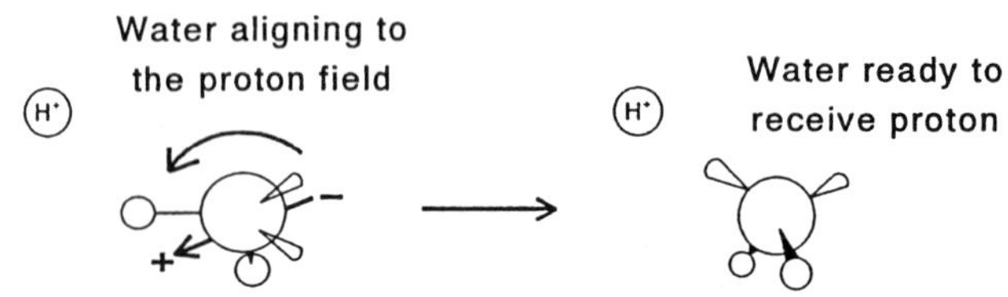

Fig. 4.133. Proton field-induced reorientation of a water molecule. (After Lyons, 1996.)

goes to a maximum numerical value. Hence, as the proton approaches a water turned away from it, it swings it around and with it, its appropriately oriented orbital, as shown in Fig. 4.133.

One last matter remains. What has been said could be a description for the self-diffusion of the proton, where the movements are random. In conduction, there has to be a *preferential* movement in the direction of the electric field applied to the cell. CBL calculated this motion also; it is a small correction, but causes a preferential drift toward the negative electrode.

4.11.7. How Well Does the Field-Induced Water Reorientation Theory Conform with the Experimental Facts?

Can the CBL theory predict experimental values? The answer is a resounding yes, and the results justify printing the lengthy story of an advance made in 1956. The rate of this field-induced water reorientation was faster than the rate of the spontaneous thermal rotation, but turned out to be much slower than the proton tunneling rate. Thus, it is the *field-induced rotation* of water that determines the overall rate of proton transfer and the rate of proton migration through aqueous solutions. According to the theory, the estimated value of the proton mobility is 28×10^{-4} and that observed experimentally is 36×10^{-4} cm^2 s^{-1} V^{-1}!

The anomalous decrease in the heat of activation with an increase in temperature also follows from the model. An increase in temperature causes increased disorder in the water structure, and consequently there are on average, fewer H bonds to break when water molecules reorient. Since the water reorientation increases, the heat of activation becomes smaller.

The decrease in anomalous mobility of the proton in the presence of added alcohol solvents (see Section 4.11.2) could also be explained: the larger size of the alcohol makes its reorientation more difficult than that for water and causes a fall in proton mobility by the tunneling and solvent-oriented method.

Finally, the stringent test for the CBL model was given by the 1.4 ratio of the mobilities of hydrogen and deuterium. Therefore, the correct calculation of this ratio—as the model indeed makes possible—is strong evidence in favor of a mechanism involving field-assisted water reorientation as the rate-determining step in proton tunneling.[58]

4.11.8. Proton Mobility in Ice

If extra confirmation of the CBL model is needed, it is given by interpretation of the fact that in pure ice the proton's *mobility* (not its *conductance*) is approximately 100 times greater than it is in water at the same temperature. Does this means that water molecules turn faster in ice than in water? Intuitively, the reverse might be expected.

However, the CBL model provides the answer. In ice, the concentration of the proton is much less than that in water. Eventually, water molecules in ice do rotate into the correct position, even without the help of an oncoming proton, so that they offer an inviting orbital to any oncoming proton. With so few protons in ice, the waters rotate spontaneously in time for the occasional oncoming proton. Every time a proton is ready to jump, the waiting orbital is there. It is like an all-green traffic light on a main thoroughfare, early in the morning when traffic is light. The former need for water rotation to be spurred by proton attraction is no longer the rate-determining step, and tunneling alone calls the shots. This tunneling is a much faster process than that of water rotation, so that it leads to higher mobilities of protons in ice than in the water or aqueous solutions in which the water rotation is rate determining.

Further Reading

Seminal

1. J. Bernal and R. E. Fowler, "The Structure of Water and Proton Conduction," *J. Chem. Phys.* **1**: 515 (1933).
2. B. E. Conway, J. O'M. Bockris, and H. Lynton, "Solvent Orientation Model for Proton Transfer," *J. Chem. Phys.* **24**: 834 (1956).
3. B. E. Conway and J. O'M. Bockris, "Proton Conduction in Ice," *J. Chem. Phys.* **28**: 354 (1958).
4. T. Erdey Gruz and S. Langyel, "Proton Transfer in Solution," in *Modern Aspects of Electrochemistry, No. 12*, J. O'M Bockris and B. E. Conway, eds., p. 349, Plenum, New York (1978).

Review

1. K. D. Krener, "Proton Conductivity," *Chem. Mat.* **8**: 610 (1996).

[58]Any proton-conduction mechanism in which the rate-determining step involves the breaking of hydrogen bonds would succeed equally well in explaining the heat of activation and alcohol anomalies, but not the u_{H^+}/u_{D^+}.

Papers

1. V. Barone, L. Orlandini, and C. Adams, *Int. J. Quantum Chem.* **56**: 697 (1995).
2. S. Consta and R. Kapral, *J. Chem. Phys.* **104**: 4581 (1996).
3. R. Pommes and B. Roux, *J. Phys. Chem.* **100**: 2519 (1996).
4. J. T. Hynes and D. Borgis, *J. Phys. Chem.* **100**: 1118 (1996).

APPENDIX 4.1. THE MEAN SQUARE DISTANCE TRAVELED BY A RANDOM-WALKING PARTICLE

In the one-dimensional random-walk problem, the expression for $<x^2>$ is found by mathematical induction as follows. Consider that after $N - 1$ steps, the sailor has progressed a distance x_{N-1}. If he takes one more step, the distance x_N from the origin will be either

$$x_N = x_{N-1} + l \tag{A4.1.1}$$

or

$$x_N = x_{N-1} - l \tag{A4.1.2}$$

Squaring both sides of Eqs. (A4.1.1) and (A4.1.2), one obtains

$$x_N^2 = x_{N-1}^2 + l^2 + 2x_{N-1}l \tag{A4.1.3}$$

and

$$x_N^2 = x_{N-1}^2 + l^2 - 2x_{N-1}l \tag{A4.1.4}$$

The average of these two possibilities must be

$$x_N^2 = x_{N-1}^2 + l^2 \tag{A4.1.5}$$

This is the result for x_N^2 when the distance traveled after $N - 1$ steps is exactly x_{N-1}. In general, however, one can only expect, for the value of the square of the distance at the $(N - 1)$th step, an averaged value $<x_{N-1}^2>$, in which case one must write

$$<x_N^2> = <x_{N-1}^2> + l^2 \tag{A4.1.6}$$

At the start of the random walk, i.e., after zero steps, progress is given by

$$<x_0^2> = 0 \tag{A4.1.7}$$

After one step it is

$$<x_1^2> = 1l^2 = l^2 \tag{A4.1.8}$$

After two steps, from Eq. (A4.1.6), one has

$$<x_2^2> = <x_1^2> + l^2 \tag{A4.1.9}$$

and, using Eq. (A4.1.8),

$$<x_2^2> = l^2 + l^2 = 2l^2 \tag{A4.1.10}$$

Similarly,

$$\begin{aligned} <x_3^2> &= <x_2^2> + l^2 \\ &= 2l^2 + l^2 \\ &= 3l^2 \end{aligned} \tag{A4.1.11}$$

Hence, in general,

$$<x_N^2> = Nl^2 \tag{A4.1.12}$$

This equation has been derived for a one-dimensional random walk, but it can be shown to be valid for three-dimensional random flights, too.

The mean square distance $<x^2>$ that a particle travels depends upon the time of travel in the following manner. The number of steps N obviously increases with time and is proportional to it, i.e.,

$$N = kt \tag{A4.1.13}$$

where k is the constant of proportionality. Hence, by combining Eqs. (A4.1.12) and (A4.1.13),

$$<x^2> = ktl^2 \tag{A4.1.14}$$

which may be written

$$<x^2> = \alpha t \tag{A4.1.15}$$

where α is a proportionality constant to be evaluated in the Einstein–Smoluchowski equation.

APPENDIX 4.2. THE LAPLACE TRANSFORM OF A CONSTANT

The Laplace transform $\overline{\alpha}$ of a constant α is by definition (4.33) given by

$$\overline{\alpha} = \int_0^\infty e^{-pz} \alpha \, dz$$

$$\overline{\alpha} = \alpha \left[\frac{e^{-pz}}{-p} \right]_0^\infty$$

$$= \alpha \left[\frac{1}{e^{pz}(-p)} \right]_{z=\infty} - \alpha \left[\frac{1}{e^{pz}(-p)} \right]_{z=0}$$

$$= 0 - \left(\frac{\alpha}{-p} \right)$$

$$= \frac{\alpha}{p}$$

Hence, the Laplace transform of a constant is equal to that same constant divided by p.

APPENDIX 4.3. THE DERIVATION OF EQUATIONS (4.279) AND (4.280)

According to notation [*cf.* Eqs. (4.273) and (4.274)],

$$p_+ = q_+ L_{++} + q_- L_{-+} \tag{A4.3.1}$$

and

$$p_- = q_+ L_{+-} + q_- L_{--} \tag{A4.3.2}$$

Hence, one can carry out the following expansions

$$\frac{p_+}{p_+ q_+ + p_- q_-} = \frac{q_+ L_{++} + q_- L_{-+}}{q_+(q_+ L_{++} + q_- L_{-+}) + q_-(q_+ L_{+-} + q_- L_{--})}$$

$$= \frac{q_+ L_{++} + q_- L_{-+}}{q_+(q_+ L_{++} + q_- L_{+-}) + q_-(q_+ L_{-+} + q_- L_{--})}$$

$$= \frac{q_+ L_{++} \frac{d\varphi}{dx} + q_- L_{-+} \frac{d\varphi}{dx}}{q_+ \left(q_+ L_{++} \frac{d\varphi}{dx} + q_- L_{+-} \frac{d\varphi}{dx} \right) + q_- \left(q_+ L_{-+} \frac{d\varphi}{dx} + q_- L_{--} \frac{d\varphi}{dx} \right)} \tag{A4.3.3}$$

It has been stated [*cf.* Eq. (4.269)] that

$$J_+ = L_{++} \vec{F}_+ + L_{+-} \vec{F}_- \tag{A4.3.4}$$

and that [*cf.* Eqs. (4.276) and (4.277)]

$$\vec{F}_+ = \frac{d\mu_+}{dx} + q_+ \frac{d\psi}{dx} \tag{A4.3.5}$$

and

$$\vec{F}_- = \frac{d\mu_-}{dx} + q_- \frac{d\varphi}{dx} \tag{A4.3.6}$$

Hence, substituting for $\vec{F}_+$ and $\vec{F}_-$ in Eq. (A4.3.4) and setting $d\mu/dx = 0$, one has

$$(J_+)_{d\mu/dx=0} = q_+ L_{++} \frac{d\varphi}{dx} + q_- L_{+-} \frac{d\varphi}{dx} \tag{A4.3.7}$$

Similarly,

$$(J_-)_{d\mu/dx=0} = q_+ L_{-+} \frac{d\varphi}{dx} + q_- L_{--} \frac{d\varphi}{dx} \tag{A4.3.8}$$

In terms of these expressions, Eq. (A4.3.3) becomes

$$\frac{p_+}{p_+ q_+ + p_- q_-} = \left(\frac{J_+}{q_+ J_+ + q_- J_-} \right)_{d\mu/dx=0}$$

$$= \frac{1}{q_+} \left(\frac{q_+ J_+}{q_+ J_+ + q_- J_-} \right)_{d\mu/dx=0} \tag{A4.3.9}$$

By notation,

$$q_+ = z_+ F \quad \text{and} \quad q_- = z_- F \tag{A4.3.10}$$

and therefore Eq. (A4.3.9) can be rewritten as

$$\frac{p_+}{p_+q_+ + p_-q_-} = \frac{1}{z_+F}\left(\frac{z_+FJ_+}{z_+FJ_+ + z_-FJ_-}\right)_{d\mu/dx=0} \tag{A4.3.11}$$

Furthermore, according to the relation between current density and flux,

$$i_+ = z_+FJ_+ \tag{A4.3.12}$$

and

$$i_- = z_-FJ_- \tag{A4.3.13}$$

Using these relations in Eq. (A4.3.11), one has

$$\frac{p_+}{p_+q_+ + p_-q_-} = \frac{1}{z_+F}\left(\frac{i_+}{i_+ + i_-}\right)_{d\mu/dx=0} \tag{A4.3.14}$$

By definition, however [*cf.* Eq. (4.247)],

$$t_+ = \left(\frac{i_+}{i_+ + i_-}\right)_{d\mu/dx=0} \tag{A4.3.15}$$

By combining Eqs. (A4.3.14) and (A4.3.15), the result is

$$\frac{p_+}{p_+q_+ + p_-q_-} = \frac{t_+}{z_+F} \tag{A4.3.16}$$

Similarly, it can be shown that

$$\frac{p_-}{p_+q_+ + p_-q_-} = \frac{t_-}{z_-F} \tag{A4.3.17}$$

APPENDIX 4.4. THE DERIVATION OF EQUATION (4.354)

One can rewrite Eq. (4.353), namely,

$$\Lambda_{\theta=0} = \Lambda^0 - (A + B\Lambda^0)(1 - \theta)^{1/2}c_a^{1/2} \tag{A4.4.1}$$

in the form

$$\Lambda_{\theta=0} = \Lambda^0 \left[1 - \frac{1}{\Lambda^0} mc_a^{1/2}(1-\theta)^{1/2} \right] \tag{A4.4.2}$$

where for conciseness the symbol m is used instead of $(A + B\Lambda^0)$.

It has been shown, however [*cf.* Eq. (4.349)], that

$$1 - \theta = \frac{\Lambda}{\Lambda_{\theta=0}} \tag{A4.4.3}$$

If this relation is used in Eq. (A4.4.2), one gets

$$\Lambda_{\theta=0} = \Lambda^0 \left[1 - \frac{mc_a^{1/2}}{\Lambda^0} \left(\frac{\Lambda}{\Lambda_{\theta=0}} \right)^{1/2} \right] \tag{A.4.4.4}$$

and substituting for $\Lambda_{\theta=0}$ in the right-hand side of Eq (A4.4.2), the result is

$$\Lambda_{\theta=0} = \Lambda^0 \left\{ 1 - \frac{m(\Lambda c_a)^{1/2}}{\Lambda^{0\,3/2}} \frac{1}{[1 - (mc_a^{1/2}/\Lambda^0)(1-\theta)^{1/2}]^{1/2}} \right\} \tag{A4.4.5}$$

One has again been left with $(1 - \theta)^{1/2}$ on the right-hand side and thus one again substitutes from Eq. (A4.4.2). This process of substitution can be repeated *ad infinitum* to give the result

$$\Lambda_{\theta=0} = \Lambda^0(1 - z\{1 - z[1 - z(\cdot\cdot\cdot)^{-1/2}]^{-1/2}\}^{-1/2}) \tag{A4.4.6}$$

$$= \Lambda^0 Z \tag{A4.4.7}$$

where

$$z = 1 - z\{1 - z[1 - z(\cdot\cdot\cdot)^{-1/2}]^{-1/2}\}^{-1/2} \tag{A4.4.8}$$

with

$$z = \frac{m(\Lambda c_a)^{1/2}}{\Lambda^{0\,3/2}} = \frac{(A + B\Lambda^0)c_a^{1/2}\Lambda^{1/2}}{\Lambda^{0\,3/2}} \tag{A4.4.9}$$

EXERCISES

1. The unit flux has been used in an attempt to simplify the solution of the partial-derivative equation of Fick's second law. (a) Calculate, for a univalent ion, what current density this flux will cause. Is this current density achievable

in a real experiment? (b) 1 mA/cm^2 is a current density typically used in electrochemistry laboratories. What is the flux in this case? (Xu)

2. Calculate the concentration gradient of a univalent ion in 0.1 M solution at 25 °C when the electric field is 10^5 V cm^{-1} and the current density at the electrode is 1 mA cm^{-2}. (Kim)

3. Estimate the electrophoretic velocity of a sodium ion in 0.01 M NaCl solution under an electric field of 0.1 V cm^{-1}. The viscosity of the solution is 0.00895 poise. (Kim).

4. Ohm's law is generally applicable to electrolytes in solution. A theory suggests that the observed current depends on the difference of two exponential terms, i.e., $(e^{AX} - e^{-AX})$, where A is $zFl/2RT$. The term l is the distance between "sites" in diffusion, z is the charge on the ion, and X is the applied field. Calculate the applied field in V cm^{-1} at which Ohm's law breaks down.

5. If a current of 10 A is passed between two electrodes, each of 10 cm^2, and one is depositing Cu metal from a $CuSO_4$ solution, calculate the thickness of this deposit after 6 h.

6. In an instantaneous-pulse experiment, the electrode material is radioactive and hence detectable by a Geiger counter. As the pulse is realized with an electronic device generating a current of 10 A on a 0.1-cm^2 electrode for 0.1 s, with a Geiger counter placed 1 cm from the electrode, register the trace of the radioactive univalent ion at 450 s after the pulse. Calculate the limiting sensitivity of the instrument. Suppose the diffusion coefficient of the ion is the typical value of 10^{-9} m^2 s^{-1}. (Xu)

7. Estimate the diffusion coefficient of Na^+ and Cl^- in water at 298 K from the equivalent conductivity at infinite dilution, Λ^0 (NaCl) = 126.46 S cm^2 eq^{-1} and the cation transport number t^0 (Na^+) = 0.396. (Herbert)

8. In experiments involving radiotracer measurements of diffusion in molten salt, the Stokes–Einstein equation has been found to be roughly applicable. For a series of ions, in molten salts it was found that the product $D\eta T^{-1} = 10^{-9}$ dyn mol^{-1}. From this information, find whether the best form of the coefficient in this expression for this case is nearer to 6 or 4.

9. For aqueous solutions at room temperature, the order of magnitude of the diffusion coefficient of most of the common simple ions (Na^+, K^+, Mg^{2+}, ClO_4^-) is 10^{-5} cm^2 s^{-1}. Suppose now that a capillary tube containing a solution of an electrolyte is brought into contact vertically and very gently with a capillary of pure water; about how far would the electrolyte diffuse into the capillary of water in 24 hr?

10. It is easy to understand that protons must have an abnormal mechanism of transport through the solution. Thus, protons in aqueous solution exist almost entirely as H_3O^+ ions yet their conductivity is several times higher than one would expect if they transported primarily by rolling along as H_3O^+. Imaginatively discuss the alternatives. Explore what is meant by "transfer to a neighboring water molecule." Can the proton come into this neighbor from any direction? Why does it prefer to drift, particularly in one direction?

11. Write equations representing each of Fick's laws. Exemplify the type of situation in which each law applies. A glucose solution ($D_{glucose} = 0.67 \times 10^{-5}$ cm^2 s^{-1}) diffuses across a region of 0.2 cm in length and 4.8 cm^2 in area. On one side, the concentration is 4 g dm^{-3} and on the other, effectively zero. Calculate the rate of diffusion of the sugar across the boundary in g s^{-1}.

12. At first the results arising from the Einstein–Smoluchowski equation ($\langle x^2 \rangle = 2Dt$) may seem difficult to understand. Thus, the diffusion considered in the equation is random. Nevertheless, the equation tells us that there is net movement in one direction arising from this random motion. Furthermore, it allows us to calculate how far the diffusion front has traveled. Is there something curious about randomly moving particles covering distances in one direction? Comment constructively on this apparently anomalous situation.

13. Show that the diffusion coefficient (D) is independent of the concentration for dilute (ideal) solutions, using the example of planar steady-state diffusion. Explain the meaning of the negative sign that is usually inserted into Fick's first law of steady-state diffusion. (Bock)

14. A definition for specific ion conductivity frequently cited in electrochemical literature is $\sigma = E_i n_i u_i z_i e_0$, where i is ionic species, n_i the number of the ion in a unit volume, u_i the conventional mobility, z_i the valence state, e_0 the elemental charge, and E the electric field. Derive this definition. (Xu)

15. When it comes to practical application, the actual conductance (the inverse of resistance R) instead of specific conductivity is the important variable. This is the reason why polymer electrolytes have drawn so much attention as a potential component of alkali-metal batteries although their specific conductivities are usually low (~10^{-5} S cm^{-1}) compared with those at nonaqueous electrolytes (~10^{-2} S cm^{-1}). Calculate the conductance of 1.0 M $LiSO_3CF_3$ in poly(ethylene oxide) and propylene carbonate, respectively. The former is fabricated into a film of 10 Tm thickness and the latter is soaked with porous separator of 1 mm thickness. (Xu)

16. A conductance cell containing 0.01 potassium chloride was found to have a resistance of 2573 Ohms at 25 °C. The same cell when filled with a solution of 0.2 N acetic acid had a resistance of 5085 Ohms. Calculate (a) the cell constant, (b) the specific resistances of the potassium chloride and acetic acid solutions,

(c) the conductance ratio of 0.2 N acetic acid, utilizing the following data at 25 °C: $\lambda^0_{H^+} = 349.82\ \Omega^{-1}\ cm^2\ eq^{-1}$; $\lambda^0_{CH_3\text{-}COO^-} = 40.9\ \Omega^{-1}\ cm^2\ eq^{-1}$; specific conductivity of 0.01 N KCl = 0.0014 $\Omega^{-1}\ cm^{-1}$. (Constantinescu)

17. A 0.2 N solution of sodium chloride was found to have a specific conductivity of $1.75 \times 10^{-2}\ \Omega^{-1}\ cm^{-1}$ at 18 °C; the transport number of the cation in this solution is 0.385. Calculate the equivalent conductance of the sodium and chloride ions. (Constantinescu)

18. A conductance cell having a constant k of 2.485 cm^{-1} is filled with 0.01 N potassium chloride solution at 25°C; the value of Λ for this solution is 141.2 S $cm^2\ eq^{-1}$. If the specific conductance, σ, of the water employed as solvent is $1.0 \times 10^{-6}\ S\ cm^{-1}$, what is the measured resistance of the cell containing the solution? (Constantinescu)

19. The ionic conductivity at infinite dilution of a divalent copper ion is $\lambda^0\ (Cu^{2+}) = 55\ S\ cm^2\ eq^{-1}$ and its ionic radius, $r_i = 72$ pm. Calculate the primary solvation number of Cu^{2+} taking r_W = 138 pm for the radius of the water molecule. (Herbert)

20. Calculate the equivalent conductivity of a 0.1 M NaCl solution. The diffusion coefficient of Na^+ is $1.334 \times 10^{-5}\ cm^2\ s^{-1}$ and that of Cl^{-1} is $2.032 \times 10^{-5}\ cm^2\ s^{-1}$. (Kim)

21. Calculate the conductivity of NaI in acetone (η = 0.00316 poise) when the radii for Na^+ and I^- are 260 and 300 pm, respectively. (Kim)

22. Calculate the conventional mobility of a sodium ion in 0.01 M NaCl. The viscosity of the solution is 0.00895 poise and the Stokes radius of a sodium ion is 260 pm. (Kim)

23. In acetonitrile (η = 0.00345 poise), the equivalent conductivity for very dilute solutions of KI is 198.2 at 25 °C. Calculate the equivalent conductance of KI in a similar concentration range in acetophenone. (η = 0.0028 poise)

24. Calculate the specific and equivalent conductivity of LiBr using the data in the text.

25. A student has to determine the equivalent conductivity at infinite dilution for KCl, NaCl, KNO_3, and $NaNO_3$ solutions and the transport numbers of the ions in these solutions. He managed to determine only λ^0 (KNO_3), λ^0($NaNO_3$), t^0_+ (Na^+/NaCl) and t^0_+ (K^+/KCl) and wrote them in a table:

	$NaNO_3$	KNO_3	NaCl	KCl
λ^0/S $cm^2\ eq^{-1}$	121.4	144.9	—	—
t^0_+	—	—	0.396	0.490
t^0_-	—	—	—	—

Assuming that the determined values are correct, help him fill in the blanks in the table without doing any further experiments. (Herbert)

26. Given the transport number of Ca^{2+} in $CaCl_2$, $t(Ca^{2+}/CaCl_2) = 0.438$ and of K^+ in KCl, $t(K^+/KCl) = 0.490$, calculate the transport number of Ca^{2+} in a solution containing both 0.001 *M* $CaCl_2$ and 0.01 *M* KCl, neglecting the variation of *t* and λ^0 with concentration. (Herbert)

27. A current of 5 mA flows through a 2-mm inner-diameter glass tube filled with 1 *N* $CuSO_4$ solution in the anode compartment and with a $Cu(CH_3COO)_2$ solution in the cathode compartment. The interface created between the two solutions moves 6.05 mm toward the anode in 10 min. Calculate the transport number of the sulfate ion in this solution. (Herbert)

28. In a 1:1 electrolyte, measurements showed diffusion coefficients of 5.1×10^{-5} $cm^2\ s^{-1}$ for the cation and 2.8×10^{-5} $cm^2\ s^{-1}$ for the anion. Calculate the transport number of the anion.

29. Calculate the absolute mobility of a sodium ion when the drift velocity is 5.2×10^{-5} cm s^{-1} under an electrical field of 0. 1 V cm^{-1}. (Kim)

30. Calculate the conventional mobility of a sodium ion in aqueous solution using the diffusion coefficient of 1.334×10^{-5} $cm^2\ s^{-1}$ at 25 °C. (Kim)

31. Calculate the radius of the solvated sodium ion in aqueous solution when the absolute mobility of the ion is 3.24×10^8 cm s^{-1} dyn^{-1}. The viscosity of the solution is 0.01 poise. (Kim)

32. The mobility of a cation is 5×10^{-4} $cm^2\ V^{-1}\ s^{-1}$ and that of the accompanying anion is 2×10^{-4} $cm^2\ V^{-1}\ s^{-1}$. Calculate the specific conductivity of a 2×10^{-3} *M* solution.

33. A tube of glass 60 cm high is closed at one end, blue copper sulfate hydrate crystals are placed at the bottom, and water is introduced. Calculate how far a blue color will have spread upward after 1 min, 1 day, 1 week, and 1 month.

34. Ions are pumped into a system electrochemically. At $t = 0$, a short burst of dissolution of an electrode is caused, giving rise to n_{total} ions, which then begin to diffuse away from the source. Seek in the text the appropriate equation by which one may know the number of ions at a distance *x* and time *t*. This is a plane-source problem. Thus, Cu ions could be dissolved from a Cu plate filling the end of a tube of solution. The question is how many ions would have diffused 11 cm in 300 s? $D_{CuSO_4} = 0.82 \times 10^{-5}$ $cm^2\ s^{-1}$.

35. Imagine, now, a horizontal tube of dilute KCl solution. Exactly in the middle is a thin slab of smooth, solid copper sulfate covered with an insoluble protective plastic layer, suddenly removed at $t = 0$. At *x* (which can be positive for the

right-hand side of the $CuSO_4$ slab and negative for the left) equals zero, the $CuSO_4$ begins to dissolve and diffuse in both directions. Draw a qualitative schematic diagram, the concentration of $CuSO_4$ (taken to be unity at $x = 0$) on the ordinate and the distance x in either direction on the abscissa. On the diagram, plot the $CuSO_4$ concentration (a) for a very long time, (b) for a day, and (c) for an hour.

36. The ionic mobilities given in tables are around 10^{-4} cm^2 V^{-1} s^{-1}. What would be the corresponding order of magnitude for the absolute mobility (the velocity under an accelerating force of 1 dyn s^{-1}).

37. A saturated solution of silver chloride, when placed in a conductance cell with a constant $k = 0.180$ cm^{-1}, has a resistance of 67.953 kΩ at 25 °C. The resistance of the water used as solvent was found to be 212.180 kΩ in the same cell. Calculate the solubility S of the salt at 25 °C assuming it to be completely dissociated in its saturated solution in water. (Constantinescu)

38. Utilize the calculated values of the thickness of the ionic atmosphere R^{-1} in 0.1 N solutions of a univalent electrolyte in (a) nitrobenzene, (b) ethyl alcohol, and (c) ethylene dichloride to calculate the relaxation times of the ionic atmospheres. (Constantinescu)

39. Estimate the time for an ionic cloud to relax around a sodium ion in 0.1 M NaCl when the drift velocity is 5.2×10^{-5} cm s^{-1} under an electrical field of 0.1 V cm^{-1}. (Kim)

40. In the text, a discussion of what happens to an ionic atmosphere when the ion at its center is discharged gives rise to an equation for the relaxation time of the ion atmosphere (as it disperses). Find such an expression. Apply it to find the time the ionic atmosphere takes to relax around Na^+ ions in a 0.01 M NaCl solution when the diffusion coefficient of Na^+ is 1.93×10^{-5} cm^2 s^{-1}.

41. The diffusion coefficient of an ion in water is 1.5×10^{-5} cm^2 s^{-1}. It seems reasonable to take the distance between two steps in diffusion as roughly the diameter of a water molecule (320 pm). With this assumption, calculate the rate constant in s^{-2} for the ion's diffusion.

42. Assume that a solution (100 ml) containing Fe^{3+} is reduced at a constant current density, (j), of 100 mA cm^{-2} employing planar electrodes of 10-cm^2 area. Calculate the time after which the concentration of Fe^{3+} ($c_{\text{initial}} = 10^{-2}$ M) would have decreased by 10%. (Bock)

43. An investigator wants to study the Debye effect of diluted NaCl solution at room temperature but has no clue about what frequency range he should look at. Please help him. The diffusion coefficient of 0.001 M NaCl solution is 1.5×10^{-9} m^2 s^{-1}. (Xu)

44. Calculate the junction potentials for the following situations. (a) 0.1 *M* HCl/0.01 *M* HCl, $t_+ = 0.83$, (b) 0.1 *M* KCl/0.01 *M* KCl, $t_+ = 0.49$. (Kim)

45. From data in the text, calculate the degree of association in NaCl in a 2 *M* solution.

PROBLEMS

1. In the experiment described in Exercise 6 it was found that at a certain time the Geiger counter registered a maximum ion flux, i.e., the intensity of the radiation has a maximum with respect to time. It was also found that by placing the Geiger counter farther away from the electrode, the time at which the maximum occurs becomes longer, and the peak intensity of the maximum decreases rapidly. Justify this observation and evaluate its usefulness in experimentally measuring diffusion coefficients of ions. (Xu)

2. Does the ion valence affect the statement that the ion diffusion coefficient can be considered a constant? Take electrolytes of the z:z type, for example, 1:1 and 2:2, and compare their diffusion coefficient variation over the concentration range of 0.1 to 0.01 mol dm^{-3}. (Xu)

3. The Einstein–Smoluchowski equation, $\langle x^2 \rangle = 2Dt$, gives a measure of the mean-square displacements of a diffusing particle in a time t. There $\langle x^2 \rangle$ is the mean-square distance traveled by most of the ions. Common observation using dyes or scents shows that diffusion of some particles occurs far ahead of the diffusion front represented by the $\langle x^2 \rangle = 2Dt$ equation. Determine the distance of this Einstein–Smoluchowski diffusion front for a colored ion diffusing into a solution for 24 hr ($D = 3.8 \times 10^{-5}$ $cm^2\ s^{-1}$). Determine for the same solution how far the farthest 1% of the total diffused material diffused in the same time. Discuss how it is possible that one detects perfume across the space of a room in (say) 30 s.

4. In a molten salt solution of $CdCl_2$ in KCl, radiotracer measurements of the diffusion coefficient of Cd at 470 °C showed the heat of activation to be 23.0 kJ mol^{-1}. A rough calculation of the entropy of activation showed this to be small, about 41.8 J $mol^{-1}\ K^{-1}$. When the composition of the melt is 66% KCl, the diffusion coefficient is 1.3×10^{-5} $cm^2\ s^{-1}$. Use these data to examine which of the two models for transport in liquids—jumping from site to site or shuffling along—is favored.

5. There are several ways of expressing ionic mobility. According to one of them, the absolute mobility, u_{abs}, is the velocity of an ion under an applied force of 1 dyne. The conventional mobility, u_{conv}, on the other hand, is the velocity under the force exerted on an ion by its interaction with an electric field of 1 V cm^{-1}. Deduce the relation between u_{conv} and u_{abs}.

6. The self-diffusion coefficients of Cl^- and Na^+ in molten sodium chloride are, respectively, $33 \times 10^{-4} \exp(-8500/RT)$ and $8 \times 10^{-4} \exp(-4000/RT)$ $cm^2 s^{-1}$. (a) Use the Nernst–Einstein equation to calculate the equivalent conductivity of the molten liquid at 935°C. (b) Compare the value obtained with the value actually measured, 40% less. Insofar as the two values are significantly different, explain this by some kind of structural hypothesis.

7. It is normal to think that positive cations are reduced and deposited at negative cathodes in electrolysis and negative anions react at anodes to be oxidized. Although this is indeed the norm, there are a number of cases where negative anions undergo reaction at negative cathodes. A well-known example is that of the chromate anion, which is the entity from which chromium metal plates out. Consider the phenomenon in terms of the Nernst–Planck equation. Using a 2:2 electrolyte, $D = 5 \times 10^{-5}$ $cm^2 s^{-1}$, $c = 1.5 \times 10^{-3}$ mol cm^{-3}, and $X = 1 \times 10^4$ V cm^{-1}, calculate the needed $\frac{dc}{dx}$ to make anion deposition at a cathode possible.

8. Blum has developed an MSA approach to conductance (Equation 4.328). It applies well in representing conductance as a function of concentration. Conversely, it neglects established characteristics of electrolytes, such as their hydration (which changes with concentration and hence affects the ionic mobility) and association (which has been measured spectroscopically and calculated theoretically to be substantial in the concentration region worked on by Blum). Examine Blum's equation in this text. Compare his treatment with that of Lee and Cheaton. Does parametrization play a part in explaining why Blum gets the right results with a model that neglects established aspects of the structure of the moving particles?

9. The equivalent conductivities of KCl and $MgCl_2$ aqueous solutions at 25 °C were estimated as 146.95 and 124.11 (S^{-1} cm^2 eq^{-1}), respectively. Calculate the molar and the specific conductivities when the concentrations of both solutions were 10^{-3} g-eq per 1000 cm^3. What would be the measured resistance of these two solutions when two planar Pt electrodes of 2-cm^2 area and 0.5 cm apart are employed? Measurements of the specific conductivity and hence of the solution resistance are usually carried out under a small ac field. Explain why a small ac field is used. (Bock)

10. Walden's empirical rule states that the product of the equivalent conductivity and the viscosity of the solvent should be constant at a given temperature. Explain the data in Table P.1, which were obtained for NaI solutions in several different solvents at 25 °C. Calculate the radius of the moving entity in acetone, applying Walden's rule. (Bock)

11. (a) Estimate the concentration of the supporting electrolyte (i.e., KCl) which must be added to a 10^{-6} M HCl solution in order to study the diffusion of protons using the data in Table P.2. (Bock)

TABLE P.1

Solvent	4O (esu C cm^{-1} eq^{-1})
Ethanol	0.754
Acetone	0.820
Isobutanol	0.725

(b) Explain the differences between the cation transference numbers listed in Table P.3.

12. According to Faraday's laws of electrolysis, an amount of electricity (i.e., number of electrons) will cause the equivalent weight of an ion in solution to react at the electrode. In a very simple case, one might envisage the deposition of Ag^+ (needing 1 mole of electrons per mole of Ag) to deposit at the negative electrode or cathode of an electrolytic cell. Correspondingly, at the other electrode, one might imagine an anodic oxidation to be occurring so that 1 mole of Fe^{2+} (say) would be oxidized to 1 mole of Fe^{3+}. At each electrode, the same number of electrons would be transferred, and the same number of moles of reactant affected. This sounds simple and expected. However, there is an apparent problem. To react at the electrodes, ions have to be transported through the solution to the interface at the electrode at a sufficient rate. This rate is a fraction of the current given by the transport number. All would be well if each transport number were exactly 0.5. However, this is not the case because transport numbers vary greatly. In extreme cases, for very large ions, they tend to be zero. Explain, with equations and diagrams, how Faraday's laws can still be obeyed.

13. It is desired to know the transport number of protons in trifluoromethanesulfonic acid. The actual measurements made were of tritium self-diffusion and the result

TABLE P.2

Cation Mobilities (*u*) Estimated at 25 °C for 0.1 *M* Solutions

Species	u [cm^2 V^{-1} S^{-1}]
H^+	$33.71 \times {}^{-4}$
K^+	6×10^{-4}

TABLE P.3

t_+^0 Measured at 25 °C, in Aqueous Solution

Salt	t_+^0
KCl	0.4905
KBr	0.4847
KI	0.4887

obtained for 5 *M* CF_3SO_2OH at 80 °C was 2.13×10^{-5} cm^2 s^{-1}. The relative values of the mobilities of u_H and u_T are (according to a modeling hypothesis) $\sqrt{3} = 1.73$. As far as $D_{CF_3SO_3^-}$ is concerned, this can be obtained from the Stokes–Einstein equation. The necessary viscosity data are shown in Figure P4.1. Calculate the transport number of H^+ in CF_3SO_2OH at 80 °C. (The radius of $CF_3SO_2O^-$ can be obtained from models.)

14. Ohm's law implies that the equivalent conductivity is independent of the strength of the applied electric field. This is certainly so for a very wide variety of applied fields, 1 to 10^5 V cm^{-1}, in fact. However, Wien showed that (with appropriate precaution taken against heating of the solution, etc.), the equivalent conductivity of electrolytes undergoes a substantial increase at about 10^6 V cm^{-1}. By appropriate consideration of the ionic atmosphere and its time of relaxation, show that a credible model to explain the above is that the high applied field

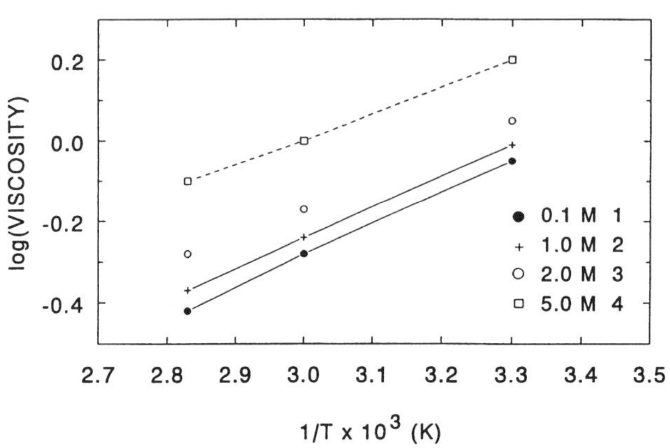

Fig. P.4.1. Plot of log η vs. $1/T \times 10^3$ K for CF_3SO_3H at different concentrations.

TABLE P.4

Ion	u_{conv} ($cm^2\ V^{-1}\ s^{-1}$)
H^+	3.625×10^{-3}
K^+	7.619×10^{-4}
Cl^-	7.912×10^{-4}
NO_3^-	7.404×10^{-4}

causes the central ion to travel so fast that in fact the ionic atmosphere does not have time to form around the ion as it travels through the solution.

15. (a) Derive and plot the relations for variation in ion concentration at the surface of the electrode, under conditions of constant flux and instantaneous pulse, respectively. (b) In the constant flux-induced diffusion, the time when the ion concentration at the electrode surface reduces to zero is called the *transition time,* and is designated as τ. Derive an expression of τ and comment on its physical significance. (Assume that in the constant flux experiment the concentration change is only caused by diffusion; i.e., the contribution of ion migration to concentration change is suppressed and therefore negligible.) (Xu)

16. Calculate the junction potentials for the following situations. (a) 0.1 *M* HCl/0.1 *M* KCl, (b) 0.1 *M* HCl/0.01 *M* KNO_3. Refer to Table P.4. (Kim)

17. A two-compartment electrochemical cell contains NaCl in one compartment and KCl in the other. The compartments are separated by a porous partition. Concentrations of both the electrolytes are equal. If Λ_{NaCl} and Λ_{KCl} are the equivalent conductivities of the two solutions, show that the liquid junction potential E_L is given by

$$E_L = \frac{RT}{F} \ln \frac{\Lambda_{NaCl}}{\Lambda_{KCl}}$$

TABLE P.5

Transference Numbers (t_+) Measured at 25 °C for 0.1 *M* Solutions

Salt	t_+
HCl	0.83
KCl	0.49

18. (a) What are the liquid–liquid junction potentials for a cell consisting of HCl (0.1 *M*) in contact with KCl when (a) the concentration of KCl is 0.1 *M* and (b) a saturated KCl solution (i.e., 4.2 *M*) is employed? Refer to Table P.5.

    (b) Discuss any practical advantages of selecting KCl and/or changing the concentration of the electrolyte in one half-cell. Also consider the factors that could introduce deviations between the calculated and measured values of E_j. (Bock)

19. In the study of nonaqueous electrolytes, the ion-pair effect is a severe factor affecting ion conduction. The degree of association of salts in nonaqueous solvents (or the solubilizing ability of the different solvents toward the salt) is often estimated by comparing the Walden product, that is, $\Lambda\eta$. Justify this method and explain what hypothesis is included and how it holds. (Xu)

20. The Einstein–Smoluchowski equation, $\langle x^2 \rangle = 2Dt$, is a phenomenological equation derived for diffusion along one coordinate. (For example, after the release of a barrier, along a tube containing a liquid.) However, it also applies to any medium. Suppose, now, that metal ions, (e.g., Pt) are deposited on a Pd substrate. Calculate how far the Pt would diffuse into the Pd in 6 weeks. (The diffusion coefficient of Pt into Pd can be estimated from other data as 9×10^{-14} at 295 K.)

MICRO RESEARCH PROBLEMS

1. (a) In an electrochemical analysis experiment, a univalent ion in a 10^{-3} *M* solution with a large amount of indifferent electrolyte is constantly oxidized at the electrode with a current density of 1.0 mA cm^{-2}. Calculate at 0.05 s after the constant current is switched on, the ion's concentration at 10^{-9} m, 5 10^{-9} m, and 20 10^{-9} m from the electrode, respectively. The diffusion coefficient of the ion is 10^{-9} m^2 s^{-1} at 250°C. Use the error function table in the text if necessary. (b) Calculate the concentration at 25 10^{-9} m, 40 10^{-9} m, and 100 10^{-9} m, at 0.5 s after the constant flux is switched on. Compare the results obtained at 0.05 s. (c) Based on the above results and results of Problem 15, qualitatively draw the three-dimensional distribution of the ion with respect to both time *t* and distance *x*. (Xu)

2. From several points of view (for example, in battery technology), it is extremely important to have solutions of maximum specific conductivity. Basically, the specific conductivity increases proportionally to the concentration of ions. On this simplistic ground, it would seem important to have as high a concentration as possible up to the solubility limits. However (see the work of Haymet described in Chapter 3), increasing concentration leads to an increase of ionic association which decreases the concentration of conducting entities. Also, as

the concentration increases, the mobility decreases. By investigating the material in the text and in books of data, propose a solution (aqueous or nonaqueous, including a room-temperature molten salt) that would have a maximum conductivity at 25°C. Estimate its value. Do not omit considering the effect of viscosity and the insights of the Stokes–Einstein equation on the diffusion coefficient and hence (through the Nernst–Einstein equation) on mobility and conductivity.

CHAPTER 5

IONIC LIQUIDS

5.1. INTRODUCTION

5.1.1. The Limiting Case of Zero Solvent: Pure Electrolytes

Modern electrochemistry is concerned not only with systems based on aqueous solutions but also with solvent-free systems. Indeed, it is in such systems that many important electrochemical processes are carried out, such as the production of metals (aluminum, sodium, and magnesium) and the development of high-energy-density batteries.

The rationale behind the use of (and the search for) media other than water will be restated (see also Section 4.8). In aqueous media, electrode reactions involving hydrogen ions and hydroxyl ions may compete with and even supplant the desired electrochemical process (as in the deposition of magnesium from aqueous solutions). Furthermore, in technologies based on the conversion of chemical energy into electrical energy and vice versa, the desired rate of conversion may be limited by the conductivity of the solution: when working with pure electrolytes (as in eutectics such as those formed by LiCl-KCl), conductivity is never limiting.

Some of the difficulties associated with carrying out processes in aqueous solutions, particularly the undesired competition from hydrogen and oxygen evolution, can be sometimes overcome by using nonaqueous solvents consisting usually of organic substances (e.g., acetonitrile) to which are added a solute that dissociates in that solvent. However, often this is not a good approach just because of the low specific conductivities of such solutions (see Section 4.8.5) and their tendencies to absorb water from the surroundings.

So the question arises: Why have a solvent at all? This limiting case of an aqueous or a nonaqueous ionic solution from which all the solvent is removed is a *pure liquid electrolyte.* Conceptually, this definition is accurate. Operationally, however, if one

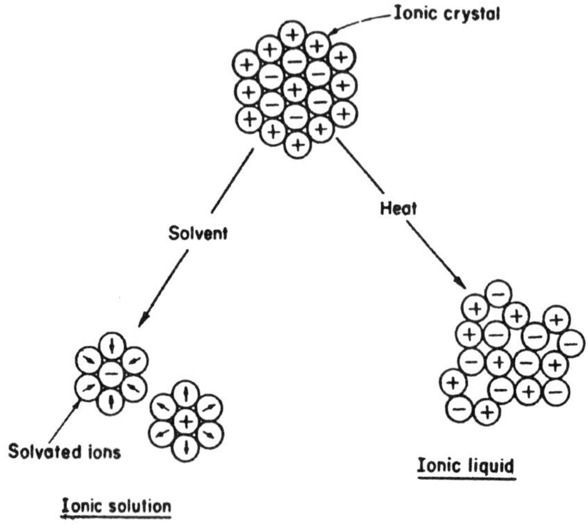

Fig. 5.1. An ionic crystal can be dismantled either by the action of a solvent or by the action of heat.

removes solvent molecules from a solution by evaporation, for example, one is left with ionic crystals, pure *solid* electrolyte. A further step in conversion from this solid to the pure liquid form is necessary.

5.1.2. Thermal Loosening of an Ionic Lattice

The process of dissolution of a true electrolyte was described in Chapter 2. The basic picture is that the ions in an erstwhile-rigid ionic lattice succumb to the strong attraction[1] of the solvent molecules and follow them into solution, executing a random walk there as free, stable solvated ions. The result is an ionic solution that has the ability to conduct electricity by means of the preferential drift of ions in the direction determined by the applied electric field and the charge on the ion. The disassembly of the ionic lattice was achieved by the solvent overcoming the Coulombic cohesive forces holding together the ions in the regular arrangement called a *lattice* (Fig. 5.1).

A solvent, however, is not the only agency that can cause an ionic lattice to fall to pieces. Heat energy also can overcome the cohesive forces and disrupt the ordered arrangement of ions in a crystal (Fig. 5.1). This process of melting results in a pure liquid electrolyte, a system having a conductance several orders of magnitude larger than that of the corresponding solid (Table 5.1).

[1]In the case of aqueous solutions, the forces are essentially ion–dipole and ion–quadrupole in character.

TABLE 5.1

Specific Conductivities of Solid and Molten NaCl

	Specific conductivity (s cm^{-1})
Solid NaCl	1×10^{-3} at 1073 K
Molten NaCl[a]	3.9 at 1173 K

[a]Melting point of NaCl is 1074 K.

5.1.3. Some Differentiating Features of Ionic Liquids (Pure Liquid Electrolytes)

A common type of ionic lattice is that of a crystalline salt. One such ionic lattice encountered in everyday life is sodium chloride. Molten sodium chloride is a typical liquid electrolyte and displays the characteristics of many liquid electrolytes.[2]

An appreciation of the properties of liquid electrolytes can be gained by a comparison between molten ice (water) and molten sodium chloride (Table 5.2). Both liquids are clear and colorless. Their viscosities, thermal conductivities, and surface tensions near their melting points are not very different.

In fact, one can go further and make the following statement: Molten salts look like water and not far above their melting points have viscosities, thermal conductivities, and surface tensions on the same orders of magnitude as those of water. In general, however, and with the important exception of some $AlCl_3$–complex organic systems, most fused salts are stable as liquids only at relatively high temperatures (500 to 1300 K) (Table 5.3).

One can quote exceptions to these generalizations. The tetraalkylammonium salts as a class are liquid at temperatures below 300 K. There are liquid electrolytes—produced from dissolving $AlCl_3$ into some complex organics—which are liquid at room temperature (Tables 5.3 and 5.4). Above the normal range of 300–1300 K is another set of molten electrolytes, the molten silicates, borates, and phosphates, for which the characteristic temperature range is 1300–2300 K (Tables 5.5 and 5.6).

5.1.4. Liquid Electrolytes Are Ionic Liquids

The crucial difference between molten salts and molten ice lies in the values of the specific conductivity (Table 5.7). Fused salts have about 10^8 times greater specific conductivity than fused ice.

The temptation to ascribe the high conductance of fused salts to conduction by electrons must be rejected. The conductivity of a molten salt is high compared to that of water; but it is ten thousand times lower than that of a liquid metal, such as mercury (Table 5.8).

---

[2]The terms *pure liquid electrolyte, ionic liquid, fused salt,* and *molten salt* are used synonymously.

TABLE 5.2

Comparison of Some Properties of Water and Molten NaCl

	Water 298 K	Molten NaCl 1123 K
Viscosity (millipoise)	8.95	2.5
Refractive index	1.332	1.408
Diffusion coefficient ions ($cm^2\ s^{-1}$)	Na^+ 3×10^{-5}	Na^+ 1.53×10^{-4}
		Cl^- 0.83×10^{-4}
Surface tension (dyn cm^{-1})	72	111.8
Density	1.00	1.539

TABLE 5.3

Melting Points of Some Inorganic Salts

Salt	Melting Point (K)	Salt	Melting Point (K)
$LiClO_4$	400		
$LiAlCl_4$	418		
$AgNO_3$	483	$PbCl_2$	774
$HgBr_2$	511	$CdCl_2$	841
$LiNO_3$	527	LiCl	883
$ZnCl_2$	548	CaCl	919
$HgCl_2$	550	NaI	924
$NaNO_3$	583	$MgCl_2$	987
KNO_3	610	KCl	1049
$PbBr_2$	646	NaCl	1081
AgBr	707	Na_2CO_3	1131
$AlCl_3$-1-methyl-3-ethyl imidazalonium and similar systems			room temperature

TABLE 5.4

Melting Points of Some Tetraalkylammonium Salts

Salt	Melting Point (K)
Tetramethylammonium bromide	503
Tetrabutylammonium iodide	417
Tetrapropylammonium iodide	553
($AlCl_3$-1-butyl-pyridinium chloride	< 303 (according to composition))

TABLE 5.5
Some Properties of Molten Oxides near the Melting Point

Molten oxide	Temp. (K)	Density ($g\ cm^{-3}$)	Surface Tension ($dyn\ cm^{-1}$)	Viscosity (cP)	Specific Conductivity ($s\ cm^{-1}$)
Li_2O-SiO_2	1523	2.07	354	2.88×10^2	5.5 (2023 K)
Li_2O-1½SiO_2	1373	2.13	331	5.02×10^3	
Li_2O-2SiO_2	1373	2.16	319	1.78×10^4	2.5 (2023 K)
Na_2O-SiO_2	1373	2.23	300	1.19×10^3	4.8 (2023 K)
Na_2O-2SiO_2	1173	2.28	289	3.33×10^5	2.1 (2023 K)
Na_2O-3SiO_2	1173	2.23	282	1.99×10^6	
K_2O-2SiO_2	1373	2.20	220	1.08×10^5	1.5 (2023 K)
CaO-SiO_2	1823	—	400	2.73×10^2	0.8 (2023 K)

Fused salts conduct by means of the preferential drift of ions in the direction of applied electric fields. They are in fact ionic liquids, that is, liquids containing only ions, the ions being free or associated (see Section 5.4).

Another class of ionic liquids consists of the *molten oxides*. These are highly conducting liquids formed by the addition of a *metal* oxide (e.g., Li_2O) to a *non-metal oxide* (e.g., SiO_2). These systems melt at much higher temperatures than the molten salts. Some properties of the molten oxides are shown in Table 5.5. To develop a perspective on the properties of liquid electrolytes, some properties of water, liquid sodium, an aqueous solution of NaCl, fused NaCl, and a mixture of fused Na_2O and SiO_2 are shown in Table 5.6.

5.1.5. Fundamental Problems in Pure Liquid Electrolytes

In dealing with aqueous and nonaqueous solutions of electrolytes, the procedure was first to seek a picture of the time-averaged structure of the electrolytic solution and second to understand the basic laws of ionic movements. The picture that emerged was of ions and solvent molecules interacting together to form solvated ions; of ions interacting with each other to form ionic clouds and associated ion pairs or complexes; and of all these entities executing an aimless random walk at equilibrium, which becomes a directed drift under a concentration gradient or an external electric field. The problems in pure liquid electrolytes are analogous, though more difficult to treat mathematically because of the increased energy of interactions between the ions (compared with those in aqueous solution) arising from the short distance between them.

The first problem can be defined as follows: What idealized model could best replicate a solvent-free system of charged particles forming a highly conducting ionic liquid? In the case of the aqueous solution, it was easy to understand the drift of ions at the behest of the applied electric field. Positive and negative ions separated by large

TABLE 5.6
Comparison of the Properties of Various Liquids

Properties	Water at 298 K	Liquid Sodium at m.p.	1 *M* Sodium Chloride Soln. at 298 K	Liquid Sodium Chloride at m.p.	Liquid Sodium Silicate near the m.p.
Melting point (K)	273.15	370.9	269.8	1074.1	1361.1
Vapor pressure (Pa)	3125.8	1.295×10^{-5}	2210.5 (293 K) 4052.6 (303 K)	45.4	
Molar volume (cm^3)	18.07	24.76	17.80	30.47	55.36
Density ($g\ cm^{-3}$)	0.997	0.927	1.0369	1.5555	2.250, 1473 K
Compressibility ($10^6\ cm^2\ atm^{-1}$)	46.3055 isoth.	18.88 isoth.	40.08 isoth.	29.08 isoth.	59.579 adiabatic, 1473 K
Diffusion coefficient ($cm^2\ s^{-1}$)	3.0×10^{-5}	2.344×10^{-7}	$D_{Na^+} = 1.25 \times 10^{-5}$ $D_{Cl^-} = 1.77 \times 10^{-5}$	$D_{Na^+} = 1.53 \times 10^{-4}$ $D_{Cl^-} = 0.83 \times 10^{-4}$ 1123 K	
Surface tension ($dyn\ cm^{-1}$)	71.97	192.2	74.3	113.3	294, 1473 K
Viscosity (cP)	0.895	0.690	1.0582	1.67	980, 1373 K
Specific electric conductance ($s\ cm^{-1}$)	4.0×10^{-8}	1.04×10^{5}	$0.101\ mol\ cm^{-3}$	3.58	4.8, 2023 K
Refractive index	1.333	1.04	1.3426	1.408, 1123 K	1.52, solid, room temp.

TABLE 5.7
Conductivities of Molten Salts and Water

Substance	Temperature (K)	Specific Conductivity ($s\ cm^{-1}$)
H_2O	291	4×10^{-8}
LiCl melt	983	6.221
NaCl melt	1181	3.903
KCl melt	1145	2.407

TABLE 5.8
Conductivities of Molten NaCl and Mercury

Substance	Temperature (K)	Specific Conductivity ($s\ cm^{-1}$)
Hg	293	1.1×10^4
NaCl melt	1181	3.903

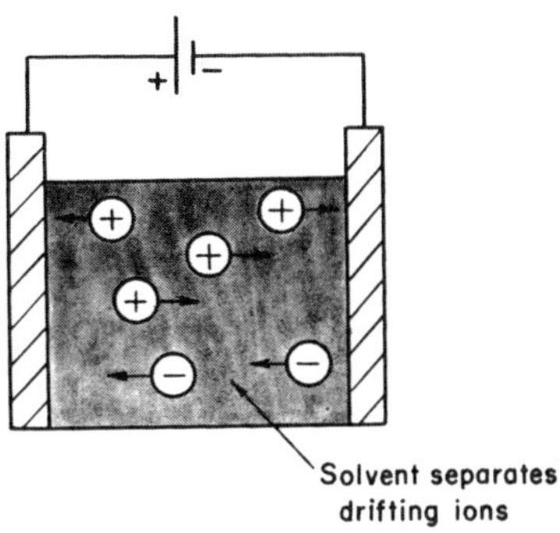

Fig. 5.2. In an aqueous solution, the solvent separates the drifting ions.

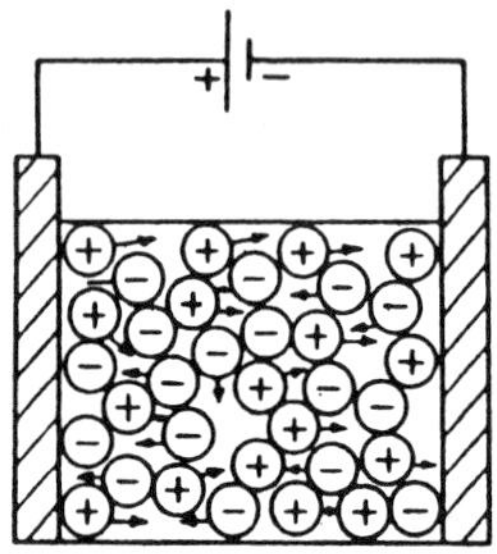

Fig. 5.3. In an ionic liquid, there is no solvent separating the drifting ions.

stretches of water drift in opposite directions (Fig. 5.2). In a pure ionic liquid, however, there is no water for the ions to swim in; the ions move about rather aimlessly among themselves (Fig. 5.3). When an external field is switched on, how is it that the ions are able to move past each other? Will not the large interionic force make them stick together, forming a poorly conducting ionic lattice? The situation appears puzzling.

What is the essential difference between the solid form and the liquid form of an ensemble of particles? This is a question that is relevant to all processes of fusion, e.g., the process of solid argon[3] melting to form a liquid. In the case of ionic liquids, the problem is more acute. One must explain the great fluidity and corresponding high conductivity in a liquid that contains only charged particles in contact.

The second problem concerns an understanding of the sharing of transport duties (e.g., the carrying of current) in pure liquid electrolytes. In aqueous solutions, it was possible to comprehend the relative movements of ions in the sense that one ionic species could drift under an electric field with greater agility and therefore transport more electricity than the other until a concentration gradient was set up and the resulting diffusion flux equalized the movements when the electrodes were reached. In fused salts, this comprehension of the transport situation is less easy to acquire. At first, it is even difficult to see how one can retain the concept of transport numbers at all when there is no reference medium (such as the water in aqueous solutions) in which ions can drift.

Third, there exists another problem, that of complex ions. In aqueous and nonaqueous solutions, it is possible to regard the ionic atmosphere as a type of incipient complex in which the mean distance between oppositely charged ions becomes smaller with increasing electrolyte concentration. Eventually the ions come sufficiently close so that the thermal forces that tend to separate them are overcome on an increasingly frequent basis by the Coulombic attraction forces so that cation and anion pairs arise, some of which remain stuck together (see Section 3.8).

[3]This is a relatively simple solid from the point of view of the forces between the uncharged particles.

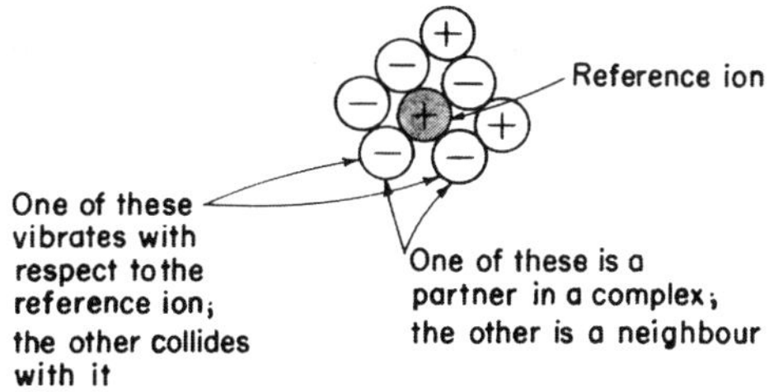

Fig. 5.4. The problem of distinguishing a neighboring ion colliding with the reference ion from a ligand (i.e., a partner in complex formation) vibrating in relation to the reference ion.

For the ionic liquids, however, without a separating solvent, the situation is different for the ions are always in contact. This absence of solvent causes conceptual problems regarding the existence of complex ions in ionic liquids. Consider a particular ion associated with another to form a vibrating complex. The ion is also in contact with, and continually jostled by, neighboring ions that are exactly like its partner in the complex (Fig. 5.4). Which is the partner and which the neighbor? Which is the vibration and which the collision? A distinction between these two types of contacts constitutes one of the problems in this field.

In aqueous solutions, the situation is clarified by the solvent. This solvent keeps the complex ions apart at mean distances, defines them as independent stable entities, and permits probing radiation (e.g., visible light) to pick them out from the surroundings (Fig. 5.5).

The concept of complex ions is therefore more subtle in ionic liquids than in aqueous solutions. There is not only the question of an objective means for identifying

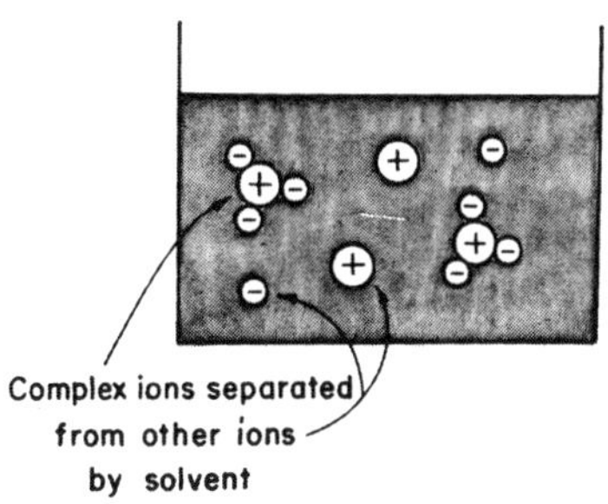

Fig. 5.5. In an aqueous solution, the complex ion is spatially separated from the other ions.

ions that can be said to be joined in some way to other ions so that the aggregate is distinctly an individual but also questions of distortion of ions in contact and the introduction of covalent bonds between them.

Further Reading

Seminal

1. W. Klemm and W. Biltz, "The Distribution of Ionic Conductivity among Molten Chlorides of the Periodic Table," *Z. Anorg. Allg. Chem.* **152**: 255, 267 (1926).
2. J. O'M. Bockris and N. E. Richards, "Free Volumes and Equations of State for Molten Electrolytes," *Proc. Roy. Soc. Lond.* **A241**: 44 (1957).
3. C. Solomons, J. H. P. Clarke, and J. O'M. Bockris, "Identification of Complex Ions in Molten Liquids," *J. Chem. Phys.* **49**: 445 (1968).
4. A. R. Ubbelohde, "Research on Molten Salts: Introduction" in *Ionic Liquids,* D. Inman and D. G. Lovering, eds., Plenum Press, New York (1981).

Review

1. J. E. Enderby, "The Structure of Molten Salts," in *Molten Salt Chemistry,* G. Mamantov and N. R. Marassi, eds., p. 115, NATO ASI Series, Reidel, Dordrecht, The Netherlands (1987).

Papers

1. I. Farnon and J. F. Stebbins, *J. Am. Chem. Soc.* **112**: 32 (1990).
2. I. Strubinizer, W. Sun, W. E. Cleland, and C. L. Hussey, *Inorg. Chem.* **29**: 993 (1990).
3. R. L. McGreevy and M. A. Howe, *Proc. Roy. Soc. Lond.* **430**: 241 (1990).
4. R. L. McGreevy and L. Pusztai, *Proc. Roy. Soc. Lond.* **241A**: 261 (1990).
5. M. L. Saboungi, D. L. Price, C. Scamehorn, and M. P. Tosi, *Europhys. Lett.* **15**: 281 (1991).
6. M. L. Saboungi, D. L. Price, C. Scamehorn, and M. P. Tosi, *Europhys. Lett.* **15**: 283 (1991).
7. Y. Toda, S. Hiroeka, Y. Katsumura, and I. Yemada, *Ind. Eng. Chem. Res.* **31**: 2010 (1992).
8. Z. Ardeniz and M. P. Tosi, *Proc. Roy. Soc. Lond.* **A437**: 85 (1992).
9. M. Oki, K. Fujishima, Y. Iwadate, and J. Mochinaga, in *Molten Salt Chemistry and Technology,* Proceedings of the Electrochemical Society, p. 9 (1993).
10. M. Mahr and K. G. Weil, in *Molten Salt Chemistry and Technology,* Proceedings of the Electrochemical Society, p. 147 (1993).
11. A. N. Yolshin and M. Yu Bryakotin, in *Molten Salt Chemistry and Technology,* Proceedings of the Electrochemical Society, p. 338 (1993).
12. M. Matsunaga, S. Hara, and K. Ogino, in *Molten Salt Chemistry and Technology,* Proceedings of the Electrochemical Society, p. 507 (1993).
13. Y. G. Boshuev and N. L. Kolesnikov, *Industrial Laboratory* **61**: 98 (1995).
14. S. Ohno, A. C. Barnes, and J. E. Enderby, *J. Phys. Cond. Matt.* **20**: 3785 (1996).
15. U. Matenaar, J. Richter, and M. D. Zeidler, *J. Magn. Reson, Series A* **122**: 72 (1996).
16. M. Abraham, M. C. Abraham, and I. Ziogas, *Electrochim. Acta* **41**: 903 (1996).

5.2. MODELS OF SIMPLE IONIC LIQUIDS

5.2.1. Experimental Basis for Model Building

One's first impression of a liquid (its fluidity, conformity to the shape of the containing vessel, etc.), would suggest that its structure has nothing to do with that of the crystal from which it was obtained by melting. If, however, a beam of monochromatic X-rays is made incident on the liquid electrolyte, the scattered beam has an interesting story to tell, as have corresponding studies on concentrated electrolytes and on molten salts carried out by means of neutron diffraction. The ions are almost at the same internuclear distances in a fused salt as in the ionic crystal, actually at a slightly lesser distance (Table 5.9). The X-ray patterns (Fig. 5.6) also indicate that in the liquid state the local order extends over a very short distance (tens of nanometers). It is as if the fused salt forgets how to continue the ordered arrangements of ions of the parent lattice, *although, curiously, the distance between individual ions becomes smaller rather than greater.*

5.2.2. The Need to Pour Empty Space into a Fused Salt

There is another important fact about the melting process. When many ion lattices are melted, *there is a 10 to 25% increase in the volume of the system* (Table 5.10). This volume increase is of fundamental importance to someone who wishes to conceptualize models for ionic liquids because one is faced with an apparent contradiction. From the increase in volume, one would think that the mean distance apart of the ions in a liquid electrolyte would be greater than in its parent crystal. On the other hand, from the fact that the ions in a fused salt are slightly closer together than in the solid lattice,

TABLE 5.9

Internuclear Distances in an Ionic Crystal and the Corresponding Fused Salt

	Distance between Oppositely Charged Ions (pm)	
Salt	Crystal, m.p.	Molten Salt
LiCl	266	247
LiBr	285	268
LiI	312	285
NaI	335	315
KCl	326	310
CsCl	357	353
CsBr	372	355
CsI	394	385

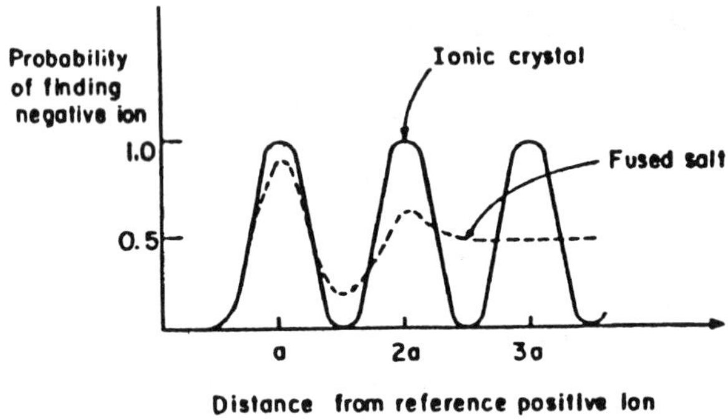

Fig. 5.6. Schematic diagram to show short-range and long-range order in an ionic crystal as opposed to only short-range order in a fused salt. In an ideal ionic crystal, if one takes a reference positive ion, there is a certainty of finding a negative ion at the lattice distance or a multiple of this distance; in a fused salt, there is a high probability of finding a negative ion one distance away; but within two or three lattice distances away, the probability becomes half, i.e., a negative ion is as likely as a positive ion. Thus, in a fused salt, there is no long-range order.

one would think that there should be a small volume decrease upon fusion.[4] How is this emptiness—which evidently gets introduced into the solid lattice on melting—to be conceptualized?

Before an answer is given to this central question, it is necessary to retrace the steps that have been taken in respect to Fig. 5.6. One examines how it was obtained because quantitative knowledge of the short-range order which does exist (dotted line in Fig. 5.6) is vital to understanding the liquidity of molten salts.

5.2.3. How to Derive Short-Range Structure in Molten Salts from Measurements Using X-ray and Neutron Diffraction

5.2.3.1. *Preliminary.* Before World War I, little was known either about the nature of X-rays (were they like waves or more like particles?) or about the

[4]In the case of some salts, the volume changes on fusion are smaller than are indicated in Table 5.10. Thus, calcium, strontium, and barium halides have volume changes that are about a fifth of the changes for the corresponding alkali halides. This is because such salts crystallize in a form that already contains plenty of open space in the solid lattice. When these open-lattice salts are melted, a smaller volume increase is needed than is the case for the space-filled lattices of the solid alkali halides.

TABLE 5.10
Volume Change on Fusion

Substance	% Increase of Volume on Fusion
NaCl	25
NaF	24
NaI	19
KCl	17
KBr	17
KI	16
RbCl	14
$CdCl_2$	20
$CdBr_2$	28
$NaNO_3$	11

arrangement of atoms in crystals. Much light was thrown on the latter question as a result of a thought experiment suggested by von Laue. He considered an apparatus known as a *diffraction grating*. This consists of a great number of parallel slits. It can be shown that if light of a specific monochromatic wavelength is incident upon a grating on which the parallel slits are a distance d apart at a certain specific angle θ, then "interference" occurs among the light waves, so that for the Bragg condition, $\lambda = 2d \sin \theta$ and no rays emerge—they have been knocked out by destructive interference with other rays.

The merit in von Laue's thought was to see that the regular rows of atoms which, in 1913 when he did his thinking, were only tentatively thought to be there, could be regarded as a diffraction grating of atomic dimensions.[5] For the diffraction experiment to work with reasonable values of the incident angle, θ, the wavelength λ must have the same order of magnitude as the width of the slits in a grating or the interatomic

[5]Von Laue's flash of insight in recognizing that a crystal lattice equals a diffraction grating is a beautiful example of the birth of a scientific idea. Since it gave rise to X-ray diffraction as a technique for obtaining knowledge of the structure of solids, it can be regarded as having great historical significance. It was the beginning of the effective study of the structure of the solid state. This creative thought is analogous to von Helmholtz's suggestion that an electrode in solution charged with excess negative or positive charges would attract to itself a monolayer of ions of opposite charge; the ensemble could be thought of as if it were the two plates of a parallel-plate condenser.

Such flashes, inspirations if you like, do not happen while one is thinking in one's study or talking with a student. They turn up like a photograph in the mind at some odd time later (Leo Szilard, for example, got some of his best ideas while soaking in the bathtub). But they don't turn up at all unless one has thought a great deal about the matter.

distance (d) in the crystals. However, X-rays can have wavelengths around 0.1 nm, which is just the right wavelength to react to a grating in which the distance the "slits" are apart is in fact the distance between ions. Von Laue boldly predicted that if a beam of X-rays were incident upon a solid crystal lattice, destructive interference and constructive augmentation would occur and a diffraction pattern[6] would result. Trial experiments succeeded and the method of X-ray diffraction for investigating structure in solids, and solid-state structural chemistry, was born.

Shortly after von Laue's suggestion and its validation, Bragg produced his equation. He showed that conditions for interference would be reached when

$$n\lambda = 2d \sin \theta \tag{5.1}$$

The terms in Bragg's law have been defined above except for n, which represents the number of the crystal planes (in succession, going downward from the surface) from which the reflected beam arises and therefore runs 1, 2, 3, ... The condition for interference is that the difference in path length of two rays of light, reflected from neighboring planes, should be an integral multiple of λ. Finding maxima and minima (non-interfered and interfered conditions) coupled with the strength of the reflection ("intensity") and the corresponding angle values of the reflected X-rays gives d [from Eq. (5.1)], the internuclear distance.

For an ordered solid, as θ is changed, the peaks representing no constructive interference repeat themselves for values of sin θ corresponding to $n = 1, 2, 3$, with a constant d; i.e., the arrangement is of long-range order. On the other hand, as has been seen, for liquids (see Fig. 5.6), the peaks fade rapidly with increasing distance from the reference plane, representing a short-range order.

5.2.3.2. *Radial Distribution Functions.* What happens when X-ray diffraction occurs in liquids? To understand this (see also Section 3.11), it is best at first to consider only a single-species liquid; one could have in mind not a binary molten salt such as liquid sodium chloride, but, say, liquid sodium.

Then one defines a quantity called a *pair correlation function,* represented by $g_{A,B}(r)$. Consider a reference particle, A. The number of particles called B that occupy a spherical shell having a radius r is

$$dn_r = 4\pi\rho_B g_{A,B}(r) r^2\, dr \tag{5.2}$$

where ρ_B is the number of B's per unit volume.

[6]A diffraction pattern is a series of bright spots alternating with dark areas that results from a plot of the intensity of the reflected beam after it passes through the object examined, as a function of the incident angle of the X-ray beam.

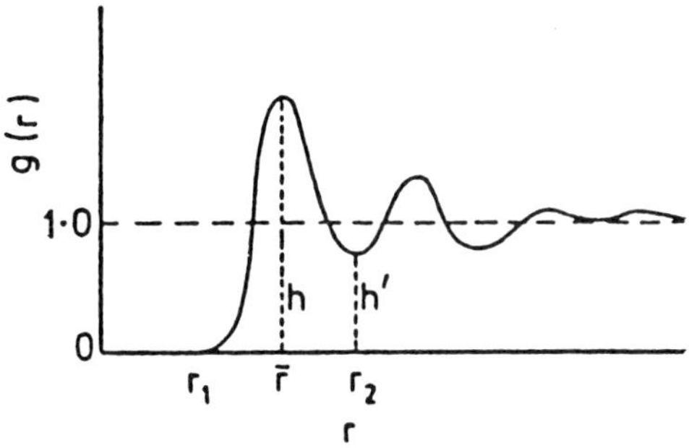

Fig. 5.7. A hypothetical radial distribution function $g(r)$ for a liquid that contains just one chemical species. (Reprinted from J. E. Enderby, in *Molten Salts, NATO ASI Science Series, Series C* **202**: 2, 1988.)

Enderby has illustrated an idealized "radial distribution function" [see Eq. (5.3)], shown in Fig. 5.7. The figure suggests that starting from the center of A, nothing is seen until a distance r_1 away, and the most probable distance to the center of the next nearest particle is $\bar{r}$. What does the distance r_2 mean? At distances greater than this, as far as A is concerned, nearest-neighbor interactions can be neglected.

From these statements, one can begin to appreciate the significance of the *radial distribution function* shown in the figure. This quantity is obtained by integrating Eq. (5.2).

$$4\pi\rho \int_0^{r_s} g_{A,B}(r) r^2 \, dr \tag{5.3}$$

If one takes $g_{A,B}(r)$ as the pair correlation function defined in Eq. (5.2), the radial distribution function represents the number of particles of B in a shell up to r_2 around A. If r_s is then r_2 (see Fig. 5.7), one can regard Eq. (5.2) as giving the *coordination number* of A in the liquid. In the example chosen for simplicity, species A is the same as species B but this of course is only true for radial distributions of monatomics, e.g., sodium. It is found in practice that in a liquid, $g_{A,B}(r)$ settles to unity by the third or fourth atom away from the reference atom A.

Radial distribution functions can be determined experimentally using diffraction (i.e., interference) experiments. X-rays or neutrons can be used. If one knows the pair correlation function $g_{A,B}(r)$ for each atom, one can work out the short-range structure in a liquid. The question is then how does one find $g_{A,B}(r)$?

5.2.4. Applying Diffraction Theory to Obtain the Pair Correlation Functions in Molten Salts

The mathematical theory of diffraction[7] is heavy stuff, and a very simplified version takes several book pages to deduce. Here, some equations from this theory will be presented to show their shape and size, and their significance to the task on hand will be explained.

The amplitude of waves scattered by nuclei is given by an equation of the form

$$\text{Amplitude} = \sum_{\alpha} f_{\alpha} \sum_{i(\alpha)} \exp[ik - r_i(\alpha)] \tag{5.4}$$

where f_α is the X-ray form factor and defines the central atom, α. In the second summation, **k** is the wave vector $2\pi/\lambda$ of the X-rays used, and $r_i(\alpha)$ is the distance from the ith nucleus of the species i. This second part deals with the phase relationships[8] of the scattered radiation as a function of distance from the given atom, α.

The intensity of the scattered light is given by the equation

$$I(\boldsymbol{k}) = N\left[\sum x_\alpha b_\alpha^2 + F(\boldsymbol{k})\right] \tag{5.5}$$

In this expression, x_α is the atom fraction of the atomic species α; b_α, the same as f_α, is the X-ray form factor; and $F(\boldsymbol{k})$ is the average of the quantities called *partial structural factors* and is given by

$$F(\boldsymbol{k}) = \sum_{\alpha} \sum_{\beta} x_\alpha x_\beta b_\alpha b_\beta [S_{\alpha\beta}(\boldsymbol{k}) - 1] \tag{5.6}$$

[7]Diffraction is sometimes described as "the bending of light around an obstacle." When light is interrupted by an object, the shadow formed is bordered by alternating white and dark bands. To observe such effects, one needs to use a point source and monochromatic light; hence diffraction effects are not observed in everyday life.

Diffraction—and the alternating dark and light bands to which it gives rise—is based on the interference of light waves that occurs when two light beams meet and annihilate or augment each other. Diffraction should not be confused with the primary phenomenon of *refraction*, which refers to the bending of light as it passes from a more dense to a less dense medium. When this occurs (and depending on the angle at which the beam from the dense medium is incident upon the less dense one), the light may pass through from one medium to the next or it may be reflected (or some of both). The change in direction of the light beam on entering the new medium can be regarded as a result of the exchange of kinetic and potential energy at the interface. Refraction is expressed in terms of the refractive index of the dense medium.

[8] In electricity theory, phase relationships refer to the relation of current to potential in an ac circuit. If the current and potential vary together in time, i.e., they reach maximum and minimum together, then they are "in phase." If the current and potential are "out of step," by, say, a quarter of a phase, they are "out of phase." In diffraction theory, the phase refers to the variation of the amplitude of the light wave with time at a given point. Two beams of radiation out of phase by a half cycle will annihilate.

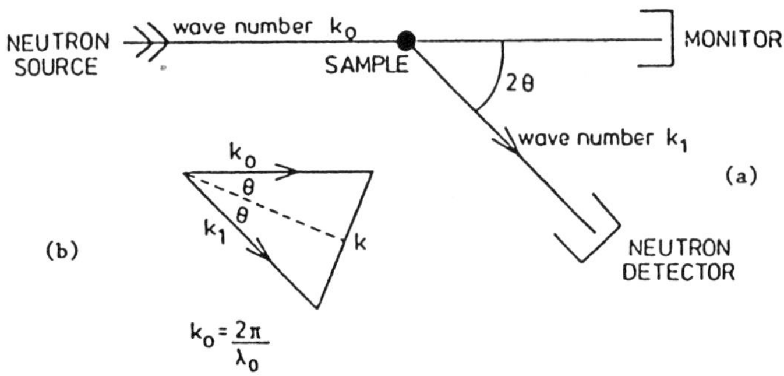

Fig. 5.8. The conventional arrangement for neutron diffraction. (Reprinted from J. E. Enderby, in *Physics and Chemistry of Aqueous Ionic Solutions*, M. C. Bellisent-Fund and G. W. Nielson, eds., *NATO ASI Series C* **207**: 131, 1987.)

The quantity symbolized by $S_{\alpha,\beta}$ is called the *structural factor.*

How would pair correlation functions and partial structural factors be written for a binary molten salt, e.g., sodium chloride? The corresponding correlation function would be given by

$$g_{\alpha,\beta} = 1 + \frac{1}{2\pi^2 \rho r} \int S_{\alpha,\beta}(h)\boldsymbol{k} \sin(\boldsymbol{k}r)\, dr \tag{5.7}$$

and the partial structural factor by

$$F(\boldsymbol{k}) = x_+^2 b_+^2 [S_{2+}(\boldsymbol{k}) - 1] + x_-^2 b_-^2 [S_{2-}(\boldsymbol{k} - 1)] + 2x_+ x_- b_+ b_- [S_{\pm}(\boldsymbol{k}) - 1] \tag{5.8}$$

These equations (also undeduced) may seem pretty fearsome. However, the quantities present in Eq. (5.5), the partial factors, can all be determined from X-ray or neutron diffraction setups. Figures 5.8 and 5.9 show at a schematic level what one does to make diffraction measurements.

The material given here then shows how measurement of the diffraction of X-rays (also neutrons, see later discussion) gives the pair correlation function, $g_{\alpha,\beta}$. It can give much more. As shown in Section 3.11, the determination of $g_{\alpha,\beta}$ allows one to calculate a number of properties of the liquid or solution. A property calculated from pair correlation functions does not involve an assumed modeling theory. Instead, the experimentally determined pair correlation functions are the basis of the calculated properties. It is as though one had "worked out"[9] the structure first and then used the knowledge of that structure to calculate the properties. Is this a "higher level" approach

[9]"Working out the structure" means determining the $g_{\alpha,\beta}$s for the various entities (e.g., α and β) present in the liquid.

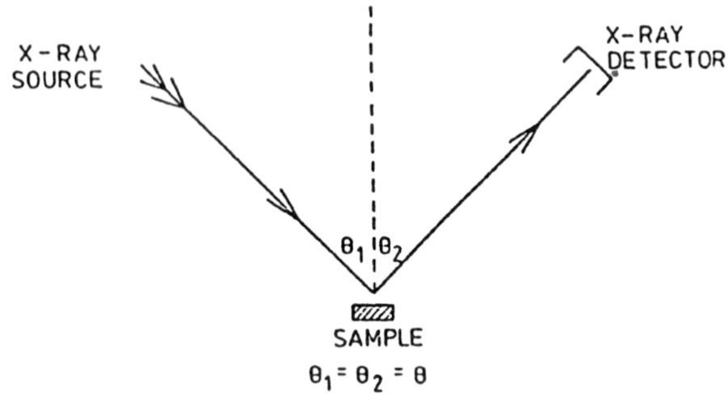

Fig. 5.9. Conventional arrangement for X-ray diffraction studies on liquids. (Reprinted from J. E. Enderby, in *Physics and Chemistry of Aqueous Ionic Solutions*, M. C. Bellisent-Fund and G. W. Nielson, eds., *NATO ASI Series C* **205**: 131, 1987.)

to the elucidation of the structure inside a liquid, rather than the alternative modeling one in which one sketches several possible pictures of the liquid and then applies uncontroversial theory to each to see the properties each model gives rise to, judging the validity of the hypothesized structure from the match of the calculated results with those of experiments?

5.2.5. Use of Neutrons in Place of X-rays in Diffraction Experiments

From the early years of this century, the only type of radiation used to find long-range order in solids and short-range order in liquids was X-rays. However, neutrons can also be used to carry out diffraction experiments, and with certain advantages, as will be seen. The trouble is that while it is fairly easy (hence, economically attractive) to use an X-ray source, one has to have a nuclear reactor handy to achieve a neutron stream. The advantages of using neutrons rather than X-rays in diffraction arise from a difference in how diffraction (i.e., interference) occurs for the two forms of radiation.

X-rays interact with matter because their electromagnetic oscillations are affected by the electrons of the material. Neutrons take no notice whatsoever of electrons when they pass through matter. They interact with the nuclei. Neutron diffraction is sensitive to the atomic number and atomic weight of the atoms constituting the substance. For example, it can distinguish easily between Fe and Co in alloys and between isotopes such as ^{35}Cl and ^{37}Cl.

Thus, the b terms in Eq. (5.6) are 11.7 for ^{35}Cl but 3.08 for ^{37}Cl, a difference much greater than that intuitively expected. Such differences are helpful in using Eq. (5.7) to give $g_{\alpha,\beta}$ and to lead to the calculation of properties (Section 5.2.3.2). Variation in the concentration of the isotopes and measurements of the patterns for a series of such

TABLE 5.11

Ratios of Ionic Radii (r_+,r_-), First Peak Positions of $g_{--}(r)$ and $g_{+-}(r)$, and Coordination Numbers (n_{+-}) for Some Molten Salts and Simple Geometrical Clusters

	r_+/r_-	r_{--}/r_{+-}	n_{+-}	Structure
$ZnCl_2$	0.457	1.621	4.3	T
$MgCl_2$	0.407	1.471	4.3	S(?)
$CaCl_2$	0.611	1.342	5.4	?
$SrCl_2$	0.691	1.311	6.9	O(?)
$BaCl_2$	0.833	1.245	7.7	C(?)
$NiCl_2$	0.426	1.610	4.7	T(?)
$NiBr_2$	0.354	1.608	4.7	T(?)
NiI_2	0.319	1.577	4.2	T(?)
Tetrahedron (T)	—	1.633	4	—
Octahedron (O)	—	1.414	6	—
Cube (C)	—	1.155	8	—
Square (S)	—	1.414	4	—

Source: Reprinted from R. L. McGreevy and L. Pusztai, *Proc. Roy. Soc. Lond.* **A430**: 241, 1990.

changes provides added data for a given salt and improves finding the unknowns. All this being the case, one has to weigh the undeniable pros of neutron diffraction against the considerable cons of having to take a team of collaborators to a nuclear reactor to obtain the neutron stream and work there day and night[10] for some days.

What about the determination of *voids* in a liquid? Determination of the short-range order may not allow one to determine the distribution (number and size) of fluctuating voids in the liquid.[11] While such voids may play a vital part in the mechanism of transport, they *are* voids and hence would hardly make much impression upon the probing radiation.

Nevertheless, neutron diffraction work in molten salts gives rise to much new knowledge of the structure of these bodies; the only caveat is that it must be used in conjunction with other kinds of measurements; the data from these measurements are used to check on the structural concepts developed.

5.2.6. Simple Binary Molten Salts in the Light of the Results of X-ray and Neutron Diffraction Work

Table 5.11 contains typical results obtained from neutron diffraction, and the pair-correlation functions for $g_{Sr-Sr}(r)$, $g_{Sr-Cl}(r)$, and $g_{Cl-Cl}(r)$ are shown in Fig. 5.10.

[10]Channels on research reactors are let to research groups for a limited total time. The cost of even a small university teaching reactor (\$10–20 million) is such that the amortization rate may be hundreds of dollars per hour.

[11]Such "volumes of nothingness" must be present to account for the large increase in volume upon fusion while at the same time the internuclear distance decreases (see Tables 5.9 and 5.10).

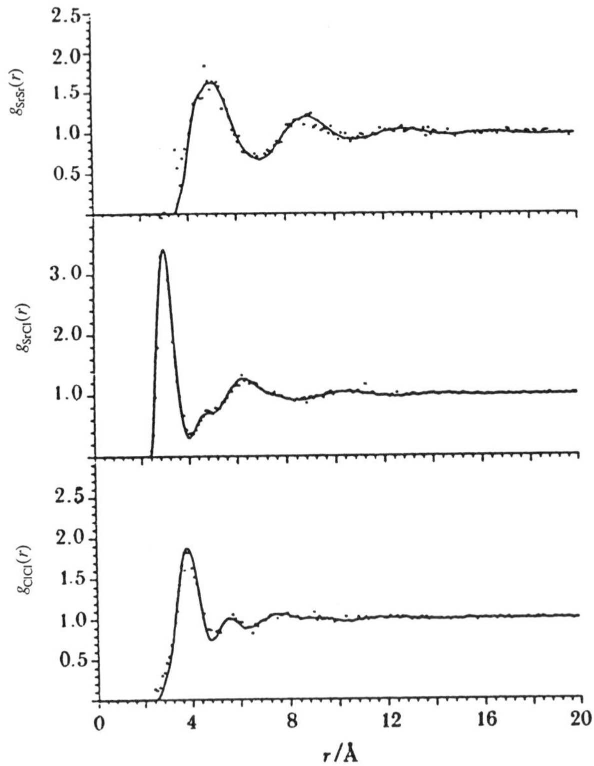

Fig. 5.10. Partial radial distribution functions for molten $SrCl_2$; solid line, experimental data; points, root mean cube fit. (Reprinted from R. L. McGreevy and L. Pusztai, *Proc. R. Soc. Lond. A* **430**: 241, 1990.)

Much neutron diffraction data of this kind are now available for molten salts. They are basic to structural knowledge of these pure electrolytes. However, the data do not play the same stellar role in determining the structure of liquid salts as they do for the solid salts because in the liquids the free space introduced on melting affects the dynamic movement of the ions and hence the liquid properties. In fact, this space is counterintuitive to the internuclear distances given by X-ray or neutron diffraction. The internuclear distances found in molten salts are smaller, not bigger, as might be thought from the increase in volume.

Data on simple molten salts can be interpreted without bringing in any orbital bond overlap; i.e., one can interpret the behavior of simple binary alkali halide molten salts in terms of ionic attraction and repulsion in a way similar to that used for the solid lattice. Molten salts containing complex anions—such as those consisting of $ZnCl_2$ and $AlCl_3$—need models that involve directed valence forces. Many results from different ways of modeling molten salts will be given in Section 5.2.8.

5.2.7. Molecular Dynamics Calculations of Molten Salt Structures

It has already been seen in Section 2.17 that computer simulation of structures in aqueous solution can give rise to calculations of some static (e.g., coordination numbers) and dynamic (e.g., diffusion coefficients) properties of ions in aqueous and nonaqueous solutions. One such computer approach is the Monte Carlo method. In this method, imaginary movements of the particles present are studied, but only those movements that *lower* the potential energy. Another technique is molecular dynamics. In this method, one takes a manageable number of atoms (only a few hundred because of the expense of the computer time) and works out their movements at femtosecond intervals by applying Newtonian mechanics to the particles under force laws in which it is imagined that only pairwise interactions count. The parameters needed to compute these movements numerically are obtained by assuming that the calculations are correct and that one needs to find the parameters that fit.

Now, these numerical simulation methods seem superior to the approaches that use alternative intuitive models. In the modeling approaches, one creates various imaginative hypotheses as to what the liquid salt might look like from inside it. Then, one calculates what one would expect from each intuited model in respect to its various properties, e.g., the diffusion coefficient. Models that predict nearest to the experimental result are taken as the best approximation to the real thing. In molecular dynamics calculations, on the other hand, one starts farther back by assuming a force law for interaction between the two kinds of particles present (K^+ and Cl^-, say) and finding out by calculation what kind of arrangement among such particles is indicated by applying such laws to the particles' movements. In the end, which approach to apply—results of models or calculation of ionic movement—depends on which is cheaper.

Experimental determinations and bench-type research may continue for a generation or two, for both methods need experimental results as a standard to judge success. As the variety of available software grows and becomes more user friendly, the choice will move toward computer simulation.

Another factor that counts in the balance is the degree to which playing with a number of intuitive, imaginative models helps thinking new thoughts. Here one comes to the question of how knowledge will be advanced in the new century. Can computers imagine and create new ideas? Or must they always be supercalculating machines, using human-derived software that instructs the calculator to work out numerical equations written by intelligent beings? If the latter, is the intuitive perception of a number of alternative possible models more likely to enhance progress? Or will molecular dynamics simulation (computational chemistry) beckon on and spark the vital new fantasies from which new paradigms arise?

5.2.8. Modeling Molten Salts

Some typical radial distribution functions for molten salts (see Section 5.2.3.2) are shown in Fig. 5.11. Those plots are made by assuming pairwise interactions

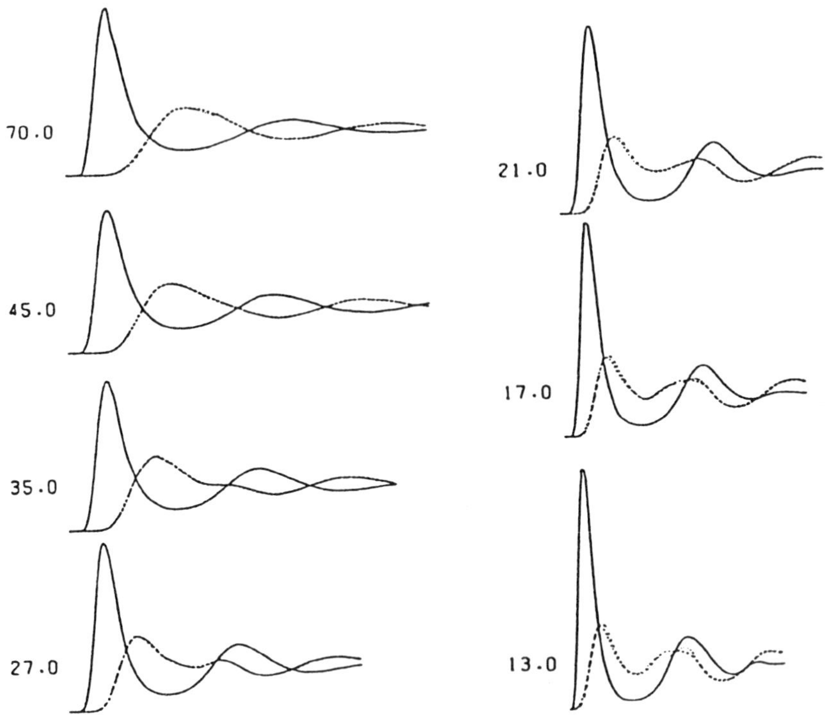

Fig. 5.11. Radial distribution functions of simulated KCl obtained by MC at T = 1700 K over a wide density range. The numbers are the molar volumes in $cm^3\ mol^{-1}$. The continuous line is g_{+-}, long dashes g_{++}, and short dashes g_{--}. (Reprinted from D. L. Price, M. L. Saboungi, W. S. Howells, and M. P. Tosi, *J. Electrochem. Soc.* **9**: 1, 1993.)

$$U_{ij} = \sum -\frac{e_i e_j}{r_{ij}} + \frac{B}{r_{ij}^{12}} \tag{5.9}$$

They naively assume that the *additivity* of such interactions (and nothing from ions outside the pairs) will allow a realistic calculation of physical properties.[12] However, a somewhat more mature view is to take into account the dipole moments induced among ions by nearest neighbors. Since this changes the ions' shapes (e.g., makes them spheroids and not spheres), the interaction equation usually assumed is not adequate.

[12]Why can patently wrong assumptions be made to give excellent agreement with experiment? Does the calibration of the parameters with experimental results in pairwise interaction equations introduce a degree of empiricism into computer simulation that suggests a lessened integrity in the calculation? Could pair potentials other than those used in Eq. (5.9) also give excellent results if their parameters were also calibrated on experimental data of a kind similar to that being calculated?

If the properties of one ion affect those of the other when they interact, an iterative approach has to be taken until constancy of intermolecular energy is obtained. This is because the dipoles induced in the ions exert counter fields upon the surrounding ions and the resulting change in shape modifies the values obtained for a system conceived as a sum of attractive and repulsive forces of unchanging spherical ions. If such a counter field can be calculated, then equations such as (5.9) can be used without the introduction of a dielectric constant in the Coulomb attractive term. Thus, dielectric constants are empirical devices that make an allowance for counter fields and forces in electrical systems.

Further Reading

Seminal

1. F. G. Edwards, J. E. Enderby, R. A. Howe, and D. I. Page, "Neutron Diffraction for Molten Salts," *J. Phys. Chem.* **8**: 3483 (1975).
2. J. Enderby, "The Structure of Molten Salts," *NATO ASI Series C* **202**: 1 (1987).
3. R. L. McGreevy and M. A. Howe, "Structure of Molten Salts," *J. Phys. Condens. Matt.* **1**: 9957 (1989).
4. R. L. McGreevy, "New Methods for Molten Electrolytes," *Nuovo Cimento* **12D**: 685 (1990).

Review

1. B. Guillot and Y. Guissani, "Towards a Theory of Coexistence and Criticality in Real Molten Salts," *Mol. Phys.* **87**: 37 (1996).

Papers

1. T. Kozlowski, *Int. J. Phys. Chem.* **100**: 95 (1996).
2. J. Ohno, A. C. Barnes, and J. E. Enderby, *J. Phys. Condens. Matt.* **8**: 3785 (1996).
3. Z. Akdeniz, G. Pastore, and H. P. Tosi, *Phys. Chem. Liq.* **32**: 191 (1996).
4. T. Koslowski, *Ber. Bunsen-Ges. Phys. Chem., Int. J. Phys. Chem.* **100**: 95 (1996).
5. L. Mouron, G. Roullet, J. J. Legendre, and G. Picard, *Computational Chemistry* **20**: 227 (1996).

5.3 MONTE CARLO SIMULATION OF MOLTEN POTASSIUM CHLORIDE

5.3.1. Introduction

The general principle behind a Monte Carlo procedure has already been described (Section 2.3.2). Woodcock and Singer were the first to make such a calculation for molten salts, and their work is the source of the following section.

In all simulations in which the interaction involves Coulomb's law, there is a difficulty in the extent to which the Coulombic terms can be summed because the number of particles used in the ensemble being considered (e.g., 216 in Woodcock and Singer's case) is too small. Thus, the Coulombic forces (which decline slowly with increasing distance) have not yet become negligible outside the "package." Some way of terminating them without spoiling the calculation had to be found. A solution to this problem, and the one that Woodcock and Singer used, is due to Ewald and lies in a purely mathematical technique by which one nonconverging series is converted into two converging series. One must bear with simplification here; a more quantitative description of Ewald's artifices would take precious space without significantly increasing the reader's comprehension of how ionic liquids work.

5.3.2. Woodcock and Singer's Model

Woodcock and Singer followed up on some considerations published earlier by Tosi and Fumi. The latter suggested that the potential that forms a suitable basis for the calculation of the pairwise addition potentials in a molten salt is suitably written as

$$U_{i,j(+)} = -\frac{z_i z_j e_0^2}{r_{i,j}} + b \exp B(\sigma_{i,j} - r) + \frac{c_{i,j}}{r^6} + \frac{d_{i,j}}{r^8} \tag{5.10}$$

This equation was first used by Born, Huggins, and Meyer and therefore bears their names. The first two terms represent, respectively, the attractive and repulsive potentials. The last two terms represent dipole-dipole and dipole-quadrupole potentials, respectively. In spite of allowing for the dipole interactions, the calculation is still a hard-sphere one, a mean spherical approximation, because the forces are not allowed to change the shape and the position of the particles. Later on, Saboungi et al.

TABLE 5.12

Parameters Used in Eq. (5.10) for KCl

ν_{ij} [a] (cm^{-8})	b [b] (10^{-12} erg)	B [c] (10^8 cm^{-1})	$c_{i,j}$ [d] (10^{-60} erg cm^6)	$d_{i,j}$ [d] (10^{-76} erg cm^8)
++2.926	0.423	2.97	−24.3	−24.0
+−3.048	0.338	2.97	−48.0	−73.0
−−3.170	0.253	2.97	−124.5	−250.0

Source: Reprinted from L. V. Woodcock and K. Singer, *Trans. Faraday Soc.* **67**: 12, 1971.
[a] Sum of ionic radii.
[b] Same value for all alkali halides.
[c] A common value for the three ion pairs in the salt.
[d] Calculated from spectroscopic data.

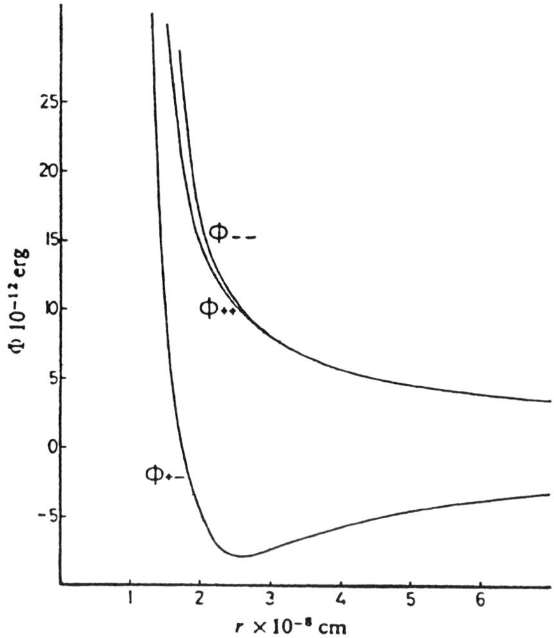

Fig. 5.12. Pair potentials for potassium chloride. (Reprinted from L. V. Woodcock and K. Singer, *Trans. Faraday Soc.* **67**: 12, 1971.)

did take into account a mutual squeezing of the ions, and some of the work of this team will be described later.

The various parameters needed in these equations come from other work and are given in Table 5.12, taken from Woodcock and Singer. The pairwise potential for potassium chloride, first calculated by Tosi and Fumi in 1964, is reproduced in Fig. 5.12. Typical computational data for an MC calculation are given in Table 5.13. The software of the day generated only 80,000 steps per hour.

5.3.3. Results First Computed by Woodcock and Singer

Changes of volume upon fusion together with compressibilities, expansivities, and specific heats calculated by Woodcock and Singer are shown in Table 5.14. What is the uncertainty in these calculations? It is around ± 0.5 cm^3 mol^{-1} and within this error the agreement of simulated and experimental results is rather good. Thus, at 1045 K for the volumes

$$V = 41.48 \text{ cm}^3 \text{ mol}^{-1}, \text{ solid: } V_f^+ = 0.68 \pm 0.03,\ V_f^- = 0.73 \pm 0.04,$$

$$V = 51.24 \text{ cm}^3 \text{ mol}^{-1}, \text{ liquid: } V_f^+ = 1.14 \pm 0.06,\ V_f^- = 1.13 \pm 0.06.$$

TABLE 5.13
Computational Data

Run	V (cm^3 mol^{-1})	T (K)	State	λ/N	$100A/\lambda$	$\Delta\phi$ (10^3 J mol^{-1})	Δp (10^3 bar)
1	41.48	1045	Solid	741	32.2	0.15	0.08
2	42.70	1045	Solid	1481	33.4	0.15	0.05
3	43.92	1045	Solid	1296	35.0	0.40	0.11
4	48.80	1045	Solid	1296	38.2	0.20	0.11
5	46.36	1045	Liquid	1296	32.9	0.20	0.17
6	48.80	1045	Liquid	1852	33.7	0.25	0.19
7	51.24	1045	Liquid	1667	35.1	0.25	0.21
8	53.68	1045	Liquid	1667	36.5	0.20	0.11
9	58.56	1045	Liquid	1481	36.9	0.15	0.07
10	48.80	1306	Liquid	1852	38.7	0.25	0.20
11	51.24	1306	Liquid	1481	39.5	0.20	0.13
12	53.68	1306	Liquid	1667	40.8	0.20	0.17
13	54.66	1306	Liquid	2407	40.7	0.10	0.07
14	58.56	1306	Liquid	1667	41.0	0.35	0.18
15	63.44	1306	Liquid	1667	43.4	0.30	0.10
16	58.56	2090	Liquid	1296	50.8	0.55	0.22
17	63.44	2090	Liquid	1296	51.8	0.25	0.19
18	68.32	2090	Liquid	1667	53.7	0.20	0.15
19	73.20	2090	Liquid	1667	54.2	0.35	0.16
20	85.40	2090	Liquid	1481	54.6	0.40	0.14
21	63.44	2874	Liquid	926	57.9	0.20	0.13
22	72.30	2874	Liquid	1667	60.1	0.25	0.15
23	85.40	2874	Liquid	1667	60.9	0.45	0.15
24	97.60	2874	Liquid	1667	61.7	0.40	0.12

Source: Reprinted from V. Woodcock and K. Singer, *Trans Faraday Soc*. **67**: 12, 1971.

Radial distribution functions are characterized by the distance of closest approach *d*, the position $r^{\max}$ and height h of the main peak, the position $r^{\min}$ of the minimum following the main peaks, and the coordination numbers *CN*, defined by

$$(CN)_m = (N-1)V^{-1}4\pi \int_0^{r_m^{\min}} r^2 g_m(r)\,dr \tag{5.11}$$

$$(CN)_u = 1/2NV^{-1}4\pi \int_0^{r_u^{\min}} r^2 g_u(r)\,dr \tag{5.12}$$

TABLE 5.14

Changes in Fusion for KCl ($T_{m.p.}$ = 1045 K)

	Solid	Liquid	Δ_{MC}	$\Delta_{expt.}$	% Changes (MC)
$<\phi_+^{DD} + \phi^{DQ}>$ 10^3 J mol^{-1}	–30.4	–25.9	4.5		–14.8
V, cm^3 mol^{-1}	41.75	50.20	8.45	8.3 9.09 7.23	20.2
κ_T, 10^{-6} bar^{-1}	10.1	36.1	26.0		257
α_p, 10^{-4} K^{-1}	1.97	3.33	1.96		99.5
β_v, bar K^{-1}	19.4	9.23	–10.2		–52.4
C_v, J K^{-1} mol^{-1}	48.4	50.5	2.1		4.3
C_p, J K^{-1} mol^{-1}	65.3	66.8	1.5	9.2 3.5	2.4

Source: Reprinted from L. V. Woodcock and K. Singer, *Trans. Faraday Soc.* **67**: 12, 1971.

$$(CN)_l = (1/2N - 1)V^{-1}4\pi \int_0^{r_l^{\min}} r^2 g_l(r)\, dr \tag{5.13}$$

On fusion, Woodcock and Singer's calculation found that the coordination number for molten KCl decreases from 5.3 to 4 in accordance with experiment. The distance of closest approach rather unexpectedly decreases as the temperature increases, while the first peak (Fig. 5.13) becomes broader and lower (less order even in the first shell).

Bockris and Richards' experimental value of the free volume for KCl (V_f = 0.68 cm^3 mol^{-1}) agrees well with that of Woodcock and Singer. The volume change on melting obtained experimentally, however, is 8.2 cm^3 mol^{-1}, which is only 3% less than the entirely acceptable calculated value obtained by Woodcock and Singer's Monte Carlo simulation. Their actual model for liquid KCl is shown in Fig. 5.14 and appears to contain "hole volume" corresponding to modelistic concepts which feature this property (Section 5.5.1).

5.3.4. A Molecular Dynamics Study of Complexing

Saboungi et al. have carried out molecular dynamics studies in the 1990s that are a great improvement on those done in the pioneering Woodcock and Singer studies. They studied, for example, complexes between NaCl and $AlCl_3$ by means of a molecular dynamics program. These systems are well known from experimental work to form $AlCl_4^-$ and other, higher complex ions, such as $Al_2Cl_7^-$.

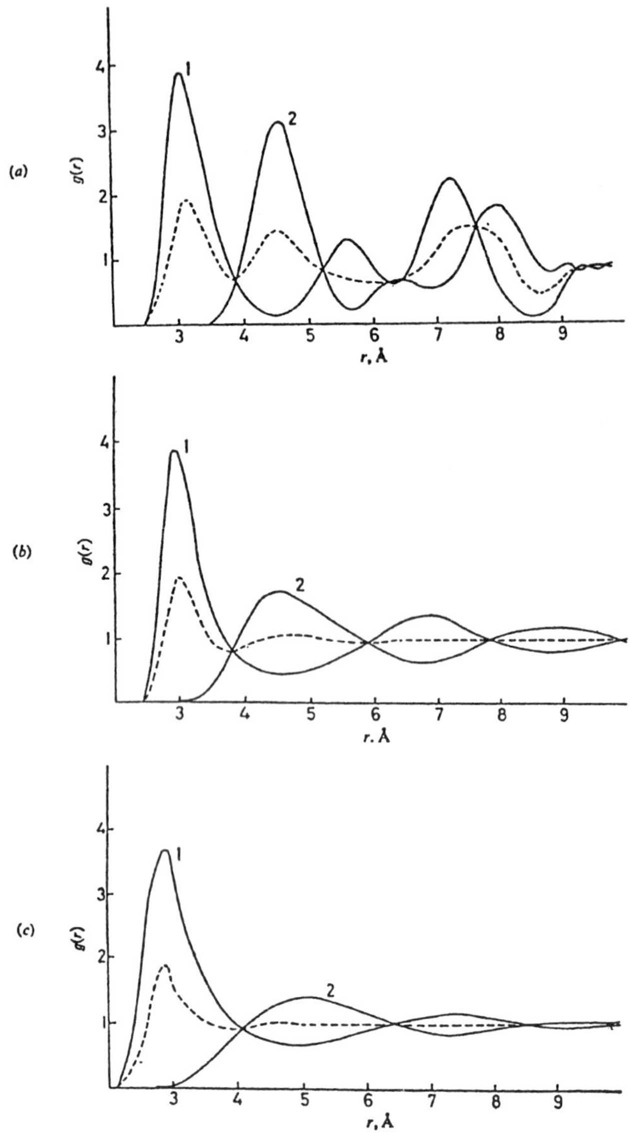

Fig. 5.13. Radial distribution functions for KCl. (a) Solid, $T = 1045$ K, $V = 41.48$ cm^3 mol^{-1}; 1, $g_u(r)$; 2, $g_1(r)$; ---, $g_e(r)$; (b) liquid, $T = 1045$ K, $V = 48.80$ cm^3 mol^{-1}; 1, $g_u(r)$; 2, $g_1(r)$; ---, $g_m(r)$; (c) liquid, $T = 2874$ K, $V = 97.60$ cm^3 mol^{-1}; 1, $g_u(r)$; 2, $g_1(r)$; ---, $g_m(r)$. (Reprinted from L. V. Woodcock and K. Singer, *Trans. Faraday Soc.* **67**: 12, 1971.)

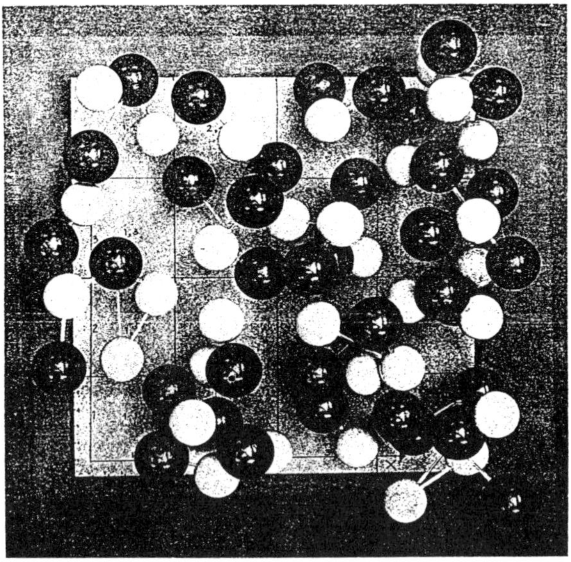

Fig. 5.14. Woodcock and Singer's representation of liquid KCl, the dark spheres representing chloride ions and white spheres representing potassium ions. (Reprinted from L. V. Woodcock and K. Singer, *Trans. Faraday Soc.* **67**: 12, 1971.)

In the molecular dynamics approaches to such systems, one of the principal novelties compared with the work on simpler systems such as KCl (Section 5.3.3) is the taking into account of anion polarizability, i.e., elimination of the hard-sphere approximation. Anions are by and large more polarizable than cations. The polarizability α is proportional to r_i^3. The radii of cations, formed by losing an outer electron, are smaller than those of the anions, which are formed by adding an electron. Hence, anions are mainly the ions affected when polarizability is to be accounted for.

When an anion feels the field of a cation X, a dipole moment is induced into the anion,

$$\mu_i = \alpha X \tag{5.14}$$

and this induced dipole moment will give rise to an additional energy of attraction between the monopole and the dipole (*cf.* the corresponding effect in the theory of hydration energy in Section 2.4.3). In addition, there will be dipole–dipole interactions.

The calculations of Saboungi et al. on these $NaCl\text{-}AlCl_3$ systems do not lack complexity. There are as many as ten terms of interaction. Thus,

$$U = u_{MM} + u_{MA} + u_{MX} + u_{AA} + u_{AX} + u_{XX}^{-} + u_{ddxx} + u_{mdmx} + u_{mdax} + u_{mdxx}^{+} \tag{5.15}$$

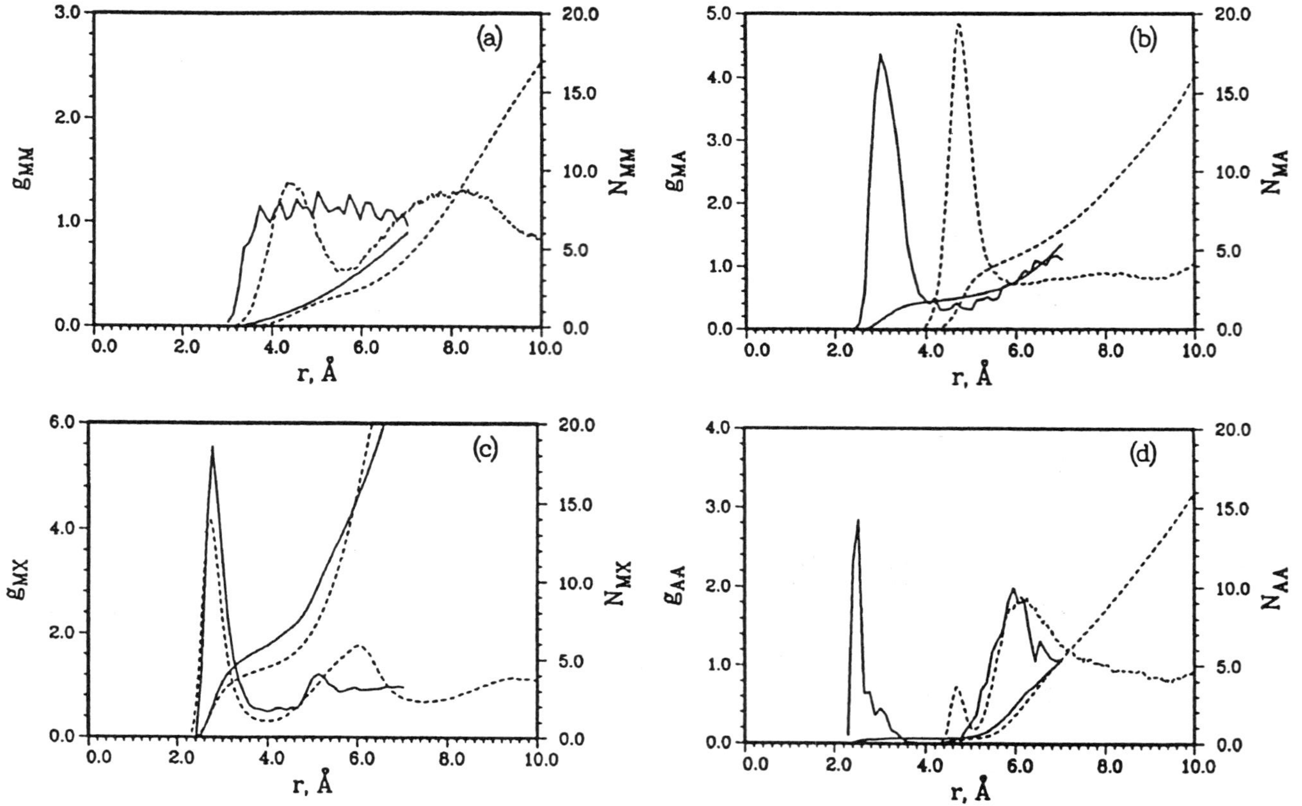

Fig. 5.15. Radial distribution functions $g_{ij}(r)$ and the radial dependence of the average coordination numbers $N_{ij}(r)$ for the molecule MAX_4. (Reprinted from M. L. Saboungi, D. L. Price, C. Scamehorn, and M. Tosi, *Europhys. Lett.* **15**: 281, 1991.)

Here, the first six terms refer to pair potential energies of interaction and the last four terms represent dipole–dipole or monopole–dipole interaction energies.

Taking the anion as polarizable not only introduces these last terms but also reduces the repulsive energies used compared with those of earlier calculations that took a "hard" anion according to the mean spherical approximation. By "hard" anion is meant one that doesn't respond to being squeezed, i.e., is not polarizable.

What do the calculated distribution coefficients look like? They are shown in Fig. 5.15. The positions of the first peaks show that the Al^{3+} cations are largely coordinated with Cl^- as $AlCl_4^-$. On the other hand, the parameter b in the repulsive potential [see Eq. (5.10)] was given values so that the calculations would indeed show up $AlCl_4^-$ ions, so their appearance is not something that fills one with awe. According to the calculations, these anions are present to the extent of 92%. The empirically enlightened calculations show that there is some indication of doubly bridged Al-Al pairs and these are assigned to Al_2Cl_6 molecules present in the melt (see Fig. 5.16). These molecular dynamics calculations (however much aided by information from prior experiments) allow many properties (e.g., the angles in entities such as NaAlCl) to be calculated (Fig. 5.16).

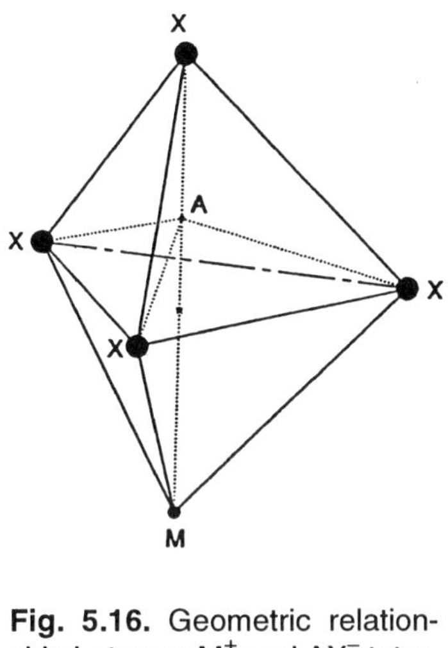

Fig. 5.16. Geometric relationship between M^+ and AX_4^- tetrahedron. The M^+ cation sits perpendicular to the center of a face of the tetrahedron. Consequently, one M-A-X triplet exists at 180° for every three triplets at 64°. (Reprinted from M. L. Saboungi, D. L. Price, C. Scamehorn, and M. Tosi, *Europhys. Lett.* **15**: 281, 1991.)

Calculations taking into account the anion polarizability in $AlCl_4^-$ reduce the approximation associated with the simple additivity of pairwise potentials in computational modeling (hard-sphere approximation) of molten salts. They predict new entities (e.g., Al_2Cl_6) and in this respect have an advantage over earlier calculations.

Further Reading

Seminal

1. N. Metropolis, A. W. Rosenblath, M. W. Rosenblath, A. H. Teller, and E. J. Teller, "The Monte Carlo Method," *J. Chem. Phys.* **21**: 1087 (1953).
2. J. O'M. Bockris, A. Pilla, and J. L. Barton, "Change of Volume in the Fusion of Salts," *J. Phys. Chem.* **64**: 507 (1960).
3. M. P. Tosi and G. Fermi, "Computer Simulation Methods for Liquid Salts," *J. Phys. Chem. Solids* **25**: 31 (1964).
4. V. Woodcock and K. Singer, "Computer Simulation for Liquid KCl," *Trans. Faraday Soc.* **67**: 12 (1971).

Reviews

1. M. P. Allen and D. Tilderley, *Computer Simulation of Liquids,* Clarendon Press, Oxford (1986).
2. J. E. Enderby, in *The Structure of Molten Salts,* G. Mamantov and R. Marassi, eds., *NATO ASI Science Series, Series C* **202**: 2 (1988).

Papers

1. H. T. J. Reijer and W. Van der Lugt, *Phys. Rev. B* **42**: 3395 (1990).
2. R. L. McGreevy and L. Pusztai, *Proc. Roy. Soc. Lond.* **A430**: 241 (1990).
3. M. L. Saboungi, D. L. Price, C. Scamehorn, and M. Tosi, *Europhys. Lett.* **15**: 281 (1991).
4. Z. Acadeniz and M. P. Tosi, *Proc. Roy. Soc. Lond.* **A437**: 75 (1992).
5. S. Itoh, I. Okada, and K. Takahashi, *Electrochemical Society, Molten Salts*, **92-16**: 88 (1992).
6. L. A. Curtiss, *Electrochemical Society, Molten Salts*, **93-9**: 30 (1993).

5.4. VARIOUS MODELING APPROACHES TO DERIVING CONCEPTUAL STRUCTURES FOR MOLTEN SALTS

5.4.1. The Hole Model: A Fused Salt Is Represented as Full of Holes as a Swiss Cheese

One of the models that can be used to approximately predict the properties of molten salts is called the *hole model.* The outstanding fact that led to this model is the large volume of fusion (10–20%) exhibited by simple salts on melting (Fig. 5.17). The basic idea of this rather artificial model is that within the liquid salt are tiny volume

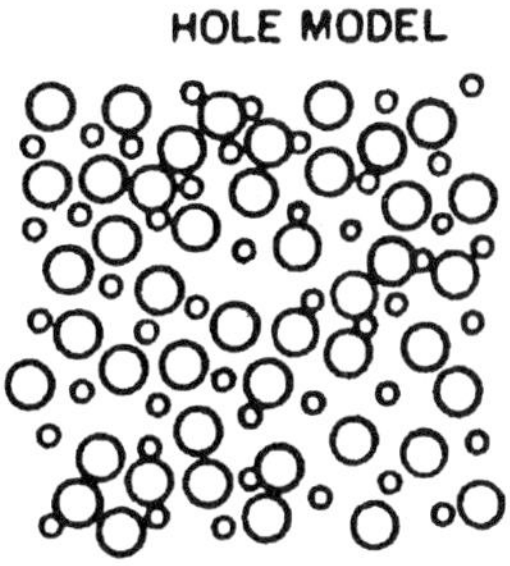

Fig. 5.17. The hole model with randomly located and variable-sized holes in the liquid.

elements, varying in size from the subatomic to about six ions, which are empty and constantly fluctuating in size. Such a model is a kind of liberation from the lattice concepts appropriate to solids and leads to equations that can explain liquid properties, some of them with reasonable numerical accuracy and without any of those "calibrating factors" that make the computer simulation approaches always agree with experiment.

How are the alleged holes in the molten salt produced? They are formed by a process analogous to the formation of a vacancy in a crystal, but in a less ordered fashion. The displacement of an ion from a lattice site in a solid produces a vacancy at its former site. In the case of the vacancy in a solid, however, the ion is removed so far from the original site that the displaced ion can be forgotten altogether. Suppose instead that at high temperatures in the course of thermal motion some of the ions constituting a cluster are displaced relative to each other but only by small amounts. Then a "hole" is produced between them (Fig. 5.18). Its size must vary in a random manner because the thermal motions that produce it are random. Further, since thermal motions occur everywhere in the liquid electrolyte, holes appear and disappear anywhere in this liquid. If one were able to label the holes with scintillating material and enlarge the signals so one could see them, the molten salt would look like a set of twinkling light sources, going on and off all over the melt.

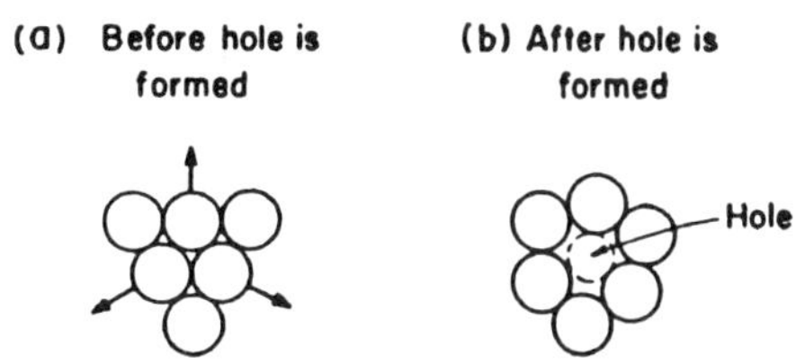

Fig. 5.18. The formation of a hole in a liquid by the relative displacement of ions in contact.

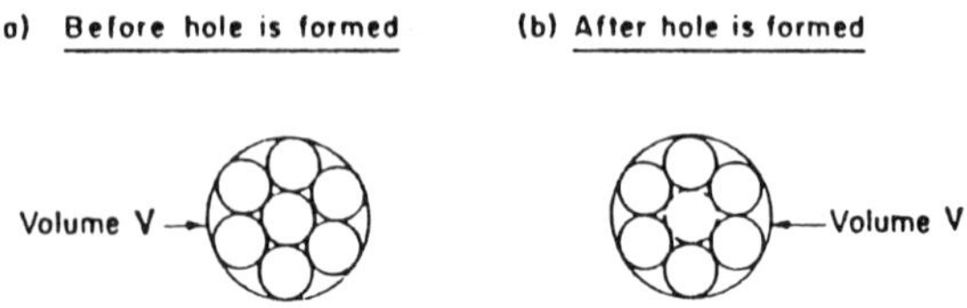

Fig. 5.19. The formation of a hole can also be looked at in terms of the number of ions occupying a volume *V*. In (a) seven particles occupy the volume *V* before the hole is formed, and, in (b), six particles occupy the same volume after hole formation.

If a liquid electrolyte (Fig. 5.19) could be sliced up and subjected to nanometer-scale photography with an exposure time of around 10^{-11} s, the photos would resemble those of cuts through a Swiss cheese. This is why, in the matter of randomness of size and location, the hole theory of liquids has been referred to in homely terms as the "Swiss cheese model." What the cheese represents, however, is the time-averaged picture of the holes. In the model, holes are continuously forming and disappearing, moving, coalescing to form larger holes, and diminishing into smaller ones.

Although this model of a liquid was suggested independently of the results obtained from computer modeling, the imagined picture of the hole model in Fig. 5.17 closely resembles the picture (Fig 5.14) from Woodcock and Singer's model derived from the Monte Carlo approach.

5.5. QUANTIFICATION OF THE HOLE MODEL FOR LIQUID ELECTROLYTES

5.5.1. An Expression for the Probability That a Hole Has a Radius between *r* and *r* + *dr*

To quantify the hole model, it is necessary to calculate a distribution function for the hole sizes. This is a plot of the number of holes per unit volume as a function of their size. As a first step toward this calculation, one can consider a particular hole in a liquid electrolyte and ask: What are the quantities (or variables) needed to describe this hole? This problem can be resolved by means of a formulation first published by Fürth in 1941.

Since a hole in a liquid can move about like an ion or other particle, the dynamic state of a hole is specified in the same way that one describes the dynamic state of a material particle. Thus, one must specify three position and three momentum coordinates: x, y, z and p_x, p_y, p_z. There is, however, an extra feature of the motion of a hole that is not possessed by material particles. This feature concerns what is called its *breathing motion* (Fig. 5.20), i.e., the contraction and expansion of the hole as its size

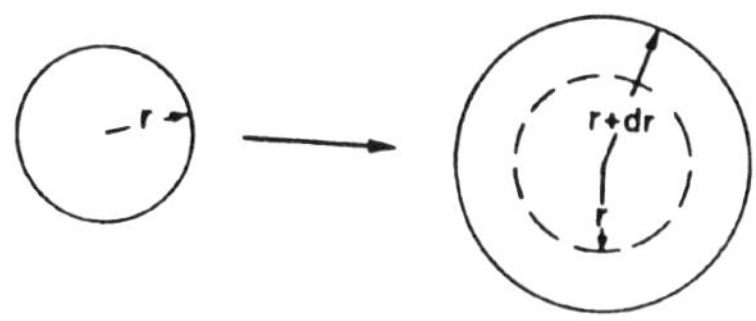

Fig. 5.20. The breathing motion of a hole involves its radial expansion.

fluctuates in the liquid. To characterize this breathing motion, it is sufficient to specify the hole radius r and the radial momentum p_r corresponding to the breathing motion. To characterize a hole completely, it is necessary to specify eight quantities: x, y, z, p_x, p_y, p_z, r, and p_r, whereas the first six only are adequate to describe the state of motion of a particle.

According to the normal equations of classical statistical mechanics, which are used to express velocities and momenta distributed in three dimensions, the probability P_r that the location of a hole is between x and $x + dx$, y and $y + dy$, z and $z + dz$; that its translational momenta lie between p_x and $p_x + dp_x$, p_y and $p_y + dp_y$, p_z and $p_z + dp_z$; that its breathing momentum is between p_r and $p_r + dp_r$; and, finally, that its radius is between r and $r + dr$, is proportional to the Boltzmann probability factor,

$$P_r\, dx\, dy\, dz\, dp_x\, dp_y\, dp_z\, dp_r\, dr \propto e^{-E/kT} \times dx\, dy\, dz\, dp_x\, dp_y\, dp_z\, dp_r\, dr \quad (5.16)$$

where E is the total energy of the hole.

Since the desired distribution function only concerns the radii (or sizes) of holes, it is sufficient to have the probability that the hole radius is between r and $r + dr$ irrespective of the location and the translational and breathing momentum of the hole. This probability $P_r\, dr$ of the hole's radius being between r and $r + dr$ is obtained from Eq. (5.16) by integrating over all possible values of the location, and of the translational and breathing momentum of the hole, i.e.,

$$P_r\, dr \propto \left(\int\int\int\int\int\int\int e^{-E/kT} \times dp_x\, dp_y\, dp_z\, dp_r\, dx\, dy\, dz\right) dr \quad (5.17)$$

However, the total energy of the hole on average does not depend upon its position; i.e., E is independent of x, y, and z. Hence,

$$P_r\, dr \propto \left(\int\int\int\int e^{-E/kT} \times dp_x\, dp_y\, dp_z\, dp_r\right)\left(\int\int\int dx\, dy\, dz\right) dr \quad (5.18)$$

Furthermore,

$$\int\int\int dx\, dy\, dz = V \quad (5.19)$$

the volume of the liquid. Thus, by incorporating this V into the proportionality constant implicitly associated with Eq. (5.18), one has

$$P_r dr \propto \left(\int\int\int\int e^{-E/kT} \times dp_x\, dp_y\, dp_z\, dp_r \right) dr \tag{5.20}$$

Now the total energy E consists of the potential energy W of the hole (i.e., the work required to form the hole) plus its kinetic energy. This kinetic energy is given by

$$\frac{p_x^2}{2m_1} + \frac{p_y^2}{2m_1} + \frac{p_z^2}{2m_1} + \frac{p_r^2}{2m_2} \tag{5.21}$$

where m_1 is the apparent mass[13] of the hole in its translational motions and m_2 is the apparent mass in its breathing motion. Hence,

$$E = W + \frac{p_x^2}{2m_1} + \frac{p_y^2}{2m_1} + \frac{p_z^2}{2m_1} + \frac{p_r^2}{2m_2} \tag{5.22}$$

Inserting this value of E into the expression (5.20) for the probability of the hole's having a radius between r and $r + dr$, one has

$$P_r\, dr \propto dr\, e^{-W/kT} \int_{-\infty}^{\infty} e^{-p_x^2/2m_1kT} dp_x \int_{-\infty}^{\infty} e^{-p_y^2/2m_1kT} dp_y \int_{-\infty}^{\infty} e^{-p_z^2/2m_1kT} dp_z \int_{-\infty}^{\infty} e^{-p_r^2/2m_2kT} dp_r \tag{5.23}$$

From the standard integral,

$$\int_{-\infty}^{\infty} e^{-ax^2} dx = \sqrt{\frac{\pi}{a}} \tag{5.24}$$

it is clear that

$$\int_{-\infty}^{\infty} e^{-p_x^2/2m_1kT} dp_x = \int_{-\infty}^{\infty} e^{-p_y^2/2m_1kT} dp_y = \int_{-\infty}^{\infty} e^{-p_z^2/2m_1kT} dp_z = (2\pi m_1 kT)^{1/2} \tag{5.25}$$

---

[13]Any entity that moves displays the property of inertia, i.e., resistance to a change in its state of rest or uniform motion. That is, the entity has a mass. If the entity is not material (a hole is a region where in fact there is no material), one refers to an apparent mass. Holes in semiconductors have apparent masses like holes in liquids. The inertia of the hole arises as a result of the displacement of the liquid around the hole as it moves, which gives rise to dissipation of energy (Appendix 5.1).

and

$$\int_{-\infty}^{\infty} e^{-p_r^2/2m_2kT} dp_r = (2\pi m_2 kT)^{1/2} \tag{5.26}$$

By using these values of the integrals in Eq. (5.23), the result is

$$P_r\, dr \propto (2\pi kT)^2 m_1^{3/2} m_2^{1/2} e^{-W/kT}\, dr \tag{5.27}$$

It can be shown, however (see Appendix 5.1), that

$$m_1 = \frac{2}{3}\pi r^3 \rho \tag{5.28}$$

and

$$m_2 = 4\pi r^3 \rho \tag{5.29}$$

After taking all the quantities that are radius independent into the proportionality constant A, one has by combining Eqs. (5.27), (5.28), and (5.29) that

$$P_r\, dr = A r^6 e^{-W/kT}\, dr \tag{5.30}$$

The evaluation of the constant is achieved through the following plausible argument. The probability that a hole has some radius must be unity (i.e., is certain by definition). Equation (5.30) expresses the probability of the radius of the hole lying between r and $r + dr$. Similarly, one can write down the probabilities of the radius' being between r_1 and $r_1 + dr$, between r_2 and $r_2 + dr$, etc. If all these probabilities for r from zero to infinity are summed up (or integrated), then the sum must be unity, i.e.,

$$\int_0^{\infty} P_r\, d_r = \int_0^{\infty} A r^6 e^{-W/kT}\, dr \tag{5.31}$$

However, to carry out this integration, one must know whether the work of hole formation W is a function of r; i.e., one must understand what determines the work of formation (or the potential energy) of a hole of radius r (Fig. 5.21).

5.5.2. An Ingenious Approach to Determine the Work of Forming a Void of Any Size in a Liquid

A remarkably simple way of calculating the work of hole formation was found by Fürth, who treated holes in liquids, the sizes of which are thermally distributed, in an article published in the *Proceedings of the Cambridge Philosophical Society.* This is an erudite university journal whose readers are mainly members of the university's

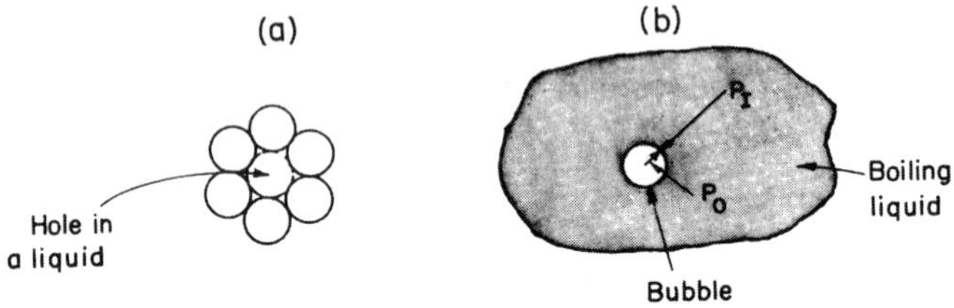

Fig. 5.21. The basis of the hole model of Fürth is the analogy between (a) a hole in a liquid and (b) a bubble in a liquid. An inward pressure P_I and an outward pressure P_O act on the bubble surface.

colleges, and this conservative choice of publication medium delayed recognition of the virtues of the model from 1941 until the 1970s, well after Monte Carlo calculations had begun. In Fürth's model, a hole in a fused salt is considered to simulate a bubble in a liquid (Fig. 5.21). The surrounding liquid exerts a hydrostatic pressure P_I on the bubble surface. Inside the bubble, however, there is vapor, which exerts an outward pressure P_O on the surface. The net pressure is therefore $P_I - P_O$. Furthermore, surface tension operates in the direction of reducing the surface area and therefore the surface energy of the bubble.

The total work required to increase the bubble size consists of two parts, the volume work $(P_I - P_O)V_H$ and the surface work γA, where $P_I - P_O$, V_O, γ, and A are the net pressure on the bubble surface, the increase of volume, the surface tension, and the increase of surface area of the bubble, respectively. Thus, the work done in making a bubble grow to a size having a radius r is

$$W = (P_I - P_O)\frac{4}{3}\pi r^3 + 4\pi r^2 \gamma \tag{5.32}$$

The first term, i.e., the volume term, is negligible compared to the second or surface term for bubbles of less than about 10^{-5} cm in diameter. The work of bubble formation reduces to

$$W = 4\pi r^2 \gamma \tag{5.33}$$

This expression can also be obtained from the general equation (5.32) by setting $P_I = P_O$. This equality between P_I and P_O represents the condition that the liquid is boiling. The analogy between a hole and a bubble consists therefore in assuming that the work of hole formation is given by the expression for the work of bubble formation in a liquid. Let the ability of the model to replicate experimental results be the criterion of the acceptability of Eq. (5.33).

5.5.3. The Distribution Function for the Sizes of the Holes in a Liquid Electrolyte

Now that an expression for the work of hole formation has been obtained, it can be inserted into Eq. (5.31), which must be integrated to evaluate the constant A. One has

$$\int_0^\infty P_r\,dr = 1 = \int_0^\infty A r^6 e^{-ar^2}\,dr \tag{5.34}$$

where

$$a = \frac{4\pi\gamma}{kT} \tag{5.35}$$

To carry out the integration, the following standard formula is used

$$\int_0^\infty r^{2n} e^{-ar^2}\,dr = \frac{1\times 3\times 5\ldots(2n-1)}{2^{n+1}a^n}\sqrt{\frac{\pi}{a}} \tag{5.36}$$

where n is a positive integer. The integral in Eq. (5.34) is consistent with the standard formula if one substitutes $n = 3$, and therefore,

$$\int_0^\infty P_r dr = 1 = A\int_0^\infty r^{2\times 3} e^{-ar^2}\,dr = A\frac{1\times 3\times 5}{2^4 a^3}\frac{\pi^{1/2}}{a^{1/2}} = A\frac{15\pi^{1/2}}{16}\frac{1}{a^{7/2}} \tag{5.37}$$

The constant in Eq (5.34) is given by

$$A = \frac{16}{15\pi^{1/2}}a^{7/2} = \frac{16}{15\pi^{1/2}}\left(\frac{4\pi\gamma}{kT}\right)^{7/2} \tag{5.38}$$

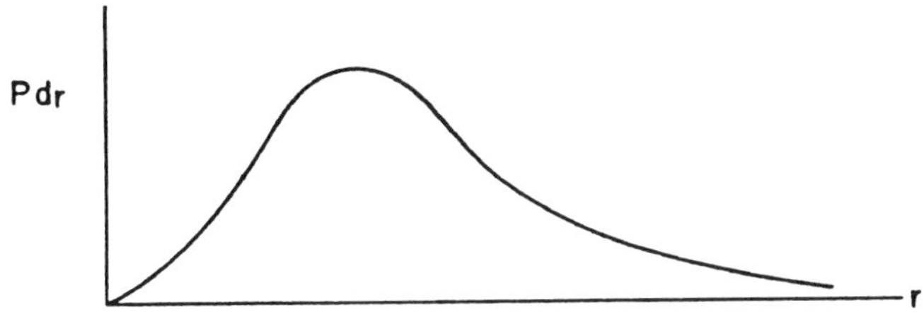

Fig. 5.22. How the probability $P_r\,dr$ that a hole has a radius between r and $r + dr$ varies with r.

and the expression (5.30) for the probability of the existence of a hole of a radius between r and $r + dr$ becomes

$$P_r\,dr = \frac{16}{15\pi^{1/2}} a^{7/2} r^6 e^{-ar^2}\,dr \tag{5.39}$$

This is the distribution function which was said to be the goal at the beginning of Eq. (5.17). From it, the average hole volume and radius will shortly be seen to be obtainable (see Fig. 5.22).

5.5.4. What Is the Average Size of a Hole in the Fürth Model?

The average radius $<r>$ of a hole can be obtained from Eq. (5.39) by multiplying the probability of the hole radius being between r and $r + dr$ by the radius of the hole and integrating this product over all possible values of r. This is the general method of obtaining average values of a quantity of which the probability is known. Thus, the average hole radius is

$$<r> = \int_0^\infty rP_r\,dr = \int_0^\infty r\frac{16}{15\pi^{1/2}} a^{7/2} r^6 e^{-ar^2}\,dr = \frac{16}{15\pi^{1/2}} a^{7/2}\int_0^\infty r^7 e^{-ar^2}\,dr \tag{5.40}$$

The integral in Eq. (5.40) can be evaluated by using the substitution $t = ar^2$,

$$\int_0^\infty r^7 e^{-ar^2}\,dr \quad \xrightarrow[t^3/a^3 = r^6]{t = ar^2,\ dt/2a = r\,dr} \quad \frac{1}{2a^4}\int_0^\infty t^3 e^{-t}\,dt \tag{5.41}$$

which leads to

$$<r> = \frac{8}{15\pi^{1/2}}\frac{1}{a^{1/2}}\int_0^\infty t^3 e^{-t}\,dt \tag{5.42}$$

The integral $\int_0^\infty t^3 e^{-t}\,dt$ is the gamma function $\Gamma(3 + 1)$ (Appendix 5.2) and is equal to 3! from

$$\Gamma(x + 1) = \int_0^\infty t^x e^{-t}\,dt = x! \tag{5.43}$$

Hence, Eq. (5.42) becomes

$$<r> = \frac{8}{15\pi^{1/2}}\left(\frac{kT}{\gamma}\right)^{1/2}\frac{1}{(4\pi)^{1/2}} \times 3 \times 2 \times 1$$

$$= \frac{8}{5\pi}\left(\frac{kT}{\gamma}\right)^{1/2}$$

$$= 0.51\left(\frac{kT}{\gamma}\right)^{1/2} \tag{5.44}$$

It follows that the average surface area of a hole calculated by this procedure is

$$4\pi{<}r{>}^2 = 3.3\,\frac{kT}{\gamma} \tag{5.45}$$

Then from Eqs. (5.44) and (5.45), one obtains, clearly:

$$\frac{4\pi{<}r{>}^2}{{<}r{>}^2} \equiv \frac{3.5kT/\gamma}{[0.51(kT/\gamma)^{1/2}]^2} = 12.57 \tag{5.46}$$

What typical values of mean hole radius does Eq. (5.44) yield (Table 5.15)? By using the macroscopic surface-tension value, Eq. (5.44) shows that the average radius of a hole in molten KCl at 1173 K is 190 pm. The mean ionic radius, however, is 160 pm.

A typical hole therefore can accommodate an ion in a possible movement into a hole. This result is remarkable because of the process by which it has been attained. One began by considering that a liquid electrolyte was a liquid continuum interspersed by holes of random size and location imagined to be forming and collapsing as in a boiling liquid. Thus, the work of hole formation was taken to be equal to the work of expanding the surface area of a bubble in a boiling liquid. With the use of this expression and simple probability arguments, the average hole radius was calculated by taking the integral of Eq. (5.40) from zero to infinity.

TABLE 5.15

Mean Hole Radius for Various Molten Salts at 1173 K

Molten Salt	Surface Tension, (dyn cm^{-1})	Mean Hole Volume (nm^3)	Mean Hole Radius (nm)
NaCl	107.1	0.032	0.170
NaBr	90.5	0.0417	0.190
KCl	89.5	0.0423	0.190
KBr	77.3	0.0527	0.210
CsCl	72.7	0.0518	0.220
Nal	66.4	0.0662	0.230
KI	60.3	0.0766	0.230

At a fixed temperature, the only parameter determining the mean hole size is the surface tension. Though one is aiming at a microscopic (structural) explanation of the behavior of ionic liquids, one goes ahead and uses the macroscopic value of surface tension. The mean hole radius then turns out to have the same order of magnitude as the mean radius of ions comprising the liquid.

A remarkable conclusion can be drawn by looking at Table 5.15. One sees that Fürth's theory of holes in liquids—liquids imagined to be nearly on the boil so that $(P_I - P_O) = 0$—indicates that the holes arising from the model are molecular in size. Why is this a remarkable conclusion? It is because no props (e.g., dependence on some measurement of volume increase on melting) which would ensure that the right hole size would turn up have been used. In fact, hitherto, nothing has been said about molecules, ions, or any structure. It might have been thought to be a long shot—taking the bubble in a boiling liquid and seeing in such a concept any molecular reality. Look back at Table 5.15. The molecular size of the holes arising from the theory is an established *fact*. The conclusion: the model is worth investigating further and seeing how its application works out when one comes to the interpretation of transport data (Section 4.5.2). At this stage, it is sufficient to be properly surprised: the idea of bubbles (holes) in liquids gives rise to the average size, which is that of the ions concerned. Later on (Section 5.6) it will be seen just how this result offers a possible mechanism for transport, e.g., viscous flow and diffusion.

The theory of an indication of molecular-sized holes supports the idea that the model reflects some aspects of reality in liquids. How can bubbles in liquids be used as the basis of a calculation of liquid properties? The answer is given by the degree to which such an approach predicts facts, for example, the compressibility (Table 5.16). The fact that it indicates the calculated size of holes seems to suggest a diffusion model in which critical acts may be the formation of the holes and the jumping into them of neighboring ions (but see Section 5.4.1).

These indications that one of the modeling approaches to molten salts looks remarkably promising does not mean that it is the final word in making models of molten salts. It is an imaginative portrayal, but there are some molten salts the properties of which are not covered at all by any theory of holes distributed like bubbles in boiling liquids. One of these undealt-with properties—the property of *supercooling liquids,* which continue to retain their properties below the normal melting point—must be looked into before one returns to a modeling interpretation of viscous flow, diffusion, and conductance (see Chapter 4).

5.5.5. Glass-Forming Molten Salts

A number of molten salt systems [e.g., the simple ionic system $Ca(NO_3)_2$-KNO_3], have the property of being able to be *supercooled*, i.e., to remain liquid at temperatures below the melting point down to a final temperature. This is called the *glass transition temperature*, and at this temperature the salts form what is called "a glass." This glass is only apparently solid. It is a highly disordered substance in which a liquid structure

TABLE 5.16

Comparison of Calculated Isothermal Compressibilities with the Experimental Values (Hole Theory)

Salt	Temp. (K)	$10^{12}\beta$, calc. ($cm^2\ dyn^{-1}$)	$10^{12}\beta$, obs. ($cm^2\ dyn^{-1}$)
LiCl	887	20.8	19.4
	1073	30.5	24.7
	1173	39.1	28.6
	1273	47.9	33.0
NaCl	1073	27.8	28.7
	1173	36.9	22.8
	1273	49.6	40.0
KCl	1045	27.6	36.2
	1073	30.2	38.4
	1173	42.3	45.7
	1273	56.7	54.7
CsCl	915	20.2	38.0
	1073	39.0	51.8
	1173	57.0	62.6
	1273	72.5	76.3
$CdCl_2$	873	31.0	29.8
	973	37.3	33.1
	1073	46.9	36.9
NaBr	1020	32.8	31.6
	1073	40.1	33.6
	1173	47.4	38.6
	1273	59.5	44.9
KBr	1008	28.0	39.8
	1073	34.5	43.8
	1173	51.3	52.1
	1273	69.9	62.1
CsBr	909	56.3	49.1
	1073	76.0	67.1
	1173	97.5	82.7
	1273	130.5	103.1

—no long-range order—is *frozen* into what appears to be a solid. Although such systems were studied in the 1960s and earlier, they subsequently received a seminal and sustained contribution from the Australian physical chemist Angell, who has been the driving force behind the discovery of much of modern knowledge of the supercooled state.

Mathematical treatment of molten salts that supercool was first carried out by Cohen and Turnbull. The principal idea of the hole theory—that diffusion involves ions that wait for a void to turn up before jumping into it—is maintained. However, Cohen and Turnbull introduced into their model a property called the *free volume*, V_f. What is meant by this "free volume"? It is the amount of space in addition to that, V_0, filled by matter in a closely packed liquid. Cohen and Turnbull proposed that the free volume is linearly related to temperature

$$V_f = V - V_0 = V_m \alpha (T - T_0) \tag{5.47}$$

where T_0 is the temperature at which the free volume becomes zero. Cohen and Turnbull named the temperature at which a supercooled liquid becomes a glass the *glass transition temperature.*

To express the probability that the free volume occasionally opens up to form a hole, Cohen and Turnbull first defined a factor γ, which allows for the partial filling of the expanded free volume to the size of a hole. It can vary between $1/2 < \gamma < 1$. A value of $\gamma = 1$ means that the holes are empty, and a value of $\gamma = 1/2$ that they are half filled.

The authors rejected the normal thermal probability term, involving $e^{-E/RT}$. They used a statistical argument concerning the number of ways it is possible to mix free spaces with ions and found thereby the probability P_r of finding a void volume V^* as a fraction of the free volume. The expression for this probability comes to

$$P_r = e^{-\gamma V^* / V_f} \tag{5.48}$$

Then the most elementary expression for a diffusion coefficient would be:

$$D = lv \tag{5.49}$$

where l is the distance covered in one jump and v the velocity of the particle. This primitive expression would be true if there were a void always available and each jump were in the same direction. Since there are three distance coordinates, each having two directions (i.e., six directions), and since the coming-into-being of a hole has the probability given by Eq. (5.48), then,

$$D = \frac{1}{6} v l e^{-\gamma V^* / V_f} \tag{5.50}$$

Substituting V_f from Eq. (5.47),

$$D = \frac{1}{6} v l e^{-\frac{\gamma V^*}{\alpha V_m (T - T_0)}} = A e^{-\frac{B}{T - T_0}} \tag{5.51}$$

where

$$B = \frac{\gamma V^*}{\alpha V_m} \tag{5.52}$$

Cohen and Turnbull's model is oriented to liquids that form glasses. At the glass transition temperature (i.e., at $T = T_0$), the diffusion coefficient becomes zero, which is a rational consequence of what is thought to be going on: the supercooled liquid finally becomes a glass in which D is effectively zero.

A difficulty might face the worker who wishes to apply Cohen and Turnbull's theory to transport phenomena in molten salts not only near the glass transition temperature but also above the normal melting point (see Section 5.6.2.2). Experimental evidence shows that the heat of activation of diffusion and of conductance for viscous flows is related to the normal melting point of the substance concerned

$$E_D^{\neq} = 3.74RT_{\text{m.p.}} \tag{5.53}$$

Accepting this on faith for the moment, then B from Eq. (5.51) can be written as

$$B = \frac{\gamma}{\alpha} \frac{V^*}{V_m} = 3.74RT_{\text{m.p.}} \tag{5.54}$$

The order of magnitude of V_f is 1 $\text{cm}^3\ \text{mol}^{-1}$. If one identifies V^* with a void into which an ion jumps, then in transport phenomena this must be around 10 $\text{cm}^3\ \text{mol}^{-1}$, a typical molar volume for simple monatomic ions. Moreover, near to the melting point, $T = T_{\text{m.p.}}$ and $\gamma \propto 0.2$.

This result increases the credibility of the Cohen and Turnbull view: the meaning of $\gamma = 0.2$ is that the "hole" is 20% full, and this seems consistent with the picture of an ion-sized space filled 20% of the time with an ion.

Further Reading

Seminal

1. R. Fürth, "On the Theory of the Liquid State, I. The Statistical Treatment of the Thermodynamics of Liquids by the Theory of Holes," *Proc. Cambridge Phil. Soc.* **37**: 252 (1941).
2. R. Fürth, "On the Theory of the Liquid State, II. The Hole Theory of the Viscous Flow of Liquids," *Proc. Cambridge Phil. Soc.* **37**: 281 (1941).
3. M. H. Cohen and D. Turnbull, "Molecular Transport in Liquids and Glasses," *J. Chem. Phys.* **31**: 1164 (1959).
4. A. F. M. Barton and R. J. Speedy, "Simultaneous Conductance and Volume Measurements on Molten Salts at High Pressure," *J. Chem. Soc. Faraday Trans.* **71**: 506 (1974).

Reviews

1. C. A. Angell, "Transport and Relaxation Processes in Molten Salts," *NATO ASI Series C* **202**: 123 (1987).
2. G. Mamantov, C. Hussey, and R. Marassi, eds., *An Introduction to the Electrochemistry of Molten Salts*, Wiley, New York (1991).

Papers

1. Y. Shirakawa, S. Tamaki, M. Saito, H. Masatoshi, and S. Harab, *J. Non-Cryst. Solids* **117**: 638 (1990).
2. W. Freyland, *J. Non-Cryst. Solids* **117**: 613 (1990).
3. R. L. McGreevy, *Nuovo Cimento* **12D** (4–5): 685 (1990).
4. T. Nakamura and M. Itoh, *J. Electrochem. Soc.* **137**: 1166 (1990).
5. M. L. Saboungi and D. L. Price, in *Proc. Int. Symp. Molten Salts,* Electrochemical Society, p. 8 (1990).
6. M. Abraham and I. Zloges, *J. Am. Chem. Soc.* **113**: 8583 (1991).
7. M. Noel, R. Allendoerfer, and R. A. Osteryoung, *J. Phys. Chem.* **96**: 239 (1992).
8. R. J. Speedy, F. X. Prielmeier, T. Vardag, E. W. Lang, and H. D. Ludemann, *J. Electrochem. Soc.* **139**: 2128 (1992).
9. C. A. Angell, C. Alba, A. Arzimanoglou, and R. Bohmer, *AIP Proc.* **256**: 3 (1992).
10. S. Deki, H. Twabuki, A. Kacinami, and Y. Kanagi, *Proc. Electrochem. Soc.* **93–9**: 252 (1993).
11. S. Itoh, Y. Hiwatari, and H. Miwagawa, *J. Non-Cryst. Solids* **156**: 159 (1993).
12. C. A. Angell, C. Lia, and E. Sanchez, *Nature* **362**: 137 (1993).
13. C. A. Angell, P. H. Poole, and J. Shao, *Nuovo Cimento* **16**: 993 (1994).
14. C. A. Angell, *Proc. Natl. Acad. Sci. U.S.A.* **92**: 6675 (1995).
15. C. A. Angell, *Science* **267**: 1924 (1995).
16. M. G. McClin and C. A. Angell, *J. Phys. Chem.* **100**: 1181 (1996).

5.6. MORE MODELING ASPECTS OF TRANSPORT PHENOMENA IN LIQUID ELECTROLYTES

5.6.1. Simplifying Features of Transport in Fused Salts

An important characteristic of liquid ionic systems is that they lack an inert solvent; they are *pure electrolytes.* Owing to this characteristic, some aspects of transport phenomena in pure molten salts are simpler than similar phenomena in aqueous solutions.

Thus, there is no concentration variable to be taken into account in the consideration of transport phenomena in a pure liquid electrolyte. Hence, there cannot be a concentration gradient in a *pure fused salt,* and [because of Fick's first law; see Eq. (4.16)] without a concentration gradient there cannot be pure diffusion. In an *aqueous solution,* on the other hand, it is possible to have a concentration gradient for the solute and thus have diffusion in the normal sense.

Another consequence of the absence of a solvent in a pure liquid electrolyte is that the mean ion–ion interaction field as a function of distance within the liquid is constant. In solutions of ions in a solvent, however, the extent of ion–ion interaction is a variable quantity. It depends on the amount of solvent dissolving a given quantity of ionic solute, i.e., on the solute concentration.

5.6.2. Diffusion in Fused Salts

5.6.2.1. *Self-Diffusion in Pure Liquid Electrolytes May Be Revealed by Introducing Isotopes.* In the absence of a solvent, it is meaningless to consider a pure liquid electrolyte (e.g., NaCl) as having different amounts of NaCl in different regions. Let the possibility be momentarily considered that the system could be made to have more ions of one species (e.g., Na^+) in one region than in another. However, this is impossible over significant time spans because any attempt of a single ionic species to accumulate in one region and decrease in another is promptly stopped by the electric field that develops as a consequence of the separation of charges. Overall electroneutrality must prevail; i.e., there can be no congregation of an ionic species in one part of the liquid.

Fortunately, electroneutrality only requires that the total positive charge in a certain region be equal to the total negative charge. Suppose therefore that in a liquid sodium chloride electrolyte, a certain percentage of the Na^+ ions are replaced by a radioactive isotope of sodium. There is no difference between the ^{22}Na and ^{23}Na as far as the principle of electroneutrality is concerned; it is only required that the number of Na^+ ions plus the number of tagged Na^{*+} ions are equal to the total number of Cl^- ions (Fig. 5.23). However, the labeled Na^{*+} ions and the nonradioactive Na^+ are completely different entities from the point of view of a counter; only the former produce the scintillations.

Herein lies a method of examining the diffusion of ions in pure ionic liquids that differs from the diffusion of ions in aqueous solution, which is governed by Fick's law. One takes a pure liquid electrolyte, say NaCl, and brings it into contact with a melt containing the same salt but with a certain proportion of radioactive ions, say, NaCl with radioactive Na^{*+} ions. There is a negligible concentration gradient for Na^+ ions, but a concentration gradient for the tracer Na^{*+} ions has been created. Diffusion of the tracer begins (Fig. 5.24).

If a capillary containing an inactive melt is suitably introduced into a large reservoir of tracer-containing melt at $t = 0$, then diffusion of the tracer into the capillary starts (Fig. 5.25). At time t, the experiment can be terminated by withdrawing the capillary from the reservoir. The total amount of tracer in the capillary can be measured by a detector of the radioactivity. From the study of the diffusion problem and the experimentally determined average tracer concentration in the capillary, the diffusion coefficient of the Na^{*+} ions can then be calculated.

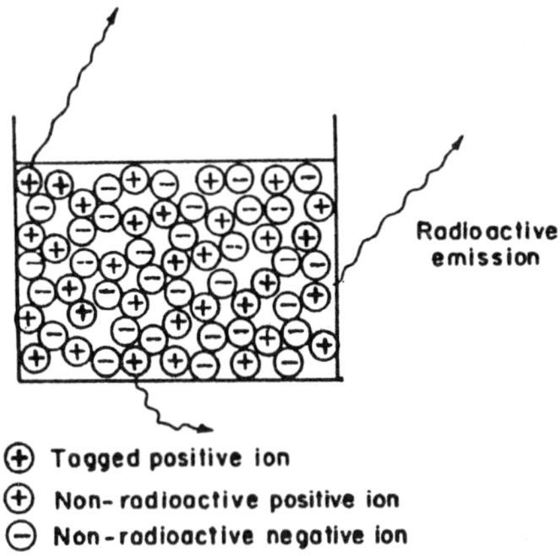

Fig. 5.23. The principle of electroneutrality is satisfied if the number of tagged positive ions plus the number of nonradioactive positive ions is equal to the total number of negative ions.

Since the tracer ions (e.g., Na^{*+}) diffuse among particles (e.g., Na^+) that are chemically just like themselves, one often refers to the phenomenon as *self*-diffusion (*tracer* diffusion is a more explanatory term) and to the diffusion coefficient thus determined as the *self*-diffusion coefficient.

5.6.2.2. *Results of Self-Diffusion Experiments.* Self-diffusion coefficient studies with fused salts really began to gather momentum after radioisotopes became

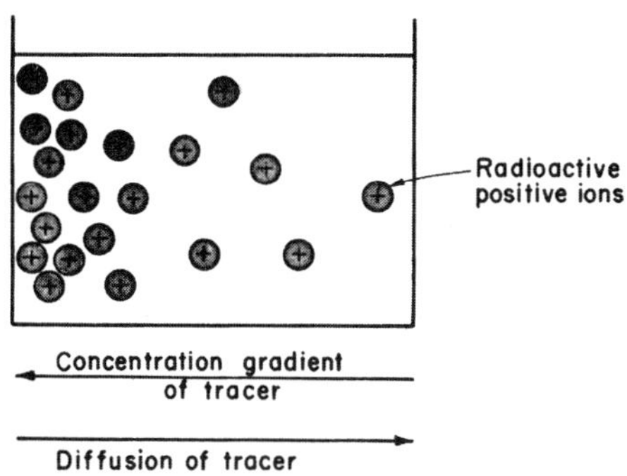

Fig. 5.24. The existence of a concentration gradient for tracer ions produces diffusion of the tracer, i.e., tracer diffusion.

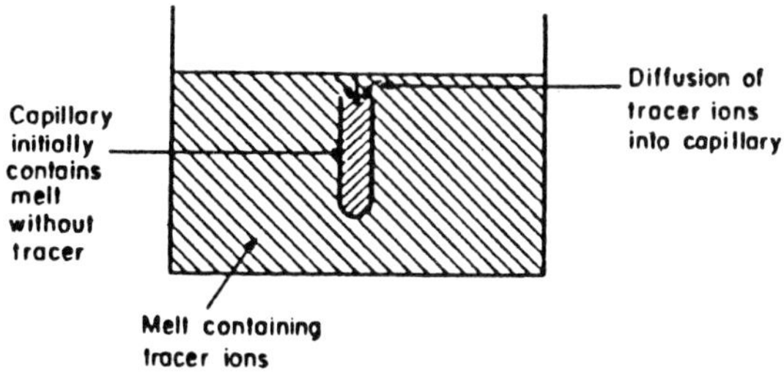

Fig. 5.25. A schematic of an experiment to study tracer diffusion. A capillary containing inactive melt is dipped into a reservoir of melt containing tracer ions. Tracer ions diffuse into the capillary.

widely available, i.e., after about 1950. Some of the data for comparing the diffusion coefficients of typical inert gases, room-temperature liquids, metals, and molten salts are presented in Tables 5.17, 5.18, and 5.19.

It can be seen that the diffusion coefficients of these liquid electrolytes (near their melting points) are of the same order of magnitude ($\sim 10^{-5}$ cm^2 s^{-1}) as for liquid inert gases, liquid metals, and normal room-temperature liquids. This fact suggests that the mechanism of diffusion is basically the same in all simple liquids, i.e., liquids in which the particles do not associate into pairs, triplets, or network structures (see Sections 3.8 and 4.8.8). The order of magnitude of the diffusion coefficient has evidently more to do with the liquid state than with the chemical nature of the liquid for in the case of the corresponding solid substances, the diffusion coefficient ranges (Table 5.19) over more than four orders of magnitude (10^{-7} to 10^{-11} cm^2 s^{-1}).

An expected feature of the results on tracer diffusion is that the diffusion coefficient varies with temperature. The temperature dependence observed experimentally can be expressed in the type of equation exhibited by virtually all transport, or rate, phenomena.

TABLE 5.17

Self-Diffusion Coefficients of Various Types of Substance

Type of Substance	Example	Temperature (K)	Diffusion Coefficient (cm^2 s^{-1})
Liquid inert gas	Ar	100	3.70×10^{-5}
Room-temperature liquid	CCl_4	298	1.41×10^{-5}
Liquid metal	Zn	693	2.07×10^{-5}
Molten salt	NaCl	1113	D_{Na^+} 9.6×10^{-5} D_{Cl^-} 6.7×10^{-5}

TABLE 5.18

Energies of Activation and Preexponential Factors for Self-Diffusion in Molten Group I and II Chlorides

Molten Salt/Tracer	$10^3 \times D_0^{\neq}$ ($cm^2\ s^{-1}$)	$E_D^{\neq}$ (kJ mol^{-1})	Temp. Range (K)
$NaCl/^{22}Na$	2.1	29.87 ± 1.05	1093–1293
$NaCl/^{36}Cl$	1.9	31.09 ± 3.51	1099–1308
$KCl/^{42}K$	1.8	28.78 ± 2.13	1071–1256
$KCl/^{36}Cl$	1.8	29.83 ± 2.05	1067–1260
$CaCl_2/^{45}Ca$	0.38	25.65 ± 2.76	1056–1277
$CaCl_2/^{36}Cl$	1.9	37.07 ± 4.02	1060–1292
$SrCl_2/^{89}Sr$	0.21	22.51 ± 4.06	1194–1393
$SrCl_2/^{36}Cl$	0.77	28.79 ± 3.01	1185–1430
$BaCl_2/^{140}Ba$	0.64	37.49 ± 5.15	1267–1480
$BaCl_2/^{36}Cl$	2.0	39.66 ± 4.27	1266–1476
$CdCl_2/^{115}Cd$	1.1	28.62 ± 2.59	880–1079
$CdCl_2/^{36}Cl$	1.1	28.45 ± 4.22	880–1075

$$D = D_0 \exp\left(-\frac{E_D^{\neq}}{RT}\right) \tag{5.55}$$

where D_0 is found to depend little on substance and temperature, and $E_D^{\neq}$ is the activation energy for self-diffusion (Fig. 5.26). Some preexponential factors and the corresponding energies of activation for diffusion are given in Table 5.18.

In some liquids, a deviation (Fig. 5.27) occurs from the straight-line log D versus $1/T$ plots expected on the basis of the empirical exponential law for the diffusion coefficient (Eq. 5.55). An example of such a deviating liquid electrolyte is molten $ZnCl_2$, but in the case of this substance, structural changes have been noted with increasing temperature. This seems to be a reasonable explanation for the deviation from the straight-line log D versus $1/T$ plot.

TABLE 5.19

Tracer Diffusion Coefficients of Crystalline Substances near the Melting Point

Substance	Tracer	D ($cm^2\ s^{-1}$)	Temperature (K)
Na	^{22}Na	1.7×10^{-7}	370
Ag	^{110}Ag	2.8×10^{-8}	1173
NaCl	^{22}Na	4.0×10^{-9}	1000
PbS	ThB	1.4×10^{-9}	1316
Pb	ThB	5.5×10^{-10}	597
PbI_2	ThB	7.7×10^{-11}	588

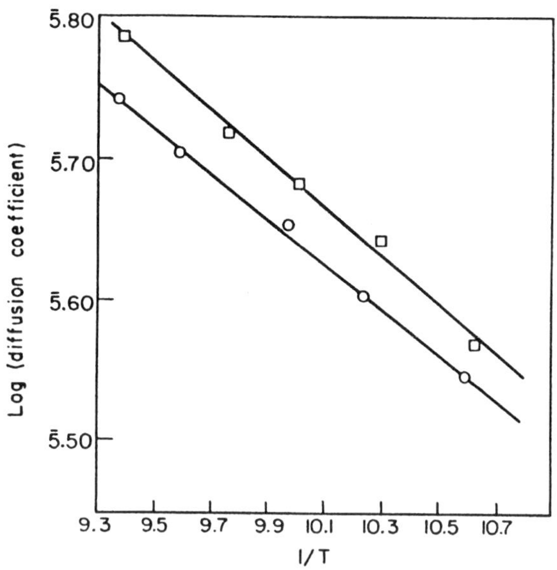

Fig. 5.26. The straight-line plot of log D versus $1/T$ observed in the case of the diffusion of ^{134}Cs (○) and ^{36}Cl (□) in molten CsCl.

The activation energy for self-diffusion is usually a constant, independent of temperature. It is, however, characteristic of the particular liquid electrolyte. The dependence of the activation energy for self-diffusion on the nature of the fused salt was experimentally found by Nanis and Bockris in 1963 to be expressible in the simple relation of Fig. 5.28, given by the equation

$$E_D^{\neq} = 3.74RT_{\text{m.p.}} \tag{5.53}$$

where $T_{\text{m.p.}}$ is the melting point.

It is important to emphasize that this *same relation is valid for liquids in general*, including the liquid inert gases, organic liquids, and the liquid metals, a remarkable fact and one which some liquid theorists have had difficulty explaining (though see Section 5.7.5). Equation (5.53) does not apply, however, to associated liquids, or to those in which there is widespread intermolecular binding throughout the liquid, such as water or molten silicates, borates, and phosphates (Section 3.8).

5.6.3. Viscosity of Molten Salts

An examination of the properties of viscous flow of molten liquids shows that viscosity varies with temperature in a way quite similar to that of self-diffusion. For

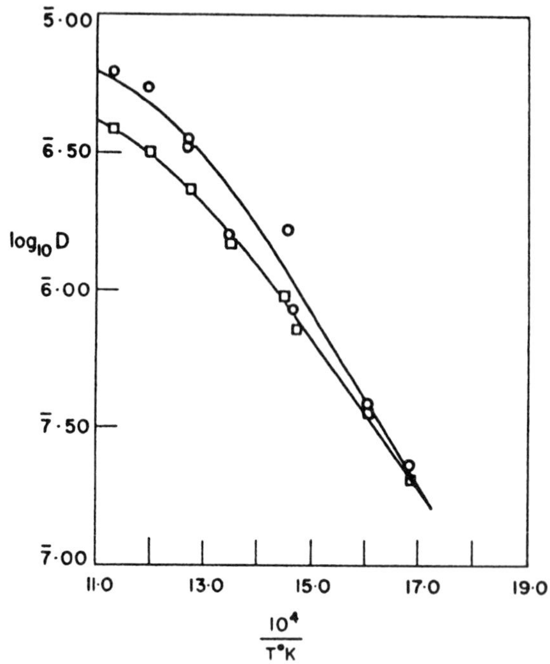

Fig. 5.27. An example of a log D versus $1/T$ plot that is not a straight line. The curve is for the diffusion of ^{65}Zn (○) and ^{36}Cl (□) in molten $ZnCl_2$.

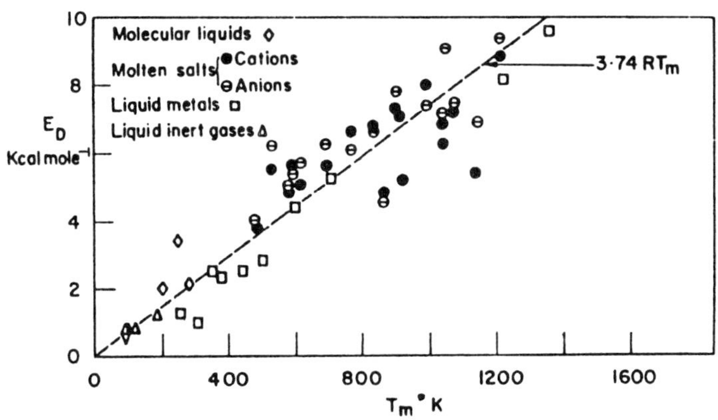

Fig. 5.28. The dependence of the experimental energy of activation for self-diffusion on the melting point (1 cal = 4.184 J).

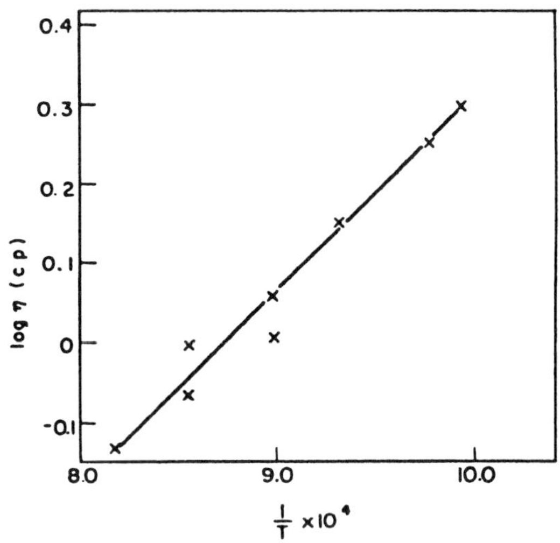

Fig. 5.29. The straight-line plot of log η versus $1/T$ for viscosity of a molten salt.

simple, unassociated liquids, the temperature dependence is given by an empirical equation

$$\eta = \eta_0 e^{+E_\eta^{\neq}/RT} \tag{5.56}$$

where η_0 is a constant analogous to D_0, and $E_\eta^{\neq}$ is the energy of activation for viscous flow (Fig. 5.29 and Tables 5.20 and 5.21).

TABLE 5.20

Viscosities of Fused Salts

Fused Salt	Temperature (K)	Viscosity (cP)
$CdCl_2$	873	2.31
$CdBr_2$	873	2.61
$PbBr_2$	823	2.98
$PbCl_2$	873	2.75
AgCl	873	1.66
AgBr	873	2.27
NaCl	1173	1.05
NaBr	1028	1.43
KCl	1046	1.51
KBr	1003	1.57

TABLE 5.21

Energies of Activation for Viscous Flow for Simple Molten Electrolytes

Molten Salt	$E_\eta^{\neq}$ (kJ mol^{-1})	Molten Salt	$E_\eta^{\neq}$ (kJ mol^{-1})
LiCl	36.82	KOH	25.52
LiBr	25.10	AgCl	12.13
LiI	18.41	AgBr	12.97
$LiNO_3$	17.57	AgI	24.27
NaCl	38.07	$AgNO_3$	12.97
NaBr	33.43	$TlNO_3$	15.06
NaI	30.96	CuCl	17.57
$NaNO_3$	16.74	$MgCl_2$	27.20
$NaNO_2$	16.74	$CaCl_2$	39.75
NaOH	23.01	$BaCl_2$	37.66
NaCNS	24.27	$PbCl_2$	28.03
KCl	32.64	$PbBr_2$	25.94
KBr	31.38	$CdCl_2$	16.74
KI	38.49	$CdBr_2$	18.83
KNO_3	15.48	NH_4NO_3	19.25

This expression is formally analogous to Eq. (5.51) for the dependence of the self-diffusion coefficient upon temperature. For simple liquid electrolytes, the experimental activation energy for viscous flow is given by an expression (Fig. 5.30) identical to that for self-diffusion, i.e.,

$$E_\eta^{\neq} = 3.74RT_{\text{m.p.}} \tag{5.57}$$

The fact that this empirical law applies so widely clearly is trying to tell us something. It is that the basic factors determining viscous flow and self-diffusion are the same for all liquids that do not have to break bonds before they undergo transport.

Simple ionic liquids have viscosities in the range of 1 to 5 centipoises (cP). However, when there is an association of ions into aggregates, as, for example, in $ZnCl_2$ near the melting point, the viscous force-resisting flow of the melt increases above that noted for simple liquids. Such complex ionic liquids are discussed later (Section 5.8).

5.6.4. Validity of the Stokes–Einstein Relation in Ionic Liquids

All transport processes (viscous flow, diffusion, conduction of electricity) involve ionic movements and ionic drift in a preferred direction; they must therefore be interrelated. A relationship between the phenomena of diffusion and viscosity is contained in the Stokes–Einstein equation (4.179).

TABLE 5.22
The $D\eta/T$ of Some Molten Salts for Testing the Stokes–Einstein Relation

Molten Salt	Tracer	Temperature (K)	$D_i \times 10^5$ ($cm^2\ s^{-1}$)	$\frac{D_i\eta}{T} \times 10^9$ (dyn deg^{-1})
NaCl	^{6}Li	1180	13.2	1.09
NaCl	^{22}Na	1180	10.2	0.85
NaCl	^{42}K	1180	9.7	0.81
NaCl	^{86}Rb	1180	9.2	0.76
NaCl	^{134}Cs	1180	8.9	0.74
KCl	^{42}K	1150	8.9	0.63
NaI	^{22}Na	1026	9.5	1.20
$CaCl_2$	^{45}Ca	1154	2.6	0.79
$SrCl_2$	^{89}Sr	1260	2.4	0.70
$BaCl_2$	^{140}Ba	1356	2.4	0.69
$CdCl_2$	^{115}Cd	925	2.7	0.58
$PbCl_2$	^{210}Pb	851	1.5	0.59
$LiNO_3$	^{6}Li	581	2.1	1.77
$NaNO_3$	^{22}Na	638	2.6	0.88
KNO_3	^{42}K	667	2.1	0.68
$AgNO_3$	^{110}Ag	534	1.5	0.80

$$D = \frac{kT}{6\pi r\eta} \tag{5.58}$$

This equation was deduced in Section 4.4.8. It is of interest to inquire here about its degree of applicability to ionic liquids, i.e., fused salts. To make a test, the experimental values of the self-diffusion coefficient D^* and the viscosity η are used in conjunction with the known crystal radii of the ions. The product $D^*\eta/T$ has been tabulated in Table 5.22, and the plot of $D^*\eta/T$ versus $1/r$ is presented in Fig. 5.31, where the line of slope $k/6\pi$ corresponds to exact agreement with the Stokes–Einstein relation.[14]

Looking at Fig. 5.32, it can be seen that there is a fairly significant fit. The anions, particularly those of the group II halides, are not very consistent with the Stokes–Einstein relation. However, their poorer fit is offset by the better Stokes–Einstein behavior of the cations. The relatively good fit of the cations tempts one to conclude that there is a particular reason for the deviations of the anions. Some attempts have been made

[14]The essential applicability of this phenomenological equation is clearly shown by using the numerical comparison of $D\eta/T = k/6\pi r$. The right-hand side is 0.7×10^9 for $r = 300$ pm, and the mean of the experimental values is 0.6×10^9, which is not bad!

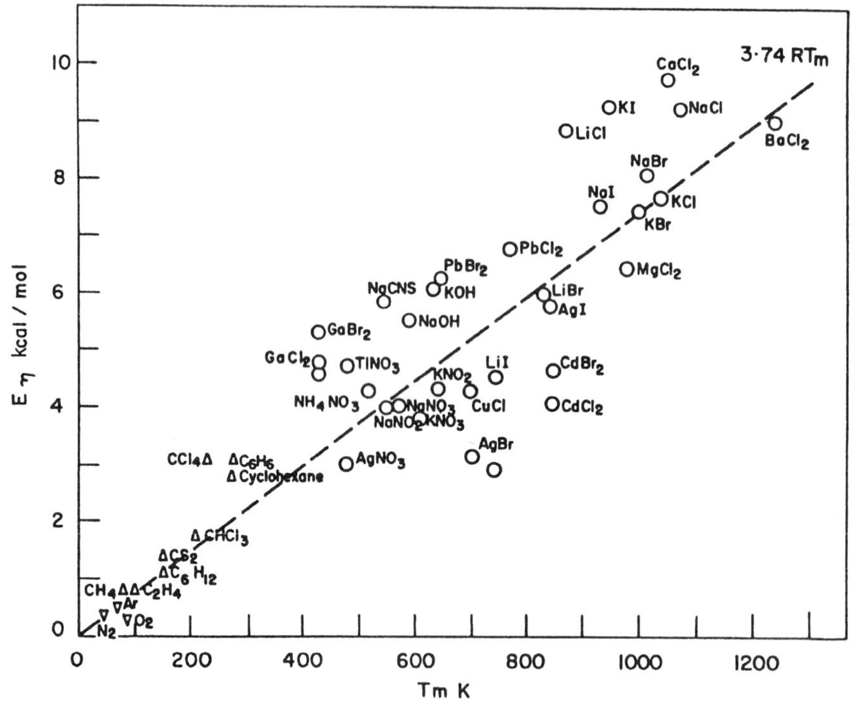

Fig. 5.30. The dependence of the experimental energy of activation for viscous flow on the melting point.

to elucidate this reason. For instance, it has been suggested that since anions are by and large larger than cations, they require greater local rearrangements at a site before they can jump into it, i.e., greater entropies of activation.

The Stokes–Einstein relation is based on Stokes' law in hydrodynamics according to which the viscous force experienced by a large sphere moving in an incompressible continuum is $6\pi r\eta v$. What Fig. 5.31 tells one is that, even though ions do not move in a continuum but among particles that are of approximately the same dimensions as the ions themselves, Stokes' law still holds! In view of the great dissimilarity of an ion in a structured medium and a sphere in an incompressible continuum, the rough applicability (in fused salts) of the Stokes–Einstein equation is somewhat unexpected and very useful. Of course, it is entirely consistent with the agreement between the heats of activation for viscous flow and self-diffusion. Each must evidently be concerned with the same rate-determining step, mechanism, and heat of activation.

5.6.5. Conductivity of Pure Liquid Electrolytes

The electrical conductance of molten salts is the easiest transport property to measure. In addition, knowledge of the order of magnitude of the equivalent conduc-

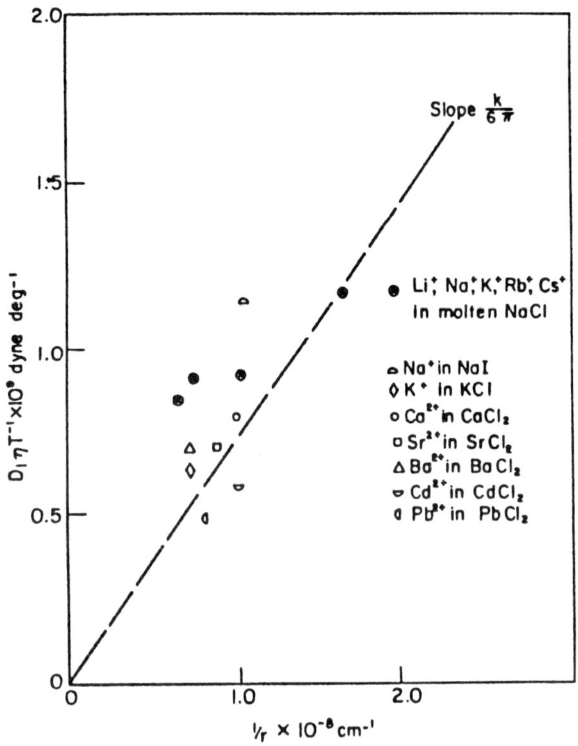

Fig. 5.31. When $D\eta/T$ is plotted against $1/r$, a straight-line of slope $k/6\pi$ should be obtained if the Stokes–Einstein relation is applicable to molten salts. The experimental points are indicated in the figure to show the degree of applicability of the Stokes–Einstein relation to molten salts.

tivity of a pure substance was used as a criterion for the nature of the bonding present. For these reasons, the electrical conductance of ionic liquids has been the subject of numerous studies.

The equivalent conductivities of some of the fused chlorides are given in Table 5.23, where the substances have been arranged according to the Periodic Table. The heavy line zigzagging across the table separates the ionic from the covalent chlorides. This structural difference is shown up sharply in the orders of magnitude of the equivalent conductivities.

Two further correlations emerge from Table 5.23. First, the equivalent conductivity decreases with increasing size of the cation (Table 5.24); second, there is a decrease in equivalent conductivity in going from the monovalent to the divalent and

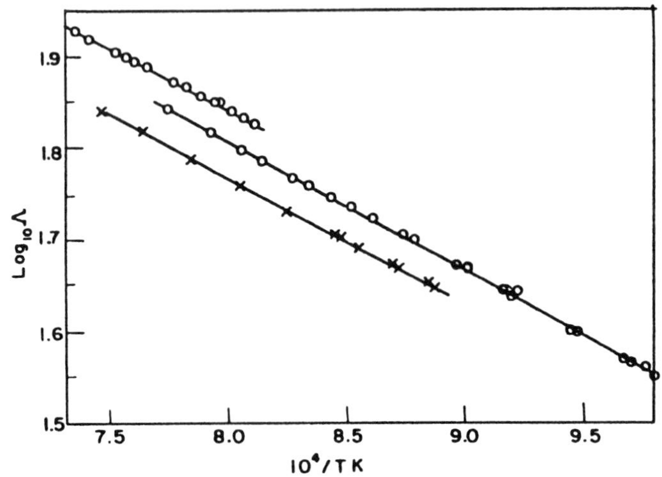

Fig. 5.32. The straight-line plot of log Λ versus $1/T$.

then to the trivalent chlorides (Table 5.25), probably because of an increase in covalent character in this order.

As with the other transport properties, the specific (or equivalent) conductivity of fused salts varies with temperature.[15] For most pure liquid electrolytes, the experimental log Λ versus $1/T$ plots are essentially linear (Fig. 5.32). This implies the usual exponential dependence of a transport property upon temperature

$$\Lambda = \Lambda_0 e^{-E^{\neq}_{\Lambda}/RT} \tag{5.59}$$

For some substances, the plots are slightly curved. In these cases, structural changes (e.g., the breaking up of polymer networks such as those observed with SiO_2) occur with change of temperature.

When the activation energies for conduction are computed from the log Λ versus $1/T$ plots, it is seen (Table 5.26) that they are a little lower than the activation energies for viscous flow and self-diffusion, i.e.,

$$E^{\neq}_{\Lambda} < E^{\neq}_{D} \approx E^{\neq}_{\eta} \tag{5.60}$$

but follow the same pattern; they are proportional to the melting point temperature.

[15]A convenient means of comparing different salts is to use "corresponding temperatures"; usually 1.05 or 1.10 times the value of the melting point in kelvins is used for this purpose.

TABLE 5.23

Equivalent Conductivities of Molten Chlorides (Ohm^{-1} cm^2 equiv.$^{-1}$)

HCl $\sim 10^{-6}$					
LiCl 166	$BeCl_2$ 0.086	BCl_3 0	CCl_4 0		
NaCl 133.5	$MgCl_2$ 28.8	$AlCl_3$ 15×10^{-6}	$SiCl_4$ 0	PCl_5 0	
KCl 103.5	$CaCl_2$ 51.9	$ScCl_3$ 15	$TiCl_4$ 0	VCl_4 0	
RbCl 78.2	$SrCl_2$ 55.7	YCl_3 9.5	$ZrCl_4$	$NbCl_5$ $x = 2 \times 10^{-7}$	$MoCl_5$ $x = 1.8 \times 10^{-6}$
CsCl 66.7	$BaCl_2$ 64.6	$LaCl_3$ 29.0	$HfCl_4$	$TaCl_5$ $x = 3 \times 10^{-7}$	WCl_6 $x = 2 \times 10^{-6}$
			$ThCl_4$ 16		UCl_4 $x = 0.34$

TABLE 5.24

Dependence of Equivalent Conductivity upon Cationic Radius

Molten Salt	Radius of Cation (pm)	Equivalent Conductivity
LiCl	68	183
NaCl	94	150
KCl	133	120
RbCl	147	94
CsCl	167	86

TABLE 5.25

Dependence of Equivalent Conductivity upon Valence of Cation

Molten Salt	Valence of Cation	Equivalent Conductivity
NaCl	1	150
$MgCl_2$	2	35
$AlCl_3$	3	15.1
$SiCl_4$	4	<0.1

TABLE 5.26

Experimental Energies of Activation for Various Transport Process

Molten Salt	$E_{\Lambda}^{\neq}$	$E_D^{\neq}$ (kJ mol^{-1})	$E_{\eta}^{\neq}$
LiCl	8.62	—	36.4
NaCl	12.22	30.5	38.1
$NaNO_3$	13.05	20.9	16.7
KCl	14.06	29.3	32.6
KNO_3	13.18	23.4	15.1
$CdCl_2$	9.20	28.5	16.7

5.6.6. The Nernst–Einstein Relation in Ionic Liquids

5.6.6.1. *Degree of Applicability.* Just as the Stokes–Einstein equation gives the relation between the transport of momentum (viscous flow) and the transport of matter (diffusion), the connection between the transport processes of diffusion and conduction leads to the Nernst–Einstein equation (see Section 4.4.9). For 1:1 electrolytes this is

$$\Lambda = \frac{F^2}{RT}(D_+ + D_-) \tag{5.61}$$

A more general expression of this same equation is

$$\Lambda = \frac{F^2}{RT}\sum_i z_i D_i \tag{5.62}$$

The Nernst–Einstein relation can be tested by using the experimentally determined tracer-diffusion coefficients D_i to calculate the equivalent conductivity Λ and then comparing this theoretical value with the experimentally observed Λ. It is found that the values of Λ calculated by Eq. (5.61) are distinctly greater (by ~10 to 50%) than the measured values (see Table 5.27 and Fig. 5.33). Thus there *are* deviations from the Nernst–Einstein equation and this is strange because its deduction is phenomenological.[16]

[16]A phenomenological deduction is one that follows from general common sense or logic and involves only very general laws. It does not involve detailed models. So if experimental phenomena don't agree with predictions made on their basis, the only conclusion is that the conditions implied for the applicability of the laws do not apply to the case at hand. In this case, the expectation that the conductivity measured should agree with that obtained from diffusion data via the Nernst–Einstein equation requires that diffusion and conduction at least involve the *same particles* (e.g., Na^+ and Cl^-). If in diffusion an extra particle comes into the picture that is not effective in conduction, then the Nernst–Einstein law will not apply.

TABLE 5.27

Test of the Nernst–Einstein Relation for Equivalent Conductivity of Molten NaCl

	Equivalent Conductivity			
	1093 K	1143 K	1193 K	1293 K
Observed	138	147	155	171
Calculated from Eq. (5.61)	159	177	198	240

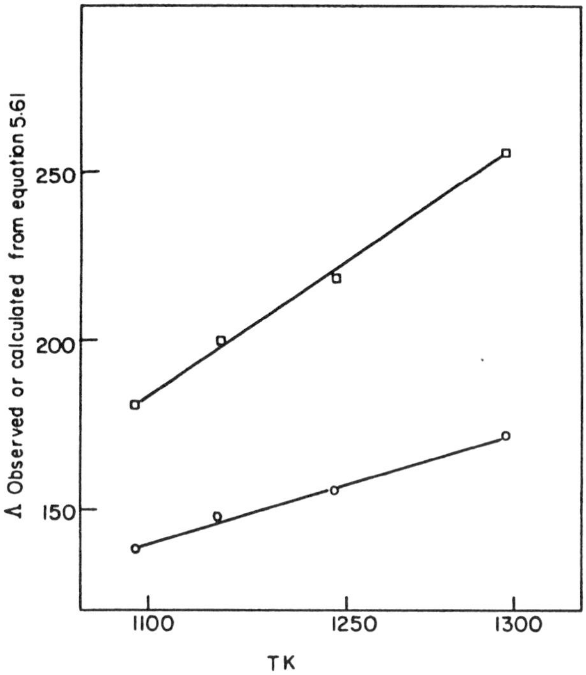

Fig. 5.33. Plot to show deviations from the Nernst–Einstein equation; (○) observed equivalent conductivity of molten NaCl and (□) calculated from Eq. (5.61).

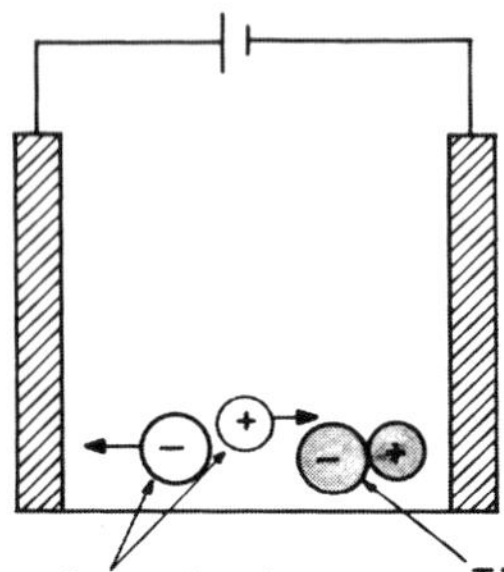

Fig. 5.34. An entity formed by the temporary or permanent association of a pair of oppositely charged ions is electrically neutral and therefore does not migrate under an electric field.

5.6.6.2. *Possible Molecular Mechanisms for Nernst–Einstein Deviations.* The observed conductivity is always found to be less than that calculated from the sum of the diffusion coefficients (Table 5.27), i.e., from the Nernst–Einstein relation [Eq. (5.61)]. Conductive transport depends only on the charged species because it is only *charged* particles that respond to an external field. If therefore two species of opposite charge unite, either permanently or temporarily, to give an uncharged entity, they will not contribute to the conduction flux (Fig. 5.34). They will, however, contribute to the diffusion flux. There will therefore be a certain amount of currentless diffusion, and the conductivity calculated from the sum of the diffusion coefficients will exceed the observed value. *Currentless diffusion will lead to a deviation from the Nernst–Einstein relation.*

It was suggested by Borucka et al. that a *permanent* association of positive and negative ions is not a necessary basis for a breakdown of the Nernst–Einstein equation. The only requirement is that diffusion should occur partly through the displacement of entities which have (momentarily, during jumps) a zero net charge and thus do not contribute to conduction. The entity may be, for instance, a pair of oppositely charged ions, in which case the diffusive displacement occurs by a coordinated movement of such a pair of ions into a paired vacancy (Fig. 5.35), i.e., a vacancy large enough to accept a positive and a negative ion at the same time. The pair of oppositely charged ions that jumps into a "paired vacancy" is neutral as a whole and therefore such coordinated jumps do not play a part in the conduction process, which is determined only by the separate, uncoordinated movements of single ions.

Thus, the experimentally observed diffusive flux of either of the ionic species is made up of two contributions—the diffusive flux occurring through the independent

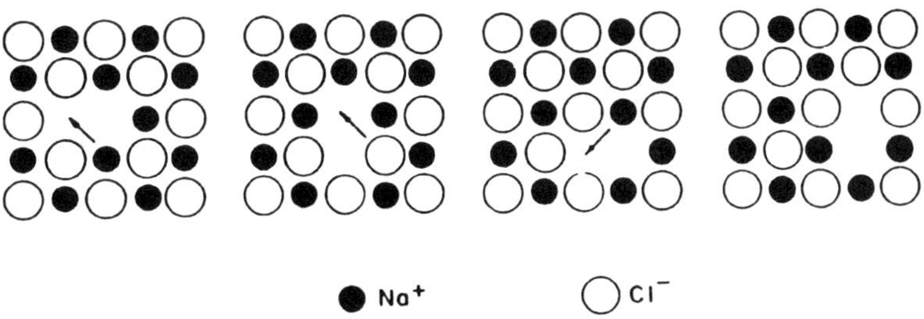

Fig. 5.35. Schematic diagrams to indicate how diffusive displacement can occur through a coordinated movement of a pair of ions into a paired vacancy.

jumps of ions and that occurring through paired jumps. Taking the example of the diffusion of Na^+ in an NaCl melt, one has[17]

$$J'_{Na^+} = J_{Na^+,ind} + J_{NaCl} \tag{5.63}$$

where J'_{Na^+} is the experimentally observed flux of Na^+ (primes refer here to experimental quantities), $J_{Na^+,ind}$ is the diffusion of Na^+ by independent jumps, and J_{NaCl} is the flux due to coordinated jumps of Na^+ and Cl^- ions into paired vacancies

$$J'_{Na^+} = D'_{Na^+} \frac{dc_{Na^+}}{dx} \tag{5.64}$$

$$J_{Na^+,ind} = D_{Na^+,ind} \frac{dc_{Na^+}}{dx} \tag{5.65}$$

and

$$J_{NaCl} = D_{NaCl} \frac{dc_{Na^+}}{dx} \tag{5.66}$$

Adding Eqs. (5.65) and (5.66), one has

[17]The subscript NaCl must not be taken to mean that there are entities in the melt that might be considered "molecules" of sodium chloride. The NaCl does not refer to Na^+ and Cl^- ions that are bound together like an ion pair in aqueous solution; rather, it refers to a pair of Na^+ and Cl^- ions that undergo a coordinated jump into a paired vacancy during the short time for which they momentarily exist in contact. They do not contribute to the conductance because their jumps are directed not by the externally applied field but by the presence of a paired vacancy that exists before the ions jump as a pair.

$$J_{\mathrm{Na}^+,\mathrm{ind}} + J_{\mathrm{NaCl}} = (D_{\mathrm{Na}^+,\mathrm{ind}} + D_{\mathrm{NaCl}}) \frac{dc_{\mathrm{Na}^+}}{dx} \tag{5.67}$$

However, from Eqs. (5.63) and (5.64)

$$J_{\mathrm{Na}^+,\mathrm{ind}} + J_{\mathrm{NaCl}} = J'_{\mathrm{Na}^+} = D'_{\mathrm{Na}^+} \frac{dc_{\mathrm{Na}^+}}{dx} \tag{5.68}$$

Hence,

$$D'_{\mathrm{Na}^+} = D_{\mathrm{Na}^+,\mathrm{ind}} + D_{\mathrm{NaCl}} \tag{5.69}$$

Similarly,

$$D_{\mathrm{Cl}^-} = D_{\mathrm{Cl}^-,\mathrm{ind}} + D_{\mathrm{NaCl}} \tag{5.70}$$

On adding, it is clear that

$$(D_{\mathrm{Na}^+,\mathrm{ind}} + D_{\mathrm{Cl}^-,\mathrm{ind}}) = (D_{\mathrm{Na}^+} + D'_{\mathrm{Cl}^-}) - 2D_{\mathrm{NaCl}} \tag{5.71}$$

The Na^+ and Cl^- ions that make coordinated jumps into paired vacancies, i.e., the NaCl species, contribute to diffusion but not to conduction since such a coordinated pair is effectively neutral. Hence, the Nernst–Einstein equation is only applicable to the ions that jump independently, i.e.,

$$D_{\mathrm{Na}^+,\mathrm{ind}} + D_{\mathrm{Cl}^-,\mathrm{ind}} = \frac{RT}{zF^2} \Lambda' \tag{5.72}$$

where Λ' is the experimentally observed equivalent conductivity of molten NaCl. Making use of Eqs. (5.72) and (5.71), one has

$$\frac{RT}{zF^2} \Lambda' = (D'_{\mathrm{Na}} + D'_{\mathrm{Cl}^-}) - 2D_{\mathrm{NaCl}} \tag{5.73}$$

or

$$\Lambda' = \frac{zF^2}{RT} (D'_{\mathrm{Na}^+} + D'_{\mathrm{Cl}^-}) - \frac{2zF^2}{RT} D_{\mathrm{NaCl}} \tag{5.74}$$

The first term on the right-hand side corresponds to the value of the equivalent conductivity that would be calculated on the basis of the experimentally observed diffusion coefficients. Using the symbol Λ_{calc} for this calculated value, i.e.,

$$\Lambda_{\mathrm{calc}} = \frac{zF^2}{RT} (D'_{\mathrm{Na}^+} + D'_{\mathrm{Cl}^-}) \tag{5.75}$$

one has

$$\Lambda' = \Lambda_{\text{calc}} - \frac{2zF^2}{RT} D_{\text{NaCl}} \tag{5.76}$$

which shows that the experimental value of the equivalent conductivity is always less than that calculated from a Nernst–Einstein equation based on experimental diffusion coefficients. This is what is observed (Table 5.27).

5.6.7. Transport Numbers in Pure Liquid Electrolytes

5.6.7.1. *Transport Numbers in Fused Salts.* The concept and determination of transport numbers in pure liquid electrolytes is one of the most interesting, and most confusing, aspects of the electrochemistry of fused salts.

The concept has been referred to in Section 4.5.2. The transport number t_i of an ionic species i is the quantitative answer to the question: What fraction of the total current $I[= Ei_i]$ passing through an electrolyte is transported by the particular ionic species i? In symbols [see Eq. (4.234)]:

$$t_i = \frac{i_i}{I} = \frac{i_i}{\sum i_i} \tag{5.77}$$

In the case of z:z-valent salts, the transport number is simply given by

$$t_i = \frac{u_i}{\sum u_i} \tag{5.78}$$

where u_i is the mobility of the ion concerned, e.g., that of Li^+ or Br^-. The coordinate system with which these mobilities are measured is considered later on. For a pure liquid electrolyte consisting of one cationic and one anionic species,

$$t_+ = \frac{u_+}{u_+ + u_-} \quad \text{and} \quad t_- = \frac{u_-}{u_+ + u_-} \tag{5.79}$$

It was seen (Section 2.4) that in aqueous solutions, the solvent could not be relegated to the status of an unobtrusive background. The solvent molecules, by entering into the solvation sheaths of ions, participated in their drift. Thus, in addition to the flows of the positive and negative ions, there was a flux of the solvent. This complication of solvent flux is absent in pure ionic liquids. There is, however, an interesting effect when a current is passed through a fused salt.

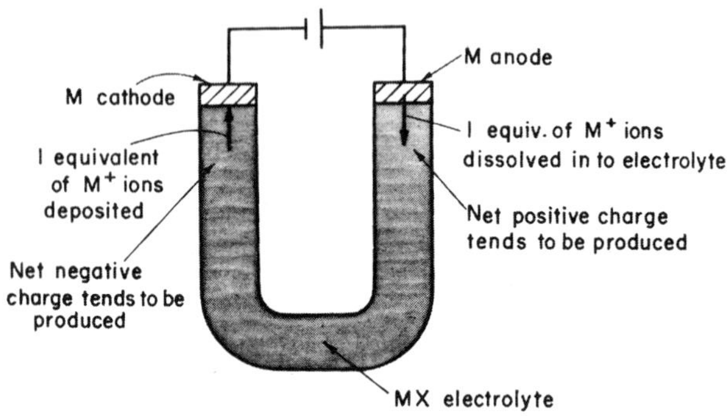

Fig. 5.36. Schematic U-tube setup with M electrodes and MX electrolyte. When 1 F of electricity is passed through the system, one equivalent of M^+ ions is deposited at the M cathode and one equivalent of ions is produced at the M anode. Hence, a negative charge tends to be produced near the cathode and a positive charge near the anode.

Consider that a fused salt MX is taken as the ionic conductor in a U tube and two M electrodes are introduced into the system as shown in Fig. 5.36. Let the consequences of the passage of 1 faraday (F) of electricity be analyzed. Near the cathode, one equivalent of M^+ ions will be removed from the system by deposition on the cathode, and near the anode one equivalent of M^+ ions will be "pumped" into the system. Since one equivalent of M^+ has been added and another equivalent has been removed, the total quantity of M^+ ions in the system is unchanged (Table 5.28).

Is the system perturbed by the passage of a faraday of charge? Yes, because near the cathode, one equivalent of M^+ ions has been removed, which has created a local excess of negative charge. This local unbalance of electroneutrality creates a local electric field.[18] A similar argument can be used for the anode region.

How do the ions of the liquid electrolyte respond to this perturbation, i.e., this creation of local fields? The ions start drifting under the influence of the fields so that the initial state of electroneutrality and zero field is restored. How do the positive and negative ions share this responsibility of moving to annul the unbalance of charges—more anions than cations near the cathode and vice versa? It should be noted (Fig. 5.37) that the original

[18]This unbalance of electroneutrality and creation of field should not be confused with that arising from the presence of the electrode, which causes an anisotropy in the forces on the particles in the electrode–electrolyte interphase region. That anisotropy *also* produces an unbalance of electroneutrality and an electric double layer (Chapter 6) with a field across the interface, but it occurs only within the first few tens of nanometers of the surface.

TABLE 5.28

Changes in MX at Electrodes of Metal M in Transport Experiment

	Anode Compartment	Cathode Compartment
Electrode process per faraday passed	1 g-eq M^+ dissolved	1 g-eq M^+ deposited
Move out of compartment	t_+ g-eq M^+	t_- g-eq X^-
Move into compartment	t_- g-eq X^-	t_+ g-eq M^+
Net change of M^+	$1 - t_+$ g-eq gained $= t_-$	$1 - t_+$ g-eq lost $= t_-$
Net change of MX	t_- g-eq gained	t_- g-eq lost
Mass change of MX in case of molten salt	$t_- M_{MX}$ g gained, where M_{MX} is eq. wt. of MX	$t_- M_{MX}$ g lost, where M_{MX} is eq. wt. of MX
Concentration change in case of aqueous solution	t_-/V_A g-eq dm^{-3} increase, where V_A is vol. of anode compartment (dm^{-3})	t_-/V_C g-eq dm^{-3} decrease, where V_C is vol. of cathode compartment (dm^{-3})

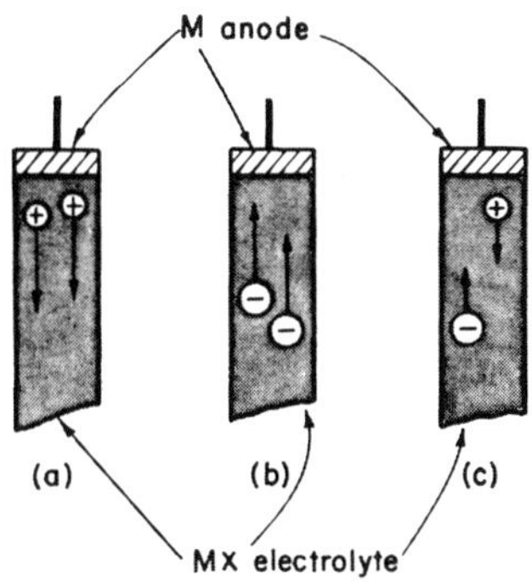

Fig. 5.37. The tendency for electroneutrality to be upset near electrodes is avoided in one of three ways. For example, near the anode, where positive charge tends to be produced, (a) positive ions can migrate away from the anode, (b) negative ions can migrate toward the anode, and (c) both (a) and (b) processes can occur to various extents.

electroneutral situation can be restored by (1) only cations moving in the anode-to-cathode direction; (2) only anions moving in the cathode-to-anode direction; and (3) both cations and anions moving in opposite directions to different extents. However, these possibilities represent different values of the transport numbers, which are the fractions of the total field-induced ionic drift arising from the various species.

5.6.7.2. *Measurement of Transport Numbers in Liquid Electrolytes.* Let t_+ and t_- be the transport numbers of the M^+ and X^- ions of the fused salt. The changes in the numbers of equivalents of M^+ and X^- near the two electrodes are shown in Table 5.28, which is based on Fig. 5.36. The analysis of the changes leads to an interesting result. The passage of 1 F of charge is equivalent to transferring t_- equivalents of the whole fused salt MX from the cathode region to the anode region (Fig. 5.38).

In the case of aqueous solutions, the ever-plentiful solvent could absorb this t_- equivalents of MX and register the transfer as a concentration change of magnitude t_-/V equivalents per liter, where V is the volume of the compartment. On the other hand, a pure molten salt has no concentration variable. Hence, the transfer leads to an increase in mass near the anode.

In molten salts, therefore, it is the change in *mass* in a compartment that reveals transport numbers; in aqueous solutions, it was the change in *concentration*. However, unless it is performed properly, the experiment provides information only on the change in mass, not on the transport property.

What future has this mass increase? Left alone, the mass increase is short-lived and the transport experiment fails. This is because gravitational flow of the molten salt from the anode to cathode tends to equalize the amounts of MX in the two tubes (Fig. 5.39). It causes the liquid level in both tubes to be the same and wipes out the change in level that the ionic movements tend to make.

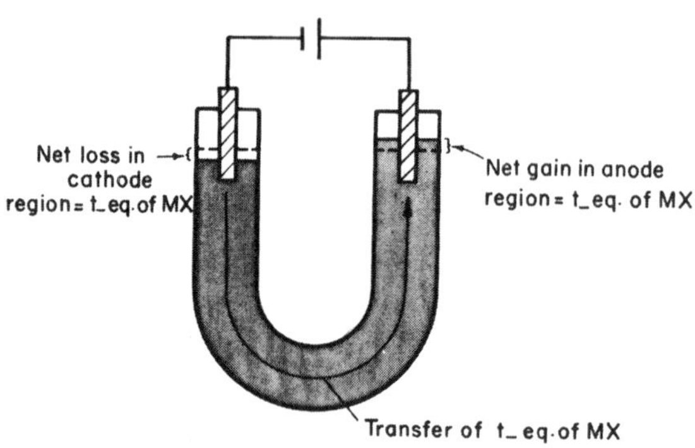

Fig. 5.38. As a result of the passage of 1 F of electricity, t_- equivalents of MX electrolyte are transferred from the cathode region (where the electrolyte level falls) to the anode region (where the level rises).

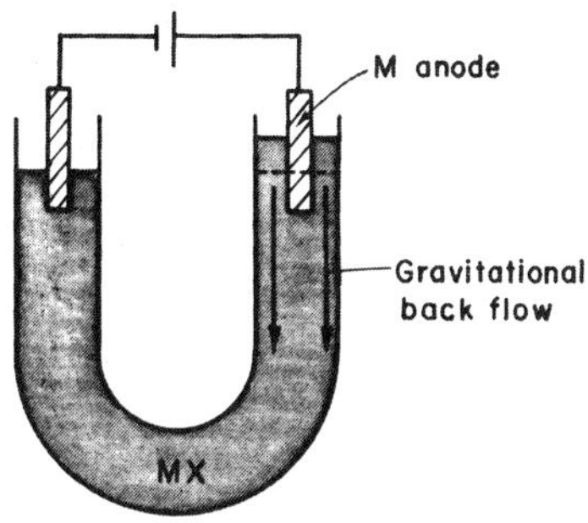

Fig. 5.39. The difference in electrolyte levels produced by the passage of 1 F of electricity leads to a gravitational flow of the electrolyte.

The first step in determining transport numbers in pure ionic liquids is to prevent the gravitational flow from masking the transfer of electrolyte. If the hydrodynamic backflow that gravity causes cannot be prevented, it must at least be taken into account. The general procedure is to minimize the gravitational flow by interposing a membrane between anode and cathode (Fig. 5.40). However, there are serious objections to the use of a membrane, owing to hydrodynamic interferences between this and the moving liquid.

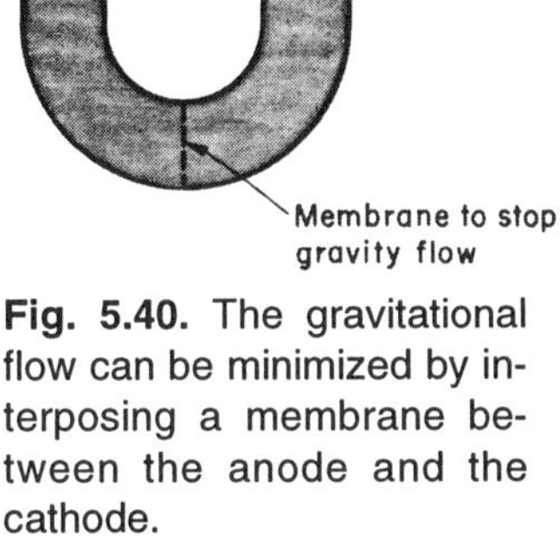

Fig. 5.40. The gravitational flow can be minimized by interposing a membrane between the anode and the cathode.

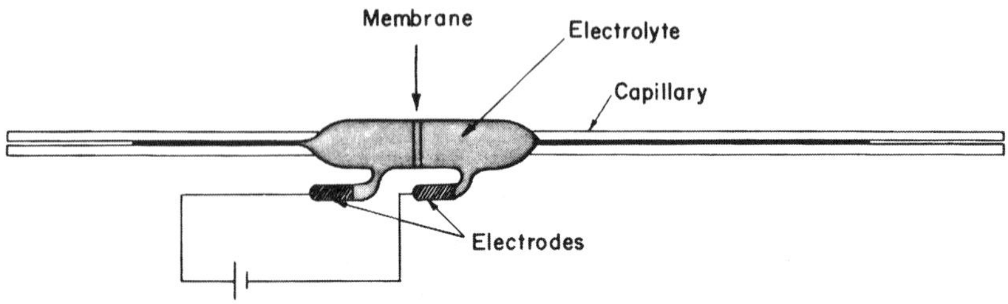

Fig. 5.41. A simple arrangement by which gravitational flow is avoided. The displacement of the electrolyte from the cathode to the anode region occurs at one level. The change in position of the melt in the capillary indicates the amount of electrolyte displaced.

It is also possible to open out the U tube and make the whole liquid "lie down" so that the movement of the fused salt occurs at one level and not against gravity. The amount of salt entering the anode region is then indicated, for example, by a sliver of molten metal pushed along by the movement of the salt in the capillary (Fig. 5.41). This method is also subject to difficulties, for the movement of salts in capillary tubes may not be smooth but is sometimes jerky.

This experiment directly demonstrates that *when electricity is passed through a fused salt, there is a movement of the salt as a whole.* In other words, the mass center of the liquid electrolyte moves. Now, the ions also are drifting with certain mobilities, i.e., velocities under unit field. But velocities with respect to what? One must define a coordinate system, or frame of reference, in relation to which the velocities (distances traversed in unit time) are reckoned. Though the laws of physics are independent of the choice of the coordinate system—the principle of relativity—all coordinate systems are not equally convenient. In fused salt it has been found convenient to use the mass center of the moving liquid electrolyte as the frame of reference, even though this choice, while providing a simple basis for computations, suffers from difficulties.

Even the elementary presentation given here makes it clear that transport-number measurements in fused salts are based on the transfer of the fused salt from the anode to the cathode compartment. The quantities measured are weight changes, the motion of indicator bubbles, the volume changes, etc. Some basic experimental setups shown in Fig. 5.42 include the apparatuses of Duke and Laity, Bloom, Hussey, and other pioneers in this field.

The migration of the electrolyte from the anode to the cathode compartment can also be followed by using radioactive tracers and tracking their drift. Since isotopic analysis methods are sensitive to trace concentrations, there is no need to wait for the electrolyte migration to be large enough for visual detection. The results of some transport-number measurements are given in Table 5.29.

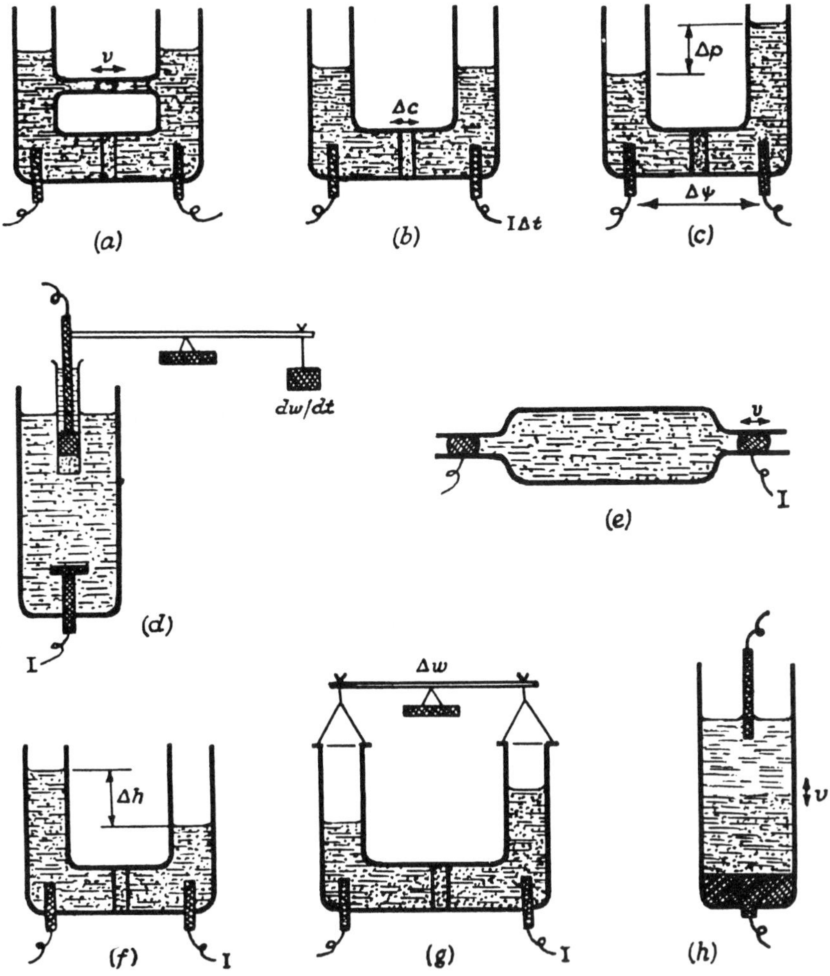

Fig. 5.42. Schematic diagrams of methods of determining transport numbers: (a) Measure velocity of the bubble; (b) measure transfer of the tracer; (c) measure the potential difference due to pressure difference; (d) measure the change in weight; (e) measure the transport of liquid metal electrodes; (f) measure the steady-state level; (g) measure the change in weight; (h) measure the moving boundary.

5.6.7.3. *Radiotracer Method of Calculating Transport Numbers in Molten Salts.* In the discussion of the applicability of the Nernst–Einstein equation to fused salts, it was pointed out that the deviations could be ascribed to the paired jump of ions resulting in a currentless diffusion. With fused NaCl as an example, it has been shown that there is a simple relation between the experimentally determined equivalent

TABLE 5.29

Some Transport-Number Results for Fused Salts

Molten Salt	Transport Number of Cation
$LiNO_3$	0.84
$NaNO_3$	0.71
$AgNO_3$	0.72
NaCl	0.87
KCl	0.77
AgCl	0.54

conductivity Λ' and the experimental diffusion coefficients of Na^+ and Cl^- as indicated by radiotracer Na^+ and Cl^-. The relation is

$$\Lambda' = \frac{zF^2}{RT}(D'_{Na^+} + D'_{Cl^-}) - \frac{2zF^2}{RT} D_{NaCl} \tag{5.74}$$

From this expression, it is clear that one can determine D_{NaCl}. Knowing D_{NaCl}, one can obtain the diffusion coefficients $D_{Na^+,ind}$ and $D_{Cl^-,ind}$, of the independently jumping Na^+ and Cl^- ions from the relations (5.69) and (5.70), i.e.,

$$D_{Na^+,ind} = D'_{Na^+} - D_{NaCl} \tag{5.80}$$

and

$$D_{Cl^-,ind} = D'_{Cl^-} - D_{NaCl} \tag{5.81}$$

Further, by using the Einstein relation [Eq. (4.172)] and the relation between absolute and conventional mobilities, one has

$$(u_{conv})_{Na^+} = z_{Na^+}e_0(\bar{u}_{abs})_{Na^+} = z_{Na^+}e_0 \frac{D_{Na^+,ind}}{kT} = \frac{z_+F}{RT} D_{Na^+,ind} \tag{5.82}$$

and similarly,

$$(u_{conv})_{Cl^-} = \frac{z_-F}{RT} D_{Cl^-,ind} \tag{5.83}$$

With these values of mobilities, the transport numbers can easily be calculated from the standard formulas

TABLE 5.30

Comparison of the Transport Number of Na^+ in Molten NaCl Calculated from Equation (5.84) with Measured Values

	Transport Number of Na^+ in NaCl
Calculated from Eq. (5.84)	0.71
Measured values	0.62
	0.87

$$t_+ = \frac{u_+}{u_+ + u_-} \quad \text{and} \quad t_- = \frac{u_-}{u_+ + u_-} \tag{5.79}$$

which, in the case of NaCl ($z_+ = z_- = 1$), reduces to

$$t_{Na^+} = \frac{D_{Na^+,ind}}{D_{Na^+,ind} + D_{Cl^-,ind}} \quad \text{and} \quad t_{Cl^-} = \frac{D_{Cl^-,ind}}{D_{Na^+,ind} + D_{Cl^-,ind}} \tag{5.84}$$

A comparison between transport numbers calculated in this way and those obtained by some of the experimental methods used is shown in Table 5.30.

Further Reading

Seminal

1. W. Klemm and W. Biltz, "Distribution of the Property of Ionic Conductance among Molten Salts on Liquid Halides in the Periodic Table," *J. Inorg. Chem.* **152**: 255 (1926).
2. F. R. Duke, R. W. Laity, and B. Owens, "Transport Numbers in Fused Salts," *J. Electrochem. Soc.* **104**: 299 (1957).
3. A. Borucka, J. O'M. Bockris, and J. A. Kitchener, "The Nernst–Einstein Equation in Molten Salts," *Proc. Roy. Soc. Lond.* **A241**: 554 (1957).
4. G. J. Janz, C. Solomons, and H. J. Gardner, "Diffusion and Transport in Molten Salts," *Chem. Rev.* **58**: 241 (1958).
5. L. Nanis and J. O'M. Bockris, "Self-Diffusion: Heats of Activation as a Function of Melting Temperature," *J. Phys. Chem.* **67**: 2865 (1963).
6. H. Bloom, *The Chemistry of Molten Salts*, W. A. Benjamin, New York (1967).

Reviews

1. C. L. Hussey, "Transport in and Transport Numbers in Molten Salts," in *Molten Salt Chemistry*, G. Mamantov and R. Marassi, eds., *NATO ASI Series C* **202**: 141 (1987).
2. C. A. Angell, "Transport and Relaxation Processes in Molten Salts," in *Molten Salt Chemistry*, G. Mamantov and R. Marassi, eds., *NATO ASI Series C* **202**: 123 (1987).
3. S. Smedley, *Interpretation of Ionic Conductivity in Liquids*, Plenum Press, New York (1990).

Papers

1. S. I. Vavilov, *J. Non-Cryst. Solids* **123**: 34 (1990).
2. D. R. Chang, *Langmuir* **66**: 11332 (1990).
3. Y. Shirakawa and S. Tamaki, *J. Non-Cryst. Solids* **117**: 638 (1990).
4. D. G. Leaist, *Electrochim. Acta* **36**: 309 (1991).
5. K. Igarashi, *J. Electrochem. Soc.* **138**: 3588 (1991).
6. H. Rajabu, S. K. Ratke, and O. T. Furland, *Proc. Electrochem. Soc.* **16**: 595 (1992).
7. F. Lanlelme, A. Barhoun, and J. Chavelet, *J. Electrochem. Soc.* **140**: 324 (1993).
8. M. Poupait, C. S. Valesquez, and K. Hasseb, *J. Am. Chem. Soc.* **116**: 1165 (1994).
9. W. Wang and John Newman, *J. Electrochem. Soc.* **142**: 761 (1995).
10. C. Larive, M. Lin, and B. J. Piersma, *J. Phys. Chem.* **99**: 12409 (1995).
11. V. A. Payne, J. H. Xu, M. Fursyth, M. A. Ratner, D. F. Shriver, and S. W. Deleuw, *Electrochim. Acta* **40**: 2087 (1995).
12. C. Cametti, *J. Phys. Chem.* **100**: 7148 (1996).

5.7. USING A HOLE MODEL TO UNDERSTAND TRANSPORT PROCESSES IN SIMPLE IONIC LIQUIDS

5.7.1. A Simple Approach: Holes in Molten Salts and Transport Processes

Some facts about transport processes in molten salts have been mentioned (Section 5.6). Whether a hole model (Section 5.4) can provide an interpretation of these must now be examined. First it is necessary to cast the model into a form suitable for the prediction of transport properties. The starting point is the molecular-kinetic expression (Appendix 5.3) for the viscosity η of a fluid, i.e.,

$$\eta = 2nm\langle v \rangle l \qquad (5.85)$$

where n and m are the number per unit volume and the mass of the particles of the fluid, $\langle v \rangle$ is the mean velocity of the particles, and l is their mean free path.

The quantity l is linked to the model for viscous flow in fluids. According to this picture (Fig. 5.43), a fluid in motion is considered to consist of layers lying parallel to the direction of flow. (The slipping and sliding of these layers against each other provides the macroscopic explanation of viscosity.) When particles jump between neighboring layers, there is momentum transfer between these layers, the cause of viscous drag (Fig. 5.44). In this picture, the symbol $\langle v \rangle$ is taken to represent the component of the average velocity of the particles in a direction normal to the layers.

Irrespective of whether the fluid is in motion, the particles constituting the fluid continuously execute random motion. The particles of a *flowing* fluid have a drift superimposed upon this random walk. It is by means of the random walk of the particles from one layer to another that the momentum transfer between layers is

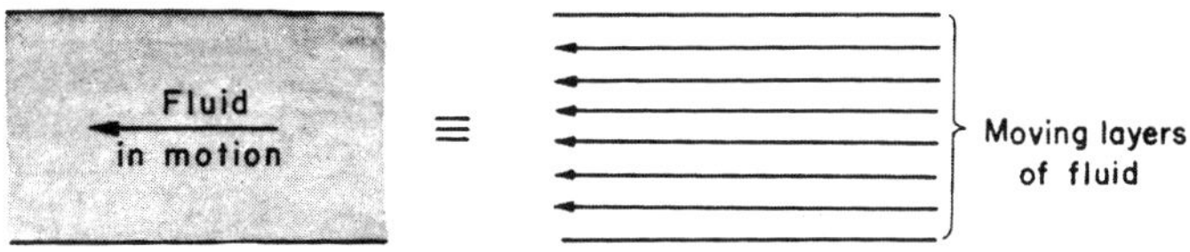

Fig. 5.43. A fluid in motion is considered equivalent to moving layers of fluid, the layers lying parallel to the flow direction.

carried on. This momentum transfer is visible to the observer as the *viscosity* of the fluid.

Holes also move. As argued earlier (Section 5.5.1), anything that moves at finite velocities must have an inertial resistance to motion, i.e., a mass (see Appendix 5.1). Although it may continue to be surprising, holes have masses and moving holes have momenta.

Thus, according to the hole theory, the random walk of holes between adjacent layers results in momentum transfer and therefore viscous drag in a moving fused salt (Fig. 5.45). On the basis of this model, the expression for the viscosity of an ionic liquid is

$$\eta = 2n_h m_h \langle v_h \rangle l_h \tag{5.86}$$

where n_h and m_h are the number per unit volume and the apparent mass for translational motion of the holes.

The velocity component $\langle v_h \rangle$ is given by the ratio of l_h, the mean distance between collisions (i.e., the mean free path), to τ, the mean time between collisions,

$$\langle v_h \rangle = \frac{l_h}{\tau} \quad \text{or} \quad \langle v_h \rangle l_h = \langle v_h \rangle^2 \tau \tag{5.87}$$

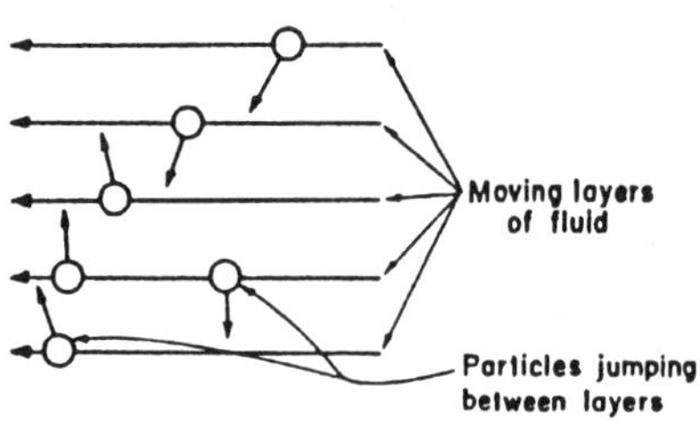

Fig. 5.44. Viscous forces are considered to arise from the momentum transferred between moving fluid layers when particles jump from one layer to another.

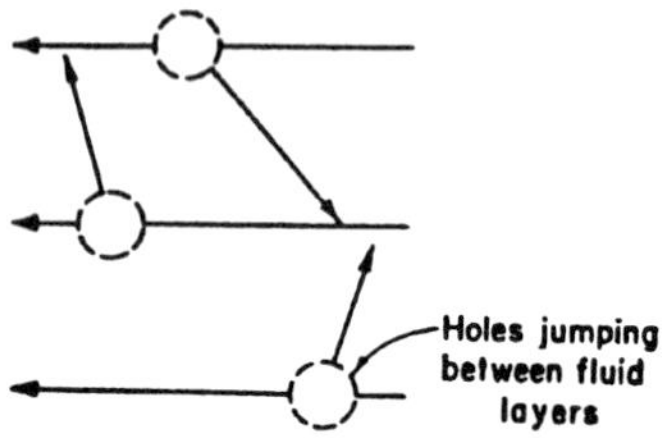

Fig. 5.45. According to the hole model, viscous drag arises from the momentum transferred between moving fluid layers when holes jump from one layer to another.

The viscosity can be written as follows

$$\eta = 2n_h \tau (m_h \langle v_h \rangle^2) \tag{5.88}$$

The theorem of the equipartition of energy can now be applied to the one-dimensional motion referred to by $\langle v_h \rangle$

$$\frac{1}{2} m_h \langle v_h \rangle^2 \approx \frac{1}{2} kT \tag{5.89}$$

and using this approximate relation, one has

$$\eta = 2n_h kT\tau \tag{5.90}$$

5.7.2. What Is the Mean Lifetime of Holes in the Molten Salt Model?

The parameter τ now invites consideration. In the gas phase, τ is the mean time between collisions. What is the significance of τ in an ionic liquid?

In a liquid, in the present model, τ would be the mean lifetime of a hole, i.e., the average time between the creation and destruction of a hole through thermal fluctuation. To calculate this, one may use the formula for the number of particles escaping from the surface of a body per unit time per unit area into empty space, i.e.,

$$a = c\left(\frac{kT}{2\pi m}\right)^{1/2} e^{-A/RT} \tag{5.91}$$

where c is the number of particles per unit volume, m is the mass of a particle, and A is the work necessary for a mole of particles lying on the surface of the hole to be released into its interior.

In a time t, $4\pi\langle r_h\rangle^2 at$ particles escape from the exterior into a spherical hole of radius $\langle r_h\rangle$. The hole will be filled by these particles if this number is equal to the number of particles in a sphere of radius $\langle r_h\rangle$, that is, $(4/3)\pi\langle r_h\rangle^3 c$. Then the time for destruction of the hole is

$$\tau = \frac{1}{3}\frac{\langle r_h\rangle c}{a} \tag{5.92}$$

Obviously, this is also the time for hole formation, and the lifetime, from Eqs. (5.91) and (5.92), is

$$\tau = \frac{\langle r_h\rangle}{3}\left(\frac{2\pi m}{kT}\right)^{1/2} e^{A/RT} \tag{5.93}$$

This is the mean lifetime of a hole. This expression is consistent with the idea that the hole theory represents a Swiss-cheese sort of model of a liquid with holes of different sizes [for $\langle r_h\rangle$ is the *mean* radius for holes varying in size according to Eq. (5.44)]. The holes keep on opening and shutting, and the mean time they are open is given by Eq. (5.93).

Before one leaves expression (5.93), it is well to note the innocent acceptance with which A has been treated. It is the heat term associated with getting a hole "unmade," with collapsing the hole, the negative of the work of forming the hole. However, it has not yet been said how this will be calculated, and what terms go into this. Such a calculation will be one test that will be made of the hole theory in Section 5.7.6.

5.7.3. Viscosity in Terms of the "Flow of Holes"

By inserting the expression for the mean lifetime of a hole [Eq. (5.93)] into Eq. (5.90), one obtains the hole-theory expression for viscosity. Thus,

$$\begin{aligned}\eta &= 2n_h kT\left[\frac{1}{3}\langle r_h\rangle\left(\frac{2\pi m}{kT}\right)^{1/2} e^{A/RT}\right]\\ &= \frac{2}{3}n_h\langle r_h\rangle(2\pi mkT)^{1/2}e^{A/RT}\end{aligned} \tag{5.94}$$

There are two quantities on the right-hand side of Eq. (5.94) which need discussion. They are n_h, the number of holes per unit volume of the liquid, and A, which occurs in the Boltzmann factor $\exp(-A/RT)$, for the probability of a successful filling of a hole. These quantities are discussed in terms of the closely related quantity of diffusion.

5.7.4. The Diffusion Coefficient from the Hole Model

Now that the viscous flow properties of an ionic liquid have been discussed, the next task is to derive an expression for the diffusion coefficient. The present modeling interpretation of the elementary act of a transport process consists of hole formation followed by a particle jumping into the hole. The focus in this elementary act has hitherto been the center of the hole.

What is the situation at the original site of the jumping ion? Alternatively stated: What has happened, as a consequence of the jump process, at the point where the ion was before it jumped? At this prejump site, there has been precisely that moving away of particles from a point that corresponds to hole formation (Fig. 5.46).

Thus, when a particle jumps, it leaves behind a hole. So then, instead of saying that a transport process occurs by particles hopping along, one could equally well say that the transport processes occur by *holes* moving. The concept is commonplace in semiconductor theory, where the movement of electrons in the conduction band is taken as being equivalent to a movement of so-called "holes" in the valence band. It has in fact already been assumed at the start of the viscosity treatment (Section 5.7.1) that the viscous flow of fused salts can be discussed in terms of the momentum transferred between liquid layers by moving holes.

Hence, when diffusion of particles occurs, there is a corresponding diffusion of holes. Instead of treating *ionic* diffusion as a separate subject, therefore, one can consider *hole* diffusion and write the Stokes–Einstein relation (Section 4.4.8) for the diffusion coefficient of holes

$$D = \frac{kT}{6\pi \langle r_h \rangle \eta} \tag{5.95}$$

The hole-theory expression for viscosity is known. It is Eq. (5.94). Let this be introduced into Stokes–Einstein relation [Eq. (5.95)]. Using Eq. (5.44) the result is

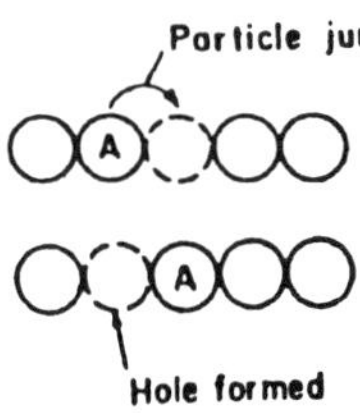

Fig. 5.46. When a particle jumps into a hole, there is the formation of a hole at the prejump site of the particle.

$$D = \frac{kT}{4\pi\langle r_h\rangle^2} \frac{(2\pi mkT)^{-1/2}}{n_h} e^{-A/RT}$$

$$= \frac{1}{4\pi(0.51)^2} \frac{\gamma}{n_h} (2\pi mkT)^{-1/2} e^{-A/RT}$$

$$= 0.31 \frac{\gamma}{n_h} (2\pi mkT)^{-1/2} e^{-A/RT} \tag{5.96}$$

The number of holes per unit volume can be expressed in terms of the known volume expansion of a mole of the liquid at the melting point ΔV_T divided by the mean hole volume $\langle V_h\rangle$ and reduced to the number per cubic centimeter by dividing by the molar volume of the liquid at T, V_T. Hence,

$$D = 0.31\gamma \frac{\langle V_h\rangle V_T}{\Delta V_T} (2\pi mkT)^{-1/2} e^{-A/RT} \tag{5.97}$$

Utilizing the value of the hole volume derived by substituting for $\langle r_h\rangle$ from Eq. (5.44) in $\langle V_h\rangle = (4/3)\pi\langle r_h\rangle^3$

$$D = \frac{0.17kTV_T}{(2\pi m)^{1/2}\Delta V\gamma^{1/2}} e^{-A/RT} \tag{5.98}$$

The assumed identity of the rates of hole diffusion and ionic diffusion is recalled. Thus, the final expression for the diffusion coefficient of ions in a fused salt is the same as that for holes, i.e., Eq. (5.98).

The first point to note about this expression, apart from the fact that it is of the form of the experimentally observed variation of D with temperature (i.e., $D = D_0 e^{-E_D/RT}$), is that the energy of activation for self-diffusion of cations and anions should be the same. This is what is observed in Table 5.31.

What is meant by the term "Arrhenius activation energy?" This term arose from the work of the early gas kineticists, who wrote equations of the type

$$\text{Rate} = Ze^{-E/RT} \tag{5.99}$$

and considered Z to have a negligible temperature dependence. Such an assumption would of course give

$$E = -R \frac{\partial \ln D}{\partial(1/T)} \tag{5.100}$$

TABLE 5.31

Energies of Activation of Cation and Anion for Self-Diffusion in Simple Molten Electrolytes

	$E_D^{\neq}$ (kJ mol^{-1})	
Molten Salt	Cation	Anion
NaCl	30 ± 1	31 ± 3
KCl	29 ± 2	30 ± 2
$CaCl_2$	26 ± 3	37 ± 4
$SrCl_2$	23 ± 4	29 ± 3
$BaCl_2$	38 ± 5	40 ± 4
$CdCl_2$	28 ± 2	29 ± 3

and it is this coefficient that is often identified with the "energy of activation." Therefore, if one wants to calculate what a theory gives for this, one has to take into account whatever temperature dependence is possessed by the preexponential factor in the theory. If one knew (as one hopes to know later on in this chapter) a theoretical expression for A of Eq. (5.98), one would have to calculate a theoretical value for E from Eq. (5.100) (including the effect of the temperature dependence of the preexponential) and compare its theoretical value with the experimental value of E calculated from Eq. (5.100). From Eq. (5.98),

$$-R\frac{\partial \ln D}{\partial(1/T)} = A + RT - RT^2\left(\frac{1}{\Delta V_T}\frac{d\Delta V_T}{dT} + \frac{1}{2\gamma}\frac{d\gamma}{dT} - \frac{1}{V_T}\frac{dV_T}{dT}\right) \tag{5.101}$$

In the following sections, a value of A will be calculated. Used in the right-hand side of Eq. (5.101), it gives there the theoretical prediction of the experimental energy of activation, i.e., the left-hand side of Eq. (5.101) [*cf.* Eq. (5.100)].

5.7.5. Which Theoretical Representation of the Transport Process in Molten Salts Can Rationalize the Relation $E^{\neq} = 3.74RT_{\text{m.p.}}$?

It has been pointed out (Sections 5.6.2.2 and 5.6.3) that the heats of activation for viscous flow and for self-diffusion are given by the empirical generalization $E_\eta^{\neq} = E_D^{\neq} = 3.74RT_{\text{m.p.}}$, where $T_{\text{m.p.}}$ is the melting point in kelvins. Some of the data that support this statement are plotted in Figs. 5.28 and 5.30. The empirical law is applicable to all nonassociated liquids, not only ionic liquids. An empirical generalization that encompasses the rare gases, organic liquids, the molten salts, and the molten metals is indeed a challenge for theories of the liquid state, and hence also for fundamental electrochemists interested in pure liquid electrolytes.

Several approaches to the theory of liquids can be distinguished (see Section 5.2). One may express the properties of a liquid in terms of a distribution function (as given in Section 5.2.3.2), an expression that indicates the probability of finding particles at a distance r from a central reference ion. One may be able to calculate the distribution function itself from molecular dynamics. Alternatively, one can make educated guesses about the scenarios inside the liquid, i.e., one agrees to temporarily assume a number of competing models of the structure, and goes on to develop the mathematical consequences of the various assumptions. The results of predictions from alternative models can then be compared with experiments and one may decide upon the most experiment-consistent model and use it as a working hypothesis to calculate other properties of the liquid model.

Several differing simple models of molten salts do indeed give reasonably close calculations of equilibrium properties, e.g., compressibility and surface tension. What these models do not do, however, is to quantitatively rationalize the data on the temperature dependence of conductance, viscous flow, and self-diffusion. The discovery by Nanis and Richards of the fact that simple liquids have heats of activation for all three properties given approximately by $3.74RT_{\text{m.p.}}$ presents a clear and challenging target for testing models of liquids.

5.7.6. An Attempt to Rationalize $E_D^{\neq} = E_\eta^{\neq} = 3.74RT_{\text{m.p.}}$

To find what Fürth's hole theory predicts for the heat of activation in viscous flow, it is necessary to attempt to use it to calculate the term A in, e.g., Eq. (5.98). The meaning of A has already been defined in Section 5.7.2. It is the work done in transferring a mole of particles from the surroundings of a hole into its interior.

An assumption will now be added to the model. It is that near or at the melting point, a hole is annihilated by the "evaporation" into it of *one* particle; i.e., just one particle fills it. There is no violation of physical sense in this assertion, for use of Eq. (5.44) shows that the size of the holes predicted by the hole theory is near that of the ions which are assumed in the model to jump into them. Correspondingly, the work done to annihilate a hole is numerically equal to the work done in forming a surface of radius r, namely, $4\pi\langle r\rangle^2\gamma$, where γ is the surface tension. Hence, if n_T particles must jump into a hole to annihilate it at temperature T

$$\frac{n_T A_T}{N_A} = 4\pi\langle r_T\rangle^2\gamma_T \tag{5.102}$$

where A_T is the term A (*per mole*, not molecule) at the temperature T. With the assumption made ($n_{\text{m.p.}} = 1$),

$$A_{\text{m.p.}} = N_A 4\pi\langle r_{\text{m.p.}}\rangle^2\gamma_{\text{m.p.}} \tag{5.103}$$

What of n_T at Ts other than the melting point, at which temperature n_T has been assumed to be unity? From Eq. (5.44), the hole volume—and hence its surface area—should increase as T increases (and γ decreases, as it does with an increase in T). Of course, ions surround the hole and it seems reasonable to assume that as the hole volume increases, the number of ions that surround it will increase, and thus the number that is needed to fill it (thus causing the work $4\pi<r_T>^2\gamma_T$) will also increase.

Let it be assumed that the difference (ΔV_T) between the volume of the liquid salt and that of the corresponding solid is due only to holes. Then the number of holes per mole of salt at temperature T is

$$\frac{\Delta V_T}{<v_{h,T}>} \tag{5.104}$$

where $<v_{h,T}>$ is obtained from Eq. (5.44).

Thus, the number of ions per hole at T is

$$\frac{2N_A}{\Delta V_T/<v_{h,T}>} \tag{5.105}$$

Hence,

$$\frac{\text{Ions per hole at } T}{\text{Ions per hole at m.p.}} = \frac{\Delta V_{\text{m.p.}}}{\Delta V_T}\,\frac{<v_{h,T}>}{<v_{h,\text{m.p.}}>} \tag{5.106}$$

Let it be assumed that the number of ions that will be needed to fill the hole at any temperature would be proportional to the number of ions per hole in the liquid. Thus,

$$\frac{n_T}{n_{\text{m.p.}}} = \frac{\Delta V_{\text{m.p.}} v_{h,T}}{\Delta V_T v_{h,\text{m.p.}}} \tag{5.107}$$

However,

$$n_T A_T = 4\pi N_A <r_T>^2 \gamma_T \tag{5.108}$$

$$A_T = \frac{4\pi N_A <r_T>^2 \gamma_T \Delta V_T v_{h,\text{m.p.}}}{\Delta V_{\text{m.p.}} v_{h,T}} \tag{5.109}$$

with $n_{\text{m.p.}} = 1$.

Thus [and with Eq. (5.44)]

$$\frac{A_T}{A_{\text{m.p.}}} = \frac{T_{\text{m.p.}}^{1/2}}{T_T^{1/2}} \frac{\Delta V_T}{\Delta V_{\text{m.p.}}} \frac{\gamma_T^{3/2}}{\gamma_{\text{m.p.}}^{3/2}} \tag{5.110}$$

One can obtain numerical values for the term on the right. It has been calculated from experimental data for some 14 simple molten salts, and if one restricts the range of experimental data used to about 200 K above the melting point, it is found that

$$\frac{T_{\text{m.p.}}^{1/2}}{T_T^{1/2}} \left(\frac{\gamma_T}{\gamma_{\text{m.p.}}}\right)^{3/2} \frac{\Delta V_T}{\Delta V_{\text{m.p.}}} \simeq 1 \tag{5.111}$$

Under such circumstances, from Eqs. (5.110) and (5.111),

$$A_T = A_{\text{m.p.}} \tag{5.112}$$

However,

$$A_{\text{m.p.}} = 4\pi \langle r_{\text{m.p.}} \rangle^2 \gamma_{\text{m.p.}} \tag{5.113}$$

From Eq. (5.44)

$$A_{\text{m.p.}} = 4\pi(0.51)^2 kT_{\text{m.p.}} = 3.3kT_{\text{m.p.}} \text{ per ion} = 3.3RT_{\text{m.p.}} \text{ per g-ion} \tag{5.114}$$

From Eq. (5.112),

$$A_T = A_{\text{m.p.}} = 3.3RT_{\text{m.p.}} \tag{5.115}$$

This result may be compared with the empirical heat of activation by substituting it in Eq. (5.101). The term on the left of this equation, the experimental values, are found to be about $3.7RT_{\text{m.p.}}$. Using $A = 3.3RT_{\text{m.p.}}$, for which the derivation has been given here, on the right of Eq. (5.101), one obtains agreement between observed and calculated values of the heat of activation to within about 5–10%. The theory of holes is thus able to give some approximate, numerical account of the heat of activation in the transport of simple liquids above the melting point.

5.7.7. How Consistent with Experimental Values Is the Hole Model for Simple Molten Salts?

The Swiss-cheese model approach is consistent with the X-ray data, which show that the distance apart of ions remains constant or decreases on melting while the volume increases ~20%. In this respect, it is more consistent with experiment than some other models which upon melting involve expansion of the "cell" in which each ion spends most of its life (so that the internuclear distance would increase).

An example of the ability of this Fürth hole model to reproduce experimental data numerically without previous appeal to experimental values of similar systems is shown in Table 5.32, which gives a comparison of experimental expansivities with the values that the hole theory yields. An interesting aspect of the evidence supporting the usefulness of this model is the relation of the (cell) free volume (Fig. 5.47) to the volume of the expansion of melting. This free volume, in the sense referred to here, is

TABLE 5.32

Comparison of Calculated Expansivities with the Experimental Values

Salt	Temp. (°C)	$10^4\alpha$, Calc. (°C^{-1})	$10^4\alpha$, Obs. (°C^{-1})
LiCl	614		
	800	4.0	3.0
	900		
	1000	4.0	3.3
NaCl	800	3.3	3.6
	900		
	1000	3.6	3.9
KCl	772		
	800	2.8	3.9
	900		
	1000	3.2	4.2
CsCl	642		
	800	4.2	4.1
	900		
	1000	4.4	4.5
$CdCl_2$	600	2.7	2.4
	700	2.7	2.4
	800	2.9	2.5
NaBr	747		
	800	3.2	3.1
	900		
	1000	4.0	3.4
KBr	735	2.5	3.7
	800	2.6	3.8
	900		
	1000	3.0	4.1
CsBr	636	4.0	4.4
	800	4.2	4.6
	900		
	1000	4.4	5.1

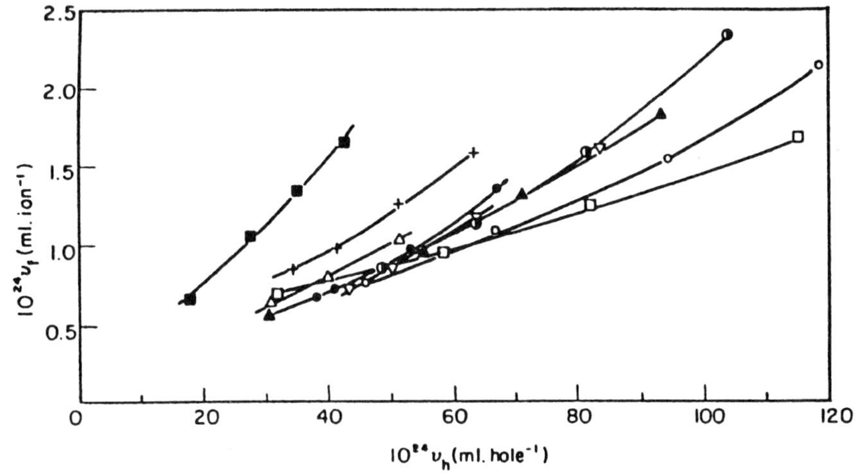

Fig. 5.47. Plot of the free volume per ion against the average hole volume; (■) LiCl, (△) NaCl, (●) KCl, (▲) CsCl, (+) NaBr, (▽) KBr, (◑) CsBr, (□) NaI, (○) KI.

the space that is free to each atom on the average. This relation is shown in Fig. 5.47. The continuous increase of the free volume with the hole volume is what would be expected with a model in which, for between one in five atoms, there is a neighboring hole, so that when a vibrating atom comes into contact with this space, its free volume (= cell free volume and part of the volume of the hole) is increased. Thus, the free volume due to the space within cells is related also to the hole volume, the volume injected on melting, because the average ion's freedom to move is increased by the presence of holes. This is just what Fig. 5.47 confirms.

One must not give the impression that Fürth's theory of holes in molten salts is more than an attempt to see what can be done in the matter of replicating experimental values without the use of experimental data relied on by competing models. It is the model that gives the greatest degree of numerical agreement for several properties, particularly those of transport. It is, on the other hand, a very crude model indeed. It attempts to deal with the problem in a curious, perhaps ingenious, way by using an analogy between holes in an actual molten salt and bubbles in a near-boiling liquid. One might at first not take it seriously, particularly when one finds that it eliminates the term $(P_I - P_O)V$ by assuming that the liquid is near boiling when $P_I = P_O$, but one should see merit in the ability of the model to predict experimental data without relying on values obtained from previous experimental determinations, as is done in Monte Carlo and molecular dynamics approaches. On the other hand, the theory involves some far-fetched imagery, for example, that *one* particle or thereabouts evaporates to annihilate a hole.

Further consideration shows that a liquid molten salt bears more than one resemblance to predictions of the hole theory. Ions do indeed tend to cling together in clusters so that the internuclear distance does not increase to allow for the observed large increase in volume on melting. Between these clusters of ions are gaps and cavities of varying sizes, undergoing rapid changes in size. Some models similar to the hole model seem unavoidable *if one is to attain consistency with the increase of volume on melting but lack of increase of the internuclear distance in the process*. No model of simple molten salts should be considered valid except a model that replicates such unbending facts, and after that it is largely a matter of how to describe the space introduced, varying in size and lifetime, in terms of physical chemistry, which is the challenge and the answer to the question: "Why is a liquid so fluid while a solid is so rigid?"

5.7.8. Ions May Jump into Holes to Transport Themselves: Can They Also Shuffle About?

Constant-Volume Measurements: Big Jumps? The theories representing trans- port in molten liquids which have been the subjects of Section 5.6 are all consistent with the one general idea as to how transport works in molten salts. This is that ions vibrate in their cells surrounded on all sides by counterions for a relatively long time, but when the opportunity arrives, they dart off into a nearby vacancy of some kind. Only Fürth's theory is clear about what they dart to—an opened-up neighboring cavity similar to those in the Woodcock and Singer model of Fig. 5.14, which then gets 90% filled up by the ion's arrival.

Then the rate at which transport, viscous flow, diffusion, and conduction occur is controlled by either the rate at which the opportunities for escape occur or the ease with which the ion jumps into the new "open structure." Of course, these statements apply only to molten salts such as sodium chloride, simple molten salts as they are called, and those for which the log D versus $1/T$ line is straight (Fig. 5.49).[19] If the molten salt forms complexes (e.g., $ZnCl_4^{2-}$, which is formed in $NaCl$-$ZnCl_2$), then it is rather different; the control of transport rate in these substances will be discussed in a later section.

It is desirable to give some evidence that supports the idea that the ions await some rearrangement that allows the transporting ion to have a less coordinated existence for the brief moment of movement. If one determines the diffusion coefficient (or conductance, or viscosity) at constant pressure, then both these processes—the making of the cavity and the jumping into it—are components of the heat of activation (see Figs. 5.48 and 5.49). This can be written as

---

[19]Many molten salts apart from the archetypal NaCl show a straight line for log D – $1/T$.

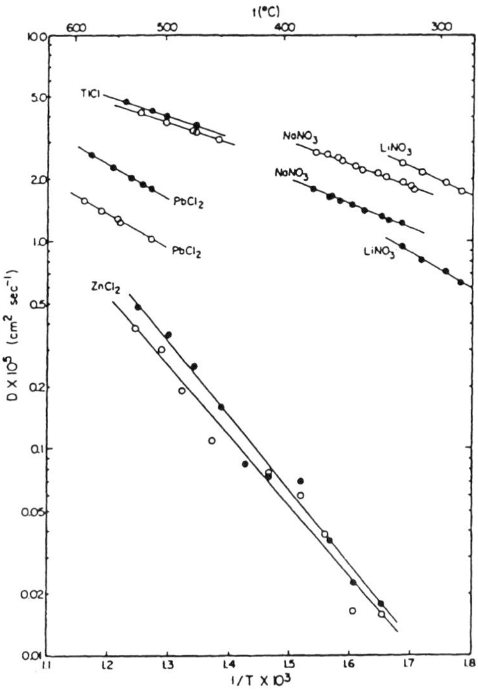

Fig. 5.48. Arrehnius plots of fused salt self-diffusion coefficients; (●) anions; (○) cations. [$LiNO_3$ and $NaNO_3$, Dworkin et al. (1960); TlCl and $PbCl_2$, Angell and Tomlinson (1965); $ZnCl_2$, Sjoblom and Behn (1968).]

$$-R\left(\frac{\partial \ln D}{\partial(1/T)}\right)_P = \Delta H_H^{\neq} + \Delta H_J^{\neq} \tag{5.116}$$

where the suffixes H and J represent, respectively, the process of hole formation and that for jumping into it.

One may also keep the temperature constant but vary the external hydrostatic pressure. Doing this at various temperatures, one obtains relations such as those shown in Fig. 5.50. Thus, knowing the expansivity and compressibility of the molten salt concerned, one can figure out what values of log D have the same volume, though at different temperatures (see Figs. 5.51 and 5.52).

At constant volume there can be no change in volume as the temperature increases, a hypothetical and artificial state. Under these hypothetical conditions, changes of transport with temperature cannot be affected by any corresponding increase in the number of holes with temperature for if more holes were to increase with an increase of volume, the volume of the solution would increase. Hence, under such (hypotheti-

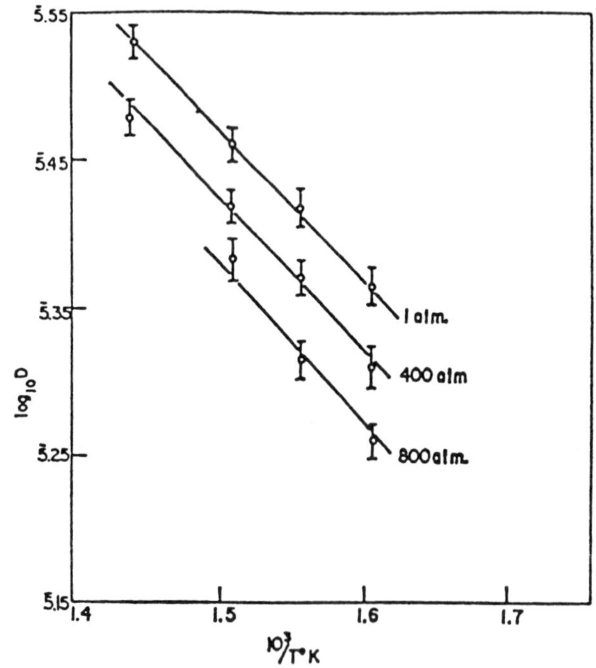

Fig. 5.49. Activation energy for diffusion at constant pressure (^{134}Cs ion in molten $NaNO_3$).

cal) conditions of constant volume, the energy for hole formation would no longer influence the rate of diffusion. For this reason, it is reasonable to write

$$-\left(\frac{\partial \ln D}{(1/T)}\right)_V = \Delta H_J^{\neq} \tag{5.117}$$

Knowing the values of $(\partial \ln D)/[\partial(1/T)]$ both at constant pressure but also at constant volume, one can separate out the heats of activation: the one for the formation of holes, and the one for jumping into holes. Some of the values obtained for ordinary liquids, molten metals, and molten salts are shown in Table 5.33.[20]

In rate theory terms, the rate constant for a happening is given by

$$\vec{k} = \frac{kT}{h} e^{-\Delta H^{\neq}/RT} e^{\Delta S^{\neq}/R} \tag{5.118}$$

[20]If $\Delta H_J^{\neq}/\Delta H_H^{\neq} < 0.3$, the "make hole, then jump in" model makes sense. It makes less sense if $\Delta H_J^{\neq}/\Delta H_H^{\neq} > 0.3$. Holes are there, all right, but they could be simply geographic features of the structure.

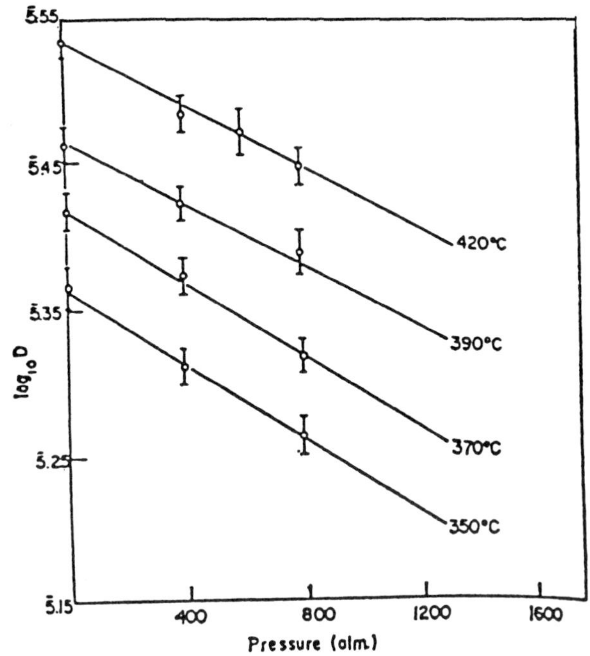

Fig. 5.50. Effect of pressure on the diffusion of ^{134}Cs ion in molten $NaNO_3$.

so that the lifetime of an event (e.g., the formation of a cavity) is

$$\tau = \frac{h}{kT} e^{\Delta H^{\neq}/RT} e^{-\Delta S^{\neq}/R} \tag{5.119}$$

Knowing $\Delta H^{\neq}$ and estimating $\Delta S^{\neq}$, the entropy of activation for its formation, one can calculate how long the hole will last.[21] Thus, the values experimentally found for $\Delta H^{\neq}_H$ and $\Delta H^{\neq}_J$ show that the hole lasts longer than the time for jumping

---

[21]There is a quantity referred to in transport calculations called the *velocity autocorrelation function* (see Section 4.2.19). When applied to the velocity of particles in liquids, it refers to the time needed for a particle to be free of the influence of the previous movement of particles (i.e., uncorrelated). For KCl at 1045 K, the value calculated by Smedley and Woodcock by means of a simulation gave 3×10^{-13} s for the autocorrelation function—about one-tenth of the time for a jump calculated by a hole model (see Table 5.33) for $NaNO_3$.

This result is more consistent with a "shuffle-along" (Swallin) model than the "wait-for-a-cavity-and-leap-into-it" model. The weight of evidence (particularly the $E^{\neq} = 3.74\, RT_{\text{m.p.}}$ law) is in the other direction and one must then ask if neglect of the 20% volume increase on melting (with a decrease of internuclear distance) has invalidated the significance of the results in Smedley and Woodcock's calculation. To support this suggestion, one may point to the wrong sign arising from such models in calculating deviations from the Nernst–Einstein equation.

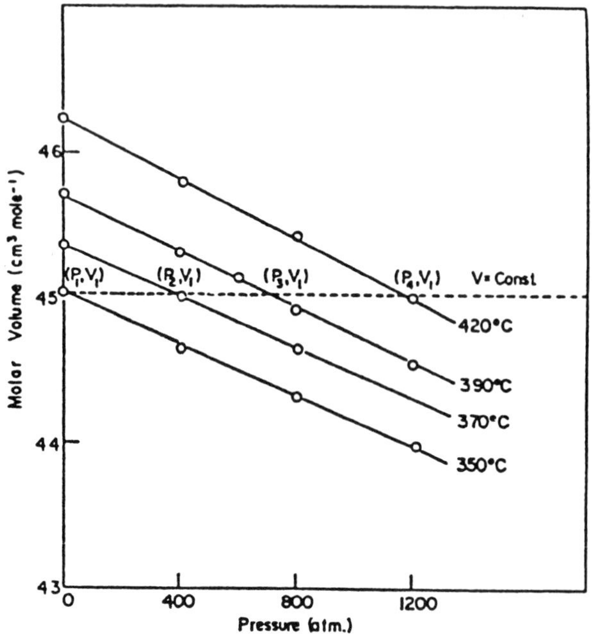

Fig. 5.51. Evaluation of constant-volume conditions (molten $NaNO_3$). Standard of reference: 1 atm, 623 K (P_1, V_1).

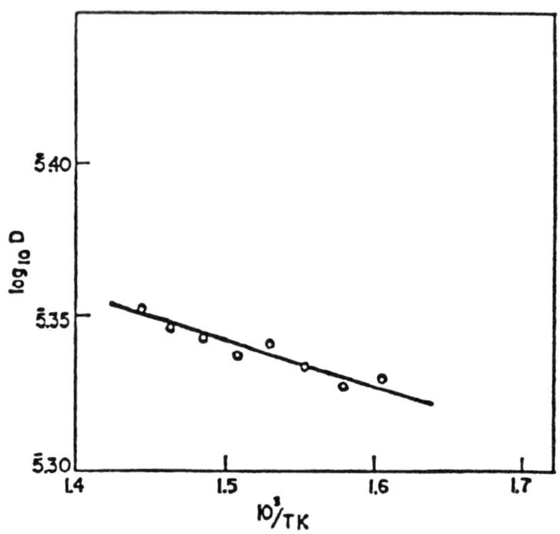

Fig. 5.52. Activation energy of "jumping" (^{22}Na ion in molten $NaNO_3$).

TABLE 5.33

Comparison of Heats of Activation at Constant Volume and Constant Pressure

Liquid	m.p. (K)	$\Delta H^{\neq}_H$ (kJ mol^{-1})	$\Delta H^{\neq}_J$ (kJ mol^{-1})	$\Delta H^{\neq}_J/\Delta H^{\neq}_H$
CCl_4	250.3	13.8	4.6	0.33
C_6H_6	278.6	16.7	2.9	0.25
Hg	234.2	4.2	4.0	0.96
Ga	302.9	4.6	4.1	0.89
$NaNO_3$	580.0	18.0	3.3	0.18

($\Delta H^{\neq}_H >> \Delta H^{\neq}_J$). This is just a confirmatory calculation and it could not be any other way if the idea of an ion waiting for the right juxtaposition of things—so it has space to move elsewhere—represents microscopic reality, at least for the systems for which the above result holds.

5.7.9. Swallin's Model of Small Jumps

Many years ago, Swallin suggested another idea as a model for transport in molten metals and molten salts.[22] In his view, the free space that occurs when a liquid melts does not play a part in the mechanism of transport, diffusion, conductance, and viscous flow. This occurs, he suggested, by means of microjumps, a movement not unlike the gyrations of a person in a large football crowd trying to get out to his seat in the front row. Gaps in the crowd are too small to aid his motion, so that the only way is to shuffle slowly forward, pushing and being pushed. Swallin's[23] suggestion did not sit well with researchers studying molten salt at the time of its publication, however, because it was accompanied by the following equation

$$\Delta V^{\neq} = \frac{1}{6}\pi x_0^2 l \tag{5.120}$$

where l is the jump distance and x_0 is the average distance between ion centers. This $\Delta V^{\neq}$ is the volume difference between the ion in its activated state—the quintessential situation of movement in the rate-determining happening—and the initial state, and it is determined easily at constant temperature from the equation

[22] The fact that (see Fig. 5.28) the temperature dependence of rare-gas liquids, ordinary liquids, liquid metals, and molten salts obeys the equation $E^{\neq} = 3.74\,RT_{\text{m.p.}}$ may mean they all have similar mechanisms for transport.

[23] Swallin's 1959 suggestion was made for metals. Here it may be more applicable than it seems to be to molten salts. Thus, in Table 5.33, it is seen that $\Delta H^{\neq}_H/\Delta H^{\neq}_J \approx 3$ for the ordinary liquids and molten salt quoted but is close to 1.1 for metals. This signifies that the transport rate in a metal is determined less by hole formation and influenced significantly by jumping.

$$\left(\frac{\partial \ln D}{\partial P}\right)_T = -\frac{\Delta V^{\neq}}{RT} \tag{5.121}$$

It can be seen that knowing $\Delta V^{\neq}$, one can calculate l, the jump distance. For $NaNO_3$ at 620 K, this is equal to 380 pm, which is too big by far to be called a microjump as required in Swellin's shuffling, though it is just the right kind of number for a jump into the conveniently opening neighboring cavity, hole, or vacant space.

On the other hand, Swallin's idea has not remained unused. One of the early simulation calculations on molten salt structures—that by Alder and Einwohner—found that a jump distance much smaller than the diameter of an ion fitted the simulation and therefore fitted Swallin's microjump model. However, this would not be consistent with the $\Delta V^{\neq}$ values, for if the microjump distance is l, then $\Delta V^{\neq} = (4/3)\pi\,(l/2)^3$ and as l in the microjump model would be around 10–20 pm, $\Delta V^{\neq}$ would be about 2×10^{-3} cm^3 mol^{-1}, compared with a measured $\Delta V^{\neq}$ of about 10 cm^3 mol^{-1} (e.g., for $NaNO_3$ at 620 K).

Work by Rice and Allnutt[24] for molten salts posed another threat to the "making-holes-and-jumping-into-them" model. In fact, their work has been seen by some to pump life into the aging Swallin theory. The simulation they made was based on calculation of the distribution functions $g_{\pm}$ (see Section 5.5.3), but this could be its Achilles heel because the function thus calculated neglects the large change in volume that occurs in salts such as NaCl when they melt. This is like trying to play Hamlet with the Elsinore castle in the backdrop, but a large live African elephant walking across the stage—it takes no notice of the main point (internuclear distance decreases on melting but volume increases 11–28%). Tables 5.33 and 5.34 contain the values for $\Delta H_H^{\neq}$ and $\Delta H_J^{\neq}$ and Table 5.35 contains those of $\Delta V^{\neq}$.

At first it was thought that the work of hole making was as much as ten times greater than that of jumping, but later on the two values were found to be not so far apart, indicating that the difficulty of jumping can be competitive with that of hole making; the ion is less eager to avail itself of the conveniently opened up neighboring hole than had been thought.

So what of Swallin's shuffling and his idea of microjumps? There is no need to *abandon* microjumping even though the heat of activation for all nonassociated, noncomplexed liquids follows the law $E^{\neq} = 3.74RT_{\text{m.p.}}$, suggesting a unified mechanism of transport for metals, organic liquids, and molten salts. Perhaps in some liquids two kinds of steps contribute to transport in parallel.

It is only in the types of liquids that fit into the relation shown in Fig. 5.28 (i.e., the linear relation of $E_D^{\neq}$ to $T_{\text{m.p.}}$) that hole *formation* seems to be the rate-controlling

[24]The Rice–Allnutt model may be understood by taking the ions that move to be analogous to a man who enters a room full of partying people. His aim is to make it to the bar. There is no gap in the crowd present. He "dives in" and essentially *jiggles* his way forward. This is a similar picture to the "shuffling along" view presented in the 1950s by Swallin for diffusion in metals.

TABLE 5.34

Diffusion Parameters at Constant Pressure and Constant Volume

Diffusion Parameters	^{22}Na in $NaNO_3$	^{134}Cs in $NaNO_3$	^{134}Cs in $CaNO_3$
		Constant pressure	
E (kcal mol^{-1})[a]	4.30 ± 0.30	4.69 ± 0.21	6.47 ± 0.32
$D_0 \times 10^3$ [b]	0.55 ± 0.13	1.24 ± 0.40	1.79 ± 0.25
		Constant volume	
E (kcal mol^{-1})	0.78 ± 0.18	1.30 ± 0.17	1.83 ± 0.26
$D_0 \times 10^5$ [b]	3.9 ± 0.95	8.3 ± 2.3	6.9 ± 1.1
E(constant volume)/E(constant pressure)	0.18	0.27	0.28

[a] 1 cal = 4.184 J.
[b] D_0 is the preexponential factor.

event in transport. Take the case of the transport of protons in water (Section 4.11.6). They dart about, continuously jumping from water to water, tunneling through the barriers in between, with a slight tendency to go more in one direction than the other if there is an applied electric field or a concentration gradient. On the other hand, about one-fifth of the H^+ ions also move along as H_3O^+ moves along. Here, therefore, two mechanisms contribute to transport. There is nothing against assuming (according to Rice and Allnutt's calculations, which neglected the distributed cavities in liquids) that *some* movements of ions occur by a shoulder-to-shoulder pushing microjump shuffle. While the evidence for $E^{\neq} = 3.74RT_{\text{m.p.}}$ holds, and the $\Delta V^{\neq}$s are about the size calculated for a hole (if $\Delta H_H^{\neq} > \Delta H_J^{\neq}$), models involving cavity making and jumping into it seem more consistent with experiments than those that feature shuffling.

Further Reading

Seminal

1. H. Eyring, "A Hole Theory of Liquids," *J. Chem. Phys.* **4:** 283 (1936).

TABLE 5.35

Activation Volumes for the Diffusion Process

	^{22}Na in $NaNO_3$	^{134}Cs in $NaNO_3$	^{134}Ca in $CaNO_3$
$\Delta V^{\neq}$ (cm^3 mol^{-1})	10.7 ± 1.7	14.9 ± 1.4	18.2 ± 2.2
V_h (cm^3 mol^{-1})	9.8 ± 1.0	9.8 ± 1.0	15.6 ± 1.6
$\Delta V_J^{\neq}$ (cm^3 mol^{-1})	0.9	5.1	2.6

2. R. Fürth, "Transport Theory and Holes in Liquid," *Proc. Cambridge Phil. Soc.* **252:** 276, 281 (1941).
3. M. Nagarajan, L. Nanis, and J. O'M. Bockris, "Diffusion of Sodium 22 in Molten Sodium Nitrate," *J. Phys. Chem.* **68:** 2726 (1964).
4. S. R. Richards and J. O'M. Bockris, "Relation of Heats of Activation to the Melting Point in All Non-Associated Liquids," *J. Phys. Chem.* **69:** 671 (1965).

Papers

1. Y. Tada, S. Hiraoka, and Y. Katsumura, *Ind. Eng. Chem. Res.* **31:** 2010 (1992).
2. C. K. Larive, M. F. Lin, B. J. Piersma, and W. R. Carper, *J. Phys. Chem.* **99:** 12409 (1995).
3. M. Watanabe, S. Yamada, and N. Ogata, *Electrochim. Acta* **40:** 2285 (1995).
4. M. Ma and K. E. Johnson, *Can. J. Chem.* **73:** 593 (1995).
5. M. Abraham, M. C. Abraham, and I. Ziogas, *Electrochim. Acta* **41:** 903 (1996).

5.8. MIXTURES OF SIMPLE IONIC LIQUIDS: COMPLEX FORMATION

5.8.1. Nonideal Behavior of Mixtures

A measure of understanding has been gained on the structure and transport properties of simple ionic liquids. In practice, however, mixtures of simple liquid electrolytes are more important than pure systems such as liquid sodium chloride. One reason for their importance is that mixtures have lower melting points and hence provide the advantages of molten salts,[25] but with a lessening of the difficulties caused by high temperatures. What happens when two ionic liquids, for example, $CdCl_2$ and KCl, are mixed together?

Consider, for instance, the electrical conductance of fused $CdCl_2$ and KCl mixtures. If the equivalent conductivity of the mixtures (at a fixed temperature) were given by a simple additivity relation, then a linear variation of equivalent conductivity with the mole fraction of KCl should be observed (dashed line in Fig. 5.53). The straight line should run from the equivalent conductivity of pure liquid $CdCl_2$ at a particular temperature to that of pure liquid KCl at the same temperature. Some binary mixtures of single ionic liquids do indeed exhibit the simple additivity implied by the dashed line of Fig. 5.53.

---

[25]Advantages include no competing hydrogen or oxygen evolution during electrode reactions and an $e^{-E/RT}$ that gives a greater velocity in any rate process. The drawbacks are corrosion and the extra precautions that must be taken to avoid the breakdown of equipment. These threaten high-temperature experiments and make those above 2000 K extremely difficult to carry out. Such difficulties are greatly reduced by using the room temperature molten salts. However, their organic nature often leads to great dissymetry in ion size between cation and anion.

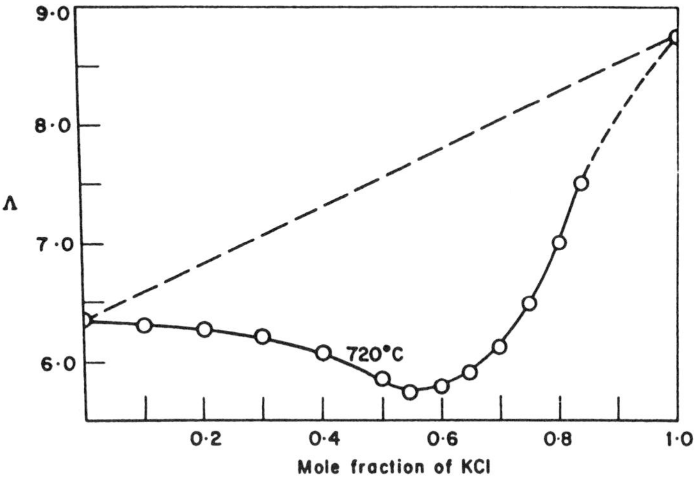

Fig. 5.53. The variation of the observed equivalent conductivity of $CdCl_2$–KCl mixtures as a function of the mole fraction of KCl. The dashed line corresponds to the variation that would be given by the additivity behavior.

There are, however, many systems in which deviations occur from a simple additive law for conductance. The system of $CdCl_2$ and KCl is a case in point (full line in Fig. 5.53). A minimum in the conductivity curve is observed. What is the significance of this minimum?

5.8.2. Interactions Lead to Nonideal Behavior

The situation is reminiscent of some happenings in aqueous solutions. At infinite dilution of, say KCl, the properties (e.g., equivalent conductivity) due to the K^+ and Cl^- ions are additive (see Section 4.3.10). This is ideal behavior (see Section 3.4.1). With increasing concentration, however, there is a departure from ideality: the equivalent conductivities are not simply additive.

Nonideality in aqueous solutions (see Chapter 3) was ascribed to Coulombic attraction between K^+ and Cl^- ions, and the ion–ion interaction theories were evolved for aqueous solutions. The electrostatic attraction between a pair of oppositely charged ions could overwhelm thermal jostling and result in the formation of ion pairs (see Section 3.8).

One can resort to similar explanations for departures from ideality in mixtures of simple ionic liquids. However, there are specific differences between the situation in fused salts and that in aqueous solutions. In pure liquid KCl, there is no concentration variable and therefore fused KCl has a single value of equivalent conductivity (at a particular temperature). The mean distance between the K^+ and Cl^- ions cannot be altered, as it can be in aqueous solutions, by interposing varying amounts of solvent because there is no solvent. Hence, the equivalent conductivity of pure liquid KCl embodies the effects of all the possible interactions for the temperature concerned. The thermodynamics of mixtures of molten salts has been intensively studied by Bloom, by Kleppa, and by Blander.

The interactions that one proposes to account for the deviations from ideality in mixtures of ionic liquids are interactions between the ions of one component of the mixture considered as a solvent and the ions of the other component that is added. In the case of KCl added to pure $CdCl_2$, one can consider, for example, the interactions between Cd^{2+} ions and Cl^- ions, and this interaction can be more than simply electrostatic, which is attraction without preferred direction. It may also involve *directed valence forces*.

5.8.3. Complex Ions in Fused Salts

It is intended here to discuss nonideality arising from complex ion formation. In the case of mixtures of pure liquid electrolytes, however, the idea of complex ion formation raises some conceptual problems.

Consider complex ion formation in the $CdCl_2$-KCl system, and let it be assumed for the moment that a $CdCl_3^-$ complex ion is formed. If such complex ions were formed in an aqueous solution of $CdCl_2$ and KCl, they would exist as little islands separated from other ions by large expanses of water. In fused salts, there are no oceans of solvent separating the ions. Thus, a Cd^{2+} ion would constantly be coming into contact on all sides with chloride ions, and yet one singles out three of these Cl^- ions and says that they are part of (or belong to) a $CdCl_3^-$ complex ion (Fig. 5.54). It appears that in the absence of the separateness possible in aqueous solutions, the concept of complex ions in molten salts is suspect. As will be argued later, however, what is dubious turns out to be not the concept but the comparison of complex formation in fused salts with complex formation in aqueous solutions.

It is more fruitful to compare complex formation in ionic liquids with the phenomenon of hydration of ions in aqueous solution (Chapter 2). It will be recalled that though an ion was seen as constantly nudged by the water molecules of the surrounding medium, a certain number of the water molecules—the "hydration

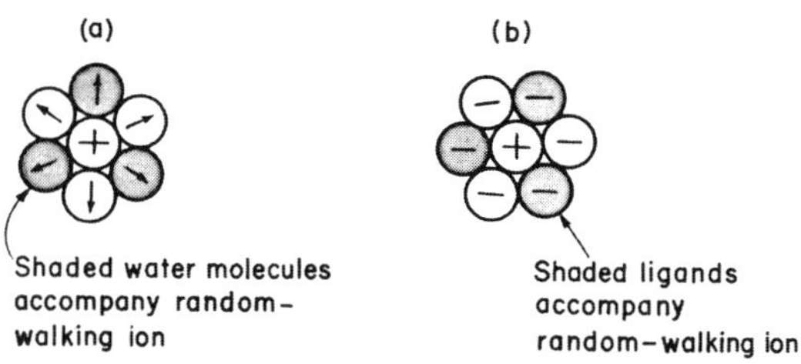

Fig. 5.54. The similarity between (a) a solvated ion in an aqueous solution and (b) a complexed ion in a mixture of ionic liquids.

number"—were considered to stay with the ion in its movement through the solution. The criterion by which the two kinds of water were distinguished was that all those water molecules that temporarily surrendered their translational degrees of freedom to the ion and participated in its random walk constituted its *solvation* sheath.

Similarly, a *complexed* ion is an entity in which a certain number of ligand ions (e.g., three chloride ions in a $CdCl_3^-$ complex) participate in the random walk of the ion (i.e., the Cd^{2+} ion in the $CdCl_3^-$ complex). The other Cl^- ions only undergo promiscuous contacts with the Cd^{2+} ion of the complex, not long-term affairs. The implication is that a complex ion (i.e., the ion and its ligands) is an entity with a lifetime that is at least several orders of magnitude longer than the time required for a single vibration.

From the standpoint of this comparison (Fig. 5.54), it is seen that the concept of a complex ion in a molten salt is at least as tenable as that of an ion with a primary solvation sheath (Section 2.4) in aqueous solutions. What experimental evidence exists for complex ions in fused salt mixtures? To anwer this question, one must discuss some results of investigating the structure of mixtures of simple ionic liquids.

5.8.4. An Electrochemical Approach to Evaluating the Identity of Complex Ions in Molten Salt Mixtures

Here one measures a quantity called the *transition time* at an electrode (see Chapter 7). One switches on a constant current through an electrode. This current is such that the electrode reactions occur fast enough and the activation overpotential is avoided. Under such conditions, an interval of time can be measured (the *transition time*, τ) at which ions in the interfacial layer in contact with the electrode exhaust because they are being removed onto the electrode faster than the rate at which they are being supplied from the solution. Then the electrode potential undergoes a dramatic change in the electrode search for other ions to act as a sink for the electrons pouring into it, which corresponds to the constant current strength chosen.

Then it is easy to show (Inman and Bockris, 1961) that under these circumstances

$$E = E_{\tau/4} + \frac{RT}{nF} \ln \frac{\tau^{1/2} - t^{1/2}}{t^{1/2}} \tag{5.122}$$

Here t is the time elapsed since the switch on of the current, and $E_{\tau/4}$ is the potential reached at one-fourth of the transition time τ described earlier.

The key point here is that $E_{\tau/4}$ is independent of the concentration of the species exchanging electrons at the electrode if there are no complexes formed. Conversely, it is empirically found that if complexes *are* present, $E_{\tau/4}$ varies with the concentration.

To illustrate how one might use this method to find the concentration of the complexes, consider a molten salt system such as $Cd(NO_3)_2$ dissolved in a eutectic of $NaNO_3$-KNO_3 at about 530 K. There are no complexes between Cd^{2+} and NO_3^- in the system of $Cd(NO_3)_2$-$NaNO_3$-KNO_3. Correspondingly, it is found that, $E_{\tau/4}$ is independent of the concentration of Cd^{2+} in this system.

What happens to $E_{\tau/4}$ when one adds to the nitrate system just described some ion that complexes Cd, e.g., Cl^-? Then it is found that $E_{\tau/4}$ changes with the concentration of added Cl^-, in contrast to its constancy in the absence of the Cl^- ligand. Now, $\Delta E_{\tau/4} = E^0_{\tau/4} - E_{\tau/4}(c)$, where $E^0_{\tau/4}$ is the quarter-time potential under standard conditions, and $E_{\tau/4}(c)$ is the value of $E_{\tau/4}$ in the presence of Cl^--containing complexes of Cd formed by adding, e.g., varying concentrations of KCl to the $Cd(NO_3)_2$-$NaNO_3$-KNO_3 eutectic.

DeFord and Hume derived expressions to treat the type of situations sketched above and a version of their equation is

$$F_0 = \left(\frac{\tau_0}{\tau_c}\right)^{1/2} \exp\left(\frac{nF}{RT}\Delta E_{\tau/4}\right) = \sum_{j=0}^{j=m} \beta_j c^j_{X^-} \tag{5.123}$$

Two symbols require definitions; c_{X^-} is the ligand concentration (the Cl^-, for example) and m is the maximum number of ligand ions associated with Cd, that is 3, as in $CdCl_3^-$; other terms have their usual meaning.

Further functions are defined as follows (F_0 is the experimentally determinable quantity):

$$F_1 = \frac{F_0 - \beta_0}{c_{X^-}} \sum_{j=1}^{j=m} \beta_j c^{j-1}_{X^-} \tag{5.124}$$

$$F_2 = \frac{F_1 - \beta_1}{c_{X^-}} \sum_{j=2}^{j=m} \beta_j c^{j-2}_{X^-} \tag{5.125}$$

$$F_3 = \frac{F_2 - \beta_2}{c_{X^-}} \sum_{j=3}^{j=m} \beta_j c^{j-3}_{X^-} \tag{5.126}$$

when $j = 0$, $\beta_0 = K_0$. β_1, β_2, etc., are related to the successive formation constants K_1, K_2, etc., by

$$\beta_1 = K_1 = \frac{[MX^{(n-1)+}]}{[M^{n+}]\,[X^-]} \tag{5.127}$$

TABLE 5.36

Successive Formation Constants for Complexes of Cd^{2+} with X^- in $NaNO_3$ + KNO_3 (Eutectic at 530 K)

Ligand	$K_0 - \beta_0$	$K_1 - \beta_1$	K_2	K_3	K_4	$C_{Cd(NO_3)_r}$ (mol kg^{-1})
Cl^-	1	100 ± 25	7 ± 3	35 ± 12	0	0.98×10^{-3}
Br^-	1	100 ± 50	65 ± 33	8 ± 3	0	0.92×10^{-3}
I^-	1	500 ± 250	60 ± 36	53 ± 22	78 ± 26	0.82×10^{-3}

Source: Reprinted from D. Inman and J. O'M. Bockris, *Trans. Faraday Soc.* **57**: 2308, 1961.

$$\beta_2 = K_1 K_2 = \frac{[MX_2^{(n-2)+}]}{[M^{n+}]\,[X^-]^2} \tag{5.128}$$

$$\beta_3 = K_1 K_2 K_3 = \frac{[MX_3^{(n-3)+}]}{[M^{n+}]\,[X^-]^3} \tag{5.129}$$

By making suitable plots, it is possible to obtain the formation constants $K_1, K_2, \ldots$ of the complexes and then to calculate from the equations given above the formulas of the complexes present. For example, if F_0 is plotted against c_{X^-} the extrapolation to $c_{X^-} = 0$ should equal β_1 ($= K_1$) and the limiting slope at $c_{X^-} = 0$ should equal β_2, and so on, up the series. The F against c_{X^-} plot for the next-to-last complex present in the system should be a straight line with a slope equal to the overall formation constant for the complex of the highest coordination number, and the intercept at the c_{X^-} plot for the last complex should be a straight line parallel to the c_{X^-} axis. The calculated formation constants are given in Table 5.36.

The way this plots out is shown in Fig. 5.55. The figure illustrates a lesson that is easily understandable: the dominant complex changes with the concentration of the ligand.

5.8.5. Can One Determine the Lifetime of Complex Ions in Molten Salts?

Ideas on complex ions in molten salts tend to vary with the time at which they were published. In the first half of the century, there seemed no doubt that complex ions in molten salts were distinct entities and, it was implied, they were permanent. Later, there was doubt as to our ability to identify complex ions in molten salts. Thus, it was argued, there is no difficulty in accepting the existence of discrete ions in aqueous solutions because each ion is a separate entity, and there are many solvent

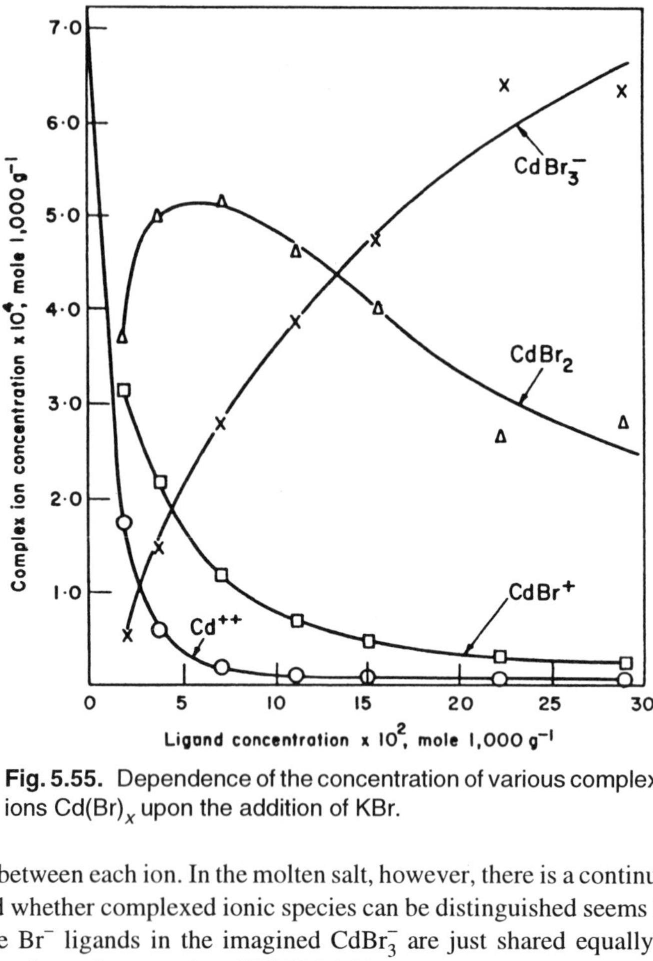

Fig. 5.55. Dependence of the concentration of various complex ions $Cd(Br)_x$ upon the addition of KBr.

molecules between each ion. In the molten salt, however, there is a continuum of ionic entities and whether complexed ionic species can be distinguished seems less certain; perhaps the Br^- ligands in the imagined $CdBr_3^-$ are just shared equally among the solvated ions in a solvent such as KNO_3^--$LiNO_3$.

These two views can be described in terms of a time τ, representing the time an ion such as Cd^{2+} remains bonded to a ligand, such as Br^-. In the two extreme views just given, τ would be minutes or even hours for long-lived complexes and zero for the concept that distinguishable complex ions in molten salts do not exist.

Such problems can be tackled by spectroscopic means, as shown later. Raman spectra, in particular, would indicate new lines having characteristic frequencies when Br^- is added to $Cd(NO_3)_2$ in KNO_3^--$LiNO_3$, and in the preceding section it has been shown that an analysis of the variations of the electrode potential of $Cd(NO_3)_2$ in KNO_3^--$LiNO_3$ with Cl^- addition has given reason to believe in complex ions in the cases quoted. However, there is a nifty electrochemical method that allows one to also obtain the *lifetime* of the individual ions and hence remove doubt as to the real existence of complex ions in molten salts.

If one applies a constant current density to an inert electrode such as Pt dipped into a molten salt consisting of a solvent (e.g., KNO_3^--$LiNO_3$) with a solute of $Cd(NO_3)_2$, it is found that the transition time varies with the strength of the electric current passing across the electrode.

The "transition time" was described briefly in Section 5.8.4, but it is described in more detail in Chapter 7. When one switches on a cathodic current (electrons ejected from the electrode to the solution) at an electrode, at first there are plenty of ions in the region near the electrode and plenty of current can flow. However, there comes a time τ, the transition time, when the solution near the electrode "runs out" of ions, and the electrode potential then undergoes a significant change.

The value of τ for simple situations (no complexing in solution) is given by

$$\tau^{1/2} = \frac{\pi n^2 F^2 D_i c_0^2}{4j^2} \tag{5.130}$$

where j is the current density, n is the number of electrons in the overall reaction, F is the faraday, D_i is the diffusion coefficient of the ion diffusing to the electrode, and c_0 is the concentration of, e.g., Cd^{2+} ions in solution.

It can be readily seen from Eq. (5.130) that $j\tau^{1/2}$ should be a constant for c_0 independent of the applied current, and Fig. 5.56 shows that for the system indicated in the annotation to the figure, it is not. Why not?

Figure 5.56 contains two linear sections. Thus, $j\tau^{1/2}$ (which would be constant if all were simple) decreases as the current density increases. This is consistent with a homogeneous reaction occurring in the bulk of the solution. As the current strength increases, the dissociation of the complex becomes increasingly unable to keep up with the electrode's need for Cd^{2+} (c_0 is kept constant). For this reason, τ sinks with an increase of current density and hence $j\tau^{1/2}$ decreases.

It is difficult to think of an interpretation consistent with these facts except some sequence of the kind:

$$CdBr_3^- \underset{\overleftarrow{k}_3}{\overset{\vec{k}_3}{\rightleftharpoons}} CdBr_2 \underset{\overleftarrow{k}_2}{\overset{\vec{k}_2}{\rightleftharpoons}} CdBr^+ \underset{\overleftarrow{k}_1}{\overset{\vec{k}_1}{\rightleftharpoons}} Cd^{2+} \tag{5.131}$$

One of these reactions must involve a slow, rate-determining step that prevents the complex from dissociating rapidly enough to make up the Cd^{2+} quantities that the electrode removes from the interfacial region around it. Thus, in Fig. 5.56 for the lower line (with higher current densities) the story is qualitatively the same. However, the mathematical treatment goes into a different approximation at higher current densities with the slope change.

It is possible to show that a two-sloped graph such as that shown in Fig. 5.56 can be treated in such a way that the rate-controlling reaction in the above series can be identified as the slow dissociation of $CdBr_2$ and the corresponding lifetime of the entity involved evaluated, also. The lifetime determined for the $CdBr_2$ was 0.3 s in the nitrate

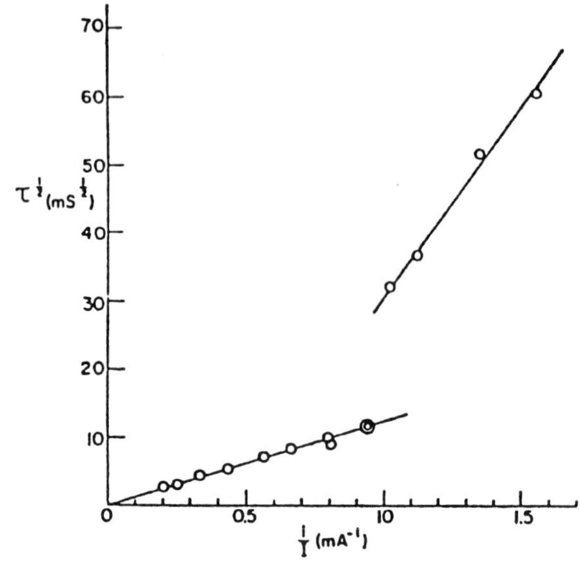

Fig. 5.56. $\tau^{1/2}$ against $1/I$ for $Cd^{2+} = 4.82 \times 10^{-3}$ mol kg^{-1}. $KBr = 9.21 \times 10^{-2}$ mol kg^{-1}. (Reprinted from S. Srinivasan, D. Inman, A. K. N. Reddy, and J. O'M. Bockris, *J. Electroanal. Chem.* **5**: 476, 1965.)

eutectic considered. All this goes to show that there can be indeed complexed entities in molten salts and they live for quite long times,[26] although they are by no means permanent.

5.9. SPECTROSCOPIC METHODS APPLIED TO MOLTEN SALTS

Cryolite is Na_3AlF_6; in the process for the extraction of Al, Al_2O_3, which is obtained from the corresponding hydrate in the mineral bauxite,[27] is added to molten

[26]Long in comparison with some molecular complexes, which have lifetimes on the order of as little as 10^{-10} s.

[27]Bauxite is most easily available to U.S. companies from Venezuela and islands in the Caribbean; there is also a lot in Australia. When bauxite ceases to be economically obtainable from nearby sources (after the year 2000), Al will be extracted from clay (sodium aluminum silicate), which is one of the more abundant minerals on earth and occurs in all countries. $AlCl_3$ would be formed by chlorination of the clay and this compound would be added to a KCl-LiCl eutectic around 770 K. Electrolysis at this temperature would deposit solid Al. Because of the very large amount of Al available from clay, and because of its light weight, Al is tending to replace Fe even in automotive manufacture.

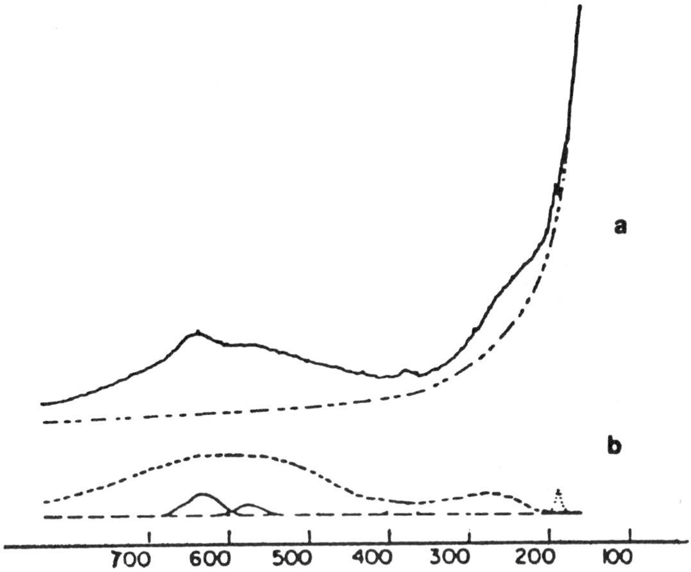

Fig. 5.57. (a) Spectrum of molten cryolite at 1380 K (— halation). (b) Analysis of the spectrum in Raman scattering (——), thermal radiation (- - - -), and instrumental background radiation (---). (Reprinted from C. Solomons, J. Clarke, and J. O'M. Bockris, *J. Chem. Phys.* **49**: 445, 1968.)

cryolite in such a way that Al deposits at the negative electrode of an electrolytic cell. This is the method by which Al is obtained throughout the world. It is the number one electrochemical process in terms of money invested.

In the study of this system, it is necessary to study the constitution of cryolite, what complex ions are present in it, if any. The work is difficult because the temperature is on the order of 1270 K, and cryolite is an aggressive substance, particularly after Al_2O_3 has been added; it attacks most materials, even refractory material such as SiO_2. Making up a cell that is both not attacked by the corrosive cryolite and has a transparent window for radiation to enter and leave is a challenge. The difficulty was first overcome in 1968 by Solomons et al., who used BN as a refractory not attacked by cryolite. In a typical Raman experiment, a focused laser beam enters into one hole in a BN crucible and the scattered light (see also Section 2.11), which carries information on the structure of the particles, emerges through another hole in the crucible. Surface tension keeps the liquid cryolite and alumina from escaping through the holes.

The Raman spectrum of cryolite is shown in Fig. 5.57. The features remaining after deconvolution are two peaks, one at a wavenumber of 633 and one at 577 cm^{-1}.

Both these peaks are polarized.[28] The relatively high frequency and polarized nature of these bands is consistent with the totally symmetric stretching frequency of Al–F in one of the two anions known to constitute cryolite—AlF_4^- (tetrahedral) and AlF_6^{3-} (octahedral).

In solid cryolite there is a band at 554 cm^{-1} (wavenumbers)[29] that has been identified as originating in AlF_6^{3-}. Is the 575-cm^{-1} band in the melt due to the same ion? The more intense band (see Fig. 5.57) at 633 cm^{-1} would be expected to be from a complex of lesser coordination number and this fits AlF_4^-.[30]

What fits the observations reported is the equilibrium

$$AlF_6^{3-} \rightleftharpoons AlF_4^- + 2F^- \tag{5.132}$$

The mole fractions of AlF_6^{3-} and AlF_4^- are determined from the relative values of the peaks in the spectrum as 0.07 and 0.3, respectively; i.e., the AlF_6^{3-} that dominates the solid at room temperature dissociates to a considerable extent in the liquid at 1270 K to the simpler complex.

When Al_2O_3 is added to molten cryolite, another ion, $AlOF_2^-$, also forms and may (surprisingly, because it is an anion) take part in the final reaction at the negative electrode[31] that produces Al.

---

[28]In Raman spectroscopy (see Section 2.11), one speaks of the polarization and depolarization of the scattered light. The depolarization ratio of a line is the ratio of the intensities within the scattered light polarized perpendicular (⊥) and parallel (| |) to the plane of polarization of the incident light. The depolarization ratio, ρ, is defined as $\rho \equiv I_\perp / I_\parallel$. There are several possibilities for the value of this ratio for the emergent light. It can be nonpolarized ($\rho = 1$), or retain its initial polarization ($I_\perp = 0$ and $\rho = 0$). In practice, a Raman line is counted as depolarized if $\rho \approx 0.75$ and polarized if $\rho < 0.75$. Of course, to observe this, one has to look at the Raman scattered light through a polarizing filter, which enables us to find the ratio of ⊥ to | | light. The importance of measuring the polarization of the scattered light is that its interpretation tells us about the symmetry of the scattering source, i.e., gives information on the nature of the ions present.

[29]*Wavenumber* is defined as $1/\lambda$, where λ is the wavelength. However, $\nu\lambda = c$, where ν is frequency and c is the velocity of light. A higher wavenumber means a lower λ and a lower λ means a higher frequency.

[30]Now for an oscillator, $\nu = (2\pi)^{-1}\sqrt{k/\mu}$, where k is the force constant of the vibration and μ is the reduced mass of the oscillator, e.g., Al-F. A lower coordination number means a higher k, and hence ν, because of lessened repulsion between the ligands and diminished screening of the nuclear charge on the cation.

[31]Readers should not be overly shocked at the idea that negatively charged anions such as $AlOF_2^-$ may react at cathodes, which are the *negative* electrodes in a cell. Electrodeposition at negative electrodes from negative anions is quite common (Cr plating of car bumpers occurs from CrO_4^{2-}). Although the anions are electrostatically repelled at a cathode, there may be a driving force as a result of a diffusion gradient.

This can be seen from the Nernst–Planck equation of Section 4.4.15. Here there are two terms. One contains a *potential* gradient, $\partial\psi/\partial x$, and the other a concentration gradient, $\partial c/\partial x$. For a negative anion approaching a negatively charged cathode, there is repulsion. However, transport *to* the electrode and deposition is still possible if the concentration gradient term (which tends to impel the anion *to* the cathode) dominates over the potential gradient term.

5.9.1. Raman Studies of Al Complexes in Low-Temperature "Molten" Systems

$AlCl_3$ complexes with inorganic ions such as Cl^- to form $AlCl_4^-$ and $Al_2Cl_7^{2-}$, both of which register in Raman spectra. $SnCl_2$ might be considered a fruitful partner for $AlCl_3$. Solomons and Clarke found pure $SnCl_2$ to be polymerized when liquid, and it is highly viscous. The obvious question is: What happens when $AlCl_3$ is added to $SnCl_2$ (both being in the liquid state)? This seems an ideal sort of question for study by a Raman spectroscopist since Al–Cl and Sn–Cl both show Raman-active bands.

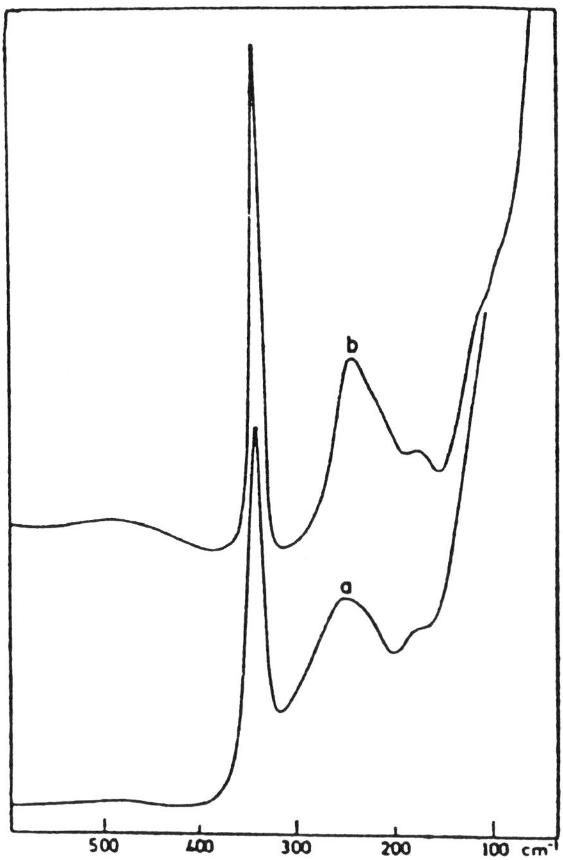

Fig. 5.58. Raman spectra of equimolar $SnCl_2$–AlC_{l3} mixtures: (a) spectrum of the liquid (T = 500 K) and (b) spectrum of the glass (T = 298 K). (Reprinted from B. P. Gilbert, F. Taulelle, and B. Tremillon, *J. Raman Spectros.* **19**: 1, 1988.)

Figure 5.58 shows the spectra for the equimolar liquid and the corresponding glass. Features that are not in the spectrum of either $SnCl_2$ or liquid $AlCl_3$ before they are brought together are observed at ~250 cm^{-1}. This wavenumber is known to be characteristic of Sn–Cl bonds. When the melt is cooled to form a glass, the 250-cm^{-1} band breaks up to show a band at 248 cm^{-1} and one at 219 cm^{-1}. The 248-cm^{-1} band is probably partly depolymerized $SnCl_2$,

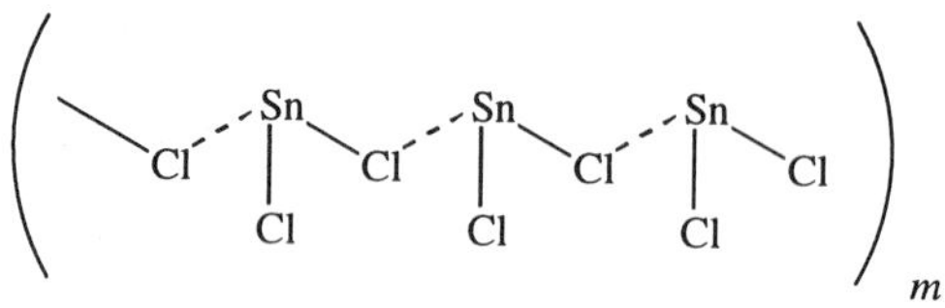

The other band at 219 cm^{-1} will be characteristic of a new Sn–Cl bond and this could come from the breakdown of the $SnCl_2$ polymer, perhaps in a new complex, $Sn(AlCl_4)_2$, either as

C_{3v}:

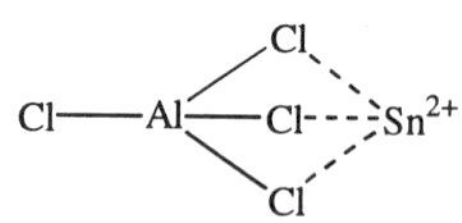

or as

C_{2v}:

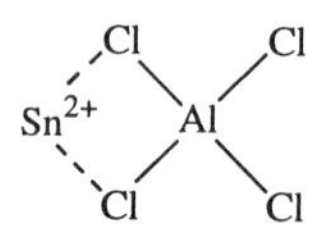

The second structure is preferred because it brings about a lesser symmetry in the complex ion and this decreased symmetry is found to lead to an explanation of one of the characteristics of the spectra, the fact that they exhibit polarization, i.e., have differences in the spectra when the polarized light incident on the sample is in either the parallel or the vertical plane, respectively. Thus, polarization of the scattered Raman radiation is expected from the C_{2v} but not from the C_{3v} complex ion.

5.9.2. Other Raman Studies of Molten Salts

Some idea of how Raman spectroscopy works—how light from nonelastic scattering on molecules contains information on the vibratory state of the bonds therein—has been given in Section 2.11. Raman spectroscopy can be used to obtain information on the structure of ions in molten salts, as has been shown in the last three sections. Here, two further molten salt systems that contain complexes and that have been subjected to Raman spectroscopy are described. The first one concerns melts of zinc chloride hydrate.

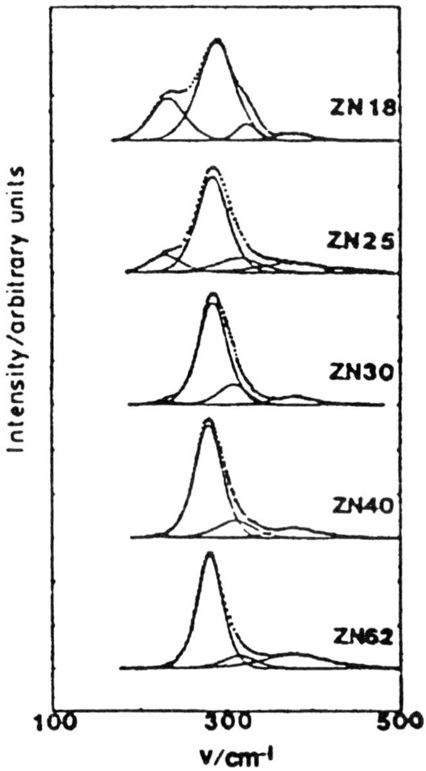

Fig. 5.59. Background-corrected Raman spectra for $ZnCl_2$ hydrate at 298 K. Solid lines show the peak components obtained by the least-squares fits. (Reprinted from T. Pom, T. Yamaguchi, S. Hayaschi, and H. Ohtaki, *J. Am. Chem. Soc.* **93**: 2620, 1989.)

The associated spectra and Raman frequencies are shown in Fig. 5.59 from the work of Yamaguchi et al. The predominant band at 280–289 cm^{-1} is attributable to the Zn–Cl vibration in $(ZnCl_4)^{2-}$. There is also a peak at 227–234 cm^{-1} for solutions containing less than 40% Zn. The same peak is found in anhydrous $ZnCl_2$ melts and is probably due to $(ZnCl_4)^{2-}$ in aggregates (i.e., a number of ions joined together by Cl atoms). Polarized light studies on the symmetry and structure do not seem to have been done.

Another system studied by Raman spectroscopy concerns molten salts at room temperature, which usually involve organic compounds. The system consists of

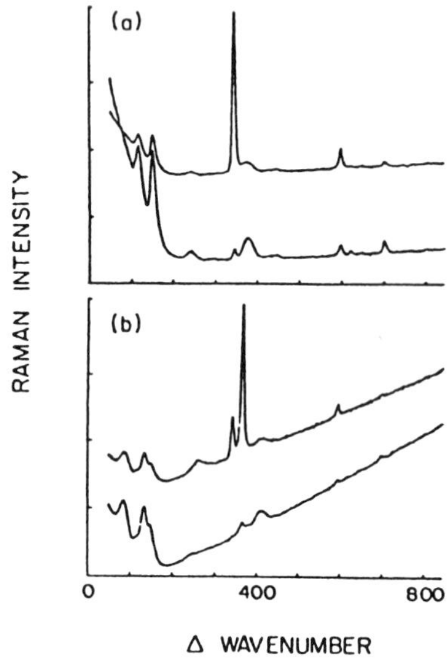

Fig. 5.60. Raman spectra for $GaCl_3$:MEIC melts at room temperature: (a) $n = 0.35$ molar fraction $GaCl_3$ and (b) $n = 0.64$ molar fraction $GaCl_3$. For both, the upper spectrum is vertically polarized, $I_{\parallel}$, and the lower one is horizontally polarized $I_{\perp}$. The intensity scales are different for the two polarizations. (Reprinted from S. P. Wicelinski, R. J. Gale, S. D. Williams, and G. Mamantov, *Spectrochim. Acta* **45A**: 759, 1989.)

gallium chloride in 1-methyl-3-ethylimidazolium chloride (MEIC) at ambient temperatures. This system was examined by Mamantov et al. using an exciting wavelength of 514.5 nm and some spectra are shown in Fig. 5.60.

Table 5.37 contains data on the chlorogallates. The peak at 376 cm^{-1} corresponds to $GaCl_4^-$ but when the molar fraction of $GaCl_3 > 0.5$, one would expect

$$GaCl_4^- + GaCl_3 \rightleftharpoons GaCl_7^- \tag{5.133}$$

The latter ion would have peaks as shown in the table; there is a weak peak at ~365 cm^{-1} that may indicate the presence of $Ga_2Cl_7^-$.

TABLE 5.37

Raman Vibrational Frequencies of Known Chlorogallate Species

($GaCl_4^-$)	($Ga_2Cl_7^-$)	($Ga_2Cl_7^-$)	(Ga_2Cl_6)	(Ga_2Cl_6)
	120 s	90 s	63 vs	98 vs
153 s	140 s	96 s	110 sh, w	115 w
343 s, p	266 w, p	107 m	166 m	168 m
370 sh, w	316 w, p	141 m	267 w	244 m
	366 s, p	284 m	320 m	330 m
	393 sh, m	365 vs	340 s	345 m
		407 m	411 s, p	405 s
			464 w	456 m
($GaCl_3$:CsCl melt) 45 mol% $GaCl_3$ 460 °C	($GaCl_3$:CsCl melt) 70 mol% $GaCl_3$ 275 °C	(KGa_2Cl_7, solid)	($GaCl_3$, 275 °C)	($GaCl_3$, solid)

Source: From S. P. Wicelinski, R. Z. Gale, S. D. Willams, and G. Mamantov, *Spectrochim. Acta* **45A**: 759, 1989.

Notes: s, strong; m, medium; w, weak; sh, shoulder; v, very strong; p, polarized.

5.9.3. Raman Spectra in Molten $CdCl_2$-KCl

Tanaka et al. investigated liquid $CdCl_2$-KCl with special reference to the complexes formed in the liquid state at from 750–980°. They used Hg blue light ($\lambda = 435.8$ nm) and green light ($\lambda = 546.1$ nm) as exciting sources. Four peaks were recorded and the strongest ones are shown in Table 5.38.

The interpretation of what these peaks mean in terms of structure was made by Tanaka et al. by comparing the numerical values of peaks shown in Table 5.38 with values obtained for the corresponding solids where the structure is known by means of the interpretation of X-ray diffraction patterns. Polarized light in the exciting source was used in the investigation and indicated the degree of symmetry in the structure of the ion being observed.

Over the range of compositions of 33–66 mol% $CdCl_2$ in $CdCl_2$-KCl, no change was observed in the Raman results; i.e., over this midrange of compositions there is a stable complex. The peaks and polarization data were compatible with the complex $CdCl_3^-$ in pyramidal structure (tetrahedral and planar structures had also been possibilities).

5.9.4. Nuclear Magnetic Resonance and Other Spectroscopic Methods Applied to Molten Salts

When a certain radio frequency ($\approx 10^7$ s^{-1}) is applied to substances which themselves have been placed in magnetic field of about 10^4 G (or 1 T), absorption of the applied frequency occurs. The origin of this absorption lies in the spin of the protons

TABLE 5.38

Frequency of Raman Peak with Strong Intensity Observed from the Molten $CdCl_2$-KCl System

Salt	Mol% KCl	Temperature (°C) (Max. Deviation ±8)	Frequency [$\Delta\nu$ (cm^{-1})]	Intensity[a]	Exciting Radiation Hg_e – 4358 Å Hg_c – 5461 Å
$2CdCl_2$-KCl	33·3	482–548	255 ± 5	M	e, c
		482–548	257 ± 2	M	
$3CdCl_2$-2KCl	40·0	605–708	260 ± 2	M	e, c
		455	254 ± 2	S	
$CdCl_2$-KCl	50·0	482–708	262 ± 2	S	e, c
		482–548	257 ± 2	S	
$2CdCl_2$-3KCl	60·0	605–708	260 ± 2	S	e, c
$CdCl_2$-2KCl	66·6	482–660	259 ± 2	S	e, c

Source: M. Tanaka, K. Balasubramanyam, and J. O'M. Bockris, *Electrochim. Acta* **8**: 621, 1964.
[a]M, medium; S, strong.

and neutrons in the nucleus of the atoms of the material. Contrary to the dogma of an earlier physics, the properties of nuclei are not immune to their surroundings, i.e., to the doings of electrons outside the nucleus, and hence to the chemical properties of the substance and to the environment. Thus, a study of these frequencies of nuclear magnetic resonance can give information on the properties of the systems being irradiated.

Instead of varying the frequency of the exciting radiation, as in other kinds of spectroscopies, and finding maxima at which absorption occurs, the usual thing with NMR is to keep the frequency of the incident radiation applied to the substance the same and vary the strength of the magnetic field H applied to it. The entity being measured is the *absorption* (A) of the applied radiation as a function of the magnetic field strength. Maxima and submaxima are observed, and the values of these (which measure absorption in the nucleus) as well as the width of the spectral band can be analyzed in a way that gives rise to information on the chemical structure of the substance being irradiated.

As an example, in liquid BeF_2-LiF, one can interpret the characteristics of the NMR absorption as being due to the existence of BeF_4^-. The ion is stable to 820 K. However, no evidence of the ion's rotation is seen, and a probable interpretation of this is that BeF_4^- groups are bonded into bigger structures, which prevent rotation of individual units in the structure.

The radiation applied to materials induces an oscillatory extra magnetic field in the nucleus and one thing which this does is alter the energy distribution among the protons there. The change is not much but it reestablishes itself to the earlier equilibrium values when the incident radiation is removed. It turns out that the time taken for

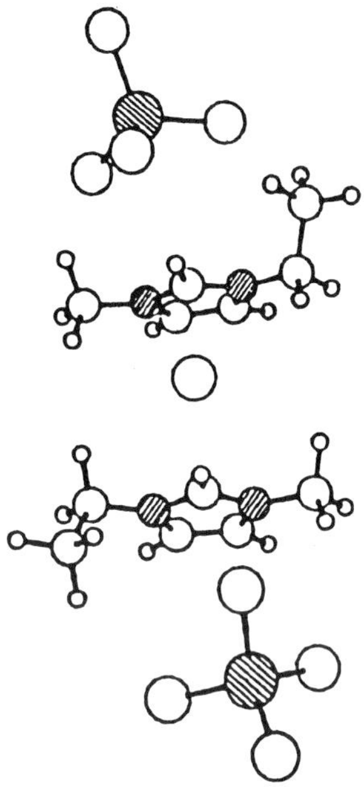

Fig. 5.61. Fully optimized configuration of the $AlCl_4^-$-MEI^+-Cl^--MEI^+-$AlCl_4^-$ system. (Reprinted from D. Dymek, N. E. Heimer, J. W. Rowang, and J. S. Wilkes, *J. Am. Chem. Soc.* **110**: 2722, 1988.)

what is named the *spin–lattice relaxation* (τ) is between 1 and 100 s and this is a convenient sort of time to measure. Study of τ can give information, e.g., on the time of relaxation in diffusional movements in salts containing ^{7}Li, ^{23}Na, and ^{87}Pb.

The study of ions containing aluminum in the liquid state can be done via NMR very conveniently, particularly since their complexes with certain large organic molecules are stable at room temperature. One may bring $AlCl_3$ into contact with, e.g., pyridine or imidazoline and the result is a number of new materials that melt at or near room temperature to form true solvent-free liquid electrolytes (see Section 5.12). The cation may be pyridinium or imidazolonium, and analysis of the degree of absorption in the Al nucleus as a function of the applied magnetic field strength can be used to determine the structure, e.g., of the Al-containing ion. ^{27}Al is the atom for which nuclear resonance is being observed, and it turns out that it gives two signals. The positions of these peaks are consistent with one peak being due to isolated $AlCl_4^-$ groups and the other to $Al_2Cl_7^-$ dimers (Fig. 5.61).

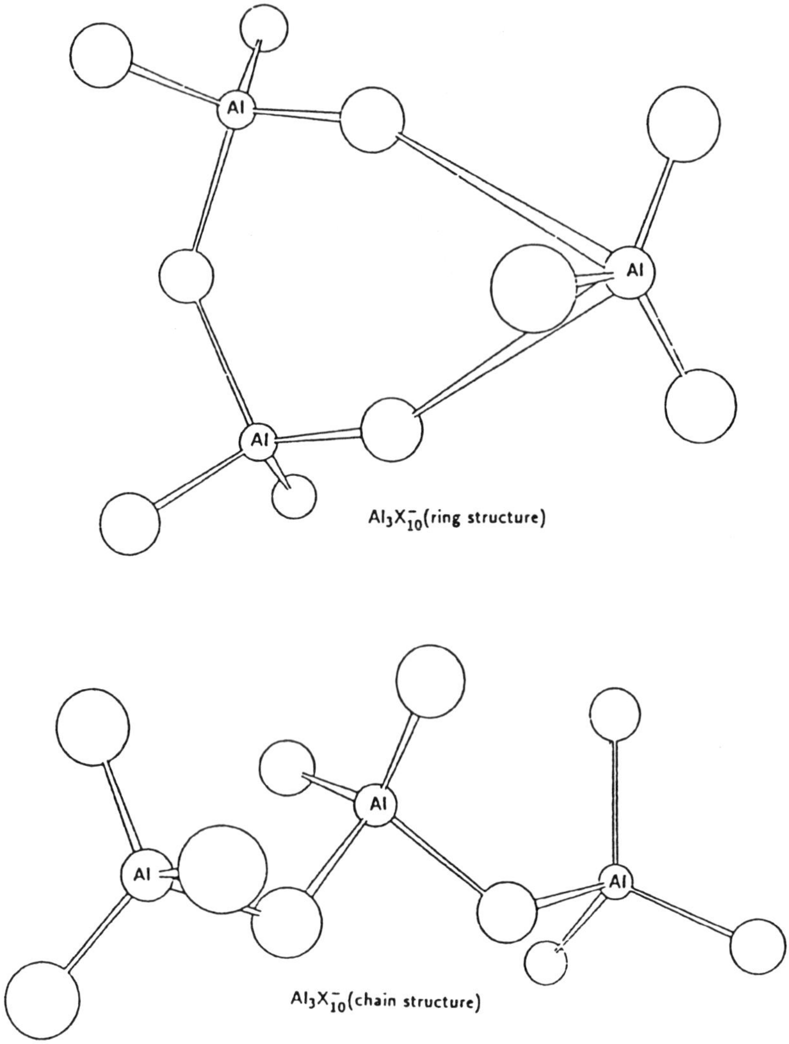

Fig. 5.62. Ring and chain structures of $Al_3X_{10}^-$. (Reprinted from M. Blander, E. Bierwagen, K. G. Calkin, L. A. Curtiss, D. L. Price, and M. L. Saboungi, *J. Chem. Phys.* **97**: 2733, 1992.)

Electron paramagnetic resonance (EPR) and neutron diffraction can also be used to study molten salts. An example of the former is a study of the motion of large organics [2,2,6,6-tetramethylpiperidine-l-oxyl (tempo) and 4-amino tempo, or tempamine] dissolved in room-temperature molten salts, e.g., 1-ethyl-3-methylimidazalonium chloride (ImCl) and $AlCl_3$. One can learn about the dynamics of movement in these systems from such studies.

Neutron diffraction has been applied to the chloroaluminate melts to determine the shape and structure of a number of anions there. They turn out to have chain and ring structures in the higher members similar to those in liquid silicates and borates (Fig. 5.62).

Further Reading

Seminal

1. D. DeFord and D. N. Hume, "The Determination of Consecutive Formation Constants of Complex Ions from Polarographic Data," *J. Am. Chem. Soc.* **73**: 5321 (1951).
2. D. Inman and J. O'M. Bockris, "Complex Ions in Molten Salts: A Galvanostatic Study," *Trans. Faraday Soc.* **57**: 2308 (1961).
3. S. Srinivasan, D. Inman, A. K. N. Reddy, and J. O'M. Bockris, "The Lifetime of Complex Ions in Ionic Liquids: an Electrode Kinetic Study," *J. Electroanal. Chem.* **5**: 476 (1963).
4. M. Tanaka, K. Balasubramanyam, and J. O'M. Bockris, "Raman Spectrum of the $CdCl_2$-KCl System," *Electrochim. Acta* **8**: 621 (1963).
5. C. Solomons, J. Clarke, and J. O'M. Bockris, "Identification of the Complex Ions in Liquid Cryolite," *J. Chem. Phys.* **49**: 445 (1968).

Review

1. J. F. Stebbins, "Nuclear Magnetic Resonance at High Temperatures," *Chem. Rev.* **91**: 1353 (1991).

Papers

1. C. L. Hussey, I-Wen Sun, S. Strubinger, and P. A. Bernard, *J. Electrochem. Soc.* **137**: 2515 (1990).
2. C. L. Hussey, I-Wen Sun, S. Strubiner, and P. A. Bernard, *J. Electrochem. Soc.* **137**: 2515 (1990).
3. M. Blander, E. Bierwagen, K. G. Calkin, L. A. Curtiss, D. L. Price, and B. L. Saboungi, *J. Chem. Phys.* **97**: 2733 (1992).
4. S. Takahashi, M. L. Saboungi, R. J. Klinger, M. J. Chen, and J. W. Rathke, *Electrochem. Soc. Proc.* **92-16**: 345 (1992).
5. E. A. Pavlatou and G. N. Papatheodorou, *Electrochem. Soc. Proc.* **92-16**: 72 (1992).
6. D. L. Price, M. L. Saboungi, S. Hashimito, and S. C. Moss, *Electrochem. Soc. Proc.* **92-16**: 14 (1992).
7. M. Oki, K. Fukushima, Y. Iwadata, and J. Mochinaga, *Electrochemical Society, Molten Salts* **93-9**: 9 (1993).
8. D. L. Price, M. L. Saboungi, W. S. Howell, and M. P. Tosi, *Electrochemical Society, Molten Salts* **93-9**: 1 (1993).
9. J. F. Stebbins, S. Sen, and A. M. George, *J. Non-Cryst. Solids* **193**: 298 (1995).

10. S. Das, G. M. Bejun, J. P. Young, and G. Mamantov, *J. Raman Spectrosc.* **26**: 929 (1995).

5.10. ELECTRONIC CONDUCTANCE OF ALKALI METALS DISSOLVED IN ALKALI HALIDES

5.10.1. Facts and a Mild Amount of Theory

Bronstein and Bredig were the discoverers (1958) of the unexpected fact that 1% concentrations of alkali metals dissolved in molten salts show electronic conductivity. In the K-KCl system, the conductance was found to increase more than linearly with concentration while in the Na-NaCl system, it increases less than linearly.

The equation for the specific conductivity (see Section 4.3.6) is

$$\kappa = FN_e u_e \tag{5.134}$$

where N_e is the number of moles of electrons per cubic centimeter, u_e is the mobility in $\text{cm}^2\ \text{V}^{-1}\ \text{s}^{-1}$, and F is the electrical charge on 1 g-ion (96,500 C). Equation (5.134) yields values for the mobilities. In the Na system, the mobility was found to be equal to $0.4\ \text{cm}^2\ \text{V}^{-1}\ \text{s}^{-1}$, and in the K system, $0.1\ \text{cm}^2\ \text{V}^{-1}\ \text{s}^{-1}$. When the Na is 0.01 M in the Na system, the concentration of electrons is about 10^{19} electrons cm^{-3}.

These values of mobilities of 0.1 and $0.4\ \text{cm}^2\ \text{V}^{-1}\ \text{s}^{-1}$ obtained at 1070 K at first appear to be 250 times more than those for the corresponding ions in aqueous solution at 300 K ($\sim 5 \times 10^{-4}\ \text{cm}^2\ \text{V}^{-1}\ \text{s}^{-1}$). It would hardly be reasonable to compare mobilities at different temperatures. If one recalculates the ionic mobility for a hypothetical situation of an alkali ion in an aqueous solution at 1070 K, the value for the mobility in aqueous solution should increase from about 5×10^{-4} to about $0.2\ \text{cm}^2\ \text{V}^{-1}\ \text{s}^{-1}$. Hence, when the correction for the temperature difference *is* accounted for, the electron's mobility in the molten salt is not very different from that of an ion in the corresponding ion in an aqueous solution.

What about the lifetime of the electrons that conduct? One can find the lifetime from a simple phenomenological theory of conductivity. Thus, the equation of motion for the electron's movement under an electric field in the molten salt is

$$m_e \frac{du}{dt} = -Xe_0 - Cu \tag{5.135}$$

Here e_0 is taken as negative. The field X accelerates the electron and there is a retarding force Cu which, because of the Stokes–Einstein equation (Section 4.4.8), one assumes to be proportional to velocity.

Integrating Eq. (5.135), and taking u at $t = 0$ to be equal to zero, one gets

$$u_e = -\frac{Xe_0\tau}{m_e}(1 - e^{-t/\tau}) \tag{5.136}$$

where t is the time after the last collision and τ is the time at which the mobility is $[1 - (1/e)] = 1 - (1/2.7183) = 0.64$ of the final value (as the electron accelerates). Thus, the value at $t = \infty$, the steady-state value, is

$$u = -\frac{Xe_0\,\tau}{m_e} \tag{5.137}$$

In order to obtain some idea of the order of magnitude of τ, it is reasonable to take u as the mean of the values for the two salts described, i.e., $\frac{1}{2}(0.1 + 0.4) = 0.25$. X is $1\ \text{V cm}^{-1}$, $m_e = 9 \times 10^{-28}$ g, and $e_0 = 1.6 \times 10^{-19}$ C.

Thus,

$$\tau = 1.5 \times 10^{-9}\ \text{s} \tag{5.138}$$

which is a typical lifetime for an electron in a solid.

The meaning of τ is the average time the electron has to move from one ion before interacting again with another one. Hence, the distance traveled in one direction between collisions when the applied field is $1\ \text{V cm}^{-1}$ is only $\sim 0.25 \times 1.50 \times 10^{-9} = 4 \times 10^{-10}$ cm = 0.004 nm, a surprisingly small distance, although it is consistent with the short tunneling distance found in the following model analysis.

5.10.2. A Model for Electronic Conductance in Molten Salts

The fact that a concentration of about 1 mol% of an alkali metal in a molten salt system can cause a considerable specific conductance demands some kind of explanation. At 1%, the electrons are about 2 nm apart, on the limit for tunneling site to site. What is the mechanism of their easy passage through the molten salt?

Emi and Bockris suggested a model for this phenomenon that bears some resemblance to the Conway–Bockris–Linton (CBL) theory of the mobility of protons in solution. Here (Section 4.11.6), protons tunnel from their positions attached to a given water to another water when—under the influence of the proton's field—this latter has rotated sufficiently to offer an orbital in which to receive a jumping proton.

There are three steps in the corresponding calculation for electrons. In the first, one finds the energy of an electron with respect to its zero potential energy *in vacuo*, both the energy in the filled state in the atom and in the empty state on a cation (each at equilibrium, i.e., having an energy at a minimum of the potential-energy–displacement relations). The next step is to allow the atom and ion to be displaced from their equilibrium positions to such an extent that the potential energy of the filled and empty electronic states in the atom and ion, respectively, become equal, which is the condition for radiationless transfer.

TABLE 5.39
Mobility of Electrons[a]

Na-NaCl						K-KCl					
1121 K			1166 K			1091 K			1133 K		
X_M	μ_{obs}	μ_{calc}	X_M	μ_{obs}	μ_{calc}	X_M	μ_{obs}	μ_{calc}	X_M	μ_{obs}	μ_{calc}
1.10	34	45	0.81	34	50	1.03	31	32	—	—	—
1.47	43	45	1.87	40	50	2.65	34	32	—	—	—
2.93	29	45	3.84	28	50	3.61	55	32	—	—	—
3.16	32	45	4.25	33	50	7.24	87	32	4.23	58	36

[a]Unit of mobility: $(cm^2\ V^{-1}\ s^{-1}) \times 10^3$; X_M = mol% of metal atoms.

Gamow-type expressions (see Section 4.11.5) can be used for the calculations of the probability of tunneling (from atom to cation) and thereafter it is relatively simple to calculate the electron mobility. This is given by the concentration times the vibration frequency times the probability of tunneling from atom to ion.

According to this theory (Table 5.39), the mobility is not affected by the concentration of electrons, because at 1% they are too far apart to interact. No change in

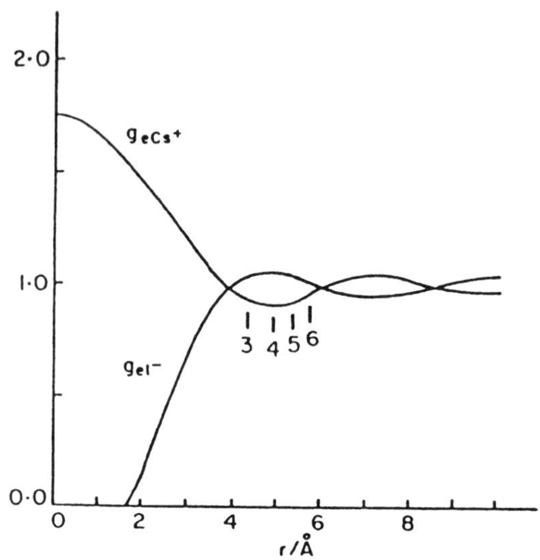

Fig. 5.63. Electron–Cs^+ and electron–I^- radial pair correlation functions. (Reprinted from C. Malescio, *Mol. Phys.* **69**: 895, 1990.)

mobility was detected as a result of a temperature change of 40 degrees for Na-NaCl (experimental; Table 5.39).

The radial distribution function of electrons in Cs has been calculated and is shown in Fig. 5.63. The numbers given on the figure are the coordination numbers of the electron by Cs^+ in terms of the distances from the center of the Cs^+ ion. They are calculated from

$$N(r) = 4\pi \int_0^r g_r r^2 \, dr \tag{5.139}$$

where g_r is the distribution function for the electron in Cs and r is the radius of the electron orbit in the outermost shell. Values ranging from 3 to 6 agree with those from corresponding neutron diffraction evidence.

Molecular dynamics calculations in metal–molten salt systems which could lead to diffusion and conduction values were first made by Parinello and Rahman in 1984. They have been used particularly by Malescio to examine the degree of delocalization of the electrons, which increases with the radii of the metal atoms.

Further Reading

Seminal

1. H. R. Bronstein and M. A. Bredig, "The Electrical Conductivity of Solutions of Alkali Metals in Their Molten Halides," *J. Am. Chem. Soc.* **82**: 2077 (1958).
2. T. Emi and J. O'M. Bockris, "Electronic Conductivity in Ionic Liquids," *Electrochim. Acta* **16**: 2081 (1971).

Papers

1. G. Malescio, *Mol. Phys.* **69**: 895 (1990).
2. G. Malescio, *Nuovo Cimento* **13D**: 1031 (1991).
3. G. M. Harberg and J. J. Egan, *Proc. Electrochem. Soc.* **16**: 22 (1992).
4. J. Lin and J. C. Poignet, *J. Appl. Electrochem.* **22**: 1111 (1992).
5. J. Bouteillon, M. Jaferian, J. C. Poignet, and A. Reydat, *J. Electrochem. Soc.* **139**: 1 (1992).
6. D. Naltland, T. Reuj, and W. Freyland, *J. Chem. Phys.* **98**: 4429 (1993).
7. J. C. Gabriel, J. Bouteillon, and J. C. Poignet, *J. Electrochem. Soc.* **141**: 2286 (1994).
8. L. Arurault, J. Bouteillon, and J. C. Poignet, *J. Electrochem. Soc.* **142**: 16 (1994).

5.11. MOLTEN SALTS AS REACTION MEDIA

Molten salts can be good media in which to carry out chemical reactions. The rate of all reactions increases exponentially with temperature. A *liquid* medium causes a higher rate of reaction to occur in a solute compared with that in a gas at the same temperature. Why is this? The situation needs thought. In the gaseous state, reactants experience only a fleeting contact when they collide, which is often too brief a contact to reach thermal activation and a successful product formation. Thus, gas molecules fly around at about 10^5 cm s^{-1} at room temperature. When they collide, the actual time of contact is less than 10^{-12} s.

A different situation is observed in liquids. Here, the time one reactant spends next to another is much longer than that in the gas phase. For this reason, the two particles can enlarge the possibilities from those of fleeting acquaintance to the more productive ones of prolonged contact, leading much more often to permanent association. One can determine the order of magnitude of the time of contact (see Section 4.2.18) as

$$D = kl^2 \tag{5.140}$$

where D is the diffusion coefficient, k is the rate constant for diffusion, and l is the distance a particle covers in one jump of its movement in diffusion. From the above equation,

$$\frac{1}{k} = \tau = \frac{l^2}{D} \tag{5.141}$$

Thus, τ is the residence time, the time between "hops," the time the two reactant particles have to decide whether to react. Near the melting point of a molten salt, the diffusion coefficient in solutes is on the order of 10^{-5} cm^2 s^{-1}. With l chosen as 3×10^{-8} cm (a typical value of the distance between sites within the molten salt structure), one obtains $\sim 10^{-10}$ s for the residence time, which is about 100 times longer than that in the gas phase at the same temperature and hence there is a hundredfold greater chance to react.

However, there is another reason why the molten salt is often a more effective medium for carrying out a reaction quickly. Reaction rates are proportional to $e^{-E_a/RT}$, where E_a is the energy of activation of the reaction. Assume $E_a = 10^5$ J mol^{-1}. Then, if one compares the rates at 300 and 600 K, the reaction rate is 10^8 higher at the higher temperature if the rate-determining step in the reaction remains the same.

Consider, for example, a dissolved organic molecule, RH, reacting with dissolved O_2 to give CO_2 and H_2O. If the reaction at 300 K occurs at a rate v_1, that at 600 K should occur at a greatly increased rate. Could this be achieved by heating the dissolved materials in an aqueous solution? Of course not! For unless one uses a pressure vessel (with the added expense of having one made), the aqueous solution cannot be heated much above 373 K before the solution boils. On the other hand, molten salts are available over the whole temperature range—from room temperatures with the $AlCl_3$ complexes in organics such as imidazoline—to molten silicates at 2000 K.

A good example of the success of a molten salt reactor is the work carried out by Guang H. Lin at Texas A&M University in 1997. This has led to a new method for the complete consumption of carbonaceous material at low cost. Lin introduced a mixture of paper, wood, and grass in pellets into a molten salt eutectic of KNO_3, and

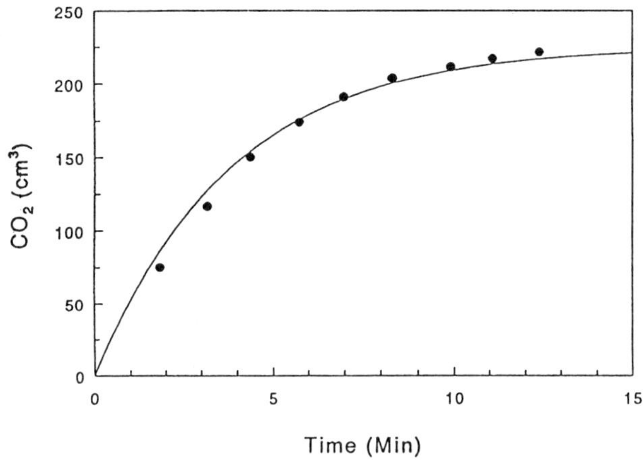

Fig. 5.64. Production of CO_2 as a function of time in a molten salt reactor of KNO_3 and $LiNO_3$.

$LiNO_3$ at temperatures that ranged from 670 to 870 K. At the same time, he contrived to have a stream of minute bubbles of O_2 at 670 K flow through the liquid electrolyte. The organic material (incompletely dissolved) reacts very rapidly to give CO_2 and H_2O. A plot of some of Lin's results[32] is shown in Fig. 5.64.

The oxygen stream can be merely air. Lin showed that the method can be extended to consume rubber. Scaled up, Lin's work *could be the solution to most waste disposal problems* (including consumption of the solid sludge left over from the modern treatment of sewage; rubber tire disposal; and that of superdangerous organics such as Agent Orange). The point is that the combustion is so complete and so quick that there are no noxious effluents to reach the air.

Further Reading

Seminal

1. J. E. Gordon, *Application of Fused Salts in Organic Chemistry,* in *Techniques and Methods of Organic Chemistry*, Vol. 1, Marcel Dekker, New York (1969).
2. L. Lessing, "Sewage Disposal in Molten Salts," *Fortune* 138 (July 1973).
3. J. W. Tomlinson, "High-Temperature Electrolytes," in *Electrochemistry: The Past Thirty and the Next Thirty Years*, H. Bloom and F. Gutmann, eds., Plenum Press, New York (1977).

Papers

1. L. Kaba, D. Hitchens, and J. O'M. Bockris, *J. Electrochem. Soc.* **137**: 5 (1990).

[32]The key to success is to keep the pellets from rising to the top of the molten salt and floating there. Lin contrived to inject the pellets beneath a cage of metal gauze where they remain trapped—keeping them in contact with the molten salt and O_2 until consumed.

2. A. Berrukov, O. Deryabina, L. Maharinsky, N. Halterinsky, and A. Berlin, *Int. J. Polym. Mater.* **14**: 101 (1990).
3. E. Sada, H. Kumazawa, and M. Kudsy, *Ind. Chem. Res.* **31**: 612 (1992).
4. U. Gat, J. R. Engel, and H. L. Dodds, *Nuc. Technol.* **100**: 390 (1992).
5. C. Tennakoon, R. Bhardwaj, and J. O'M. Bockris, *J. Appl. Electrochem.* **26**: 18 (1996).

5.12. THE NEW ROOM-TEMPERATURE LIQUID ELECTROLYTES

At the beginning of this chapter it was pointed out that aqueous and many nonaqueous electrochemical systems suffer from the small size of the potential range in which solutes dissolved in them can be examined. This is because (for pH = 0, say) if the potential of an electrode immersed therein is more *negative* than 0.00 V on the normal hydrogen scale (see Section 4.8.3), the water itself in the solution begins to decompose to form H_2. On the other hand, at a potential more *positive* than 1.23 V on the same scale, the aqueous solvent tends to decompose to form O_2.

It is true that this small window of 1.2 V is extendable in both potential directions, particularly on the positive side because the phenomenon of overpotential (Chapter 7) is especially strong there and the potential that has to be applied to the positive electrode to get a significant current density may be as high as 1.8 V. Nevertheless, it has been felt for decades that systems were needed in which one could make the potential of electrodes more positive (thus releasing a greater power in oxidation) and also more negative (greater power of reduction) than is currently possible because of the solvent decomposition problem in aqueous solutions.

Indeed, the possibility of using molten salts to extend the potential window in which electrochemical reactions can occur has been one of the driving forces behind the need to know about the liquid electrolytes described in this chapter. Thus, for liquid NaCl, significant decomposition does not occur till c. 3.0 V have been applied across an electrochemical cell containing liquid NaCl. This gives twice the electrochemical window, which in aqueous systems is as low as 1.2 V. However, the sacrifice has been that one had to work at more than 1120 K, with all the attendant experimental difficulties that work at high temperatures brings.[33]

It has long been known that liquid electrolyte systems that melted in the low hundreds of degrees were available in systems of metal chlorides and $AlCl_3$, and that some tetraalkylammonium salts melted at < 373 K. Hurley and Wier in 1951 showed that a 2:1 mixture of some complex organic chlorides with $AlCl_3$ gave liquid electrolytes at room temperatures. The discovery remained undeveloped for more than 25

[33]There are a good many of these, although they are less at temperatures below, say 770 K, when glass vessels can still be used, and they increase exponentially as the temperature is increased. They include the difficulty that the containing vessels tend to dissolve in the liquid electrolyte solvent, the evaporation and decomposition thereof, the need to take precautions in experimental design to achieve a uniform temperature, the troubles of extensive thermal insulation, cracking of the refractory vessels, etc.

years. However, in 1977, Halena Chum, writing with Koch and Osteryoung,† found that such systems (to be precise: 1-Methyl-3-ethylimidazolonium chloride in union with $AlCl_3$) could be made into liquid electrolyte solvents, and that many such systems had melting points near room temperature. Such systems have been under intensive examination in the '80s and '90s, largely by American electrochemists, among whom Osteryoung, Wilkes, and Hussey have—each with his own team—led most of the contributions. Although the chemistries (including the redox properties) of many of these systems have been researched, their applications (e.g., to energy storage systems) are in the early development stage and promise rich yields. The electrochemical windows are often above 3 V and occasionally extend even to 6 V!

5.12.1. Reaction Equilibria in Low-Melting Point Liquid Electrolytes

Redox reactions are subject to examination in solvent systems such as 1-butylpyridinium chloride with $AlCl_3$, which melts at 308 K (correspondingly, 1-3-dialklyimidazolonium chloride). These molecules are

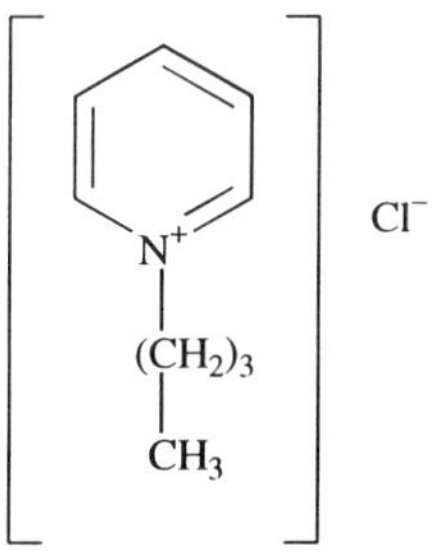

1-butylpyridinium chloride

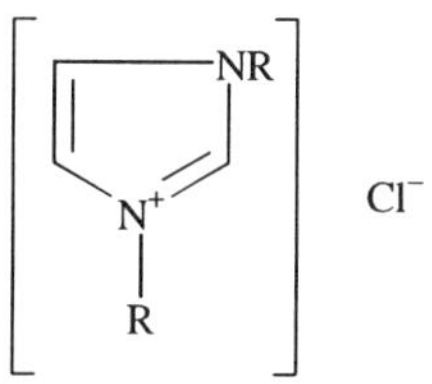

1,3-dialkylimidazolonium chloride

Many treatments of these solvent systems are in terms of acid–base equilibria, and the basis is to regard Cl^- as a Lewis base.[34] In these terms, the heptachloroaluminate is a strong Lewis acid.

---

†Robert Osteryoung is picked out here for recognition because—apart from his pioneering work on low temperature molten salts—he is well known for his early work on pulse techniques (Chap. 8). He was the first to develop computers to control electrochemical experiments. Professor Osteryoung is a Head of Chemistry at North Carolina State University where unlike some great researchers, he is well known for his success as an able administrator.

[34]In Lewis's view, a base *donated* an electron pair and an acid accepted an electron pair. For example, consider the reaction

$$R_3N + BF_3 \rightarrow R_3N - BF_3$$

The R_3N is the base and the BF_3 the acid.

This definition of acids and bases is broader than that of Brønsted, which had an acid as something that *gave* protons and a base as something that *accepted* protons. Thus, Lewis's definition includes reactions involving H^+ ions, oxide ions, and solvent interactions.

In one of the low-melting liquid electrolyte systems containing $AlCl_3$ one has a changing acid/base character, depending on the ratio of $AlCl_3$ to the organic partner.

BASE	**ACID**
$AlCl_3/RCl < 1$	$AlCl_3/RCl > 1$
R^+, Cl^- and $AlCl_4^-$	$Al_2Cl_7^-$

In systems of organic chlorides such as R–Cl with $AlCl_3$, when the $AlCl_3$ is first added (in low concentrations), the Al is largely in the form of $AlCl_4^-$, but as the $AlCl_3$ increases in relative concentration, the system becomes acid in the Lewis sense, and the dominating constituent is the ion $Al_2Cl_7^-$:

$$2\,AlCl_4^- \rightleftharpoons Al_2Cl_7^- + Cl^- \qquad (5.142)$$

To obtain the equilibrium constant in these systems, one can use electrochemical cells such as those described in Section 3.4.8. For example, measurements that involve Al electrodes have a K value of 3.8×10^{-13} at 308 K and the reaction is displaced to the right in a system formed with butylpyridinium chloride.

Attractive interactions occur between acids such as R^+, $Al_2Cl_7^-$, $AlCl_4^-$, and Cl^-. A sign of this is the increase in viscosity.

Among the kind of probable structures are (Im^+ = imidazolonium),

$$Cl^- - Im^+ - Cl^-$$

$$Cl^- - Im^+ - AlCl_4^-$$

$$AlCl_4^- - Im^+ - Al_2Cl_7^-$$

and NMR measurements are consistent with these structures.

5.12.2. Electrochemical Windows in Low-Temperature Liquid Electrolytes

Whereas 1.2 V is the fundamental electrochemical window in aqueous solutions, more than 3 V is available in some of the currently discussed systems. As a concrete example, Al can be electrochemically deposited from $Al_2Cl_7^-$ at –0.4 V with an Al electrode taken as reference. The evolution of Cl_2 occurs at +2.5 V against the same reference electrode. Thus, the window is 2.5 – (–0.4) = 2.9 V.

5.12.3. Organic Solutes in Liquid Electrolytes at Low Temperatures

A number of organic solutes undergo reaction with $AlCl_3$. For example, aromatic amines dissolve in the room-temperature molten electrolyte butylpyridinium chloride. Triphenylamine (TPA) shows two stages of oxidation upon polarographic examination. The first reaction involves the formation of the radical cation. On the other hand, in a neutral or acid melt, the behavior is different and probably

$$TPA \rightarrow TPA^{+} + e$$

followed by

$$TPA^{+} \rightleftharpoons TPB + H^{+}$$

$$TPB \rightarrow TPB^{2+} + 2\,e^{-} \tag{5.143}$$

5.12.4. Aryl and Alkyl Quaternary Onium Salts

Although researches on room-temperature systems of $AlCl_3$ and butylpyridinium chloride have contributed a new vista to the physical chemistry and electrochemistry of molten salts, the systems carry with them the disadvantage that few of the materials mentioned are commercially available. Correspondingly, molten systems containing the easily available pyridinium cation have to pay a penalty in that pyridinium is easy to reduce.

On the other hand, *onium* salts based in N (or S or P) are commercially available in abundance. Extended by introducing a variety of alkyl and aryl groups, they present a large variety of properties, being all liquid electrolytes.

TABLE 5.40

Sources of Quaternary and Ternary Onium Compounds

Compound	Source	Purity (%)
Trimethylphenylammonium chloride	Kodak	98
Triethylphenylammonium iodide	Kodak	99
Trimethylbenzylammonium chloride	Sherex (in water)	60–65
Triethylbenzylammonium chloride	Kodak	99
Dimethylethylbenzylammonium chloride	Sherex	N.A.
Dimethylphenylbenzylammonium chloride	Pfaltz and Bauer	N.A.
Methyldiphenylsulfonium tetrafluoroborate	Lancaster Synthesis	98
Ethyldiphenylsulfonium tetrafluoroborate	Lancaster Synthesis	98
Trimethylphenylphosphonium iodide	K and K	N.A.
Tetraethylammonium chloride	Kodak	98
Tetrabutylammonium chloride	Kodak	92
Tetrapentylammonium chloride	Kodak	97
Tetrahexylammonium chloride	Kodak	—
Tetraheptylammonium chloride	Kodak	99
Trioctylpropylammonium chloride	Kodak	90
Tetrabutylammonium perchlorate	Kodak	99

Source: Reprinted from S. D. Jones and G. E. Blomgren, *J. Electrochem. Soc.* **136**: 424, 1989.
[a]N.A. = not available

TABLE 5.41

Melting Points of Various Ammonium Chlorides

Chloride	Melting Point (K)
Tetramethylammonium chloride	693
Tetrapentylammonium chloride	414
Tetrahexylammonium chloride	385
Tetraheptylammonium chloride	264
Trimethylphenylammonium chloride	510(subl.)

Source: Reprinted from S. D. Jones and G. E. Blomgren, *J. Electrochem. Soc.* **136**: 424, 1989.

TABLE 5.42

Room-Temperature Conductivities of Mixtures of Various Ammonium Chlorides and Aluminum Chloride in a 1:2 Mole Ratio

Chloride	Conductivity (mS cm^{-1})
Trimethylphenylammonium chloride	3.40
Triethylphenylammonium chloride	1.29
Trimethylbenzylammonium chloride	1.46
Triethylbenzylammonium chloride	0.32
Dimethylethylbenzylammonium chloride	0.63
Dimethylphenylbenzylammonium chloride	Very low

Source: Reprinted from S. D. Jones and G. E. Blomgren, *J. Electrochem. Soc.* **136**: 424, 1989.

TABLE 5.43

Behavior of Sulfonium and Phosphonium Compounds with Aluminum Chloride at a 1:2 Mole Ratio

Compound	Observation
Methyldiphenylsulfonium tetrafluoroborate	Liquid at room temperature; conductivity = 0.76 mS cm^{-1} at 300 K
Ethyldiphenylsulfonium tetrafluoroborate	Melting point slightly above room temperature; conductivity = 2.1 mS cm^{-1} at 308 K
Trimethylphenylphosphonium iodide	Melting point slightly above room temperature; conductivity = 3.0 mS cm^{-1} at 308 K

Source: Reprinted from S. D. Jones and G. E. Blomgren, *J. Electrochem. Soc.* **136**: 424, 1989.

The breadth of the compounds available can be gauged from Table 5.40. Some of the remarkable properties of these compounds are exhibited in Tables 5.41 to 5.43, taken here from the pioneering work of Jones and Blomgren in 1989. Melting points go down as the complexity increases. Tetraheptylammonium chloride melts at 264 K! However, the more complex salts have high viscosity and low conductivity. Conductivity is reduced by the presence of aryl groups. Mixed with $AlCl_3$, liquid eutectics at 198 K can be obtained.

There is probably an interaction between the π electrons of the arylammonium cations and the $Al_2Cl_7^-$ anions. This would account for the lowering of the melting point by these salts. Furthermore, these π bonds may well be *delocalized*, again helping to lower the melting point. The smaller organic groups provoke lower viscosity and hence (Walden's rule) a higher conductance.

The electrochemical window obtainable with the low-temperature liquid onium systems is about 3 V, which is about the same as that with high-temperature liquid NaCl. NMR measurements of these low-temperature electrolytes can be informative. For example, $EtAlCl_2$-containing melts can be examined by registering the NMR characteristics of probe groups (e.g., ^{13}C and ^{47}Al). The NMR relaxation provides information on interactions in systems such as MEI^+ groups and various properties of $EtAlCl_2$. It is found that MEICl-$AlCl_3$ exhibits interaction of the MEI^+ ring and $EtAlCl_2$. Even greater electrochemical windows, up to 4 V, are available, for example, with an alkyl-substituted aromatic heterocyclic cation and trifluoromethane sulfonate.

5.12.5. The Proton in Low-Temperature Molten Salts

H^+ is regarded as a contaminant in molten salt work. It arises from the pervasive presence of water vapor. It can be studied, e.g., with $AlCl_3$-1-ethyl-3-methylimidazolonium chloride (ImCl). The methods used have been Fourier transform infrared (FTIR) and NMR spectroscopies and standard electrochemical techniques.

The proton is found to exist in these melts as HCl and HCl_2^-. The equilibrium between these two forms can be studied by the methods stated (particularly NMR) and turns out to be much in favor of the anion. If the concentration of HCl is above 0.5 in mole fraction, various complex ions ($H_nCl_{n+1}^-$, $n > 2$) begin to form. Thus, it turns out that H^+ can be taken up in these salts, but only in the form of complex anions.

Further Reading

Seminal

1. F. Hurley and J. P. Wier, "Electrodeposition of Metals from Fused Quaternary Ammonium Salts," *J. Electrochem. Soc.* **98**: 203 (1951).
2. H. L. Chum, V. R. Koch, L. L. Miller, and R. A. Osteryoung, "An Electrochemical Scrutinity of Organometallic Iron Complexes and Hexamethylbenzene in a Room Temperature Molten Salt," *J. Am. Chem. Soc.* **97**: 3265 (1975).

Review

1. R. A. Osteryoung, "Organic Chloraluminates Ambient Temperature," in *Molten Salts*, G. Mamantov, ed., *NATO ASI Series C* **202**: 329 (1987).

Papers

1. J. Jeng, R. D. Allendorfer, and R. A. Osteryoung, *J. Phys. Chem.* **96**: 3531 (1992).
2. J. L. E. Campbell and K. E. Johnson, *Proc. Electrochem. Soc.* **92-16**: 317 (1992).
3. P. C. Truelove and R. A. Osteryoung, *Proc. Electrochem. Soc.* **92-16**: 303 (1992).
4. P. C. Truelove and R. T. Carlin, *Proc. Electrochem. Soc.* **93-9**: 62 (1993).
5. R. T. Carlin and T. Sullivan, *J. Electrochem. Soc.* **139**: 144 (1992).
6. T. L. Riechel and J. S. Wilkes, *Proc. Electrochem. Soc.* **92-16**: 351 (1992).
7. T. L. Riechel and J. S. Wilkes, *J. Electrochem. Soc.* **140**: 3104 (1993).
8. R. A. Mantz, R. C. Truelove, K. T. Carlin, and R. A. Osteryoung, *Inorg. Chem.* **34**: 3846 (1995).
9. W. J. Gau and I. W. Sun, *J. Electrochem. Soc.* **143**: 914 (1996).
10. E. Hondrogiannis, *J. Electrochem. Soc.* **142**: 1758 (1995).
11. W. R. Carper, *Inorg. Chim. Acta* **238**: 115 (1995).
12. K. E. Johnson, *Can. J. Chem.* **73**: 593 (1995).

5.13. MIXTURES OF LIQUID OXIDE ELECTROLYTES

5.13.1. The Liquid Oxides

Fused salts (and mixtures of fused salts) are not the only type of liquid electrolytes. Mention has already been made of *fused oxides* and in particular mixtures of fused oxides. A typical fused oxide system is the result of intimately mixing a nonmetallic oxide (SiO_2, GeO_2, B_2O_3, P_2O_5, etc.) and a metallic oxide (Li_2O, Na_2O, K_2O, MgO, CaO, SrO, BaO, Al_2O_3, etc.) and then melting the mixture. The system can be represented by the general formula $M_xO_y - R_pO_q$, where M is the metallic element and R is the nonmetallic element.

Why give these liquids special consideration? Are not the concepts developed for understanding molten salts adequate for understanding molten oxides? The essential features of fused salts emerge from models of the liquid state. There is no doubt that the fluidity of molten salts demands a model with plenty of free space, and a model based on density fluctuations that are constantly occurring in all parts of the liquid seems about the best way to think of the inside of a molten liquid. Is the same dependence on the opening up of temporary vacancies an adequate basis for explaining the behavior of the fused oxides?

5.13.2. Pure Fused Nonmetallic Oxides Form Network Structures Like Liquid Water

Some of the special features of molten oxides must now be described for it is these features that do not permit the hole model of ionic liquids to be applied to fused oxides

TABLE 5.44
Specific Conductivity of Water and Molten SiO_2 and NaCl near the Melting Point

Substance	κ (S cm^{-1})	Temp (K)
SiO_2	7.7×10^{-4}	2073
NaCl	3.6	1074
H_2O	4×10^{-8}	291

in the same way it is applied to molten salts. The first interesting feature of molten silica, is that its conductivity is more like that of water (i.e., molten ice) than that of fused NaCl (Table 5.44).

The dissimilarity in the conductivities of liquid NaCl, on the one hand, and liquid water and liquid silica, on the other, is of fundamental importance. When NaCl is fused, the ionic lattice (the three-dimensional periodic arrangement of ions) is broken down (see Section 5.1.2) and one obtains an *ionic liquid*. When ice is melted, the tetrahedrally directed hydrogen bonding involved in the crystal structure of ice is partially retained. Thus, water is not a collection of separate water molecules but an association (based on hydrogen bonding) of water molecules in a three-dimensional network. The network, however, does not extend indefinitely. There is a periodicity and only short-range order, implying a certain degree of bond breaking. It is this network structure that is responsible for the small mole fraction of free ions (H^+ and OH^-) in water, in contrast to the almost total absence of any ion association (into pairs, complexes, etc.) in liquid NaCl. This great difference in the concentration of charge carriers is responsible for a difference in the specific conductivities of liquid NaCl (high charge carrier concentration) and liquid water (very low charge carrier concentration) that is several orders of magnitude.

The specific conductivities of water and of fused silica are both very low. This suggests that the structures of crystalline water [Fig. 5.65(a)] and crystalline silica [Fig. 5.65(b)] have much in common. Each oxygen atom in ice is surrounded tetrahedrally by four other oxygens, the oxygen–oxygen bonding occurring by a hydrogen bridge (the hydrogen bond). In crystalline silica, there are SiO_4 tetrahedra occurring through an oxygen bridge. The different forms of ice and the different forms of silica (Fig. 5.66) correspond to different arrangements of the tetrahedra in space.

It is reasonable therefore to consider that fused silica resembles liquid water. Just as liquid water retains from the parent structure (ice) the three-dimensional network but not the long-range periodicity of the network, one would expect that liquid silica also retains the continuity of the tetrahedra, i.e., the space network, but loses much of the periodicity and long-range order that are the essence of the crystalline state. This model of fused silica, based on keeping the extension of the network but losing the translational symmetry of crystalline silica, implies a low concentration of charge

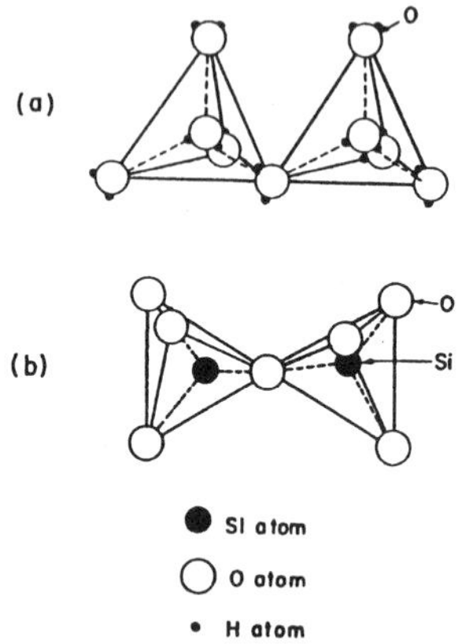

Fig. 5.65. The similarity between the basic building blocks of (a) ice and (b) crystalline silica structures.

carriers in pure liquid silica and therefore low conductivity in comparison with a molten salt (see Tables 5.44 and 5.45).

5.13.3. Why Does Fused Silica Have a Much Higher Viscosity Than Do Liquid Water and the Fused Salts?

It has just been argued that the conductivities of simple ionic liquids, on the one hand, and liquid silica and water, on the other, are vastly different because a fused salt is an unassociated liquid (it consists of individual particles) whereas both molten silica and water are associated liquids with network structures. What is the situation with regard to the viscosities of fused salts, water, and fused silica? Experiments indicate that whereas water and fused NaCl have similar viscosities not far above the melting points of ice and solid salt, respectively, fused silica is a highly viscous liquid (Table 5.46). Here then is an interesting problem.

One successful theory of transport processes in liquids are based on elementary acts, each act consisting of two steps: (1) holes are formed and (2) particles jump into these holes (see Section 5.7.4). For fused salts and other nonassociated liquids, this theory was successful in explaining the movements and drift of particles although it clashed with molecular dynamics calculations that seemed to favor a shuffle-along mechanism for transport. The *mean* volume of a hole is determined by the surface tension as follows [*cf.* Eq. (5.44)]:

$$\langle v_h \rangle = 1.6\left(\frac{kT}{\gamma}\right)^{3/2} \tag{5.144}$$

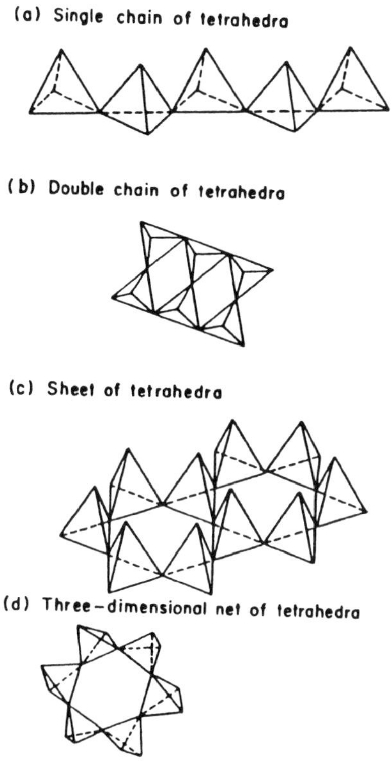

Fig. 5.66. Forms of silicates resulting from different ways of linking up SiO_4 tetrahedra: (a) single chain of tetrahedra; (b) double chain of tetrahedra, as in asbestos; (c) sheets of tetrahedra, as in clay, mica, and talc; and (d) networks of tetrahedra, as in ultramarine.

from which it turns out that in *unassociated fused salts*, the size of the holes is roughly equal to the size of individual ions. In those simple liquids, holes can receive ions of the fused salt which jump into them. Furthermore, in simple ionic liquids, the free energy of activation for the jumping of ions into holes is much less than that of the free energy for forming the holes. Once the hole is formed in a fused salt, the jump into the hole is three to ten times easier than forming the hole.

The surface tension of fused silica is only about three times that of fused sodium chloride. Hence [see Eq. (5.144)] in fused silica also there would be many holes of atomic dimensions, as for fused ionic liquids.

TABLE 5.45
Electric Conductance in the Liquid Silicates

Cation	Composition $XM_xO - YSiO_2$, $X:Y$	$\kappa_{1750\,°C}$ (S cm^{-1})	$\Lambda_{1750\,°C}$	ΔH^* (kcal g^{-1} ion^{-1})[a]
K^+	1:2	1.5	71.8	8.2
	1:1	2.4	82.7	8.0
Na^+	1:2	2.1	83.3	12.0
	1:1	4.8	126.0	13.5
Li^+	1:2	2.5	77.8	11.6
	1:1	5.5	109.0	10.6
	2:1	23.2	332.0	9.6
Ba^{2+}	1:2	0.18	6.4	33.2
	1:1	0.60	16.2	17.5
	2:1	1.32	29.9	9.0
Sr^{2+}	1:2	0.21	7.7	36.0
	1:1	0.63	15.7	26.7
	2:1	1.4	26.8	17.0
Ca^{2+}	1:2	0.31	11.4	30.0
	1:1	0.83	18.4	20.0
	2:1	1.15	18.8	20.0
Mn^{2+}	1:2	0.55	18.2	24.0
	1:1	1.8	35.1	16.0
	2:1	6.3	85.5	12.0
Fe^{2+}	1:1	1.82	44.0	15.0
Mg^{2+}	1:2	0.23	6.5	34.0
Mg^{2+}	1:1	0.72	12.2	24.0
	2:1	2.15	24.7	17.0
Al^{3+}	10 wt%	3.10^{-3}	0.20	22.0
Ti^{4+}	10 wt%	6.10^{-4}	0.05	35.7

[a] 1 cal = 4.184 J

TABLE 5.46
Viscosities of Fused Silica, Water, and Fused NaCl

Substance	Temp. (K)	Viscosity (poise)
Liquid SiO_2	1993	3×10^6
Water	298	9×10^{-3}
Fused NaCl	1123	13×10^{-3}

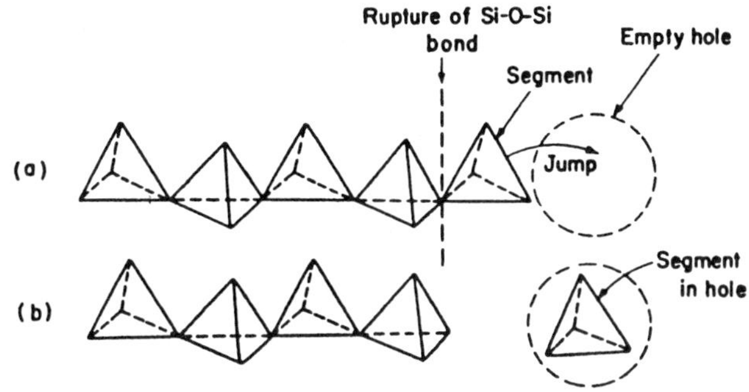

Fig. 5.67. Schematic diagram to show a segment of an SiO_4 chain breaking off and jumping into an empty hole: (a) before jump and (b) after jump.

In simple ionic liquids (e.g., NaCl), the holes are atom sized; *so are the jumping ions*. Jumping is easier than making holes. What particles can jump into holes in fused silica? Obviously, large chunks of a silicate network (for suggestions, see later discussion) cannot jump into holes that are about the same size as those that receive Na^+ and Cl^-, etc. How then can jump-dependent transport processes occur? There may be a way. Small segments (one to a few atoms in size) can break off from the network and these pieces (segments) can do the jumping (Fig. 5.67).

A comparison between transport processes in simple fused salts and in fused SiO_2 is interesting. In molten salts, the jump was easier than making the hole but nevertheless many voids of different sizes were shown to form in the liquid. The rate of the process was controlled more by the rate of hole formation than the easier jumping. However, with molten silica, the balance of influences is different. Holes are as easily formed as in the ionic liquids because the rate of hole formation is controlled by the vibrations of the atoms relative to each other, but it is difficult to produce individual small particles because this would involve rupturing strong Si–O–Si bonds holding the network together (Table 5.47). Therefore, in the silicates, the rate-determining process is the bond-rupture step. While in simple ionic liquids the experimental activation energy[35] for a transport process, such as a viscous flow, is determined predominantly by the enthalpy of hole formation, in associated liquids with network structures it is determined entirely by the energy required to break the bonds of the network to produce a "flow unit" sufficiently small to jump into the relatively easily made hole.

[35]It will be recalled that it was decided that the quantity obtained from the slope of the log η versus $1/T$ curve would be termed an *energy* of activation irrespective of whether it pertains to constant-pressure or constant-volume conditions, though under the former it is an enthalpy and under the latter, it is an energy.

TABLE 5.47

Comparison between Transport Processes in Simple Fused Salts and in Liquid Silica

	System	
	Fused Salt	Liquid Silica
Essence of situation	Particles waiting to jump into holes	Holes waiting for small enough segments of networks
Rate-determining step	Hole formation	Rupture of bonds between segments and network
Energy of activation for transport process (approx.)	$\Delta H_{\text{hole formation}}$	$\Delta H_{\text{bond rupture}}$

In this difference of the rate-controlling mechanism for flow lies the answer to the next question. Why is the viscosity of associated water similar to that of molten NaCl, which has no network? The difference in the viscosity behavior of water and of fused silica lies in the ease with which segments can be broken off the two networks. The Si–O–Si chemical bonds are much more difficult to rupture than the O–H–O hydrogen bonds (Table 5.48). Thus, flow units—probably individual H_2O molecules—are so easily produced in water that the holes do not have much of a wait; an ease of flow, high fluidity or low viscosity, results. This is not the case with fused silica because of the much higher bond-breaking energy, and a high viscosity results.

Some support for the idea that the viscous-flow properties of associated liquids such as liquid silica and water are determined by the step of bond breaking rather than that of hole formation comes from the experimental plots of log η versus $1/T$. These plots suggest a slight trend away from linearity (Fig. 5.68), which is not the case for fused salts (Fig. 5.29). For water also, the plot is curved slightly, with the experimental energy of activation for viscous flow $E_{\eta,p}$ decreasing with increasing temperature. The explanation for this phenomenon is as follows. Because the energy of activation for viscous flow depends upon the breaking of bonds and because, according to the Boltzmann distribution, the fraction of broken bonds increases with temperature, the

TABLE 5.48

Heat of Dissociation of Si–O and O–H Bonds

Bond	Heat of dissociation (kcal mol^{-1})[a]
Si–O–Si	104
O–H–O, hydrogen bond	5

[a]1 cal = 4.184 J.

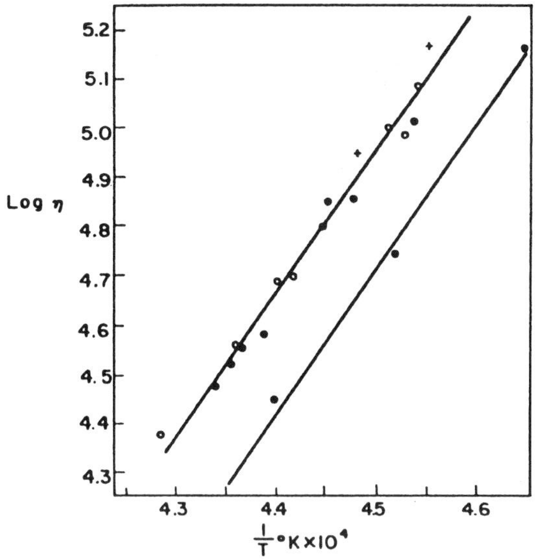

Fig. 5.68. Two independent sets of data for the difficultly determinable viscosity η of liquid SiO_2. Both suggest a slight tendency to curve in the sense that the energy of activation becomes higher at the lower temperatures.

fraction required to be broken by the shearing force in viscous flow decreases with an increase in temperature and, correspondingly, there is a decreasing energy of activation with increasing temperature.

To summarize: Unlike fused salts, mixtures of fused oxides are *associated liquids*, with extensive bonding between the individual molecules or ions. In fused oxides, hole formation occurs but it is not the step that determines the rate of transport processes. It is the rate of production of individual small jumping units that controls them. This conclusion makes it essential to know what (possibly different) entities are present in fused oxides and what are the *kinetic* entities. In simple fused salts, the jumping particles are already present (they are the ions themselves); the principal problem is the structure of the empty space or free volume or holes, and the properties of these holes. In molten oxides, the main problem is to understand the structure of the macrolattices or particle assemblies from which small particles break off as the flow units of transport.

5.13.4. Solvent Properties of Fused Nonmetallic Oxides

If fused silica is a three-dimensional, aperiodic network, all the atoms are to some extent joined together, i.e., the liquid is a giant molecule. Can ions dissolve in such a

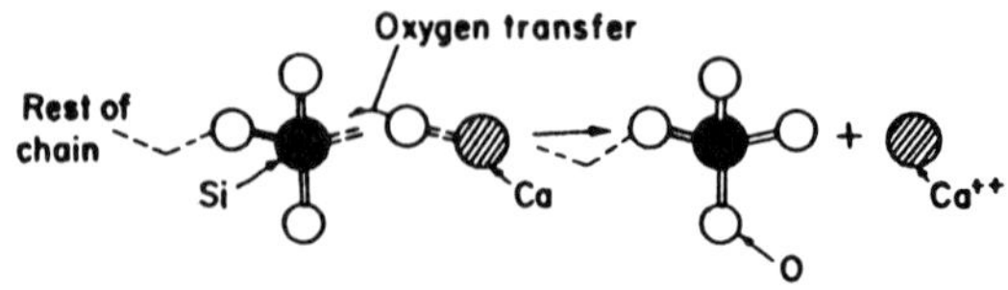

Fig. 5.69. The interaction between a metal oxide and a silicon atom of the silica network.

structure? Water also is a network structure like SiO_2, and ions dissolve in water. Hence, liquid SiO_2 may well be expected to have solvent properties leading to the production of ionic solutions.

Water, it may be recalled (Chapters 2 and 4), has two modes of solvent action, depending on the nature of the added electrolyte. The water can contact an ionic crystal (e.g., NaCl), detach the ions from the lattice through the operation of ion–dipole (or ion–quadrupole) forces, and convert them to hydrated ions (Chapter 2).

The water can also interact *chemically* with a potential electrolyte (e.g., CH_3COOH). The hydrogen atom forming part of the hydroxyl group of the organic acid does not differentiate between the oxygen atoms of the water network and that of the OH group. The hydrogen of the OH group detaches itself from the organic acid. Two ions are thus formed: a hydrogen bonded to a water molecule from the solvent and an organic anion. This mode of solvent action is a proton-transfer or acid–base reaction.

The type of solvent action that fused nonmetallic oxides have on metallic oxides may be likened to the second type of dissolution process, i.e., proton-transfer reactions. The process may be pictured as follows. The oxygens cannot discriminate between the metal ions M^+ (of the metallic oxide), with which they have been associated in the lattice of a metal oxide before dissolution, and the oxygen atoms of the SiO_4 tetrahedra contained in the solvent—fused silica. The oxygen atoms sometimes therefore leave the metal ions and associate with those of the tetrahedra. Dissolution has occurred with a type of oxygen-transfer reaction (see Fig. 5.69).

There is a further analogy between the solvent actions exercised by water and a fused nonmetallic oxide. Just as water dissolves an electrolyte at the price of having its structure disturbed, so also the reaction resulting from the addition of a metallic oxide to a fused nonmetallic oxide like silica is equivalent to a bond rupture between the SiO_4 tetrahedra (Fig. 5.70). Solvent action occurs in fused oxide systems along with a certain breakdown of the network structures present in the pure liquid solvent (e.g., in pure liquid silica).

5.13.5. Ionic Additions to the Liquid-Silica Network: Glasses

An interesting aspect of molten oxide electrolytes may be mentioned at this point. Some liquids can appear to be solids, i.e., some solids are really liquids of such high

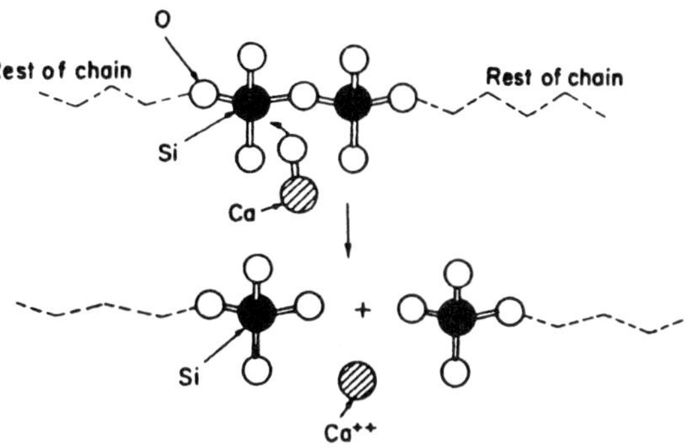

Fig. 5.70. The oxygen-transfer reaction leads to a rupture between the SiO_4 tetrahedra.

viscosity that no significant flow occurs over tens or hundreds of years. These solidlike liquids are called *glasses*. The structure breaking that has just been described is an aspect of the basic mechanism behind the formation of glasses, which might be regarded as "cold molten silicates."

When ions with a relatively large radius and small peripheral field are added to liquid silica, they produce structure breaking in the network. This is shown in a one-dimensional way in Figs. 5.69 and 5.70. With an increase in the number of ruptures in the network, there is an increase in the number of "free" or "dangling" ends of the ruptured network. The network becomes increasingly *distorted* with an increase in the mole fraction of the metallic oxide present.

If the "broken-down network" is at a sufficiently high temperature, it is known as a *liquid silicate*. Such a system results, for example, from adding an alkali oxide (e.g., Na_2O) in low concentration to SiO_2. At these temperatures the system can be distinctly a *liquid*, though the viscosity near the melting point may be, for example, 427 poises (P) (at 1820 K), whereas that of water at its melting point is about 0.01 P. When the temperature is dropped, the liquid silicate attempts to "freeze," but, to do this, the long-range order of the crystalline silicate has to be reestablished. The establishment of order, however, requires rearrangement of the structure; i.e., the kinetic units of the broken-down network must move into the sites corresponding to order. Were these particles, or kinetic units, simple, they would be agile, i.e., their movements would be easy, the viscosity would be low, and the restoration of crystalline order would be accomplished almost immediately. A crystalline solid with a sharp melting point would result.

However, this reorganization is precisely what is not quickly accomplished by the entities in the broken-down networks in liquid silicates. The entities[36] resulting from

[36]The nature of these large anionic entities that exist in glass-forming silicates is discussed in Section 5.13.8.

the rupturing of three-dimensional networks when metal oxides are added are large and sluggish. They cannot get into line in time; the viscosity is too high. The loss of thermal energy during cooling catches them still out of position as far as the regular three-dimensional arrangement of the crystalline silicate is concerned. Then it is too late, for as the temperature drops still further, they are still less likely to be able to get back into line before the structure forming about them is too inhibiting to allow further movement. The loss of thermal energy freezes the structure of the liquid silicate—a glass is formed. It is a "frozen liquid," i.e., a liquid that has been supercooled to such a high viscosity that it seems to have the essential requirement of a solid—absence of flow. A beam of X-rays, however, would reveal an essential characteristic of the liquid state, namely, the absence of long-range periodicity.

If, however, sufficient time is allowed (e.g., a few hundred years), a sufficient number of the units of the broken-down network will get back into line. Long-range order will be partly reestablished; the glass will "deglassify" or devitrify, as it is called, and will often split up and break.

How is the liquid-silicate network affected by the addition of various types of ions in the production of the peculiar and complicated kind of pure electrolyte, a glass? It is the answer to this structural question that provides the basis for the understanding of the *glassy* state.

5.13.6. The Extent of Structure Breaking of Three-Dimensional Network Lattices and Its Dependence on the Concentration of Metal Ions Added to the Oxide

The extent of breakdown of the network structures present in a pure liquid solvent can be viewed in the following way. The SiO_4 tetrahedron is accepted as the basic structural unit in a mixture of metallic oxide and fused silica. However, are the tetrahedra linked together at all and, if so, how are they linked together? What is the number of links per silicon? Water molecules are the basic structural unit in an aqueous solution, but extensive linkage and intermolecular bonding occurs in pure water. How is this linkage in molten SiO_2 affected by the presence of new ions dissolved in it (Section 5.1)?

In a fused-oxide system, the metal oxide (e.g., Na_2O) is the structure breaker; in aqueous solutions, the electrolyte (e.g., NaCl) is the structure breaker. Does the extent of structural breakdown of the continuous Si–O network present in pure silica before the addition of MO depend on the concentration of MO? The extent of breakdown must indeed depend on the concentration of the structure breaker; then one would expect that properties that depend on the size and nature of the structures present would also be concentration dependent.

In fused-oxide systems, a simple way of expressing the concentration of the metal oxide M_xO_y in the fused nonmetallic oxide R_pO_q is sometimes used. This involves the so-called "O/R ratio." The O/R ratio is related to the mole fraction of the metallic

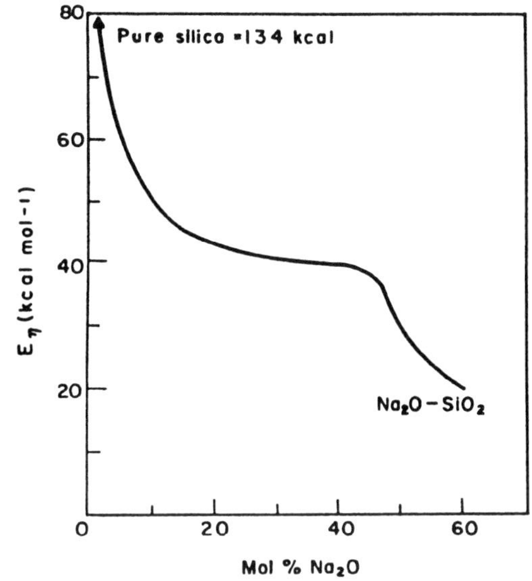

Fig. 5.71. The variation of the energy of activation for viscous flow in an $Na_2O + SiO_2$ melt as a function of the mole percent of Na_2O. (1 cal = 4.184 J)

oxide. For example, an O/Si ratio of 4 in a system of Li_2O and SiO_2 is obtained when the Li_2O has a mole fraction of 66% (i.e., $2Li_2O + SiO_2$ has four Os to one Si).

One way of probing the sizes of structures present in fused-oxide systems and their variation with the mole fraction of MO added to the nonmetallic oxide is through the variation of the ease of flow with composition. The viscosity of the system must be keenly sensitive to the size and nature of the kinetic entities present because it is these entities that must make the jumps from site to site involved in viscous flow.

Experimental results on the variation of the energy of activation for viscous flow, E_η, as a function of the mole percent of the metal oxide are shown in Fig. 5.71. The basic feature of the results appears to be a very high (~630 kJ mol^{-1} or 150 kcal mol^{-1}) energy of activation for viscous flow of the pure nonmetallic oxide and then a rapid fall with addition of the metallic oxide, whereupon there is a leveling off to a value of about 167 kJ mol^{-1} (40 kcal mol^{-1}), which remains relatively unchanged between about 10 and 50 mol% of Na_2O in SiO_2. This behavior can be used (Section 5.13.7) as a touchstone in deciding between alternative models for the structural changes accompanying changes in metal-ion concentration.

The structural theories presented will be in terms of the liquid silicates because most research on molten oxides has been done with them, but one can extend the basic structural ideas to fused-oxide systems involving metal oxides dissolved in B_2O_3 and

P_2O_5 and probably to most liquid electrolytes in which there are largely continuous network structures at very low concentrations of M_xO_y.

5.13.7. Molecular and Network Models of Liquid Silicates

A naive view of events after the addition of metallic oxides to molten silica is to think of different uncharged molecular species, the species changing with the mole fraction of the metallic oxide. This view has to be given up quickly because studies of *mixtures* of M_xO_y and SiO_2 show that these systems are highly conducting and therefore are rich in charge carriers (Table 5.45). One has to suggest ionic, rather than molecular, structures.

A second attempt at interpreting the structure of liquid silicates starts with a consideration of the curve of E_η versus the mol% of M_xO_y (Fig. 5.71). It is in terms of the gradual breakdown of the three-dimensional network of fused silica. Just as there is *thermal* bond breaking on going from crystalline to fused silica, one can consider that with the addition of, say, Na_2O, the additional O atoms cause Si–O–Si bonds in the originally continuous network of SiO_2 to break. This produces structure breaking to various composition-dependent degrees. A mole fraction of 66% for M_2O implies, as already stated, an O/Si ratio of 4 and must therefore be considered a composition at which the only anions are SiO_4^{4-} ions.[37]

What model can be suggested for the corresponding structural changes "inside" the fused-oxide system for the M_2O mole-fraction range from 0 to 66%? From the fact that there is such a sharp fall in the energy of activation for viscous flow between zero and about 10 mol% of M_2O (Fig. 5.71), one must think of a radical change (over this composition range) in the difficulty of causing a kinetic entity or flow unit to break off from the rest of the structure and jump from site to site in the random walk that is the basis of diffusion (Section 5.7.4). The model must of course contain the explanation of the generation of more free ions to account for an increase in conductivity with an increasing amount of MO. The anions postulated as dominant for a given O/R ratio must meet some exacting requirements. Thus (1) they must have formulas consistent with the overall O/R ratio; (2) they must have a total charge that compensates for the total charge of the cations and thereby ensures overall electroneutrality; and (3) they must be shaped in a way consistent with the bond angles, particularly the R–O–R angle, shown from X-ray data in the corresponding solids.

The broad approach used by Zachariasen in the *network* model of liquid silicates is to break down the network present in the pure fused nonmetallic oxide thus

---

[37]Correspondingly, for M_2O > 66 mol%, there are oxygen atoms in excess of the ability of the Si atoms present to coordinate them (O/Si ratio > 4). Hence, liquids with such compositions probably contain SiO_4^{4-} and O^{2-} entities in addition to the M^+ cations present.

TABLE 5.49

Network Theory of Liquid Silicate Structure

Range of Composition (mol% M_2O)	Silicate Entities Present
0	Continuous 3-D network of SiO_4 tetrahedra with small degree of thermal bond breaking
0–33	Essential 3-D network of SiO_4 tetrahedra with number of broken bonds equal to number of added O atoms from M_2O; end of 3-D boundary at 33%
33	"Infinite" 2-D sheets of SiO_4 tetrahedra; M^+ ions and O^- ions between sheets
33–50	Region of sheets and some chains of tetrahedra
50	Chains of infinite length
50–60	Chains of decreasing length
66	SiO_4^{4-}

$$\text{Three-dimensional network with some thermal bond breaking (e.g., } SiO_2) \underset{\text{added}}{\overset{M_2O}{\longrightarrow}} \text{two-dimensional sheets}$$

$$\underset{\text{added}}{\overset{\text{more } M_2O}{\longrightarrow}} \text{one-dimensional chains} \underset{\text{more } M_2O}{\overset{\text{still}}{\longrightarrow}} \text{simple } SiO_4^{4-} \text{ monomers}$$

The details of the network theory of liquid-silicate structures, which was first suggested to explain the glassy state, are presented in Table 5.49. The chief defect of this model is that it argues for very large changes in the heat of activation for viscous flow in the composition range of 33 to 66%. However, this is not what is indicated in Fig. 5.71, where the large change comes after 10% oxide. This is because in the network model the size and shape of the kinetic unit—the jumping entity—is supposed to undergo a radical change in the composition range of 33 to 66%. Thus (Table 5.49), sheets are being broken up into chains. The kinetic unit of flow would therefore be expected, in this model, to change radically in size over *this* composition range, and this diminution in size of the flowing unit would be expected to make the heat of activation for viscous flow strongly dependent on composition in *this* composition range (33 to 66% M_2O). In contrast to the expectation, however, the E_η changes by only 25% over the composition range of 10 to 50% M_xO_y (Fig. 5.71), whereas from 0 to 10% the change in energy of activation is about 200% (Fig. 5.71).

Another inconsistency of the earlier network model concerns the implications that it has for phenomena in the composition region around 10% M_xO_y. This is an important composition region. Experimentally, whether one measures the composition dependence of the heat of activation for viscous flow, of expansivity, of compressibility, or

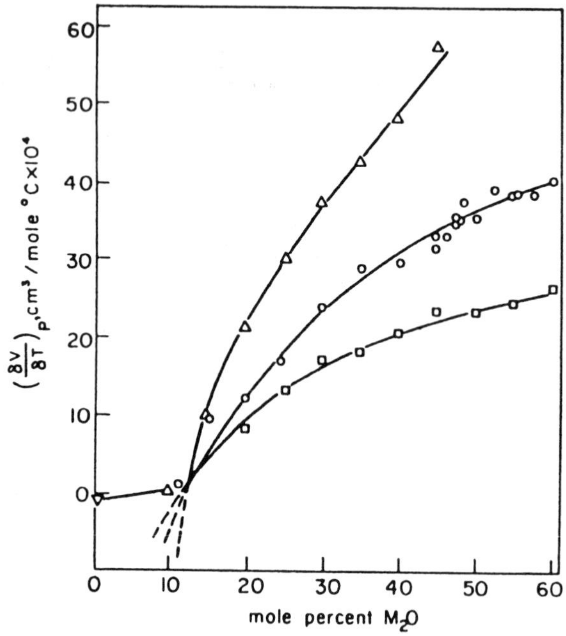

Fig. 5.72. The sharp change in the expansivity of M_2O-SiO_2 melts around the 10 mol% of M_2O composition; (△) $K_2O + SiO_2$, (○) $Na_2O + SiO_2$, (□) $Li_2O + SiO_2$, (▽) SiO_2.

of other properties, in all cases there is always a sharp and significant change (Fig. 5.72) around the 10% M_xO_y composition, which indicates a radical structural change at this point. However, this composition has no special significance at all, according to the former network model. Thus, although the network model served well in an earlier stage of the development of the theory of glasses, one must reconsider the situation and develop a model that corresponds more closely, in the expectations it produces, to the experimental facts.

5.13.8. Liquid Silicates Contain Large Discrete Polyanions

Consider the situation as one *decreases* the O/R ratio, i.e., decreases the mole percent of the metallic oxide M_xO_y. Between 100 and 66% there is little need for special modeling because the quadrivalency of silicon and the requirements of stoichiometry demand that the ionic species present be monomers of SiO_4^{4-} tetrahedra. It is in the composition range of 66 to 10% M_xO_y that the network model fails (see Section 5.13.7) in the face of facts.

What are the requirements of a satisfactory structural model for ionic liquids in this composition range? First, since transport-number determinations show that the conduction is essentially cationic, the anions must be large in size compared with the cations (see Sections 4.4.8 and 4.5). Second, the marked change in properties (e.g., expansivity) occurring at 10% M_xO_y indicates radical structural changes in the liquid in the region of this composition. From the sharp rise of the heat of activation for the flow process to such high values (toward 420 kJ mol^{-1} at compositions below 10 mol% M_xO_y) with decreasing mole percent of M_xO_y in this composition region, one suspects that the structural change that is the origin of all the sudden changes near 10 mol% M_2O must involve sudden aggregation of the Si-containing structural units into networks. The difficulty of breaking bonds to get a flowing entity out of the network and into another site explains the very sluggish character of the liquid at 10 or less mol% of M_xO_y. The potential flow unit (perhaps here SiO_2 itself) has to break off four bonds to flow. Finally, from 66 to 10% *there must be no major changes in the type of structure*, except some increase in the *size* of the entities, because there is only a small increase in the heat of activation for viscous flow over this composition region (*cf.* Fig. 5.71). This relative constancy of the heat of activation for flow over this composition region means that the various structural units present can become the kinetic entities of flow over this region without great change of the energy involved; i.e., the flow units present over the composition range must be similar.

The construction of a model (Table 5.50) can therefore start from the SiO_4^{4-} ions present in the *orthosilicate* composition (66% M_xO_y). With a decrease in the molar fraction of M_xO_y, the size of the anions must increase to maintain the stoichiometry. One can consider that there is a joining up, or polymerization, of the tetrahedral SiO_4^{4-} monomers. For example, the dimer $Si_2O_7^{6-}$ (Fig. 5.73) could be obtained thus

TABLE 5.50

Discrete-Polyanion Model of Liquid Silicates

Range of Composition (mol% M_2O)	Type of Silicate Entities Present
0	Continuous 3-dimensional networks of SiO_4 tetrahedra with some thermal bond breaking and a fraction of SiO_2 molecules
0–10	Essentially SiO_4 network with number of broken bonds approximately equal to number of added O atoms from M_2O, having fraction of SiO_2 molecules and radicals containing M^+
10–33	Discrete silicate polyanions based upon a six-membered ring ($Si_6O_{15}^{6-}$)
33–55	Mixture of discrete polyanions based on $Si_3O_9^{6-}$ and $Si_6O_{15}^{6-}$ or $Si_4O_{12}^{8-}$ and $Si_8O_{20}^{8-}$
~55–66	Chains of general form $Si_nO_{3n+1}^{(2n+2)-}$, e.g., $Si_2O_7^{6-}$
66–100	SiO_4^{4-} + O^{2-} ions

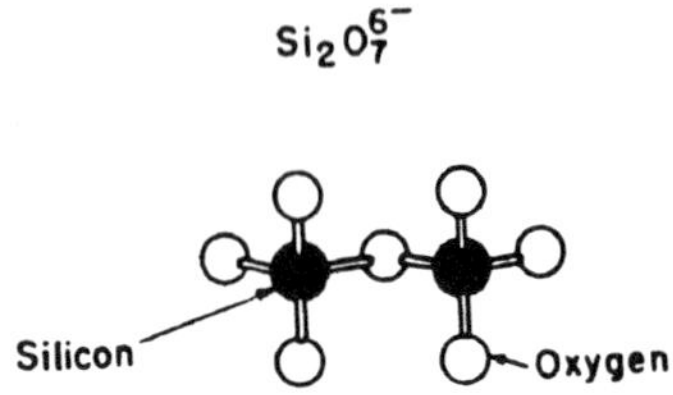

Fig. 5.73. The dimer ion $Si_2O_7^{6-}$.

$$SiO_4^{4-} + SiO_4^{4-} \xrightarrow{-O} O_3^{3-}Si - O - SiO_3^{3-}$$

This chemical change into polyanions is the concept of the *discrete-polyanion* model (Bockris and Lowe, 1954) for the structure of mixtures of liquid oxides corresponding to compositions greater than 10% M_xO_y and less than 66%.

At the outset, it does not seem easy to derive the structure of the polyanions predominant for each composition of the liquid oxides. However, several criteria can be used to suggest their structure. Electroneutrality must be maintained for all compositions; i.e., the total charge on the polyanion group per mole must equal the total cationic charge per mole for a given composition. Since the cationic charge per mole must decrease with decreasing M_xO_y mole percent, the negative charge on the polyanions per mole equivalent of silica must also decrease. It follows that the size of the polymerized anions must increase as the molar fraction of M_xO_y decreases.

After a dimer, i.e., $Si_2O_7^{6-}$, is formed, the next likely anionic entity to appear as the M_xO_y/SiO_2 ratio falls might be expected to be the trimer

$$O_3^{3-}\,Si - O - SiO_3^{3-} + SiO_4^{4-} \rightarrow O_3^{3-}\,Si - O - Si(O^-)_2 - O - SiO_3^{3-}$$

Following the trimer, a polymeric anion with four units may be invoked to satisfy the requirements of electroneutrality and stoichiometry, etc., the general formula being $Si_nO_{3n+1}^{(2n+2)-}$. On this basis, however, when a composition of 50% M_xO_y is reached, i.e., when the mole fraction of M_xO_y is equal to that of SiO_2, the Si/O ratio is 1:3. However, from the general formula for the chain anion, i.e., $Si_nO_{3n+1}^{(2n+2)-}$, it is clear that O/Si = 3 when $(3n + 1)/n = 3$, i.e., when $n \rightarrow \infty$. Near 50% M_xO_y, an attempt to satisfy electroneutrality and stoichiometry by assuming that linear chain anions (extensions of the dimers and trimers) are formed would imply a large increase in the energy of activation for viscous flow in the composition range of 55 to 50 mol% of M_xO_y because here the linear polymer would rapidly approach a great length. However, no such sharp increase in the heat of activation for viscous flow is observed experimentally in the range of 55 to 50 mol% of M_xO_y (see Fig. 5.71). Hence, the composition range in the liquid oxides in which linear anions can be made consistent with the flow data is relatively small—between 66 and somewhat greater than 50 mol% of M_xO_y. The linear

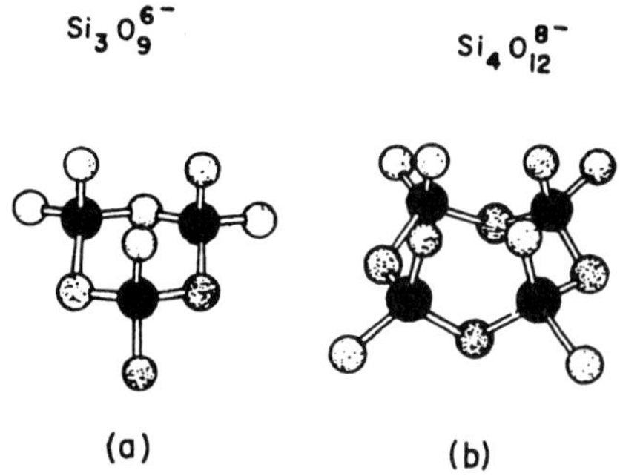

Fig. 5.74. The proposed ring anions: (a) $Si_3O_9^{6-}$ and (b) $Si_4O_{12}^{8-}$.

anion must be given up as a dominant anionic constituent before the metal oxide composition has dropped to 50 mol%.

An alternative anionic structure near a 50% composition that satisfies stoichiometric and electroneutrality considerations is provided by *ring formation.* If, in the composition range of 55 to 50% M_xO_y, the linear anionic chains (which are assumed to exist at compositions between some 50 and 60 mol% M_xO_y) link up their ends to form rings such as $Si_3O_9^{6-}$ or $Si_4O_{12}^{8-}$ (Fig. 5.74), then such ring anions satisfy the criteria of the O/Si ratio, electroneutrality, and also the Si–O–Si valence angle that X-ray data leads one to expect. Thus, the $Si_3O_9^{6-}$ anion corresponds to an O/Si ratio of 3, and if one is considering a 50% CaO system, the charge per ring anion is 6–, and the charge on the three calcium ions required to give $3CaO/3SiO_2$ is 6+. Further, the $Si_3O_9^{6-}$ anion is not very much larger than the $Si_2O_7^{6-}$ dimer and hence there would not be any large increase in the heat of activation for viscous flow. With regard to the Si–O–Si bond angle, in the $Si_3O_9^{6-}$ and $Si_4O_{12}^{8-}$ ions, it is near that observed for the corresponding solids, i.e., the minerals wollastonite and poryphrite, respectively, which are known to contain $Si_6O_9^{6-}$ and $Si_4O_{12}^{8-}$.

Further structural changes between 50 and 30% M_xO_y are based on the $Si_3O_9^{6-}$ and $Si_4O_{12}^{8-}$ ring system. At the 33% compositions, polymers $Si_6O_{15}^{6-}$ and $Si_8O_{20}^{8-}$ (Fig. 5.75) can be postulated as arising from dimerization of the ring anions $Si_3O_9^{6-}$ and $Si_4O_{12}^{8-}$ (Fig. 5.74). As the M_xO_y concentration is continuously reduced, further polymerization of the rings can be speculatively assumed. For example, at $M_xO_y/3SiO_2$ when M_xO_y is 25%, the six-membered ring would have the formula $Si_9O_{21}^{6-}$ and would consist of three rings polymerized together (Fig. 5.76) (Table 5.50).

Ring stability might be expected to lessen with increase in size and increasing proportion of SiO_2 because the silicate polyanions that correspond to compositions approaching 10 mol% of M_xO_y would be very long. The critical 10% composition at which there is a radical change in many properties may be explained as that composition in the region of which a *discrete* polymerized anion type of structure becomes

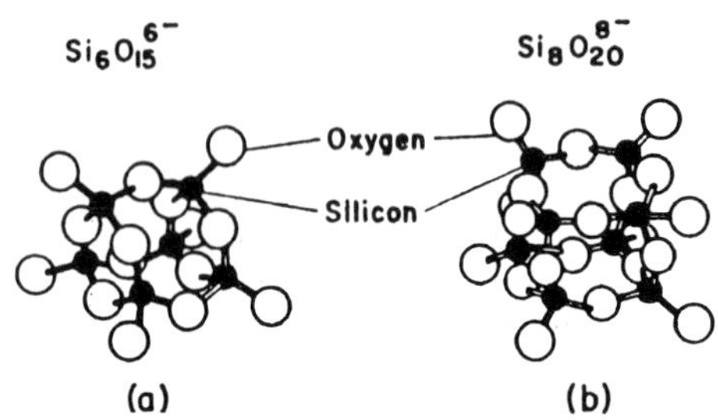

Fig. 5.75. The proposed large ring anions: (a) $Si_6O_{15}^{6-}$ and (b) $Si_8O_{20}^{8-}$.

unstable (because of size) and rearrangement to the random three-dimensional network of silica occurs. That is, a changeover in structural type occurs to what is fundamentally the SiO_2 network, with some bond rupture due to the metal cations. The very large increase in the heat of activation for flow that takes place at this composition (Fig. 5.71) would be consistent with this suggestion, as would also the sudden fall in expansivity.

These ideas about the discrete-polyanion model for liquid-silicate structures are summarized in Table 5.50. As in most models, the description is highly idealized. Thus, all the silicate anions may not have the Si–O–Si angle of the crystalline state; only the *mean* angle may have the crystalline value. Furthermore, the discrete anion suggested for a particular composition is intended to be that which is *dominant*, though not exclusive. Mixtures of polymerized anions may be present at a given composition, the proportions of which vary with composition.

The discrete-polyanion model is a speculative one because no direct proof of the existence of its ring-polymeric anions, for example, is available. It provides a much more consistent qualitative account of facts concerning the behavior of liquid silicates than does the network model. It predicts the observed marked changes in properties near 10% M_2O (Fig. 5.72), the relatively small variation in E_η over the concentration range of 10 to 50% (Fig. 5.71), etc. The suggestions for the structure of the anions

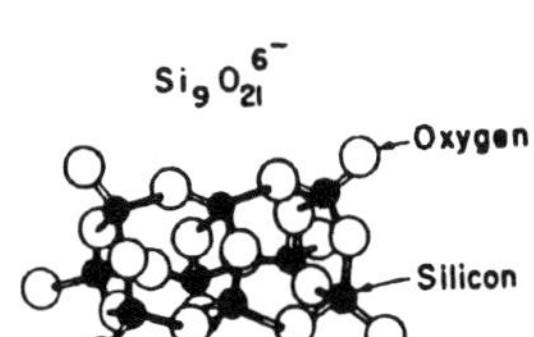

Fig. 5.76. The proposed six-membered ring anion $Si_9O_{21}^{6-}$.

receive some indirect support from solid-state structural analyses of certain mineral silicates.

5.13.9. The "Iceberg" Model

There are some facts, however, which cannot be easily explained in terms of the discrete-polyanion model. First, the partial molar volume of SiO_2, which is related to the size of the SiO_2-containing entities, is relatively constant from 0 to 33 mol% M_2O (Fig. 5.77). On the basis of the discrete-polyanion model, the critical change at 10% M_xO_y involving the breakdown of three-dimensional networks and the formation of discrete polyanions would require a decrease in molar volume of SiO_2 at about 10% M_xO_y for the following reasons. The SiO_2 structure is a particularly open one and has a large molar volume; in contrast, a structure with discrete polyanions would involve a closing up of some of the open SiO_2 volume and a decrease in partial molar volume compared with the SiO_2 networks. Second, it is a fact that in certain ranges of composition (e.g., 12 to 33% M_2O) M_2O and SiO_2 are not completely *miscible*. The two liquids consist of an SiO_2 phase and a metal-rich phase. The discrete-polyanion model cannot accommodate this phenomenon.

It was suggested by Tomlinson, White, and Bockris (1958) that in the composition range of 12 to 33% M_2O, two structures are present. One is similar to that which exists at 33% in the discrete-polyanion model. The other is a structure corresponding to glassy or vitreous silica, i.e., fused silica with the randomized three-dimensional

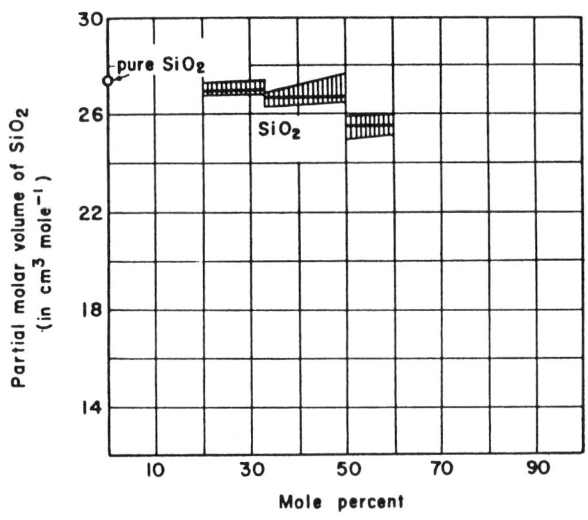

Fig. 5.77. The negligible change in the partial molar volume of an Li_2O-SiO_2 melt over the range of 0 to 33 mol% of Li_2O.

networks frozen in. The vitreous silica is in the form of "islets" or "icebergs"—hence the name, the *iceberg model* of liquid-silicate structure. These icebergs are similar to the clusters that occur in liquid water (Section 5.13.2). The submicroscopic networks may be pictured as continually breaking down and re-forming. Microphase regions of $M_xO_y/2SiO_2$ (the 33% structure) occur in the form of thin films separating the SiO_2-rich icebergs; hence there is the possibility of a separation of the liquid into two phases, one rich in SiO_2 and the other rich in M_xO_y. Since most of the SiO_2 is present in the icebergs, the almost constant partial molar volume of SiO_2 is rationalized.

An estimate of the order of magnitude of the iceberg size can be made. For 12% M_2O, the radius of an (assumed spherical) iceberg is about 1.9 nm, and at 33% M_2O, the iceberg of the iceberg model becomes identical in size with the discrete polyanion of the discrete-polyanion model, which has a radius of about 0.6 nm.

5.13.10. Icebergs As Well as Polyanions

In the iceberg model, the structure of the medium on a microscale is heterogeneous. Flow processes would involve slip between the iceberg and the film. No Si–O–Si bonds need be broken. At present, it seems that both the discrete-polyanion model and the iceberg model probably contribute to the structure of liquid silicates. In a sense, the iceberg model is the most complete model because it involves the discrete polyanions as well as the SiO_2 entities called icebergs.

5.13.11. Spectroscopic Evidence for the Existence of Various Groups, Including Anionic Polymers, in Liquid Silicates and Aluminates

The structure of liquid silicates suggested in the last few sections was presented as a reasonable *interpretation* of conductance, viscous flow, and density measurements and the variation of the heats of activation with composition, for example. What evidence for the chains, rings, and icebergs is given from spectral approaches such as Raman and NMR? If all is well in the earlier interpretation, it should be possible to see evidence of the structures in the spectral peaks.

The earliest workers to obtain this evidence were Furokawa et al., and then a little later, work was carried out by McMillan. These researchers found that in systems based on Na_2O-SiO_2, the metasilicate and the disilicate had two peaks, one that showed up at 950 cm^{-1} and one at 1150 cm^{-1}, respectively. What molecular entities should these peaks represent? This kind of question is grist for the mill among spectroscopists, who often argue a great deal about the significance of the blips and "shoulders" on their curves. In this system they have come up with a decisive but unambitious answer—each peak represents the vibrations of the Si–O bond and the peak for this varies with composition. Such a conclusion, of course, does not indicate the structural units present.

Evidence of various structures of the silicate anions shows up from these Raman and IR studies. The results are shown with the attendant peaks in Table 5.51.

Although these anions are simpler than the structures deduced on the basis of transport measurements, they agree with the structure (see Section 5.6). Chains are present, as are O^{2-} and SiO_2. However, up to 1997, no peaks have yet been registered that are characteristic of the ring structures known to exist in the corresponding solids, the suggested presence of which in the liquid silicates fits stoichiometry, bond angles, and the behavior of the heat of activation for viscous flow as a function of composition (Fig. 5.71).

Farnan and Stebbins performed NMR measurements on simple molten silicates in 1990 and managed to pull more out of their data than other workers in this area.

TABLE 5.51

Compositional Ranges (Expressed as Bulk Melt Nonbridging Oxygens/Si) of Coexisting Structural Units (within Analytical Uncertainty) for Melts on Metal Oxide–Silica Systems

Nonbridging Oxygens/Si Range	Coexisting Anionic Structural Units
	$Na_2O\text{-}SiO_2$
>0–0.05	SiO_2, $Si_2O_5^{2-}$
0.05–1.0	SiO_2, $Si_2O_5^{2-}$, SiO_3^{2-}
1.0–1.4	$Si_2O_5^{2-}$, SiO_3^{2-}, SiO_4^{4-}
1.4–2.0	$Si_2O_5^{2-}$, SiO_3^{2-}, $Si_2O_7^{6-}$, SiO_4^{4-}
	$BaO\text{-}SiO_2$
>0–0.2[2]	SiO_2, $Si_2O_5^{2-}$
0.2–1.0	SiO_2, $Si_2O_5^{2-}$, SiO_3^{2-}
1.0–1.1	SiO_2, $Si_2O_5^{2-}$, SiO_3^{2-}, SiO_4^{4-}
1.1–2.4[2]	$Si_2O_5^{2-}$, SiO_3^{2-}, $Si_2O_7^{6-}$, SiO_4^{4-}
>2.4[2]	SiO_3^{2-}, $Si_2O_7^{6-}$, SiO_4^{4-}
	$CaO\text{-}SiO_2$
>0–0.3[2]	SiO_2, $Si_2O_5^{2-}$
0.3–1.0	SiO_2, $Si_2O_5^{2-}$, SiO_3^{2-}
1.0–2.2	$Si_2O_5^{2-}$, SiO_3^{2-}, SiO_4^{4-}
2.2–3.0	SiO_3^{2-}, $Si_2O_7^{6-}$, SiO_4^{4-}
>3.0[3]	SiO_3^{2-}, $Si_2O_7^{6-}$, SiO_4^{4-}, O^{2-}

Source: From B. O. Mysen, *Structure and Properties of the Silicate Melts*, New York, Elsevier (1988).

They measured the lifetime of the SiO_4^{4-} tetrahedra and found it to be ~10^{-6} s. What happens after an ion dies? Lifetime implies an exchange (perhaps by means of O^-) between SiO_4^{4-} and another ion. By making such measurements on quenched samples as a function of temperature, it is possible to get the heat of activation for the critical step in viscous flow. Remarkably, the values for Li_2O-SiO_2 turn out to be 176 kJ mol^{-1}, which is within 8–13 kJ mol^{-1} of those measured earlier for viscous flow in K_2O-SiO_2. This fits well with the model for viscous flow (see Fig. 5.71). According to this model, an entity has to break off from the silicate anion (i.e., exchange to another structure) before the critical unit (which must be the size of a hole) can move into the cavity, thus giving consistency to the idea of ions of limited lifetime. The dependence on temperature of the NMR peaks for $2Li_2O$-SiO_2 is shown in Fig. 5.78.

Up until the late 1990s the spectroscopic confirmation of the liquid silicates of the model suggested in Sections 5.13.7 and 5.13.9 was limited to the ions

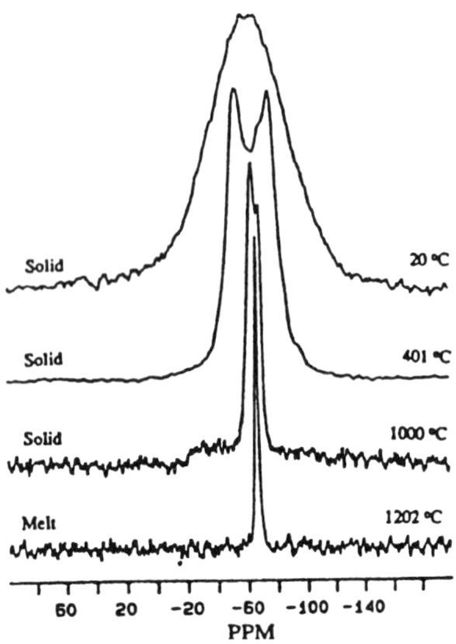

Fig. 5.78. Variable-temperature ^{29}Si NMR spectra of the lithium orthosilicate/metasilicate mixture. Typically, narrow spectra required 64 pulses and broader spectra 500 pulses. The line broadening used in processing each spectrum was less than 10% of the total line width. (Reprinted from I. Farnan and J. F. Stebbins, *J. Am. Chem. Soc.* **112**: 32, 1990.)

SiO_4^{4-}, SiO_3^{2-}, $Si_2O_5^{2-}$, and $Si_2O_7^{6-}$, etc. However, the ring anions also suggested are well confirmed in the corresponding solid silicates.

The degree of certainty in the structures deduced as a result of modeling reasoning from the facts of transport, partial molar volumes, etc., are better than one might think. However, one needs to consider two important facts to be able to pick up the proper structures; these are the necessity for electroneutrality and the fixed O–Si–O angle. Thus, for a certain O/Si ratio, there must be a certain number of charges per ion to match the concentration of cations that the ratio determines. The other matter concerns the O–Si–O bond angle, which is fixed at 120°.

Taking these two facts into account effectively decreases the number of structures possible at a given O/Si value. They are those given in Table 5.50. Now, if there is molecular silica present (H_2O molecules in the complex water structure), as seems to be indicated by the small change of $\overline{V}_{SiO_2}$ over the composition range 0–33 mol% Li_2O in the Li_2O-SiO_2 system, molecular SiO_2 may indeed be added to the structure present in the range of 12–33 mol% in M_xO_y–SiO_2, and this would decrease the size of the corresponding polyanion, probably to $Si_4O_{12}^{6-}$.

A system of interest among the silicates is Al_2O_3-SiO_2, partly because of the relevance of knowledge of the structure of the corresponding glasses and liquids to geochemistry and materials science. Using ^{27}Al and ^{29}Si, Poe et al. found it possible to obtain the coordination number of O for the molecules cited above. Using IR spectra, they were able to get information on the structure, leading to the idea of considering polyhedra involving Al, which tends to replace Si as the main ion to which O coordinates in Al_2O_3-SiO_2 structures. Poe et al. identified AlO_4, AlO_5, and AlO_6 (polyhedra) with peaks at 700–900, 600–700, and 500–600 cm^{-1}, respectively. How are these polyhedra related to each other? It seems likely that exchange of O is the active process:

$$AlO_4 \rightleftharpoons AlO_5 \rightleftharpoons AlO_6$$

The critical variables are the O/Al ratio and the molecular size, and hence the representative anion formulated must depend on these parameters just as it does for the M_xO_y–SiO_2 systems (see Section 5.13.8).

The change in composition with the ions presented was determined by shifts of the NMR peaks. The liquid samples were splat-quenched—dropped into a cold body in such a way that rapid solidification occurred, easing the measurements which then could be made at room temperature. There is a danger in this method since the quenching may not be fast enough, so that some change in structure from that at a given composition and temperature may occur before the quenching ends.

5.13.12. Fused Oxide Systems and the Structure of Planet Earth

We live on an apparently solid earth, but the solidified crust is only some 60 km thick. As one moves down into it, the temperature increases from about 300 K on the surface toward about 1000 K at the bottom of the solid crust where the magma begins. From here, for the next 6000 km (i.e., to the center of the earth) the temperature is thought to remain in the 1000–2000 K range but the *pressure* increases linearly. The mantle down to the iron core is made up of silicates of varying kinds which are liquefied from the solid rock not only by the high temperatures but also by the increasing pressure. Do the structures presented in the last few sections exist in the earth's interior? What is the effect of pressure upon the structure of liquid silicates?

These questions cannot be answered unless one knows the pressure inside the earth as the depth increases. Knowledge of this is incomplete and it clearly must be obtained somewhat indirectly (e.g., from the study of the pattern of seismic distur-

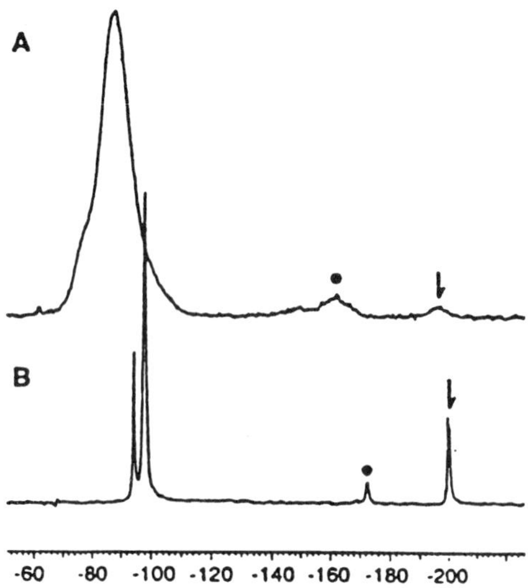

Fig. 5.79. ^{29}Si NMR spectra of $Na_2Si_2O_3$ samples quenched from 8 GPa: (a) 1700 K, glassy and (b) 1420 K, crystalline. Peaks marked with dots are spinning side bands; peaks marked with arrows are attributed to ^{29}Si. Exponential line broadenings of about 20% of the peak widths (20 and 100 Hz) were applied to enhance signal-to-noise ratios. Scales for all spectra are in parts per million relative to tetramethylsilane (TMS). (Reprinted from X. Sue, J. F. Stebbins, M. Kanzaki, and R. D. Tronnes, *Science* **245**: 963, 1993.)

bances, etc.). However, one can also use the average density of the earth (6.7) and ask what is the force per unit area at the bottom of a vertical column 1000 km deep. This force per unit area turns out to be on the order of 6×10^{10} Pa.

Making a physiochemical measurement at ~10^{11} Pa and ~2000 K is clearly very difficult indeed. How could it be achieved? One approach is to pressurize a silicate such as $Na_2Si_2O_5$ at 1700 K (glassy state) and then freeze it very rapidly so that the structure corresponding to the high pressure and temperature remains frozen in the sample. Then NMR measurements can be made at room pressure and temperature while the structure from the high pressure and temperature remains present. What can be seen (Fig. 5.79)?

The main difference that occurs at these high pressures (like those deep in the earth) is that the coordination number of Si begins to change. Throughout the material described in this chapter so far, the central assumption has been that Si is four-coordinated with O, as in SiO_4^{2-} but also in the polymer ions assumed for O/Si < 4. Now, as seen from the arrow points in Fig. 5.79, some amount of six-coordinated Si exists as well as a trace of octahedral structures.

Does this mean that deep in the magma liquid silicates would have a different structure from that described on the basis of measurements at atmospheric pressure? Probably! We are far from knowing what the detailed changes are, or being able to say, for example, that the polymer ions in Fig. 5.75 are now smaller in size. However, at least according to Xue et al. (1992), deep in the earth there is a more random distribution of bridging and nonbridging oxygens than that observed in solid and liquid silicates under conditions at the earth's surface.

5.13.13. Fused Oxide Systems in Metallurgy: Slags

Knowledge of what goes on inside fused oxide systems is important not only as a basis for future advances in glass technology but also for metallurgical processes. Consider, for example, one of the most basic processes in industry, the manufacture of iron in a blast furnace (Fig. 5.80). Iron ore, coke, and flux (essentially limestone and dolomite) are fed into the top of the furance. Compressed air fed in through openings in the bottom of the furnace converts the carbon in coke to carbon monoxide, which reduces the iron oxide to iron. Molten iron collects at the bottom. On top of the molten metal is a layer of molten material called *slag*.

What is slag? A typical chemical analysis (Table 5.52) shows that it consists mainly of silica (SiO_2), alumina (Al_2O_3), lime (CaO), and magnesia (MgO)—in fact, precisely the kinds of substances (i.e., nonmetallic oxides such as SiO_2 and metallic oxides such as CaO) the structure of which is being discussed here. (Slags, in fact, can be regarded as molten glasses.) The constituents of the slags are present in the ores and in coke.

Successful operation of the furnace and production of an iron with the desired composition (and hence metallurgical properties) depend so much on making the slag

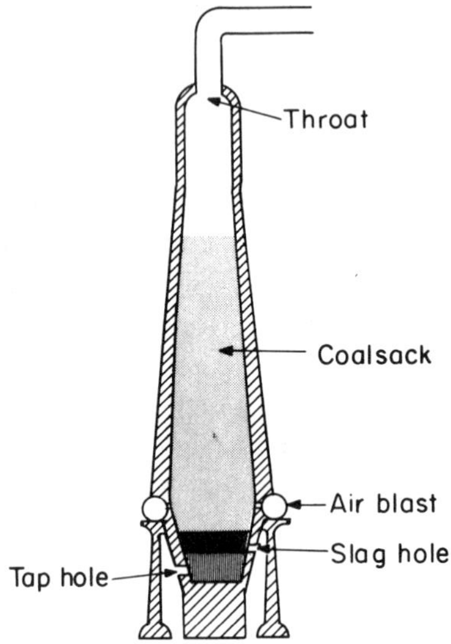

Fig. 5.80. Schematic of a blast furnace.

with the right composition by controlling the raw materials fed into the furnace that it is sometimes said: "You don't make iron in the blast furnace, you make slag!"

Can one explain this importance of the slag? Measurements of conductance as a function of temperature and of transport number indicate that the slag is an ionic conductor (liquid electrolyte). In the metal–slag interface, one has the classic situation (Fig. 5.81) of a metal (i.e., iron) in contact with an electrolyte (i.e., the molten oxide electrolyte, slag), with all the attendant possibilities of corrosion of the metal. Corrosion of metals is usually a wasteful process, but here the current-balancing partial electrodic reactions that make up a corrosion situation are indeed the very factors that control the equilibrium of various components (e.g., S^{2-}) between slag and metal and hence the properties of the metal, which depend greatly on its trace impurities. For example,

TABLE 5.52
Analysis of a Typical Slag

Constituents	Fe	Mn	SiO_2	Al_2O_3	CaO	MgO	S
Percentage	0.34	1.18	34.67	14.58	44.78	3.21	1.36

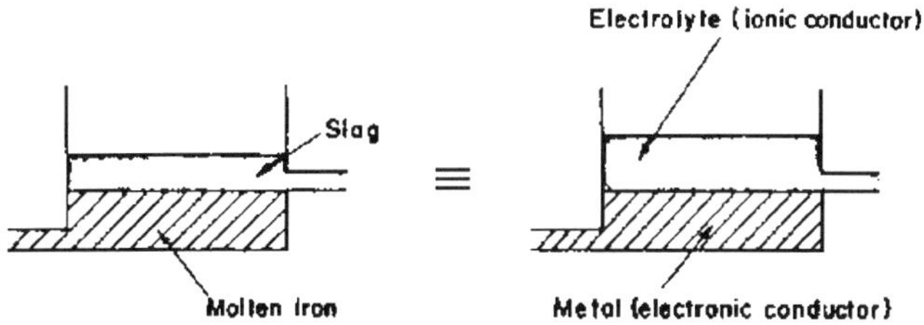

Fig. 5.81. Molten metal in contact with slag is electrochemically equivalent to a metal in contact with an electrolyte.

$$S[\text{in metal}] + 2\ e \rightarrow S^{2-}[\text{in slag}]$$

$$Fe \rightarrow Fe^{2+} + 2\ e$$

The quality of the metal in a blast furnace is thus determined largely by electrochemical reactions at the slag–metal interface. Making good steel and varying its properties at will depends on making good slag first. Future developments in steel making depend on having electrochemists controlling the electrical potential at the slag–metal interface.

Further Reading

Seminal

1. W. Zachariasen, "Chain and Sheet Structure of Glasses," *J. Am. Chem. Soc.* **54**: 3841 (1932).
2. J. O'M. Bockris and D. C. Lowe, "Viscosity and the Structure of Liquid Silicates," *Proc. Roy. Soc. Lond.* **A226**: 423 (1954).
3. J. O'M. Bockris, J. D. MacKenzie, and J. A. Kitchener, "Viscous Flow in Silica and Binary Liquid Silicates," *Trans. Faraday Soc.* **51**: 1734 (1955).
4. J. W. Tomlinson, J. L. White, and J. O'M. Bockris, "The Structure of the Liquid Silicates; Partial Molar Volumes and Expansivities," *Trans. Faraday Soc.* **52**: 299 (1956).

Review

1. B. O. Mysen, *Structure and Properties of the Silicate Melts*, Elsevier, New York (1988).

Papers

1. J. E. Stebbins and I. Farnan, *Science* **255**: 586 (1992).
2. B. T. Poe, P. E. McMillan, C. A. Angell, and R. K. Sato, *Chem. Geol.* **96**: 333 (1992).
3. Y. Kawasita, J. Dong, T. Tsuzuki, Y. Ohmassi, M. Yao, H. Endo, H. Hoshimo, and M. Inni, *J. Non-Cryst. Solids* **156**: 756 (1993).
4. C. Scamehorn and C. A. Angell, *Geochim. Cosmochim. Acta* **55**: 721 (1991).
5. X. Xui, J. F. Stebbins, M. Kenza, and R. G. Tronnes, *Science* **245**: 962 (1996).

APPENDIX 5.1. THE EFFECTIVE MASS OF A HOLE

The pressure gradient in a fluid in the direction x as a result of an instantaneous velocity u in that direction can be expressed as

$$\frac{dP}{dx} = \rho X - \rho\frac{du}{dt} - \left(u\frac{du}{dx} + v\frac{du}{dy} + w\frac{du}{dz}\right) + \eta\left(\frac{d^2u}{dx^2} + \frac{d^2u}{dy^2} + \frac{d^2u}{dz^2}\right) + \frac{\eta}{3}\frac{d}{dx}\left(\frac{du}{dx} + \frac{dv}{dy} + \frac{dw}{dz}\right) \tag{A5.1.1}$$

Similar equations exist for the pressure gradients along the other two mutually perpendicular axes, y and z. In these equations, P is pressure; ρ is the density of the fluid, and η is its viscosity; u, v, and w are the instantaneous fluid velocities at the points x, y, and z in the directions of the three coordinate axes; and X is the component of the accelerating force in the x direction.

Stokes has shown that in cases where the motion is small and the fluid is incompressible and homogeneous, etc., these equations can be simplified to a set of three equations of the form

$$\frac{dP}{dx} = \eta\left(\frac{d^2u}{dx^2} + \frac{d^2u}{dy^2} + \frac{d^2u}{dz^2}\right) - \rho\frac{du}{dt} \tag{A5.1.2}$$

plus an equation of continuity

$$\frac{du}{dx} + \frac{dv}{dy} + \frac{dw}{dz} = 0 \tag{A5.1.3}$$

Solution of Eqs. (A5.1.2) and (A5.1.3) in spherical coordinates leads to a general expression of the form

$$\text{Force} = 2\pi a\int_0^{\pi}(-P_r\cos\theta + T_\theta\sin\theta)_a\sin\theta\times d\theta \tag{A5.1.4}$$

for the force of the fluid acting on the surface ($r = a$) of a hollow sphere oscillating in it with a velocity v given by $ce^{i\omega t}$, where P_r and T_θ are the instantaneous normal and tangential pressures at the points r and θ, c is the velocity at time $t = 0$, and ω is the frequency of oscillation. The term in T_θ can be ignored for present purposes since it corresponds to a viscous force acting on the sphere owing to its motion through the liquid.

Inserting the appropriate expressions for P_r, ignoring the terms arising from the viscosity of the liquid since there can be no viscous slip between a liquid and a hole, and proceeding through a number of algebraic stages yields

$$\Gamma\left(\frac{1}{2}\right)=\int_0^\infty e^{-t}t^{-1/2}\,dt$$

Put $t = x^2$, in which case

$$dt = 2x\,dx$$

$$t^{-1/2}\,dt = 2\,dx$$

and

$$\Gamma\left(\frac{1}{2}\right)=2\int_0^\infty e^{-x^2}\,dx$$

Using Eq. (5.24), i.e.,

$$\int_0^\infty e^{-ax^2}\,dx=\frac{1}{2}\sqrt{\frac{\pi}{a}}$$

one has

$$\Gamma\left(\frac{1}{2}\right)=\sqrt{\pi}$$

3.
$$\Gamma(n+1)=\int_0^\infty e^{-t}t^n\,dt$$

Integrating by parts,

$$\int_0^\infty \underset{u}{t^n}\underset{dv}{e^{-t}\,dt} = -[\underset{u}{t^n}\underset{v}{e^{-t}}]_0^\infty + n\int_0^\infty \underset{v}{e^{-t}}\,\underset{du}{t^{n-1}\,dt}$$

$$= n\Gamma(n)$$

Hence,

$$\Gamma(n+1)=n\Gamma(n)$$

APPENDIX 5.3. THE KINETIC THEORY EXPRESSION FOR THE VISCOSITY OF A FLUID

Consider three parallel layers of fluid, T, M, B (Fig. 5.82), moving with velocities $v + (\partial v/\partial z)\lambda$, v, and $v - (\partial v/\partial z)\lambda$, respectively, where z is the direction normal to the

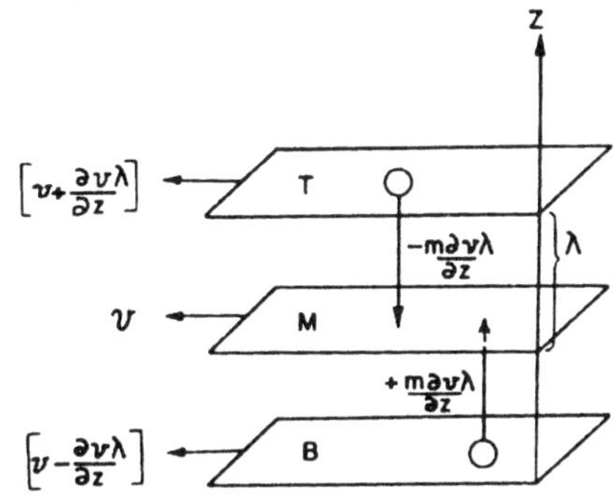

Fig. 5.82. Viscous forces arise from the transfer of momentum between adjacent layers in fluid.

planes and λ is the mean free path of the particles populating the layers, i.e., the mean distance traveled by the particles without undergoing collisions. In the direction of motion of the moving layers, the momenta of the particles traveling in the T, M, and B layers is $m[v + (\partial v/\partial z)\lambda]$, mv, and $m[v - (\partial v/\partial z)\lambda]$, respectively.

When a particle jumps from the T to the M layer, the net momentum gained by the M layer is $mv - m(v + (\partial v/\partial z)\lambda) = -m(\partial v/\partial z)\lambda$. If $<\omega>$ is the mean velocity of particles in the direction normal to the layers, then, in 1 s, all particles within a distance $<\omega>$ will reach the M plane. If one considers that there are n particles per cubic centimeter of the fluid and the area of the M layer is A cm^2, then $n<\omega> A$ particles make the $T \rightarrow M$ crossing per second, transporting a momentum per second of $-[n<\omega>Am(\partial v/\partial z)\lambda]$ in the downward direction.

When a particle jumps from the B to the M layer, the net momentum gained by the M layer is $mv - m[v - (\partial v/\partial z)\lambda] = +m(\partial v/\partial z)\lambda$, i.e., the momentum transported per particle in the *downward* direction is $-m(\partial v/\partial z)\lambda$. Hence, the momentum transferred per second in the downward direction owing to $B \rightarrow M$ jumps is $-[n<\omega>Am(\partial v/\partial z)\lambda]$.

Adding the momentum transferred owing to $T \rightarrow M$ and $B \rightarrow M$ jumps, it is clear that the momentum transferred per second in the downward direction, i.e., the rate of change of momentum, is $-2n<\omega>m\lambda[A(\partial v/\partial z)]$. This rate of change of momentum is equal to a force (Newton's second law of motion). Thus, the viscous force F_η is given by

$$F_\eta = -(2n<\omega>m\lambda)A\,\frac{\partial v}{\partial z} \tag{A5.3.1}$$

However, according to Newton's law of viscosity, the viscous force is proportional to the area of the layers and to the velocity gradient, and the proportionality constant is the viscosity, i.e.,

16. Explain the value of constant-volume and constant-pressure measurements of transport rates. What can one conclude from utilizing the values of each?

17. The determination of transport numbers in aqueous electrolytes is relatively easy (Chapter 3), but in molten salts it poses difficulties of concept, which in turn demand specialized apparatus. Explain why direct determination is difficult. Would it not be better to abandon the direct approach and use the approximate applicability of the Nernst–Einstein equation, relying on self-diffusion determinations? Any counter considerations?

18. Calculate the transport numbers of the cation and the anion in molten CsCl at 943 K. The experimental equivalent conductivity of the fused salt is 67.7 $ohms^{-1}$ cm^2 equiv. The observed diffusion coefficients of Cs^+ and Cl^- ions in molten CsCl are 3.5×10^{-5} and 3.8×10^{-5} cm^2 s^{-1}, respectively. (Contractor)

19. Use the text to find an equation giving the "lifetime" of a hole in a molten salt. Calculate it for KCl using data in the text.

20. Calculate the relaxation time for an electron conducting in a metal–molten salt mixture. (The mobility for such systems is about 0.2 cm^2 V^{-1} s^{-1}.)

21. For a molten salt mixture in which the reaction at an electrode involves one electron, $D_{solute} = 5.4 \times 10^{-5}$ cm^2 s^{-1} and $c_0 = 0.1$ mol dm^{-3}. Electrolysis is occurring in constant-current pulses of 4 A cm^{-2}. Calculate the time at which the reactant in the interface is exhausted.

22. In Raman spectroscopy of cryolite, a higher value of ν favors a lower coordination number. Why?

23. In the text of Chapter 5, there are X-ray data on the internuclear distance in the solid and liquid forms of the alkali halides. In all cases, the internuclear distance *contracts* on melting. Find the mean contraction in percent. What kind of structure in the liquid could be consistent with your finding?

24. In Chapter 5, there are data on the increase in volume of the solid lattice when it becomes liquid. Work out the average increase in volume in percent and compare it with the average contraction of the internuclear distance on melting (see the previous problem). What kind of structure of molten salts does this suggest?

25. Why is it that neutrons are preferred to X-rays in carrying out diffraction experiments with, e.g., molten salts? What is the disadvantage (in practice) of using neutrons compared with X-rays?

26. Sketch a few of the typical anionic polymer ions likely to be present at each composition range in molten oxides. What new ions are formed if O/Si > 4? Which ions have been confirmed to be present by spectroscopic methods?

27. State, in less than 50 words, the essential principles behind the Monte Carlo and molecular dynamics methods of calculating the numerical values of phenomena in liquids. Why is it that such methods need prior experimental determinations in nearby systems?

28. Some of the ambient-temperature molten salts are made up from certain alkyl ammonium salts or, alternatively, a mixture of $AlCl_3$ with organics such as imidazolonium chloride. They have two strong advantages over traditional molten salts with melting points several hundred degrees above room temperature: their great ease of handling and the very large electrochemical window that they allow.

    From the information given in the chapter, suggest up to six solvents (systems) that would allow the electrochemical oxidation of complex organics such as polymerized isoprene (rubber) or even Teflon (polymerized tetrafluoroethylene) at less than 373 K.

29. From the Cohen-Turnbull configurational entropy model, prove that the temperature dependence of relaxation time τ and viscosity η of super-cooled liquids are both non-Arrhenius, i.e., of the $\exp[BT_0/(T - T_0)]$ type, where T_0 is a characteristic temperature. (Xu)

30. $Ca(NO_3)_2$-KNO_3 (CKN) is a well-known molten salt that easily vitrifies upon cooling. An attempt to ascertain the fragility of this system was made on a CKN sample with a glass transition temperature of 350 K. This sample was heated up to 390 K and its dielectric relaxation time measured by an impedance bridge as 10^{-2} s. Classify this ionic liquid. (Xu)

31. Assume that the electrical conductivity of CaO is determined primarily by the diffusion of Ca^{2+} ions. Estimate the mobility of cations at 2070 K. The diffusion coefficient of Ca^{2+} ions in CaO at this temperature is 10^{-14} m^2 s^{-1}. CaO has an NaCl structure with $a = 0.452$ nm. Account for your observation. (Contractor)

32. The equivalent conductivity of molten salts depends upon the cationic radius. Plot the equivalent conductivities of molten salts of monovalent cations against the corresponding cationic radii. Comment on this linear dependence. (Contractor)

33. The diffusion coefficient of tracer Cl^- ion in molten NaCl is 4.2×10^{-5} cm^2 s^{-1} at 1290 K and 6.7×10^{-5} cm^2 s^{-1} at 1110 K. Calculate the values of the activation energy and the preexponential factor. (Contractor)

34. Given that the equivalent conductivity of molten NaCl is 153 ohm^{-1} cm^2 $equiv^{-1}$ at 1193 K and that the self-diffusion coefficients of Na^+ and Cl^- ions in molten NaCl are 9.6×10^{-5} and 6.7×10^{-5} cm^2 s^{-1}, respectively, evaluate the Faraday constant. (Contractor)

of K^+ and F^- is 0.131 nm.) Then calculate the contribution of the paired vacancies to the diffusion and find out how much greater the Nernst–Einstein equation would indicate the conductance to be than it really is.

11. When a metal such as Na is dissolved in a molten salt such as NaCl, it is found that 1 mol% of the metal gives rise to significant electronic conductance. Utilize the equation $\kappa = FNu$, where N is the number of moles of Na per cubic centimeter and u is the mobility (0.4 $cm^2\ V^{-1}\ s^{-1}$). What would be the average distance apart of the Na atoms? On the basis of this distance and an approximation square-well model, calculate the probability of electron tunneling between a K atom and a K^+ ion in molten KCl and thereby the mobility.

    Is the order of magnitude of your result consistent with the observed mobility? If not, suggest an alternative model for electronic conductance of alkali metals in alkali metal salts.

12. Using the distribution function, make a plot of probability that a hole has a radius $\langle r_h \rangle$ in molten sodium chloride at 1170 K. The surface tension of molten sodium chloride at 1170 K is 107.1 dyn cm^{-1}, and the mean hole radius is 1.7×10^{-8} cm. (Contractor)

13. In the Fürth hole model for molten salts, the primary attraction is that it allows a rationalization of the empirical expression $E^{\neq} = 3.74RT_{\text{m.p.}}$. In this model, fluctuations of the structure allow openings (holes) to occur and to exist for a short time. The mean hole size turns out to be about the size of ions in the molten salt. For the distribution function of the theory (the probability of having a hole of any size), calculate the probability of finding a hole two times the average (thereby allowing paired-vacancy diffusion), compared with that of finding the most probable hole size.

14. In the liquid oxides of the type M_xO_y–SiO_2, there are certain limitations on the absolute structures. The O–Si–O angle may be assumed to be between 90° and 120°; the saturated valence of Si is 4 and that of O is 2. Electroneutrality means that the total charges on the metal ions (e.g., Na^+, Ca^{2+}, and Al^{3+}) must be equal in number to the corresponding charges on the silicate anions. Finally, it is always possible from the known composition, in mol%, to determine the value of O/Si, e.g., 4 in Ca_2SiO_4. On the basis of these "givens," calculate the likely structure of a liquid oxide for which O/Si is 2, 3, 4, and 5.

15. The heats of activation for flow in simple molten salts are generally <40 kJ mol^{-1}. In the liquid silicates, the corresponding heat of activation is 5 to 10 times higher than that for CaO_2, Na_2O <50% in SiO_2. Why is there this very significant difference? Is there evidence that connects transport properties in NaCl-type molten salts—the movement into gaps or vacancies in the struc-

ture—to a probable rate-determining step in the flow of liquid oxides (i.e., liquid silicates, borates, and phosphates)?

16. There are several ways by which the structure of pure liquid electrolytes may be examined. A leading way currently is by means of neutron diffraction, a method that acts by registering the interferences caused by the reflection of neutrons and gives the internuclear distance. Examine this method with respect to its ability to contribute to conclusions stimulated by the two most outstanding facts relevant to the structure of molten salts. These are that when the salt melts and expands, the ions get nearer together rather than farther apart, and that the degree of expansion on melting, particularly for the alkali halides, is about 20% of their volume in the solid. To what kind of structures do these striking facts point? Can neutron diffraction measure empty space?

17. Ring and chain structures of $Al_3X_{10}^-$ were deduced by Blander and Saboungi in 1992 and are given in diagrams in the text. Compare the structure shown there (Al^{3+} the coordinating ion) with those deduced by MacKenzie and Lowe for the liquid silicates in 1955 (also given in the text). Differences? Similarities? Why is the working temperature range for the liquid silicates (>1600 K) different from the room-temperature systems studied by Blander and Saboungi?

18. A pairwise potential widely used in both Monte Carlo and molecular dynamics computations is given as

$$\phi_{ij}(r) = A_{ij}\frac{z_i z_j}{r} + B_{ij}\exp[C_{ij}(\sigma_{ij} - r)] + D_{ij}r^{-6} + E_{ij}r^{-8}$$

which describes the potential as a function of distance between two ions i and j; z_i and z_j are charges on i and j, respectively, while σ_{ij} is the size parameter of the ion pair (normally the sum of the crystallographic radii of i and j). A_{ij}, B_{ij}, C_{ij}, D_{ij}, and E_{ij} are constants estimated from the studies on the crystal of the corresponding salt.

(a) Identify the term that dominates the attraction between a pair of oppositely charged ions in long range and the term that prevents these two ions from "falling into each other."

(b) Which parameter in the second term determines the steepness of the repulsion felt by these two ions once their size parameters and center-to-center distances have been fixed?

(c) In molten silicate, the Si–Si equilibrium distance is ca. 0.32 nm. Determine by calculation whether the force due to the second term or the Coulombic-like charge repulsion dominates. What does the result suggest concerning the stability of the silicate network? ($B_{ij} = 0.19 \times 10^{-19}$ J; $C_{ij} = 3.44 \times 10^{10}$ m^{-1}; $r_{Si} = 1.31 \times 10^{-10}$ m.)

SUPPLEMENTARY REFERENCES HELPFUL IN THE SOLUTION OF THE PROBLEMS

Adler, B. J., and Wainwright, T. E., *J. Chem. Phys.* **27**: 1208 (1959).

Angell, C. A., Cheesman, P. A., and Tamaddon, S., **218**: 885 (1982).

Bernal, J. D., and Fowler, R. H., *J. Chem. Phys.* **1**: 515 (1933).

Bjerrum, N., *Kgl. Danske Videnskab. Selskab.* **7**: 9 (1926).

Blum, E. L., *Inorg. Chem.* **24**: 139 (1977).

Bockris, J. O'M., *Quart. Rev. Chem. Soc.* **44**: 151 (1948).

Bockris, J. O'M., *Trans. Faraday Soc.* **45**: 989 (1949).

Bockris, J. O'M., and Lowe, D., *J. Electrochem. Soc.* **102**: 609 (1954).

Böhmer, R., Sanchez, E., and Angell, C. A., *J. Phys. Chem.* **96**: 9090 (1992).

Card, D. N., and Valleau, J. P., *J. Chem. Phys.* **52**: 6232 (1970).

Conway, B. E., *J. Electrochem. Soc.* **129**: 138 (1982).

Davies, C. W., *Ionic Association*, Butterworth, London (1962).

Fajans, K., and Johnson, O., *J. Am. Chem. Soc.* **64**: 668 (1942).

Fawcett, W. R., and Tikanen, A. C., *J. Phys. Chem.* **100**: 4251 (1996).

Frank, B., and Wen, J., *J. Electrochem. Soc.* **129**: 138 (1982).

Friedman, H. L., *Ionic Solution*, Interscience, New York (1962).

Friedman, H. L., *NATO ASI Series C: Mathematical and Physical Sciences* **205**: 61 (1987).

Friedman, H. L., Raineri, E. O., and Wodd, O. M. D., *Chem. Scr.* **29A**: 49 (1989).

Gurney, R., *J. Chem. Phys.* **19**: 1499 (1953).

Inman, D., and Bockris, J. O'M., *Trans. Faraday Soc.* **57**: 2308 (1961).

Kebarle, P., and Godbole, E. W., *J. Chem. Phys.* **39**: 1131 (1968).

Ling, C. S., and Drost-Hansen, W., *Adv. Chem. Soc.* **129**: 75 (1975).

Metropolis, N., *J. Chem. Phys.* **21**: 1087 (1953).

Ramanathan, K. V., and Pinto, R., *J. Electrochem. Soc.* **134**: 165 (1987).

Rasaiah, J. C., *NATO ASI Series B: Physics* **193**: 135 (1987).

Rasaiah, J. C., and Friedman, H. L., *J. Chem. Phys.* **48**: 2742 (1968).

Rossky, P. J., and Friedman, H. L., *J. Chem. Phys.* **72**: 5694 (1980).

Soper, A. K., Nielson, G. W., Enderby, J. E., and Howe, R. A., *J. Phys. Soc.* **10**: 1793 (1977).

Tanaka, M., Tanaka, J., and Nagakura, S., *Bull. Chem. Soc. Japan* **39**: 766 (1965).

Tomlinson, J. W., White, A., and Bockris, J. O'M., *Trans. Faraday Soc.* **54**: 1822 (1958).

Wertheim, G. K., *Phys. Rev.* **30**: 4343 (1984).

Xue, X., Stebbins, J. F., Kanzaki, M., and Tronnes, R. G., *Science* **245**: 963 (1990).